T0181728

Lecture Notes in Computer Science 12239

More information about this series at http://www.springer.com/series/7409

Xingming Sun · Jinwei Wang ·
Elisa Bertino (Eds.)

Artificial Intelligence and Security

6th International Conference, ICAIS 2020
Hohhot, China, July 17–20, 2020
Proceedings, Part I

 Springer

Editors
Xingming Sun 🆔
Nanjing University of Information Science
Nanjing, China

Jinwei Wang 🆔
Nanjing University of Information Science
Nanjing, China

Elisa Bertino 🆔
Purdue University
West Lafayette, IN, USA

ISSN 0302-9743 ISSN 1611-3349 (electronic)
Lecture Notes in Computer Science
ISBN 978-3-030-57883-1 ISBN 978-3-030-57884-8 (eBook)
https://doi.org/10.1007/978-3-030-57884-8

LNCS Sublibrary: SL3 – Information Systems and Applications, incl. Internet/Web, and HCI

This Springer imprint is published by the registered company Springer Nature Switzerland AG
The registered company address is: Gewerbestrasse 11, 6330 Cham, Switzerland

Preface

The 6th International Conference on Artificial Intelligence and Security (ICAIS 2020), formerly called the International Conference on Cloud Computing and Security (ICCCS), was held during July 17–20, 2020, in Hohhot, China. Over the past five years, ICAIS has become a leading conference for researchers and engineers to share their latest results of research, development, and applications in the fields of artificial intelligence and information security.

We used the Microsoft Conference Management Toolkits (CMT) system to manage the submission and review processes of ICAIS 2020. We received 1,064 submissions from 20 countries and regions, including Canada, Italy, Ireland, Japan, Russia, France, Australia, South Korea, South Africa, Iraq, Kazakhstan, Indonesia, Vietnam, Ghana, China, Taiwan, Macao, the USA, and the UK. The submissions cover the areas of artificial intelligence, big data, cloud computing and security, information hiding, IoT security, multimedia forensics, encryption, cybersecurity, and so on. We thank our Technical Program Committee (TPC) members and external reviewers for their efforts in reviewing papers and providing valuable comments to the authors. From the total of 1,064 submissions, and based on at least three reviews per submission, the program chairs decided to accept 142 papers, yielding an acceptance rate of 13%. The volume of the conference proceedings contains all the regular, poster, and workshop papers.

The conference program was enriched by a series of keynote presentations, and the keynote speakers included: Xiang-Yang Li, University of Science and Technology of China, China; Hai Jin, Huazhong University of Science and Technology (HUST), China; and Jie Tang, Tsinghua University, China. We thank them for their wonderful speeches.

There were 56 workshops organized in ICAIS 2020 which covered all the hot topics in artificial intelligence and security. We would like to take this moment to express our sincere appreciation for the contribution of all the workshop chairs and their participants. We would like to extend our sincere thanks to all authors who submitted papers to ICAIS 2020 and to all TPC members. It was a truly great experience to work with such talented and hard-working researchers. We also appreciate the external reviewers for assisting the TPC members in their particular areas of expertise. Moreover, we want to thank our sponsors: Nanjing University of Information Science and Technology, New York University, ACM China, Michigan State University, University of Central Arkansas, Université Bretagne Sud, National Natural Science Foundation of China, Tech Science Press, Nanjing Normal University, Inner Mongolia University, and Northeastern State University.

May 2020

Xingming Sun
Jinwei Wang
Elisa Bertino

Organization

General Chairs

Yun Q. Shi	New Jersey Institute of Technology, USA
Mauro Barni	University of Siena, Italy
Elisa Bertino	Purdue University, USA
Guanglai Gao	Inner Mongolia University, China
Xingming Sun	Nanjing University of Information Science and Technology, China

Technical Program Chairs

Aniello Castiglione	University of Salerno, Italy
Yunbiao Guo	China Information Technology Security Evaluation Center, China
Suzanne K. McIntosh	New York University, USA
Jinwei Wang	Nanjing University of Information Science and Technology, China
Q. M. Jonathan Wu	University of Windsor, Canada

Publication Chair

Zhaoqing Pan	Nanjing University of Information Science and Technology, China

Workshop Chair

Baowei Wang	Nanjing University of Information Science and Technology, China

Organization Chairs

Zhangjie Fu	Nanjing University of Information Science and Technology, China
Xiaorui Zhang	Nanjing University of Information Science and Technology, China
Wuyungerile Li	Inner Mongolia University, China

Technical Program Committee Members

Saeed Arif	University of Algeria, Algeria
Anthony Ayodele	University of Maryland, USA

Zhifeng Bao	Royal Melbourne Institute of Technology University, Australia
Zhiping Cai	National University of Defense Technology, China
Ning Cao	Qingdao Binhai University, China
Paolina Centonze	Iona College, USA
Chin-chen Chang	Feng Chia University, Taiwan, China
Han-Chieh Chao	Taiwan Dong Hwa University, Taiwan, China
Bing Chen	Nanjing University of Aeronautics and Astronautics, China
Hanhua Chen	Huazhong University of Science and Technology, China
Xiaofeng Chen	Xidian University, China
Jieren Cheng	Hainan University, China
Lianhua Chi	IBM Research Center, Australia
Kim-Kwang Raymond Choo	The University of Texas at San Antonio, USA
Ilyong Chung	Chosun University, South Korea
Robert H. Deng	Singapore Management University, Singapore
Jintai Ding	University of Cincinnati, USA
Xinwen Fu	University of Central Florida, USA
Zhangjie Fu	Nanjing University of Information Science and Technology, China
Moncef Gabbouj	Tampere University of Technology, Finland
Ruili Geng	Spectral MD, USA
Song Guo	Hong Kong Polytechnic University, Hong Kong, China
Mohammad Mehedi Hassan	King Saud University, Saudi Arabia
Russell Higgs	University College Dublin, Ireland
Dinh Thai Hoang	University Technology Sydney, Australia
Wien Hong	Sun Yat-sen University, China
Chih-Hsien Hsia	National Ilan University, Taiwan, China
Robert Hsu	Chung Hua University, Taiwan, China
Xinyi Huang	Fujian Normal University, China
Yongfeng Huang	Tsinghua University, China
Zhiqiu Huang	Nanjing University of Aeronautics and Astronautics, China
Patrick C. K. Hung	Ontario Tech University, Canada
Farookh Hussain	University of Technology Sydney, Australia
Genlin Ji	Nanjing Normal University, China
Hai Jin	Huazhong University of Science and Technology, China
Sam Tak Wu Kwong	City University of Hong Kong, Hong Kong, China
Chin-Feng Lai	National Cheng Kung University, Taiwan, China
Loukas Lazos	University of Arizona, USA
Sungyoung Lee	Kyung Hee University, South Korea
Chengcheng Li	University of Cincinnati, USA
Feifei Li	Utah State University, USA

Arun Kumar Sangaiah   VIT University, India
Di Shang   Long Island University, USA
Victor S. Sheng   University of Central Arkansas, USA
Zheng-guo Sheng   University of Sussex, UK
Robert Simon Sherratt   University of Reading, UK
Yun Q. Shi   New Jersey Institute of Technology, USA
Frank Y. Shih   New Jersey Institute of Technology, USA
Biao Song   King Saud University, Saudi Arabia
Guang Sun   Hunan University of Finance and Economics, China
Jianguo Sun   Harbin University of Engineering, China
Krzysztof Szczypiorski   Warsaw University of Technology, Poland
Tsuyoshi Takagi   Kyushu University, Japan
Shanyu Tang   University of West London, UK
Jing Tian   National University of Singapore, Singapore
Yoshito Tobe   Aoyang University, Japan
Cezhong Tong   Washington University in St. Louis, USA
Pengjun Wan   Illinois Institute of Technology, USA
Cai-Zhuang Wang   Ames Laboratory, USA
Ding Wang   Peking University, China
Guiling Wang   New Jersey Institute of Technology, USA
Honggang Wang   University of Massachusetts-Dartmouth, USA
Jian Wang   Nanjing University of Aeronautics and Astronautics, China
Jie Wang   University of Massachusetts Lowell, USA
Jin Wang   Changsha University of Science and Technology, China
Liangmin Wang   Jiangsu University, China
Ruili Wang   Massey University, New Zealand
Xiaojun Wang   Dublin City University, Ireland
Xiaokang Wang   St. Francis Xavier University, Canada
Zhaoxia Wang   A*STAR, Singapore
Sheng Wen   Swinburne University of Technology, Australia
Jian Weng   Jinan University, China
Edward Wong   New York University, USA
Eric Wong   The University of Texas at Dallas, USA
Shaoen Wu   Ball State University, USA
Shuangkui Xia   Beijing Institute of Electronics Technology and Application, China
Lingyun Xiang   Changsha University of Science and Technology, China
Yang Xiang   Deakin University, Australia
Yang Xiao   The University of Alabama, USA
Haoran Xie   The Education University of Hong Kong, Hong Kong, China
Naixue Xiong   Northeastern State University, USA
Wei Qi Yan   Auckland University of Technology, New Zealand

Aimin Yang	Guangdong University of Foreign Studies, China
Ching-Nung Yang	Taiwan Dong Hwa University, Taiwan, China
Chunfang Yang	Zhengzhou Science and Technology Institute, China
Fan Yang	University of Maryland, USA
Guomin Yang	University of Wollongong, Australia
Qing Yang	University of North Texas, USA
Yimin Yang	Lakehead University, Canada
Ming Yin	Purdue University, USA
Shaodi You	The Australian National University, Australia
Kun-Ming Yu	Chung Hua University, Taiwan, China
Weiming Zhang	University of Science and Technology of China, China
Xinpeng Zhang	Fudan University, China
Yan Zhang	Simula Research Laboratory, Norway
Yanchun Zhang	Victoria University, Australia
Yao Zhao	Beijing Jiaotong University, China

Organization Committee Members

Xianyi Chen	Nanjing University of Information Science and Technology, China
Yadang Chen	Nanjing University of Information Science and Technology, China
Beijing Chen	Nanjing University of Information Science and Technology, China
Baoqi Huang	Inner Mongolia University, China
Bing Jia	Inner Mongolia University, China
Jielin Jiang	Nanjing University of Information Science and Technology, China
Zilong Jin	Nanjing University of Information Science and Technology, China
Yan Kong	Nanjing University of Information Science and Technology, China
Yiwei Li	Columbia University, USA
Yuling Liu	Hunan University, China
Zhiguo Qu	Nanjing University of Information Science and Technology, China
Huiyu Sun	New York University, USA
Le Sun	Nanjing University of Information Science and Technology, China
Jian Su	Nanjing University of Information Science and Technology, China
Qing Tian	Nanjing University of Information Science and Technology, China
Yuan Tian	King Saud University, Saudi Arabia
Qi Wang	Nanjing University of Information Science and Technology, China

Lingyun Xiang   Changsha University of Science and Technology,
                               China
Zhihua Xia   Nanjing University of Information Science
                               and Technology, China
Lizhi Xiong   Nanjing University of Information Science
                               and Technology, China
Leiming Yan   Nanjing University of Information Science
                               and Technology, China
Li Yu   Nanjing University of Information Science
                               and Technology, China
Zhili Zhou   Nanjing University of Information Science
                               and Technology, China

Contents – Part I

Internet of Things

Contents – Part II

Information Processing

Artificial Intelligence

Method of Multi-feature Fusion Based on Attention Mechanism in Malicious Software Detection

Yabo Wang[✉] and Shuning Xu

School of Software Technology, Zhengzhou University, Zhengzhou, China
zzuwangyb@126.com

Abstract. Malicious software is designed to destroy or occupy the resources of the target computer, which seriously violates the legitimate interests of users.

Currently, methods based on static detection have certain limitations to the malicious samples of system call confusion. The existing dynamic detection methods mainly extract features from the local system Application Programming Interface (API) sequence dynamically invoked, and combine them with Random Forests and N-grams, which have limited accuracy for detection results. This paper proposes a weight generation algorithm based on Attention mechanism and multi-feature fusion approach, combined with the advantages of Convolutional Neural Networks (CNN) and Gated Recurrent Unit (GRU) algorithms to learn local features of the API sequence and dependencies and relations among API sequences. The experiment tested eight of the most common types of malware. Experimental results show that the proposed method shows a better work than traditional malware detection model in the research of malware detection based on system API call sequences.

Keywords: Malware detection · Attention · Multi-feature fusion

1 Introduction

With the rapid development of computer technology and the increasing popularity of related applications, malicious software technology continues to bring forth the new through the old. At present, malware has become an important issue that threatens the security of computer systems and the development of networks. Therefore, how to identify unknown malware has become a new challenge. In the field of traditional malware detection, the detection method must obtain the signature of the malware before it can be detected. This shortcoming increases the probability of a computer infecting new malware and makes it more difficult to detect malware. With the rise of artificial intelligence, machine learning has been used in the field of malicious software detection, in addition, good detection results are obtained [1–3]. To some extent, malware detection technology combined with machine learning can improve the generalization ability and the recognition rate of malicious samples.

© Springer Nature Switzerland AG 2020
X. Sun et al. (Eds.): ICAIS 2020, LNCS 12239, pp. 3–13, 2020.
https://doi.org/10.1007/978-3-030-57884-8_1

Detection and classification of malware based on machine learning is the main research direction at present. Different behavioral characteristics are used as input to the statistical model, including static disassembly results, and API call sequences processed using N-grams. Machine learning algorithms commonly used in malware detection research based on system API call sequences are logistic regression and random forest. Yang proposed to use the improved random forests classification model to detect and classify the resulting feature vectors [2]. Dahl used dynamic API sequences and processed them using N-grams, then, through the use of the association analysis algorithms to classify them based on confidence coefficient [3]. Most of the above methods can't take good care of the relationship between sequences, fail to make reasonable use of API parameters, lose a lot of original information, and it is difficult to verify the accuracy and generalization of the model based on other datasets [5, 6].

In recent years, malicious sample detection based on deep learning has become a hot spot. Dahl used neural network methods to classify malicious samples under large-scale data sets [3]. Saxe used forward neural networks to classify static analysis results [4], however, the call sequences for dynamic analysis is not considered. After the rise of forward neural networks, deep learning model Convolutional Neural Network (CNN), Recurrent Neural Network (RNN) and their improved versions became the focus of malicious sample detection. However, none of the above methods can properly take into account local features of the API sequence and dependencies and relations among API sequences. the proposed method is used in the research of malware detection based on system API call sequences, meanwhile, combined with the advantages of CNN and RNN algorithms to learn local features of the API sequence and dependencies and relations among API sequences, GRU, LSTM and attention-based algorithms are combined using a multi-feature fusion approach, which proposes a multi-feature fusion malware detection method based on attention mechanism [9, 10].

The paper is organized as follows. Section 2 discusses Attention, CNN and GRU in detail. The model structure of the multi-feature fusion malware detection method based on attention mechanism proposed in this paper is discussed in Sect. 3. Source of experiment ideas and analysis of experimental results, conclusion based on the experiment and the prospect of the application of this model are shown in Sects. 4 and 5 respectively.

2 Related Works

2.1 Malicious Software API Sequence

The Application programming interface (API) is an interface function for the Windows operating system to provide system services to users in the dynamic link library, running in user mode or kernel mode. The API running in kernel mode is the Native API, which is an interface function of the kernel-level system service in the dynamic link library, that is quite different from the API in the user mode. The Native API call sequence can reflect the characteristics of the application at the kernel level, therefore can be used as a data source for anomaly detection. The extraction of dynamic API call sequences is an automated dynamic analysis process that provides true and complete API call sequences by dynamically running and monitoring each PE file in a real machine analysis environment. Since the call to the Application Programming Interface (API) of the portable executable file can reflect the behavior information of the file, it is one of the most effective features used as a smart feature detection method based on dynamic features [1, 4–6, 11].

2.2 Attention Mechanism

The Attention mechanism was firstly used in tasks such as dialogue systems or machine translation, mainly inspired by people's attention, generally, people only focus on the key information that can solve the problem when reading. The model processing mechanism assigns more attention to key parts and allocates less attention to noncritical parts through reasonable distribution of attention [7]. The Attention mechanism can imitate the way people think, the principle is to calculate the attention distribution probability of each element in the input sequence for the current output y in the Decoder stage [16], so that for each output, a unique corresponding intermediate semantic code C can be calculated. In this way, the model assigns different weights to different inputs, highlights data with large weights, and weakens data with small weights, thereby improving the classification effect [8] (Fig. 1).

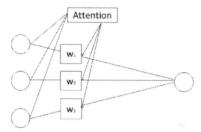

Fig. 1. Attention mechanism assigns weights to features

2.3 Convolutional Neural Networks (CNN)

Convolutional neural networks (CNN) were firstly used in computer vision, and CNN model was firstly proposed by Yoon Kim [12]. CNN is applied to the text classification task, and multiple kernel with different sizes are used to extract the key information in sentences, so as to better capture the local correlation [17]. Convolutional layers have been successfully applied to sequence classification and text classification problems. CNN mainly learns the local features of the input through the convolutional layer and the pooling layer, extracting and retaining the important information of the representation. CNN can achieve the desired results without excessive pre-processing work, significantly reducing the reliance on feature engineering [11]. The CNN mainly consists of an input layer, a convolutional layer, a pooling layer and a fully connected layer. For the sake of CNN used in the field of natural language processing and sequence processing, the input layer is a vector representation of the vocabulary [15].

For a given sentence of length n, the representation of the input layer matrix is as shown in formula (1).

$$E \in R^{n*m} \tag{1}$$

Where m is the word vector dimension. The convolution layer convolves the input matrix using convolution kernels of different sizes, extracts local features of the input data, and obtains the feature vector of the text, as shown in formula (2).

$$c = f(W * x + b) \tag{2}$$

Where x is the word embedding matrix, W is the weight matrix, b is the bias amount and f is the convolution kernel activation function.

The feature vector obtained by the convolution operation can be further down sampled by the pooling layer to extract the most important feature information in the sequence. The pooled output is used as an input to the fully connected layer to classify the sequence data.

2.4 Gated Recurrent Unit (GRU)

The Gated Recurrent Unit (GRU) is an improvement to the Recurrent Neural Network. GRU effectively solves the gradient explosion and gradient dispersion of the Recurrent Neural Network (RNN) during the training process by introducing the update gate and the reset gate. Compared with LSTM, GRU streamlines the network structure, reduces model parameters, and improves model training speed [18]. In the sequence data processing task, the GRU network can learn the long-term dependencies of words in sentences, which can also better characterize and model the text. The GRU network not only memorizes the important feature information in the stored sentence through the memory unit, but also can forget the unimportant information at the same time. Each neuron in the GRU network includes 1 memory unit and 2 gate units [13, 14].

At time t, for a given input x_t, the hidden layer output of the GRU is h_t, and the specific calculation process is as shown in formulas (3)–(6).

$$z_t = \sigma(W^{(z)} * x_t + U^{(z)} * h_{t-1}) \tag{3}$$

$$rt = \sigma(W^{(r)} * x_t + U^{(r)} * h_{t-1}) \tag{4}$$

$$h'_t = \tanh(W * x_t + r_t \odot U^{(r)} h_{t-1}) \tag{5}$$

$$h_t = z_t \odot h_{t-1} + (1 - z_t) \odot h'_t \tag{6}$$

Where W is the weight matrix connecting the two layers, σ and $tanh$ are the activation functions, and z and r are the update gate and the reset gate respectively.

The improvement of GRU based on LSTM is shown in Fig. 2 below.

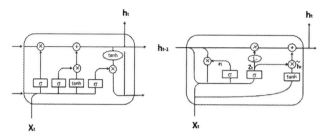

Fig. 2. The improvement of GRU based on LSTM

3   The Malware Detection Model

In order to extract more effective features to identify malicious programs, This paper proposes a weight generation algorithm based on attention mechanism (ATT-CNN-GRU) as shown in Fig. 3, Combined with the advantages of CNN and RNN algorithms to learn local features of the API sequence and dependencies and relations among API sequences. The model mainly includes Embedding layer, Multi-channel Convolutional layer, GRU layer, Model fusion layer and Attention layer. This section mainly introduces the malware detection model proposed in this paper.

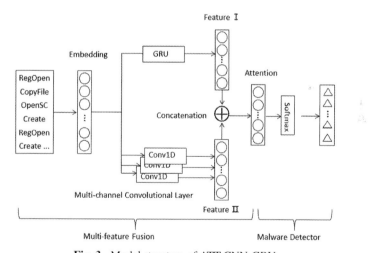

Fig. 3. Model structure of ATT-CNN-GRU

3.1   Model Introduction

The model starts with an embedding layer that maps the API sequence to an embedded vector of any size as an input to the model. Embedded vectors can often improve the performance of large neural networks in complex situations.

This is followed by a multi-channel Convolutional layer and a GRU layer. The multi-channel convolutional layer uses different convolution kernels to extract sequence feature information of different granularities, which can extract the contextual information of the local API sequence more efficiently. GRU is a recurrent neural network layer that can learn the correlation between embedded vectors. This paper uses both GRU and multi-channel convolutional layers to extract features from the sequence.

The feature fusion layer is responsible for merging the features extracted by the GRU and the multi-channel convolutional layer into the attention layer, and the attention layer assigns weights to different features, which can better grasp the important information in the API sequence, and finally send the output into the classifier, get malware detection results.

4   Experiment

4.1   Introduction to Experimental Data

The experimental data comes from the file (Windows executable program) through the sandbox program simulation run API command sequence, all windows binary executable program. It is divided into Normal Software (NS), Ransomware (RW), Mine, DDoS, Worm, Infectious Virus, Backdoor (BD) and Trojans. The data size of each category is shown in the Table 1 below.

Table 1. Malware types, size of dataset

Type	Size
Normal Software (NS)	35.8%
Ransomware (RM)	3.6%
Mine	8.6%
DDoS	5.9%
Worm	0.7%
Infectious virus	38.9%
Backdoor (BD)	3.7%
Trojan	10.7%

4.2   Experimental Environment

This paper uses the deep learning framework of Keras and TensorFlow to build and train the new model of this paper. The model evaluation method used is F1score, and in the experiment, data are divided into training set and verification set according to the proportion of 80% and 20%. Then, the trained model is used to predict the test set data. This paper configures the model with the following parameters.

- Embedding Layer: 256
- Maxlen: 6000
- Kernel Size: (3,4,5)
- GRU: 64
- Dropout regularization: 0.3
- Batch Size: 32
- Epoch: 50

4.3 Experimental Result

This paper trains our model using Adaptive Moment Estimation (Adam) with batch size of 32 samples and 50 epochs. Adam algorithm simultaneously has gained the advantages of the Adaptive Gradient Algorithm (AdaGrad) and Root Mean Square Prop (RMSProp) algorithms at the same time. Adam not only calculates the adaptive parameter learning rate based on the first-order moment mean as the RMSProp algorithm, but also makes full use of the second-order moment mean of the gradient. In this paper, the cross entropy loss is defined as the loss value of the model. Set the learning rate to 0.1 times of the current value for every 10 epochs, and then observe the state of the model training process. Figure 4 shows the curve of the loss and F1 Score value with the number of iteration steps during the training.

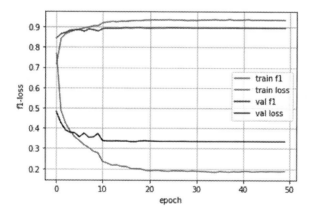

Fig. 4. Loss and F1 score change with epoch

It can be seen that the F1 score of the verification set is steadily increased in the first 10 epochs, and in the last 40 epochs, the F1 score of the verification set shows a small amplitude oscillation, and finally tends to be stable, the model classification effect is optimal at this moment, F1 score and the accuracy rates are 0.896 and 89.4%, respectively.

4.4 Comparative Test

In order to verify the validity of the proposed model, this paper evaluates the malware detection model. On the same data set, and compared with the previous behavioral malware detection model. In the past, the best results of classification models based on malware detection from OS system API call sequences were Logistic Regression (LR) and Random Forests (RF). LR and RF were used with N-grams feature extraction and Term Frequency Inverse Document Frequency (TF-IDF) feature transformation. This paper also performed comparative experiments with the depth models LSTM, respectively. The experimental results of several methods are shown in the Table 2 below.

Table 2. Average F1 score and average accuracy for ATT-CNN-GRU, logistic, random forests, and LSTM

Model	F1 score	Accuracy
Logistic	0.783	0.795
Random forests	0.871	0.874
LSTM	0.722	0.718
ATT-CNN-GRU	**0.896**	**0.894**

As can be seen from Table 2, the ATT-CNN-GRU proposed in this paper is compared to the previous LSTM, Logistic, and Random Forests. The average F1 score increased by 17.4%, 11.3%, and 2.25%, respectively, and the recognition accuracy increased by 17.6%, 9.9%, and 2%, separately.

Considering the different totals of various malwares, this paper examines the accuracy of individual identification of different malware by several different models. The result is shown in the following Table 3.

Table 3. The accuracy of individual identification of different malware

	NS	RM	Mine	DDoS	Worm	Virus	BD	Trojan
Logistic	0.960	0.908	0.851	0.816	0	0.925	0.305	0.670
Random forests	**0.991**	**0.980**	0.921	0.833	0.118	0.962	0.646	0.767
LSTM	0.918	0.871	0.840	0.815	0	0.855	0.448	0.576
ATT-CNN-GRU	0.990	0.975	**0.960**	**0.873**	**0.222**	**0.976**	**0.718**	**0.788**

Based on the results of individual identification of various types of malware, it can be seen that the accuracy of the new model proposed in the identification of multiple malware is higher than that of the previous methods. The fusion of multiple features makes the model more feasible to learn something than the previous methods. and the experimental result is plotted as shown below (Fig. 5).

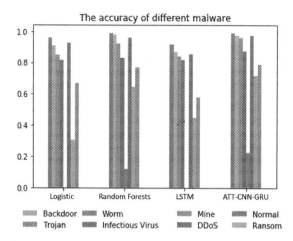

Fig. 5. The accuracy of individual identification of different malware

Considering the endless stream of malware in the real scene, this paper controls the number of malware types, compares the detection capabilities of different models in the case of increasing software types, and observes the general applicability of the model. The experimental results are shown in the Fig. 6 below.

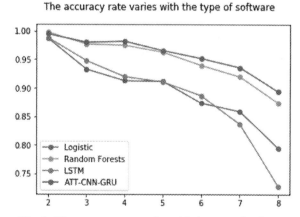

Fig. 6. The accuracy rate varies with the type of software

It can be seen that the model detection method proposed in this paper still maintains a good detection effect in the case of increasing types of malware. Compared with the case where the accuracy of the previous model decreases with the increase of the varieties, the model proposed in this paper can improve the detection effect better under the condition of the increasing number of malware.

5 Conclusion

This paper proposes a multi-feature fusion model based on attention mechanism for malware detection, combined with the advantages of CNN and GRU algorithms to learn local features of the API sequence and dependencies and relations among API sequences. The features extracted by different channels are merged to model the system API call sequences. This paper introduces such a attention mechanism that allows the model to pay more attention to those parts that have a greater impact on the judgment of emotional polarity. the experiment tests eight of the most common types of malware, Experimental results show that the proposed method used in the research of malware detection based on system API call sequences, works better than the traditional malware detection model in the field of malicious program detection.

In the next step, consider improving the attention mechanism to promote the stability and generalization ability of the model. The method of integrating multi-features based on attention mechanism not only can be applied outside the field of malware detection, but also can be applied to other fields of sequence data classification. This has certain reference significance for the classification problems widely existed in machine learning.

References

1. Vinod, P., Zemmari, A., Conti, M.: A machine learning based approach to detect malicious android apps using discriminant system calls. Future Gener. Comput. Syst. **94**, 333–350 (2019)
2. Yang, H., Xu, J.: Android malware detection based on improved random forest algorithm. J. Commun. **38**(04), 8–16 (2017)
3. Dahl, G.E., Stokes, J.W., Deng, L., et al.: Large-scale malware classification using random projections and neural networks. In: 38th IEEE International Conference on Acoustics, Speech and Signal Processing, pp. 3422–3426. IEEE Press, New York (2013)
4. Saxe, J., Berlin, K.: Deep neural network based malware detection using two dimensional binary program features. In: 10th International Conference on Malicious and Unwanted Software, pp. 11–20. IEEE Press, New York (2015)
5. Sami, A., Yadegari, B., Rahimi, H., Peiravian, N., Hashemi, S., Hamze, A.: Malware detection based on mining API calls. In: 25th Proceedings of the 2010 ACM Symposium on Applied Computing, pp. 1020–1025. ACM Press, New York (2010)
6. Alazab, M., Venkataraman, S., Watters, P.: Towards understanding malware behaviour by the extraction of API calls. In: 2nd 2010 Second Cybercrime and Trustworthy Computing Workshop, pp. 52–59. IEEE Press, New York (2010)
7. Vaswani, A., Shazeer, N., Parmar, N., et al.: Attention is all you need. In: Advances in Neural Information Processing Systems, pp. 5998–6008 (2017)
8. Bahdanau, D., Chorowski, J., Serdyuk, D., Brakel, P., Bengio, Y.: End-to-end attention-based large vocabulary speech recognition. In: 13th 2016 IEEE International Conference on Acoustics, Speech and Signal Processing (ICASSP), pp. 4945–4949. IEEE Press, New York (2016)
9. Cho, K., Van Merrienboer, B., Gulcehre, C., Bahdanau, D., Bougares, F., Schwenk, H., et al.: Learning phrase representations using RNN encoder-decoder for statistical machine translation. In: Conference on Empirical Methods in Natural Language Processing (2014)
10. Dai, Y., Li, H., Qian, Y., Yang, R., Zheng, M.: SMASH: a malware detection method based on multi-feature ensemble learning. IEEE Access **7**, 112588–112597 (2019)

11. Yakura, H., Shinozaki, S., Nishimura, R., Oyama, Y., Sakuma, J.: Malware analysis of imaged binary samples by convolutional neural network with attention mechanism. In: Proceedings of the Eighth ACM Conference on Data and Application Security and Privacy, pp. 127–134. ACM Press, New York (2018)

12. Kim, Y.: Convolutional neural networks for sentence classification. In: Conference on Empirical Methods in Natural Language Processing (EMNLP), pp. 1746–1751 (2014)

13. Cho, K., Van Merrienboer, B., Gulcehre, C., Bahdanau, D., Bougares, F., Schwenk, H., et al.: Learning phrase representations using RNN encoder-decoder for statistical machine translation. In: Conference on Empirical Methods in Natural Language Processing(EMNLP) (2014)

14. Gulcehre, C., Cho, K., Pascanu, R., Bengio, Y.: Learned-norm pooling for deep Feedforward and recurrent neural networks. In: Calders, T., Esposito, F., Hüllermeier, E., Meo, R. (eds.) ECML PKDD 2014. LNCS (LNAI), vol. 8724, pp. 530–546. Springer, Heidelberg (2014). https://doi.org/10.1007/978-3-662-44848-9_34

15. Shin, H.C., Roth, H.R., Gao, M., et al.: Deep convolutional neural networks for computer-aided detection: CNN architectures, dataset characteristics and transfer learning. IEEE Trans. Med. Imaging 35(5), 1285–1298 (2016)

16. Ling, H., Wu, J., Li, P., et al.: Attention-aware network with latent semantic analysis for clothing invariant gait recognition. Comput. Mater. Continua 60(3), 1041–1054 (2019)

17. Xiong, Z., Shen, Q., Wang, Y., et al.: Paragraph vector representation based on word to vector and CNN learning. Comput. Mater. Continua 55(2), 213–227 (2018)

18. Qiu, J., et al.: Dependency-based local attention approach to neural machine translation. Comput. Mater. Continua 59(2), 547–562 (2019)

Computing Sentence Embedding by Merging Syntactic Parsing Tree and Word Embedding

Yong Wang[1] , Maosheng Zhong[1,2(✉)] , Lan Tao[2] , and Shuixiu Wu[1]

[1] Jiangxi Normal University, Nanchang Ziyang Avenue No. 99,
Nanchang 330022, China
zhongmaosheng@sina.com
[2] East China Jiao Tong University, Nanchang Shuanggang East Street No. 808,
Nanchang 330013, China

Abstract. Recent progress in using deep learning for training word embedding has motivated us to explore the research of semantic representation in long texts, such as sentences, paragraphs and chapters. The existing methods typically use word weights and word vectors to calculate sentence embedding. However, these methods lose the word order and the syntactic structure information of sentences. This paper proposes a method for sentence embedding based on the results of syntactic parsing tree and word vectors. We propose the SynTree-WordVec method for deriving sentence embedding, which merges word vectors and the syntactic structure from the Stanford parser. The experimental results show the potential to solve the shortcomings of existing methods. Compared to the traditional sentence embedding weighting method, our method achieves better or comparable performance on various text similarity tasks, especially with the low dimension of the data set.

Keywords: Semantic representation · Word vector · Syntactic parse · Sentence embedding

1 Introduction

Natural Language Processing (NLP) research has acquired a significant breakthrough with the rapid development of deep learning. The text embedding representations learned from neural network are basic building blocks for NLP and Information Re-trieval (IR). Researchers compute word embedding which capture the similarities between words by using diverse methods [6,8,15,18]. Meanwhile researchers generally use the semantic combination for the sake of obtaining the semantic representation of sentences or documents. German mathematician Gottlob Frege proposed that the semantics of a sentence was determined by the

Supported by the National Natural Science Foundation of China (NSFC) under Grant No.61877031, No.61876074.

X. Sun et al. (Eds.): ICAIS 2020, LNCS 12239, pp. 14–25, 2020.
https://doi.org/10.1007/978-3-030-57884-8_2

semantics of its constituent parts and the combination of them [9]. And mostly previous work has shown that the semantic combination is a good choice for deriving sentences or documents embedding. Hermann summarizes commonly used semantic combination methods, such as linear weighting, matrix multiplication and tensor multiplication [10]. Furthermore, recent work represent sentences or documents by training neural network [12,22,24,27]. [11] proposed a recurrent convolutional neural network that requires fewer parameters and captures context information as much as possible to preserve long distance word order. Particularly, [5] proposed the smooth inverse frequency (SIF) for calculating word weight and merged them with word vectors to represent the sentence. The weighting method performs well and is quite better or better than the complex supervision method, but unfortunately the word order is lost. To overcome the weakness of existing weighting methods, we consider that syntactic parsing tree can not only derive synthetic structures but also derive word order, which is a better way to represent synthetic information.

Therefore, we propose a method for sentence embedding based on the results of syntactic parsing tree and word vector, named SynTree-WordVec. The traditional way to seize syntactic structure information is to construct a parsing tree by a parser [13]. And we choose the Stanford Parser for creating syntactic parsing tree. Then merge the parsing tree and word vectors referring to the order of post-order traversal (LRD) in tree to derive sentence embedding. Our proposed method can preserve the word order and syntactic structure in sentence embedding, such that our method have the potential to overcome the weakness of existing methods. Ultimately, we achieve significant improvements on the textual similarity tasks over the most datasets compared to SIF [5] and traditional TF-IDF.

In this work, we compute sentence embedding by weighting methods on the basic of parsing tree and word embedding. The innovation of our proposed method is to consider syntactic information and word embedding. Compared to the traditional sentence embedding weighting method, our method achieves better or comparable performance on various text similarity tasks, especially with the low dimension of the data set.

2 Related Work

2.1 Sentence Embedding

Previous works combined word embedding by using operations on vectors and matrices to derive phrase or sentence embedding [7,16,17,25]. The results showed that operating the vectors by coordinate-wise multiplication achieved a very well performance in the binary operations studied. And [19–21] focused on the distributed representations of phrases and sentences. They typically required parsing and the result was shown to work for sentence-level representations. And in [11], the recurrent convolutional neural network that synthesized the advantages of RNN and CNN represented the text precisely. Another approach raised up an unsupervised algorithm to learn the distributed representations of sentences or documents [12]. Their experiments indicated that their method was competitive with state-of-the-art methods.

More recently, [5] explored the smooth inverse frequency (SIF) as word weight to compute sentence embedding. The word weight of word w in sentence is SIF. The definition of SIF [5] is

$$SIF = \frac{a}{a + p(w)} \tag{1}$$

where a was scalar hyperparameter and $p(w)$ was estimated marginal probabilities of word w.

Inspired by the weighting method SIF, we merge the weights which are transformed from syntactic parsing tree and the word vectors to calculate sentence embedding.

2.2   Syntactic Parsing

Syntactic parsing is one of the core technologies of natural language processing, and the cornerstone of understanding language deeply. The task of syntactic parsing is to identify the syntactic components contained in the sentence and the relationship between these components, generally using parsing trees to represent the results of syntactic parsing [23].

In the paper, we prefer Stanford Parser which is simple to operate and provides various useful tools. We apply the results of syntactic parsing to sentence embedding computation because of the distinctive semantic representation. On the other hand, the structure of parsing tree can preserve the word order in sentence.

Our method combines the word vectors and syntactic structure to derive sentence embedding which may capture more semantics representation in sentences.

3   SynTree-WordVec Method

3.1   Main Idea

In view of the shortcomings of existing methods, we compute the sentence embedding through merging the syntactic parsing tree from Stanford Parser and word vectors following to the order of post-order traversal (LRD) in binary tree. The order of merging nodes in parsing tree ensure that our method is able to keep the word order in sentences. We regard part-of-speech tags, phrase tags and clause tags in the parsing tree as model parameters namely word weights. Then the method learns iteratively the parameter till model gets the optimal performance.

3.2   SynTree-WordVec's Framework

We consider that sentence s's parsing tree is composed of one Parent Node ($Pnode$, namely the root node of parsing tree), n Child Nodes ($Cnode_i(i =$

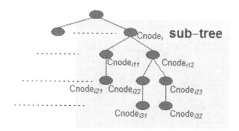

Fig. 1. The structure of parsing tree

$1, 2, ..., n))$. W_{node} is the weight of each node and $SUMWeight_s$ is the sum of the whole weights.

In Fig. 1, the dashed area is one sub-tree of the whole parsing tree. The parent node of the sub-tree is $Cnode_i$. The $Cnode_i$'s vector V_{Cnode_i} is merged by its child nodes' represented as formula (2).

$$V_{Cnode_i} = W_{Cnode_i} \cdot (V_{Cnode_{i11}} + V_{Cnode_{i12}})$$
$$= W_{Cnode_i} \cdot \sum_{k=1}^{2} V_{Cnode_{i1k}} \tag{2}$$

In the same way, the child nodes' vectors are merged by their child nodes' as well. The vector of every node in parsing tree can compute by using formula (2) with different value of k (namely the number of each node's child nodes).

Therefore, we can recursively calculate all the child nodes' vectors to derive the parent node's vector V_{Pnode}. Then divide V_{Pnode} by $SUMWeight_s$ for normalization shown in formula (3).

$$V_{Pnode} = \frac{W_{Pnode} \cdot \sum_{k=1}^{i} V_{Cnode_k}}{SUMWeight_s} \tag{3}$$

where i means that the parent node has i child nodes.

In detail, we choose the example sentence *I love you* to elaborate the computation steps. We run the Stanford Parser to get parsing tree as shown in Fig. 2. Each leaf node in tree is a word of sentence so that its vector is v_w, tag $ROOT$ is the root node of the tree, tag S stands for simple declarative clause, tag NP means noun phrase, and VP means verb phrase, tag PRP represents pronoun and tag VBP represents verb, non-3rd person singular present.

In Fig. 2, the 1-th node S has two child nodes NP and VP, so V_S^1 is

$$V_S^1 = W_S \cdot (V_{NP}^2 + V_{VP}^3) \tag{4}$$

where W_S is the weight of node S. In the same way, the 2-th node NP has one child node PRP and 3-th node VP has two child nodes VBP and NP. So, V_{NP}^2, V_{VP}^3 are calculated as below.

$$V_{NP}^2 = W_{NP} \cdot V_{PRP}^4 \tag{5}$$

18   Y. Wang et al.

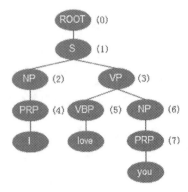

Fig. 2. The parsing tree of example sentence *I love you*

$$V_{VP}^3 = W_{VP} \cdot (V_{VBP}^5 + V_{NP}^6) \tag{6}$$

Similarly, the 4-th node PRP is computed by formula (7). The rest nodes are computed by the same way.

$$V_{PRP}^4 = W_{PRP} \cdot v_I \tag{7}$$

Thus, we can recursively merge all the vectors of nodes from bottom to top till to get root node's vector V_{ROOT}^0. Namely,

$$
\begin{aligned}
V_{ROOT}^0 = V_S^1 &= W_S \cdot (V_{NP}^2 + V_{VP}^3) \\
&= W_S \cdot (W_{NP} \cdot V_{PRP}^4 + W_{VP} \cdot (V_{VBP}^5 + V_{NP}^6)) \\
&= W_S \cdot (W_{NP} \cdot W_{PRP} \cdot v_I + W_{VP} \cdot \\
&\quad (W_{VBP} \cdot v_{love} + W_{NP} \cdot W_{PRP} \cdot v_{you}))
\end{aligned}
\tag{8}
$$

Ultimately, divide V_{ROOT}^0 by $SUMWeight_{V_{Iloveyou}}$ for normalization, where

$$SUMWeight_{V_{Iloveyou}} = 2W_{PRP} + W_{VBP} + 2W_{NP} + W_{VP} + W_S \tag{9}$$

So sentence $V_{Iloveyou}$'s embedding is calculated as shown in below formula (10).

$$
\begin{aligned}
V_{Iloveyou} &= \frac{V_{ROOT}^0}{SUMWeight_{V_{Iloveyou}}} \\
&= \frac{W_S \cdot (W_{NP} \cdot W_{PRP} \cdot v_I + W_{VP} \cdot (W_{VBP} \cdot v_{love} + W_{NP} \cdot W_{PRP} \cdot v_{you}))}{SUMWeight_{V_{Iloveyou}}}
\end{aligned}
\tag{10}
$$

3.3   Computation Steps of SynTree-WordVec

We integrate syntactic structure into sentence embedding by allocating weights to the part-of-speech tag, phrase tag and clause tag, which is capable of capturing syntactic information. The order of merging nodes in tree is similar to the order of post-order traversal (LRD) in binary tree so we can also obtain the word order in sentence. Ultimately, the sentence embedding is computed by subtracting the projection to their first principal component.

The steps of SynTree-WordVec method are summarized in following Table 1.

Table 1. Computation Steps of SynTree-WordVec Method

SynTree-WordVec Method	
Input:	Sentence s's parsing tree $\{T_s : s \in Dataset\}$, word embedding $\{v_w : w \in WordVector\}$, nodes' weights $\{W_{nod} : nod \in T_s\}$
Output:	Sentence embedding $\{V_s : s \in Dataset\}$
Steps:	1: for each sentence s in Dataset do: for each node in T_s do: $V_{nod} \leftarrow W_{nod} \cdot \sum_{k=1}^{i} V_{Cnode_k}$, where nod has i child Nodes. end for; $V_s = \dfrac{W_{Pnode} \cdot \sum_{k=1}^{i} V_{Cnode_k}}{SUMWeight_{T_s}}$, where $SUMWeight_{T_s} = \sum_{nod \in T_s} W_{nod}$, end for. 2: Computing the first principal component μ of $\{V_s : s \in D\}$. 3: for all sentences s in D do: $V_s \leftarrow V_s - \mu\mu^T V_s$

3.4 Parameters Learning

Assuming that there are n (n is 60% of number of sentences pairs for all datasets) pairs of training sentences (S_1^n, S_2^n), and their vectors are $(V_{S_1}^n, V_{S_2}^n)$, the pre-dicted score of similarity between sentences pair (S_1^n, S_2^n) is $(V_{S_1}^n, V_{S_2}^n)$ and the ground-truth score is $GTScore(V_{S_1}^n, V_{S_2}^n)$.

In generally, the target function is minimizing the difference $\Delta Score$ between pre-dicted scores and ground-truth scores shown in formula (11).

$$Min(\Delta Score) = Min(GTScore(V_{S_1}^n, V_{S_2}^n) - PScore(V_{S_1}^n, V_{S_2}^n)) \qquad (11)$$

And people will calculate all the combinations of parameters to obtain the optimal distribution of parameters. And there are 36 part-of-speech tags, 22 phrase tags and 9 clause tags in parsing results. All the part-of-speech tags combine in different order during adjusting weights, which could be reach 36!=3.721041 combinations amazingly. If we iteratively run the model in all the weights combinations, it will cost huge running time.

Therefore, referring to the Expectation Maximization Algorithm (EM), we keep the rest of the weights intact when adjusting the weight, then we can obtain the highest Pearson correlation coefficient while the weight values a appropriate value $W_{a_{best}}^t$ ($t = 1, 2, 3, ..., 36, a = 1, 2, 3, ...$). And we use X_n, Y_n to succinctly represent $PScore(V_{S_1}^n, V_{S_2}^n)$, $GTScore(V_{S_1}^n, V_{S_2}^n)$ shown in below.

$$X_n = PScore(V_{S_1}^n, V_{S_2}^n), Y_n = GTScore(V_{S_1}^n, V_{S_2}^n) \qquad (12)$$

And the Pearson correlation coefficient of X_n and Y_n is

$$Pearson(X_n, Y_n) = \frac{\sum\limits_{i=1}^{n}(X_i - \bar{X})(Y_i - \bar{Y})}{(\sqrt{\sum\limits_{i=1}^{n}(X_i - \bar{X})^2})(\sqrt{\sum\limits_{i=1}^{n}(Y_i - \bar{Y})^2})} \qquad (13)$$

where $W_a^t(t = 1, 2, 3, ..., 36, a = 1, 2, 3, ...)$, the Pearson correlation coefficient is represented as $P^{X_n, Y_n}(W_a^t)$.

So, the target is adjusting weights to derive the biggest $P^{X_n, Y_n}(W_a^t)$. While adjusting the first parameter,

$$a_{best} = \arg\max_{a=1} \sum P^{X_n, Y_n}(W_a^1) \qquad (14)$$

where $W_{a+1}^1 = W_a^1 + epoch$.

Namely $W_a^1 = W_{a_{best}}^1$, we derive the best Pearson correlation coefficient. Then, we continue to adjust the second parameter in the same way. An iteration of parameters learning are finished till 36 parameters are done. According to the experimental demand, the parameters learning may compute for several iterations till gain the optimal performance. Efficiently, we save a lot of time in parameter learning.

4   Experiment and Analysis

4.1   Datasets and Evaluation Criterion

The datasets: 22 textual similarity datasets including all the datasets from SemEval semantic textual similarity (STS) tasks (2012–2015) [1–4], the SemEval 2015 Twitter task [26] and the SemEval 2014 Semantic Relatedness task [14].

We run our SynTree-WordVec method on all these datasets to predict the similarity between pairs of sentences. The evaluation criterion is the Pearson's coefficient between the predicted scores and the ground-truth scores.

4.2   Weighting Parameter's Distribution

We firstly consider the weight of part-of-speech during experiment. And the weights of phrase and clause structure will be thought over in future work.

Normally, the important components of a sentence are verbs, nouns, adjectives, adverbs, or other meaningful part-of-speech in human subjective expression. Similarly, according to the optimal parameters distribution derived by parameters learning, the weights of verbs, nouns, adjectives and adverbs dominate the proportion as well. The detailed interpretation and distribution of the part-of-speech tags are shown in Table 2, 3.

From Table 2 and Table 3, the weights of verbs, nouns, adjectives and adverbs take the most proportion especially verbs. In syntactic structure, usually, noun or verbnoun phrase is the subject, verb is predicate, the object generally is

Table 2. The interpretation of part-of-speech tags

Part-of-speech	Components
Verb (VB)	VB (base form), VBD (past tense), VBN (past participle), VBP,VBZ (non-3rd and 3-rd person singular present), VBG (gerund or present participle) and MD (Modal)
Noun (NN)	NN, NNS (singular or mass, plural), NNP, NNPS (Proper noun, singular and plural)
Adjective (JJ)	JJ (Adjective), JJR (comparative), JJS (superlative)
Pronoun (PRON)	PRP (Personal pronoun), PRP$ (Possessive pronoun), WP (wh-pronoun), WP$ (Possessive wh-pronoun)
Cardinal number (CD)	CD (Cardinal number)
Existential (EX)	Existential (EX)
To (TO)	To (TO)
Determiner (DT)	DT (Determiner), PDT (Predeterminer), WDT (wh-determiner)
Other	CC (Coordinating conjunction), FW (Foreign word), IN (Preposition or subordinating conjunction), LS (List item marker), SYM (Symbol), UH (Interjection), POS (Possessive ending), RP (Particle)

adjective, adverb or pronoun, and clause commonly is a complete sentence which is introduced by the wh-pronoun, wh-determiner, wh-adverb or other introducing words such as "that". Considering the characteristic of sentence structure, the main components of sentence are subject, predicate, object and the same as in clause. Hence, the weights of part-of-speech contained by the main components get bigger so that the sentence embedding may obtain more significant semantics. As normally, a sentence can omit the subject but predicate can't, in addition, just one verb can be regard as a sentence. Thus, the weight of verb is the biggest which indicates that verb plays a vital role in syntactic structure.

As for the part-of-speech in low proportion, such as the part-of-speech in *Other*, this kind of words may carry less semantics in sentence, some words even don't have any information just useful for the sentence structure. However, in

Table 3. Distribution of part-of-speech weights

Part-of-speech	Proportion	Part-of-speech	Proportion
Verb (VB)	30.84%	Cardinal number (CD)	5.26%
Noun (NN)	17.53%	Existential (EX)	4.72%
Adverb (RB)	13.59%	To (TO)	3.81%
Adjective (JJ)	10.49%	Determiner (DT)	1.74%
Pronoun (PRON)	9.54%	Other	2.47%

some ways, the meaningless words can be meaningful while combine with other words. So, their weights occupy the small proportion instead of 0.

4.3   Parameters Learning of Part-of-speech

During the parameters learning, we record the Pearson's coefficient (namely *Score* in following figures) and weights at each step. Here are the tendency charts of weights and *Score* while adjust the *VB*.

From Fig. 3–6, *Score* will reach the peak with changes of parameter which means that parameter gets the optimum while *Score* reaches the maximum. The maximum *Score* and optimum weights of *VB* in each parameter learning iteration are shown in Table 4.

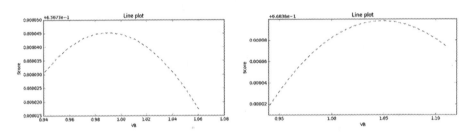

Fig. 3. First adjustment (VB)   **Fig. 4.** Second adjustment (VB)

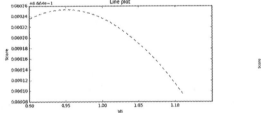
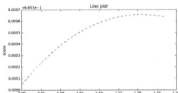

Fig. 5. Third adjustment (VB)   **Fig. 6.** Fourth adjustment (VB)

Table 4. The weights of VB and the value of *Score* (Pearson's r×100)

Iteration	W_{VB}	$Score^{best}$
1	0.991	65.68
2	1.051	66.85
3	0.951	66.67
4	1.211	66.58

4.4   Comparison of Experimental Results

We compare our SynTree-WordVec method with SIF and traditional TF-IDF by using same word vector in different dimensions. Here are results in Table 5.

Table 5. Experimental results (Pearson's r×100) on textual similarity tasks by using different dimensions of word vector

	glove.6B.50d				*glove.6B.100d*		
Tasks	SIF	TF-IDF	Ours	Tasks	SIF	TF-IDF	Ours
STS2012	50.76	52.01	**52.91**	STS2012	53.75	55.10	**56.13**
STS2013	50.76	53.47	**53.63**	STS2013	53.90	**56.91**	55.93
STS2014	63.98	65.59	**65.78**	STS2014	66.61	**68.31**	68.18
STS2015	65.00	**66.37**	66.11	STS2015	68.79	**70.50**	69.57
Sick2014	68.55	68.02	**69.56**	Sick2014	70.02	69.52	**70.76**
Twitter2015	33.15	34.22	**40.68**	Twitter2015	38.14	39.46	**43.28**
	glove.6B.200d				*glove.6B.200d*		
Tasks	SIF	TF-IDF	Ours	Tasks	SIF	TF-IDF	Ours
STS2012	55.65	56.90	**57.57**	STS2012	56.72	57.82	**57.83**
STS2013	56.73	**59.38**	58.00	STS2013	56.82	**59.61**	58.26
STS2014	68.40	**70.06**	69.35	STS2014	69.20	**70.69**	69.78
STS2015	71.46	**73.04**	71.33	STS2015	72.48	**74.02**	71.78
Sick2014	70.22	71.27	**71.76**	Sick2014	71.45	70.15	**71.73**
Twitter2015	41.60	40.44	**45.11**	Twitter2015	40.84	41.96	**46.19**

The scores in boldface are the highest over different word vectors. Our proposed method beats the SIF on all the datasets, furthermore, gets better or comparable performance to the TF-IDF especially using 50 dimensions of word vector shown in Table 5. This demonstrates the merits of our method. We emphasize that the syntactic structure plays a significant role in sentence embedding.

We also compare the influence of different word vector dimensions on the results of sentence embedding. As shown in Table 5, with the increase of word vector's dimensions, experimental results don't get improved greatly. Nevertheless, the computing time increases largely. Hence, the dimension of word vector can't be as high as possible due to the huge time cost.

Extraordinarily, our method performs much better than SIF and TF-IDF on Twit-ter2015 over the whole dimensions shown in Table 5. We analyse that the sentences in Twitter2015 mostly are in complete form that the parser performs better.

5   Conclusions and Future Work

The paper proposes a method of sentence embedding based on the results of syntactic parsing tree and word vectors. The order of merging nodes in parsing tree ensure that our method is able to keep the word order in sentences. And we regard the tags of parsing tree as parameters of weights, which captures the syntactic information in sentence embedding. So our proposed method has the potential to overcome the weakness of existing methods. What's more, it is fast to compute and learn parameters, furthermore achieves better or comparable performance than SIF and traditional TF-IDF on various textual similarity tasks. In addition, there are also some problems in our method should be improved. For example, parameters learning should be optimized such as designing an unsupervised learning target function to get better performance.

References

1. Agirre, E., et al.: Semeval-2015 task 2: Semantic textual similarity, english, spanish and pilot on interpretability. In: Proceedings of the 9th International Workshop on Semantic Evaluation (SemEval 2015), pp. 252–263 (2015)
2. Agirre, E., et al.: Semeval-2014 task 10: Multilingual semantic textual similarity. In: Proceedings of the 8th International Workshop on Semantic Evaluation (SemEval 2014), pp. 81–91 (2014)
3. Agirre, E., Cer, D., Diab, M., Gonzalez-Agirre, A., Guo, W.: *sem 2013 shared task: Semantic textual similarity. In: Second Joint Conference on Lexical and Computational Semantics (*SEM), Volume 1: Proceedings of the Main Conference and the Shared Task: Semantic Textual Similarity, pp. 32–43 (2013)
4. Agirre, E., Diab, M., Cer, D., Gonzalez-Agirre, A.: Semeval-2012 task 6: A pilot on semantic textual similarity. In: Proceedings of the First Joint Conference on Lexical and Computational Semantics-Volume 1: Proceedings of the Main Conference and the Shared Task, and Volume 2: Proceedings of the Sixth International Workshop on Semantic Evaluation, pp. 385–393. Association for Computational Linguistics (2012)
5. Arora, S., Liang, Y., Ma, T.: A simple but tough-to-beat baseline for sentence embeddings (2016)
6. Bengio, Y., Ducharme, R., Vincent, P., Jauvin, C.: A neural probabilistic language model. J. Mach. Learn. Res. **3**(Feb), 1137–1155 (2003)
7. Blacoe, W., Lapata, M.: A comparison of vector-based representations for semantic composition. In: Proceedings of the 2012 Joint Conference on Empirical Methods in Natural Language Processing and Computational Natural Language Learning, pp. 546–556. Association for Computational Linguistics (2012)
8. Collobert, R., Weston, J.: A unified architecture for natural language processing. In: International Conference on Machine Learning (2008)
9. Frege, G.: Über begriff und gegenstand (1892)
10. Hermann, K.M.: Distributed representations for compositional semantics (2014). arXiv preprint arXiv:1411.3146
11. Lai, S.: Word and document embeddings based on neural network approaches (2016). arXiv preprint arXiv:1611.05962

12. Le, Q., Mikolov, T.: Distributed representations of sentences and documents. In: International Conference on Machine Learning, pp. 1188–1196 (2014)
13. Manning, C.D., Manning, C.D., Schütze, H.: Foundations of Statistical Natural Language Processing. MIT press, Cambridge (1999)
14. Marelli, M., Bentivogli, L., Baroni, M., Bernardi, R., Menini, S., Zamparelli, R.: Semeval-2014 task 1: Evaluation of compositional distributional semantic models on full sentences through semantic relatedness and textual entailment. In: Proceedings of the 8th International Workshop on Semantic Evaluation (SemEval 2014), pp. 1–8 (2014)
15. Mikolov, T., Sutskever, I., Chen, K., Corrado, G.S., Dean, J.: Distributed representations of words and phrases and their compositionality. In: Advances in Neural Information Processing Systems, pp. 3111–3119 (2013)
16. Mitchell, J., Lapata, M.: Vector-based models of semantic composition. In: proceedings of ACL-08: HLT, pp. 236–244 (2008)
17. Mitchell, J., Lapata, M.: Composition in distributional models of semantics. Cogn. Sci. $34(8)$, 1388–1429 (2010)
18. Pennington, J., Socher, R., Manning, C.: Glove: Global vectors for word representation. In: Proceedings of the 2014 Conference on Empirical Methods in Natural Language Processing (EMNLP), pp. 1532–1543 (2014)
19. Socher, R., Huang, E.H., Pennin, J., Manning, C.D., Ng, A.Y.: Dynamic pooling and unfolding recursive autoencoders for paraphrase detection. In: Advances in Neural Information Processing Systems, pp. 801–809 (2011)
20. Socher, R., Lin, C.C., Manning, C., Ng, A.Y.: Parsing natural scenes and natural language with recursive neural networks. In: Proceedings of the 28th International Conference on Machine Learning (ICML-11), pp. 129–136 (2011)
21. Socher, R., et al.: Recursive deep models for semantic compositionality over a sentiment treebank. In: Proceedings of the 2013 Conference on Empirical Methods in Natural Language Processing, pp. 1631–1642 (2013)
22. Song, M., Zhao, X., Liu, Y., Zhao, Z.: Text sentiment analysis based on convolutional neural network and bidirectional LSTM model. In: Zhou, Q., Miao, Q., Wang, H., Xie, W., Wang, Y., Lu, Z. (eds.) ICPCSEE 2018. CCIS, vol. 902, pp. 55–68. Springer, Singapore (2018). https://doi.org/10.1007/978-981-13-2206-8_6
23. Wu, W., Zhou, J., Qu, W.: A survey of syntactic parsing based on statistical learning. J. Chin. Inf. Process. $27(3)$, 9–19 (2013)
24. Xiong, Z., Shen, Q., Wang, Y., Zhu, C.: Paragraph vector representation based on word to vector and CNN learning. CMC Comput. Mater. Contin 55, 213–227 (2018)
25. Xiong, Z., Shen, Q., Xiong, Y., Wang, Y., Li, W.: New generation model of word vector representation based on CBOW or skip-gram. CMC-Comput. Mater. Continua $60(1)$, 259–273 (2019)
26. Xu, W., Callison-Burch, C., Dolan, B.: Semeval-2015 task 1: Paraphrase and semantic similarity in twitter (PIT). In: Proceedings of the 9th International Workshop on Semantic Evaluation (SemEval 2015), pp. 1–11 (2015)
27. Zhang, X., Lu, W., Li, F., Peng, X., Zhang, R.: Deep feature fusion model for sentence semantic matching. Mater. Continua Comput. $61(2)$, 601–616 (2019)

An Overview of Cross-Language Information Retrieval

Liang Zhang[1,2] and Xiaobing Zhao[1,2(✉)]

[1] School of Information Engineer, Minzu University of China, Beijing 100081, China
cunyu1943@qq.com, nmzxb_cn@163.com
[2] Minority Languages Branch, National Language Resource and Monitoring
Research Center, Beijing 100081, China

Abstract. With the development of the Internet, various language resources are becoming more and more abundant. Users are no longer satisfied with obtaining information expressed only in their mother tongue, and there is a growing need to retrieve documents in other languages. It demands the advancement of technologies to eliminate the communication barriers among languages. Cross-language information retrieval (CLIR), as a sub-field of information retrieval, can meet the purpose of users to retrieve information in other languages which are different from the queries, thus greatly improving the ability of users to retrieve multi-language documents. This paper introduces the translation techniques such as query translation, document translation and machine translation used in cross-language information retrieval, and analyzes their advantages and disadvantages. Besides, we also discuss the main challenges in cross-language information retrieval. We also introduce some of the latest algorithms and techniques, such as word embedding and query expansion. Finally, we summarize the overview and make a further outlook about CLIR.

Keywords: Cross-language information retrieval · Translation techniques · Word embedding

1 Introduction

In the first, there are only several mainstream languages such as English, Chinese and French on the Internet. With the development of the Internet and the popularization of smart devices [44], there is more and more information in different languages. The internet has become an indispensable information acquisition method [46] and the amount of textual information that people need to process daily is increasing [48]. How to obtain information become a problem, especially for resources which are described in different languages. To solve the problem, Cross-Language Information Retrieval (CLIR) are in strong demand.

As a sub-field of Information Retrieval (IR), CLIR provides a method for users to retrieve relevant information stored in target language different from

© Springer Nature Switzerland AG 2020
X. Sun et al. (Eds.): ICAIS 2020, LNCS 12239, pp. 26–37, 2020.
https://doi.org/10.1007/978-3-030-57884-8_3

the source language of user's given queries. According to [14], CLIR offers a solution that enables users to retrieve documents in target language by using queries in source language [34], it eliminates the differences between the two languages, and reduces the cost. [36].

The earliest research on CLIR started in 1970s, which can be traced back to the international online retrieval experiment originated by [40]. In 1997, CLIR experiments were conducted officially in TREC-6 (Text Retrieval Conference) [42]. In 1999, the NTCIR series of workshops organized by the National Institute for Information of Japan started, they focused on the work of Asian languages in addition to English. In 2000, CLIR experiments on European languages started in CLEF (Cross-Language Experiment Forum), they are the first experiments dealt with English, German, French and Italian documents using queries in Dutch, English, Italian and so on. FIRE (Forum for Information Retrieval Evaluation) started in 2008 and it concentrated on the IR and CLIR technologies about India series. Apart from these academic conferences and forums, search engine providers such as Google, Yahoo and Baidu have started to include CLIR in there product development line, which provides the landing conditions for CLIR.

The structure of the paper is as follows: Section 2, we analysis the query translation, document translation and pivotal translation approaches respectively. Section 3 describes the current challenges in CLIR. Section 4 introduces some new algorithms and technologies in CLIR. Section 5 is a summary of the overview and the outlook on the future of CLIR.

2   Approaches in CLIR

There are mainly the following processing steps during processing. First, we take the queries in source language as the inputs. Then, we use a series of approaches to process the queries and get the retrieving result documents in the target language. Finally, the documents obtained in the previous step are fed back to the user as the outputs. The related approaches are mainly query translation, document translation and pivotal translation [17], and query translation is the most used widely. Now, let we introduce and compare these approaches.

2.1   Query Translation Approach

The key in CLIR is eliminating the barrier between the source language and the target language and enabling users to retrieve the documents in target language by the queries in source language. Currently, the query translation is still in the leading position [30,38,45], the structure is shown in Fig. 1 [49]. The core of query translation is that when given a source language, we first translate the queries from source language to target language, then use the translated queries to retrieve the documents. In this way, the CLIR problems are converted into monolingual retrieval problems. Compared with other approaches, there are the following advantages and disadvantages: a) The demand of grammar are less strict, its purpose is to enable the system to match queries and documents rather than making the user understand fully. **b)** Less computing resources. The need of time and space are both relatively low. **c)** Insufficient context information. Query usually is a single word or a short phrase, when we translate the queries, the lack of context information may result in the results are not the desired. **d)** Once the results of translation are poor, it will lead to further retrieval errors [20].

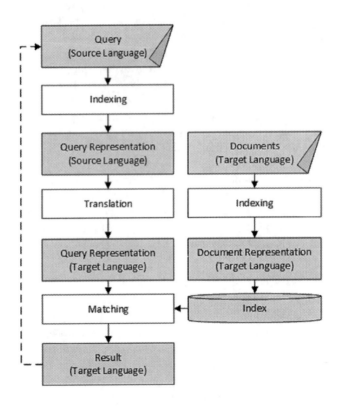

Fig. 1. Query translation.

In CLIR, query translation play a import role and it can be divided into dictionary-based approach, corpus-based approach and machine translation approach.

Dictionary-Based Translation Approach. The basis of dictionary-based translation approach is to represent the queries in source language with a word or a group of words in target language [19], then match them in the documents, finally output the matches as the result. And there are the following problems exist: **a)** Ambiguity. It's mainly caused by polysemy and homonymy, and there are two different types. (1) With the evolution of language, a word has different meanings, such as "neck", it often means "the part of man or animal joining the head to the body", but is also means "the part of the garment". (2) Because the same pronunciation or spelling of the words, the words are from different etymology and even different languages. Such as "ball", which means both "any object in the shape of a sphere" and "a large formal occasion where people dance", and the later meaning is from old French. Due to the ambiguity of the queries, it may fail while matching the relevant documents in target language [1,18,23,35]. **b)** Out-of-vocabulary, OOV. There is no doubt the accuracy of translation through dictionaries is high, but the coverage of the dictionary is limited, it is impossible to find the all corresponding between the queries and documents. The constructing of a dictionary is time-consuming and laborious. Nowadays, the update speed of dictionary can not keep up with the explosion of information, it is difficult to construct a complete dictionary. Once a new word is added into the queries, the corresponding translation can not be found in the dictionary. In this case, the method is no longer useful and affects the performance of system [18,32].

Corporal-Based Translation Approach. Corpus is a collection of language materials that we use in our daily life. The original purpose of using corpus is to extract relevant statistical information, and then construct a dictionary [22,33]. However, due to the high cost of constructing a dictionary and with the rise of statistical methods, researchers begin to give up the indirect approach from corpus to dictionary and use corpus directly. Compared with the dictionary-based approach, there are the following advantages: **a)** We can get the translation knowledge from corpus directly and there is no need to set rules and build dictionary by ourselves, so the cost is lower than that of dictionary-based approach. **b)** The knowledge is derived from large-scale corpus, and the final translation result will be closer to the expression form of the target language. **c)** There in no need to build bilingual dictionaries or semantic resources in advance [8].

Corpus can be divided into parallel corpus and comparable. A parallel corpus requires that the two languages are aligned (sentence level or paragraph level) by translating the text in source language into the text in target language. The method improve the accuracy, but there are still some factors that affect the accuracy [7]. **a)** The corpus is often downloaded from the Internet directly, which is difficult to align between texts. **b)** It is often impossible to translate the materials according to strict rules. **c)** The corpus is from different fields and difficult to unify.

Parallel corpus is more friendly to the language with abundant corpus resources. However, comparable corpora are more suitable for the texts between different language with the same topic but not translate strictly. Although the texts are not aligned strictly, the content is comparable as long as the topic elements are the same [29], [41] conduct a series of experiments based on the strategy and collect a set of comparable texts. Compared with the method based on parallel corpus, the translation result is not accurate enough, but when the available resources are insufficient, the method based on comparable corpus prove to be useful [3,13,28,41].

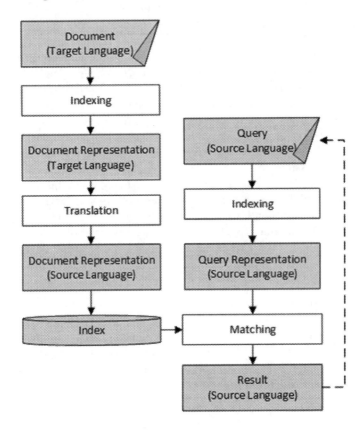

Fig. 2. Document translation.

Machine Translation Approach. The purpose of machine translation is to use the context to translate the queries from the source language to the target language. Compared with the former approaches and combined with the current growth rate of information in the Internet, the machine translation approach seems to be a better choice [25]. There are the following advantages and disadvantages: a) When facing with large texts, machine translation approach can save more time. b) It can translate the whole sentence when translating and combine more context to reduce the occurrence of ambiguity. c) On the one hand, the results of machine translation are often less than ideal when faced with grammatical problems such as polysemous words, complex syntactic structures, and pronouns. d) On the other hand, the cost of building a machine translation system is high. And the long time required in the translation process also has an impact on retrieval performance.

2.2 Document Translation

If you want to use the queries in source language to retrieve the documents in target language and get the result, document translation is one of the best choices [4,9], the structure is shown in Fig. 2 [49]. When using document translation, all documents are translated from target language into source language. The translation engineering is so huge and the labor cost is so high that the machine translation approach is adopted. Therefore, offline translation is performed in advance, and then index the translated document before the retrieval.

Compared with query translation, the advantages and disadvantages of document translation are also obvious. Query translation performs poor when translating short sentences because of ambiguity. There are following advantages over the query translation approach [17]: **a)** Document translation can provide more context and find out the exact meaning of queries more accurately according to the context semantics. **b)** There is a little influence on the results when an error occurs during translation, because the similarity is obtained by weighting the entire document. **c)** The documents are indexed before retrieval, because of the offline translation, it needs less waiting time.

However, there are still the following problems: **a)** More powerful computational support is needed when indexing the translation results. **b)** Document translation needs higher performance requirements for machine translation, but the accuracy of machine translation is still unsatisfactory [24]. **c)** It is necessary to translate the documents through offline translation in advance, so huge amount of work are needed. And it is also a burden for the computer memory.

2.3 Pivotal Translation

When the source language and the target language are both low resource languages, the translation of them is not always feasible due to limited resources. At this point, translation can be carried out by means of pivot language. There are two general cases: one is to translate one of them into the pivot language, and then use the pivot language and the rest of them for translation. The second is to translate both of them into the pivot language. This is a method that combines query and document translation to obtain better performance, as shown in Fig. 3 [49]. Generally, we combine some bilingual dictionaries and monolingual resources such as monolingual (Word Net) [11] ontology and corpus [37] to extract translation knowledge, and generate optimal results [47].

3 Challenges in CLIR

Each of the approaches above has created challenges to the CLIR. In CLIR, the problems are more serious than that in monolingual retrieval. How to solve these problems has become the most difficult at present. In query translation, due to insufficient context, the translation accuracy is not high, which reduces the accuracy of the retrieval. In document translation, there is more context and the accuracy is much higher. However, it is necessary to translate the documents from the target language into the source language in advance, which is higher in cost. The problems in pivotal translation are more serious. The pivotal translation is a combination of query translation and document translation. Through the accumulation of problems in different sub-modules, the final result will be further deteriorated. As mentioned earlier, most of the methods in CLIR are translation methods, so the problems are mainly concentrated in the following aspects:

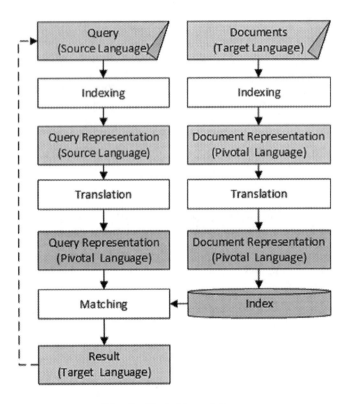

Fig. 3. Pivotal translation.

3.1 Language Resources

Any CLIR experiment requires the support of large-scale language resources such as bilingual dictionaries, parallel corpora and comparable corpora. At present, most of the popular resources in the network are related to the mainstream languages, and there are very few corpus resources for the low-resource languages. Moreover, the quality of available corpus resources is often uneven, which requires further processing before it can be applied to model training. Therefore, it is urgent to build high-quality language resources.

3.2 Ambiguity

When a word is a polysemy, ambiguity may occur [6]. For example, "bank" can be translated as "a financial institution that accepts deposits and channels the money into lending activities" or "sloping land (especially the slope beside a body of water)". When translating them into other languages, because of the polysemy, we needs to combine more context to improve accuracy. Ambiguity in monolingual retrieval is mainly at the semantic level, while in CLIR, it is at the semantic and lexical levels. In contrast, ambiguity is more common in CLIR than in monolingual information retrieval [10].

4 The Latest Algorithms and Technologies in CLIR

4.1 Word Embedding

First, let us learn something about Word Embedding. The study that representing the words as continuous real-valued vectors can date back as far as the mid-1980s [39]. [26] proposed a neural network architecture that leans word representation by predicting neighbouring words. There are two main methods to learn the distributed word representation. One is the Continuous Bag-of-Words (CBOW) model, which learn the word in the middle by combining the representations of the words in the context. The other is the Skip-Gram model, which uses the word in the middle to predict the words in the context. In addition, [31] also proposed an algorithm for learning word embedding in an unsupervised way. The purpose is to learn the word embedding so that the dot product of any pair of words is equal to the logarithm of the co-occurrence probability, then constructing the global vector by combining the global matrix decomposition and local context window.

[12,27] obtain word embedding through Word2Vec, the former constructs the translation matrix by linear regression and then converts the source language vector into the target language space. The latter trains word embedding of the two languages respectively, and maps the word embedding into the same space through Canonical Correlation Analysis (CCA).

[2,43] conduct CLIR experiments with word embedding, [43] is based on the document-level aligned bilingual corpus, and uses Skip-Gram model to obtain the cross-language word embedding composed of query embedding and document embedding, and then calculates the cosine similarity between them and sorts them according to the similarity. The method reduces the dependence on resources such as parallel corpus or bilingual dictionary to acquire high-quality bilingual word embedding, and paves the way for learning bilingual word embedding from comparable corpus. [2] leverages the CBOW model to capture the words in the context of a specific word in the source language, and then interprets these words as translations of the target language to obtain word embedding of the language pairs. Then use the dictionary of word translation pairs to map the relationship between the two language, which alleviates the difficulty of obtaining low-resource language data. Finally, they experiment with FIRE

2008 and 2012 datasets for Hindi and English CLIR, the result shows that the approach based on word embedding outperforms the dictionary-based approach by 70%.

4.2   Query Expansion

The most used in query expansion is a local analysis approach called correlation feedback, which is mainly divided into explicit feedback, implicit feedback and pseudo-correlation feedback [16]. The first two require the direct or indirect participation of users, and the final result depends on the subjective behavior of users, it is difficult to realize because of privacy. However, pseudo-correlation feedback does not require any participation of users. It extracts relevant words directly that are in the front position and can express the user's query intention, and then carries out the secondary retrieval. The advantage of pseudo-correlation feedback enables users to build a multi-language queries based on the original document and then integrate them into the cross-language information retrieval [15].

Query expansion is introduced to optimize the mismatches between queries and documents [5]. [21] present a query expansion approach based on word embedding, they use CBOW model [26] to train the corpus and then select the words that are relevant to the query semantics (because the similarity between embedding corresponds to semantic similarity). Then they conduct two sets of experiments with these words. The first is to extend the original query sets with these words, the result not only outperform the original query approach, but also better than the word2vec-based translation model [50]. The second is to integrate these words into a model based on pseudo-correlated feedback, the result shows better than the first one.

5   Conclusion and Future Work

The former studies on CLIR are mainly focused on the mainstream languages, however, with the development of Internet, there are various information in different languages, the demand for CLIR is also growing. CLIR has attracted more and more researchers' attention and made some progress, but there are still many problems to worth further discussion. The overview introduces the most used approaches and the latest algorithms and technologies, and then analyzes the challenges in CLIR. The conclusion shows that query translation is more convenient than document translation, and document translation relies on the development of machine translation technology. Its cost is so high but the accuracy of document translation is higher. The approach of pivotal translation combines the former two approaches, but the effect is poor due to the problems exist in the former two approaches. With the popularity of neural network and deep learning, word embedding and query expansion have become the hot topics. In the future, with the improvement of computer, the expansion of high-quality language resources and various new algorithms, we believe that the new approaches integrate these elements together will become powerful tools. At the same time, we are also looking forward to applying the technologies into practice.

References

1. Ballesteros, L., Croft, B.: Dictionary methods for cross-lingual information retrieval. In: Wagner, R.R., Thoma, H. (eds.) DEXA 1996. LNCS, vol. 1134, pp. 791–801. Springer, Heidelberg (1996). https://doi.org/10.1007/BFb0034731
2. Bhattacharya, P., Goyal, P., Sarkar, S.: Using word embeddings for query translation for hindi to english cross language information retrieval. Computación y Sistemas **20**(3), 435–447 (2016)
3. Braschler, M., Schäuble, P.: Using corpus-based approaches in a system for multilingual information retrieval. Inf. Retrieval **3**(3), 273–284 (2000)
4. Buckley, C., Singhal, A., Mitra, M., Salton, G.: New retrieval approaches using smart: Trec 4. In: Proceedings of the Fourth Text REtrieval Conference (TREC-4), pp. 25–48 (1995)
5. Carpineto, C., Romano, G.: A survey of automatic query expansion in information retrieval. ACM Comput. Surv. (CSUR) **44**(1), 1 (2012)
6. Chandra, G., Dwivedi, S.K.: A literature survey on various approaches of word sense disambiguation. In: 2014 2nd International Symposium on Computational and Business Intelligence, pp. 106–109. IEEE (2014)
7. Chen, J., Nie, J.Y.: Automatic construction of parallel English-Chinese corpus for cross-language information retrieval. In: Proceedings of the Sixth Conference on Applied natural language processing, pp. 21–28. Association for Computational Linguistics (2000)
8. Chengzhi, L.S.Z.: Survey of multilingual document representation. Data Anal. Knowl. Discovery **26**(6), 33–41 (2010)
9. Croft, W.B., Turtle, H.R., Lewis, D.D.: The use of phrases and structured queries in information retrieval. In: Proceedings of the 14th Annual International ACM SIGIR Conference on Research and Development in Information Retrieval, pp. 32–45. ACM (1991)
10. Diekema, A.R.: Translation events in cross-language information retrieval: lexical ambiguity, lexical holes, vocabulary mismatch, and correct translations (2003)
11. Dini, L., Peters, W., Liebwald, D., Schweighofer, E., Mommers, L., Voermans, W.: Cross-lingual legal information retrieval using a wordnet architecture. In: Proceedings of the 10th International Conference on Artificial Intelligence and Law, pp. 163–167. ACM (2005)
12. Faruqui, M., Dyer, C.: Improving vector space word representations using multilingual correlation. In: Proceedings of the 14th Conference of the European Chapter of the Association for Computational Linguistics, pp. 462–471 (2014)
13. Franz, M., McCarley, J.S., Roukos, S.: Ad hoc and Multilingual Information Retrieval at IBM, pp. 157–168. NIST special publication SP (1999)
14. Gaillard, B., Bouraoui, J.L., De Neef, E.G., Boualem, M.: Query expansion for cross language information retrieval improvement. In: 2010 Fourth International Conference on Research Challenges in Information Science (RCIS), pp. 337–342. IEEE (2010)
15. Gao, L.: Research on the Query Expansion of Tibetan-Chinese Cross-language Information Retrieval Based on Topic Model. Master's thesis, Minzu University of China (2017)

16. Hua-yun, Y., Qi-ping, L., Liang-jun, X.: Survey on relevance feedback for information retrieval. Appl. Res. Comput. **26**(1), 11–14 (2009)
17. Huang, G., Wang, M., Hao, Y.: A novel cross language information retrieval model based on interlingua semantics. J. Chin. Inf. Process. **23**(2), 77–82 (2009)
18. Hull, D.A., Grefenstette, G.: Querying across languages: a dictionary-based approach to multilingual information retrieval. In: Proceedings of the 19th Annual International ACM SIGIR Conference on Research and Development in Information Retrieval, pp. 49–57. Citeseer (1996)
19. Jun-Lin, Z., Wei-Min, Q., Lin, D., Yu-Fang, S.: State-of-the-art:cross-language information retrieval. Comput. Sci. **31**(7), 16–19 (2004)
20. Kayode, K., Ayetiran, E.: Survey on cross-lingual information retrieval. Int. J. Sci. Eng. Res. **9**, 484–491 (2018)
21. Kuzi, S., Shtok, A., Kurland, O.: Query expansion using word embeddings. In: Proceedings of the 25th ACM International on Conference on Information and Knowledge Management, pp. 1929–1932. ACM (2016)
22. Landauer, T.K.: Fully automatic cross-language document retrieval using latent semantic indexing. In: Proceedings of the Sixth Annual Conference of the UW Centre for the New Oxford English Dictionary and Text Research (1990)
23. Lyons, J.: Language and Linguistics. Cambridge University Press, Cambridge (1981)
24. Ma, L., Lai, W., Zhao, X.: Mongolian-chinese cross-language query expansion based on cross-language word vectors. J. Chin. Inf. Process. **33**(6), 27–34 (2019)
25. McCarley, J.S.: Should we translate the documents or the queries in cross-language information retrieval? In: Proceedings of the 37th Annual Meeting of the Association for Computational Linguistics on Computational Linguistics, pp. 208–214. Association for Computational Linguistics (1999)
26. Mikolov, T., Chen, K., Corrado, G., Dean, J.: Efficient estimation of word representations in vector space (2013). arXiv preprint arXiv:1301.3781
27. Mikolov, T., Le, Q.V., Sutskever, I.: Exploiting similarities among languages for machine translation (2013). arXiv preprint arXiv:1309.4168
28. Moulinier, I., Molina-Salgado, H.: Thomson legal and regulatory experiments for CLEF 2002. In: Peters, C., Braschler, M., Gonzalo, J., Kluck, M. (eds.) CLEF 2002. LNCS, vol. 2785, pp. 155–163. Springer, Heidelberg (2003). https://doi.org/10.1007/978-3-540-45237-9_12
29. Nie, J.: Cross-language Information Retrieval: Synthesis Lectures on Human Language Technologies. Morgan & Claypool Publishers, San Rafael (2010)
30. Oard, D.W., He, D., Wang, J.: User-assisted query translation for interactive cross-language information retrieval. Inf. Process. Manage. **44**(1), 181–211 (2008)
31. Pennington, J., Socher, R., Manning, C.: Glove: Global vectors for word representation. In: Proceedings of the 2014 Conference on Empirical Methods in Natural Language Processing (EMNLP), pp. 1532–1543 (2014)
32. Pfeifer, U., Poersch, T., Fuhr, N.: Retrieval effectiveness of proper name search methods. Inf. Process. Manage. **32**(6), 667–679 (1996)
33. Picchi, E., Peters, C.: Cross-language information retrieval: A system for comparable corpus querying. In: Cross-Language Information Retrieval, pp. 81–92. Springer, Boston (1998)
34. Pigur, V.: Multilanguage information-retrieval systems: Integration levels and language support. Autom. Documentation Math. Linguis. **13**(1), 36–46 (1979)
35. Pirkola, A., et al.: The effects of query structure and dictionary setups in dictionary-based cross-language information retrieval. In: Sigir 1998, pp. 55–63 (1998)

36. Qu, P., Li, L., Zhang, L.: A review of advanced topics in informationretrieval. Library and Information Service **3** (2008)
37. Qu, Y., Grefenstette, G., Evans, D.A.: Resolving translation ambiguity using monolingual corpora. In: Peters, C., Braschler, M., Gonzalo, J., Kluck, M. (eds.) CLEF 2002. LNCS, vol. 2785, pp. 223–241. Springer, Heidelberg (2003). https://doi.org/10.1007/978-3-540-45237-9_19
38. Raju, B.N., Raju, M.B., Satyanarayana, K.: Translation approaches in cross language information retrieval. In: International Conference on Computing and Communication Technologies, pp. 1–4. IEEE (2014)
39. Rumelhart, D.E., Hinton, G.E., Williams, R.J., et al.: Learning representations by back-propagating errors. Cogn. Model. **5**(3), 1 (1988)
40. Salton, G.: Experiments in Multi-lingual Information Retrieval. Cornell University, Technical report (1972)
41. araic Sheridan, P., Ballerini, J.P.: Experiments in multilingual information retrieval using the spider system. In: Proceedings of the 19th Annual International ACM SIGIR Conference on Research and Development in Information Retrieval, pp. 58–65. Citeseer (1996)
42. Voorhees, E.M., Harman, D.: Overview of the sixth text retrieval conference (TREC-6). Inf. Process. Manage. **36**(1), 3–35 (2000)
43. Vulić, I., Moens, M.F.: Monolingual and cross-lingual information retrieval models based on (bilingual) word embeddings. In: Proceedings of the 38th International ACM SIGIR Conference on Research and Development in Information Retrieval, pp. 363–372. ACM (2015)
44. Wang, M., Niu, S., Gao, Z.: A novel scene text recognition method based on deep learning. Comput. Mater. Continua **60**, 781–794 (2019)
45. Wu, D., He, D.: A study of query translation using google machine translation system. In: 2010 International Conference on Computational Intelligence and Software Engineering, pp. 1–4. IEEE (2010)
46. Xu, F., Zhang, X., Yang, Z.: Investigation on the Chinese text sentiment analysis based on convolutional neural networks in deep learning. Comput. Mater. Continua **58**, 697–709 (2019)
47. Xu, H., Wang, H.: Research on query translation method in cross-language information retrieval. Digit. Libr. Forum **4**, 41–46 (2009)
48. Yin, L., Meng, X., Li, J., Sun, J.: Relation extraction for massive news texts. Comput. Mater. Continua **58**, 275–285 (2019)
49. Zhou, D., Truran, M., Brailsford, T., Wade, V., Ashman, H.: Translation techniques in cross-language information retrieval. ACM Comput. Surv. (CSUR) **45**(1), 1 (2012)
50. Zuccon, G., Koopman, B., Bruza, P., Azzopardi, L.: Integrating and evaluating neural word embeddings in information retrieval. In: Proceedings of the 20th Australasian Document Computing Symposium, p. 12. ACM (2015)

A Formal Method for Safety Time Series Simulation of Aerospace

Ti Zhou[1(✉)], Jinying Wang[2], and Yi Miao[1]

[1] Beijing Aerospace Control Center, Beijing 100094, China
tzhou@nudt.edu.cn
[2] Beijing Institute of Tracking and Telecommunications Technology,
Beijing 100094, China

Abstract. To develop software of spacecrafts' telemetry and command, it's needed to research extensible and reusable simulation system. This paper studies time series simulation of aerospace, presents the general model based on CCS-like, gives the calculi method of model, and proves the termination of computation. This paper analyzes the time process in the simulation of spacecrafts, simplifies the general model, designs the framework of the temporal configuration files and the control flow of temporal actions in programs, gives two key algorithms which implement this simplified model in simulation system, and implements the framework in a simulation system. And then, it describes the computation process by the example of the concrete temporal process of the spacecraft's simulation. Finally, this paper gives the conclusion and point out the field in which this model can be used in the future.

Keywords: CCS-like · Aerospace · Time series simulation ·
Computing termination · Algorithms

1 Background

With the development of aerospace industry in China, there are so many aerospace crafts which are launched and operated. The safety of spacecraft is very important, so we need to find a method to simulate its actions before the launch. Because they are used only once, different spacecrafts are produced by different companies who define the temporal control processes of spacecrafts in their own way. To simulate real-world business exactly, it's needed to research the temporal processes of spacecraft in the simulation system and simulate special sequence actions in these processes. Temporal actions without conditions can be triggered by temporal actions defined in configuration files, but others with conditions, such as the process of space rendezvous and docking, are written usually in software or model. If these processes are changed, the software or model is needed to be modified. The modifications of software may introduce logic mistakes into system, and raise the risk during the running of software. According to the need of software engineer, the modified software has to be

X. Sun et al. (Eds.): ICAIS 2020, LNCS 12239, pp. 38–48, 2020.
https://doi.org/10.1007/978-3-030-57884-8_4

tested by regression testing, such as unit testing and component testing. If the sequences of temporal actions are changed, it will lead to the poor efficiency of software development and the increase of labor cost. Therefore, the sequences of temporal actions need to be departed from the code of software. The execution logic which is relatively fixed need to be solidified in the software, and the sequences of temporal actions which may be changed are defined in the configuration files. When the sequences are modified, only the configuration files need to be changed. Without updating the code and testing software, the simulation system can be used in simulating the running of spacecrafts quickly.

When the pre-conditions are satisfied, temporal simulation will trigger a series of continuous temporal actions usually and reaches final states. For example, in the space rendezvous and docking, when many pre-conditions are all satisfied, the docking process begins, such as touching and landing, etc. until its completion. Therefore, we need to research the framework of software in cases where many pre-conditions are satisfied, temporal actions are done one by one, and the final states are reached.

[3] introduces the calculus of communicating systems which is as an abstract code that corresponds to a real program whose primitives are reduced to simple send and receive on channels. [4] gives a simple type system for direct extension of both the λ and the π calculi, and provides a programming notation for higher-order concurrency. [5] introduces a radically different approach to parallel simulation for gate level design, aimed at completely eliminating the communication and synchronization overhead between simulators. [6] investigates matching systems for CCS, the π-calculus, and so on. [7] presents the feasibility of employing the temporal and fuzzy inferences in power system fault diagnosis, and proposes alarm message preprocessing and a temporal and fuzzy Cause-Effect network(TFCEN). [8] proposes a methodology using a formal B-method approach, which supports formal stepwise development with refinement to model component-based systems in view of interaction among components. [9] derives a matrix technique which allows determination of whether or not a given rule could have generated a specified time series, but this method is not suitable for configuration description for time series of aerospace. CCS is suitable for describing sequential process, but there is no work to solve time series simulation with CCS. Therefore, this paper will use this method to describe the sequence of time series simulation.

This paper will research a usual model of temporal actions in software and direct the development of software by this model. The structure of this paper is described as follows: Sect. 2 designs a temporal model on CCS-like, gives the definitions in this model, such as temporal action and reasoning rule, introduces the instantaneous fixpoint of temporal sequences, and proves the termination of the computation of the fixpoint. Section 3 discusses the implementation of this model in the simulation system, and the simplified algorithms in realization. Section 4 presents the experiment and takes an example of the exection of temporal process in real world business. Finally, this paper sums up the work, and points out the future development.

2   Temporal Model

R. Milner gives the calculus of communicating systems as the semantic specification of concurrent systems. CCS is one of mathematics methods which is used to describe the concurrent programs. By restricting and extending CCS, time-restricted processes can be defined in formal method. The number of processes of the simulation of aerospace is limited, and the messages translated belong the finite set. Therefore, the time model of the simulation of aerospace can be defined by restricting and extending CCS. This section will model the time controls by CCS-like and prove the termination of computation in this model.

Definition 1 (Message). *Programmed instruction, tele-control commands and self-defined tags are messages, noted by* **Mes**. *The received messages are noted by C_1, C_2, $\cdots$. The sent messages are noted by $\overline{C_1}, \overline{C_2}, \cdots$. Programmed control commands and tele-control commands are finite, and self-defined tags are denumerable, so* **Mes** *is a denumerable set.*

Definition 2 (Temporal action). *Action a is executed at the time t from a special time clock, noted by $< t, a >$. The set of all temporal actions is noted by T_A.*

Definition 3 (Temporal action sequence). *The sequence of temporal actions noted by $< t_1, a_1 > . < t_2, a_2 > . \cdots . < t_n, a_n >$, which means the action a_{i+1} will be executed in t_{i+1} seconds after the execution of a_i, where $1 \leq i < n$.*

Definition 4 (Satisfied Message Set). *A message set is satisfied if all messages in it are sent by processes.*

Definition 5 (Temporal process term). *The temporal process term is $P = \mathcal{C}_G.TA.\overline{C_A}$, where*
(1) $\mathcal{C}_G = C_{G_1}. \cdots .C_{G_n}, C_{G_i}(i = 1, \cdots, n)$: the i-th received message set $C_{G_i} \subseteq Mes$ is unsorted and finite, and is called as guard conditions. The satisfied computation of $C_{G_{i+1}}$ is after that the messages of C_{G_i} have been received. C_{G_i} is noted by M where $C_{G_i} = \{M\}$.
(2) TA: the temporal action sequence as $< t_1, a_1 > . < t_2, a_2 > . \cdots . < t_m, a_m >$ $(m \in N)$. If $m = 0$, then the sequence is empty, noted by ε. The action a_i may be commands, evens(such as extending solar panel), and so on.
(3) $\overline{C_A}$: the finite sent message set, and $\overline{C_A} \subseteq Mes$ contains the messages which will be sent after the action is executed. $\overline{C_A}$ is noted by $\overline{M}$ where $\overline{C_A} = \{M\}$.

The temporal process term describes the guard conditions sequence and the post conditions of a temporal action sequence. When the guard conditions $C_{G_i}(i = 1, \cdots, n)$ in the temporal process term are satisfied one by one, the pre-conditions of the action sequence are satisfied. And then, the actions are fired by time such that the action a_k is executed after delaying time t_k. When the last action a_m is executed, the messages in the send message set are sent in the network, and fire other temporal actions. $\mathcal{C}_G$ or $\overline{C_A}$ is noted by $\emptyset$ if it is empty.

Definition 6 (Temporal process). *Temporal process is constructed as follows:*
(1) the empty process **0** *is a temporal process;*
(2) if **P** *is a temporal process term,* **P** *is a temporal process;*
(3) if **P₁** *and* **P₂** *are temporal processes,* **P₁|P₂** *is a temporal process;*
(4) a temporal process can be constructed by using (1) ∼ (3) finitely.

The derivation rules of temporal process are following as Table 1.

Table 1. The derivation rules of temporal process.

Rule Name	Expression	
Zero process	$\dfrac{P_1 \mid 0}{P_1}$	
Commutation	$\dfrac{P_1 \mid P_2}{P_2 \mid P_1}$	
Association	$\dfrac{(P_1 \mid P_2) \mid P_3}{P_1 \mid (P_2 \mid P_3)}$	
Guard departing	$\dfrac{C_{G_1} \cdots (C_{G_i} \textbf{ or } C'_{G_i}) \cdots C_{G_n} .TA.C_A}{(C_{G_1} \cdots C_{G_i} \cdots C_{G_n} .TA.C_A) \mid (C_{G_1} \cdots C'_{G_i} \cdots C_{G_n} .TA.C_A)}$	
Action departing 1	$\dfrac{C_G.<t_1,a_1>.\cdots.<t_m,a_m>.\overline{C_A}}{(C_G.<t_1,a_1>.\overline{S_1}) \mid (S_1.<t_2,a_2>.\overline{S_2}) \mid \cdots \mid (S_{m-1}.<t_m,a_m>.\overline{C_A})}$	$m \in N,\ S_i$ is a new tag.
Action departing 2	$\dfrac{C_G.<t_1,a_1>.\cdots.<t_m,a_m>.\overline{C_A}}{(C_G.<t_1,a_1>.\emptyset) \mid \cdots \mid (C_G.<\sum_{i=1}^{m-1} t_i,a_{m-1}>.\emptyset) \mid (C_G.<\sum_{i=1}^{m} t_i,a_m>.\overline{C_A})}$	$m \in N, 1 \le k \le m$
Send Message departing	$\dfrac{\overline{C_A}}{\overline{C_1} \mid \cdots \mid \overline{C_n}}$	$C_A = \{C_1, \cdots, C_n\}$

Zero process law presents that the empty process **0** can be eliminated. Commutation law presents that the execution sequence of concurrent processes are independent. Guard departing law can simplify the guard conditions, so the guard conditions in which there is at least one satisfied are translated into the guard conditions which are satisfied one by one. By departing many single temporal actions from a complex temporal action, action departing laws simplify the implementation of the temporal model.

Let Q_t be the message set produced in clock t. $\mathcal{A}_t \subseteq \{< t,a > .\overline{C_A} \mid < t,a >\in \mathcal{T}_A, \overline{C_A} \subseteq \textbf{Mes}\}$, in which actions will be executed, is the action set produced in clock t. Q_t^i is the i-th message set computed in clock t, $Q^0 = \emptyset$. The will-be executed action set $\mathcal{A}_t^i$ is the i-th computation result set in clock t. And then, the deriving rules of temporal process $[P]Q\mathcal{A}t$ is defined as follows:

1. if $P = \varepsilon$, then $[P]Q\mathcal{A}t = \varepsilon$;
2. let $P = C_{G_1}.\cdots.C_{G_n}.TA.\overline{C_A}, TA =< t_1, a_1 > .\cdots. < t_m, a_m >$,
   (a) if $n > 0$, then $[P]Q\mathcal{A}t = C_{G_1} \setminus Q.C_{G_2}.\cdots.C_{G_n}.TA.\overline{C_A}$;
   (b) if $n \le 0$, then $[P]Q\mathcal{A}t = \varepsilon, \mathcal{A} = \mathcal{A} \cup \{< t + \sum_{i=1}^{k} t_i, a_k > .\emptyset, < t + \sum_{i=1}^{m} t_i, a_m > .\overline{C_A} \mid 1 \le k < m\}, Q = Q \cup \{a_1, \cdots, a_m\} \cup \overline{C_A}$;
3. if $P = P_1 \mid P_2$, then $[P]Q\mathcal{A}t = [P_1 \mid P_2]Q\mathcal{A}t = [P_1]Q\mathcal{A}t \mid [P_2]Q\mathcal{A}t$.

Let P be a temporal process model, time clock is $t(t \geq 1)$, define:

$\mathcal{A}_0 = \{\};$

$Proc_0(P) = P;$

$Proc_t^{(0)}(P) = Proc_{t-1}(P);$

$Q_t^{(0)} = \{M | M$ is the new message set produced in time interval $[t-1, t)\};$

$\mathcal{A}_t^{(0)} = \mathcal{A}_{t-1};$

$Proc_t^{(i+1)}(P) = [Proc_t^i(P)](Q \leftarrow Q_t^{(i)})(\mathcal{A} \leftarrow \mathcal{A}_t^{(i)})t;$

$Q_t^{(i+1)} = Q \cup \{a | < t_a, a > \in \mathcal{A}, \text{ and } t_a \leq t\}$, where both Q and $\mathcal{A}$ are the results of computation of $Proc_t^{(i+1)}(P);$

$\mathcal{A}_t^{(i+1)} = \{< t_0, a > \in \mathcal{A} | t_0 > t\}$, where $\mathcal{A}$ is the result of computation of $Proc_t^{(i+1)}(P);$

$Proc_t(P) = \bigcup_{i=0}^{\infty} Proc_t^{(i)}(P).$

$Proc_t(P)$ is defined as the instaneous fixpoint of P in clock t.

By the definition, this model can be used to describe the complex control of process. For example, the process $C_1. < t_1, a_1 >< t_2, a_2 > .\overline{C_2} | C_2.C_3. < t_3, a_3 > .C_4$ describes that the action a_1 will be executed after t_1 clock if C_1 is satisfied. And then, the action a_2 will be executed after t_2 clock if a_1 is executed. Finally, the message set C_2 is sent. After the process on the right receives the message set C_2, it checks whether all the message in C_3 are received. Let t be the time when C_3 is satisfied, the action a_3 is executed in the clock $t + t_3$. At last, the message C_4 is sent.

Theorem 1. *Let P be a temporal process. The computation of deriving rules of $[P]Q\mathcal{A}t$ is terminated.*

Proof. inductive proof is used about the length of P.

1. If $P = \varepsilon$, then $[P]Q\mathcal{A}t = \varepsilon$ is terminated;
2. Let $P = C_{G_1}. \cdots .C_{G_n}.TA.\overline{C_A}$, $TA =< t_1, a_1 > \cdots < t_m, a_m >,$
   (a) if $n > 0$, then $[P]Q\mathcal{A}t = C_{G_1} \setminus Q.C_{G_2}. \cdots .C_{G_n}.TA.\overline{C_A}$, it's terminated.
   (b) if $n \leq 0$, then $[P]Q\mathcal{A}t = \varepsilon$, $\mathcal{A} = \mathcal{A} \cup \{< t + \sum_{i=1}^{k} t_i, a_k > .\emptyset,$
   $< t + \sum_{i=1}^{m} t_i, a_m > .\overline{C_A} | 1 \leq k < m\}, Q = Q \cup \{a_1, \cdots, a_m\} \cup \overline{C_A},$
   it's terminated;
3. If $P = P_1 | P_2$, then $[P]Q\mathcal{A}t = [P_1|P_2]Q\mathcal{A}t = [P_1]Q\mathcal{A}t | [P_2]Q\mathcal{A}t$. By the inductive hypothesis, both $[P_1]Q\mathcal{A}t$ and $[P_2]Q\mathcal{A}t$ are terminated, so $[P]Q\mathcal{A}t$ is terminated.

Therefore, the deriving $[P]Q\mathcal{A}t$ of temporal process P is terminated.   ∎

Theorem 2. *Let P be a temporal process. The instaneous fixpoint $Proc_t(P)$ of temporal sequences P is terminated in clock t.*

Proof. The simultaneous induction with respect to t and i is used to prove the termination of $Proc_t(P)$ and $Proc_t^{(i)}(P)$.

1. when $t = 0$, the proposition is established by the definition of $Proc_0(P) = P;$

2. Suppose it is established when $t = k$, when $t = k + 1$,

   (a) let $i = 0$, $Proc_{k+1}^{(0)}(P) = Proc_k(P)$ is terminated;

   (b) suppose it is established by $i = m$, $Proc_{k+1}^{(m)}(P)$ is terminated;
   let $i = m + 1$, $Proc_{k+1}^{(m+1)}(P) = [Proc_{k+1}^{(m)}(P)](Q \leftarrow Q_t^{(m)})(\mathcal{A} \leftarrow \mathcal{A}_t^{(m)})t$.
   From the inductive hypothesis, $Proc_{k+1}^{(m)}(P)$ is terminated, and the result
   is process P'. By Theorem 1, $[Proc_{k+1}^m(P)](Q \leftarrow Q_t^{(m)})(\mathcal{A} \leftarrow \mathcal{A}_t^{(m)})t =$
   $[P']Q'\mathcal{A}'t$ is terminated. Therefore, $Proc_{k+1}^{(m+1)}(P)$ is terminated.   ∎

Corollary 1. *Let P be a temporal process. There are finite actions in the
sequence in clock t.*

Corollary 2. *Let P be a temporal process. The messages are finite sent by P
in clock t.*

Theorem 1 guarantees that the calculus of temporal simulation in the simulation step is terminated in this way. Corollary 1 guarantees that actions are
finite in the simulation system, and that simulation can be done in finite time.

3   Software Design

It is the fact that the guard conditions need not to be ordered in constraint in
the simulation system of spacecraft, such as the flight control simulator in space
rendezvous and docking, so the model of temporal process can be simplified.
Many sequences of guard conditions can be simplified as a single set of guard
condition. To be clear and fluent in the realization logic of software, the sequence
of actions can split in many temporal processes by action departing rules. There-
fore, there is an effective min term of temporal process $P = \mathcal{C}_G. < t_a, a > .\overline{\mathcal{C}_A}$.
The program is departed from data by the temporal model running in software
and temporal processes defined in configure files. This section will describe the
implementation of the program and the definition of configure file.

The time series of simulation written in configure file is defined as follows:

$$< \text{guard conditions} >< \text{time} >< \text{action} >< \text{post conditions} >$$

When the *guard conditions* are satisfied, by passing *time*, action (such as
command/event executed) will be executed. And then, it reaches the final state
(post conditions) of execution. The temporal process is defined as the structure in Fig. 1. **PreConds** is the set of guard conditions, **PreCNo** is the count
in PreConds, **DelayTime** is the delay time, and K is the action, which can
be tele-control command, programmed instruction and self-defined command.
PostConds is the set of messages after actions are executed. In the first term,
the guard condition is *Touch*, delay time is $0\,\mathrm{s}$, the action is executing the self-
defined-command *TSN005*, and the message *TouchDone* will be sent. In the
six term, there are two guard conditions in **PreConds**. It describes that when

	PreConds			DelayTime	K		PostConds			
1	PreConds			0	TSN005		PostConds			
		PreCNo	1						PostCNo	1
		PreC	Touch						PostC	TouchDone
2	PreConds			8	TSN006		PostConds			
3	PreConds			120	TSN007		PostConds			
4	PreConds			30	TSN008		PostConds			
5	PreConds			60	TSN009		PostConds			
6	PreConds			5	TSN009		PostConds			
		PreCNo	2						PostCNo	1
		PreC {2}							PostC	Locked
			Abc Text							
			1 K1							
			2 K2							

Fig. 1. Configure file of the definitions of temporal processes

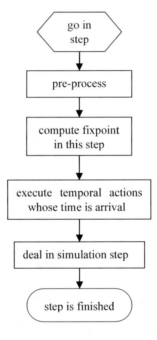

Fig. 2. The control flow graph of the computation in temporal process.

the tele-commands *K1* and *K2* are executed, the action self-defined-command *TSN009* is executed after 5 s delay. And then, the message *Locked* is sent.

The Fig. 2 gives the control flow graph of program of temporal actions in a step of simulation system. The pre-process of simulation step will save or resume the breakpoint of simulation, receive and execute the tele-commands, and execute the program commands whose time is arrival.

The time cost of the computation of the fixpoint is too high for real-time simulation, so we design the on-the-fly algorithm. In this algorithm, the fixpoint

Table 2. The algorithm of computation of fixpoint in simulation system.

Algorithm: Reduce

Input: Simulation time $Tsim$,

       the delay actions queue $\mathcal{A}$,

       new message set Q in the last simulation step.

Process:

  Loop: for each temporal process P do:

    Loop: for each message $m \in Q$ do:

1.   delete the message m from the set of guard conditions in P

2.   if $P = \emptyset. < t_a, a > .\overline{C_A}$, then

2.1. $\mathcal{A} \leftarrow \mathcal{A} \cup \{< Tsim + t_a, a > .\overline{C_A}\}$.

2.2.   set the set of guard conditions in P restored to the initial stat.

---

can be computed by the messages received in the simulation step. The messages received now can be delay to the next step to resolve. In this way, all the computations in one step can be finished in a small simulation step. The system has only a set of guard conditions, so the algorithm of the computation of fixpoint is designed as Table 2. This algorithm checks the set of guard conditions in every temporal process one by one. If the message is received between last simulation step and this one, it will be deleted from the set. When the set of guard conditions is empty, it means that all the preconditions of the action sequence are satisfied. So it is needed to compute the time when actions are executed, and add the last process to the action queue $\mathcal{A}$ in which actions will be executed at the suitable time. At the same time, the process P will be restored to the initial state so that the temporal actions can be executed in the running of software when guard conditions are satisfied again. Let the number of processes be $\#PSet = n$, the number of guard conditions in a process P be $\#C_P$, the total of guard conditions of processes be $m = \sum_{P \in PSet} \#C_P$, and the number of messages in Q be $\#Q = k$. This algorithm has two loops, and step 1 checks the union set of guard conditions in all $P \in PSet$. That means this step be $O(m * k)$. Step 2 costs $O(1)$, so the total cost of two loops is $O(n * k)$. Therefore, the cost of this algorithm is $O(m * k + n * k)$.

Because the temporal actions reached the execution time are executed after the calculus of fixpoint, all the messages produced between last simulation step and this one are used in the calculus of fixpoint. To receive new messages between this simulation step and next step, the new message set Q should be clear. And then, actions reached the execution time are executed, and messages sent are added in the new message set Q. The algorithm is presented in Table 3.

Let the number of the delay actions in the queue $\mathcal{A}$ be n. $\mathcal{A}$ is partitioned by Step 2 and 3, so the cost of time is $O(n)$. Step 4 (Loop) is executed less than n times, so this algorithm of time cost is $O(n)$.

Table 3. Algorithm of the execution of temporal sequence.

Algorithm: ExecTimeAction

Input: Simulation time $Tsim$,

the delay actions queue $\mathcal{A}$,

new message set Q in the last simulation step.

Process:

1. $Q \leftarrow \{\}$;

2. $A \leftarrow \{< a, \overline{C_A} > \mid < t_a, a > .\overline{C_A} \in \mathcal{A}$, and $t_a \leq Tsim\}$

3. $\mathcal{A} \leftarrow \mathcal{A} \setminus \{< t_a, a > .\overline{C_A} \in \mathcal{A} \mid t_a \leq Tsim\}$

4. Loop: for each action $a \in A$ do:

4.1   execute a

4.2 $Q = Q \cup \{a\} \cup \overline{C_A}$

4   Experiment

This paper implements these algorithms in Solaris operation system by C++. This section describes how to calculate the docking temporal sequence by these algorithms. Let the docking sequences in the automation and in the command be presented as follows. The column "time" represents the current time in simulation. The column "current message set Q" represents all the messages between last simulation step and current one. The column "process P after the algorithms executed" presents the process state after the algorithm are executed. The column "action set $\mathcal{A}$ after algorithm 1 executed" presents the set of actions after the algorithm 1 is executed. "Action set A waiting to be executed" presents the set in which actions produced by received messages will be executed. "action set $\mathcal{A}$ after algorithm 2 executed" presents the action set in which algorithm 2 is executed.

In the sequence 3, $T05$ is executed currently, and is added in current message set Q with message $TouchDone$. In the Sect. 2, the theoretical model computes the fixpoint iteratively. Temporal processes receive new messages generated and actions appeared at the sampling time frequently and resolve the guard conditions. And then, the fixpoint state of processes is reached. In Sect. 3 about the implementation of simulation system, current simulation step only uses the message set until time coming into step to achieve the goal of real time simulation and output results. New messages generated in this step will be used in the next step. Therefore, the action $T06$ executed and the message Locked sent in the sequence 5 couldn't be used in the resolution of guard conditions in current processes, but they take part in the resolution in the next step in sequence 6.

Table 4. The sequence of temporal process in real world business.

num	time	current message set Q	process P after the algorithms	action set A after algorithm 1 executed	action set A waiting to be executed	action set A after algorithm 2 executed.
1	0		$Touch.<0,T05>.TouchDone$ $TouchDone.<8,T06>.Locked$ $\{Cmd1,Cmd2\}.<5,T07>.Locked$			
2	...	...				...
3	t_0	$Touch$	$Touch.<0,T05>.TouchDone$ $TouchDone.<8,T06>.Locked$ $\{Cmd1,Cmd2\}.<5,T07>.Locked$	$<t_0,T05>$ $.TouchDone$	$<T05,$ $TouchDone>$	
4	t_0+t_{step}	$T05,$ $TouchDone$	$Touch.<0,T05>.TouchDone$ $TouchDone.<8,T06>.Locked$ $\{Cmd1,Cmd2\}.<5,T07>.Locked$	$<t_0+t_{step}+8,$ $T06>.Locked$		$<t_0+t_{step}+8,$ $T06>.Locked$
5	t_0+8 $+t_{step}$		$Touch.<0,T05>.TouchDone$ $TouchDone.<8,T06>.Locked$ $\{Cmd1,Cmd2\}.<5,T07>.Locked$	$<t_0+t_{step}+8,$ $T06>.Locked$	$<T06,Locked>$	
6	t_0+8+ $2*t_{step}$	$T06,$ $Locked$	$Touch.<0,T05>.TouchDone$ $TouchDone.<8,T06>.Locked$ $\{Cmd1,Cmd2\}.<5,T07>.Locked$			
7	...	...				...
8	t_1	$Cmd2$	$Touch.<0,T05>.TouchDone$ $TouchDone.<8,T06>.Locked$ $\{Cmd1\}.<5,T07>.Locked$			
9	t_2	$Cmd1$	$Touch.<0,T05>.TouchDone$ $TouchDone.<8,T06>.Locked$ $\{Cmd1,Cmd2\}.<5,T07>.Locked$	$<t_2+5,T07>$ $.Locked$		$<t_2+5,T07>$ $.Locked$
10	t_2+5 $+t_{step}$		$Touch.<0,T05>.TouchDone$ $TouchDone.<8,T06>.Locked$ $\{Cmd1,Cmd2\}.<5,T07>.Locked$		$<T07,Locked>$	

To use the set of processes repeatedly, the guard condition set will be restored to initial state like the step 2.2 of algorithm Reduce in software design when it is empty. The Table 4 above presents guard conditions restored by the way that conditions are added bottom color. Sequence 8 and 9 describe respectively that command $Cmd2$ received in clock t_1 and command $Cmd1$ received in clock t_2. And then, the action $T07$ is fired in clock t_2+5 and the message Locked is sent. Sequences from 3 to 6 describe the process simplified of two spacecrafts from touched to locked automatically. Sequences from 8 to 10 give the docking process of two spacecrafts by man-commands. From this example, this temporal model can simulate the temporal process and execution of space rendezvous and docking. When the sequence is changed, only configure file need to be modified for changing temporal sequence in simulation. In this method, both the efficiency of simulation in spacecraft mission and the maintains of temporal sequence are improved greatly.

5   Conclusion

The model of temporal process effectively separates the control flow in software from the temporal sequence, so that the software module can be applied in the multi-mission simulation system without modification, enhancing the universality of the software and reducing the software maintenance and testing work. With definition of different temporal actions in the configuration file, the simulation flexibility is increased and the efficiency in simulation preparation is improved. At the same time, the work in this paper promotes the software development work and makes software application more loosely coupled, so that any person who uses software can complete complex control in space simulation by configuring temporal actions without understanding the software development and design. The temporal sequence model given in this paper is not only suitable for the software in space simulation, but also suitable for other software with loose coupling of temporal actions and software control, such as the software in which there are more time series control cases. Future work will be directed to promoting the use of this temporal model to other timing control software systems.

References

1. Lippman, S.B., Lajoie, J.: C++ Primer, 4th edn. Addison-Wesley Professional, Boston (2005)
2. Cormen, T.H., Leiserson, C.E., Rivest, R.L., Stein, C.: Introduction to Algorithms, 3rd edn. The MIT Press, Cambridge (2009)
3. Milner, R.: Communication and Concurrency. Prentice-Hall, New Jersey (1989)
4. Boudol, G.: The π-calculus in direct style. High. Order Symbolic Comput. **11**(2), 177–208 (1998)
5. Kim, D., Ciesielski, S.K. et al.: Temporal parallel gate-level timing simulation. In: IEEE International High Level Design Validation & Test Workshop. IEEE (2008)
6. Haagensen, B., Maffeis, S., Phillips, I.: Matching Systems for Concurrent Calculi. Electron. Notes Theor. Comput. Sci. **194**(2), 85–99 (2008)
7. Fei, Z., Yong, Z., Liu, Y., et al.: A temporal and fuzzy logic inference based method for power system fault diagnosis. J. North China Electric Power Univ. **41**(1), 7–14 (2014)
8. Khan, A.I, Alam, M.M., Noor-ul-Qayyum, et al.: Validation of component based software development model using formal B-method. Int. J. Comput. Appl. **67**(9), 24–39 (2013)
9. Voorhees, B.H.: Time Series Simulation: Computational Analysis Of One-Dimensional Cellular Automata. World Scientific, pp. 208–219 (1995)
10. Tang, X., Juan, X., Duan, B.: A memory-efficient simulation method of Grover's search algorithm. Comput. Mater. Continua **57**(2), 307–319 (2018)
11. Liu, L., Li, W., Jia, H.: Method of time series similarity measurement based on dynamic time warping. Comput. Mater. Continua **57**(1), 97–106 (2018)
12. Vanegas-Useche, L.V., Abdel-Wahab, M.M., Parker, G.A.: Determination of the normal contact stiffness and integration time step for the finite element modeling of bristle-surface interaction. Comput. Mater. Continua **56**(1), 169–184 (2018)

Tuple Measure Model Based on CFI-Apriori Algorithm

Qing-Qing Wu[1], Xing-Shuo An[2(✉)], and Yan-yan Zhang[3]

[1] University of Science and Technology Beijing, Beijing 100083, People's Republic of China
[2] North China Institute of Computing Technology, Beijing 100083, People's Republic of China
axsdhh@163.com
[3] Beijing Economic and Technological Development Area, Beijing 100176, People's Republic of China

Abstract. Aiming at the loss of valuable information in the traditional Apriori algorithm. A new improved algorithm CFI-Apriori (Candidate frequent item sets-Apriori) is proposed to solve this problem, A tuple measurement model based on the algorithm is created. Compared with traditional Apriori algorithm, CFI-Apriori algorithm stores infrequent candidate item sets and creates two tuple measurement models based on probability weight and information entropy. The measurement models could screen out high-value target tuples to provide a basis for efficient use of data. Simulation results show that the improved algorithm and tuple measure model can utilize data efficiently, which is more efficient than traditional methods.

Keywords: CFI-Apriori algorithm · Tuple measurement model · Candidate sets

1 Introduction

In recent years, with the development of social network and the continuous innovation of information technology, network data presents a large number, complex structure, difficult to trace and other characteristics. Therefore, how to effectively dig out valuable information from massive data is the research focus in the era of big data [1]. Data mining is the process of automatically discovering useful information in large data stores. And from the original data mining hidden, useful, not yet discovered information and knowledge, to explore an effective data mining knowledge and rules [2]. Association rules are used to find out the relationship between data information in the process of data mining and to find the meaningful relationship hidden in large data sets. The most classical association rule algorithm is Apriori algorithm [3], The association rules of the algorithm are generated in iterative form. The traditional meaning is to discover the degree of association between transactions, so as to provide better decisions [4]. In order to improve the efficiency of Apriori algorithm, many literatures have also improved the algorithm. Such as using Database mapping, Transaction compression [5, 6], Boolean matrix, Maximum item mining [7, 8], Distributed [9], Plot number [10, 11], Recommendation system [12] etc. However, these improved algorithms are based on reducing the I/O overhead in the mining process, and do not consider how to effectively utilize

X. Sun et al. (Eds.): ICAIS 2020, LNCS 12239, pp. 49–61, 2020.
https://doi.org/10.1007/978-3-030-57884-8_5

the hidden value information of the infrequent item sets in the iterative process. How to effectively improve Apriori, extracting the hidden value in the data is a research topic worthy of attention [13]. Therefore, an improved Apriori algorithm is proposed to mine valuable infrequent itemsets in the dataset, and two tuple metric models are proposed to calculate the availability and integrity of the tuples [14].

The main contributions are as follows

1) The CFI-Apriori is proposed, which temporarily stores the infrequent candidate set generated by each iteration process, to mine the hidden data information.
2) A tuple measurement model based on the summation of probability weights is proposed to screen valuable data information, dig out the potential value of information, and increase the availability of data judgment.
3) A tuple measurement model based on information entropy is proposed to find out high-value target information, and complete the integrity data analysis.

2   CFI-Apriori Algorithm

2.1   Disadvantages of Apriori Algorithm

Apriori algorithm is a classic data mining algorithm proposed by RAgrawa [15]. The advantage is that it is easy to find associations between data. But the Apriori algorithm is pre-validated. and the process of finding frequent terms needs to follow two properties: 1). Connection properties 2). Pruning properties:

According to the first two properties of Apriori, Frequent item support is also required to meet the minimum support and other characteristics, which can lead to multiple traversal of the data table, resulting in the following problems:

1) In order to obtain the data set with frequent item N, Apriori algorithm needs to calculate the data table for N times. When the data sample size is large, the requirement for frequent item N is also large, so redundant candidate set information will be generated by multiple calculations of the data table.
2) Candidate set is generated in the iteration process of Apriori algorithm, which will occupy a large amount of memory. The classical Apriori algorithm does not take advantage of frequent candidate sets that do not meet the requirements, resulting in the loss of some information. This data mining process needs to be redone when the user has different needs and wants to work again.

2.2   CFI-Apriori Algorithm Working Process

The working process of CFI-Apriori algorithm is that the user obtains the data from the original database by using the mining idea and filters to obtain the sub-data table. Organize it into a transaction table for the CFI-Apriori algorithm. By setting rules such as minimum support and storage to obtain the potential value information of candidate set. The algorithm flow is as Fig 1.

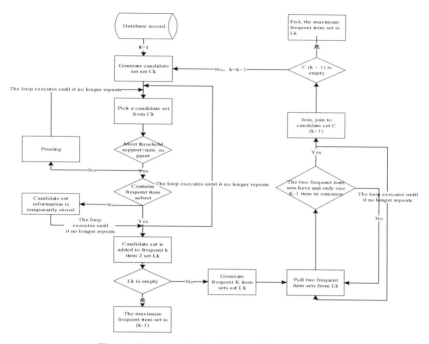

Fig. 1. CFI-Apriori algorithm working flow chart

CFI-Apriori algorithm:

1) Scan the source database and generate the transaction table corresponding to the data, in which the transaction table code TID represents the record of all the general series of behaviors during the transaction. ITEMS are used to represent a transaction record of all the specific details of an activity.

2) Set the minimum support, scan the transaction table, and set the number of elements in the item set $k = 1$.

3) Scan the transaction table, calculate the support degree of each transaction, and generate the set C_k of candidate frequent k item set. Select the candidate item set whose support degree is greater than or equal to the minimum support degree from the set C_k, and generate the set L_k of frequent k-item set. If L_k is empty, ends. If L_k is not null, execution continues. When $k > 1$, the set of infrequent K items less than the minimum support is stored in memory. When $k = 1$, traverse the candidate set C_1 and count its support, delete the candidate frequent 1 item set less than the minimum support, and find the frequent item L_1 satisfying the minimum support.

4) Two frequent item sets are taken from the set L_k of frequent item sets. If the two frequent item sets have and only $k - 1$ items are the same, the two frequent item sets are connected to obtain candidate frequent $(k + 1)$ item sets.

5) Repeat step 4) until the obtained candidate frequent $(k + 1)$ item set C_{k+1} is no longer repeated. If C_{k+1} is not empty, $k = k + 1$, and return the execution step (3).

The overall working diagram is as Fig. 2:

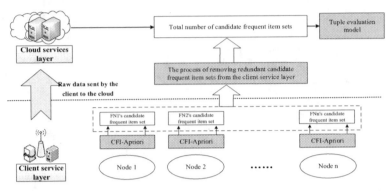

Fig. 2. Working diagram

When CFI Apriori algorithm is used to transfer frequent item sets and temporary non frequent item sets to Cloud server. Different item sets can be measured according to the value of data. Since the data value is related to the number of item sets, the basis of measurement is set by the order of magnitude of item sets and user requirements.

3 Establishment of Tuple Measurement Model Based on CFI-Apriori Algorithm

3.1 Tuple Measurement Model

In this paper, the measurement of data quality [15] refers to the availability and integrity of data. Different metrics can be used to derive the corresponding tuple metric model based on CFI-Apriori algorithm. It can measure the value of hidden data. The cloud server generates a corresponding transaction table according to the source data; each row in the transaction table represents a tuple, and the tuple is composed of transactions; The specific measurement process is shown in Fig. 3.

3.2 Tuple Measurement Model Based on Sum of Probability Weights

If the data set obtained by the CFI-Apriori algorithm is D, The number of frequent i items and infrequent i items in the transaction table tuple D_j is x_i. The ratio of x_i to the number of combinations of i items in the tuple D_j is the item set probability $p(x_i)$. Where D_j represents the jth row tuple, The transaction table tuple information set D_j is composed of different transaction item data. For example, $D_3 = \{A, B, C, E\}$. For example, $D_3 = \{A, B, C, E\}$, D_3 contains frequent itemsets and infrequent itemsets of different item levels, such as 2-item frequent set $\{A, B\}$ and 3-item infrequent set $\{A, C, E\}$. and x_i represents the number of i item sets. The probability of the item set is:

$$P(x_i) = \frac{x_i}{C_m^i}(i = 1 \cdots m) \tag{1}$$

m represents the number of transactions of the tuple, and C_m^i represents the number of combinations of tuple i items.

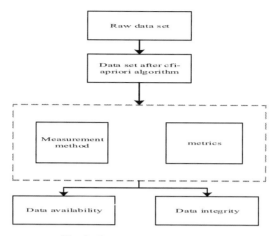

Fig. 3. Data measurement model

the mathematical probability matrix model A of data set D can be expressed as:

$$A = \begin{pmatrix} X \\ P(X) \end{pmatrix} = \begin{pmatrix} x_1 & x_2 & x_3 & \cdots & x_i & \cdots & x_n \\ p(x_1) & p(x_2) & p(x_3) & & p(x_i) & & p(x_n) \end{pmatrix} \tag{2}$$

Where n represents the number of transactions contained in the largest item set in tuple D_j.

Determining the weight $w(x_i)$ of x_i, where $w(x_i)$ is determined by $p(x_i)$ together with the user demand; Then the corresponding probability weight matrix B can be expressed as:

$$B = \begin{pmatrix} X \\ W(X) \end{pmatrix} = \begin{pmatrix} x_1 & x_2 & x_3 & \cdots & x_i & \cdots & x_n \\ w(x_1) & w(x_2) & w(x_3) & & w(x_i) & & w(n) \end{pmatrix} \tag{3}$$

Determining the value Q_j of the tuple D_j from $p(x_i)$ and $w(x_i)$. A tuple measurement model using weights and probabilities can be derived:

$$Q_j = \sum_{i=1}^{i=n} p(x_i) * w(x_i) \tag{4}$$

$$= A * B^T$$

Among them, Q_j represents the value of the tuple D_j, which is availability. When Q_j is larger, it is proved that the larger the target data contained in the tuple D_j, and the availability is higher. In practice, it may be determined whether the value Q_j is greater than a preset value threshold, and if so, the tuple D_j is highly utilized, and the tuple D_j is extracted. In addition, w_i can be adjusted according to the importance and demand of the data, thereby improving Q_j.

3.3 Tuple Measurement Model Based on Information Entropy

Information entropy is an indicator used to measure the uncertainty of a sample set [15]. This paper uses information entropy to evaluate the uncertainty of transaction table tuples, and then to judge the integrity of the data. That is to say that whether the data of the hidden value information of the infrequent item is utilized, and the degree of utilization. List a transaction table tuple, if there are many different items in the tuple with frequent items and infrequent items, such as frequent 2 items, infrequent 2 items, frequent 3 items, infrequent 3 items, the sample is rather messy and uncertain; If the sample set contains only frequent items, the sample is determined. The higher the uncertainty of information entropy, the more infrequent items the sample set contains and the higher the sample integrity.

Based on the information entropy-based tuple measurement model, the information entropy is introduced into the algorithm. The transaction tuple has various sets according to the CFI-Apriori algorithm rule, and each item in each set is regarded as an event, and the number of occurrences of an event is related to the probability. Frequent term probability is P_k The number of frequent k items in the transaction table tuple D_j is x_k. The ratio of x_k to the number of combinations of k items in the tuple D_j is the item set probability P_k. Where D_j represents the jth row tuple, x_i represents the number of i item sets, The probability of frequent items is:

$$P_k = \frac{x_k}{C_m^k}(k = 1 \cdots m)$$ (5)

m represents the number of transactions of the tuple, and C_m^k represents the number of combinations of tuple k items.

The transaction table tuple item set $X = \{x_1, x_2, x_3 \ldots x_n\}$, the probability of the frequent item set is $P = \{p_1, p_2, p_3, \ldots p_n,\}$, the frequent item sets of the transaction table tuple The amount of information for a transaction is:

$$I = -P_k \log_2 P_k$$ (6)

The information amount I of each set of tuples is known by (formula (4)), and the information entropy of the transaction tuple is judged by using Eq. (5).

$$\text{Ent}(D_j) = -\sum_{k=1}^{|y|} P_k \log_2 P_k$$ (7)

Where $|y|$ represents the number of frequent item sets in the tuple D_j, and $\text{Ent}(D_j)$ is the information entropy of the obtained tuple set. $\text{Ent}(D_j)$ is used to determine the uncertainty of each tuple in the transaction table. The smaller the entropy value, the lower the uncertainty of the tuple, that is to say that the less information of the infrequent item set. When the information entropy value is 0, it means that the tuple contains only frequent item sets and does not contain hidden information. Conversely, the higher the information entropy, the higher the uncertainty of the data, and the more infrequent item set information is contained. In practice, it is determined whether the information entropy is greater than a preset information entropy threshold, and if so, the uncertainty of the tuple D_j is large, and the tuple D_j is extracted.

4   Simulation

4.1   CFI-Apriori Instance Analysis

In order to better demonstrate the working process of CFI-Apriori, this section creates a transaction table D to demonstrate its verification process.

Database D is as shown in Table 1, setting minsuppprt = 2

Table 1. Transaction table

TID	Items
TD1	A, B, C, D
TD2	B, C, E
TD3	A, B, C, E
TD4	B, E

Scanning the transaction Table 1, according to the nature of the algorithm, the following table is obtained. Where table a represents the frequent candidate set obtained after the traversal, and b table is the frequent item set L_1 obtained according to the association rule.

Table a candidate 1 - item set

Items	support
{A}	2
{B}	4
{C}	3
{D}	1
{E}	3

Delete item for support<2 →

Table b frequent 1 - item set

Items	support
{A}	2
{B}	4
{C}	3
{E}	3

Fig. 4. Candidate 1-item set and frequent 1-item set

According to the CFI-Apriori algorithm rule, the frequent candidate sets that do not satisfy the minimum support degree are temporarily stored from the 2 item set, as shown in the Fig. 5:

Table a Frequently 1 - Item Set

Items	support
{A}	2
{B}	4
{C}	3
{D}	1
{E}	3

Connection gets 2 items set

⟶

Table b frequent 1-item set Find the 2 frequent

2 items collection
{A,B}
{A,C}
{A,E}
{B,C}
{B,E}
{C,E}

item sets and 2 temporary item sets are obtained from the candidate 2 item sets, as shown in the table :

Table c frequent 1-item set

2 items collection	support
{A,B}	2
{A,C}	2
{A,E}	1
{B,C}	2
{B,E}	3
{C,E}	2

Delete the item of Support<2

⟶

Temporary storage

Table d candidate 2-item set

2 items collection	Support
{A,B}	2
{A,C}	2
{B,C}	2
{B,E}	3
{C,E}	2

Table e Temporary storage of 2 infrequent items

2 items collection
{A,E}

Fig. 5. Candidate 2-item set and frequent 2-item set

According to the following method of frequent 3- items, the frequent n- item set and the n item temporary storage set are calculated successively.

4.2 Tuple Measurement Weight Setting and Case Analysis

The experiment takes a tuple measurement model based on the sum of probability weights as an example. In the example herein, The weight $w(x_i)$ of the item set is defined only according to the ratio of the item set probability value $p(x_i)$ to the total probability value of the transaction item tuple.

$$w(x_i) = \frac{P_i}{\sum_1^n P_i}(i = 1 \cdots n) \tag{8}$$

Take transaction Table 1 as an example. The specific operation process is as shown in the figure:

The probability of the transaction tuple set obtained by Eq. (1) is shown in Table 2. Take the TD1 transaction tuple as an example, Probability $p(x_i)$ of each item set is shown in Eq. (1). As can be seen from Figs. 4, 5 and 6, the TD1 tuple contains four elements, One of these sets is {A}, {B}, {C},So the probability of one term set is 3/4;The combination of two items in the TD1 tuple has $C_4^2 = 6$, of which the two items include, the two frequent sets {A, B}, {A, C}, {B, C}, there is no infrequent binomial set, Therefore,

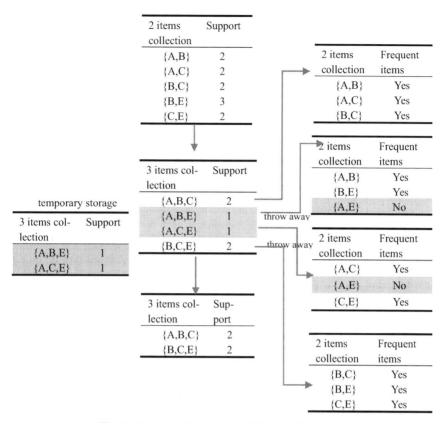

Fig. 6. Frequent 3-item set and infrequent 3-item set

the two items have a probability 1/2. The probability sets of the remaining transaction tuples are calculated accordingly.

Table 2. Transaction tuple probability

Tid	1 items collection $P(X_1)$	2 items collection $P(X_2)$	3 items collection $P(X_3)$
TD1	3/4	1/2	1/4
TD2	1	1	1
TD3	1	1	1
TD4	1	1	0

The weight calculation in this paper is shown in Eq. (8). For example, the set of the TD1 transaction tuple is 3/4, and the weight ratio of the TD1 transaction tuple is 1/2, According to this calculation, the probability weights of other transaction items can be obtained in Table 3.

Table 3. Transaction table weights

Weight	1 items collection $P(X_1)$	2 items collection $P(X_2)$	3 items collection $P(X_3)$
TD1	1/2	1/3	1/6
TD2	1/3	1/3	1/3
TD3	1/3	1/3	1/3
TD4	1/2	1/2	0

From Eq. (4), the probability weight summation number Q_j of the transaction tuple TD1 can be calculated to be 0.875.

4.3 Experimental Data and Analysis of Tuple Measurement Model Based on Sum of Probability Weights

On the one hand, the comparative analysis analyzes the tuple metric model based on the sum of probability weights in the CFI-Apriori and Apriori algorithm environments to obtain Q_i, and the corresponding probability value and weight of the Apriori algorithm are the same as those of the 4.2 section. On the other hand, compare data availability analysis in the case of different transaction tuples, determine the value of the transaction item tuple, where the frequent item minimum support is set to Support = 2.

The set probability $p(x_i)$ and the probability weight $w(x_i)$ calculated by the tuple metric model based on the sum of probability weights are obtained by Eqs. (1) and (8), As shown in Table 4 and Table 5:

Table 4. CFI-Apriori algorithm transaction table data

Tid	1 items collection $P(X_1)$	$w(x_1)$	2 items collection $P(X_2)$	$w(x_2)$	3 items collection $P(X_3)$	$w(x_3)$	Q_i
TD1	3/4	1/2	1/2	1/3	1/4	1/6	0.58
TD2	1	1/3	1	1/3	1	1/3	1
TD3	1	1/3	1	1/3	1	1/3	1
TD4	1	1/2	1	1/2	0	0	1

It can be seen from the experimental results in Fig. 7 that under the same transaction tuple, the sum of the probability weights of the CFI-Apriori algorithm is slightly higher than the sum of the probability weights of the Apriori algorithm. It shows that the CFI-Apriori algorithm has better measurement value, and the availability of the algorithm is higher, which makes the overall judgment more accurate. At the same time, by comparing the Q_i values, it can be seen that the tuple measurements based on the sum of the probability weights are larger in the CFI-Apriori algorithm, D_2, D_3, D_4 has a larger Q_i, and have higher target measurement values; In the Apriori algorithm, D_2 and D_4 a has a

Table 5. Apriori algorithm transaction table data

Tid	1 items collection P(X_1)	$w(x_1)$	2 items collection P(X_2)	$w(x_2)$	3 items collection P(X_3)	$w(x_3)$	Q_i
TD1	3/4	1/2	1/2	1/3	1/4	1/6	0.58
TD2	1	1/3	1	1/3	1	1/3	1
TD3	1	3/7	5/6	5/14	1/2	3/14	0.918
TD4	1	1/2	1	1/2	0	0	1

larger Q_i and have higher target measurement value, but CFI-Apriori has a much larger tuple metric value range and more valuable information.

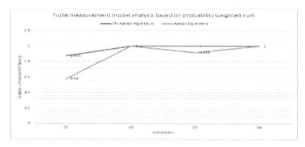

Fig. 7. Comparative analysis of tuple measurement models based on sum of probability weights

4.4 Simulation Analysis of Correlation Data Tuple Measurement Model Based on Information Entropy

The probabilistic value P_k calculation method of the tuple measurement model based on information entropy is shown in Eq. (5). Since the transaction tuple combination of the Apriori algorithm only contains frequent item sets, the information entropy measured by the measurement model is always 0, so the comparison between CFI-Apriori and Apriori cannot be performed. Finally, Table 6 is available.

Table 6. Transaction data of the CFI-Apriori algorithm

Tid	1 items collection	2 items collection	3 items collection	$Ent(D_j)$
TD1	3/4	1/2	1/4	1.31
TD2	1	1	1	0
TD3	1	5/6	1/2	0.719
TD4	1	1	0	0

From the experimental results in Fig. 7, it can be seen from the comparison of $Ent(D_j)$ values that the entropy-based tuple measurement model in the CFI-Apriori algorithm, D_1, D_3, has a larger $Ent(D_j)$, indicating higher uncertainty. It contains more infrequent item sets, has higher target value and is more comprehensive and complete (Fig. 8).

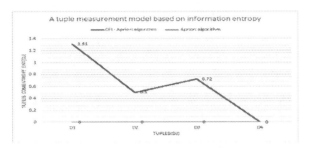

Fig. 8. Comparative analysis of tuple measurement models based on sum of probability weights

5   Conclusion

In this paper, the CFI-Apriori algorithm is proposed by Apriori algorithm. The algorithm can store the hidden valuable infrequent information. A tuple data metric model based on probability weight summation and a tuple metric model based on information entropy are established for the algorithm. The tuple measurement model based on the sum of probability weights is mainly to improve the availability of data, and the weight of the probability is added to set the overall judgment. The tuple measurement model based on information entropy is mainly to measure the uncertainty of the data, and consider the influence of more sudden factors or increased infrequent items on the overall data judgment. and more comprehensive judgment. And both measurement models can filter out high-value target tuples, providing a basis for efficient use of data.

References

1. Lin, H., Yang, S.: A cloud-based energy data mining information agent system based on big data analysis technology. Microelectron. Reliab. **97**, 66–78 (2019)
2. Jingtao, S., Lin, F., Zhou, X., Xing, L.: Steiner tree based optimal resource caching scheme in fog computing. China Commun. **12**(8), 161–168 (2015)
3. Xiao, D., Liang, J., Ma, Q., Xiang, Y., Zhang, Y.: High capacity data hiding in encrypted image based on compressive sensing for nonequivalent resources. Comput. Mater. Continua **58**(1), 1–13 (2019)
4. Xia, Z., Lu, L., Qiu, T., Shim, H.J., Chen, X., Jeon, B.: A privacy-preserving image retrieval based on AC-Coefficients and color histograms in cloud environment. Comput. Mater. Continua **58**(1), 27–43 (2019)
5. Yuan, X.: An improved Apriori algorithm for mining association rules. In: AIP Conference Proceedings. AIP Publishing (2017). 1820(1): 080005

6. Singh, H., Ram, D.J.S.: Sodhi improving efficiency of Apriori algorithm using transaction reduction. Int. J. Sci. Res. Publ. **3**(1), 1–4 (2013)
7. Yu, H., Wen, J., Wang, H., et al.: An improved Apriori algorithm based on the Boolean matrix and Hadoop. Procedia Eng. **15**, 1827–1831 (2011)
8. Narvekar, M., Syed, S.F.: An optimized algorithm for association rule mining using FP tree. Procedia Comput. Sci. **45**, 101–110 (2015)
9. Rathee, S., Kashyap, A.: Adaptive-Miner: an efficient distributed association rule mining algorithm on Spark. J. Big Data **5**(1), 6 (2018)
10. Park, J., Chen, M., Yu, P.: Using a hash-based method with transaction trimming for mining association rules. IEEE Trans. Knowl. Data Eng. **9**(5), 813–825 (1997)
11. Kuramochi, M., Karypis, G.: Frequent subgraph discovery. In: Proceedings 2001 IEEE International Conference on Data Mining, pp. 313–320. IEEE (2001)
12. AlZu'bi, S., Hawashin, B., ElBes, M., et al.: A novel recommender system based on Apriori algorithm for requirements engineering. In: 2018 Fifth International Conference on Social Networks Analysis, Management and Security (SNAMS), 323–327. IEEE (2018)
13. Lin, F., Zhou, Y., An, X., You, I., Choo, K.-K.R.: Fair resource allocation in an intrusion-detection system for edge computing: ensuring the security of internet of things devices. IEEE Consum. Electron. Mag. **7**(6), 45–50 (2018). https://doi.org/10.1109/MCE.2018.2851723
14. Liu, Yu., Yan, X., Liu, G., Hu, B., Yang, Z.: A novel multi-hop algorithm for wireless network with unevenly distributed nodes. Comput. Mater. Continua **58**(1), 79–100 (2019)
15. Lin, F., Zhou, Y., You, I., Lin, J., An, X., Lü, X.: Content recommendation algorithm for intelligent navigator in fog computing based IoT environment. IEEE Access **7**, 53677–53686 (2019)

Deep Learning Video Action Recognition Method Based on Key Frame Algorithm

Li Tan[⊠], Yanyan Song, Zihao Ma, Xinyue Lv, and Xu Dong

Beijing Technology and Business University, Beijing, China
tanli@th.btbu.edu.cn

Abstract. In order to solve the problem of action recognition in short video and capture the key information of video, this paper first proposes a KGAF-means method for key frame extraction. The KGAF-means method is based on the clustering principle and combines the K-means algorithm with the artificial fish swarm algorithm to realize the key frame sequence extraction. Based on the extracted key frame sequence, the RGB image and the optical flow image are separately extracted by the improved dual-stream variable convolution network. Then, using the cascading method, the image feature vector and the optical flow feature vector are fused to obtain the fused feature vector for action recognition. The selected data set is the Charades data set. The experimental results show that the mAP value of the method is 22.9 on the public dataset Charades. And the results show that the proposed method has better robustness than other network models and improves the short video action recognition effect.

Keywords: Key frame · K-means algorithm · Artificial fish swarm algorithm · Action recognition · Video understanding

1 Introduction

In recent years, with the improvement of China's comprehensive national strength and the maturity of Internet communication technologies, the desires and consumption demands of Internet users for Internet resources are increasing. The Internet and its functions have gradually replaced traditional letters and entertainment projects. Among them, video has become the most vivid way of transmitting Internet information because of the diversity of information. Among them, the chic product of short video is especially worse. Compared with the traditional video, the short video shortens the duration of the video, increases original and personalized content in video, and brings more experience to users in limited fragmentation time. Therefore, the short video related industry has developed rapidly in recent years. Behind the development of the short video industry, there is a huge amount of short video data and regulatory pressure on platform video regulators for short video resources.

The video content itself has the rich image, audio and text information. In the face of viewers with different needs, video can effectively present its diversified and distinctive features to meet the experience of different users. And the videos watched by the same

© Springer Nature Switzerland AG 2020
X. Sun et al. (Eds.): ICAIS 2020, LNCS 12239, pp. 62–73, 2020.
https://doi.org/10.1007/978-3-030-57884-8_6

user often contain a lot of information and resources related to the action of individual users. Then, as a video platform, the most important and most appealing content of the short video is extracted, and the accurate content and advertisement push are specifically targeted to the individual user. This not only helps to meet the user's viewing needs but also enhances the user's stickiness and improves the user's cohesiveness. At the same time, with the rapid development of the short video industry, the huge short video information is more or less mixed with all kinds of bad information, which is not only bad for the physical and mental development of young people but also will affect the social atmosphere to some extent. The traditional means of relying on manpower to supervise these video content is not only time-consuming and laborious, but also causes unnecessary troubles because the trial criteria for different information cannot be unified. At the same time, it is very difficult for manpower to accurately screen out key information from an increasingly large amount of video information. Then, if the key frames of the video can be effectively extracted, the video platform supervisor can easily retrieve the video information and grasp the main content, and the information delivery becomes simpler and more accurate. The video key frame extraction algorithm has a strong practical significance.

The short video key frame extraction technique refers to a technique for effectively replacing the original short video with a limited sequence of still images. This technology is not only the basis of video summarizing, video retrieval and video review, but also one of the key technologies of video information mining. It has an irreplaceable role in video resource processing. The excellent short video key frame extraction technology [1–4] can greatly reduce the computational complexity of massive video data processing, and also improve the efficiency of people to quickly identify and query information from video data, and avoid unnecessary human and material waste. The main contributions of this paper include the following sections:

1) In order to solve the problem that the K-means algorithm is sensitive to the initial value, and can initially determine the number of initial cluster centers, a KGAF-means clustering method that can automatically determine the number of categories and portray the key frames more deeply and accurately is proposed. The average gray value and image sequence number of the gray image of each frame of the video are taken as artificial fish, and the improved artificial fish swarm algorithm is used for pre-clustering processing, thereby obtaining a controllable K value range. Therefore the final key frame can be obtained because the best K value is selected by evaluation.

2) In order to measure the accuracy of the improved video key frame algorithm, this paper uses the improved dual-stream variable convolution network to extract the key frame sequences of the dataset and compare it with other network models to analyze its action recognition accuracy.

2 Related Work

2.1 Key Frame Extraction

The key frame extraction method based on cluster analysis is built on the deep learning clustering algorithm. By finding the specific similarity measure between key frames, the picture objects are grouped, and similar video frames are divided into the same cluster. Then, from each of the finalized clusters, a suitable frame is selected as a representative key frame according to a certain method. Liu and Hao [5] proposed a key frame extraction algorithm for an improved hierarchical clustering algorithm, which calculates the information entropy of image blocks as a measure of key frames in clusters. Furini et al. [6] used the color feature of the video frame as the difference of the inter-frame information during clustering to more accurately divide the video frame. Zhuang et al. [7] used the most classical k-means clustering algorithm to divide the frames in the shot as samples, and then selected the center of the final cluster as the key frame. This method is simple to calculate, can adapt to different data processing types, and has a good clustering effect. However, more adequate pre-processing of the raw data is required, and it is highly dependent on the initial predetermined accurate cluster center value. The clustering method mentioned above needs to determine the number of cluster centers in advance. To improve this problem, Wu et al. [8] proposed a method of using the word bag model to describe the visual features of a video. Thereafter, clustering of the video image frames is performed according to the visual features depicted to select key frames.

2.2   Action Recognition

With the great success of deep learning in the field of image processing [11, 13, 20–22], more and more people are extending the deep learning method to video applications. Video is essentially a 3D modality compared to an image, which contains appearance information and motion information. How to extract effective spatiotemporal features to represent video information has become the mainstream research direction. Simonyan et al. [18] proposed a classic two-stream network. A two-stream network divides video information into spatial and temporal information. The spatial information is obtained by a spatial stream with RGB input, and the time information is obtained by the stacked optical stream. The action recognition result is obtained through the late fusion of the dual stream information. However, the calculation of optical flow is quite time consuming and not suitable for real-time processing. Ullah et al. [19] proposes a new motion recognition method for processing video data by using a convolutional neural network and a deep bidirectional LSTM network. First, depth features are extracted from every six frames of the video, and then the sequence information between the frame features is learned using a deep two-way network. The method is capable of learning long-term sequences and processing lengthy videos by analyzing features of specific time intervals. In order to improve the representation of key frames, Zhao Hong et al. [17] proposed a key frame sequence optimization method and performed action recognition based on this sequence. Firstly, based on the 3D human skeleton feature, the K-means clustering algorithm is used to extract the key frames in the video action sequence. Then, the second optimization is performed according to the position in the sequence of the key frame to extract the optimal key frame, which solves the problem of key frame sequence redundancy in the traditional method. Finally, the video is identified by the Convolutional Neural Network (CNN) classifier according to the optimal key frame.

3 Method

The video key frame extraction method based on cluster analysis is based on the principle of the clustering algorithm. After the video is framed, each independent frame image is treated as a data object, and then clustering operations are performed by using different calculation indexes. Finally, the video frame closest to the cluster center theory is selected as the key frame. The method extracts video key frames by dividing the video frames and calculating the correlation of video frames. Compared with the existing key frame extraction methods [9, 12, 15, 16], the fidelity of this method is more excellent.

3.1 KGAF-Means Algorithm Model

The artificial fish swarm algorithm is a process of judging the overall data through individual data and then obtaining the overall optimal value. The ability of the algorithm to find the optimal solution does not converge to the local optimal value. At the same time, it can quickly obtain feasible solutions for various problems and has good robustness. The key frame extraction algorithm based on K-means is a process of clustering the central values of the entire image frame by clustering the individual image frames. The two algorithms are highly corresponding with each other. Therefore, the KGAF-means algorithm combining artificial fish swarm algorithm and K-means algorithm is proposed, which can effectively solve the problem of sensitive initial cluster center selection in K-means algorithm and obtain more accurate initial K value. On the other hand, it can improve the computational efficiency of the artificial fish swarm algorithm.

Artificial Fish Position Coding Structure. When the artificial fish swarm algorithm performs optimization processing on the fitness function in the visual field space, each artificial fish is regarded as a potential solution. For the key frame extraction process based on the K-means algorithm, the data object to be processed is each picture frame obtained by framing the video according to RGB encoding. Therefore, a mapping relationship can be constructed between the artificial fish and the picture frame of the video. Then, for each artificial fish position coding structure, by optically framing each picture of the video frame, after obtaining each picture of the video, the average gray level of the gray picture is used as the intermediate mapping amount. The average of the gray levels represents the overall brightness of the image. The artificial fish coding structure corresponding to each picture is a data group formed by the image sequence number and the average gray value of the corresponding image grayscale image.

Fitness Function. The fitness function is combined with the clustering algorithm, and the adjustment algorithm is in the following form:

$$f(x) = \frac{a}{\sum\limits_{i=1}^{k}\sum\limits_{j=1}^{n} \|C_i - X_i\|^2} \tag{1}$$

where a is the established function, C_i is the i-th cluster center, and X_i is the i-th picture data.

After the fitness function is adjusted, the square sum of the errors of the clusters can be calculated for the picture and the cluster center according to the initial picture clustering. The greater the fitness, the smaller the square of the clustering error of this artificial fish.

KGAF-Means Algorithm Implementation. The KGAF-means algorithm combined with K-means and artificial fish population is as follows:

1) Set the fish size, field of view, step size, maximum number of iterations, and congestion factor;
2) Initialize the fish population parameters, calculate the fitness function of all artificial fish, and record the optimal individual state;
3) Perform an evaluation function on the behavior of each artificial fish, and select the behavior function with the most fitness for execution;
4) Change the position of the artificial fish, calculate the fitness of the current position, and update the optimal state;
5) Determine whether the maximum number of iterations is reached or the optimization criteria are reached. If the judgment is valid, the number of categories is output. Otherwise, return to step 3);
6) The number of clusters of the output of the improved artificial fish swarm algorithm is used as the initial cluster center number K of the K-means algorithm;
7) For the number of cluster centers from 2 to K, the first K frame of the video key frame is selected as the initial cluster center, and the first clustering algorithm is performed to calculate the Euclidean distance between the cluster center and each data;
8) Dividing the picture frame object into classes belonging to the cluster center according to the distance;
9) Then calculate the cluster center in each cluster to perform a new cluster center update;
10) Determine whether the clustering center moves within the threshold or reaches the maximum number of iterations. If yes, save the result; otherwise, return to step 8) to continue execution;
11) Calculate the error mean of the cluster center obtained by different K values, select the K with the best optimization result as the final selection data, update the cluster center, and extract it as a key frame sequence;

The algorithm flow chart is shown in Fig. 1,

Evaluation Criteria. The recall rate R and the precision rate P are the evaluation criteria of the clustering result. The recall rate is the rate used to measure missing picture frames in the key frames. The precision rate is used to measure the accuracy of key frames result.

$$R = \frac{N_C}{N_C + N_M} \times 100\% \tag{2}$$

$$P = \frac{N_C}{N_C + N_F} \times 100\% \tag{3}$$

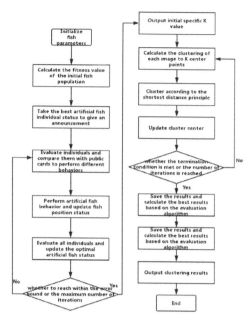

Fig. 1. KGAF-means algorithm flow chart

where N_C, N_M and N_F respectively represent the number of key frames belonging to the correct value in the video, the number of picture frames that are missed, and the number of picture frames that are mistaken for the key frame.

3.2 Dual-Stream Variable Convolution Network

In order to further verify the rationality of the key frame extraction of the algorithm, this paper uses VGGNet16 network as the basic network to construct a dual-stream variable convolution network to evaluate the rationality of the algorithm and analyze the effect of video action recognition.

Network Architecture. In the experimental part of constructing neural network, the classical two-stream network and variable convolution network are combined to extract the features of the key frame sequence RGB image and optical flow image [10, 14]. The image features and optical flow features are fused in a cascading manner. Action recognition is performed after the fused feature vector is obtained. The workflow is shown in the Fig. 2.

Action Feature Extraction. The feature extraction algorithm of dual-stream variable convolution network constructed in this paper is as follows,

1) Input the RGB feature key frame image of each video of the video dataset, and extract image feature vectors for each image;

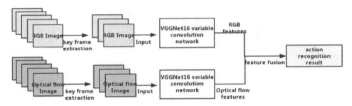

Fig. 2. Work flow chart

2) Calculate the image feature vector of the selected video dataset;
3) Enter the optical stream key frame image for each video of the video set, and extract the feature vector of the image for each image;
4) Calculate the optical flow eigenvector of the selected video dataset;
5) The image feature vector and the optical flow feature vector are fused by a cascading method. The fused feature vector is obtained by using the feature weighted fusion method. Let the weight of the image feature vector be W_i and the weight of the optical flow feature vector be W_f. Set each feature value to have a weight range of 0.1–0.9 and $W_i + W_f = 1$. Then the fusion feature calculation formula is:

$$MIX = W_i M_i + W_f M_f \qquad (4)$$

Where MIX denotes a fusion feature vector, and M_i and M_f are image feature vectors and optical flow feature vectors. W_i and W_f are the weights of the image feature vector and the optical flow feature vector.

4   Experimental Results

4.1   Experimental Parameter Setting

Key Frame Extraction Settings. In order to more intuitively reflect the clustering effect of the new algorithm, this paper extracts four animation videos from a platform to build a small video dataset. Each animation video is 3 min long, and the parameters of the artificial fish swarm algorithm are set as follows: $N = 60$, $step = 0.005$, $\delta = 8$, $visual = 0.001$, $try\_number = 40$, the maximum number of iterations is 100. In order to verify the validity of the proposed algorithm, the maximum number of iterations of the clustering algorithm is 300, and the KGAF-means algorithm is compared with the key frame extraction algorithm based on the difference between frames.

Action Recognition Parameter Settings. In this paper, 8561 videos in the standard data set of charade_v1 are selected as the source data set, and the RGB color frame-by-frame image and the optical stream frame-by-frame image of different 1000 videos are selected as the picture training set for the improved dual-stream variable convolution network training. At the same time, the picture of the key frame sequence obtained by the KGAF-means algorithm is converted into gz file as the data set for data input operation, and the fully connected layer feature of the VGGNet16 dual-stream variable convolution network is extracted as the feature vector for calculation. The recognition rate of the feature information of the key frame sequence image under the key frame extraction algorithm is analyzed.

Since the constructed neural network needs to fuse the image feature information and the optical flow feature information, this experiment sets different parameters to select the optimal weight coefficient. As known in Sect. 3.2, the image feature weight is W_i and the optical flow feature weight is W_f. For the selection of two parameter values, this paper chooses to compare the weights by preset weights. 10 videos were randomly selected from charade_v1 to make a training set and a test set. Different weight parameters were set, and classification training was carried out in the VGGNet16 dual-stream variable convolutional neural network model. The setting of each feature weight parameter was tested 20 times and the cross-validation method was adopted.

4.2 Experimental Results

The inter-frame difference method is performed on video 1, and the obtained key frame picture sequence is as follows: 20, 61, 97, 156, 234, 287, 356, 392, 457, 521, 554, 620, 654. The KGAF-Means algorithm is executed on video1, and the key frame picture sequence obtained is as follows: 32, 74, 192, 248, 289, 374, 402, 469, 545, 619, 678, 732. The specific visual effect chart comparison results are shown in the Fig. 3.

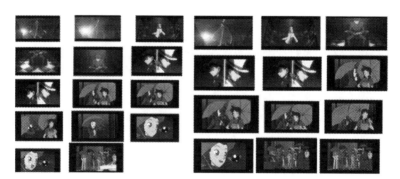

Fig. 3. Video1 key frame extraction effect comparison chart

The left side of Fig. 3 is the 14 key frames detected by the inter-frame difference extraction algorithm. The right side of Fig. 3 is the 12 key frames detected by the KGAF-Means algorithm. There are 2 frames of redundancy in the left sequence. According to the content of the original video, the scene in the animation is switched 6 times, and there are 7 scenes in total, wherein the inter-frame difference algorithm has redundant frames for the third scene and the sixth scene. From the above results, it can be seen that the KGAF-Means algorithm is more accurate for video expression content.

The inter-frame difference method is performed on video 2, and the obtained key frame picture sequence is as follows: 4, 10, 13, 19, 46, 213, 260, 354, 394, 423, 545,

630, 645, 674, 707. The KGAF-Means algorithm is executed on video2, and the key frame picture sequence obtained is as follows: 31, 150, 192, 248, 289, 374, 402, 469, 545, 619, 678, 732. The specific visual effects are compared as shown in the Fig. 4.

Fig. 4. Video2 key frame extraction effect comparison chart

The left side of Fig. 4 is the 15 key frames detected by the inter-frame difference extraction algorithm. The right side of Fig. 4 is the 14 key frames detected by the KGAF-means algorithm. There are 5 frames of redundancy in the left sequence and 2 frames of redundancy in the right sequence. It can be found from the content of the original video that the scene in the animation adds a large number of light and shadow changes. And the light change effect causes the image processed by the KGAF-means optical stream to be able to segment and distinguish the image well, and the inter-frame difference method is affected by the optical flow characteristics, resulting in a large redundant frame error.

After the key frame extraction algorithm is applied to the video of the four data sets, since the concept of key frames is not real, the key frame itself is only an abstract concept that people use to extract the main content of the video, so when compared with the inter-frame difference algorithm, the results of the KGAF-means algorithm and the inter-frame difference algorithm are used as evaluation criteria to calculate the number of missing frames and redundant frames. The statistical results are shown in Table 1.

Table 1. Comparison of different key frame algorithm extraction results.

Video	Frames	Key frames		Missed frame		Redundant frame	
		inter	KGAF	inter	KGAF	inter	KGAF
1	750	14	12	1	0	3	0
2	750	15	14	1	1	5	2
3	750	14	12	0	0	3	1
4	750	13	11	1	0	2	0

From the comparison results in Table 1, it can be known that the KGAF-means algorithm has a lower false detection rate and a higher fidelity rate. Compared with the inter-frame difference method, the KGAS-MEANS algorithm, which combines the artificial fish swarm algorithm and the K-MEANS algorithm, can express the main content of the video more completely and accurately.

After key frame extraction in the RGB image dataset, different depth network models are trained and tested to calculate the behavioral information recognition rate. The comparison results are shown in Table 2.

Table 2. Behavioral recognition results of RGB image datasets on different networks.

Method	TrainPrec1	TrainPrec5	ValPrec1	ValPrec5	mAP
VGG16	13.331	40.280	7.339	24.272	0.185
GoogLeNet	17.384	53.359	5.473	20.388	0.156
Our methods	**20.925**	**62.149**	**8.715**	**26.947**	**0.254**

After the key frame extraction is performed in the RGB image dataset and the optical flow image dataset, the feature weighted fusion method is used to fuse the RGB and flow features. Compared with other network models, the results are shown in Table 3.

Table 3. Comparison of action recognition results under different network models.

Method	Modality	mAP
2-Stream+LSTM	RGB+Flow	17.8
Asyn-TF	RGB+Flow	22.4
Our methods	RGB+Flow	**22.9**

It can be seen from the experimental results in Table 2 and Table 3 that the improved key frame extraction algorithm can effectively provide the feature information of the video. At the same time, based on the key frame sequence, the improved dual-stream variable convolution network can effectively improve the video action recognition result and has better robustness.

5   Conclusion

In this paper, the key frame extraction algorithm KGAF-means based on cluster analysis effectively realizes the capture of short video key information by integrating K-means algorithm and artificial fish swarm algorithm and greatly reduces the computational complexity of massive video data processing. Based on key frame extraction, the improved dual-stream variable convolution network improves the efficiency of short video action recognition. The experimental results show that the mAP value of the method is 22.9 on the public dataset Charades, which is higher than other network models, which proves the validity of the model.

Acknowledgments. This research was funded by Beijing Natural Science Foundation-Haidian Primitive Innovation Joint Fund grant number (L182007) and the National Natural Science Foundation of China grant number (61702020).

References

1. Jiang, P., Qin, X.L.: Adaptive video key frame extraction based on visual attention model. J. Image Graph. **14**(8), 1650–1655 (2009)
2. Liu, T., Zhang, H.J., Qi, F.: A novel video key-frame-extraction algorithm based on perceived motion energymodel. IEEE Trans. Circuits Syst. Video Technol. **13**(10), 1006–1013 (2003)
3. Zhang, H.J., Wu, J., Zhong, D., et al.: An integrated system for content-based video retrieval and browsing. Pattern Recogn. **30**(4), 643–658 (1997)
4. Lai, J.L., Yi, Y.: Key frame extraction based on visual attention model. J. Vis. Commun. Image Represent. **23**(1), 114–125 (2012)
5. Liu, H., Hao, H.: Key frame extraction based on improved hierarchical clustering algorithm. In: 11th International Conference on IEEE Fuzzy Systems and Knowledge Discovery (FSKD), pp. 793–797 (2014)
6. Furini, M., Geraci, F., Montangero, M., et al.: STIMO: still and moving video storyboard for the web scenario. Multimedia Tools Appl. **46**(1), 47 (2010)
7. Zhuang, Y., Rui, Y., Huang, T.S., et al.: Adaptive key frame extraction using unsupervised clustering. In: 1998 International Conference on Image Processing, ICIP 98. Proceedings, vol. 1, pp. 866–870. IEEE (1998)
8. Wu, J., Zhong, S., Jiang, J., et al.: A novel clustering method for static video summarization. Multimedia Tools Appl. **76**(7), 9625–9641 (2017)
9. Hannane, R., Elboushaki, A., Afdel, K., et al.: An efficient method for video shot boundary detection and keyframe extraction using SIFT-point distribution histogram. Int. J. Multimedia Inf. Retrieval **5**(2), 89–104 (2016)
10. Heidari, S., Abutalib, M.M., Alkhambashi, M., Farouk, A., Naseri, M.: A new general model for quantum image histogram (QIH). Quantum Inf. Process. **18**(6), 175 (2019)
11. Hao, H.Q., Wang, Y.L.: Moving Target Detection Based on Interframe Difference and Pyramid Optical Flow Method[J]. Video Engineering **40**(07), 134–138 (2016)
12. Liu, T., Zhang, H.J., Qi, F.: A novel video key-frame-extraction algorithm based on perceived motion energy model. IEEE Trans. Circuits Syst. Video Technol. **13**(10), 1006–1013 (2006)
13. Nedzvedz, O.V., Ablameyko, S.V., Gurevich, I.B., Yashina, V.V.: A new method for automating the investigation of stem cell populations based on the analysis of the integral optical flow of a video sequence. Pattern Recogn. Image Anal. **27**(3), 599–609 (2017). https://doi.org/10.1134/S1054661817030221
14. Verma, N.K., Singh, S.: Generation of future image frames using optical flow (2013)
15. Chakraborty, S., Tickoo, O., Iyer, R.: Adaptive keyframe selection for video summarization. In: 2015 IEEE Winter Conference on Applications of Computer Vision (WACV), pp. 702–709. IEEE (2015)
16. Li, C., Wu, Y., Yu, S.S., et al.: Motion-focusing key frame extraction and video summarization for lane surveillance system. In: Proceedings of IEEE International Conference on Image Processing, pp. 4329–4332 (2010)

17. Zhao, H., Xuan, S.B.: Key frame optimization and behavior recognition of human motion video. J. Graph. **39**(03), 463–469 (2018)
18. Simonyan, K., Zisserman, A.: Two-stream convolutional networks for action recognition in videos (2014)
19. Ullah, A., Ahmad, J., Muhammad, K., et al.: Action recognition in video sequences using deep bi-directional LSTM with CNN features. IEEE Access **99**, 1 (2017)
20. Shengting, W., Liu, Y., Wang, J., Li, Q.: Sentiment analysis method based on kmeans and online transfer learning. Comput. Mater. Continua **60**(3), 1207–1222 (2019)
21. Ling, T., Chong, L., Jingming, X., Jun, C.: Application of self-organizing feature map neural network based on k-means clustering in network intrusion detection. Comput. Mater. Continua **61**(1), 275–288 (2019)
22. Xianyu, W., Luo, C., Zhang, Q., Zhou, J., Yang, H., Li, Y.: Text detection and recognition for natural scene images using deep convolutional neural networks. Comput. Mater. Continua **61**(1), 289–300 (2019)

Optimization and Deployment of Vehicle Trajectory Prediction Scheme Based on Real-Time ANPR Traffic Big Data

Zhe Long and Zuping Zhang[✉]

School of Computer Science and Engineering, Central South University, Changsha 410083, Hunan, People's Republic of China
zpzhang@csu.edu.cn

Abstract. Path prediction is an important issue in traffic management. In big data environment, the real-time ANPR (Automatic Number Plate Recognition) data can provide evident support for path prediction. Taking advantage of the real-time ANPR data, this study adopts and modifies a trajectory prediction model based on order-k Markov model, calculating a supplementary probability matrix to compensate for the deficiency of the prediction model by exploring the hot spot area. This model produces a five-step prediction procedure which takes at most 8 s, exhibiting the efficiency of data processing. Meanwhile, the model shows the stability when dealing with updating data which has a 20 million growth on a daily basis, and displays the compatibility in dynamically accommodating itself into the periodic trajectory changes. By implementing the deployment mode of prediction algorithm with real-time big data, the fruits of this study can meet the requirement of the traffic management. Although an improved scheme for random trajectory prediction is added to the model, the accuracy of random trajectory prediction is still far from perfect. Therefore, the study will further optimize the prediction accuracy of the order-k Markov model by adding more factors into it and upgrading it into a higher order one.

Keywords: Path prediction · Order-k Markov model · Real-time ANPR data · Traffic big data

1 Introduction

With the gradual improvement of traffic pavement facilities, trajectory prediction has played an increasingly significant role in traffic management. Trajectory Pattern Tree (TPT) schemes [1], partial periodic pattern data-mining [2, 3] and Machine Learning (ML) schemes [4] have an important impact in this field.

Generally, these schemes consist of two steps, namely processing and querying. However, with the increase of tracking data, the ones mentioned above show their defects in various aspects, such as time complexity, spatial complexity, and timeliness. Actually, in the real production environment, there are three steps to achieve these algorithms, namely reading, processing and querying. The extra one reading means to prepare the

© Springer Nature Switzerland AG 2020
X. Sun et al. (Eds.): ICAIS 2020, LNCS 12239, pp. 74–85, 2020.
https://doi.org/10.1007/978-3-030-57884-8_7

data source or query from the dataset. In this paper, we use real-time ANPR [5] data as an experimental source and take the Order-k Markov Model for prediction.

Section 2 introduces the previous research in path prediction and the application of the Markov model in traffic management. Section 3 analyses the real-time ANPR data that has made tradition schemes worked unsatisfactorily. Section 4 describes the major issues and business needs of the management department, and then puts forward an Order-k Markov Model to deal with the singular trajectory problem, proposing a feasible solution to deal with real-time traffic big data.

2 Related Work

2.1 Path Prediction Schemes

The trajectory prediction method has undergone tremendous changes owing to the development of computer technology and hardware technology. This development is driven by the selection of different data sources, the existence of road network structure and the proposal of a new algorithm.

Earlier algorithms were mainly based on the moving state of the current moving target, such as speed, direction, and timestamp, simulating the moving position of the target in the next moment through mathematical functions. For example, Tao et al. [6] fitted a matrix function in two-dimensional plane coordinates to predict the moving trend of moving objects. The recursive motion function (RMF) takes into account the recent historical moving position of moving objects and has a good prediction accuracy for smooth moving, but the problem is also obvious, that is, it cannot predict the sudden change of direction. This problem is a typical one involved in predicting trajectory by mathematical function simulation, which is called the fork dilemma, that is, the direction choice at this time depends more on the driver than on the vehicle status.

In order to alleviate the effect of fork dilemma, Jeung et al. [1] proposed an improved scheme and a hybrid prediction model. When the moving object satisfies the motion conditions, this model uses accurate motion mode prediction, and when it is not satisfied, the model uses motion function completion. However, this method still does not solve the prediction problem when the direction of the motion function changes abruptly.

A method called frequent pattern mining was used in trajectory prediction by comparing all historical trajectories. Take Brilingaitė et al. [7] as an instance. Their method has a good effect on periodic trajectories, but the disadvantage is that trajectory comparison takes too much time when the data set increase. In 2007, Kim et al. [8] proposed a trajectory prediction method for known destinations, but in reality, the destination is unknown.

Machine learning method [4] is also attempted to be used for trajectory prediction. The algorithm divides the historical trajectory into training tuples and uses decision tree for training. It has good accuracy when there is a small amount of data. However, when the amount of data increases, especially when the number of prediction points increases, the parameters of the learning model need to be adjusted accordingly.

2.2 Markov Model

Markov model [9] is a statistical model, based on the relations of data. A Markov model must satisfy the existence of state transition chains between data. This chain is the so-called Markov chain and consists of a sequence of random variables (States). And the current state is only related to the most recent state in the historical state which is expressed as:

$$P(s_{x+1}|s_x \ldots s_1) = P(s_{x+1}|s_x) \tag{1}$$

This model is widely used in many natural language processing applications such as speech recognition, automatic part-of-speech tagging, and probabilistic grammar. In traffic problems, Markov chain also has effective applications, such as traffic speed forecasting [10], and traffic accident warning [11].

In the aspect of trajectory prediction, Markov model also has good performance. Yu et al. [12] proposed a hybrid Markov model to solve the problem that the probability matrix of order-k Markov model was too large. On the premise of ensuring the accuracy of prediction, the storage space of the probability matrix was optimized. Hasan et al. [13] presented a probabilistic modeling approach to predict human next activity time and location. But like many of these experiments, the data set used in this experiment contained fewer locations.

2.3 Existing Problems

In the previous work, the optimization of the model mainly focuses on the prediction accuracy, time complexity, and spatial complexity, and has achieved good results. But there are still several problems that need to be considered.

1. Fork dilemma is the common problem in the method of predicting trajectory by mathematical function simulation. Considering only the current state of the vehicle, it cannot reflect the purpose of the vehicle, that is, it cannot predict whether the vehicle turns at the intersection.
2. The real vehicle trajectory data sets are quite large. The traditional model has achieved considerable benefits in the accuracy of modeling a small number of vehicles in the experimental environment. But in reality, the number of vehicles is millions and the number of trajectories is billions, so trajectory prediction in an acceptable time has become a problem worthy of study.
3. In related work, data sets are static, which have not been expanded as time goes by. Most of the algorithms are validated by experiments on static data sets, resulting in that these algorithms may be incompatible with dynamic data sets.

Markov model has a simple structure and a fairly fast way of statistical calculation because its statistical calculation method is suitable for parallel computing in the big data environment. In this paper, a prediction scheme of the singular trajectory using Markov model is proposed, and a scheme of maintaining the probability matrix of Markov model in the big data environment is put forward.

3 Data Source Analysis

3.1 ANPR Data Source

In various traffic pavement data, ANPR data are the main data source of traffic big data and widely used in traffic management because of the convenient installation of camera sets and the compatibility of roads. Other data source such as Bluetooth detection data, GPS location data [14] and VANET [15] have a common disadvantage that not all the vehicles have the discoverable Bluetooth device or GPS device. Therefore, these data cannot fully reflect traffic conditions and vehicles behavior.

Table 1 gives some records of ANPR data that are captured and recognized by the cameras set in the intersections. Every record contains the number plate (NP), the plate color (PC), intersection and timestamp, which can be expressed as r = {NP, PC, Intersection, timestamp}. The encrypted NP and PC show the message of a unique recognized vehicle, while the NP is marked as 'Unidentified' to represent the unrecognized one. The intersection in the record combines location code with direction. The timestamp is the real time of vehicles passing through the intersection.

Table 1. ANPR data format (with Unidentified)

NP	PC	Intersection	Timestamp
AF**12	02	0123****8902_4	2018-**-27 00:00:37
AB**M3	02	0123****8902_4	2018-**-27 00:02:28
AA**29	01	0980****1050_1	2018-**-27 09:12:56
Unidentified	41	0980****2050_2	2018-**-27 10:08:56

3.2 Quantity and Quality of Data Set

The ANPR data are generated from the photography and recognition by the intersection camera facility. Generally, the data are captured from the camera, identified by the facility, transmitted to the relational database (Oracle), and finally extracted to the big data computing platform, which takes only 11 s. At least two devices have been installed on the entrance of each intersection to record the data from at least two directions. At present, 4886 sets of such equipment have been deployed in Changsha, covering 1573 intersections. And it is planned to raise the coverage in the foreseeable future. In such a large equipment base, records are generated every minute and every second and stored in the database. The quantity of total and incremental ANPR records is quite large. For example, in Fig. 1, the data have been selected for 6 months (from July 1st, 2018 to January 1st, 2019) as below. It shows that the data increase by 23 million per day and 43 billion in 6 months.

Figure 2 shows the number of vehicles of ANPR records in one month. About 90% vehicles have less than 748 times of passing records and the average number of

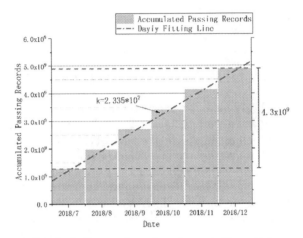

Fig. 1. Daily growth number of records (6 months)

these records is 168. In Fig. 2, we exclude some inactive vehicles of which the number of records is less than 10 times. These excluded records generally come from false recognition. In fact, the number of false recognition records for single vehicle is small, but the vehicles in this kind are of large quantity.

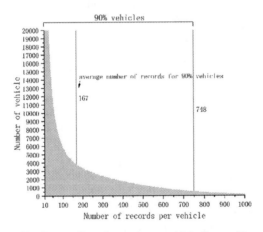

Fig. 2. Number of records per vehicle (1 month)

Through deep data retrieval, we find more recognition errors in the data. In March 2019, the system successfully identified 9.6 million different number plates, while the current number of registered vehicles in Changsha is only 3.4 million. Even it covers foreign vehicles, the amount of recognition data is seriously inconsistent with the facts. After further retrieval, 4.4 million of the 9.6 million identified plates had been identified only once by the system. Careful comparisons show that most of the causes of false recognition are due to light problems, leading to system errors in recognizing the body structure or body advertisements as license plates. Therefore, when choosing algorithms and models, we must have considerable fault tolerance to deal with this kind of data.

4 Prediction Model with Real-Time Big Data

4.1 Order-K Markov Model

It can be assumed that a vehicle trajectory is like $L = c_1 c_2 \ldots c_n$. We define that this vehicle has just pass a cross $c_x = \{Intersection, timestamp\}$ which will be a random variable so called current state in the Markov chain. So, Markov chain assumption will be:

$$P(c_{x+1}|c_x \ldots c_i) = P(c_{x+1}|c_x) \tag{2}$$

But vehicle activity regularity often has its subjective destination and follows individual driving habits and periodic activity regularity. Order-k Markov chain [16] will be more suitable for this situation. There will be a pre-trajectory $L_{pre} = c_1 c_2 \ldots c_x$. Then:

$$P(c_{x+1}|c_x \ldots c_1) = P(c_{x+1}|c_x \ldots c_k) \tag{3}$$

Obviously, we can get the transition probability from a lot of statistics. Among them, $N_{c_{x+1} c_x \ldots c_k}$ is the number of trajectories.

$$P(c_{x+1}|c_x \ldots c_k) = \frac{N_{c_{x+1}c_x \ldots c_k}}{\sum_{i=1}^{n} N_{c_i c_x \ldots c_k}} \tag{4}$$

The predicted result will be c_X:

$$P(c_X|c_x \ldots c_k) = \max_{i=1..n}(P(c_i|c_x \ldots c_k)) \tag{5}$$

Actually, the second and third large transition probabilities are not important, so we can only keep the max result in the matrix which we need and called probability result matrix.

$$P_{max}(c_x \ldots c_k) = c_X \tag{6}$$

In practice, for every car, we need to collect the record and build this model based on the data.

$$P_{max}(NP, c_x \ldots c_k) = c_X \tag{7}$$

According to this model above we can predict every vehicle's periodic activity regularity except random activities.

4.2 Model Improvement

Random trajectory behavior has always been a difficult problem in the absence of suffi-cient information. Random trajectory occurs in the following two situations. One is that the current trajectory of a vehicle can partly match the historical trajectory of it and will finally arrive at a historical place. The other is that the current trajectory of this vehicle cannot be matched its historical trajectory, and the next path selection is related to the road traffic condition and the driver's purpose.

For situation one, we have partially improved the Markov model. We collect the intersections c_{hs} around the key business districts, enterprises and institutions in the city as hot spots, take the stop points as trajectories breakpoints and use the trajectories covering these hot spots as data sources for calculating the Markov probability matrix. Generally, hot spots are trajectory destinations in the above trajectories. In the model calculation (4) and (5), $L_{pre} = c_1 c_2 \ldots c_{hs} \ldots c_x$. Then get the result matrix blow.

$$P_{max}(HS, c_x \ldots c_k) = c_X \tag{8}$$

For situation two, we build another matrix with all the records called total transition probability matrix to cope with this unique situation. We count every record as $N_{c_{x+1} c_x \ldots c_k}$. It means every car in this intersection will make the same choice.

$$P_{max}(ALL, c_x \ldots c_k) = c_X \tag{9}$$

This method dealing with two situations cannot ideally solve the problem of random destination, but it can approach the problem of the random route under the condition of periodic destination, such as the routes of taxi and private cars traveling at random place to important hub sites, which can get better prediction results.

4.3 Model Maintenance

For the real-time response, we can only use offline modeling and build the querying model as simple and fast as possible. Therefore, based on the order-k Markov model, this paper proposes a feasible modeling scheme, the order-2 Markov model, coping with this situation. Using the order-2 Markov model to predict the trajectory requires two latest vehicle's passing records. In order to increase the querying speed of historical trajectories, we maintain a short-term historical trajectory table of vehicles and keep the latest two passing records. As Fig. 3 shows, when there is a prediction demand, we will query the short-term historical trajectory table instead of the full trajectories.

Considering the parking state, when the interval between two trajectories exceeds 30 min, a virtual parking point is inserted to the trajectory and participates in the construction of the Markov probability matrix.

The main problem of offline modeling is how to update the order-2 Markov proba-bility matrix efficiently. In the actual trajectory prediction process, we will use the clear license plate number as the input parameter.

In this model, the original trajectory data is stored in Oracle and converted to Hive in real time and the calculation is processed in Hive. Initially, we do the initialization and set $k = 2$.

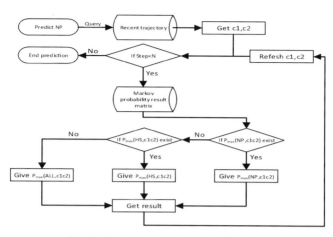

Fig. 3. Querying and prediction flow chart

① Query data $c_1 c_2 \dots c_n$
② If $t_i - t_{i-1} > 30min$ then add a parking point c_r, so $L = c_1 c_2 \dots c_{i-1} c_r c_i \dots c_n$
③ Count $N_{c_{x+1} c_x \dots c_k}$ and store in Hive
④ Calculate $P(c_{x+1}|c_x \dots c_k)$ and store in Hive
⑤ Extract $P_{max}(NP, c_x \dots c_k)$ and $P_{max}(ALL, c_x \dots c_k)$ and store in HBase

By taking step 1, we can accelerate the querying speed via using the analysis window function Lead in Hive [17]. Step 2 is to add a parking point for actual prediction demand. In step 3, 4 and 5, when we set $k = 2$ and calculate six-months vehicle trajectory records, the final result takes up 138G of space in Hive and the calculate process takes about 15 min.

The main problem in processing real-time data is how to calculate and maintain the probability matrix effectively. As mentioned above, the daily data increment is about 23 million. Here we use hierarchical and step-by-step calculation, updating the values of different layers and reducing the amount of data in the final results to the largest extent. The final result is hashed and stored in HBase [18].

① Query new daily data $c_1 c_2 \dots c_n$
② If $t_i - t_{i-1} > 30min$ then add a parking point c_r, so $L = c_1 c_2 \dots c_{i-1} c_r c_i \dots c_n$
③ Count $N'_{c_{x+1} c_x \dots c_k}$ and store in Hive as temporary data
④ Update $N_{c_{x+1} c_x \dots c_k} = N_{c_{x+1} c_x \dots c_k} + N'_{c_{x+1} c_x \dots c_k}$ and recalculate $P(c_{x+1}|c_x \dots c_k)$
⑤ If $P(c_{x+1}|c_x \dots c_k) > P(P_{max}(NP, c_x \dots c_k)|c_x \dots c_k)$ then update $P_{max}(NP, c_x \dots c_k) = c_{x+1}$

Inserting a full probability matrix into HBase is time-consuming, but the matrix updates fairly fast. In the final step, we set up a Restful service to respond to the predicted requirements. For a five-step prediction requirement, this model will take about 8 s to produce a complete prediction path.

4.4 Model Implementation Results

We randomly select a private car whose NP is AD**40. Its historical trajectories are showed in Fig. 4, within which some intersections are not marked. L1 and L2 are its historical trajectories, while L1 is its periodic trajectory. They are all counted in the probability matrix, and the crosses on L1 will be counted in the probability result matrix. Here this vehicle has a singular trajectory L3, and this trajectory has never been counted in the probability result matrix $P_{max}(\text{AD} * * 40, c_x \ldots c_k)$. The prediction results are shown in Table 2.

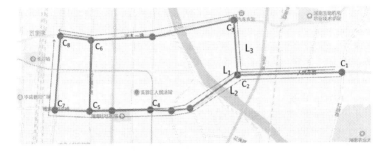

Fig. 4. Historical trajectories of a private car

Table 2. Some prediction result

Current trajectory	P_{NP}	P_{HS}	P_{ALL}	Result
C_2C_1	C_4	C_4	C_4	C_4
C_5C_4	C_6	C_7	C_7	C_6
C_3C_2	Null	C_5	C_5	C_5
C_6C_3	Null	C_8	C_8	C_8
C_3C_{stop}	Null	Null	C_5	C_5
C_8C_6	C_{stop}	C_{stop}	Not in Fig. 4	C_{stop}

In Table 2, the trajectory C_2C_1, C_5C_4 are periodic trajectory, so the model gets the result $P_{max}(\text{AD} * * 40, c_2c_1) = c_4$, $P_{max}(\text{AD} * * 40, c_5c_4) = c_6$. When there is a singular trajectory C_3C_2, the model gives C_5 as a result because $P_{max}(\text{AD} * * 40, c_3c_2) = \text{null}$ and $P_{max}(\text{HS}, c_3c_2) = c_5$. If this vehicle drives to C_3 and stopped, its current trajectory is C_3C_{stop}, then the model will give C_5 as a result, because $P_{max}(\text{AD} * * 40, c_3c_{stop}) = \text{null}$ and $P_{max}(\text{HS}, c_3c_{stop}) = \text{null}$.

This model also has some fault tolerance. If C_2 equipment loses the trajectory data, the current trajectory will be C_3C_1, the model will give C_6, because C_3 and C_6 are on the trunk road, and there are many trajectories counted in $P_{max}(\text{ALL}, c_3c_1)$.

In order to test the accuracy of the model, we randomly select 1000 vehicles, and 5 trajectories whose length is longer than 10 for each vehicle. Altogether the ANPR data we select include 73 279 records. The prediction process and prediction accuracy results show in Fig. 5 and Fig. 6

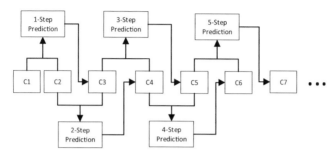

Fig. 5. Prediction process

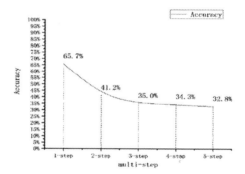

Fig. 6. Multi-step prediction accuracy

Figure 6 shows that the prediction accuracy is unsatisfactory compared with previous studies, because the real-time ANPR data have a lot of recognition errors including wrong recognition, missed capture, delayed submission and so on. These will generate untrusted path which will accelerate accumulation error. Although the model has some fault tolerance, these errors transform a large number of periodic trajectories into singular trajectories, which makes the prediction more difficult. However, it can also be seen from Fig. 6 that the model has exhibited a stable status as for the accuracy in predicting the periodic trajectories, because there is little cumulative prediction error and the prediction accuracy tends to be stable with the progress of multi-step prediction.

5 Conclusion

Compared with other related works using GPS data for the experiment, the main work of this paper is to design and implement a trajectory prediction model based on order-k Markov model by using real-time ANPR data in a real production environment, and considering the actual needs of the traffic management department.

Processing massive trajectory data is the focus of this paper. From November 2018 to April 2019, the model has been employing about 2 billion trajectory data for the latest three months. Moreover, the model maintains the data in the probability result matrix which occupies 138G of space. Under these conditions, a five-step prediction procedure will take only 8 s, exhibiting the efficiency of data processing. At the same time, the model can still deal with updating data which has a 20 million growth on a daily basis, showing the stability of this model. For the vehicles with periodic trajectory mode, such as commuter buses, private cars during rush hours, the trajectory prediction accuracy can meet the requirements of traffic management. And when the periodic trajectory changes, the corresponding prediction matrix can be dynamically adjusted, displaying the compatibility of this model.

The model proposed in this paper is mainly devoted to dealing with real-time data modeling and continuous data maintenance. Considering the huge growth of traffic passing records, a high-order Markov prediction matrix with simple structure and high efficiency is used to build the prediction model, and a scheme of continuous updating and maintenance of the prediction matrix is deployed. The scheme meets the real-time prediction and the needs of the instructor department Although an improved scheme for random trajectory prediction is added to the model, the accuracy of random trajectory prediction is still far from perfect. On the other hand, there are a lot of recognition errors in the real-time ANPR data. We can't eliminate these errors effectively and accurately, so that the prediction accuracy is further reduced. Moreover, due to the properties of the Markov chain, the model cannot accurately reflect the purpose of the vehicle, even though it is extended to order-k Markov model. Additionally, the influence of travel time on trajectory prediction is not considered in the application of this model.

We will further optimize the prediction accuracy of the order-k Markov model, and add more factors such as travel time, travel flow, and weather to the Markov model. At the same time, we will study whether we can upgrade the order-2 Markov model into a higher-order one under the premise of satisfying the space demand.

References

1. Jeung, H., Liu, Q., Shen, H.T., Zhou, X.: A hybrid prediction model for moving objects. In: Proceedings - International Conference on Data Engineering, pp. 70–79 (2008)
2. Kang, J., Yong, H.S.S.: A frequent pattern based prediction model for moving objects. IJCSNS **10**, 200 (2010)
3. Zhang, Y., Ji, G., Zhao, B., Sheng, B.: An algorithm for mining gradual moving object clusters pattern from trajectory streams. Comput. Mater. Continua **59**, 885–901 (2019)
4. Anagnostopoulos, T., Anagnostopoulos, C.B., Hadjiefthymiades, S., Kalousis, A., Kyriakakos, M.: Path prediction through data mining. In: 2007 IEEE International Conference on Pervasive Services, ICPS, pp. 128–135 (2007)

5. Zhu, M., Liu, C., Wang, J., Wang, X., Han, Y.: 2016 IEEE International Conference on Services Computing A Service-Friendly Approach to Discover Traveling Companions based on ANPR Data Stream (2016)
6. Tao, Y., Faloutsos, C., Papadias, D., Liu, B.: Prediction and indexing of moving objects with unknown motion patterns, p. 611 (2004)
7. Brilingaitė, A., Jensen, C.S.: Online route prediction for automotive. In: Proceedings of the 13th World Congress and Exhibition on Intelligent Transport Systems and Services, pp. 1–8 (2006)
8. Kim, S.-W., Won, J.-I., Kim, J.-D., Shin, M., Lee, J., Kim, H.: Path prediction of moving objects on road networks through analyzing past trajectories. In: Apolloni, B., Howlett, R.J., Jain, L. (eds.) KES 2007. LNCS (LNAI), vol. 4692, pp. 379–389. Springer, Heidelberg (2007). https://doi.org/10.1007/978-3-540-74819-9_47
9. Bacha, R.E.L., Zin, T.T.: A Markov Chain Approach to Big Data Ranking Systems. 3–4 (2017)
10. Qi, Y., Ishak, S.: A hidden Markov model for short term prediction of traffic conditions on freeways. Transp. Res. Part C Emerg. Technol. 43, 95–111 (2014)
11. Xiong, X., Chen, L., Liang, J.: A new framework of vehicle collision prediction by combining SVM and HMM. IEEE Trans. Intell. Transp. Syst. 19, 699–710 (2018)
12. Yu, X.G., Liu, Y.H., Wei, D., Ting, M.: Hybrid markov models used for path prediction. In: Proceedings - International Conference on Computer Communications and Networks, ICCCN, pp. 374–379 (2006)
13. Hasan, S., Ukkusuri, S.V.: Incomplete check-in data : a semi-markov continuous-time bayesian network model, pp. 1–12 (2017)
14. Sun, H., McIntosh, S.: Analyzing cross-domain transportation big data of New York City with semi-supervised and active learning. Comput. Mater. Continua 57, 1–9 (2018)
15. Mohammed, B., Naouel, D.: An efficient greedy traffic aware routing scheme for internet of vehicles. Comput. Mat. Continua 60, 959–972 (2019)
16. Yang, J., Xu, J., Xu, M., Zheng, N., Chen, Y.: Predicting next location using a variable order Markov model, pp. 37–42 (2014)
17. Song, J., He, H., Thomas, R., Bao, Y., Yu, G.: Haery : a Hadoop based query system on accumulative and high-dimensional data model for big data, 4347 (2019)
18. Chang, B.R., Tsai, H., Chen, C., Huang, C., Hsu, H.: Implementation of secondary index on cloud computing NoSQL database in big data environment 2015 (2015)

Ear Recognition Based on Gabor-SIFT

Ying Tian[(⊠)], Huiwen Dong, and Libing Wang

School of Computer Science and Software Engineering, University of Science
and Technology Liaoning, Anshan 114051, China
astianying@126.com

Abstract. Scale invariant feature transform is a local point features extraction method. It can find those feature vectors in different scale space which are invariant for scale changes and rotations, and are flexible for illumination variations and affine transformations. The paper chooses SIFT to extract key points of ear images. Then the features of key points are extracted with the local multi-scale analysis feature of the Gabor wavelet. In this way, every key point is represented by a series of multi-scale and multi-orientation Gabor filter coefficients. Finally Ear recognition based on these feature is carried out with Euclidean distance as similarity measurement. Experimental results show that proposed method can effectively extract ear feature points, and obtain high recognition rate by using few feature points. It is robust to rigid changes, illumination and rotations changes of ear image, provides a new approach to the research for ear recognition.

Keywords: Ear recognition · SIFT · Gabor transform

1 Introduction

Ear recognition as a branch of biometric technology, its research and application prospects have caused people's concern. The human ears have rich and stable structure that is preserved from childhood into old age [1, 2]. The ear image, compared with the face image, has a small image area, the color relatively stable, not affected by the make-up and facial expressions and aging, glasses, covering with hair. The ear is considered as an alternative to be used separately or in combination with the face as it is comparatively less affected by such changes. However, its smaller size and often the presence of nearby hair and ear-rings make it very challenging to be used for non-interactive biometric applications.

The existing feature extraction methods of the human ear can basically be divided into two categories: the overall expression based on the global feature vector and based on the separation local characteristics. In the former, the whole person ear is a global description, using statistical or algebraic methods to extract global features. The mainstream is the PCA [3], DCT [4] and 2DLDA [5] and so on. These methods generally require a large amount of calculation and a group of experimental parameters related to the specific environment. Furthermore, although these algebraic methods can grasp the difference between the images [6], they are inherently sensitive to pose, illumination, scale and occlusion, and their robustness is poor. And basing on these methods inevitably encounter

© Springer Nature Switzerland AG 2020
X. Sun et al. (Eds.): ICAIS 2020, LNCS 12239, pp. 86–94, 2020.
https://doi.org/10.1007/978-3-030-57884-8_8

promotional problem that the performance of the algorithm relies heavily on the degree of similarity between the test set and the training set data distribution. You can not conduct effective training if impossible to be recognized by multiple samples. In the latter, the extracted local feature is discrete and its component generally comprises ear contour or internal groove formed curve, the feature point and angle, etc., such as using the outer ear edge [7], characterized in point, auricle anatomy, ear shape. The combination of features and inner ear structure characteristics [8]. These methods are generally small amount of calculation, intuitive sense of physical characteristics. But edge features and point features have direction and scale differences. If scale and rotation angle changes in the image, the extracted edge will produce errors, thereby increasing the false rejection rate [9].

In recent years, the gray image [10] Gabor transform is becoming one of the mainstream ideas. This is mainly because the Gabor can be well simulated cerebral cortex in single-cell receptive fields outline capture prominent visual properties, in particular Gabor can extract the specific area of the image inside the multi-scale, multi-direction spatial frequency domain characteristics, such as microscope enlarges gradation changes, which means that it can obtain the best localization in the time domain and frequency domain. First 2D Gabor transform is applied to the field of computer vision by Daugman [11], and followed in fingerprint and palm print identification and so on. The Gabor decomposition makes the data dimension a significant increase, and especially when the image size is too large. The dimension must be reduced to avoid disaster.

SIFT [12] is proposed by David Lowe as a point feature technology. It is a local feature extraction algorithm to find the extreme points in the scale space. Extracted feature is invariant to image scale changes and rotation, but also has strong adaptability to illumination changes and image deformation, and is stable to changes of perspective affine transformation, and the noise.

In this paper, SIFT algorithm and Gabor transform is combined to obtain feature vectors. First SIFT algorithm is used to extract stabile feature points of ear image, and then the feature point as the center, we calculate the feature string which is formed by multi-scale and multi-direction Gabor transform features. The feature string is treated as a key point descriptor to form a feature vector, using the Euclidean distance to identification. Experiments show that proposed method can effectively extract ear feature points, and obtain high recognition rate by using few feature points.

2 Feature Point Detection Based on SIFT Algorithm

We tried several feature point detection methods in 2D gray ear image, including the Harris corner detector [13], curve-shaped structure detector [14]. However, the extraction feature points by these methods do not have a scale and rotation invariance. In 2008, Bustard [15] and Tian Ying [16] proposed the ear recognition method based on SIFT features. SIFT algorithm consists of four steps:

1) Scale space extreme point detection: The keypoint is detected in the scale space to keep Characteristic scale invariance. The SIFT algorithm filters images to get scale space image using a Gaussian function, two-dimensional Gaussian function is defined as followed:

$$G(x, y, \sigma) = \frac{1}{2\pi\sigma^2} e^{-(x^2+y^2)/2\sigma^2} \qquad (1)$$

The scale space representation of the image with Gaussian kernel convolution is formula (2):

$$L(x, y, \sigma) = G(x, y, \sigma) * I(x, y) \qquad (2)$$

Wherein, (x, y) represents a pixel position of the image, and σ called the scale factor. A pyramid of images is created by convolving the original image by a set of Difference-of-Gaussian (DoG) kernels. The difference of the Gaussian function is increased by a factor in each step. The SIFT descriptor detects feature points efficiently through a staged filtering approach that identifies stable points in the scale space of the resulting image pyramid. Local feature points are extracted through selecting the candidates for feature points by searching peaks in the scale-space from a DoG function.

1) Precise localizing extreme point position: Three-dimensional quadratic function through the preparation and precise determination of location and scale of key points, while removing the low contrast and unstable critical points in order to enhance the stability of match and improve noise immunity. Figure 1 is some ear images their size 400×300, each piece can probably extract about 100 key points.

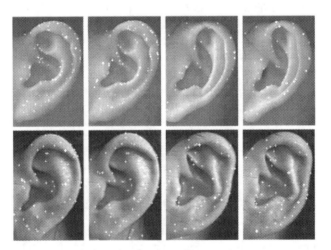

Fig. 1. Extracted ear feature points

2) Specify the direction for key point: One or more orientations are assigned to each key point location based on local image gradient directions. All future operations are performed on image data that has been transformed relative to the assigned orientation, scale, and location for each feature, providing invariance to these transformations.

3) Feature point descriptor structure: First rotation of axes that as the key points direction ensure rotation invariance. The key point as the center then select the 16×16

windows, and divided into 16 sub-regions of 4×4 pixels. For each pixel within a sub-region, SIFT adds the pixels gradient vector to a histogram of gradient directions by quantizing each orientation to one of 8 directions and weighting the contribution of each vector by its magnitude. Calculating the cumulative value add for each gradient direction and form a seed point 16 sub-regions of 4×4.

3 Ear Recognition Algorithm Based on Gabor-SIFT

Our local feature point description algorithm has the same input with standard SIFT descriptor. Namely: the location, scale and the main direction of key point. First on a given scale, rotate axis as the direction of the key point to ensure rotation invariance. Then the key point for the center, calculate multi-scale, multi-directional Gabor transform characteristic features of the formation of the string as a key point descriptor.

3.1 Extract Ear Partial Multi-scale Features by Gabor

Taking into account the excellent characteristics of previously described Gabor transform, we use Gabor function to extract the characteristics of the input key points attribute. 2D form of the Gabor filter is defined as a constraint plane wave with a Gaussian envelope function:

$$\psi_{\overrightarrow{k_j}}(\overrightarrow{x}) = \frac{\overrightarrow{k_j}^2}{\sigma^2} e^{-\frac{\overrightarrow{k_j}^2 \overrightarrow{x}^2}{2\sigma^2}} [e^{i\overrightarrow{k_j}\overrightarrow{x}} - e^{-\sigma^2/2}] \tag{3}$$

Here: $\sigma: \overrightarrow{x} = (x, y)$; $\overrightarrow{k_j} = \begin{bmatrix} k_{jx} \\ k_{jy} \end{bmatrix} = \begin{bmatrix} k_v \cos\varphi_\mu \\ k_v \sin\varphi_\mu \end{bmatrix}$; $k_v = 2^{(v+2)/2}\pi$; $\varphi_\mu = \mu\frac{\pi}{8}$;

σ is Gaussian standard deviation, v is the scale parameter, μ is direction parameters. The first part decides Gabor nuclear shock part, and the second is the compensation the DC component to eliminate nuclear function in response to the dependence of the absolute value of the brightness of the image changes. Parameter $\overrightarrow{k}$ controls the width of the Gaussian window, the shock part of the wavelength and direction, parameter σ determines the ratio between the width of the window and the wavelength.

Gabor feature of keypoints in the ear image is generated by image points and Gabor filter convolution. $f(x, y)$ represents the distribution of the gradation of the ear image, then $f(x, y)$ and the Gabor filter convolution can be defined as formula (4):

$$G(x, y, v, \mu) = f(x, y) * \psi_{\overrightarrow{k_j}}(\overrightarrow{x}) \tag{4}$$

A different set of Gabor filter can be obtained by setting different scale parameters v and direction parameters μ. Thinking of the speed and effectiveness, the paper uses three different scales, the four different directions, i.e. $v \in \{0, 1, 2\}$, $\mu \in \{0, 1, 2, 3\}$, so that $3 \times 4 = 12$ different Gabor filters can be generated.

The key points extracted by SIFT method in ear images do convolution with each Gabor filter. Each of the key point has 12 complex outputs. In order to improve the speed of the convolution operation, Fast Fourier Transform is used. The 12 complex responses are on behalf of the key point of the local multi-scale features.

3.2 Gabor-SIFT Feature Descriptor Structure

In fact, in the vicinity of edge, the real part and the imaginary part of the Gabor transform will produce shocks, rather than a smooth peak response, and thus it is not conducive to the matching of the identification phase. Therefore, we only retain the Gabor response amplitude and reflect the characteristics of the energy spectrum of the characteristic point. Figure 2 is the different amplitude patterns of an ear image, in which each Gabor transform is calculated pixel by pixel. Due to limited space, the amplitude spectra is only given when the scale parameter $v = 0$ and the direction parameter $\mu \in \{0, 1, 2, 3\}$.

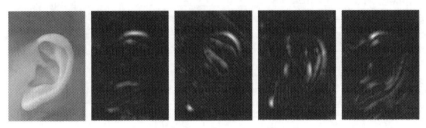

Fig. 2. Different amplitude patterns of an ear image (scale is: 0) leftmost is original image, rightward in turn is: μ from 0 to 3

The 12 amplitude characteristics obtained by the position of the key points of the image by above method reflect the energy distribution characteristics of the local region to the point for the center. The 12 amplitude characteristics are cascaded together as the keypoint feature descriptor F, the descriptor of the key point in the position of $k(x, y)$ is: $F_{k(x,y)} = \{f_{v\mu,k(x,y)} | v = (0, 1, 2), \mu = (0, 1, 2, 3)\}$.

Each of keypoint in the proposed algorithm is represented as just 12-dimensional feature vector rather than 128 in the standard SIFT, thus reduce many substantial increase in computing speed.

3.3 Classification and Recognition

For a given template and test ear image, each image can be reliably detected a series of feature points and their corresponding robust feature descriptor, and then we have to do is to classify and recognize ear image based on these feature vectors.

We use Euclidean distance of the eigenvectors as similarity measurement of key points of two images. Given two features descriptor F_i and F_j, measurement ND is formula (5):

$$ND = |F_i - F_j| = \sqrt{\sum_{k=1}^{n} (f_{ik} - f_{jk})^2} \tag{5}$$

In formula (5) $n = 12$, taken a key point in test image, according to formula (5), respectively, to find out two key points from template image where their Euclidean distance is shortest, their distance denoted as ND_1 and ND_2, if the nearest distance ND_1

dividing second nearest distance ND_2 is less than a threshold, i.e. $ND_1/ND_2 < T_n$, then accept this pair of matching points. Lowering the threshold, the number of matching points will reduce, but more precise and stable. Through experiment we select $T_n =$ 0.7. In practical applications, according to security requirements adjust the threshold. Along with change of T_n, FRR (False Reject Rate) and FAR (False Accept Rate) can be changed, distribution is shown in Table 1.

Table 1. Matches precision with a threshold T_n relationship

Threshold T_n	0.5	0.6	0.7	0.8	0.9
FAR	18.6%	9.7%	6.9%	8.6%	11.8%
FRR	5.4%	3.7%	4%	4.6%	4.6%

4   Experimental Results Analysis

We use the ear image database I and II which was provided the Beijing University of Science and Technology, and our human ear image database we called database III. Image database I contains 61 people, each person has 3 images. Image database II contains 77 people each person has 4 images (2 light change and 2 angle change). Image database III contains 15 people each person has 10 images (5 light change and 5 angle change). We remove some poor quality images, and select 50 persons from image database I and II respectively, select 10 people from image database III. We select a total of 450 images as a small-scale sample library to do experiment.

We do two experiments: study of the effectiveness of the algorithm used in a controlled environment and robustness of the algorithm when light conditions and ear rotation angle change. We measure its performance according to CMS (Cumulative Match Score).

Experiment 1: Use image database I, it has no angle changes only in light intensity minor change, to compare the recognition accuracy of standard PCA method, standard SIFT methods and Gabor-SIFT method. We take the 120 dimensional eigenvectors for PCA method, so that it can retain 98.5% of the information, using Euclidean distance for similarity calculation. Take two images of image database as training samples, the other one as test sample. Compare with a 128-dimensional feature vector extracted using standard SIFT method, and 12-dimensional feature vectors using Gabor-SIFT, results are shown in Fig. 3. The figure shows that our method is effective on the ear recognition with standard PCA method and the SIFT algorithm. Rank 1 recognition rate consistent, and in Rank 5 reached 100% faster than the PCA method.

This method is less feature vector dimensions, thus, its matching speed is faster than the other two methods. Table 2 gives the contrast with SIFT algorithm and Gabor-SIFT algorithm about matching time and recognition rate in image database I. Seen from the Table 2, the match time of Gabor-SIFT method is much shorter than the SIFT method.

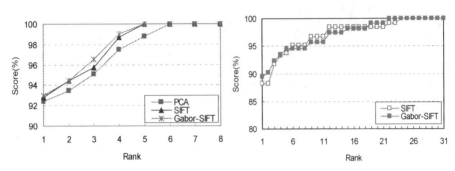

Fig. 3. CMS curve of experiment 1   **Fig. 4.** CMS curve of experiment 2

Table 2. Recognition rate and matching time on database I

Methods	Point amount	Vector dimension	Recognition rate	Matching time(s)
SIFT	80–120	128	92.6%	2.62
Gabor-SIFT	80–120	12	93.1%	0.25

Experiment 2: Test the robustness of the ear recognition method based on Gabor-SIFT algorithm when lighting conditions and angle of rotation change. We do experiments in image database II and image database III. It is found that the three methods have been worse results than experiment 1. In standard PCA method, when the image rotation, illumination and scale changed, the recognition rate is greatly affected and would drop to less than 30%, but based on SIFT and Gabor-SIFT method still maintain a high recognition rate. Therefore, the comparison with PCA method has no meaning. Figure 4 shows the robustness of this method and the standard SIFT method in different databases. Seen from the figure, ear recognition method based on Gabor-SIFT has strong robustness for changes in lighting conditions and the angle of rotation changes.

From experiments can find, a significant improvement in recognition rate would appear if we appropriate combined the standard 128-dimensional SIFT feature vector with 12-dimensional Gabor-SIFT feature:

For each detected key-point, we have created a feature vector composed by SIFT descriptor and Gabor-SIFT descriptor, and it is defined as: $F = \begin{bmatrix} \omega S \\ (1-\omega)G \end{bmatrix}$, where S is a 128-dimensional SIFT descriptor, G is a 12-dimensional description of the Gabor-SIFT vector, ω is associated weighting factor.

We calculate NDS ($n = 128$, SIFT descriptor portion) and NDG ($n = 12$, Gabor-SIFT descriptors portion) with formula (5), then the distance is: $ND = \omega NDS + (1-\omega)NDG$. We do experiment in above three images database, the recognition rate has increased in different degree. The comparison of average recognition rate is shown in Fig. 5, wherein taken $\omega = 0.5$.

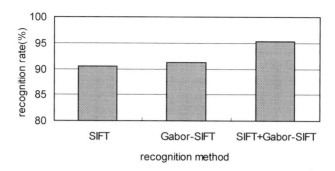

Fig. 5. The recognition rate of different method

5 Conclusion

This article describes a new outer ear recognition method based on Gabor-SIFT feature descriptor. The Gabor-SIFT characteristic describes local structural information of the image. Although it doesn't have very obvious visual significance like corner and edge feature, for image change factors it can maintain a certain invariance because of its estimated scale and direction. Experimental results show that based on Gabor-SIFT feature extraction algorithm not only can effectively solve the static image automatic ear recognition problem, but also has strong robustness, insensitive to changes in image light conditions and angle of rotation.

Acknowledgments. Thanks for the image library of Ear Recognition Laboratory at USTB.

This research was supported by (1) Foundation of Liaoning Educational Committee (Grant number 2019LNJC03); (2) Foundation of University of Science and Technology Liaoning (Grant number 2016RC06).

References

1. Iannarelli, A.: Ear Identification. Forensic Identification Series. Paramount Publishing Company, California (1989)
2. Wyawahare, M.V., Patil, P.M., Abhyankar, H.K.: Image registration techniques: an overview. Int. J. Sig. Process. Image Process. Pattern Recogn. **2**(3), 11–28 (2009)
3. Li, Y., Cao, J., Zhang, H.: Improved method of ear recognition based on tensor PCA. Comput. Eng. Appl. **25**(47), 171–174 (2011)
4. Saraswathy, K., Vaithiyanathan, D., Seshasayanan, R.: A DCT approximation with low complexity for image compression. In: Proceedings of 2013 International Conference on Communications and Signal Processing, Melmaruvathur, India, 3–5 April, pp. 465–468 (2013)
5. Lu, X.-L., Shen, H,-F., Zhao, L.-H.: Ear recognition based on 2DLDA and FSVM. Sci. Technol. Eng. **12**, 2852–2855 (2012)
6. Wu, H., Liu, Q., Liu, X.: A review on deep learning approaches to image classification and object segmentation. Comput. Mater. Continua **60**(2), 575–597 (2019)
7. Yuan, W., Tian, Y.: Ear contour detection based on edge tracking. In: Proceedings of Intelligent Control and Automation (WCICA 2006), Dalian, China, 21–23 June, pp. 10450–10453 (2006)

8. Mu, Z., Li, Y., Xu, Z., Xi, D., Qi, S.: Shape and structural feature based ear recognition. In: Proceedings of 5th Chinese Conference on Biometric Recognition, Guangzhou, China, 13–14 December, pp. 633–670 (2005)

9. Li, S., Liu, F., Liang, J., Cai, Z., Liang, Z.: Optimization of face recognition system based on azure IoT edge. Comput. Mater. Continua **61**(3), 1377–1389 (2019)

10. Wang, C., Feng, Y., Li, T., Xie, H., Kwon, G.-R.: A new encryption-then-compression scheme on gray images using the markov random field. Comput. Mater. Continua **56**(1), 107–121 (2018)

11. Daugman, J.G.: Complete discrete 2-D gabor transforms by neural networks for image analysis and compression. IEEE Trans. Acoustics Speech Sig. Process. **7**(36), 1169–1179 (1988)

12. Lowe, D.G.: Distinctive image features from scale-invariant keypoints. Int. J. Comput. Vis. **2**(60), 91–110 (2004)

13. Wen, W.: An improved algorithm for Harris multi-scale color detection. J. Chongqing Univ. Technol. (Natural Sci.) **8**(26), 94–96 (2012)

14. Steger, C.: An unbiased detector of curvilinear structures. IEEE Trans. Pattern Anal. Mach. Intell. **3**(20), 113–125 (1998)

15. Dakshina, R.K., Hunny, M.: SIFT-based ear recognition by fusion of detected keypoints from color similarity slice regions. In: Proceedings of Advances in Computational Tools for Engineering Applications, Zouk, Mosbeh, 17–19 July (2009)

16. Tian, Y., Yuan, W.: Ear recognition based on fusion of scale invariant feature transform and geometric feature. Acta Optica Sinica **8**(28), 1485–1491 (2008)

17. Mo, W., Ding, Z.: A novel template weighted match degree algorithm for optical character recognition. Int. J. Smart Home **7**(3), 261–270 (2013)

18. McConnon, G., Deravi, F., Hoque, S., Sirlantzis, K., Howells, G.: An investigation of quality aspects of noisy colour images for Iris recognition. Int. J. Sig. Process. Image Process. Pattern Recogn. **4**(3), 165–178 (2011)

Differentiated Services Oriented Auction Mechanism Design for NVM Based Edge Caching

Zhenyuan Zhang[1], Fang Liu[1(✉)], Zhenhua Cai[1], Yihong Su[1], and Weijun Li[2]

[1] Sun Yat-sen University, Guangzhou 510006, China
liufang25@mail.sysu.edu.cn
[2] Shenzhen Dapu Microelectronic Co., Ltd., Shenzhen 518000, China

Abstract. In the age of big data, contents of internet are accessed frequently from the cloud by different users, which brings heavy load pressure on network servers of service provider (SP). Dozens of researches show that this problem can be effectively solved by caching contents on the network edge. However, edge caching still faces some challenges. First, cache capacity requirements and dilemma. Most existing edge devices have small storage capacity, which can be scaled by non-volatile memories (NVMs) that are deployed in large number, but NVMs have the disadvantages of read-write asymmetry and limited writes. Second, diverse and varied user favors for content also affect the effectiveness of edge caching. Internet data explosive growth changes user preference from simple Zipf distribution to stretched exponential distribution (SED) with weakened long tail effect. The change of user's preference affects the content placement of edge cache device, making the NVM wear worse. Third, in the actual scenario of Internet services, different Internet users have different service level requirements (i.e. differentiated services). The quality of services depends on their requirements for content, and SP needs differentiated services for users. To solve the above challenges, aiming at improving SP revenue, we propose an auction mechanism, which can offer differentiated services. Compared with the baseline which mainly optimize SP revenue without taking NVM, differentiated services and varied user's preference into consideration, extensive simulations verified that our proposal not only improved user's experience and SP revenue up to 103.90% and 99.33% respectively, but also suppressed the NVM wear times similar to the baseline even when user preference distribution changes.

Keywords: NVM · Differentiated service · Auction mechanism · Edge caching

1 Introduction

In recent years, with the explosive growth of Internet mobile data, IDC has predicted that more than 50 billion devices will be connected to the Internet by

X. Sun et al. (Eds.): ICAIS 2020, LNCS 12239, pp. 95–106, 2020.
https://doi.org/10.1007/978-3-030-57884-8_9

2020, and Internet data will grow to 44ZB, and 70% of this data will need to be processed on the edge [12]. The traditional cloud computing technology has been difficult to meet quality-of-service (QoS) and quality-of-experience (QoE) due to the heavy load of mobile traffic in the network during the peak period. Mobile edge caching has become a promising paradigm to support mobile traffic recently, where service providers (SPs) prefetch popular content and cache them at the network edge [1]. When requested, the cached content can be delivered directly to the user with low latency, thereby reducing the traffic load on the backhaul network during peak times and improving the QoE.

In other words, data processing near the data producer does not send request to the cloud computing center through the network, which greatly reduces the system delay and enhances the service response capability [15]. Caching some popular contents on network edge devices can provide lower latency, but edge devices usually have less storage capacity and are far from the cloud, making a low cache hit rate and a high cost of missing [11]. In recent years, with the rapid development of non-volatile memory (NVM) storage technology research, NVM used in the edge has become a hot research area [11]. Therefore, we can introduce this kind of NVM storage media with large capacity, such as 3D XPoint [7]. Although the introduction of this high-capacity NVM storage medium can help edge devices improve their hit ratio, we still need to consider some disadvantages of NVM, such as limited write times, read/write asymmetry, and high price. Specifically, when content is cached on edge devices, we expect to reduce unnecessary write times without degrading the performance of the system, reducing the wear cost of NVM, improving the availability of the cache system. To this end, we modeled the wear cost of NVM in Sect. 2, solved it in Sects. 3 and 4, and simulated it in Sect. 5.

In addition, user favor for content also affects the effectiveness of edge caching. Due to the increasing variety of Internet data, the previous Zipf distribution is no longer applicable to describe content popularity (i.e. users' preference or valuation, willingness-to-pay for their need) [4,13]. Based on Guo et al. [4], as shown in Fig. 1, we generated users valuation within [1, 100] under different distribution after analyzing kinds of video-on-demand data sets such as Youtube-06, IFILM-06 etc and web media data sets such as ST-SVR-01, PS-CLT-04 etc. The observation shows that SED (i.e. line 2 to 9 in Fig. 1) has a weaker long tail effect than Zipf distribution. In the case of limited storage capacity of SP, the increase of content lead to the decrease of user's loyalty to a certain content [13] (i.e. the number of types of content affects the user's loyalty to a certain content), which brings increasing cache replacement times and more wear cost of NVM storage media in SP. This is also a reason why we expect to reduce NVM wear times in above.

In real Internet service scenarios, different users have different levels of service quality requirements (i.e. personalized services), such as high level video-on-demand (VoD) users pursue high video resolution and fluency while low level one can endure low delivery video quality contrarily. According to Nielsen consulting report [13], 61% of Chinese consumers choose to buy high-end products because of their superior quality. In other words, people are willing to pay more for higher service quality, so SP should provide different content interaction quality to different users. In addition, the interest and activity patterns of mobile users are predictable [16]. Therefore, in our experiment in Sect. 5, we generated users valuation under some appropriate distribution as shown in Fig. 1 and analyzed effectiveness of our caching system.

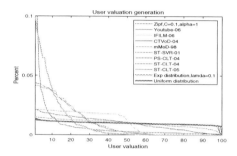

Fig. 1. Datasets analysis

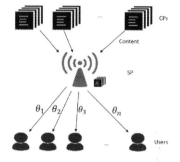

Fig. 2. An example for our system

1.1   Related Work

Li et al. proposed a hierarchical edge caching framework that introduces D2D collaboration, which greatly reduces traffic in the backhaul network and the delay in user access to popular content [9]. In a case of online request edge server service scenario, Tan et al. considered the cost of relay between multiple edge servers to reduce traffic and delay on the backhaul network, while ensuring a certain hit ratio [17]. Based on the storage capacity and computing power of 5G base station equipment in the future, Zeng et al. combined online edge caching with load balancing to unload the traffic load on traditional base station (BS) to small base station (SBS) on the edge, relieving the load pressure of traditional BS and reducing the traffic on backhaul network [20]. In order to minimize the total overheads of all the users in mobile edge computing, Wei et al. proposed a deep reinforcement learning approach for computation offloading decision reducing the latency, energy consumption and enhancing the computation efficiency [18]. There are many different cache replacement or resource allocation strategies based on different application scenarios. Jia et al. modeled the cache replacement cost and proposed a cache replacement strategy at last layer cache (LLC). By maximizing the "Cache Value" on edge hybrid storage platform consisted of NVM and DRAM, they reduced average cache replacement cost and improved performance in hybrid main memory-based mobile edge devices [8]. Guo et al. formulated the resource allocation problem and improved the incomes of RP when resources in the edge computing datacenter are limited [5]. Zhang et al. transformed the multi-dimensional cloud computing resource allocation problem into a regression or classification machine learning problem, and their algorithm reached approximated optimal solution in the terms of social welfare allocation accuracy and resource utilization [21].

1.2 Motivation

First, many related works, during the edge cache "service provider-user" modeling process [1,2,8] do not fully take the features of storage devices on edge servers into consideration and even ignore the cost caused by their features. It is worth noting that Cao et al. [1] proposed an optimal auction mechanism which boosted SP earnings without NVM feature and some other consideration, so we followed up on the work. Recently, published by Intel, the technology of 3D XPoint [10,19], which exists between DRAM and flash SSD in storage hierarchy, has optimistic development prospects and extensive usage scenarios. Due to large capacity and limit wear times of NVM, we take it for our caching system example, and optimize the expected revenue of SP in the more real scene by focusing on suppressing NVM wear times. Second, uncover by [4,12,13], with the coming age of big data, user preferences are diverse and varied. We consider that SP should collect users demand for content but not just model it by simple Zipf distribution. Third, by analyzing [1], we also conclude that SP expected revenue is greatly affected by content delivery quality θ. As we all know, differentiated services are common in Internet services, so we should analyze the change of SP benefits after differentiated services for users.

1.3 Summary of Contributions

Our main contributions are as follows:

1) Considering the change of data distribution in the era of big data, we used uniform, exponential distribution and modified uniform-to-exponential distribution for analyzing, and verifying users valuation of content.

2) Observing the real internet service scenario, we provided differentiated services to different users by different content delivery quality to further optimize the benefits of SP.

3) Further from the actual scenario, in the edge caching scenario with NVM as the storage medium, we suppressed the NVM wear consumption, and made the expected revenue of SP more realistic and available.

4) With the explosive growth of the Internet content category, user preferences are diverse and varied. User favor for content also affect the effectiveness of edge caching. We verified that our proposal improved the expected revenue of SP and the average utility of users when users valuation changes from uniform to exponential distribution.

2 System Modeling

In reality, our video service provider (SP) usually collect video content from clients over all the world. For example, when vblogers publish their videos to

Titok or Youtube, the SP should pay them some rewards. In addition, when other client users watch or download videos from the SP, the SP can earn from them and improve expected revenue or cut back on some unnecessary expenses by some measures, such as caching the most valuable content and providing differentiated services.

Of course, edge caching is not a new topic. As listed in the related working section above, it has been studied by many scholars. Among them, Cao et al. modeled caching system and carried out extensive simulations, then effectively improved profit of SP and analyzed some factors impacting it. Based on Cao's [1], we propose our edge caching system, then verify and improve it in detail in the followings.

Our system modeling can be introduced as follows. An example as illustrated in Fig. 2. Consider that our model with one SP, m CPs and n users.

In here, SP can get content i from video content provider (CP) $i, i = 1, \ldots, n$. Without loss of generality, we denote the collection of video content which interests user j as $\mathbb{S}_j \subset 1, \ldots, m, j = 1, \ldots, n$. We also denote that the collection of client users or downloaders interested in content i as $\Omega_i \subset 1, \ldots, n$. In fact, we can easily deduce $\sum_{j=1}^{n} |\mathbb{S}_j| = \sum_{i=1}^{m} |\Omega_i|$. It's worth noting that different users have different interests of video content types, however, some specific contents are in favor with them. Moreover, it cannot be ignored that not all users are equally interested in the same content (i.e. some users may have more desire for some specific content). So, if we just cache contents by ranking their request frequency or popularity, our caching system could not reap more profits from client users who have more willingness-to-pay. It is difficult to determine what types of video content should be cached in SP. But an observation showed that the interest and activity patterns of mobile data users are predictable [16]. So we generate the users valuation shown as Fig. 1 to simulate users' willingness-to-pay (depend on how much their favor and desire for this content) for their favorite content.

As mentioned above, the users' willingness-to-pay should be listed as user privacy to some extent. Frankly, we hardly collect this type of dataset. But one SP can deliver contents which users request and be charge of users in its region. SP can collect and do some statistical analysis before delivering contents to users. So, we suppose user j has true willingness-to-pay value t_j, and noise value τ_j considering user privacy.

In particular, suppose that all users have true value $\boldsymbol{t} = [t_1, t_2, \ldots, t_n]$, noise value $\boldsymbol{\tau} = [\tau_1, \tau_2, \ldots, \tau_n]$, $\boldsymbol{t}$ and $\boldsymbol{\tau}$ are independent across users, and t_j is a random variable (r.v.) over interval $\mathbb{T}_j = [\underline{a}_j, \overline{a}_j]$ with probability density function (PDF) $f_j(t_j)$ and cumulative distribution function $F_j(t_j)$. The joint valuation space of all users is denoted as $\mathbb{T} = \mathbb{T}_1 \times \cdots \times \mathbb{T}_n$ while the joint PDF is denoted as $f(\boldsymbol{t}) = \prod_{j=1}^{n} f_j(t_j)$.

The SP pays CP i for content i with fixed unit price r_i for simplicity. When the SP delivers video content to K users with same video transmission/delivery quality $\theta_k > 0, k = 1, \ldots, K$, the SP will incurs delivery cost $\sum_{k=1}^{K} h(\theta_k)$ where $h(\cdot)$ is the cost function of content delivery. On the other hand, suppose that user j receives content in $\mathbb{S}_j$ with quality θ, and his unit satisfaction level will be θt_j.

3   Problem Formulation

The optimal auction model proposed by Cao et al. [1] considers the cost of obtaining content and the cost of delivering content by SP. In the system with n users and m contents, the expected revenue of SP is defined as formula (1):

$$ER \triangleq \int_{\mathbb{T}} [\sum_{j=1}^{n} x_j(t) - \sum_{i=1}^{m} p_i(t) r_i - \sum_{i=1}^{m} p_i(t)|\Omega_i|h(\theta)] f(t) dt \qquad (1)$$

Among formula (1), the integral terms mean in the order: users payment amount, the SP access cost for content, the SP delivery cost. t means the users real value for content, $p_i(t)$ is expressed as the capacity percent of content i in SP, r_i means unit acquisition costs of content i, $h(\theta)$ means delivery cost when SP deliver content to user j with delivery quality θ, $|\Omega_i|$ means the number of users interested in content i.

In a real storage scenario, just taking Intel DC Persistent Memory Module (PMM) which is constructed by 3D XPoint technology [7] for instance, we can discover that it is expensive, but no price. Due to its high cost SP should take NVM wear into a serious consideration, and our problem formulation can be redefined as follow.

$$ER \triangleq \int_{\mathbb{T}} [\sum_{j=1}^{n} x_j(t) - \sum_{i=1}^{m} p_i(t) r_i - \eta \sum_{i=1}^{m} g_i(t) c_i - \sum_{i=1}^{m} p_i(t) \sum_{j \in \Omega_i} h(\theta_j)] f(t) dt \quad (2)$$

Above, we redefine SP expected revenue by the payment of all users(the first term), the storage cost of corresponding contents in SP (the second term), the NVM wear cost (the third term) which caused by frequent cache replacement when the age of big data is coming, and delivery cost (the fourth term).

In particular, the term $\eta \sum_i^m g_i(\mathbf{t}) \mathbf{c_i}$ means the NVM wear cost of SP due to device wear and cache replacement caused by the changing of users' favor. Here, c_i means the unit replace cost due to content i replaced. η means the device current status. (i.e. Actually # of valid cells in NVM device, η increases over time and η always greater than 0, initialize $\eta_1 = 1$).

As mentioned above, the interest and activity patterns of mobile users are predictable [16], and due to the increasing variety of Internet data, the previous Zipf distribution is no longer applicable to users' prediction of content valuation, and users valuation are gradually present as stretched exponent distribution [4,13]. Hereto, we have to compute in advance and analyze the percent change of content i in SP. Thereby, define

$$g_i(t) = p_i(x) - p_i(t) \qquad (3)$$

Here, $p_i(\mathbf{t})$ means the current proportion of content i subject to $\mathbf{t}$. $p_i(\mathbf{x})$ which is computed in advance, means the future next proportion of content i subject to $\mathbf{x}$, so $g_i(t)$ means the future next changes in proportion of content i.

4   Optimal Auction Mechanism Design

Our caching system have two phase for optimizing the ER. Next we will discuss in the following paragraph.

4.1   First Phase

In initial state of system $T = 0$, we assume $\eta_1 = 1$ and $g_i(t) = 0$ for preventing from SP expected revenue ≤ 0 such that SP will not cache any content, so ER in (2) could simply to

$$ER \triangleq \int_{\mathbb{T}} [\sum_{j=1}^{n} x_j(t) - \sum_{i=1}^{m} p_i(t)r_i - \sum_{i=1}^{m} p_i(t) \sum_{j \in \Omega_i} h(\theta_j)] f(t)dt \tag{4}$$

We can easily deduce equation (4) based on Cao et al. [1] and classify user type into low valuation and high valuation based on the Pareto principle [6].

First, we Define $c_j(t_j) \triangleq t_j - \frac{1-F_j(t_j)}{f_j(t_j)}, for \quad any \quad j = 1, \ldots, n, t_j \in \mathbb{T}_j$. the optimal $\boldsymbol{p}^*(\boldsymbol{t})$ can be given as:

$$\begin{cases} p_k^*(\boldsymbol{t}) = 1, p_i^*(\boldsymbol{t}) = 0, \forall i \neq k, \\ \quad if \quad max_{i=1,\ldots,m}\{ \sum_{j \in \Omega_i} [\theta_j c_j(t_j) - h(\theta_j)] - r_i \} > 0; \\ p_i^*(\boldsymbol{t}) = 0, \forall i = 1, \ldots, m, otherwise. \end{cases} \tag{5}$$

Next, we can get the optimal user payment $x_j^*(\boldsymbol{t})$ is

$$x_j^*(\boldsymbol{t}) = \theta_j t_j \sum_{i \in \mathbb{S}_j} p_i(\boldsymbol{t}) - \theta_j \int_{\underline{a}_j}^{t_j} p_i(\tau_j, \boldsymbol{t}_{-j})d\tau_j, \forall j = 1, \ldots, n, \forall t \in \mathbb{T}. \tag{6}$$

the solution is not be detailed here, and their proof can be found in [1].

Motivated by energy consumption in [14] and resource allocation in [1,3,20], the common property of monetary cost function is convex, so we define the content delivery cost function is

$$h(\theta_j) = \alpha \theta_j^2 \tag{7}$$

Different user j will be delivered by different θ_j according their valuation t_j.

Finally, we get the sum of delivery cost of SP is $\sum_{i=1}^{m} p_i(t) \sum_{j \in \Omega_i} h(\theta_j)$. So, the sum of expected revenue of SP is showed as formulation (4). On the other hand, the utility of user j can be formulated as:

$$u_j(\boldsymbol{\tau}, t) \triangleq [\sum_{i \in \mathbb{S}_j} p_i(\boldsymbol{\tau})] \theta_j t_j - x_j(\boldsymbol{\tau}) \tag{8}$$

4.2   Second Phase

In the middle or late stage of system operation, it is advisable to set the running time as $T = 1$, and users current valuation as t^1, and last valuation as t^0. We know that users valuation will change over time when the kinds of content is greatly increasing, as different distributions are shown in Fig. 1.

Considering NVM wear and content replacement cost of SP, so we define $Cost_w = \eta^w \sum_{i \in \Delta} g_i(t)c_i$ where η^w represents the wear status of the NVM storage device when $T = w$, and the set Δ including the content which upcoming $p(t^1)$ and $p(t^0)$. $\eta^0 = 1$ represents the state of NVM when it is first used. As the system runs, η is gradually increasing, so does NVM wear cost when cache replacement happened.

Next, we need recalculate the $p(t)$ to suppress the # of NVM wear, and the wear cost. The formula (9) is showed as follows:

$$\begin{cases} p_k^*(t) = 1, p_i^*(t) = 0, \forall i \neq k, \\ \quad if \quad max_{i=1,\ldots,m}\{ \sum_{j \in \Omega_i} [\theta_j c_j(t_j) - h(\theta)] - r_i - \sum_{i \in \Delta} [\eta^w c_i]\} > 0; \\ p_i^*(t) = 0, \forall i = 1, \ldots, m, otherwise. \end{cases} \quad (9)$$

As illustrated by the first phase of system running, so we easily deduce that $Cost_w$ is either 0 if no replace or ηc_i if content i will be replace (i.e. $i \in \Delta$).

If happened replace, the payment $x_j(t)$ will be:

$$x_j^*(t^1) = \theta_j t_j^1 \sum_{i \in \mathbb{S}_j} p_i(t^1) - \theta_j \int_{\underline{a}_j}^{t_j} p_i(\tau_j, t^1_{-j})d\tau_j + [\eta^w \sum_{i \in \Delta} \frac{g_i(t)c_i}{|\Omega_\Delta|}]\mathbf{1}(j \in \Omega_\Delta)$$

$$\forall j = 1, \ldots, n, \forall t \in \mathbb{T}. \quad (10)$$

where $\mathbf{1}(\cdot)$ is the indicator function. $\mathbf{1}(\cdot) = 1$ if user $j \in \Omega_\Delta$, $\mathbf{1}(\cdot) = 0$ if user $j \notin \Omega_\Delta$. Therefore, the utility of user j can be updated as:

$$u_j(\tau^1, t^1) \triangleq [\sum_{i \in \mathbb{S}_j} p_i(\tau^1)]\theta t_j^1 - x_j(\tau^1) \quad (11)$$

Last, a brief algorithm framework 1 is illustrated to solve the formula (2).

Algorithm 1. Solution for formula (2)

1.$T = 0$(the initial stage of system), the user valuation collected by the system is t.
2.According to formula (5), we can compute $p(t^0)$,ER, user utility u_j.
3.$T = 1$(the middle stage of system),the user valuation collected by the system is t^1.
4.According to formula(9), formula (10), we can further compute $p(t^1)$, and $x_j^*(t^1)$.
5.Substitute $p(t^1)$, $x_j^*(t^1)$ into formula (2), we can get ER, user utility u_j.
6.When $T = 2, 3, \ldots, w$, their solution are similar to steps 3–5 above.

---

5   Numerical Results

In this section, we carried out extensive numerical experiments and analyzed many improved results. In particular, we set Cao et al. [1] as baseline, and all results involving experiments are average over 10^4 independent trials.

By analyzing many datasets of web media and VoD in Fig. 1, we can discover these trend line is very likely to uniform or exponential trend line in Fig. 1, so our experiment generate users valuation under them for simplicity. We consider a caching system (Fig. 2) with $m = 3$ CPs and $n = 10$ users. Unlike the baseline, for our differentiated service, we introduce a coefficient ϵ to divide the user valuation into two categories. Specifically, we assume user j has a high valuation if $t_j \geq \underline{a}_j + \epsilon(\overline{a}_j - \underline{a}_j)$, and user j has a low valuation if $t_j < \underline{a}_j + \epsilon(\overline{a}_j - \underline{a}_j)$. The θ_j is set to 2 if user j is high level, otherwise 1. The parameters configuration is shown as Table 1 in detail.

Table 1. Parameters configuration of our proposal

Parameters	Configuration of our proposal
The number of CPs and users	$m = 3, n = 10$
The set of contents Ω_i	$\Omega_1 = \{1, 3, 4, 5, 6, 10\}$, $\Omega_2 = \{1, 3, 5, 7, 8, 9\}$, $\Omega_3 = \{1, 2, 3, 5, 9, 10\}$
The acquisition costs of contents r_i	$r_1 = 4.2036, r_2 = 1.2714, r_3 = 4.0714,$
The valuation space of user j	$\mathbb{T}_j = [\underline{a}_j, \overline{a}_j], \underline{a}_j = 1 + 0.1(j-1), \overline{a}_j = 4 + 0.1(j-1)$
The PDF of exponential distribution	$f_j(t_j) = \lambda_j e^{(-\lambda_j t_j)}, for \; t_j \geq 0, \lambda_j = \frac{1}{10+0.4(j-1)}$
Differentiated service factor ϵ	$\epsilon = 0.2$ for Exp distribution, $\epsilon = 0.5$ for uniform distribution
Content delivery quality θ	$\theta = 1$ for low valuation, $\theta = 2$ for high valuation
The content delivery cost function of user j	$h(\theta_j) = \alpha\theta_j^2, \alpha = 0.1$ and $\theta_j = 1$ temporarily

In our extensive experiments under three distribution, we compare our proposal with Cao's [1] natural modification version which takes NVM wear factor into cost. Similarly, we mainly compare four performance index to analyze these result.

First, what SP concerns most always is expected revenue (ER). In Fig. 3, we can discover our proposal is superior to baseline under these three distribution, exactly improving 82.55%, 99.33%, 92.50% under uniform distribution (Uni), exponential distribution (Exp), and modified uniform-to-exponential distribution (Uni2Exp) respectively.

Second, what user concern most always is utility (i.e. it's illustrated by formula 8 and 11). In Fig. 4, our proposal is 74.07%, 103.90%, 99.59% better than baseline.

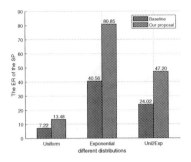

Fig. 3. The ER of the SP under different distribution

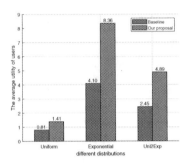

Fig. 4. The average utility of users under different distribution

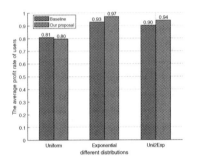

Fig. 5. The average profit of users under different distribution

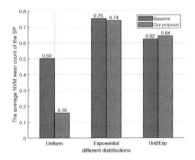

Fig. 6. the average number of NVM wear(or content replace) under different distribution

Third, in addition to the users' utility, it's also worth focusing our attention on the profit rate of users' utility (i.e. $\frac{utility}{payment}$). Owing to our realization with different service quality, SP obtain more from users in comparison to baseline which just provide $\theta = 1$ (i.e. the user payment of proposal is higher than baseline). The reason why we do that is 61% of Chinese consumers are ready to buy high-end products for their superior quality [13] and considering a realistic internet service scenario in common. So, for the sake of fairness between our proposal and baseline, we define profit rate $PR_j(t) = \frac{u_j(t)}{x_j(t)}$. In Fig. 5, we can see that our proposal's average PR is near to baseline under Uni, and 4.30%, 4.45% better under Exp and Uni2Exp respectively.

Fourth, considering the expensive cost and limited wear times (one of the determining factor of reliability) of NVM, we continue to compare our proposal with baseline on NVM wear times under three different users valuation distribution, as shown in Fig. 6. The average NVM wear times of proposal is 212.50%, 1.35% less than baseline under Uni and Exp respectively. However, it is 3.85% more than baseline under Uni2Exp. It is because there are some differences in solving for proposal between $p_i(t)$ of proposal which decided by (θ, Δ) (i.e. formula 9) and $p_i(t)$ of baseline which just decided by θ (i.e. formula 5).

Last, by analyzing Fig. 3–6, we can discover proposal is not overall better than baseline in all performance index. This is because the minimum of NVM wear times and the maximum of ER or utility do not satiable together in fact under random t. To be specific, high θ can effect $p_i(t)$. The change set of content Δ is the same.

6   Conclusions

In this paper, based on NVM storage and differentiated services scenarios, we proposed an optimal auction mechanism for edge caching. Our proposal took necessary and non-negligible aspects or costs into optimal auction mechanism in edge servers(for example, SPs). In particular, we considered that NVM will be wear heavier with internet data exponentially proliferation, and took the NVM cost into our SP operating cost. In Addition, by considering that different users should be served with different delivery of quality, our proposal not only are adapted to the more realistic scenario but further optimize SP earnings. Finally, some necessary numerical results in this paper showed and proved our proposal effectiveness and correctness.

Next, our future work will consider the problem of energy consumption of edge cache , making our work more practical. In addition, some problem of bandwidth consumption on network edge is also becoming a hot research topic and attracting SP's attention in business.

Acknowledgements. This work is supported by The National Key Research and Development Program of China (2016YFB1000302), National Natural Science Foundation of China (61832020, 61702569), and Key-Area Research and Development Program of Guang Dong Province (2019B010107001).

References

1. Cao, X., Zhang, J., Poor, H.V.: An optimal auction mechanism for mobile edge caching. In: 2018 IEEE 38th International Conference on Distributed Computing Systems (ICDCS), pp. 388–399. IEEE (2018)
2. Forecast, C.V.: Cisco visual networking index: Global mobile data traffic forecast update 2015-2020. In: Cisco Public Information, p. 9 (2016)
3. Georgiadis, L., Neely, M.J., Tassiulas, L., et al.: Resource allocation and cross-layer control in wireless networks. Found. Trends® Networking **1**(1), 1–144 (2006)
4. Guo, L., Tan, E., Chen, S., Xiao, Z., Zhang, X.: The stretched exponential distribution of internet media access patterns. In: Proceedings of the Twenty-seventh ACM Symposium on Principles of Distributed Computing, pp. 283–294. ACM (2008)
5. Guo, Y., Liu, F., Xiao, N., Chen, Z.: Task-based resource allocation bid in edge computing micro datacenter. Comput. Mater. Continua **61**(2), 777–792 (2019)
6. Hosking, J.R., Wallis, J.R.: Parameter and quantile estimation for the generalized pareto distribution. Technometrics **29**(3), 339–349 (1987)
7. Intel: 3d xpoint introduction (2019). https://www.intel.com/content/www/us/en/products/memory-storage/optane-dc-persistent-memory.html

8. Jia, G., Han, G., Du, J., Chan, S.: A maximum cache value policy in hybrid memory based edge computing for mobile devices. IEEE IoT J. **6**, 4401-4410 (2018)
9. Li, X., Wang, X., Wan, P.J., Han, Z., Leung, V.C.: Hierarchical edge caching in device-to-device aided mobile networks: Modeling, optimization, and design. IEEE J. Sel. Areas Commun. **36**(8), 1768–1785 (2018)
10. Liu, J., Chen, S.: Initial experience with 3d xpoint main memory. In: 2019 IEEE 35th International Conference on Data Engineering Workshops (ICDEW), pp. 300–305. IEEE (2019)
11. Yang, D.L., Juan, T.Y.: A survey on the storage issues in edge computing. ZTE Technol. J. **25**(3), 15–22 (2019)
12. McAuley, D., Mortier, R., Goulding, J.: The dataware manifesto. In: 2011 Third International Conference on Communication Systems and Networks (COMSNETS 2011), pp. 1–6. IEEE (2011)
13. Nielsen: Total consumer report 2019 (2019). https://www.nielsen.com/us/en/insights/report/2019/total-consumer-report2019
14. Poularakis, K., Iosifidis, G., Tassiulas, L.: A framework for mobile data offloading to leased cache-endowed small cell networks. In: 2014 IEEE 11th International Conference on Mobile Ad Hoc and Sensor Systems, pp. 327–335. IEEE (2014)
15. Shi, W., Zhang, X., Wang, Y., Zhang, Q.: Edge computing: State-of-the-art and future directions. J. Comput. Res. Dev. **56**(1), 69 (2019). https://doi.org/10.7544/issn1000-1239.2019.20180760, http://crad.ict.ac.cn/CN/abstract/article_3851.shtml
16. Tadrous, J., Eryilmaz, A.: On optimal proactive caching for mobile networks with demand uncertainties. IEEE/ACM Trans. Netw. **24**(5), 2715–2727 (2015)
17. Tan, H., Han, Z., Liu, L., Han, K., Zhao, Q., et al.: Camul: Online caching on multiple caches with relaying and bypassing. In: IEEE INFOCOM 2019-IEEE Conference on Computer Communications, pp. 244–252. IEEE (2019)
18. Wei, Y., Wang, Z., Guo, D., Yu, F.R.: Deep q-learning based computation offloading strategy for mobile edge computing. Comput. Mater. Continua **59**(1), 89–104 (2019)
19. Wu, K., Arpaci-Dusseau, A., Arpaci-Dusseau, R.: Towards an unwritten contract of intel optane SSD. In: 11th USENIX Workshop on Hot Topics in Storage and File Systems (HotStorage 19). USENIX Association, Renton (2019)
20. Zeng, Y., Huang, Y., Liu, Z., Yang, Y.: Joint online edge caching and load balancing for mobile data offloading in 5g networks. In: 2019 IEEE 39th International Conference on Distributed Computing Systems (ICDCS), pp. 923–933. IEEE (2019)
21. Zhang, J., Xie, N., Zhang, X., Yue, K., Li, W., Kumar, D.: Machine learning based resource allocation of cloud computing in auction. Comput. Mater. Continua **56**(1), 123–135 (2018)

Conjunctive Keywords Searchable Encryption Scheme Against Inside Keywords Guessing Attack from Lattice

Xiaoling Yu, Chungen Xu$^{(\boxtimes)}$, and Bennian Dou

School of Science, Nanjing University of Science and Technology,
Nanjing 210094, China
xuchung@njust.edu.cn

Abstract. Public-key encryption with conjunctive keywords search (P-ECKS) is an extension of public-key encryption with keywords search (PEKS), which can realize the efficient search of the encrypted data stored in the cloud server and keep the privacy of these data during the search phase. However, there exists an inherent security issue for the typical PEKS schemes that they cannot prevent the inside keywords guessing attack (KGA). Moreover, most PEKS schemes were constructed based on the hardness of some number theory problems which can be solved in polynomial time using the quantum computer. Thus once large-scale quantum computers are built, these PEKS schemes aren't secure in the future quantum era. To address the above issues, this paper proposes a lattice-based PECKS scheme which can resist inside KGA. Its security can be reduced to the hardness of LWE problem and ISIS problem, thus it can resist quantum computing attack. We also give a comparison with other searchable encryption schemes on the computational cost of the main algorithms and the size of related parameters.

Keywords: Searchable encryption · Lattice assumption · Conjunctive keywords search · Inside keywords guessing attack · Post-quantum secure

1 Introduction

To save the local storage and keep the privacy of data, more and more users outsource the encrypted data to the cloud server. Public-key encryption with keywords search (PEKS) is a technique that allows users to search the encrypted data. Boneh et al. [5] proposed the first PEKS scheme, which refers to three parties, the sender (data owner), the receiver (data user) and the server. The sender

The authors would like to thank the support from Fundamental Research Funds for the Central Universities (No. 30918012204), China. The authors also gratefully acknowledge the helpful comments and suggestions of other researchers, which has improved the presentation.

© Springer Nature Switzerland AG 2020
X. Sun et al. (Eds.): ICAIS 2020, LNCS 12239, pp. 107–119, 2020.
https://doi.org/10.1007/978-3-030-57884-8_10

encrypts data using the receiver's public key and then uploads the ciphertext to the cloud. When the receiver intends to query data related to some keywords, he will compute a token of keywords by his secret key as a search request and sends it to the server. In addition to the data storage, the server provides the function of data searching, where he executes a test algorithm between the ciphertext stored in the server and the search token and then returns the search results. Furthermore, to enrich the query function and improve search efficiency, public-key encryption with conjunctive keywords search (PECKS) is studied, where more than one keyword can be searched once time.

Byun et al. [7] introduced an attack for the traditional PEKS schemes, keywords guessing attack (KGA) which is resulted from the fact that the keywords space has low entropy [12]. Thus given a search token, an adversary can exhaust the ciphertexts of keywords from the keyword space using the receiver's public key and confirm the keyword in the token when a test for a guessed keyword succeeds. Under this attack, the privacy of the receiver is fragile, such as the keywords of other returned documents, and which keyword the receiver intends to search, etc. Furthermore, this attack is more convenient for a malicious server controlled frequently by untrusted third parties, where this attack is named as inside KGA.

Moreover, the security of most PEKS schemes [5,12,14,17,19,20,22] depends on the hardness of number theory problems which can be solved in polynomial time by a quantum computer. Thus once large-scale quantum computers are ever built, they will be able to break these cryptosystems currently in use. This vulnerability has prompted the research on post-quantum cryptography (PQC) which uses some alternative mathematical problems to resist quantum computing attack. Lattice-based cryptography has been studied by many researchers as one of the most vital techniques of PQC.

To address the above issues, this paper focuses on constructing a lattice-based PECKS scheme which can resist inside KGA. We adopt a special hash function to transform the keyword into a matrix. In this scheme, we employ the lattice trapdoor generation algorithm to generate the sender's and the receiver's public and secret key pairs. To achieve the inside KGA-resistance, we append a GPV signature to the ciphertext, which makes sure that the KGA attacker including a malicious server cannot forge a valid ciphertext. To achieve the conjunctive keywords search, a vector associated with each queried keyword is chosen temporarily and randomly during the generation of token, thus the generated token cannot leak the information of the queried keyword.

The contributions of this paper are as follows.

- We construct a public-key encryption with conjunctive keywords search (PECKS) scheme, whose security can be reduced to the hardness of LWE problem and ISIS problem from lattice. Thus this scheme is post-quantum secure. Moreover, our scheme can prevent the malicious server from performing KGA.
- We also show a comparison with other searchable schemes on the computational cost and the size of the related parameters. In addition to the property

of post-quantum computing, the operations involved in our scheme are mainly multiplication and addition of modular matrix as well as sample algorithm, instead of bilinear maps and exponent operation whose computational costs are expensive.

Related Works: There are some extensions for PEKS model proposed which refer to multi-keywords search [6,22,24]. To resist inside KGA, Huang et al. [12] proposed a PEKS construction using bilinear maps, where the secret key of the sender is added into ciphertext generation to prevent the server from testing casually. Chen et al. [8] and Sun et al. [18] used more than one servers to prevent any single server from conducting the test independently. There were some lattice-based PEKS schemes proposed [4,11,13,21,23,25–27]. Zhang et al. [25] proposed a PEKS scheme with a designated cloud server from lattice. Zhang et al. [26] and Xu et al. [21] constructed the identity-based searchable encryption scheme. Mao et al. [15] showed a lattice-based encryption scheme with conjunctive keywords search. To resist inside KGA, this scheme requires two servers which don't collude with each other. Kuchta et al. [13] gave an attribute-based searchable encryption scheme from lattice. Behnia et al. [4] focused on the experimental analysis on lattice-based PEKS constructions. Zhang et al. [27] first introduced the forward secure PEKS model and showed a construction on lattice.

2 Preliminaries

Notations: $[n]$ denotes a set of $\{1, 2, ..., n\}$. $\|\tilde{A}\|$ means the Gram-Schmidt norm of the matrix A, where $\tilde{A}$ is the Gram-Schmidt orthogonalization of A. Given two random variables X and Y over a countable domain S, a function $\Delta(X, Y) = \frac{1}{2} \sum_{s \in S} |X(s) - Y(s)|$ is defined to be statistical distance. If the distance $\Delta(X, Y)$ is negligible, we say X and Y are statistically close.

2.1 Lattice

Definition 1. (Lattice). *Given positive integers n, m and some linearly independent vectors $\boldsymbol{b}_i \in \mathbb{R}^m$ for $i \in [n]$, the set generated by the above vectors $\Lambda(\boldsymbol{b}_1, \cdots, \boldsymbol{b}_n) = \{\Sigma_{i=1}^n x_i \boldsymbol{b}_i | x_i \in \mathbb{Z}\}$ is a lattice.*

From the above definition, the set $\{\boldsymbol{b}_1, \cdots, \boldsymbol{b}_n\}$ is a lattice basis. m is the dimension and n is the rank. One lattice is full-rank if its dimension equals to the rank, namely, $m = n$.

Definition 2. *For positive integers n, m and a prime q, a matrix $A \in \mathbb{Z}_q^{n \times m}$ and a vector $\boldsymbol{u} \in \mathbb{Z}_q^n$, define:*

$$\Lambda_q^{\perp}(A) := \{\boldsymbol{e} \in \mathbb{Z}^m | A\boldsymbol{e} = 0 \mod q\}$$
$$\Lambda_q^{\boldsymbol{u}}(A) := \{\boldsymbol{e} \in \mathbb{Z}^m | A\boldsymbol{e} = \boldsymbol{u} \mod q\}.$$
$$\Lambda_q(A) := \{\boldsymbol{u} \in \mathbb{Z}_q^m | \exists \boldsymbol{s} \in \mathbb{Z}_q^n, \boldsymbol{u} = A^T \boldsymbol{s} \mod q\}.$$

Observe that $\Lambda_q^{\mathbf{u}}(A)$ isn't a lattice but a shift of the lattice $\Lambda_q^{\perp}(A)$, where $\Lambda_q^{\mathbf{u}}(A) = \Lambda_q^{\perp}(A) + \mathbf{t}$ for a vector $\mathbf{t} \in \Lambda_q^{\mathbf{u}}(A)$. Assuming that $T \in \mathbb{Z}^{m \times m}$ is a basis of $\Lambda_q^{\perp}(A)$, T is a basis of $\Lambda_q^{\perp}(BA)$ for a full-rank $B \in \mathbb{Z}_q^{n \times n}$.

2.2   Hardness Assumptions

Definition 3. *[1] For $\alpha \in (0,1)$ and a prime q, let Ψ_α denote a probability distribution over $\mathbb{Z}_q$ by choosing $x \in \mathbb{R}$ from the normal distribution with mean 0 and standard deviation $\alpha/\sqrt{2\pi}$, outputs $\lfloor qx \rceil$.*

Definition 4 (Learning with errors, LWE problem). *[16] For a positive integer n, set $m = m(n)$ and $\alpha \in (0,1)$ such that a prime $q = q(n) > 2$ and $\alpha q > 2\sqrt{n}$, a secret $\mathbf{s} \xleftarrow{\$} \mathbb{Z}_q^n$, define the following distributions over $\mathbb{Z}_q^{n \times m} \times \mathbb{Z}_q^m$:*

- *LWE distribution: choose uniformly a matrix $A \xleftarrow{\$} \mathbb{Z}_q^{n \times m}$, and sample $\mathbf{e} \leftarrow \Psi_\alpha^m$, output $(A, A^T \mathbf{s} + \mathbf{e}) \in \mathbb{Z}_q^{n \times m} \times \mathbb{Z}_q^m$;*
- *Uniform distribution: choose uniformly a matrix $A \xleftarrow{\$} \mathbb{Z}_q^{n \times m}$ and vector $\mathbf{x} \xleftarrow{\$} \mathbb{Z}_q^m$, output $(A, \mathbf{x}) \in \mathbb{Z}_q^{n \times m} \times \mathbb{Z}_q^m$,*

The decisional LWE problem is to distinguish between the above distributions. The LWE problem is hard, if for any probability polynomial time (PPT) adversary $\mathscr{A}$, $|Pr[\mathscr{A}(A, A^T \mathbf{s} + \mathbf{e}) = 1] - Pr[\mathscr{A}(A, \mathbf{x}) = 1]|$ is negligible on n.

Definition 5. (Inhomogeneous small integer solution, ISIS problem). *Given a matrix $A \in \mathbb{Z}_q^{n \times m}$ and a uniform vector $\mathbf{u} \in \mathbb{Z}_q^n$, a real number $\eta > 0$, the problem is to find a non-zero integer vector $\mathbf{e} \in \mathbb{Z}_q^m$ such that $A\mathbf{e} = \mathbf{u} \mod q$ and $\|\mathbf{e}\| \leqslant \eta$.*

LWE problem [16] and ISIS problem [10] have been proved as hard as approximating the worst-case Gap-SVP and SIVP with certain factors.

2.3   Lattice Algorithms

Definition 6 (Gaussian distribution). *Given parameter $\sigma \in \mathbb{R}^+$, a vector $\mathbf{c} \in \mathbb{R}^m$ and a lattice Λ, $\mathbf{D}_{\Lambda,\sigma,c}$ is a discrete gaussian distribution over Λ with a center $\mathbf{c}$ and a parameter σ, denoted by $\mathbf{D}_{\Lambda,\sigma,c} = \dfrac{\rho_{\sigma,c(x)}}{\rho_{\sigma,c(\Lambda)}}$ for $\forall x \in \Lambda$, where $\rho_{\sigma,c(\Lambda)} = \sum_{x \in \Lambda} \rho_{\sigma,c(x)}$ and $\rho_{\sigma,c(x)} = \exp(-\pi\dfrac{\|x - c\|^2}{\sigma^2})$.*

Lemma 1 (TrapGen algorithm). *[2, 3, 10] Given integers n, m, q with $q > 2$ and $m \geqslant 6n \log q$ as the input, there is a PPT algorithm TrapGen, outputs a matrix $A \in \mathbb{Z}_q^{n \times m}$ along with a basis T_A of the lattice $\Lambda^{\perp}(A)$, namely, $A \cdot T_A = 0 \mod q$, where the distribution of A is statistically close to uniform on $\mathbb{Z}_q^{n \times m}$, and $\|\tilde{T}_A\| \leqslant O(\sqrt{n \log q})$.*

Lemma 2. (Preimage sample algorithm). *Given a matrix $A \in \mathbb{Z}_q^{n \times m}$ with a basis $T_A \in \mathbb{Z}_q^{m \times m}$, a vector $\boldsymbol{u} \in \mathbb{Z}_q^n$, and a parameter $\sigma \geqslant \|\tilde{T}_A\| \cdot \omega(\sqrt{\log m})$ as the input, where $m \geqslant 2n\lceil \log q \rceil$, there is a PPT algorithm SamplePre, outputs a sample $\boldsymbol{e} \in \mathbb{Z}_q^m$ distributed in $\mathbf{D}_{\Lambda_q^u(A), \sigma}$, such that $A\boldsymbol{e} = \boldsymbol{u} \mod q$.*

Lemma 3 (SampleLeft algorithm). *[1] Given a matrix $A \in \mathbb{Z}_q^{n \times m}$ with basis T_A of lattice $\Lambda^\perp(A)$ and a matrix $A' \in \mathbb{Z}_q^{n \times m_1}$, a vector $\boldsymbol{u} \in \mathbb{Z}_q^n$, a gaussian parameter σ, there is an algorithm SampleLeft to output a short vector $\boldsymbol{e}$ sampled from a distribution statistically close to $D_{\Lambda_q^u(F_1), \sigma}$, where $F_1 = (A|A')$.*

Lemma 4. (SampleRight algorithm). *[1] Given a matrix $A \in \mathbb{Z}_q^{n \times k}$ and a matrix $B \in \mathbb{Z}_q^{n \times m}$, a random matrix $R \in \mathbb{Z}_q^{k \times m}$, a basis T_B of $\Lambda^\perp(B)$, a vector $\boldsymbol{u} \in \mathbb{Z}_q^n$, and a parameter σ, there is an algorithm SampleRight to output a vector $\boldsymbol{e}$ sampled from a distribution statistically close to $D_{\Lambda_q^u(F_2), \sigma}$, where $F_2 = (A|AR + B)$.*

Our scheme employs an injective encoding function, full-rank difference map (FRD), to transform from a keyword to a matrix. The keywords set can be expanded to $\{0,1\}^*$ using a collision-resistant hash function over $\mathbb{Z}_q^n$.

Definition 7. (FRD). *[1] Given a prime q and a positive integer n, a function $H : \mathbb{Z}_q^n \to \mathbb{Z}_q^{n \times n}$ is a full-rank difference function, if it satisfies the following conditions:*

1. *For all distinct $\boldsymbol{u}, \boldsymbol{v} \in \mathbb{Z}_q^n$, the matrix $H(\boldsymbol{u}) - H(\boldsymbol{v}) \in \mathbb{Z}_q^{n \times n}$ is full rank;*
2. *H is computable in polynomial time (in $n \log q$).*

Gentry et al. [10] constructed a signature using a one-way preimage sampleable function (PSF), which has been proved to be strongly existentially unforgeable based on the hardness of ISIS problem.

Definition 8. (GPV signature). *[10] The signature includes three algorithms, {KeyGen, Sign, Verify}.*

- **KeyGen:** *Given a security parameter λ, the signature key generation algorithm invokes the $TrapGen(n, m, q)$ to generate (A, T), where A describes a function $f_A(\boldsymbol{e}) = A\boldsymbol{e} \mod q$ with $\boldsymbol{e} \leftarrow D_{\mathbb{Z}^m, s}$ and $\|\boldsymbol{e}\| \leqslant s\sqrt{m}$. The verification key is the matrix A and the signing key is a basis T of the lattice $\Lambda^\perp(A)$.*
- **Sign:** *Given a message M and the signing key T as the input, the signing algorithm chooses randomly $r \leftarrow \{0,1\}^k$, then runs $SamplePre(A, T, H(M|r), \sigma)$ to generate a sample $\boldsymbol{e}$ as the signature of M.*
- **Verify:** *if $\boldsymbol{e}$ satisfies $A\boldsymbol{e} = H(M|r) \mod q$ and $\boldsymbol{e} \leftarrow D_{\mathbb{Z}^m, s}$ and $r \leftarrow \{0,1\}^k$, then accept. Otherwise, reject.*

Lemma 5. *[10] The above signature is strongly existentially unforgeable under a chosen-message attack if ISIS problem is hard.*

3   Syntax of Public-Key Encryption with Conjunctive Keywords Search

Overview: This section introduces the model of PECKS against inside KGA. There are some instructions as follows:

- In this paper, we focus on the encryption of keywords. For the encryption of data document, we adopt the classical symmetric encryption technique which is widely believed to be more efficient than public-key encryption schemes.
- There are l keyword fields defined for each document. One keyword consists of two parts, a field and a specific keyword. Considering an email system, the keywords fields set is {"From", "To", "Date", "Subject"}. "Field name+ Null" can be used to denote a field which has not a valid keyword. Given a keywords set of {From: Null, To: Alice, Date: 2019.01.01, Subject: Education}, documents (or emails) on "Education" will be returned which were sent to "Alice" on Jan. 1st, 2019. For simplify the notations, we do a sort for the keywords field and use the specific keyword w along with its position which denotes the location of the keyword instead of "Field name+ w".

The PECKS model involves four participants, the central authority, the sender, the receiver and the cloud server. In addition to computing the public parameters, the central authority computes the sender's and receiver's public and secret key pairs, and sends them to the sender and the receiver via a secure channel, respectively. The sender encrypts the ordered keywords set $W = \{w_1, ..., w_l\}$ of each document using his secret key and the receiver's public key, and then stores the ciphertext C_W to the server. When the receiver intends to search some documents associated with keywords set $\{w'_1, ..., w'_t\}$ along with its position set $\{j_1, ..., j_t\} \subset [l]$, he will compute a search token T_W on this keywords set. The server will execute a test algorithm to check if the queried keywords set is contained in the keywords set of document or not, that is, $(w_{j_1} = w'_1) \cap (w_{j_2} = w'_2) \cap ... \cap (w_{j_t} = w'_t)$, then he will return the queried results.

3.1   System Model

One PECKS scheme against inside KGA consists of six algorithms, $\Pi =$(**Setup, SKeyGen, RKeyGen, Enc, Token, Test**).

- $pp \leftarrow$ **Setup**(λ, l): Given the security parameter λ and the number l of keywords field in one document as the input, **Setup** algorithm outputs the public parameter pp.
- $(pk_s, sk_s) \leftarrow$ **SKeyGen**(pp): Given the public parameter pp, the key generation algorithm returns the sender's public key pk_s and secret key sk_s.
- $(pk_r, sk_r) \leftarrow$ **RKeyGen**(pp): Given the public parameter pp, the key generation algorithm returns the receiver's public key pk_r and secret key sk_r.
- $C_W \leftarrow$ **Enc**(W, pk_s, sk_s, pk_r): Given a keywords set $W = \{w_1, ..., w_l\}$, the sender's key pair and the receiver's public key as the input, the encryption algorithm outputs the ciphertext of keywords set C_W.

- $T_W \leftarrow \textbf{Token}(\{Q, I\}, sk_r, pk_r)$: Given a queried keywords set $Q = \{w_{j_1}, ...,$ $w_{j_t}\}$ along with its position set $I = \{j_1, ..., j_t\} \subset [l]$, the receiver's key pair as the input, the token generation algorithm outputs a search token T_W.
- $\{0, 1\} \leftarrow \textbf{Test}(T_W, C_W, pk_s)$: Given the token, the ciphertext, as well as public key of the sender as the input, the test algorithm outputs 1 if $S \subset W$, otherwise outputs 0.

3.2   Security Model

The security model of PECKS schemes requires that there is no PPT adversary who can distinguish the ciphertext with a random element chosen form the ciphertext set. We introduce the following game to define security model, ciphertext indistinguishability under chosen keywords attack, which refers to two parties, a challenger and an attacker. Given security parameter λ, a challenger $\mathscr{C}$ and an adversary $\mathscr{A}$ carry out the following games, denoted by $Exp_{\mathscr{A}}$.

- $\textbf{Setup}(\lambda, l)$: The adversary $\mathscr{A}$ chooses the target keywords set W^* and sends it to $\mathscr{C}$. Given security parameter λ and the number of keywords l as input, $\mathscr{C}$ outputs public parameter pp and sends them to $\mathscr{A}$.
- $\textbf{Query phase 1}$: $\mathscr{A}$ is allowed to query tokens of any keywords set $W_1, ..., W_t$ adaptively, which means the adversary can issue the q_kth query with the knowledge of $q_1, ..., q_{k-1}$ queries. Then $\mathscr{C}$ returns the token to $\mathscr{A}$.
- $\textbf{Challenge phase}$: $\mathscr{C}$ randomly choose a bit $r \in \{0, 1\}$, if $r = 0$, computes $C_W \leftarrow Enc(W^*, pk_r, sk_s)$ and returns it to $\mathscr{A}$. Otherwise, selects randomly an element C from the ciphertext space to return to $\mathscr{A}$.
- $\textbf{Query phase 2}$: $\mathscr{A}$ can continue to query tokens of any keywords set adaptively.
- $\textbf{Guess}$: $\mathscr{A}$ guesses a bit $r' \in \{0, 1\}$. If $r = r'$, return 1. Otherwise, return 0.

One PECKS scheme satisfies ciphertext indistinguishability under chosen keywords attack if the advantage $Adv_{\mathscr{A}}(\lambda) = |Pr[Exp_{\mathscr{A}}(\lambda) = 1] - \frac{1}{2}|$ is negligible for any PPT adversary $\mathscr{A}$.

4   PECKS Scheme Against Inside KGA from Lattice

4.1   Our Construction

In this section, we show a construction of conjunctive keywords searchable encryption scheme from lattice. In this scheme, we adopt the TrapGen algorithm to generate the sender's and the receiver's public and secret key pairs. During the encryption phase, we construct a GPV signature on a random bit and the part of the ciphertext, which makes sure the KGA attacker including a malicious server cannot forge a valid ciphertext used for the test algorithm. In the generation of token, the vector $\mathbf{u}_i$ associated with each queried keyword is chosen temporarily and randomly, thus the generated token cannot leak the information of the queried keyword.

- **Setup**(λ, l): For system parameters $n, m, q, \delta, \sigma, \alpha$ where q is prime, the central authority generates the public parameter $pp = (q, n, m, \delta, \sigma, \alpha, \mathbf{u}, H, H_0, \{A_i\}_{i=1}^{l}, B)$, where a vector $\mathbf{u}$ is selected randomly from the uniform distribution over $\mathbb{Z}_q^n$. A function $H : \{0,1\}^* \rightarrow \mathbb{Z}_q^{n \times n}$ is chosen from the FRD hash function set $\mathscr{H}$ and a hash function $H_0 : \mathbb{Z}_q^{2m} \times \{0,1\} \rightarrow \mathbb{Z}_q^n$ is chosen from the hash function set $\mathscr{H}_0$. $l+1$ random matrices $(B, \{A_i\}_{i=1}^{l})$ are chosen from the uniform distribution over $\mathbb{Z}_q^{n \times m}$.

- **SKeyGen**(pp): Given a public parameter pp as input, the key generation algorithm runs algorithm $TrapGen(n, m, q)$ to generate a random matrix $A_s \in \mathbb{Z}_q^{n \times m}$ with a short basis T_s of $\Lambda_q^{\perp}(A_s)$, then outputs the sender's public key $pk_s = A_s$ and secret key $sk_s = T_s$.

- **RKeyGen**(pp): Given a public parameter pp as input, the key generation algorithm runs algorithm $TrapGen(n, m, q)$ to generate a random matrix $A_r \in \mathbb{Z}_q^{n \times m}$ with a short basis T_r of $\Lambda_q^{\perp}(A_r)$, then outputs the receiver's public key $pk_r = A_r$ and secret key $sk_r = T_r$.

- **Enc**$(W = \{w_1, w_2, ..., w_l\}, pk_s, sk_s, pk_r)$: Given keywords set $W = \{w_1, ..., w_l\}$, the sender's key pair and the receiver's public key, the sender encrypts these keywords and stores the ciphertext to the server. In the phase, a random vector $\mathbf{s}$ is selected from a uniform distribution over $\mathbb{Z}_q^n$ and a single bit b is chosen from $\{0,1\}$ at random. Then the sender chooses a noise vector $\mathbf{y} \leftarrow \Psi_\alpha^m$ and $x \leftarrow \Psi_\alpha$, where Ψ_α is a distribution over $\mathbb{Z}_q$ as Definition 3. Nextly, the sender computes $C_0 = \mathbf{u}^T \mathbf{s} + x + b \lfloor \frac{q}{2} \rfloor \in \mathbb{Z}_q$. For each keyword w_i, the sender chooses a random matrix $R_i \in \{-1, 1\}^{m \times m}$, then computes $C_i = (A_r | A_i + H(w_i)B)^T \mathbf{s} + \mathbf{z}_i \in \mathbb{Z}_q^{2m}$ and employs a hash function H_0 in (C_i, b) to obtain $\mathbf{h}_i = H_0(C_i | b)$. The preimage sample algorithm as Definition 2 is invoked to obtain a preimage $\mathbf{e}_i \in \mathbb{Z}_q^m$ which satisfies $A_s \mathbf{e}_i = \mathbf{h}_i \mod q$. Finally, the sender returns $(C_0, \{C_i\}_{i=1}^{l}, \{\mathbf{e}_i\}_{i=1}^{l})$ as the ciphertext.

- **Token**$(S = \{w_{j_1}, ..., w_{j_t}\}, I = \{j_1, ..., j_t\}), pk_r, sk_r)$: For the queried keywords set Q along with the position set I, the receiver computes a token T_W using his own secret key. This algorithm firstly selects t random vecrots from $\mathbb{Z}_q^n$ such that $\sum_{k=1}^{t} \mathbf{u}_{j_k} = \mathbf{u}$. Then he computes a matrix $M_{j_k} = A_{j_k} + H(w_{j_k})B$, and invokes the SampleLeft algorithm to sample $T_{j_k} \in \mathbb{Z}_q^{2m}$ which satisfies $(A_r | A_{j_k} + H(w_{j_k})B)T_{j_k} = \mathbf{u}_{j_k} \mod q$. Finally, the receiver returns all of the queried tokens to the server.

- **Test**(T_W, C_W, pk_s): The server computes $r = C_0 - \sum_{k=1}^{t} T_{j_k}^T \cdot C_{j_k}$ to recover b, then for each queried keyword, the server computes the hash value on the second part of ciphertext C_i along with b to verify if $A_s \mathbf{e}_i = \mathbf{h}_i \mod q$ holds. When this equation holds for all of the queried keywords which indicates the token matches ciphertext, namely, the ciphertext includes the queried keywords set encapsulated in the token.

For the **Test** algorithm, we have the following result:

$$r = C_0 - \sum_{k=1}^{t} T_{j_k}^T \cdot C_{j_k}$$

$$= \mathbf{u}^T \mathbf{s} + x + b\lfloor \frac{q}{2} \rfloor - \sum_{k=1}^{t} T_{j_k}^T \cdot ((A_r | A_{j_k} + H(w_{j_k})B)^T \mathbf{s} + \begin{pmatrix} \mathbf{y} \\ R_{j_k}^T \mathbf{y} \end{pmatrix})$$

$$= b\lfloor \frac{q}{2} \rfloor + x - \sum_{k=1}^{t} T_{j_k}^T \cdot \begin{pmatrix} \mathbf{y} \\ R_{j_k}^T \mathbf{y} \end{pmatrix},$$

where $x - \sum_{k=1}^{t} T_{j_k}^T \cdot \begin{pmatrix} \mathbf{y} \\ R_{j_k}^T \mathbf{y} \end{pmatrix}$ is error term. To decrypt correctly, we need to make sure that these error terms are less than $\frac{q}{4}$. And the work in [1] gave more discussions on the security parameters.

4.2   Security Analysis

In the section, we show the security proof of the proposed scheme for ciphertext indistinguishability. Then we show an analysis on inside KGA resistance.

Lemma 6. *For any PPT adversary $\mathscr{A}$, the proposed PECKS scheme is ciphertext indistinguishability against selectively chosen keywords attack from the hardness of $(\mathbb{Z}_q, n, \Psi_\alpha)$-LWE problem.*

Proof. Let $\mathscr{A}$ be a PPT adversary to break the ciphertext indistinguishability under selectively chosen keywords of the proposed scheme. Then we construct a challenger $\mathscr{C}$ who can solve the LWE problem.

Setup: The adversary $\mathscr{A}$ chooses $W^* = (w_1^*, ..., w_m^*)$ as the traget keywords set and sends it to $\mathscr{C}$. $\mathscr{C}$ queries the LWE oracle $\mathscr{O}$ for $m + 1$ times and obtains fresh pairs $(\mathbf{u}_i, v_i) \in \mathbb{Z}_q^n \times \mathbb{Z}_q$ for $i \in \{0, 1, ..., m\}$. Then $\mathscr{C}$ prepares the system parameters pp and public key as follows:

1. Construct a random matrix $A_r = (\mathbf{u}_1, ..., \mathbf{u}_m) \in \mathbb{Z}_q^{n \times m}$ from m of the given LWE samples. Let $\mathbf{u}$ be $\mathbf{u}_0$.
2. By running *TrapGen* algorithm, the challenger $\mathscr{C}$ obtains a random matrix $B \in \mathbb{Z}_q^{n \times m}$ as well as a trapdoor T_B for $\Lambda_q^{\perp}(B)$. For each $i \in [t]$, $\mathscr{C}$ selects a random matrix $R_i^* \in \{-1, 1\}^{m \times m}$ to construct $A_i \leftarrow A_r R_i^* - H(w_i^*)B$.

Query Phase 1: $\mathscr{C}$ answers some token queries $W_1, W_2, ...W_T$, where $W_t = \{w_{t1}, ..., w_{tk}\}$ for $t \in [T]$. For each element w_{tj} in the keywords set, $\mathscr{C}$ computes $(A_r | A_{tj} + H(w_{tj})B) = (A_r | A_r R_{tj}^* + (H(w_{tj}) - H(w_i^*))B)$, where $H(w) - H(w^*)$ is a non-singular matrix and T_B is a trapdoor for $\Lambda_q^{\perp}(B')$, where $B' = (H(w_{tj}) - H(w_i^*))B$. Then $\mathscr{C}$ can answer the token query to $\mathscr{A}$ which is generated from the SampleRight$(A_r, (H(w_{tj}) - H(w_i^*))B, R_i^*, T_B, \mathbf{u}, \delta)$.

H_0 query: For an $H_0(C_2|b)$ query, $\mathscr{C}$ returns the hash value if the related input $(C_2|b)$ has been queried. Otherwise, $\mathscr{C}$ selects at random $\mathbf{h}$ from $\mathbb{Z}_q^n$ and returns it into the hash list L_0.

Challenge Phase: $\mathscr{A}$ selected randomly a bit $b^* \in \{0, 1\}$, then $\mathscr{C}$ generates the ciphertext of the target keywords as follows:

1. set $\mathbf{v}^* = (v_1, ..., v_m) \in \mathbb{Z}_q^m$, compute $C_0^* = v_0 + b^* \lfloor \frac{q}{2} \rfloor \in \mathbb{Z}_q$ and $C_i^* = (\mathbf{v}^*|(R_i^*)^T \mathbf{v}^*) \in \mathbb{Z}_q^{2m}$. Compute $h_i^* = H_0(C_i^*|b^*)$ and invokes SamplePre algorithm to generate a preimage sample $\mathbf{e}_i^*$.
2. select randomly a bit $r \in \{0, 1\}$, if $r = 0$, return $(C_0^*, \{C_i^*, \mathbf{e}_i^*\}_{i=1}^t)$ to $\mathscr{A}$, otherwise return a random $(C_0, \{C_i\}_{i=1}^t)$ from the part of ciphertext space $\mathbb{Z}_q \times \mathbb{Z}_q^{2m \times t}$, and check if $\mathbf{h}_i = H_0(C_i|b)$ for a randomly chosen $b \in \{0, 1\}$ has been queried or not, if it hasn't been queried, compute the hash value $\mathbf{h}_i = H_0(C_i|b)$ and add it into the hash list L_0. The challenger generates $\mathbf{e} \leftarrow \mathbf{D}_{\Lambda_q^u(A_s), \delta}$ by SampleDom algorithm in [10]. Finally, $\mathscr{C}$ returns $(C_0, \{C_i, \mathbf{e}_i\}_{i=1}^t)$ to the adversary.

Query Phase 1: $\mathscr{A}$ can continue to query the token except the target keywords set W^*.

Guess: $\mathscr{A}$ outputs a guess on $r \in \{0, 1\}$.

If the LWE oracle is pseudorandom, $(A_r|A_i + H(w_i^*)B) = (A_r|A_r R_i^*)$. For the random noise vector $\mathbf{y} \leftarrow \Psi_\alpha^m$, $\mathbf{v}^* = A_r^T \mathbf{s} + \mathbf{y}$. Thus

$$C_2^* = \begin{pmatrix} A_r^T \mathbf{s} + \mathbf{y} \\ (R_i^*)^T A_r^T \mathbf{s} + (R_i^*)^T \mathbf{y} \end{pmatrix} = \begin{pmatrix} A_r^T \mathbf{s} + \mathbf{y} \\ (A_r R_i^*)^T \mathbf{s} + (R_i^*)^T \mathbf{y} \end{pmatrix} = (A_r|A_r R_i^*)^T \mathbf{s} +$$

$\begin{pmatrix} \mathbf{y} \\ (R_i^*)^T \mathbf{y} \end{pmatrix}$, which is a vaid part of ciphertext for the target keyword w_i^*. From the LWE instance, we can note that $v_0 = \mathbf{u}_0^T \mathbf{s} + x$, then $C_0^* = \mathbf{u}_0^T \mathbf{s} + x + b^* \lfloor \frac{q}{2} \rfloor$ is also a part of ciphertext. When the oracle is a truly random oracle, we have v_0 and $\mathbf{v}^*$ are uniform. Then C_i^* is uniform and independent. Thus the part of ciphertext $(C_0^*, \{C_i^*\}_{i=1}^t)$ is uniform in $\mathbb{Z}_q \times \mathbb{Z}_q^{2m}$. From the lemma 2, the preimage sample $\mathbf{e}^*$ is distributed in $\mathbf{D}_{\Lambda_q^u(A_s), \delta}$. Thus if the adversary $\mathscr{A}$ can distinguish the ciphertext, then $\mathscr{C}$ can solve the LWE problem.

Lemma 7. *For any PPT adversary $\mathscr{A}$, the proposed PECKS scheme can resist inside KGA if the GPV signature as Definition 8 is strongly existentially unforgeable under a chosen-message attack.*

From the proposed scheme, the ciphertext of keywords consists of three parts, $C_0, \{C_i\}_{i=1}^l$, and $\{\mathbf{e}_i\}_{i=1}^l$. Actually, we can note that the third part of the ciphertext $\mathbf{e}_i$ is a GPV signature on C_i and a random bit b encapsulated in C_0. According to the Lemma 5, the signature is strongly existentially unforgeable under a chosen message attack and can be reduced to the hardness of ISIS problem [10]. Hence, an adversary, including malicious server, cannot forge a valid ciphertext for performing the Test algorithm, where the server has to recover a single bit b and verifies this signature, then he can obtain the return results.

4.3   Performance Comparison

This section shows a comparison with other PEKS schemes as Table 1 on the computational cost of main algorithms and the size of parameters. There are some notations as follows: H: hash operation, E: exponent operation in group, P: pairing operation, M_P: multiplication operation of group elements, T_{SL}: SampleLeft operation, T_{SP}: SamplePre operation; M: modular multiplication, m: the dimension of matrix, l_q: the size of element in $\mathbb{Z}_q$, $|G|$: the size of group elements, κ: the security parameter used in the scheme [4], l: the number of keywords, t: the number of queried keywords.

Table 1. The performance comparison with other schemes:

Schemes	[12]	[9]	[4]	Ours
PEKS	$l(H + 3E + M_P)$	$2E + lH + lP$	$3l\kappa M$	$(2l+1)M + l(2H + T_{SP})$
Token	$t(P + H + E)$	$(t-1)M_P + E$	tT_{SL}	$t(H + M + T_{SL})$
Test	$t(2P + M_P)$	$P + (t-1)M_P$	$t\kappa M$	$(t-1)M + tH$
Ciphertext size	$2l|G|$	$(l+1)|G|$	$l((2m+1)l_q + 1)$	$(3ml+1)l_q$
Token size	$t \cdot |G|$	$t \cdot |G|$	$2mtl_q$	$2mtl_q$
Hardness	DBDH/mDLIN	DBDH	LWE	LWE/ISIS
Conjunctive	✗	✓	✗	✓
Inside KGA	✓	✗	✗	✓
Post-quantum	✗	✗	✓	✓

Here we refer to two schemes [4,12] for single keywords search, so these schemes are extended to conjunctive keywords for comparison. The schemes [9,12] are constructed using bilinear maps, thus they cannnot resist quantum computers compared with [4] and our scheme. The scheme in [4] does not consider inside KGA and rich query function. And this scheme deals with the keywords bit-by-bit encrypt each keyword instead of using the hash function. Our scheme requires matrix multiplications, hash operation and the SamplePre algorithm, during the encryption phase. The Token generation algorithm needs SampleLeft algorithm except matrix multiplications, hash operation. The Test algorithm of our scheme needs some matrix multiplications and hash operation. It's widely accepted that these operations are more efficient than the bilinear maps and exponent operation.

References

1. Agrawal, S., Boneh, D., Boyen, X.: Lattice basis delegation in fixed dimension and shorter-ciphertext hierarchical IBE. In: Rabin, T. (ed.) CRYPTO 2010. LNCS, vol. 6223, pp. 98–115. Springer, Heidelberg (2010). https://doi.org/10.1007/978-3-642-14623-7_6

2. Ajtai, M.: Generating hard instances of the short basis problem. In: Proceedings of the ICALP, pp. 1–9 (1999)
3. Alwen, J., Peikert, C.: Generating shorter bases for hard random lattices. In: Proceedings of the STACS, pp. 75–86 (2009)
4. Behnia, R., Ozmen, M.O., Yavuz, A.A.: Lattice-based public key searchable encryption from experimental perspectives. IEEE Trans. Dependable Secure Comput. (2018). https://doi.org/10.1109/TDSC.2018.2867462
5. Boneh, D., Di Crescenzo, G., Ostrovsky, R., Persiano, G.: Public key encryption with keyword search. In: Cachin, C., Camenisch, J.L. (eds.) EUROCRYPT 2004. LNCS, vol. 3027, pp. 506–522. Springer, Heidelberg (2004). https://doi.org/10.1007/978-3-540-24676-3_30
6. Boneh, D., Waters, B.: Conjunctive, subset, and range queries on encrypted data. In: TCC, pp. 535–554 (2007)
7. Byun, J.W., Rhee, H.S., Park, H., Lee, D.H.: Off-line keyword guessing attacks on recent keyword search schemes over encrypted data. In: Secure Data Management, Third VLDB Workshop, SDM, pp. 75–83 (2006)
8. Chen, R., Mu, Y., Yang, G., Guo, F., Wang, X.: A new general framework for secure public key encryption with keyword search. In: ACISP, pp. 59–76 (2015)
9. Farràs, O., Ribes-González, J.: Provably secure public-key encryption with conjunctive and subset keyword search. Int. J. Inf. Secur. 18(5), 533–548 (2019). https://doi.org/10.1007/s10207-018-00426-7
10. Gentry, C., Peikert, C., Vaikuntanathan, V.: Trapdoors for hard lattices and new cryptographic constructions. In: Proceedings of the ACM Symposium on Theory of Computing, pp. 197–206 (2008)
11. Gu, C., Zheng, Y., Kang, F., Xin, D.: Keyword search over encrypted data in cloud computing from lattices in the standard model. In: Cloud Computing and Big Data - Second International Conference, CloudCom-Asia, pp. 335–343 (2015)
12. Huang, Q., Li, H.: An efficient public-key searchable encryption scheme secure against inside keyword guessing attacks. Inf. Sci. 403, 1–14 (2017)
13. Kuchta, V., Markowitch, O.: Multi-authority distributed attribute-based encryption with application to searchable encryption on lattices. In: Mycrypt. Malicious and Exploratory Cryptology, pp. 409–435 (2016)
14. Liu, Y., Peng, H., Wang, J.: Verifiable diversity ranking search over encrypted outsourced data. Comput. Mater. Continua 55(1), 37–57 (2018)
15. Mao, Y., Fu, X., Guo, C., Wu, G.: Public key encryption with conjunctive keyword search secure against keyword guessing attack from lattices. Trans. Emerg. Telecommun. Technol. 30(11), e3531 (2018). https://doi.org/10.1002/ett.3531
16. Regev, O.: On lattices, learning with errors, random linear codes, and cryptography. J. ACM 56(6), 34:1–34:40 (2009)
17. Rhee, H.S., Park, J.H., Susilo, W., Lee, D.H.: Trapdoor security in a searchable public-key encryption scheme with a designated tester. J. Syst. Softw. 83(5), 763–771 (2010)
18. Sun, L., Xu, C., Zhang, M., Chen, K., Li, H.: Secure searchable public key encryption against insider keyword guessing attacks from indistinguishability obfuscation. SCIENCE CHINA Inf. Sci. 61(3), 038106:1–038106:3 (2018)
19. Tang, Y., Lian, H., Zhao, Z., Yan, X.: A proxy re-encryption with keyword search scheme in cloud computing. Comput. Mater. Continua 56(2), 339–352 (2018)
20. Xu, L., Xu, C., Liu, Z., Wang, Y., Wang, J.: Enabling comparable search over encrypted data for IoT with privacy-preserving. Comput. Mater. Continua 60(2), 675–690 (2019)

21. Xu, L., Yuan, X., Steinfeld, R., Wang, C., Xu, C.: Multi-writer searchable encryption: an LWE-based realization and implementation. In: Asia Conference on Computer and Communications Security, AsiaCCS, pp. 122–133 (2019)
22. Yang, Y., Ma, M.: Conjunctive keyword search with designated tester and timing enabled proxy re-encryption function for e-health clouds. IEEE Trans. Inf. Forensics Secur. **11**(4), 746–759 (2016)
23. Yang, Y., Zheng, X., Chang, V., Ye, S., Tang, C.: Lattice assumption based fuzzy information retrieval scheme support multi-user for secure multimedia cloud. Multimed. Tools Appl. **77**, 1–15 (2018)
24. Zhang, B., Zhang, F.: An efficient public key encryption with conjunctive-subset keywords search. J. Netw. Comput. Appl. **34**(1), 262–267 (2011)
25. Zhang, X., Xu, C.: Trapdoor security lattice-based public-key searchable encryption with a designated cloud server. Wireless Pers. Commun. **100**(3), 907–921 (2018)
26. Zhang, X., Xu, C., Mu, L., Zhao, J.: Identity-based encryption with keyword search from lattice assumption. China Commun. **15**(4), 164–178 (2018)
27. Zhang, X., Xu, C., Wang, H., Zhang, Y., Wang, S.: FS-PEKS: lattice-based forward secure public-key encryption with keyword search for cloud-assisted industrial internet of things. IEEE Trans. Dependable Secure Comput. (2019). https://doi.org/10.1109/TDSC.2019.2914117

Detecting Bluetooth Attacks Against Smartphones by Device Status Recognition

Fan Wei[1,2(✉)]

[1] Information Security Center, State Key Laboratory of Networking and Switching Technology, Beijing University of Posts and Telecommunications, Beijing 100876, China
buptweifan@qq.com
[2] Sate Key Laboratory of Public Big Data, Guizhou University, Guiyang 550025, Guizhou, China

Abstract. Bluetooth is a universal wireless standard which is used on most smartphones. With the widespread use of Bluetooth on smartphones, Bluetooth security has received a lot of attention, and it is increasingly important to identify and block Bluetooth attacks to ensure that smartphones are free of the threat of Bluetooth attacks. Traditional attack detection techniques are generally based on traffic, and unfortunately smartphone Bluetooth traffic is extremely difficult to capture. Based on this problem, we propose a Bluetooth attack detection method based on device status recognition, which uses the response time of the smartphone to the ping in the Bluetooth L2CAP to remotely monitor the Bluetooth status of the smartphone. Due to the variety of Bluetooth attacks, we can't easily identify multiple Bluetooth attacks based on response time. For this purpose, we explored the effectiveness of a detection approach based on deep learning. Our experimental results show our algorithm can detect Bluetooth attack with a high precision and high recall.

Keywords: Attack detection · Bluetooth · Smartphone · LSTM

1 Introduction

Bluetooth is an open standard for short-range radio frequency communications. Bluetooth is a low-cost, low-power technology that provides a mechanism to create small wireless networks in a special way [1]. Bluetooth has a wide range of applications in many types of commercial and consumer devices, including smartphones, laptops, automobiles, printers, headsets, and more [2]. Nowadays, Bluetooth has been widely used in smartphones. Users can use Bluetooth module of the smartphone to transfer files, transmit audio and video, control other Bluetooth devices, and so on. Mobile phone Bluetooth has become an indispensable part of people's lives.

Due to the heavy use of Bluetooth, Bluetooth security has received extensive attention. As a wireless technology standard, Bluetooth wireless technology and related devices are vulnerable to general wireless network threats, such as denial of service attacks, man-in-the-middle attacks, eavesdropping, message tampering, etc. [3].

© Springer Nature Switzerland AG 2020
X. Sun et al. (Eds.): ICAIS 2020, LNCS 12239, pp. 120–132, 2020.
https://doi.org/10.1007/978-3-030-57884-8_11

Traditional attack detection techniques are generally based on traffic, that is, traffic behavior analysis through traffic extraction and traffic analysis techniques. The Bluetooth communication traffic of smartphones is very difficult to capture and there is no analysis. When the attacker attacks the smartphone through Bluetooth, the Bluetooth status of the smartphone will change. Therefore, we consider attack detection by detecting the change of the Bluetooth status of the smartphone. We chose to use the ping method in the Bluetooth L2CAP protocol to determine the Bluetooth status of the smartphone, so that no physical contact or observation of the smartphone is required.

Based on the above ideas, this paper proposes a way to realize Bluetooth attack detection by remotely monitoring the status of Bluetooth devices. We use the attack detection system deployed on the remote host to initiate a continuous Bluetooth ping to the smartphone, process the obtained response time into sequence data, and use the deep learning method to identify the Bluetooth status of the smartphone, thus completing the Bluetooth attack detection. We summarize our contributions as follows:

- We propose a Bluetooth attack detection by remotely monitoring the status of the Bluetooth device, which does not require Bluetooth traffic or physical contact devices.
- We introduce the deep learning algorithm LSTM into the recognition of Bluetooth device status, successfully identify multiple states of Bluetooth devices, and have high precision and recall.

This paper is organized as follows. In Sect. 2, we discuss a few closely related work. A Bluetooth background and the deep learning algorithm LSTM are provided in Sect. 3 and Sect. 4 describes the Bluetooth attack. We introduce an attack detection method based on Bluetooth device status in Sect. 5. The experimental results and discussion are provided in Sect. 6 and Sect. 7 concludes the paper.

2 Related Work

Bluetooth security has received widespread attention, Hassan et al. performed a comprehensive survey to identify major security threats in Bluetooth communication [4]. Padgette et al. provided information on the security capabilities of Bluetooth and gives recommendations to organization [5]. Some of the tools and techniques that are currently available to attackers to exploit the vulnerabilities in Bluetooth have been demonstrated [6].

Satam et al. presented an anomaly-based intrusion detection system for Bluetooth, and the system uses an n-gram based approach to characterize the normal behavior of the Bluetooth protocol [7]. A new efficient Intrusion Detection and Prevention System for Bluetooth networks which is based on the set of rules that are used to identify strange communication behavior of Bluetooth devices has been put forward [8]. Terrence J. OConnor improved upon existing techniques, which only detected only a limited set of known attacks through measuring anomalies in the power levels of the Bluetooth device [9].

Vrizlynn LL Thing proposed a solution based on anomaly detection and classification using a deep learning approach which self-learns the features necessary to detect network anomalies and is able to perform attack classification accurately [10]. Self-taught

Learning(STL), a deep learning based technique, was used on NSL-KDD – a benchmark dataset for network intrusion [11]. Fiore et al. explored the effectiveness of a detection approach based on machine learning, using the Discriminative Restricted Blotzmann Machine to combine the expressive power of generative models with good classification accuracy capabilities to infer part of its knowledge from incomplete training data [12].

3 Background

The Bluetooth specification promulgated by the Bluetooth Technology Alliance is the Bluetooth wireless communication protocol standard, which specifies the standards and requirements that Bluetooth application products should follow. According to the logic function of the Bluetooth protocol, the protocol stack is divided into three parts from bottom to top: transport protocol, intermediary protocol, application protocol [1].

The transport protocol is primarily responsible for mutually confirming the location of each other between Bluetooth devices and establishing and managing physical links between Bluetooth devices. Mainly divided into low-level transport protocols and high-level transport protocols. The underlying transport protocol includes a Bluetooth radio part, a baseband link management controller, and a link management protocol, and is responsible for the physical implementation of language, data wireless transmission, and interconnection networking between Bluetooth devices.

The high-level transport protocol includes Logical Link Control and Adaptation Protocol (L2CAP) and Host Controller Interface (HCI), which shields the underlying transmission operations such as hopping sequence selection for high-level applications, and provides high-level programs with effective and favorable data packet formats [13]. The intermediary protocol provides necessary support for high-level application protocols or programs to work on the Bluetooth logical link, and provides various standard interfaces for the application layer. This part of the protocol mainly includes RFCOMM (Radio Frequency Communication Protocol) and SDP (Service Discovery Protocol), IrDA, PPP, IP, TCP, UDP, TCS, AT instruction set, etc.

L2CAP is above the link control protocol and belongs to the data link layer. It uses protocol multiplexing, segmentation, reassembly, and group abstraction to provide connection-oriented and connectionless data services to upper-layer protocols. L2CAP allows higher layer protocols and applications to send receive data packets with a maximum length of 64k bytes, as well as flow control and retransmission for each channel through flow control and retransmission mode [14]. L2CAP provides a logical channel, called the L2CAP channel, which maps to the L2CAP logical link through the support of ACL logical transport.

As defined in the Bluetooth specification, a ping based on L2CAP can occur between two Bluetooth devices, one Bluetooth device initiates a ping request and another Bluetooth device will respond to the request. The ping can occur without any authentication, and is generally used to test the connectivity between Bluetooth devices.

Deep learning is a new research direction in the field of machine learning. Deep learning combines low-level features to form more abstract high-level representation attribute categories or features to discover distributed feature representations of data. Long-short Term Memory (LSTM) is a deep learning algorithm proposed by Hochreiter and Schmidhuber in 1997 [15]. It is a time recurrent neural network, mainly to solve the gradient disappearance and gradient explosion in long sequence training [16].

4 Bluetooth Attacks

As a Bluetooth device, smartphones are susceptible to multiple Bluetooth attacks. In the survey of Bluetooth security by Padgette et al. [3], Bluetooth and associated devices are threatened by more specific Bluetooth related attacks, such as Bluesnarfing, Bluejacking, Bluebugging, Car Whisperer, Denial of Service, Fuzzing Attacks, Pairing Eavesdropping, Secure Simple Pairing Attacks. John Paul Dunning classified the Bluetooth threats based on a framework call "A Bluetooth Threat Taxonomy" and also explained each of these classes in detail and gives a gist of attack classification in the form of a table which is given below in Table 1 [17].

For the Bluetooth devices such as smartphones, we mainly test the following four common Bluetooth attacks.

Table 1. Bluetooth attacks

Attack classification	Threats
Surveillance	Blueprinting, bt_audit, redfang, War-nibbling, Bluefish, sdptool, Bluescanner, BTScanner
Range extension	BlueSniqing, bluetooone, VeraNG
Obfuscation	Bdaddr, hciconfig, Spooftooph
Fuzzer	BluePass, Bluetooth Stack, Smasher, BlueSmack, Tanya, BlueStab
Sniffing	FTS4BT, Merlin, BlueSniff, HCIDump, Wireshark, kismet
Denial of service	Battery exhaustion, signal jamming, BlueSYN, Blueper, BlueJacking, vCardBlaster
Malware	BlueBag, Caribe, Comm Warrior
Unauthorized direct data access	Bloover, BlueBug, BlueSnarf, BlueSnarf++, BTCrack, Car Whisperer, HeloMoto, btpincrack
Man in the middle	BT-SSP-Printer-MITM, BlueSpooof, bthidproxy

Denial of Service Attacks. Like other wireless technologies, Bluetooth is susceptible to

Denial of Service Attacks. Impacts include making a device's Bluetooth interface unusable and draining the device's battery. We use the ping flood in Bluetooth L2CAP to attack the smartphone. A Bluetooth ping flood is a form of Denial of Service designed to send a high volume of packets to a target device to saturate the communication channel.

Fuzzing Attacks. Fuzzing is a software testing technique whose core idea is to generate malformed data and send it to the application and capture its anomalies. Malformed

Bluetooth packets can cause anomalies in smartphone Bluetooth, such as a Bluetooth stack crashing or even unresponsive. We mainly conduct Fuzzing Attacks against the Bluetooth protocol.

Scanning Attacks. Bluetooth scanning attacks can scan a certain range of smartphone Bluetooth to obtain the Bluetooth address, Bluetooth basic information and Bluetooth protocol information of the smartphone.

Attacks Based on Bluetooth Vulnerabilities. The Bluetooth inventory of common smartphone operating systems (Android, iOS) is in a variety of Bluetooth vulnerabilities. If the smartphone fails to install security patches or upgrade the system in time, there is a risk of being attacked by some specific Bluetooth vulnerabilities. Attackers can exploit known vulnerabilities to exploit attacks that can cause smartphone memory information to leak or even execute arbitrary code remotely.

5 Attack Detection Based on Smartphone Bluetooth Status

5.1 Attack Detection Scenario

We design the detection scenario as shown in Fig. 1. There are three devices in the detection scenario: Smartphone, Attacking host, Detection host.

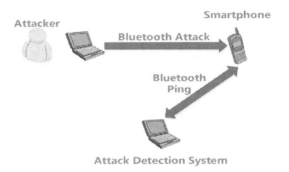

Fig. 1. Attack detection scenario.

Smartphone. Bluetooth is turned on and provides a Bluetooth address to the attacking host and detection host. We offer four smartphones, the most popular smartphone operating systems, Android and iOS. All four mobile phones are capable of responding to ping messages of the Bluetooth L2CAP protocol and are affected by Bluetooth attacks.

Attacking Host. With a Bluetooth module, it can initiate multiple Bluetooth attacks on smartphones with specified Bluetooth address.

Detection Host. It has a Bluetooth module and deploys a Bluetooth attack detection system to send Bluetooth ping packets to smartphones with specified Bluetooth address in real time.

5.2 Bluetooth Ping

Definition 1. (Application Response Time (ART)). The time during which the detection host sends a ping request packet to receive a response from the smartphone.

Definition 2. (Application Not Responding (ANT)). In some cases, the application will refuse to respond to the ping request sent by the detection host. When ANT appears, the ART value is considered 0.

Since the fast ping itself is a Bluetooth attack behavior, in order to reduce the impact of ping on the Bluetooth of the smartphone, we use slow ping [18]. The detection host will send a ping request packet to the smartphone every second and record the current ART. Figure 2 shows the changes in ART collected by the detection host under four types of Bluetooth attacks.

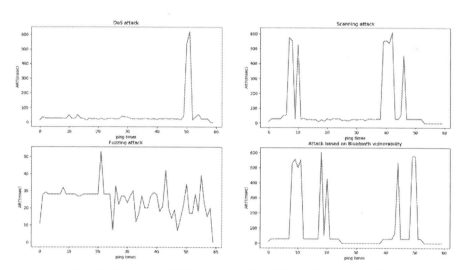

Fig. 2. Changes in ART over time under different Bluetooth attacks.

5.3 Approach of Bluetooth Attack Detection

According to the different characteristics of the acquired ART data stream, we distinguish the different states of the Bluetooth of the smartphone and realize the attack detection. However, it is difficult for us to manually extract rule features directly from the ART data stream, so we introduced the deep learning algorithm LSTM. Figure 3 shows our Bluetooth attack detection process.

5.4 Data Preprocessing

In order to facilitate data analysis, we introduce a denoising algorithm and a segmentation algorithm. The denoising algorithm (Algorithm 1) removes the noise in the ART stream data to obtain new stream data.

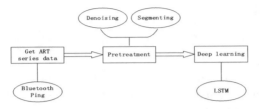

Fig. 3. The process of Bluetooth attack detection.

Algorithm 1: Denoising(A, Q)

This **Denoising** function is used to remove the noise of the ART stream data.

Input: A: the ART stream data;

              Q: threshold, exceeding this value is considered noise.

Output: B: the new ART stream data.

Denoising(A, Q):

1. N=Length(A)
2. j=0
3. **for** i **from** 0 **to** N
4.   **if** A[i] < Q:
5.   B[j] = A[i]
6.   j = j + 1
7. **return** B

After obtaining the denoised ART stream data, we use the segmentation algorithm (Algorithm 2) to segment it, and then segment the data to obtain the ART group. These sets of data will be the raw data for our feature extraction.

Algorithm 2: Segmenting(B, L)

This **Segmenting** function is used to segment the ART stream data.

Input: B: the ART stream data;

              L: the maximum length of grouping.

Output: C: the group of ART.

Segmenting(B, L):

1. N = Length(B)
2. J = 0
3. K = 0
4. **for** i **from** 0 **to** N
5.   **if** B[i] = ANT or (i − k) > L
6.   C[j] = (B[k], B[k+1], ..., B[i])
7.   j = j + 1
8.   k = i
9. **return** C

5.5 LSTM

The LSTM is an algorithm in deep learning, which is often used to process time series data. We will introduce the LSTM to detect Bluetooth attacks [19]. We simulated the attack to launch different Bluetooth attacks on the smartphone, and collect the ART stream data. The ART stream data is pre-processed and tagged according to different attack types, so we convert Bluetooth attack detection into a multi-class problem. There are five types of situation in the resulting training set, one for normal and four for attack.

The loss function is used to estimate the degree of inconsistency between the predicted value of the model and the true value. In general, the smaller the loss function, the better the robustness of the model. We choose cross entropy as the loss function of the algorithm [20]. Cross entropy loss is calculated as:

$$L\left(\widehat{y^{(i)}}, y^{(i)}\right) = -(y^{(i)} \ln \widehat{y^{(i)}} - \left(1 - y^{(i)}\right) \ln\left(1 - \widehat{y^{(i)}}\right)) \tag{1}$$

Where $\hat{y}$, y respectively represent predictive output and actual value.

We use the gradient descent method to solve the solution step by step, and obtain the minimized loss function and model reference value [21]. Below we give the gradient derivation of a single sample. First we have the following formula definition:

$$z = W^T + b \tag{2}$$

$$\sigma(z) = sigmoid(z) \tag{3}$$

$$sigmoid(z) = \frac{1}{1 + e^{-z}} \tag{4}$$

$$\sigma'(z) = \frac{e^{-z}}{\left(1 + e^{-z}\right)^2} = \frac{1}{\left(1 + e^{-z}\right)} \left(1 - \frac{1}{1 + e^{-z}}\right) = \sigma(z)(1 - \sigma(z)) \tag{5}$$

$$L(\hat{y}, y) = -\left(y \ln \hat{y} + (1 - y)\ln(1 - \hat{y})\right) \tag{6}$$

Where W, T and b respectively represent weight matrix, transpose of matrix and bias. Now let's assume a sample with two eigenvalues (x_1, x_2), then there are also two weights (w_1, w_2). Let $z = w_1 x_1 + w_2 x_2 + b$, $a = \sigma(z)$. Below we want to calculate the derivative of $L(a, y)$ for w and b to find the direction in which the loss function decreases the fastest:

$$da = \frac{\vartheta L(a, y)}{\vartheta a} = -\frac{y}{a} + \frac{1 - y}{1 - a} \tag{7}$$

$$dz = \frac{\vartheta L(a, y)}{\vartheta a} * \frac{\vartheta a}{\vartheta z} = a - y \tag{8}$$

$$dw_j = \frac{\vartheta L(a, y)}{\vartheta a} * \frac{\vartheta a}{\vartheta z} * \frac{\vartheta z}{\vartheta w_j} = (a - y)x_j \tag{9}$$

$$db = \frac{\vartheta L(a, y)}{\vartheta a} * \frac{\vartheta a}{\vartheta z} * \frac{\vartheta z}{\vartheta b} = a - y \tag{10}$$

α presents the learning rate. Then update the values of w and b, $w_j = w_j - \alpha\, dw_j$, $b = b - \alpha\, db$.

6 Experiment, Results and Discussion

6.1 Experiment

In this experiment, our detection host and attack host are both Linux operating systems. On the detection host, we use the Linux system's own l2ping command to complete the Bluetooth L2CAP ping behavior [22]. On the attack host, we use Linux Bluetooth protocol stack Bluez to simulate various Bluetooth attacks. Experiments have been performed using four smartphones listed in Table 2.

Table 2. List of tested smartphones.

Smartphone name	Operating system	Bluetooth version
Mi phone 8	Android 8.1	5.0
Google Nexus 5X	Android 7.1.2	4.2
Samsung Galaxy S4	Android 6.0.1	4.0
Apple iPhone 8	iOS 11.0.3	5.0

6.2 Results

We focus on the Google Nexus 5X because it is native Android and is susceptible to multiple Bluetooth attacks. We implement our deep learning frameworks, and send the train dataset to generate the self-learnt models. Based on our generated models, we performed the experimental evaluation using the test dataset. There are 200 sets of data for each type (normal type and four attack types) in the test set, a total of one thousand sets of data. Table 3, 4, 5, 6, 7 and 8 shows the various types of experimental data. Our classifier is a binary classifier, so when we perform attack detection, we take 200 sets of data of the same type as positive samples, and take 50 sets of data each of the other 4 categories (200 sets in total) as negative samples.

Table 3. Normal.

Actual category		Forecast category		
		Yes	No	Sum
	Yes	143	67	200
	No	21	179	200
	Sum	164	236	400

Table 4. Fuzzing attacks.

Actual category		Forecast category		
		Yes	No	Sum
	Yes	164	36	200
	No	39	161	200
	Sum	203	197	400

Table 5. Dos attacks.

Actual category		Forecast category		
		Yes	No	Sum
	Yes	131	69	200
	No	51	149	200
	Sum	182	218	400

Table 6. Scanning attacks.

Actual category		Forecast category		
		Yes	No	Sum
	Yes	135	65	200
	No	44	156	200
	Sum	179	221	400

Table 7. Attack based on Bluetooth vulnerabilities.

Actual category		Forecast category		
		Yes	No	Sum
	Yes	191	9	200
	No	11	189	200
	Sum	202	198	400

We use our Precision, Recall and F-Measure to evaluate our model [23]. F-Measure is the weighted harmonic average of Precision and Recall:

$$F = \frac{(a^2 + 1)P * R}{a^2(P + R)} \tag{11}$$

Table 8. Various types of precision, recall and F1-measure values.

	Precision	Recall	F1-measure
Normal	87.20%	71.50%	78.57%
Fuzzing attack	80.79%	82.00%	81.39%
DoS attack	71.98%	65.50%	68.59%
Scanning attack	75.42%	67.50%	71.24%
Attack based on Bluetooth vulnerabilities	94.55%	95.50%	95.02%

Where a, F, P and R respectively represent parameter, F-Measure, Precision and Recall. We use the parameter a to set to 1, which is F1 to evaluate our model. Table 7 gives the various type of P, R and F1 values.

Our model identifies higher F1 values for attacks based on Bluetooth vulnerabilities. Through the analysis of the Bluetooth vulnerability of smartphones, we know that attacks based on Bluetooth vulnerabilities often hijack the Bluetooth process of smartphones by triggering the Bluetooth vulnerability and execute arbitrary code remotely. During this time, the smartphone will not respond to the ping request of the detection host, and the detection host will get multiple ANT data, so the feature is more obvious.

6.3 Discussion

Our experiment is based on the fact that the smartphone only turns on Bluetooth and is not affected by other wireless, but the actual situation is much more complicated than this. When the smartphone is subject to other wireless interference such as WIFI, etc., the test results may change. We can extend the model's application scenarios by customizing the state, such as adding the status "Smartphone to open WIFI".

The Bluetooth devices we are testing are smartphones. The detection object can be extended to all Bluetooth devices that can respond to Bluetooth L2CAP ping. These Bluetooth devices can be Bluetooth headset, Bluetooth speaker, etc.

7   Conclusion

We propose a Bluetooth attack detection method based on device status recognition, which uses the response time of the ping of the Bluetooth L2CAP protocol to remotely detect the Bluetooth status of the smartphone, and uses the LSTM algorithm to distinguish the Bluetooth status of the smartphone. In this paper, we present four Bluetooth attack methods and simulate these four methods to attack the smartphone. Our experimental results show that our model has a high precision and recall rate for the recognition of these Bluetooth attacks.

In the future we will increase the Bluetooth status of the detection and optimize the algorithm.

Acknowledgement. This work is supported by the NSFC (Grant Nos. 61671087, 61962009, 61003287), the Fok Ying Tong Education Foundation (Grant No. 131067), the Major Scientific and Technological Special Project of Guizhou Province (Grant No. 20183001), the Foundation of State Key Laboratory of Public Big Data (Grant No. 2018BDKFJJ018), the High-quality and Cutting-edge Disciplines Construction Project for Universities in Beijing (Internet Information, Communication University of China), the Fundamental Research Funds for the Central Universities, and the Fundamental Research Funds for the Central Universities No. 2019XD-A02.

References

1. Bluetooth Specification. www.bluetooth.com/Bluetooth/Technology/Building/Specifications/Default.htm. Accessed 18 Sept 2019
2. Zhang, Q., Liang, Z., Cai, Z.: Developing a new security framework for Bluetooth low energy devices. Comput. Mater. Continua **59**(2), 457–471 (2019)
3. Sandhya, S., Devi, K.A.S.: Analysis of Bluetooth threats and v4.0 security features. In: Proceedings of the 2012 International Conference on Computing, Communication and Applications, pp. 1–4 (2012)
4. Hassan, S.S., Das Bibon, S., Hossain, M.S., Atiquzzaman, M.: Security threats in Bluetooth technology. Comput. Secur. (2017)
5. Padgette, J., Scarfone, K., Chen, L.: Guide to Bluetooth Security. NIST Special Publication (2012). http://csrc.nist.gov/
6. Cope, P., Campbell, J., Hayajneh, T.: An investigation of Bluetooth security vulnerabilities. In: 2017 IEEE 7th Annual Computing and Communication Workshop and Conference, pp. 1–7 (2017)
7. Satam, P., Satam, S., Hariri, S.: Bluetooth intrusion detection system (BIDS). In: 2018 IEEE/ACS 15th International Conference on Computer Systems and Applications (AICCSA), pp. 1–7. IEEE (2018)
8. Haataja, K.: New efficient intrusion detection and prevention system for Bluetooth networks. In: Proceedings of the ACM International Conference on Mobile, Wireless MiddleWare, Operating Systems, and Applications, Innsbruck, Austria, 12–15 February 2008
9. O'Connor, T.J.: Bluetooth Intrusion Detection (2008). http://www.lib.ncsu.edu/theses/available/etd03212008-135411/unrestricted/etd.pdf
10. Thing, V.L.L.: IEEE 802.11 network anomaly detection and attack classification: a deep learning approach. In: Proceedings IEEE Wireless Communications and Networking Conference (WCNC), pp. 1–6 (2017)
11. Niyaz, Q., Sun, W., Javaid, A.Y., Alam, M.: A deep learning approach for network intrusion detection system. In: EAI International Conference on Bio-inspired Information and Communications Technologies, pp. 21–26 (2015)
12. Fiore, U., Palmieri, F., Castiglione, A., Santis, A.D.: Network anomaly detection with the restricted Boltzmann machine. Nerocomputing **22**, 13–23 (2013)
13. Lee, J.-S., Su, Y.-W., Shen, C.-C.: A comparative study of wireless protocols: Bluetooth, UWB, ZigBee, and Wi-Fi. In: Industrial Electronics Society, IECON 2007 (2007)
14. Haartsen, J.C.: The Bluetooth radio system. IEEE Pers. Commun. **7**(1), 28–36 (2000)
15. Hochreiter, S., Schmidhuber, J.: Long short-term memory. Neural Comput. **9**(8), 1735–1780 (1997)
16. Zhaowei, Q., Cao, B., Xiaoru Wang, F., Li, P.X., Zhang, L.: Feedback LSTM network based on attention for image description generator. Comput. Mater. Continua **59**(2), 575–589 (2019)
17. Dunning, J.P.: Taming the blue beast: a survey of bluetooth-based threats. IEEE Priv. Secur. **8**, 20–27 (2010)

18. Celosia, G., Cunche, M.: Detecting smartphone state changes through a Bluetooth based timing attack. In: Conference on Security & Privacy in Wireless and Mobile Networks (WiSec). Stockholm, Sweden, pp. 154–159. ACM (2018)
19. Shen, Y., et al.: Hashtag recommendation using LSTM networks with self-attention. Comput. Mater. Continua **61**(3), 1261–1269 (2019)
20. Li, C.H., Lee, C.K.: Minimum cross entropy thresholding. Pattern Recogn. **26**(4), 617–625 (1993)
21. Burges, C., et al.: Learning to rank using gradient descent. In: Proceedings of the 22nd International Conference on Machine Learning, pp. 89–96. ACM (2005)
22. Linux man page. https://linux.die.net/man/8/l2ping. Accessed 25 Oct 2019
23. Powers, D.M.: Evaluation: From precision, recall and f-measure to ROC, informedness, markedness and correlation. J. Mach. Learn. Technol. **2**(1), 37–63 (2011)

Covered Face Recognition Based on Deep Convolution Generative Adversarial Networks

Yanru Xiao[1(✉)], Mingming Lu[1], and Zhangjie Fu[2]

[1] School of Computer Science, Central South University, ChangSha 410083, Hunan, China
3105160696@qq.com
[2] School of Computer Science and Software, Nanjing University of Information Science and Technology, Nanjing 210000, Jiangsu, China

Abstract. At present, the use of deep convolution generative adversarial networks to generate images is a new research direction in the field of machine vision. It can generate very high quality face pictures, so it plays a very important role in the field of criminal investigation. But during surveillance, suspects often wear face covers, such as glasses, masks, hats and so on. It may affect the machine's recognition of the face. The traditional method is to complete the incomplete image, that is, to imagine and complete the covered part, and then carry on the face recognition. In this paper, an image arithmetic operation based on DCGAN is proposed to process the traditional photos so and store them in the face database, so that the details of the human face can be well preserved. When face recognition is carried out, such pictures are used to assist in face recognition. The experimental results show that this method can improve the accuracy.

Keywords: Face recognition · Generative adversarial networks · Deep convolution generative adversarial networks

1 Introduction

With the rapid development of computer image processing technology, face recognition has become a hot research field. With the continuous maturity of face recognition technology, face recognition has become an important part of our lives. For example, when we unlock our mobile phones, we use face recognition; when we enter into the train station, we use face recognition; and when we pay our bills, we use the face recognition. And it also applies to criminal investigation. Monitoring equipment is a manual right-hand assistant for policemen, but not every picture taken by the monitor is available. Some suspects cover their faces, which affects the machine's recognition of faces. So how to improve the accuracy of face recognition has become a matter of great concern. Especially when wear covers such as glasses, sunglasses, masks, beards, bangs and so on, it will affect the effect of face recognition to a certain extent. In this paper, wearing glasses are taken as an example to carry out the experiment.

Compared with the traditional face recognition technology, the deep convolution generation adversarial networks has better characteristics of automatic feature extraction,

© Springer Nature Switzerland AG 2020
X. Sun et al. (Eds.): ICAIS 2020, LNCS 12239, pp. 133–141, 2020.
https://doi.org/10.1007/978-3-030-57884-8_12

and can effectively generate higher quality pictures. The images formed by the generated antagonistic model can even deceive people's eyes. In the contest between discriminator and generator, the face can be learned better, so that the recognition features can be better. The traditional mask of face usually adopts the completion of incomplete image, that is, DCGAN is used to generate the covered place, the glasses are restored to a part of the face, and then the face recognition is carried out. But the one that was restored, part of it is generated by DCGAN, which may have a great influence on the parameters of the eye, nose beam, eye corner and so on. The accuracy of the restored part is less than 80%, so it is difficult to fully meet the needs of the society.

In this paper, by using the method of reverse thinking, using the principle of DCGAN and image arithmetic operation, the image with covers is generated by "face + mask", which is stored in the database and then the face recognition is carried out. Because the image obtained by image arithmetic operation method does not change the details of the original image, the loss can be reduced in the corner of the eye, nose and beam, etc. So it can be used as an auxiliary test to improve the accuracy.

2   Generative Adversarial Networks

Before introducing the deep convolution generative adversarial networks(DCGAN), let's introduce the generative adversarial networks (GAN). When it comes to GAN, you have to mention a person, Ian Goodfellow, the father of generative adversarial networks. At that time, Ian was doing an experiment, using the traditional neural network method, but the result was that the quality of the generated picture was always not ideal. Then he came up with a way to use two neural networks at the same time, and these two neural networks are a relationship between confrontation and game, we call they are generators and discriminator. To put it simply, the generator generates a fake picture, and the discriminator identifies the picture and find the error, determine whether it is a "true picture" or "false picture", then output the results. The generator improves itself according to the feedback given by the discriminator, and then generate a picture. The new picture goes to the discriminator, which takes the time to find a deeper difference in order to distinguish the picture. In this way, many repeated games are between the two, and they will become stronger in the whole training process. This process will continue until the generator generates a picture and the discriminator determines that it is a true picture, and finally outputs.

Let's take a look at what the two models need to do. First, a set of random numbers z enter into the generator, and an image is output. Then input the image into the discriminant model, and then the discriminant model outputs a probability value to judge the truth and false. When the probability is greater than 0.5, it is considered to be true, while if it is less than 0.5, it is judged to be false. The method of training the two models is to train alternately and iteratively (Fig. 1).

For generators, it is closely related to discriminators, which can be regarded as a zero-sum game, and their cost synthesis should be zero.

Optimize discriminator:

$$\max_{D} V(D, G) = E_{x \sim p_{data}(x)}[\log(D(x))] + E_{z \sim p_z(z)}[\log(1 - D(G(z)))] \quad (1)$$

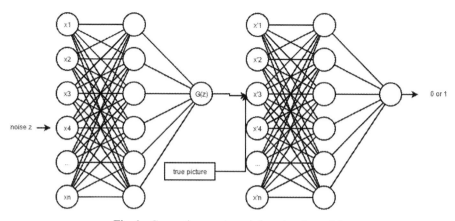

Fig. 1. Generating a antagonistic network model.

Optimize generator:

$$\min_{G} V(D, G) = E_{z \sim p_z(z)}[\log(1 - D(G(z)))] \tag{2}$$

Ian's goal formula in the original paper:

$$\min_{G} \max_{D} V(D, G) = E_{x \sim p_{data}(x)}[\log(D(x))] + E_{z \sim p_z(z)}[\log(1 - D(G(z)))] \tag{3}$$

It can be seen from the formula that the larger the V is, the better the discriminator is, and the smaller the V is, the better the generator is, thus the game relationship between the two be formed.

In this paper, an example of GAN visualization in Ian's original paper is used, in which the point line is the real data distribution and the curve is the generated data sample. The goal of GAN is to get the curve close to the point line. Picture a) is in the initial state, the difference between the point line and the curve is obvious, picture b) means it is trained, and finally as the picture d) shows [1, 2] (Fig. 2).

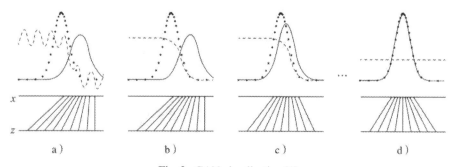

Fig. 2. GAN visualization [1]

3 Deep Convolution Generative Adversarial Networks

The first paper of DCGAN, Unsupervised Representation Learning with Deep Convolutional Generative Adversarial Networks [3], was published in 2015. On the basis of GAN, a new DCGAN framework is proposed. DCGAN has better effect and more stable training state than GAN.

First of all, in DCGAN, the traditional pooling layer is removed and replaced by convolutional layer. It uses step-length convolution for discriminator. The picture shows the spatial downsampling of the discriminator. The data matrix is 5 * 5, the filter is 3 * 3, the step size is 2 * 2, and the output matrix is 3 * 3 (Fig. 3).

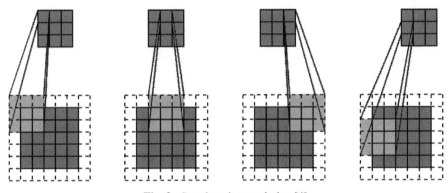

Fig. 3. Step-length convolution [4]

For the generator, the fractional step-length convolution is used. The picture shows the spatial upsampling. The input is 3 * 3 matrix and the reverse step size is 2 * 2, so 0 is filled between each point in the matrix, and 0 is filled around the point, and a 5 * 5 matrix is finally output (Fig. 4).

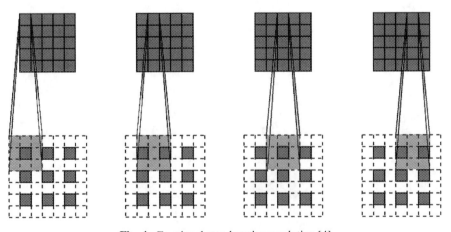

Fig. 4. Fractional step-length convolution [4]

Secondly, the fully connected hidden layer is removed.

The third point is the use of batch normalization. Because the number of network layers is large, each layer may change the output data, and batch normalization is to normalize the input of each layer, so that the data obeys the fixed data distribution.

Fourth, use the appropriate activation function. The activation function is used for nonlinear transformation in the neural network. The formula of the Sigmoid function is

$$\sigma(x) = \frac{1}{1 + e^{-x}} \tag{4}$$

This is a function with a value of $(0, 1)$. Taking $x = 0$ as the bound, when $x > 0$, the value of is closed to 1, while when $x < 0$, the function approaches 0, so it is often used as the output of 01 binary classification. But what is commonly used in practice is the Tanh function, which is a deformation of the Sigmoid function, and the formula is

$$\tanh(x) = 2\sigma(2x) - 1 \tag{5}$$

Its value is $(-1, 1)$, which solves the problem that the mean value of Sigmoid function is not 0. The ReLU (Rectified Linear Unit) function is a very popular activation function in recent years. The formula is

$$f(x) = \max(0, x) \tag{6}$$

Its advantage is that it is easier to converge than Sigmoid and Tanh in gradient decline, but it may occur in training that some neurons can never be updated. Therefore, an improved method, LeakyReLU, is proposed to take $f(x) = \alpha x$ when $x < 0$, so that the gradient will not disappear when reverse conduction when $x < 0$.

In the DCGAN architecture, the ReLU function is used in the generator, but the Tanh activation function is used in the output layer. And all layers in the discriminator use LeakyReLU.

The structure of DCGAN is shown in the figure, which shows that a 100-dimensional input data z is transformed into a 64 * 64 * 3 RGB picture after a series of fractional step convolution [3] (Fig. 5).

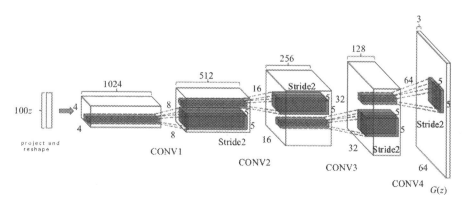

Fig. 5. DCGAN's framework [3]

4   Image Completion

Between image and statistics, we can view the image as a sample derived from a high-dimensional probability distribution. The probability distribution corresponds to the pixel of the image. Imaging that the picture we're seeing is made up of a limited number of pixels, and we can sample from these complex picture samples to estimate the probability distribution, so that the missing value of the unknown part can be complemented. Then this requires us to do two things, one to generate a false image and the other to find the best solution to complement the full image.

Generating new samples is the method of using DCGAN, which uses the game between generator and discriminator to generate high quality false pictures.

As for the image completion of the original image y, we project the original image onto the generated distribution. We use a binary mask with only 0 and 1 values, and call it M. For M, the value 1 is the part we want to retain and the value 0 is the part we need to make up. Then through the combination of contextual loss and perceptual loss, we can find a complete image which we most want to find [5].

5   Image Arithmetic Operation Based on DCGAN

5.1   Experimental Environment

Table 1 shows the operating environment. The operating system of the computer for experiment is Win10. The development environment is PyCharm 2019, and the python vision is 3.7. The experiment is based on these.

Table 1. Experimental Environment.

Condition	Environment
Operating system	Windows 10
Development environment	PyCharm 2019.2
Python version	Python 3.7

5.2   Image Processing

Through the confrontation training of generator and discriminator, the generator finally generates a distribution pG similar to the distribution pdata of true photos. The distribution of these picture features is in the deep convolution confrontation network, and the feature extraction is finally embodied in 64 * 64 * 3 convolution feature map. Based on these, this paper mainly needs to do the processing and generation of graphics.

The dataset CalebA used in this paper is an open data of the Chinese University of Hong Kong, which contains 202599 face pictures with 10177 real-life samples, and all of them have been marked with features. It is indeed very useful for experiments. In this paper, we take glasses for example, while the principle of other masks, such as hats, masks and so on, is similar from the case of wearing glasses.

CelebA affectionately marks every picture, such as Big_Lips, Eyeglasses, Wearing_Hat, and so on, and is recorded in the list_attr_celeba.txt file.

```
202599
S_o_Clock_Shadow Arched_Eyebrows Attractive Bags_Under_Eyes Bald Bangs Big_Lips Big_Nose Black_Hair Blond_Hair Blurry Brown_Hair Bushy_Eyebrows Chubby Double_Chin
000001.jpg -1  1  1 -1 -1 -1 -1 -1 -1 -1  1 -1 -1 -1 -1 -1  1  1 -1  1 -1 -1  1 -1  1  1 -1  1 -1  1 -1 -1  1
000002.jpg -1 -1 -1  1 -1 -1 -1  1 -1 -1  1  1 -1 -1 -1 -1 -1  1 -1  1  1 -1  1 -1 -1 -1 -1  1 -1 -1 -1 -1  1
000003.jpg -1 -1 -1 -1 -1 -1  1 -1 -1 -1  1 -1 -1 -1 -1 -1 -1 -1  1 -1 -1  1  1 -1 -1  1 -1 -1 -1 -1 -1 -1  1
000004.jpg -1 -1  1 -1 -1 -1 -1 -1 -1 -1 -1 -1 -1 -1 -1 -1 -1 -1 -1 -1  1 -1  1  1 -1 -1 -1  1  1 -1  1  1 -1
000005.jpg -1  1  1 -1 -1 -1  1 -1 -1 -1 -1 -1 -1 -1 -1 -1  1  1 -1  1  1 -1  1 -1 -1 -1 -1 -1  1 -1  1 -1  1
000006.jpg -1  1  1 -1 -1 -1  1 -1 -1 -1  1 -1 -1 -1 -1 -1  1 -1  1 -1  1 -1  1 -1 -1 -1 -1 -1 -1  1  1  1 -1 -1  1
000007.jpg  1 -1  1  1 -1 -1  1  1  1 -1 -1 -1 -1 -1 -1 -1 -1  1 -1 -1 -1 -1 -1  1 -1 -1  1  1 -1 -1 -1 -1  1
000008.jpg  1  1 -1  1 -1 -1  1 -1  1 -1 -1 -1 -1 -1 -1 -1 -1 -1 -1 -1  1 -1 -1  1  1 -1 -1  1 -1 -1 -1 -1 -1 -1  1
000009.jpg -1  1  1 -1  1  1 -1 -1 -1 -1 -1 -1 -1 -1 -1 -1 -1  1  1 -1  1 -1 -1 -1  1 -1 -1 -1 -1 -1 -1  1
000010.jpg -1 -1  1 -1 -1 -1 -1 -1 -1 -1 -1 -1 -1 -1 -1  1  1 -1 -1 -1  1 -1 -1 -1 -1 -1 -1 -1 -1  1 -1  1 -1  1
000011.jpg -1 -1  1 -1 -1 -1 -1  1 -1 -1 -1 -1 -1 -1 -1 -1 -1 -1 -1  1 -1 -1  1 -1 -1 -1 -1 -1 -1 -1  1 -1 -1 -1 -1 -1 -1  1
000012.jpg -1 -1  1  1 -1 -1 -1  1 -1  1 -1 -1 -1 -1 -1 -1  1  1  1 -1 -1 -1  1  1 -1 -1  1 -1  1 -1 -1 -1 -1  1
```

Fig. 6. list_attr_celeba.txt file.

According to these features, images with glasses and pictures without glasses are screened by traversal of TXT files. Of course, the format of these pictures is not standard, so they need to be processed again. In this paper, we use the target detection algorithm of AdaBoost in Opencv, and use the haar feature to obtain the face detection. After the detection, the processing image is with width = 64 and height = 64, and the image is subject to grayscale correction and noise filtering. Finally, the same person's pictures are placed in the same folder. At this step, the data set is basically processed.

5.3   Matrix Operation to Complete the Image

In the construction of neural network, the generator network adopts five layers structure, the first layer is the full connection layer, the Second, third, fourth, fifth layers are the reverse convolution layers, and the function is ReLU function. The second layer is the 1024 dimension of the reverse convolution layer, the third layer is the 512 dimension of the reverse convolution layer, the fourth layer is the 256th dimension of the reverse convolution layer, and the fifth layer is the 128 dimension of the reverse convolution layer. The discriminator is the opposite. The last layer of the five layers structure is the full connection layer, the first four layers are convolution layers, and the function is Leaky_ReLU function.

The loss function is based on the loss function pointed out earlier, and the optimization is defined by the loss function.

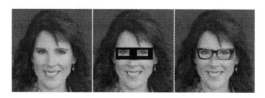

Fig. 7. Result

According to the method discussed above, the image is completed. Because we have to practice masking, keep the part of the eye when we remove the pixels. Here, under the epoch = 2000, learning rate = 0.00005, select a set of better situations.

5.4 Build the Database of Covered Face

The best benefit of this methods, compared to traditional methods of removing covers and restoring them, is that it preserves some of the details of the original face, so that the machine can make the data it gets in some places more accurate when processing some pictures. This can be used as an auxiliary method for traditional methods to obtain more accurate data by comparing the images in the database after the traditional facial restoration.

According to the automatically mended images, a face picture database with masking objects is established. As an auxiliary detection method, the accuracy of face recognition is improved by combining forward and reverse bidirectional methods.

6 Conclusion

This paper mainly introduces the GAN, DCGAN, image completion technology, then uses the original sample to generate the picture of wearing glasses, complements the picture without glasses, and puts forward the establishment of face picture library with occlusive object, so that the accuracy of face recognition can be improved by combining forward and reverse in face recognition.

References

1. Goodfellow, I.J., Pouget-Abadie, J., Mirza, M. et al.: Generative adversarial nets. In: International Conference on Neural Information Processing Systems, pp. 2672–2680. MIT Press (2014)
2. Goodfellow, I., Bengio, Y., Courville, A.: Deep Learning. MIT Press (2016)
3. Radford, A., Metz, L., Chintala, S.: Unsupervised representation learning with deep convolutional generative adversarial networks. arXiv preprint arXiv:1511.06434 (2015)
4. Dumoulin, V., Visin, F.: A guide to convolution arithmetic for deep learning. arXiv preprint arXiv:1603.07285 (2016)
5. Raymond, A.Y., Chen, C., Lim, T.Y., Schwing, A.G., Hasegawa-Johnson, M., Do, M.N.: Semantic image inpainting with deep generative models (2016)
6. Tang, X.-L., Du, Y.-M., Liu, Y.-W., Li, J.-X., Ma, Y.-W.: Image recognition with conditional deep convolutional generative adversarial networks. Zidonghua Xuebao/Acta Automatica Sinica **44**(5), 855–864 (2018)
7. Liu, S., Yu, M., Li, M., Xu, Q.: The research of virtual face based on deep convolutional generative adversarial networks using TensorFlow. Phys. A Stat. Mech. Appl. **521**, 667–680 (2019)
8. Liu, X., Xie, C., Kuang, H., Ma, X.: Face aging simulation with deep convolutional generative adversarial networks. In: Proceedings of the 2018 10th International Conference on Measuring Technology and Mechatronics Automation (ICMTMA), pp. 220–224 (2018)
9. Liu, Z., Luo, P., Wang, X., Tang, X.: Deep learning face attributes in the wild. In: Proceedings of International Conference on Computer Vision. IEEE (2015)

10. Yang, S., Luo, P., Loy, C., Tang, X.: From facial parts responses to face detection: a deep learning approach. In: Proceedings of the IEEE International Conference on Computer Vision (2015)
11. Huang, Z., Xue, W., Mao, Q., Zhan, Y.: Unsupervised domain adaptation for speech emotion recognition using PCANet. Multimedia Tools Appl. **76**(5), 6785–6799 (2016). https://doi.org/10.1007/s11042-016-3354-x
12. Nazeri, K., Ng, E., Joseph, T., et al.: EdgeConnect: generative image inpainting with adversarial edge learning (2019)
13. Li, Y., Liu, S., Yang, J. et al. Generative face completion (2017)
14. Goodfellow, I.: NIPS 2016 Tutorial: Generative Adversarial Networks. arXiv preprint arXiv: 1701.00160 (2017)
15. Goodfellow, I., Bengio, Y., Courville, A.: Deep Learning. MIT Press, Cambridge, UK (2016)
16. Arjovsky, M., Chintala, S., Bottou, L.: Wasserstein GAN (2017)
17. Berthelot, D., Schumm, T., Metz, L.: BEGAN: boundary equilibrium generative adversarial networks. arXiv preprint arXiv:1703.10717 (2017)
18. Simonyan, K., Zisserman, A.: Very deep convolutional networks for large-scale image recognition. Comput. Sci. arXiv preprint arXiv:1409.1556 (2014)
19. Li, X., Liang, Y., Zhao, M., Wang, C., Jiang, Yu.: Few-shot learning with generative adversarial networks based on WOA13 data. Comput. Mater. Continua **60**(3), 1073–1085 (2019)
20. Tu, Y., Lin, Y., Wang, J., Kim, J.U.: Semi-supervised learning with generative adversarial networks on digital signal modulation classification. Comput. Mater. Continua **55**(2), 243–254 (2018)
21. Li, S., Liu, F., Liang, J., Cai, Z., Liang, Z.: Optimization of face recognition system based on Azure IoT edge. Comput. Mater. Continua **61**(3), 1377–1389 (2019)

An Efficient Method for Generating Matrices of Quantum Logic Circuits

Zhiqiang Li$^{(\boxtimes)}$, Jiajia Hu, Xi Wu, Juan Dai, Wei Zhang, and Donghan Yang

Yangzhou University, Yangzhou 225000, China
yzqqlzq@163.com

Abstract. This paper presents an efficient method for generating matrices of quantum logic circuits. First, the truth table is generated by the operation rules of quantum gates in the quantum circuit, and then the matrix of the quantum circuit is constructed according to the mapping relationship between the truth table and the matrix. A common method is to generate a matrix by using the topological transformation rules of quantum gates, and then multiply these matrices generated by each quantum gate in the quantum circuit to construct a quantum circuit. When the scale of the quantum circuit is large, the method involves the generation and product of many large matrices, and complicated matrix multiplication, which takes a huge time cost. In contrast, our method achieves dimensionality reduction skillfully, which greatly improves the efficiency of the algorithm. Taking the GT circuit and the NCV circuit as examples, when the number of quantum lines is as large as 8, our method is hundreds of thousands of times faster than the method proposed in the previous paper.

Keywords: Quantum computing · Generate unitary matrix · Array · GT circuit · NCV circuit

1 Introduction

In the last decades, quantum computing [1] has evolved into one of the most important research area for computer scientists, physicists and for computing developers due to its widespread applications [2, 3]. Quantum gates and quantum circuits play a vital role in quantum computing. Usually, we utilize a unitary matrix to describe the quantum gate. Unitary matrix is a mathematical model which accurately reflects the mathematical properties of quantum gate, as well as quantum circuits. Because the essence functional of the quantum gate is to realize a certain displacement of input data, and quantum circuit is composed of several cascaded quantum gates, and its essence is the product of several permutation matrices, so a common method of constructing structure of the matrix of a quantum circuit is to generate the matrix by using the topological transformation rules of quantum gates, and then multiply matrix generated by each quantum gate in the quantum circuit. Some literature has simplified the quantum gate through a certain process [4], but it cannot effectively compress the matrix, which leads to the exponential increase in the time and space complexity of the algorithm.

© Springer Nature Switzerland AG 2020
X. Sun et al. (Eds.): ICAIS 2020, LNCS 12239, pp. 142–150, 2020.
https://doi.org/10.1007/978-3-030-57884-8_13

In this paper, the array is used to store the quantum state of the input qubit, and the matrix of quantum circuit is constructed by using the truth table and the transformation rules of the quantum circuit. The algorithm in this paper does not involve matrix multiplication. Compared with the traditional algorithm, the time and space complexity of the algorithm is reduced. In this paper, we use reversible circuits in NCV and GT library as the reference circuit and .tfc file storage circuit. The experimental results show that when the number of quantum wires is less than 8, the efficiency can be improved by hundreds of times, or even up to hundreds of thousands of times when the number of quantum wires is 8 and the number of gates is 637.

The remainder of this paper is structured as follows. The next section briefly reviews the basics of reversible and quantum circuits. Later in this section, it provides the information about the machine-readable format of reversible and quantum circuits. Section 3 introduces the proposed algorithm. The experimental results are presented in Sect. 4. Finally, the paper is concluded in Sect. 5.

2 Preliminaries

2.1 Reversible Gates and Circuits

A logic function $f : B^n \rightarrow B^m$ over inputs $X = \{x_1, x_2, \cdots x_n\}$ is reversible if and only if

- its number of inputs is equal to its number of outputs (i.e. n = m) and
- it maps each inputs pattern to a unique output pattern.

Otherwise, the function is termed irreversible [14]. A reversible function can be realized by a circuit comprised of a cascade of reversible gates with no fan-out and feedback [1].

Several reversible gates have been introduced, including the Toffoli gate, the Fredkin gate, and the Peres gate. A multiple control Toffoli (MCT) gate has a target line and several control lines. The target line is inverted if all the controls have value 1, otherwise the value on the target is passed through unchanged. A MCT gate with no control lines acts like the NOT gate and The MCT gate with one control line and two control lines are called CNOT and Toffoli gate respectively as shown in Fig. 1. The family of MCT gates is generalized Toffoli (GT) library. MCT gate is universal because of the fact that all reversible functions can be realized using this gate type alone.

2.2 Quantum Gates and Circuits

Quantum gates are the basic units of quantum circuits and a cascade of quantum gates constitutes the quantum circuit. A quantum circuit may be composed of an infinite number of different quantum gates. However, in practice, researchers think that circuits composed of a small number of gate types [10].

The NCV gate library was introduced by Barenco et al. [12], and contains the following set of quantum gates (Table 1):

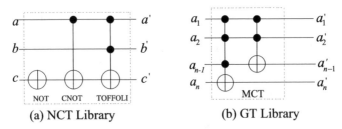

Fig. 1 (a) Fundamental gates and NCT library, (b) MCT gates and GT library

Table 1. Table captions should be placed above the tables.

Quantum gate	Schematic representation	Transformation Matrix	Features
NOT(X)	$\oplus$	$\begin{bmatrix} 0 & 1 \\ 1 & 0 \end{bmatrix}$	$X\lvert 0\rangle = \lvert 1\rangle$ $X\lvert 1\rangle = \lvert 0\rangle$
CNOT(C)	$x_1 \!-\!\bullet\!-\! y_1$ $x_0 \!-\!\oplus\!-\! y_0$	$\begin{bmatrix} 1 & 0 & 0 & 0 \\ 0 & 1 & 0 & 0 \\ 0 & 0 & 0 & 1 \\ 0 & 0 & 1 & 0 \end{bmatrix}$	$\lvert x_1\rangle\lvert x_0\rangle \xrightarrow{CNOT} \lvert x_1\rangle\lvert x_1 \oplus x_0\rangle$
CV	$x_1 \!-\!\bullet\!-\! y_1$ $x_0 \!-\!\boxed{V}\!-\! y_0$	$\begin{bmatrix} 1 & 0 & 0 & 0 \\ 0 & 1 & 0 & 0 \\ 0 & 0 & \frac{1+i}{2} & \frac{1-i}{2} \\ 0 & 0 & \frac{1+i}{2} & \frac{1-i}{2} \end{bmatrix}$	$\lvert x_1\rangle\lvert x_0\rangle \xrightarrow{CV}$ $\lvert x_1\rangle\left(x_1 V\lvert x_0\rangle + (1-x_1)\lvert x_0\rangle\right)$
CV$^+$	$x_1 \!-\!\bullet\!-\! y_1$ $x_0 \!-\!\boxed{V^+}\!-\! y_0$	$\begin{bmatrix} 1 & 0 & 0 & 0 \\ 0 & 1 & 0 & 1 \\ 0 & 0 & \frac{1-i}{2} & \frac{1+i}{2} \\ 0 & 0 & \frac{1+i}{2} & \frac{1-i}{2} \end{bmatrix}$	$\lvert x_1\rangle\lvert x_0\rangle \xrightarrow{CV^+}$ $\lvert x_1\rangle\left(x_1 V^+\lvert x_0\rangle + (1-x_1)\lvert x_0\rangle\right)$

The NCV library can realize any reversible Boolean function. Therefore, the NCV library is the universal library of quantum logic gates. Although the NCV gate library is universal, other libraries also arise wide interests as they can lead to better circuits, e.g. fewer gates. In [9], a modification to the NCV gate library is introduced based on the concept of multi-valued logic.

2.3 Machine-Readable Format: .tfc File

The .tfc file [11] is in the machine-readable format for reversible circuits. It contains detailed information about reversible circuits such as input lines, output lines, gate netlists, garbage lines and constant inputs. In short, the .tfc file provides a unified, formatted file storage method for reversible circuits that allows us to quickly read and identify reversible circuits. For more detailed introduction of .tfc file, please refer to the relevant materials.

Figure 2(b) is the .tfc file of ham3 [11], and the reader can construct the corresponding quantum circuit according to the information provided by the .tfc file, as shown in Fig. 2(a).

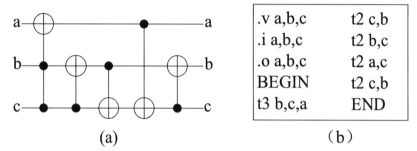

a	a	.v a,b,c	t2 c,b
b	b	.i a,b,c	t2 b,c
c	c	.o a,b,c	t2 a,c
		BEGIN	t2 c,b
		t3 b,c,a	END

(a)   （b）

Fig. 2. (a) Reversible ham3 benchmark circuit, (b) .tfc file for ham3 circuit

2.4 Using Truth Table to Generate the Matrix

Suppose the matrix of the quantum circuit is A, when the input qubits are $|x\rangle$, the output qubit are $|y\rangle$, that is to say, $A|x\rangle = |y\rangle$, where x is between 0 and $2^n - 1$. Assuming that the input vector corresponding to the ith input qubit is X_i, and the ith output vector corresponding to the output qubit is Y_i, and $N = 2^n - 1$, then

$$AX_i = \begin{bmatrix} a_{00} & \cdots & a_{0i} & \cdots & a_{0N} \\ \vdots & & \vdots & & \vdots \\ \vdots & & a_{ji} & & \vdots \\ \vdots & & \vdots & & \vdots \\ a_{N0} & \cdots & a_{Ni} & \cdots & a_{NN} \end{bmatrix} \begin{bmatrix} 0 \\ \vdots \\ 1 \\ \vdots \\ 0 \end{bmatrix} = \begin{bmatrix} a_{0i} \\ \vdots \\ a_{ji} \\ \vdots \\ a_{Ni} \end{bmatrix} = Y_i$$

$A = \begin{pmatrix} Y_0 & Y_1 & \cdots & Y_{2^n-1} \end{pmatrix}$. This proves that the matrix of the quantum circuit can be quickly constructed by the truth table.

C^2V gate is used as an example to generate the matrix of quantum circuit by truth table. The logic function of the quantum circuit is to flip the quantum qubits of the target line when input qubits of the control end is zero, otherwise, there is no operation (Table 2) (Fig. 3).

3 Preliminaries

3.1 Quantum Gate

The essential function of a quantum gate is to achieve some kind of data replacement. The logical function of the MCT gate is to invert the target qubits when the input qubits

Table 2. C^2X gate truth table

Input $x_0\ x_1\ x_2$	Input vector	Output $y_0\ y_1\ y_2$	Output vector
0 0 0	$X_0 = (1,0,0,0,0,0,0,0)^T$	0 0 0	$Y_0 = (1,0,0,0,0,0,0,0)^T$
0 0 1	$X_1 = (0,1,0,0,0,0,0,0)^T$	0 0 1	$Y_1 = (0,1,0,0,0,0,0,0)^T$
0 1 0	$X_2 = (0,0,1,0,0,0,0,0)^T$	0 1 0	$Y_2 = (0,0,1,0,0,0,0,0)^T$
0 1 1	$X_3 = (0,0,0,1,0,0,0,0)^T$	0 1 1	$Y_3 = (0,0,0,1,0,0,0,0)^T$
1 0 0	$X_4 = (0,0,0,0,1,0,0,0)^T$	1 0 0	$Y_4 = (0,0,0,0,1,0,0,0)^T$
1 0 1	$X_5 = (0,0,0,0,0,1,0,0)^T$	1 0 1	$Y_5 = (0,0,0,0,0,1,0,0)^T$
1 1 0	$X_6 = (0,0,0,0,0,0,1,0)^T$	1 1 1	$Y_6 = (0,0,0,0,0,0,0,1)^T$
1 1 1	$X_7 = (0,0,0,0,0,0,0,1)^T$	1 1 0	$Y_7 = (0,0,0,0,0,0,1,0)^T$

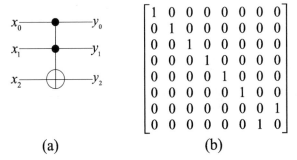

$$
\begin{bmatrix}
1 & 0 & 0 & 0 & 0 & 0 & 0 & 0 \\
0 & 1 & 0 & 0 & 0 & 0 & 0 & 0 \\
0 & 0 & 1 & 0 & 0 & 0 & 0 & 0 \\
0 & 0 & 0 & 1 & 0 & 0 & 0 & 0 \\
0 & 0 & 0 & 0 & 1 & 0 & 0 & 0 \\
0 & 0 & 0 & 0 & 0 & 1 & 0 & 0 \\
0 & 0 & 0 & 0 & 0 & 0 & 0 & 1 \\
0 & 0 & 0 & 0 & 0 & 0 & 1 & 0
\end{bmatrix}
$$

(a)   (b)

Fig. 3. C^2X gate and corresponding the matrix

of all control terminals are the same as the control qubits; if circuits with input use V 、 V^+ gate only, the value of each qubit at each stage of the circuit is restricted to one of $\{0, V_0, 1, V_1\}$ where $V_0 = \frac{1}{2}\begin{pmatrix} 1+i \\ 1-i \end{pmatrix}$ and $V_1 = \frac{1}{2}\begin{pmatrix} 1-i \\ 1+i \end{pmatrix}$. The V and V^+ operations over these four values are:

x	$V(x)$	$V^+(x)$
0	V_0	V_1
V_0	1	0
1	V_1	V_0
V_1	0	1

As shown, V is the cycle $(0 \rightarrow V_0 \rightarrow 1 \rightarrow V_1)$, and V^+ is the inverse cycle [9]. This algorithm uses $V_2()$, $V_3()$ to implement the data replacement of V gate, V^+ gate. The software provided by [11] has not yet been implemented to calculate the truth table of the NCV circuit.

3.2 Matrix Algorithm Based on Truth Table

In order to obtain the truth table of quantum circuit, first, we read the total number of variables, quantum circuit of input variable and output variable information from the .tfc file and then further processing them, at the same time quantum circuit details stored in the array, such as quantum control side door type, number, target location, control bit line stored in a two-dimensional array *Circuit*, control qubits are stored in Array *Circuit01*. The input quantum qubits are stored in a one-dimensional array *QubitArray*, and these three arrays are combined to determine whether the quantum gate operation should be carried out, and the output qubits are obtained after traversing all the quantum gates.

---

input: .tfc file

output: the matrix of the quantum circuit

1. $Circuit[QCount][CCount]$ // *Circuit* main storage .tfc file characters

2. $Circuit01[QCount][CCount]$ // *Circuit01* stores the control qubit of *Circuit*

3. **for** $input = 0$ **to** $2^n - 1$ **do** {

4.   store the binary number of *input* in array *QubitArray*

5.   **for** $i = 0$ **to** $QCount - 1$ **do**

6.   **for** $count = 2$ **to** $Circuit[i][1] + 1$ **do**

7.   **if** $QubitArray[Circuit[i][count]] = Circuit01[i][count]$ **then**

8. $targetIdx = Circuit[i][Circuit[i][1]+1] - 'a'$

9.   **if** ($Circuit[i][0] == 'T'$) **then**

10. $QubitArray[targetIdx] = 1 - QubitArray[targetIdx]$

11.   **else if** $(Circuit[i][0] == 'V')$ **then**

12. $QubitArray[targetIdx] = V_2(QubitArray[targetIdx])$

13.   **else if** $(Circuit[i][0] == 'v')$ **then**

14. $QubitArray[targetIdx] = V_3(QubitArray[targetIdx])$

15.   **else**

16.   return false

17.   **end**

18.   **end**

19.   **end**

20.   **end**

21. $Decimal \leftarrow QubitArray$ corresponding decimal number

22. $Unitary\_Matrix[Decimal][input] = 1$

23.**end**

24.return $Unitary\_Matrix$

Step 1–2: *QCount* is the number of quantum gates that constitute the quantum circuit, *CCount* is the number of characters in a line of .tfc file. Process the data between *'BEGIN'* and *'END'* in the .tfc file by line and store the data in the array *Circuit*. Determine the type of quantum gate based on the first two characters and store it in array *Circuit[i][0]*.

The algorithm in this paper use 'T' to represent MCT gate, 'V' to represent V gate, and 'v' to represent V^+ gate. The number of " ' " of ith row is stored in array $Circuit[i][1]$, that is, array $Circuit01[i][1]$ stores the number of control qubits of the corresponding quantum gate. Take ham3 in Fig. 2 as an example, $Circuit[0] = \{'t', '2', 'b', 'c', 'a'\}$, $Circuit01[0] = \{1, 1, 1, 1, 1\}$.

Step 3–4: n is the maximum number of input qubits of the circuit, traverse from 0 to $2^n - 1$ input qubits, and store the input qubits of the operation in array $QubitArray$. Assuming the input qubit is 0, $QubitArray = \{0, 0, 0\}$.

Step 5: Traverse the quantum gate in.tfc file and complete the data replacement.

Step 6–7: We will compare whether the control qubits stored in array $Circuit01$ are the same as the data in array $QubitArray$. If the input qubits and the control qubits are all the same, perform steps 8–12 to complete the operation of a quantum gate. If the corresponding value is not equal, no operation is performed, steps 8–12 are skipped.

Step 8–12: Judge the type of quantum gate according to the value of $Circuit[i][0]$. If the gate is MCT gate, then take the reverse operation of the target qubits in array $QubitArray$. If it is V, performs function V_2 on the target qubits. If it is a V^+ gate, a function V_3 is performed on the target qubits.

Step 13–14: The decimal number $Decimal$ corresponding to array $QubitArray$ is calculated and generated. $Unitary\_Matrix$ is a two-dimensional array with an assigned value of 0. The value of the $Decimal$ row $input$ column of u is changed to 1, that is, the input column of array $Unitary\_Matrix$ is changed to the tensor product of the output qubit, which is used to construct the matrix of the quantum circuit. When 0 passes all the quantum gates, the final result of $QubitArray = \{0, 0, 0\}$, it means that the value of the 0th row and 0th columns will be changed to 1.

4 Experimental Results

The result time [4] in Table 3 are obtained by using python3.7 on ASUS W518L based on the theory proposed by [4], the algorithm in this paper is implemented in C language. It can be seen from the experimental results that when the number of lines is as high as 8, the efficiency can be increased by hundreds of thousands of times.

Suppose that there are n lines in a quantum circuit which is composed of m quantum gates. When the matrix of quantum gate is multiplied to get the matrix of the quantum circuit. First, the unitary matrix of m quantum gates should be constructed by tensor product, and the unitary matrix of a quantum gate should be constructed by multiplying the tensor product of n matrixes (2×2) for $n-1$ times, that is performing $(2 \times 2)^{n-1}$ times numerical multiplication, and then multiply the unitary matrix of the quantum gates to construct the matrix of the quantum circuit, that is to perform $m*4^{n-1} + (m-1)*2^{3n}$ times numerical multiplication. The algorithm in this paper only needs to perform m times permutation of 2^n data to generate the truth table and 2^n times permutation of the target matrix to generate the final matrix, that is, it only needs to perform $m*2^n + (m-1)*2^n = (2m-1)*2^n$ times permutation of data to generate the final matrix. When the number of lines is 8, the traditional method is to multiply the value of $16793600*m - 16777216$ times. The algorithm only needs to perform $512*m + 256$ times value replacement. Obviously, the computational complexity of the former is much higher than the latter. Experimental data comparison is shown in Table 3.

Table 3. Circuit running time

Benchmark circuits	Source	Gate count	Quantum cost	Gate library	Time [4]/ms	Time [thispaper]/ms
4b15g_1	[12]	15	47	GT	117	1
mspk_4b15g_1	[12]	15	39	GT	116	1
4b15g_2	[12]	15	61	GT	118	1
mspk_4b15g_2	[12]	15	31	NCT	117	1
4b15g_3	[12]	15	53	GT	115	1
mspk_4b15g_3	[12]	15	33	NCT	117	1
4b15g_4	[12]	15	47	GT	114	1
mspk_4b15g_4	[12]	15	35	GT	116	1
4b15g_5	[12]	15	43	NCT	116	1
mspk_4b15g_5	[12]	15	29	NCT	115	1
ham7-23-81	[12]	23	81	GT	86634	16
ham7-21-65	[12]	21	65	GT	83982	15
ham7-25-49	[12]	25	49	NCT	95887	15
hwb4-11-23	[12]	11	23	NCT	87	1
hwb4-11-21	[12]	11	21	NCT	86	1
mspk-hwb4-12	[12]	12	20	NCT	94	1
mspk-hwb4-13	[12]	13	19	NCT	102	1
hwb8_637	[12]	637	16522	GT	21162278	40
p3_1	[10]	4		NCV		1
p3_6	[10]	4		NCV		1
p4_2	[10]	6		NCV		1

5   Experimental Results

In this paper, a method for rapidly generating the matrix of quantum circuit is proposed. The details of quantum gate that makes up the quantum circuit are stored in two two-dimensional arrays A and B, and the array that stores the input qubits is changed according to the value of arrays A and B, and the true table will be generated. Finally, the truth table is used to generate the matrix, thus avoiding a large number of high-dimensional matrices multiplications and a lot of tensor products. Compared with the traditional method, the rate of generating the matrix is increased by a hundredfold or even hundreds of thousands of times.

References

1. Nielsen, M.A., Chuang, I.L.: Quantum computation and quantum information. Math. Struct. Comput. Sci. **17**(6), 1115 (2007)

2. IBM Quantum Computing. http://www.ibm.com/quantum-computing/. Accessed 1 Nov 2019
3. Yiru, W.U., Chen, Y., Ahmad, H., et al.: An asymmetric controlled bidirectional quantum state transmission protocol. Comput. Mater. Continua **59**(1), 199–214 (2019)
4. Xue, X., Chen, H., Liu, Z., et al.: Divide and conquer algorithms for quantum circuit simulation. Acta Electronica Sin. **38**(02), 439–442 (2010)
5. Dou, Z., Xu, G., Chen, X., et al.: Rational non-hierarchical quantum state sharing protocol. Comput. Mater. Continua **58**(2), 335–347 (2010)
6. Li, Z., Chen, H., Xu, B., et al.: Fast algorithms for 4-qubit reversible logic circuits synthesis. In: 2008 IEEE Congress on Evolutionary Computation, CEC 2008 (2008)
7. Cheng, X., Tan, Y., Guan, Z., et al.: An optimized simplification algorithm for reversible MCT circuits. Chin. J. Quantum Electron. **34**(06), 713–720 (2007)
8. Feynman, R.: Quantum mechanical computers. Opt. News **16**(6), 1120 (1986)
9. Sasanian, Z., Wille, R., Miller, D.M.: Realizing reversible circuits using a new class of quantum gates. In: 49th Design Automation Conference (DAC). ACM/EDAC/IEEE (2012)
10. Li, Z., Chen, S., Song, X., et al.: Quantum circuit synthesis using a new quantum logic gate library of NCV quantum gates. Int. J. Theor. Phys. **56**(4), 1023–1038 (2017)
11. Bennett, A., et al.: Elementary gates for quantum computation. Phys. Rev. A At. Mol. Opt. Phys. **52**(5), 3457 (1995)
12. Reversible logic synthesis benchmark. http://webhome.cs.uvic.ca/~dmaslov/. Accessed 1 Nov 2019
13. Chen, X., Chen, H., Liu, Z., Li, Z.: A fast quantum simulation algorithm based on state vector. Acta Electronica Sin. **39**(03), 500–504 (2011)
14. Wille, R., Lye, A., Drechsler, R.: Considering nearest neighbor constraints of quantum circuits at the reversible circuit level. Quantum Inf. Process. **13**(2), 185–199 (2013). https://doi.org/10.1007/s11128-013-0642-5
15. Liu, W., Xiao, Y., Yang, J.C.N., Yu, W., Chi, L.: Privacy-preserving quantum two-party geometric intersection. Comput. Mater. Continua **60**(3), 1237–1250 (2019)

A Mobile Device Multitasking Model with Multiple Mobile Edge Computing Servers

Tao Deng[1] , Yunkai Zhao[1] , Ximing Zhang[1] , Bin Xu[1,2(✉)] , Jin Qi[1(✉)] ,
and Yuntao Ma[2]

[1] Nanjing University of Posts and Telecommunications, Nanjing 21000, China
{xubin2013,qijin}@njupt.edu.cn
[2] NanJing Pharmaceutical Co., Ltd., Nanjing 21000, China

Abstract. Mobile Edge Computing (MEC) is an emerging computing framework that meets the growing computing needs of mobile applications. However, when the user scale is too large, the MEC server may be overloaded. In order to provide satisfactory computing power, it is of great significance that how to break the performance bottleneck of the MEC server. In this paper, we propose a Mobile Device Multitasking Model in Multiple MEC Servers Scenario (MDMS). This model combines collaborative computing with MEC. Each idle mobile device is treated as a virtual MEC server, which shares computing resources with other devices in the network, thereby relieving the pressure on real MEC servers. Each mobile device can perform tasks to the maximum. Based on these, we consider the multi-task interference problem, which further improves the rationality of the model. Meanwhile, this model can run in multiple MEC server scenarios. Furthermore, taking into account that the offloading decision and the computing resource allocation are tightly coupled, a two-layer optimization method is adopted to minimize the total energy consumption of MEC system. Experimental results show the superiority of the model. Moreover, we analyse the relationship between the number of MEC servers and the energy consumption.

Keywords: Mobile edge computing · Computational offloading · Multi-user collaboration model

This paper was supported in part by Natural Science Foundation of Jiangsu Province of China under Grant BK20191381, in part by Jiangsu Planned Projects for Postdoctoral Research Funds under Grant 2019K223, in part by the National Natural Science Foundation of China under Grant 61802208, Grant 61772286, and Grant 61771258, in part by Project funded by China Postdoctoral Science Foundation under Grant 2020M671552 and Grant 2019M651923, in part by Primary Research & Development Plan of Jiangsu Province under Grant BE2019742, in part by the Natural Science Fund for Colleges and Universities in Jiangsu Province under Grant 18KJB520036, and in part by NUPTSF under Grant NY220060.

X. Sun et al. (Eds.): ICAIS 2020, LNCS 12239, pp. 151–161, 2020.
https://doi.org/10.1007/978-3-030-57884-8_14

1   Introduction

The emergence of 5G technology has promoted the development of a large number of new applications, such as artificial intelligence, natural language processing, driverless, etc. These applications require a large amount of computing resources and are sensitive to delay, which lead to higher performance requirements for operating equipment. Although mobile devices have been greatly developed, due to their physical limitations (CPU, battery capacity, etc.), its computing power is still far from satisfactory.

The emergence of Mobile Cloud Computing (MCC) technology provides an idea for solving the above problems [1]. MCC transfers computing tasks on the mobile device side to the cloud for processing, which overcomes the shortcomings of weak mobile device computing capacity and limited storage capacity. It effectively improves mobile device battery life, allowing mobile users to use more complex high-volume programs. But this method requires long distance data transmission, which leads to a high delay caused by the extra radio backhaul load. Meanwhile, data transmission results in high energy consumption of mobile terminals, which affects the user experience [2,3].

In order to provide users with good service quality, an emerging computing framework - Mobile Edge Computing (MEC) has been proposed [4]. MEC provides users with strong computing power and reduces program return delay by deploying MEC servers [5]. Compared with local computing, MEC overcomes the shortcoming of limited computing power of mobile devices. Compared with MCC, although edge cloud coverage is small and resources are few, it effectively reduces program delay. In general, MEC can perform latency-sensitive and computationally intensive tasks well. However, as more and more mobile devices access the MEC server, serious interference may occur between devices, which reduces the transmission rate, resulting in increased transmission time and a waste of energy [6]. Meanwhile, since the computing resources of the MEC server are also limited, when the scale of users expands, the MEC server may be overloaded.

In order to solve the above problems, we assume idle mobile devices on to be virtual MEC servers, which share computing resources with other devices in the network, thereby relieving the pressure on real MEC servers. So a Mobile Device Multitasking Model in Multiple MEC Servers Scenario (MDMS) is proposed in this paper. Different from the existing collaborative unloading MEC model [7–9], we consider the channel interference and delay-sensitive tasks. Moreover, a more complex scenario of multiple MEC servers is studied, there are multiple MEC servers to choose from in our hypothetical scenario. Meanwhile, mobile devices can perform their tasks to the maximum.

The main contributions of this paper are as follows:

1. Considering that a single MEC server is not enough to represent complex scenarios, a task offloading model for multiple MEC servers scenario is proposed. In addition, the relationship between the number of MEC servers and energy consumption is studied based on this model.

2. Collaborative computing is integrated into the model. The mobile device is regarded as a virtual MEC server, which can perform multiple tasks as long as it meets the delay constraints of tasks and has sufficient resources on its own. Multitasking interference is also considered, which is more in line with the reality.

3. A large number of experiments are performed on two instance suites which have 400 mobile users at most, which demonstrate the superiority of the model.

2   Related Work

In recent years, many studies focus on MEC in different scenarios. In [6], the multi-user computational offloading problem of mobile edge computing in multi-channel wireless interference environment is studied. All users upload the task data to a MEC server for calculation over five channels. Furthermore, a distributed computing offloading algorithm is designed to solve the problem which can achieve Nash equilibrium. In [10], the problem of energy-saving computing and communication resource joint allocation in the Multi-User-Single-MEC-Server (MUSMS) scenario is studied, and the greedy strategy and the SMSEF (Select Maximum Saved Energy First) algorithm are used to solve the problem. In [11], a new integrated architecture of cloud, MEC and Internet of Things is presented and a lightweight request and permission framework comes up to solve scalability problems in MUSMS scenarios. In [12], based on the multi-base station scenario, a joint mobile sensing cache and SBS density placement scheme and a hybrid computing offloading strategy are proposed and the experimental results demonstrate that the system energy consumption is effectively reduced. In [7], a multi-user collaborative offloading decision model that can accept sensitive tasks in a multi-channel wireless interference environment in a single MEC server scenario is proposed, which assumes that the collaborative device can only perform one task at most. In [8], a computing offloading system consisting of a secondary node, a user node and a MEC server is proposed to relieve the pressure of MEC servers by using idle resources. In [9], a new multi-user collaborative MEC offloading system is developed, in which 20 mobile devices and 2 MEC servers were tested. Different from the existing work [6–12], we consider channel interference. And delay sensitive tasks can be executed in MDMS model. In addition, the number of MEC servers is no longer limited to one. It is allowed to be multiple, which makes the scenario more reasonable and expands the user scale. Meanwhile, for collaborative devices, mobile devices are allowed to maximize the number of tasks they perform. In Table 1, we summarize the current work and comparison of models.

3   System Model

We propose a multi-user collaborative mobile edge computing model which consists of s MEC servers and n mobile users. Each task has $(s + n)$ assignment

Table 1. Model comparison in the filed of mobile edge computing.

Ref.	Multiple users	Multiple MEC servers	Multi-user interference	Delay-sensitive task	Cooperative computing	Mobile device multitasking
[6]	✓		✓			
[10]	✓					
[11]	✓			✓		
[12]	✓	✓		✓		
[7]	✓		✓	✓	✓	
[8]				✓	✓	
[9]	✓	✓	✓		✓	
Our work	✓	✓	✓	✓	✓	✓

modes, as shown in Fig. 1. $J = \{1, 2, \cdots, s, s+1, s+2, \cdots, s+n\}$ is to identity all base stations and devices. The first s elements in the set represent the index of base stations, and the next n elements represent n mobile device modes. All elements in J are candidate task execution modes. In addition, we assume that each mobile device user has a task that needs to be executed. So n mobile users have n tasks. A set of n tasks is denoted as $N = \{1, 2, \cdots, n\}$.

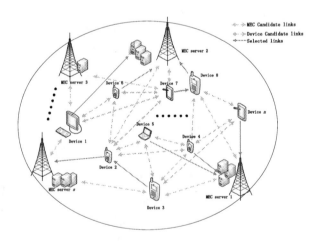

Fig. 1. Network scenario model based on multiple base stations and mobile devices

A two-tuple containing two sets $B = (U, V)$ is used to represent the complex allocation scenarios in MEC system. U is a dynamic set of integers representing the assigned task. The maximum number of elements in the set is the number of all tasks; V is a relational set that represents the allocation pattern of the assigned tasks.

In order to more specifically represent the assigned tasks and their allocation patterns, u_t represents an assigned task number, where $t \in \{1, 2, \cdots, n\}$, $u_t \in J\backslash\{1, 2, \cdots, s\}$. A quadruples $\left(C_{u_t}, D_{u_t}, B_{u_t}, T_{u_t}^{\max}\right)$ is used to describe the task u_t. C_{u_t} represents the number of CPU cycles required to execute the task u_t; D_{u_t} is the amount of data of the task u_t, including code, input parameters and so on; B_{u_t} represents the size of the resulting data generated by task u_t when the computing is completed; $T_{u_t}^{\max}$ represents the maximum delay that the task can tolerate.

$v_{ti} = <u_t, i>$ represents the task u_t is assigned to the device or the base station i to execute. So $i \in J$. However, not all candidate modes will be selected, and only a few may be used to execute. I is defined as a dynamic set storing determined execution mode i of the task. Therefore, $I \subset J$. For different i, we have the following situations:

- $i \in J\backslash\{s+1, s+2, \cdots, s+n\}$, represents that the task u_t is running in base station mode.
- $i \in J\backslash\{1, 2, \cdots, s\}$ and $i \neq u_t$ represent that the task u_t is running in collaboration mode.
- $i = u_t$, represents that the task u_t is running in local mode.

3.1 Calculation Model

Giving the required computing resources $r_{v_{ti}}$ of task u_t in the allocation mode i, the computation time of task u_t in mode i is $T_{v_{ti}}^c$, as described in Eq. (1). It is worth noting that $r_{v_{ti}} \leq f_i$, f_i represents the maximum computing power of the device or the base station itself i.

$$T_{v_{ti}}^c = \frac{C_{u_t}}{r_{v_{ti}}}, \ u_t \in U, \ i \in I \qquad (1)$$

Only the computational power consumption on mobile devices is considered, because mobile devices are usually battery-equipped and do not have long-lasting battery life. The proportion of processing-intensive tasks in their power consumption is very significant. According to [13], dynamic power consumption dominates the circuit power consumption, so the computational task energy consumption $E_{v_{ti}}^c$ can be expressed as Eq. (2).

$$E_{v_{ti}}^c = \begin{cases} k\left(r_{v_{ti}}\right)^2 C_{u_t}, & \forall u_t \in U, \ i \in J\backslash\{1, 2, \cdots, s\} \\ 0, & \forall u_t \in U, \ i \in J\backslash\{s+1, s+2, \cdots, s+n\} \end{cases} \qquad (2)$$

The constant coefficient k is used to represent effective capacitance coefficient.

3.2 Communication Model

As shown in Fig. 2, when introducing multiple MEC servers, mobile device is allowed to perform multiple tasks with its own computing resources, which makes the communication model become more reasonable.

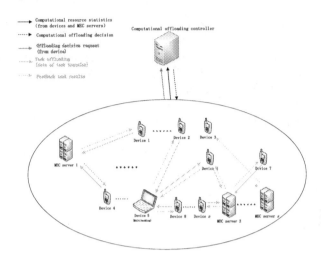

Fig. 2. Mobile device multitasking communication model in multiple MEC servers scenario

When multiple tasks are assigned to the same mode, mobile users need to upload data to the same device or MEC server, which can cause interference between users. So the maximum transmission rate $R_{v_{ti}}$ of the task u_t with selection mode i can be obtained according to the Shannon formula. However, it is worth noting that when selecting the local device mode, there is no need to transfer data. $R_{v_{ti}}$ can be got by Eq. (3).

$$
R_{v_{ti}} = \begin{cases} \infty, & u_t \in U, \ i = u_t \\ B^* \log_2 \left(1 + \dfrac{p_{u_t}^{tr} H_{v_{ti}}}{\delta^2 + \sum\limits_{h=1}^{n} p_{u_h}^{tr} H_{v_{hi}} - p_{u_t}^{tr} H_{v_{ti}}} \right), & otherwise \end{cases} \tag{3}
$$

where B is the bandwidth, $H_{v_{ti}}$ is the channel gain when mode i is selected for the task u_t. In addition, $p_{u_t}^{tr}$ is the power when the device sends the data of task u_t, and δ^2 is the noise power. Similarly, after the task data is transmitted and calculated, the result data needs to be returned to the device corresponding to task u_t. We assume that the transmission rate of the output data and the returned result data is consistent.

Then, the transmission time $T_{v_{ti}}^{tr}$ can be got by Eq. (4).

$$
T_{v_{ti}}^{tr} = \frac{D_{u_t}}{R_{v_{ti}}} + \frac{B_{u_t}}{R_{v_{ti}}}, \ u_t \in U, \ i \in I \tag{4}
$$

Obviously, the mode is local mode when $i = u_t$. So there is no need to transfer data, set $R_{v_{ti}} = \infty$, $T_{v_{ti}}^{tr} = 0$.

For the energy consumption $E_{v_{ti}}^{tr}$ generated in the transmission process, if the computing task u_t is carried out under the cooperative device mode, the device in need of assistance will generate energy consumption when uploading the data of task u_t and receiving the data of result. In addition, the equipment that helps to complete the computing task u_t will also generate energy consumption when receiving the data of task u_t and returning the data of result. The total transmission energy consumption is seen as the result of completing task u_t. $p_{u_t}^{re}$ is the power when the device receives the result of the task u_t. However, when the mode is selected as MEC server mode, this part of energy consumption is borne by MEC server, which is not included in the transmission energy consumption in this paper. Therefore, the calculation formula of $E_{v_{ti}}^{tr}$ is shown in Eq. (5).

$$
E_{v_{ti}}^{tr} = \begin{cases} \dfrac{p_{u_t}^{tr} D_{u_t}}{R_{v_{ti}}} + \dfrac{p_i^{re} D_{u_t}}{R_{v_{ti}}} & i \in J\backslash\{1,2,\cdots,s\},\ i \neq u_t \\ \quad + \dfrac{p_i^{tr} B_{u_t}}{R_{v_{ti}}} + \dfrac{p_{u_t}^{re} B_{u_t}}{R_{v_{ti}}}, \\ \dfrac{p_{u_t}^{tr} D_{u_t}}{R_{v_{ti}}} + \dfrac{p_{u_t}^{re} B_{u_t}}{R_{v_{ti}}}, & i \in J\backslash\{s+1,s+2,\cdots,s+n\} \\ 0, & i = u_t \end{cases} ,\ u_t \in U \quad (5)
$$

3.3   Optimization of Objective Expression

In the first two subsections, the computational models and communication models are quantified. In this section the main research issues on the basis of them is quantified, that is, in the scenario of multiple mobile devices and multiple MEC servers, how to jointly optimize resource allocation and offloading decisions when computing resources and communication resources are limited. The desired goal is to minimize the total energy consumption of the entire model under the time constraints of each task. The problem can be summarized as Eq. (6).

$$
P : \min_{V,r} \sum_{i \in I} \sum_{t=1}^{n} \left(E_{V_{ti}}^c + E_{V_{ti}}^{tr} \right) \tag{6}
$$

$$
s.t.\ 0 \le \sum_{t=1}^{n} \left(T_{v_{ti}}^c + T_{v_{ti}}^{tr} \right) \le T_{u_t}^{\max},\ \forall u_t \in U,\ i \in I \tag{7}
$$

$$
0 \le \sum_{t=1}^{n} r_{v_{ti}} \le f_i,\ i \in I \tag{8}
$$

$$
U = N \tag{9}
$$

There are two decision variables to be optimized in Eq. (6), namely V, r, which are offloading decision and computing resource allocation respectively. Equation (7) guarantees that each task will not exceed the maximum delay when it is executed; Eq. (8) allows each device to perform multiple tasks simultaneously without exceeding its own computing power; Eq. (9) guarantees that each task will be executed. It can be noticed that the assigned task set and the entire task set are ultimately equal.

Obviously, P is a mixed variable non-linear problem. Moreover, the offloading decision set V and the computing resource allocation matrix r are interdependent. The allocation of computing resources depends on the formulation of the offloading decision, and the pros and cons of the offloading decision change with the distribution of resources. The traditional optimization method can only solve them separately, but it destroys the strong dependence of the two, which will greatly interfere with the quality of the solution. Therefore, we adopt a bilevel optimization method [7]. The upper level optimization problem is the formulation of the offloading decision. The lower level optimization problem is the allocation of computing resources and the minimum mobile device energy consumption. The transformed problem $P1$ is shown in Eq. (10).

$$P1 : \min_{V,r} \sum_{i \in I} \sum_{t=1}^{n} \left(E_{V_t}^c + E_{V_t}^{tr} \right)$$

$$s.t. \ \arg\min_{r} \left\{ \sum_{t=1}^{n} E_{v_{ti}}^c : Eq. \ (7) \ and \ Eq. \ (8) \ are \ constraints \right\}$$

$$Eq. \ (9)$$

$$(10)$$

4   Simulation and Analysis

The purpose of the experiment in this section is to verify the validity of the MDMS model. The data set in [7] is used in the experiment, which includes information of one MEC server and two instance suits, such as channel gain, computing power, task data size, etc. Moreover, additional information about the 4 MEC servers is generated based on the above data set.

Two indicators are used to measure the quality of the results, namely the number of tasks completed(NTC) and the total energy consumption(EC). NTC records the number of tasks completed within the delay constraint after 300 iterations of the algorithm. EC is the energy consumption of the MEC system when all tasks are completed. Please note that all experiments are carried out in MATLAB R2018b using a personal computer running on a 2.20 GHz (processor) and 8 GB of RAM with an i7-8750 CPU. The solving algorithm of each model is a double-layer optimization method that integrates ant colony algorithm [7], and the algorithm parameters are consistent with the original paper. The experimental results are the average of the results obtained after 5 independent runs.

In Figs. 3 and 4, in order to verify the validity of the MDMS model, we compare the MDMS model with the following three models according to the average number of task completions of the respective tasks on the two instance suites with different scales:

Local computing (LC): The task published by a mobile device runs only on that device.

MEC server computing (MEC-C): The tasks published by mobile devices are only performed on the MEC server.

Cooperative computing (CC): The tasks published by the mobile device itself can be performed not only on the device but also on the other mobile devices in this network.

It can be seen from Figs. 3 and 4 that LC, MEC-C and CC models are unable to achieve the maximum number of tasks in all cases, but the MDMS model can complete all released tasks. In the case of the small number of tasks, ie, instance suit 1, it can be seen that the CC model results are superior to the LC model, and are substantially comparable to the MEC-C model. In fact, in the case of low task count, the device-to-device collaborative computing mode and the MEC server computing mode are both good choices. But as the number of tasks continues to increase, ie, in suite 2, the effect of the MEC-C model remains basically the same. Because the MEC server has limited computing resources. When the task scale is rapidly expanding, the MEC server has already reached the performance bottleneck so that it can no longer perform the task. The LC model and the CC model are all in a steady state. This is because the available resources of the LC model and the CC model are gradually increasing as the user scales up. However, because the CC model can maximize the resource utilization by sharing the idle resources of all devices in the entire network, the CC model is always better than the LC model. But the mobile device itself has physical limitations, so not all tasks of CC model can be completed. The MDMS model combines above three models to overcome the shortcomings of each of the above three models, and thus shows the best performance.

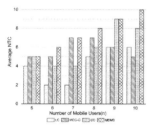

Fig. 3. Instance suite 1

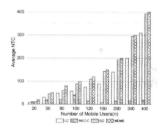

Fig. 4. Instance suite 2

In Fig. 5, MDMS-s represents the MDMS model in s MEC servers scenarios. BIJOR is a collaborative offloading model in [7]. It can only be executed in a single MEC server scenario, and the collaboration mode is a pseudo-cooperative mode, which assumes that each mobile device can only perform one task under constraints. For the sake of fairness, this paper uses the same solution method and parameter settings as in [7], and solves the model on the instance suite 2, the results are shown in Fig. 5. Obviously, the MDMS-5 model has the best performance under all tasks, and the MDMS model has better performance than the BIJOR model.

Table 2 refers to the energy consumption ratio reduced by each model compared with BIJOR, where $rate = \frac{\overline{EC}_{BIJOR} - \overline{EC}_{MDMS-s}}{\overline{EC}_{BIJOR}}$. According to Table 2, MDMS-2 has the largest improvement range for each MEC server added when

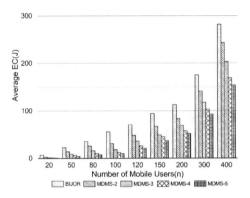

Fig. 5. The average EC of MDMS and BIJOR in instance suite 2 in a multiple MEC servers scenario

the number of tasks is 20, 100, 120, and 200, indicating that 2 is the best number of MEC servers in these cases. When the number of tasks is 50, 80, 150, 300, for each additional MEC server in the MDMS model, the energy consumption can be reduced by nearly half. While the number of MEC server reaches 3, the effect is reduced. When the number of tasks is 400, the energy consumption of the MDMS model can be reduced by nearly half with each MEC server added. This trend has been maintained when the number of MEC servers reaches 4. When the number of MEC servers reaches 5, the effect drops sharply. It can be seen that although the increasing number of MEC servers can reduce the energy consumption of the MDMS model, the number of MEC servers needs to be controlled within a certain range under different task sizes. This clearly shows the relationship between the number of MEC servers and the reduction in energy consumption on this model.

Table 2. MDMS model increases with the number of MEC servers compared to the average EC reduction ratio of BIJOR.

Task	MDMS-2	MDMS-3	MDMS-4	MDMS-5
Tasks = 20	60.39%	80.20%	85.20%	88.24%
Tasks = 50	38.59%	60.10%	71.72%	81.47%
Tasks = 80	27.31%	55.09%	70.78%	79.26%
Tasks = 100	44.21%	66.89%	77.68%	82.21%
Tasks = 120	31.74%	48.88%	63.44%	70.14%
Tasks = 150	28.79%	48.11%	51.71%	60.87%
Tasks = 200	25.85%	38.65%	49.11%	54.31%
Tasks = 300	19.36%	32.76%	41.08%	47.11%
Tasks = 400	13.71%	27.84%	40.09%	45.30%

5   Conclusion

Aiming at the bottleneck problem of MEC server in the field of edge mobile computing, we combine collaborative computing with mobile edge computing. A multi-user collaborative mobile edge computing model MDMS suitable for multi-MEC servers scenarios is proposed. The model assumes that each mobile device is a virtual MEC server. Each mobile device can perform multiple tasks while meeting delay constraints and computing resource constraints. In addition, multitasking interference issues are considered. In order to verify the validity of the MDMS model, two indicators NTC, EC and multiple models are compared on the two example suites. Meanwhile, the relationship between the number of MEC servers and the energy consumption is analyzed. In the future, we will consider the impact of users movement and the decomposability of tasks in dynamic scenarios.

References

1. Akherfi, K., Gerndt, M., Harroud, H.: Mobile cloud computing for computation offloading. Issues Chall. **14**(1), 1–16 (2018)
2. Chen, X.: Decentralized computation offloading game for mobile cloud computing. IEEE Trans. Parallel Distrib. Syst. **26**(4), 974–983 (2015)
3. Li, Y., Wang, X., Fang, W., et al.: A distributed ADMM approach for collaborative regression learning in edge computing. CMC **59**(2), 493–508 (2019)
4. Wei, Y., Wang, Z., Guo, D., et al.: Deep q-learning based computation offloading strategy for mobile edge computing. CMC **59**(1), 89–104 (2019)
5. Guo, Y., Liu, F., Xiao, N., et al.: Task-based resource allocation bid in edge computing micro datacenter. CMC **61**(2), 777–792 (2019)
6. Chen, X., Jiao, L., Li, W., Fu, X.: Efficient multi-user computation offloading for mobile-edge cloud computing. IEEE/ACM Trans. Netw. **24**(5), 2795–2808 (2016)
7. Huang, P., Wang, Y., Wang, K., et al.: A bilevel optimization approach for joint offloading decision and resource allocation in cooperative mobile edge computing. IEEE Trans. Cybern. (2019). https://doi.org/10.1109/TCYB.2019.2916728
8. Tao, Y., You, C., Zhang, P., Huang, K.: Stochastic control of computation offloading to a helper with a dynamically loaded CPU. IEEE Trans. Wirel. Commun. **18**(2), 1247–1262 (2019)
9. Ren, C., Lyu, X., Ni, W., et al.: Distributed online learning of fog computing under nonuniform device cardinality. IEEE IoT J. **6**(1), 1147–1159 (2018)
10. Wei, F., Chen, S., Zou, W.: A Greedy algorithm for task offloading in mobile edge computing system. China Commun. **15**(11), 155–163 (2018)
11. Lyu, X., et al.: Selective offloading in mobile edge computing for the green internet of things. IEEE Netw. **32**(1), 54–60 (2018)
12. Min, C., Yixue, H., Meikang, Q., et al.: Mobility-aware caching and computation offloading in 5G ultra-dense cellular networks. IEEE Netw. **16**(7), 974–987 (2016)
13. Mao, Y., You, C., Zhang, J., et al.: A survey on mobile edge computing: the communication perspective. IEEE Commun. Surv. Tutor. **19**(4), 2322–2358 (2017)

A Localization Algorithm Based on Emulating Large Bandwidth with Passive RFID Tags

Die Jiang, Liangbo Xie$^{(\boxtimes)}$, and Qing Jiang

School of Communication and Information Engineering, Chongqing University of Posts and Telecommunications, Chongqing, China
xielb@cqupt.edu.cn

Abstract. The passive Ultra High Frequency (UHF) RFID tag localization technology has attracted more and more attention in recent years, due to its advantages of low cost, simple deployment and long-distance communication. Most researches use the Received Signal Strength (RSS) of the tag backscattered signal or reference tags to realize localization. Owing to the large fluctuation of RSS and its sensitivity to antenna direction and environment, the localization accuracy is meter-level and can hardly be further improved. Reference tag based localization systems are complex to deploy and consume more resources. To solve these problems, a passive UHF RFID localization algorithm based on carrier phase and distance difference is proposed in this paper. A large virtual bandwidth is obtained by frequency hopping technology to improve the multipath resolution. Benefiting from the high multipath resolution provided by the large bandwidth, a modified Frequency Hopping Continuous Wave (FHCW) ranging algorithm is proposed to suppress the multipath effect on phase and solve the phase cycle ambiguity. And the Least Square (LS) method is adopted to realize two-dimensional localization. Simulation results demonstrate that the proposed FHCW algorithm can achieve ranging accuracy of 2 cm, and the probability is more than 97%. Under the triangle layout, the proposed localization algorithm can achieve localization error of less than 6 cm.

Keywords: Carrier phase · Virtual bandwidth · Multipath suppression · Phase cycle ambiguity

1 Introduction

With the rapid development of Internet, Internet of things, mobile communication and other technologies, people urgently need to obtain location information anytime and anywhere. In the outdoor environment, Global Positioning System (GPS), BeiDou Navigation Satellite System (BDS) [1] and other Global Navigation Satellite Systems (GNSS) can provide users with meter-level location services, which basically solves the need of people in outdoor localization. However, due to indoor multipath effect, the GNSS signal is rapidly attenuated and even cannot be detected [2], leading to a sharp drop in indoor localization accuracy. A variety of indoor localization technology came into being under such circumstance. The RFID indoor localization technology exploits radio

© Springer Nature Switzerland AG 2020
X. Sun et al. (Eds.): ICAIS 2020, LNCS 12239, pp. 162–172, 2020.
https://doi.org/10.1007/978-3-030-57884-8_15

frequency signals to automatically identify tags without direct contact, so it has been widely used [3–5].

At present, the research on RFID localization is mainly based on the Received Signal Strength (RSS) of the tag backscattered signal or with the help of reference tags. However, due to the large fluctuation of RSS and its sensitivity to antenna direction and environment, the localization accuracy can hardly be further improved. And reference tags are complex to deploy and consume more resources. Compared to the localization algorithm based on RSS or reference tags, the phase-based algorithm can achieve precise localization by obtaining accurate phase information, since phase is very sensitive to time and distance. Therefore, a passive Ultra High Frequency (UHF) RFID location algorithm based on carrier phase and distance difference is proposed in this paper.

The rest of this paper is organized as follows. The related works on localization using RFID technology are reviewed in Sect. 2. The description of system is presented in Sect. 3. The Frequency Hopping Continuous Wave (FHCW) ranging algorithm is discussed in Sect. 4, including Line-of-Sight (LOS) path identification, multipath suppression optimization algorithm, and phase cycle ambiguity solving. Section 5 introduces the Least Square (LS) localization algorithm. Section 6 presents the related simulation results. And Sect. 7 concludes this paper.

2 Related Works

Many researches have been conducted on RFID localization [6, 7]. Early proposals in this field depend on measuring the RSS, due to the requirements of no additional hardware support and additional network consumption. However, RSS suffers from poor accuracy because of its large fluctuation and sensitivity to antenna direction and environment. In order to get better localization performance, reference tag is introduced into subsequent proposals. Whereas, the deployment and density of the reference tags not only consume more resources, but also are difficult in high-precision localization.

Compared to the localization algorithm based on RSS or reference tags, the phase-based algorithm can achieve precise localization by obtaining accurate phase information, because phase is very sensitive to time and distance. A ranging method based on multi-frequency carrier phase difference is proposed [8], which uses Chinese Remainder Theorem (CRT) to solve the phase cycle ambiguity. The authors propose a uniform linear array of passive tags to implement tag localization relying on the phase information of the backscattered tag signal [9], but it consumes more resources and suffers from higher requirements because what they need to localize are tag arrays. Ma Yunfei et al. [10, 11] have done a lot of research works on passive UHF RFID indoor localization based on phase information. In [10], the authors propose to use the nonlinear backscattering mechanism of the tag and the phase information to achieve accurate 3D real-time indoor localization. However, the tag needs to be customized. Reference [11] proposes to achieve centimeter-level indoor localization based on phase information. It is much easier for researchers to reproduce, since the authors directly use off-the-shelf narrowband passive tags.

In this paper, we propose a centimeter-level passive UHF RFID localization algorithm, which relies on carrier phase for a tag ranging, and uses distance difference to achieve two-dimensional localization by LS method.

3 System Description

Since passive tags can be attached to the surface of the objects without occupying space, and the cost is low for large-area deployment, passive tag localization has a broader prospect in indoor localization. However, passive tags have an ISM (Industrial Scientific Medical) band of only 26 MHz. Even if the reader transmits signals at all frequencies in this band, its bandwidth is much smaller than that of the GHz band [12, 13] which achieves the localization accuracy of centimeters.

In order to achieve centimeter-level localization accuracy, it is necessary to meet the requirements of large bandwidth and ensure that the transmission power can drive passive tags for communication. We employ backscatter communication mechanism of passive tags to emulate a large virtual bandwidth by hopping technology [11]. Specifically, two reader simultaneously transmit continuous waves at a frequency f_p inside the ISM band and at another frequency f_c outside the ISM band, respectively. It uses high power f_p to power up and communicate with the passive tag, and uses lower power f_c to hop over time as shown in Fig. 1. Then we stitch the channels at the various frequencies to realize a large virtual bandwidth and improve multipath resolution. Such an approach still remains compliant with FCC regulations. Since the passive tag is like a mirror, both signals at f_p and f_c will be backscattered back to the reader. We use low pass filter to filter out the high-frequency signal f_p and only leave the signal information corresponding to the frequency f_c.

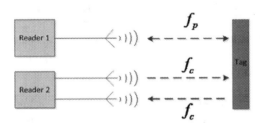

Fig. 1. System structure for ranging.

The signal transmitted by the reader to emulate the bandwidth can be denoted as

$$S_{TX}(t) = A \cos(2\pi f_c t + \varphi_0) \tag{1}$$

where A is the signal magnitude, f_c is carrier frequency, φ_0 is initial phase of the carrier.

If the tag takes Binary Amplitude Shift Keying (2ASK) modulation, the received signal is given by

$$S_{RX}(t) = \Gamma A \alpha_f \alpha_b \cos\left(2\pi f_c\left(t - \tau_f - \tau_b\right) + \varphi_0\right) + S_{IN}(t) \tag{2}$$

where Γ is reflection coefficient of the tag, α_f and α_b is attenuation coefficient of the forward link (reader-to-tag) signal and the back link (tag-to-reader) signal, respectively, τ_f and τ_b is Time of Flight (TOF) of the forward link signal and the back link signal, respectively, $S_{IN}(t)$ is interference signals, mainly including the signals only reflected by obstacles without the tag reflection and direct leaked signals from the transmitter to the receiver.

After I/Q quadrature demodulation and filtering the interference signals $S_{IN}(t)$, we can obtain the in-phase $I(\tau)$ and quadrature $Q(\tau)$ branches of the received signal, and the phase value is obtained by

$$\theta = \arctan\left(\frac{Q(\tau)}{I(\tau)}\right) \tag{3}$$

Due to the complexity of the indoor environment, RF signals bounce off different obstacles (such as ceilings, walls and furniture). As a result, the receiver obtains several copies of the signal and every copy has experienced a different TOF. Therefore, the phase obtained by (3) is not accurate and TOF calculated by the phase is also with large error. The impacts of multipath and noise on the phase should be minimized in order to obtain accurate phase and TOF.

4 FHCW Ranging Algorithm

Since the proposed localization algorithm relies on the distance parameter, we can evaluate the distance based on Ref. [14], which presents a FHCW ranging algorithm, including LOS path identification, multipath suppression and phase cycle ambiguity solving, as shown in Fig. 2.

Fig. 2. The flow diagram of FHCW ranging.

4.1 LOS Path Identification

Since the communication distance of UHF RFID system is short and generally within 10 m and the tag to be located is stationary, the channel can be regarded as a time-invariant linear channel, which can be expressed as [15]

$$H(f) = \sum_l \alpha_l e^{-j2\pi f \tau_l} = \alpha_{los} e^{-j2\pi f \frac{d_{los}}{c}} + \sum_i \alpha_i e^{-j2\pi f \frac{d_i}{c}} \tag{4}$$

where α_l and τ_l are attenuation coefficient and TOF of $l-th$ path, respectively. α_{los} and d_{los} are attenuation coefficient and length of the LOS path, respectively. α_i and d_i are attenuation coefficient and length of $i-th$ multipath, respectively. c is the speed of light.

In order to identify the LOS signal from the multipath signals and achieve ranging, we regard the first peak of Channel Impulse Response (CIR) profile as judgment standard. The multipath resolution is important to identify LOS path and inversely proportional to the signal bandwidth [13]. A larger signal bandwidth results in a higher multipath resolution, which helps to separate the LOS path from the multipath. In Sect. 3, we have obtained a virtual large bandwidth, so the bandwidth requirement of the LOS path identification can be satisfied.

By performing Inverse Discrete Fourier Transform (IDFT), we can transform the channels from the frequency domain, i.e. (4), to the time domain, and plot CIR profile as shown in Fig. 3(a). And the first peak denotes the LOS signal.

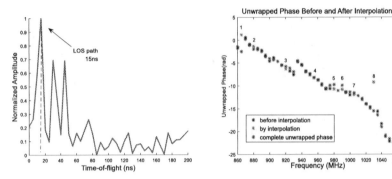

Fig. 3. (a) CIR profile at 200 MHz bandwidth; (b) Unwrapped phase before and after cubic spline interpolation method.

Since the commercial passive tag is matched with the antenna in the ISM band, the receiver may not receive the signals at all the frequencies due to interference, resulting in the wrong identification of LOS path by IDFT. We proposed inversing Non-uniform Discrete Fourier Transform (NDFT) as demonstrated in [14] to obtain the same distance estimate as shown in Fig. 3(a). But we find that the method is at the cost of complexity because we need to consider how to process the CFR data that are symmetrical with the missing data in following multipath suppression. In order to reduce the complexity, we propose that the CFR data are firstly recovered by cubic spline interpolation method, and then perform IDFT to identify LOS path in this paper. The phases corresponding to missing frequencies recovered by cubic spline interpolation method are as shown in Fig. 3(b).

Each phase in Fig. 3(b) is an overlay value affected by three multipath and noise, whose signal to noise ratio is 10 dB. For ease of analysis, we only take three multipath into consideration in this paper. It is seen that there are eight phases obtained by interpolation, which are denoted by red symbols. Most of them are very close to the theoretical values (i.e., denoted by blue symbols), but exist a little deviation due to random noise. The difference between interpolation phase values of the first and the eighth and corresponding theoretical values is 2π, which is a cycle of phase. What we need to emphasize is that the difference of 2π still can prove the interpolated result is accurate because of the periodicity of phase. In summary, Fig. 3(b) demonstrates that cubic spline interpolation method can recover missed phase information, and CFR data because it also depends on the same in-phase and quadrature branches.

4.2 Multipath Suppression Optimization Algorithm

When the phase errors caused by the system itself are eliminated by technical means, the main source of phase error is multipath and noise, which can only be suppressed. Reference [11] presents an idea that using the rough distance estimate $\tilde{d}_{los}$ of the identified LOS path performs suppression. $\tilde{d}_{los}$ is obtained from c multiplying by the TOF estimate of the LOS signal. The accurate phase obtained by suppression [11] is given by

$$
\theta_k = \angle \sum_{i=1}^{K} h_i e^{j\frac{2\pi}{c}(f_i - f_k)\tilde{d}_{los}} = \angle a_{los} e^{-j\frac{2\pi}{c} f_k d_{los}} \sum_{i=1}^{K} e^{j\frac{2\pi}{c}(i-k)\Delta f\left(\tilde{d}_{los} - d_{los}\right)}
$$

$$
+ \angle \sum_{l=1}^{L} a_l e^{-j\frac{2\pi}{c} f_k d_l} \sum_{i=1}^{K} e^{j\frac{2\pi}{c}(i-k)\Delta f\left(\tilde{d}_{los} - d_l\right)} \tag{5}
$$

where K is the total number of frequency hopping. d_{los} is the actual distance. Δf is frequency space. As described in [11], there is

$$
\left| \frac{\sum_{i=1}^{K} e^{j\frac{2\pi}{c}(i-k)\Delta f\left(\tilde{d}_{los} - d_l\right)}}{K} \right| \approx \left| \sin c\left(B\left(\tilde{d}_{los} - d_l\right)\right) \right| \ll 1 \tag{6}
$$

The equation of (6) takes the absolute value, but $\sum_{i=1}^{K} e^{j\frac{2\pi}{c}(i-k)\Delta f\left(\tilde{d}_{los} - d_l\right)}$ of (5) in fact is a complex. Therefore, (5) has no perfect suppression performance. We hence put forward an optimization algorithm by introducing an additional condition as shown by

$$
\theta_m = \angle \sum_{i=1}^{K} h_i e^{j\frac{2\pi}{c}(f_i - f_m)\tilde{d}_{los}}
$$

$$
s.t. \begin{cases} m = \dfrac{K+1}{2} \\ K \geq 3 \text{ is odd number} \end{cases} \tag{7}
$$

when K is odd number and $m = (K+1)/2$, there exists a symmetrical behavior in frequency difference $(f_i - f_m)$ as shown in Fig. 4.

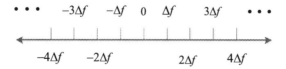

Fig. 4. Symmetrical frequency difference.

Thus, the filtering effect is further enhanced to obtain an accurate phase, and the suppressed phase is close to the phase of the LOS path as shown in Fig. 5.

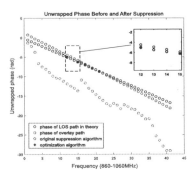

Fig. 5. Unwrapped phase before and after suppression.

We only select five suppressed phases in this paper to solve the cycle ambiguity, and the ranging accuracy of centimeter-level can be achieved. It should be noted that the cycle ambiguity is solved by combining multiple phases in next subsection and the basic number of phases is three. The more phase, the more favorable to solve cycle ambiguity. However, we find that five suppressed phases are enough to obtain centimeter ranging accuracy.

4.3 Solving Phase Cycle Ambiguity

When the measured distance exceeds one wavelength, there will be phase cycle ambiguity due to the periodicity of the phase. That is, TOF computed by θ_m is wrong at this moment, so we need to solve the phase cycle ambiguity.

The well-known method to solve cycle ambiguity is CRT [16] as described by

$$\tilde{\tau} = \tau_m + M_m/f_m = -\frac{\theta_m}{2\pi f_m} + M_m/f_m \qquad (8)$$

where M_m is the cycle of each phase, $m = 12, 15, 17, 19, 21$ in this paper.

However, the set of equations established by (8) is under-determined. Thus, the set of equations does not have a single closed-form solution. And the traditional CRT can only solve integer problems, so we propose to construct a search matrix by (8) as written by

$$A = \begin{bmatrix} \tau_{12} \ \tau_{12} + 1/f_{12} \ \tau_{12} + 2/f_{12} \cdots \tau_{12} + M_{max}/f_{12} \\ \tau_{15} \ \tau_{15} + 1/f_{15} \ \tau_{15} + 2/f_{15} \cdots \tau_{15} + M_{max}/f_{15} \\ \tau_{17} \ \tau_{17} + 1/f_{17} \ \tau_{17} + 2/f_{17} \cdots \tau_{17} + M_{max}/f_{17} \\ \tau_{19} \ \tau_{19} + 1/f_{19} \ \tau_{19} + 2/f_{19} \cdots \tau_{19} + M_{max}/f_{19} \\ \tau_{21} \ \tau_{21} + 1/f_{21} \ \tau_{21} + 2/f_{21} \cdots \tau_{21} + M_{max}/f_{21} \end{bmatrix} \qquad (9)$$

where $M_{max} = R_{max}/\lambda_{min}$, $\lambda_{min} = c/f_{max}$, R_{max} is maximum reading distance between the reader and the tag. The R_{max} can limit the traversal range of matrix A. The steps are summarized as follows:

- Selecting the first row of data of matrix A as each initial cluster center.

- Calculating the distance between all the data and the cluster centers in the matrix A, and classifying all data to form several clusters according to the nearest distance principle.
- Calculating the variance of each cluster, and the cluster with the smallest variance demonstrates that the data in the cluster have the smallest deviation from each other.
- Selecting the mean of the cluster with the smallest variance as the TOF estimate $\tilde{\tau}$ of the LOS path.

After the suppression optimization algorithm and the phase cycle ambiguity solving algorithm, we can obtain ranging accuracy of 2 cm by $\hat{d} = \tilde{\tau} \times c$, and the probability is more than 97%.

5 LS Localization Algorithm

In the bi-static dislocated backscatter link, the reader transmitter and receiver are separated by an electrical distance. Thus, it is easier to spatial layout and localization. Under the premise that the transmitter and receiver are synchronized, we can use distance difference and LS method to locate the target tag. Assume that $\hat{d}_i = \hat{r}_0 + \hat{r}_i (i = 1, 2, 3)$ is ranging estimate in Sect. 4, where $\hat{r}_0$ is the distance between the reader transmitter and the target tag, $\hat{r}_i$ is the distance between the target tag and the reader receivers. The distance differences are given by

$$\begin{cases} \hat{d}_2 - \hat{d}_1 = \hat{r}_{21} \\ \hat{d}_3 - \hat{d}_1 = \hat{r}_{31} \end{cases} \tag{10}$$

Then we can obtain distance error equations by

$$\begin{cases} \sqrt{(x_2 - x)^2 + (y_2 - y)^2} - \sqrt{(x_1 - x)^2 + (y_1 - y)^2} - \hat{r}_{21} = e_{21} \\ \sqrt{(x_3 - x)^2 + (y_3 - y)^2} - \sqrt{(x_1 - x)^2 + (y_1 - y)^2} - \hat{r}_{31} = e_{31} \end{cases} \tag{11}$$

where (x, y) is the coordinate of the target tag, $(x_i, y_i)(i = 1, 2, 3)$ is the coordinate of one receiver.

Assume that the initial coordinate of the target tag is (x_0, y_0), then the nonlinear equations represented by (11) are linearized by Taylor series, i.e.

$$\begin{cases} A_2\delta_x + B_2\delta_y - k_2 = e_{21} \\ A_3\delta_x + B_3\delta_y - k_3 = e_{31} \end{cases} \tag{12}$$

where $A_i = \dfrac{x_0 - x_i}{\sqrt{(x_0 - x_i)^2 + (y_0 - y_i)^2}} - \dfrac{x_0 - x_1}{\sqrt{(x_0 - x_1)^2 + (y_0 - y_1)^2}}$,

$$B_i = \frac{y_0 - y_i}{\sqrt{(x_0 - x_i)^2 + (y_0 - y_i)^2}} - \frac{y_0 - y_1}{\sqrt{(x_0 - x_1)^2 + (y_0 - y_1)^2}},$$

$$k_i = \hat{r}_{i1} - \left(\sqrt{(x_0 - x_i)^2 + (y_0 - y_i)^2} - \sqrt{(x_0 - x_1)^2 + (y_0 - y_1)^2} \right),$$

$$\delta_x = x - x_0, \ \delta_y = y - y_0.$$

A vector can be given by

$$W\delta - K = E \tag{13}$$

where $W = \begin{pmatrix} A_2 \ B_2 \\ A_3 \ B_3 \end{pmatrix}$, $\delta = \begin{pmatrix} \delta_x \\ \delta_y \end{pmatrix}$, $K = \begin{pmatrix} k_2 \\ k_3 \end{pmatrix}$, $E = \begin{pmatrix} e_2 \\ e_3 \end{pmatrix}$.

By minimizing the sum of squared errors, we can obtain

$$\hat{\delta} = (W^T W)^{-1} W^T K \tag{14}$$

Thus, $\hat{\delta}_x$ and $\hat{\delta}_y$ are obtained. Then take $x_0' = x_0 + \hat{\delta}_x$ and $y_0' = y_0 + \hat{\delta}_y$, and repeat the above process until $\hat{\delta}_x$ and $\hat{\delta}_y$ are small enough. That is, set a threshold ε and when $\left|\hat{\delta}_x\right| + \left|\hat{\delta}_y\right| < \varepsilon$, (x_0', y_0') is position estimate of the target tag.

When Non-Line-of-Sight (NLOS) errors are not considered, the localization accuracy will depend on two key factors, including the geometric layout of the three receivers and each ranging accuracy. The geometric layout of the receivers can be denoted by Geometric Dilution of Precision (GDOP). When $G = \begin{bmatrix} A_2 \ B_2 \\ A_3 \ B_3 \end{bmatrix}$, thus GDOP can be given by

$$GDOP = (q_{11} + q_{22})^{\frac{1}{2}} \tag{15}$$

where q_{11} and q_{22} are from $Q = (G^T G)^{-1} = \begin{bmatrix} q_{11} \ q_{12} \\ q_{21} \ q_{22} \end{bmatrix}$.

6 Simulation Results

We perform intensive emulation experiments and prove that the presented approach can realize centimeter accuracy on ranging and localization. In our emulation, the bandwidth is 200 MHz and the hopping space is 5 MHz, so the multipath resolution is about 5 ns and the corresponding distance resolution is around 1.5 m. We assume the receiver obtained four copies of the transmitted signal in indoor environment, including one LOS path and three NLOS paths. Through 1000 simulation experiments, it is shown that the proposed algorithm can achieve the ranging accuracy of 2 cm, and the probability is more than 97% as plotted by the red line in Fig. 6, while the probability is only 50% before suppression plotted by the blue line.

The ranging accuracy has been analyzed above, and the localization accuracy is also related to the relative position of the target tag and the receivers. In order to clearly demonstrate the influence of GDOP on the localization error, we assume that no errors are added in the process and only the position of one receiver changes randomly to represent different GDOP, while the other two receivers are not moving in the horizontal plane.

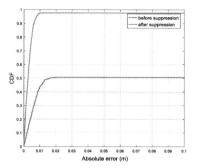

Fig. 6. The CDFs of ranging error.

Suppose that the receivers are set up in an area of 8 m × 8 m, the errors of $\hat{d}_i$ are within 2 cm, the actual coordinate of the target tag is set to $(4, 4)$, and the coordinates of the three receivers are set to $(0,0)$, $(8, 0)$ and $(x, 8)$, respectively, where x randomly changes within $[0, 8]$. We only take nine value of x to show errors of $x/y/localization$ and variation of GDOP as plotted in Fig. 7(a).

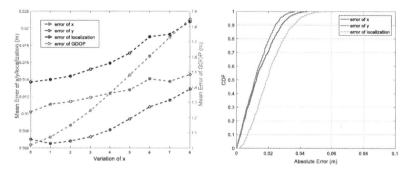

Fig. 7. (a) Curves of mean error of $x/y/localization$; (b) The CDFs of localization error.

We can observe that as x changes, the mean errors of $x/y/localization$ and GDOP are gradually increasing, respectively in Fig. 7(a). The bigger the mean of GDOP is, the bigger localization error is. When $x = 8$, the localization error is biggest but is less than 6 cm as shown in Fig. 7(b). In summary, the proposed algorithm can achieve centimeter-level accuracy of ranging and localization.

7   Conclusion

This paper presents a FHCW algorithm based on carrier phase that enables centimeter-level accuracy on ranging. For the purpose of two-dimensional localization, we rely on the distance differences and take the LS method. The main contributions of this paper are to take the solution of missing data into account and propose the improved multipath suppression algorithm. We believe the proposed algorithm can provide a new idea for RFID indoor localization technology.

Acknowledgements. This work was supported partly by the National Natural Science Foundation of China (No. 61704015), and General program of Chongqing Natural Science Foundation (special program for the fundamental and frontier research) (No. cstc2019jcyj-msxmX0108).

References

1. Xie, J., Liu, T.: Research on technical development of BeiDou navigation satellite system. In: Sun, J., Jiao, W., Wu, H., Shi, C. (eds.) China Satellite Navigation Conference (CSNC) 2013 Proceedings. LNEE, vol. 243, pp. 197–209. Springer, Heidelberg (2013). https://doi.org/10. 1007/978-3-642-37398-5_19
2. Cummins, C., Orr, R., O'Connor, H.: Global Positioning Systems (GPS) and microtechnology sensors in team sports: a systematic review. Sports Med. **43**, 1025–1042 (2013)
3. Wang, X., Zhang, M., Lu, Z.: A frame breaking based hybrid algorithm for UHF RFID anti-collision. Comput. Mater. Continua **59**(3), 873–883(2019)
4. Huang, Z., et al.: A physical layer algorithm for estimation of number of tags in UHF RFID anti-collision design. Comput. Mater. Continua **61**(1), 399–408 (2019)
5. Chen, H., Liu, K., Ma, C., Han, Y., Su, J.: A novel time-aware frame adjustment strategy for RFID anti-collision. Comput. Mater. Continua **57**(2), 195–204 (2018)
6. Shao, S., Burkholder, R.J.: Passive UHF RFID tag localization using reader antenna spatial diversity, pp. 1–4. IEEE (2012)
7. Zhang, Z., Lu, Z., Saakian, V.: Item-level indoor localization with passive UHF RFID based on tag interaction analysis. IEEE Trans. Industr. Electron. **61**, 2122–2135 (2014)
8. Li, X., Zhang, Y., Amin, M.G.: Multifrequency-based range estimation of RFID tags, pp. 147–154. IEEE (2009)
9. Scherhaufl, M., Pichler, M., Stelzer, A.: UHF RFID localization based on phase evaluation of passive tag arrays. IEEE Trans. Instrum. Meas. **64**, 913–922 (2015)
10. Ma, Y., Hui, X., Kan, E.C.: 3D real-time indoor localization via broadband nonlinear backscatter in passive devices with centimeter precision. In: International Conference on Mobile Computing & Networking (2016)
11. Ma, Y., Selby, N., Adib, F.: Minding the billions: ultra-wideband localization for deployed RFID tags. In: Proceedings of the 23rd Annual International Conference on Mobile Computing and Networking, pp. 248–260 (2017)
12. Levanon, N.: Radar Principles. Wiley-Interscience, New York (1988)
13. Adib, F., Kabelac, Z., Katabi, D.: 3D tracking via body radio reflections. In: Usenix NSDI, pp. 317–329 (2014)
14. Xie, L., Jiang, D., Fu, X., Jiang, Q.: Centimeter-level localization algorithm with RFID passive tags. In: Jia, M., Guo, Q., Meng, W. (eds.) WiSATS 2019. LNICSSITE, vol. 280, pp. 160–170. Springer, Cham (2019). https://doi.org/10.1007/978-3-030-19153-5_15
15. Tse, D., Vishwanath, P.: Fundamentals of Wireless Communications. Cambridge University Press (2005)
16. Zhao, P., Ouyang, X., Peng, H.: A phase delay estimation algorithm of frequency hopping signal based on Chinese Reminder Theorem. JEIT **40**, 656–662 (2018)

An Efficient Approach for Selecting
a Secure Relay Node in Fog Computing
Social Ties Network

Mingxi Yin[1], Jinliang Yu[1], Shanshan Tu[1(✉)], Muhammad Waqas[1,2],
Sadaqat Ur Rehman[1,3], Ghulam Abbas[2], and Ziaul Haq Abbas[2]

[1] Beijing Key Laboratory of Trusted Computing, School of Computing,
Beijing University of Technology, Beijing 100124, China
yinmx.3@gmail.com, sstu@bjut.edu.cn, engr.waqas2079@gmail.com,
engr.sidkhan@gmail.com
[2] Telecommunications and Networking (TelCon) Research Lab, Ghulam Ishaq Khan
Institute of Engineering Sciences and Technology, Topi 23460, KPK, Pakistan
{abbasg,ziaul.h.abbas}@giki.edu.pk
[3] College of Science and Engineering, Hammad Bin Khalifa University, Doha, Qatar

Abstract. With the development of the Internet of Things technology
and the advent of the 5G era, cloud computing is difficult to meet the
requirements of low latency, high reliability, and data security for various application services. The fog computing model newly proposed by
researchers in recent years can increase the task processing and data analysis capabilities of network edge devices, so it can reduce the computing
load of cloud computing devices and improve the efficiency of data operations. Although fog computing can largely solve existing cloud computing
problems, secure access and privacy protection are still a very important
and urgent problem to be solved. Aiming at the problem of selecting a
relay node to generate a security key under the fog computing model
under the fog computing model, this paper proposes a secure relay node
selection method based on game theory. In this paper, a fog computing
model based on social relationships is constructed. Under this model, a
secure relay node selection method based on game theory is designed to
realize the selection of relay nodes when a communication link generates
a security key. Finally, this paper analyzes the selection efficiency from
the two dimensions of total equipment and dynamic changes of terminal equipment. Simulation results show that the method proposed in this
paper can quickly select secure relay nodes in the social connection-based
fog computing model.

Keywords: Fog computing · Social ties · Game theory · Physical
layer security

1 Introduction

Cloud computing, which is characterized by centralized big data processing, has
strong computing and storage capabilities. This centralized processing method

© Springer Nature Switzerland AG 2020
X. Sun et al. (Eds.): ICAIS 2020, LNCS 12239, pp. 173–183, 2020.
https://doi.org/10.1007/978-3-030-57884-8_16

can save overhead and reduce losses [1]. With the development of the Internet of Things, the number of Internet-connected devices is growing rapidly. These devices include personal computers, laptops, tablets, smartphones, tablets and other embedded devices. These things have different sizes, functions, processing and computing capabilities. And supports different types of applications [2], according to the forecast of the Cisco Internet Business Solutions Group [3], by 2019, 45% of the data created by iot will be stored, processed, analyzed, and located near the network or the network Processing at the edge, 50 billion devices will be connected to the Internet by 2020. Some IoT applications may require very short response times, some may involve private data, and some may generate large amounts of data, which are all heavy loads on the network. Cloud computing is not efficient enough to support these applications [4]. 5G is now commercially available. Processing some computing tasks at the edge of the network can reduce latency and energy consumption, reduce the burden on cloud computing centers, and in addition, users' privacy will be better protected. Therefore, fog computing can be an important extension of cloud computing. A new computing model will bring new security and privacy challenges. Among them, the mutual trust of IoT devices is important for establishing a secure environment and maintaining the security of IoT services. Reliability plays a very important role [5]. In the fog computing model, IoT devices communicate through wireless networks. The open communication environment makes wireless transmission more vulnerable to malicious attacks than wired communications, such as wiretapping attacks [6], denial of service attacks [7], and spoofing attacks [8], Man-in-the-middle attack [9]. The wireless network generally adopts the OSI protocol, which includes an application layer, a transport layer, a network layer, and a MAC layer. Each layer will select an appropriate security protection scheme according to its characteristics. At present, due to the multi-layered structure of the fog computing model, the devices are in a dynamic change, making it difficult for traditional security technologies to solve the problem of secure communication between devices.

In wireless communication, the encryption technology used by the upper layer of the wireless network is currently the most commonly used security method. Take symmetric encryption technology as an example. The public key is shared by both ends of the communication. If there are no private keys at both ends of the communication, a secure channel is required Key exchange. At this time, physical layer methods can be used for key distribution and location privacy instead of secure channels. Physical layer security makes it more difficult for eavesdroppers to decipher the information transmitted by a communication. In [10], Wyner studied the discrete memoryless eavesdropping channel consisting of the source and destination of the channel and the eavesdropper, and proved that as long as the channel capacity of the main link from the channel source to the destination is greater than the capacity of the eavesdropping channel, Achieve very secure channel transmission. In [11], Leung extended Wyner's research results to Gaussian channels, in which he proposed the concept of security capabilities. The difference between the channel capacity of the main link and the channel capacity of the eavesdropping link is equal to the security capability. In [12], according to Wyner's method, Maurer proposed a method in which the main link can also communicate securely when the eavesdropper channel is superior to the main link. This method enables the two communicating parties to generate security without sharing the key. Compared with traditional key generation algorithms, this method does not need to perform key distribution, and can directly implement physical layer security through keys. The physical layer key is generated by using the reciprocity of the random fading channel, and using closely related wireless channel characteristics, such as the received signal strength or channel impulse response, as a shared random source to generate the shared key. When an attacker is more than half a wavelength away from a legitimate user, the attacker will not be able to obtain channel measurements for irrelevant communication links, and therefore cannot infer that a large amount of information about the key is generated. The physical layer key generation does not require strong computing power and has the potential to achieve the security of information theory, because the confidentiality of the generated key does not depend on the strength of the calculation, but on the physical laws of the wireless fading channel [13]. In [14], Ali proposed a new protocol for generating shared secret keys. In [15], Thai proposed a physical layer key generation scheme for multi-antenna legal nodes.

At present, the global data traffic is increasing exponentially, and people's social circles on the Internet are becoming more and more perfect. The interaction between users and users on social software can to some extent reflect the social ties between users. Such social ties are certain To a certain degree, it can represent the degree of trust between users. In [16], [17], Waqas proposed to add social relationships in the D2D communication model, and select relay nodes through user social ties to improve the key generation rate. In summary, this article will establish a fog computing model with social ties to reduce the security risks of fog computing, use game theory to speed up the selection of relay nodes to reduce communication delays, and have higher efficiency of relay nodes selection in a dynamic environment.

2   Model Overview

2.1   System Model

The social relationship-based fog computing system model constructed in this paper is shown in Fig. 1. The model has three layers from top to bottom. The first layer is the cloud computing centers,We use $C = \{c_1, c_2, \cdots, c_n\}$ to represent cloud nodes set, the second layer is the fog nodes,We use $F = \{f_1, f_2, \cdots, f_n\}$ to represent fog nodes set, and the third time is the terminal devices,We use $T = \{t_1, t_2, \cdots, t_n\}$ to represent terminal devices set. The nodes and the cloud computing centers are connected through a wired network, and the fog node and the terminal device are connected via a wireless network. The terminal device transmits the computing task to the fog node. As a network edge device, the fog node can preprocess the data and transmit it to the cloud computing center or

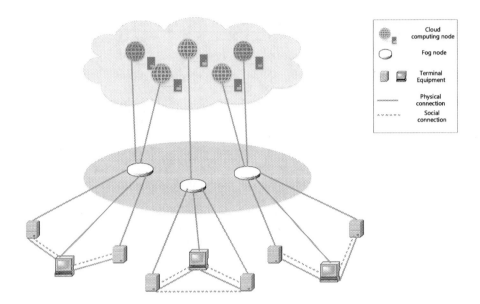

Fig. 1. Fog calculation network model based on social social ties, the end-users connected to fog node nearby, the for computing nodes are link with cloud computing node. The solid blue line means there is a physical connection between them, and the blue dotted line means there is a social connection between them.

terminal device. In the fog computing model based on social relationships, we use $D = \{d_1, d_2\}$ to represent the devices at both ends of the communication link, and $R = \{r_1, r_2\}$ to represent the relay nodes. Both ends of the communication link are terminal devices, or one end is a terminal device and the other end is a fog node, so the physical connection is between the terminal device, or the physical connection is between the terminal device and the fog node. In social relations, there are three states. The first is social trust, which means that potential relay node pairs have social links with devices at both ends of the communication link. The second is social untrust, which means that potential relay node pairs do not have social links to devices at both ends of the communication link. The third is a mix of social trust and social untrust, which means that one relay node has a social link with the devices at both ends of the communication link, while the other relay node has no social link with the devices at both ends of the communication link. We use $\delta_{(i_1,i_2,j_1,j_2)}$ to represent the physical connection, when $\delta_{(i_1,i_2,j_1,j_2)} = 1$, this indicates that there is a physical connection between the communication nodes. We use $\alpha_{(i_1,i_2,j_1,j_2)} = 1$ to indicates that there are social trust connections between the communication nodes and relay nodes, $\beta_{(i_1,i_2,j_1,j_2)} = 1$ to indicates that there are social untrust connections between the communication nodes and relay nodes, and $\gamma_{(i_1,i_2,j_1,j_2)} = 1$ to indicates that there are mix of social trust and social untrust connections between the communication nodes and relay nodes. The states are concluded as follow formula (1)

$$Mode = \begin{cases} \delta_{(i_1,i_2,j_1,j_2)} \times \alpha_{(i_1,i_2,j_1,j_2)} = 1 & \text{Trust} \\ \delta_{(i_1,i_2,j_1,j_2)} \times \beta_{(i_1,i_2,j_1,j_2)} = 1 & \text{Untrust} \\ \delta_{(i_1,i_2,j_1,j_2)} \times \gamma_{(i_1,i_2,j_1,j_2)} = 1 & \text{Mixed} \end{cases} \tag{1}$$

2.2 Key Generation Model Based on Social Ties

This paper uses the communication key generation model proposed by [16]. The specific model is shown in Fig. 2. d_1 and d_2 are the communication nodes, r_1 and r_2 are the relay nodes that assists the communication link in generating keys. And then, we use traditional key generation methods, the process of secret key generation between communication nodes and relay consists of the following three steps.

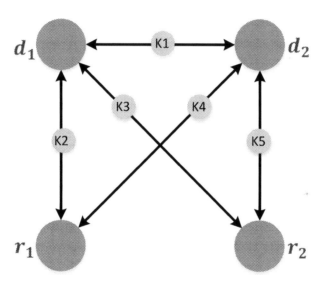

Fig. 2. The key generation process with relay node in different modes. The model is chosen in different modes depends on whether the relay node is in a social network or not. It will decide the way to generate the main key with K1, K2, K3, K4, and K5.

1. First step: Channel estimation. In the key generation model, nodes at both ends of the communication link send training signals to each other and quantify the training signals.
2. Second step: Keys generation. Generate the initial key among the communication node and the relay node based on the quantized information of the training signals.
3. Third step: Key Agreement Based on Social Ties.

In [16], the author derived the key generation rate when there are no relay nodes and the key generation rate when there are relay nodes involved in generating secure keys. The key generation rate R_{direct} and R_{relay} are shown below:

$$R_{direct} = \frac{1}{T_c} log_2(1 + \frac{\sigma_i^4 p^2 T_c^2}{4(\sigma_n^4 + \sigma_n^2 \sigma_i^2 pT_c)})$$ (2)

$$R_{relay} = \frac{1}{T_c} log_2\left(\left(1 + \frac{\sigma_i^4 p^2 T_c^2}{8(2\sigma_n^4 + \sigma_n^2 \sigma_i^2 pT_c)}\right) \prod_{i=1}^{2} \prod_{j=1}^{2} \left(1 + \frac{\sigma_j^4 p^2 T_c^2}{8(\sigma_n^4 + \sigma_n^2 \sigma_i^2 pT_c)}\right)\right)$$ (3)

T_c^2 is 2.4. Coherence time in the GHz band. p is the signal transmission power of the communication nodes and the relay nodes. The additive Gaussian white noise corresponding to the link channel of the communication node obeys the normal distribution with variance σ_n^2. The channel gain of the communication node and the channel gain of the relay node obey the normal distribution of variance σ_i^2 and σ_j^2 respectively.

When the communication nodes and the relay nodes have finished generation of keys, the keys need to be agreed based on social ties. For security reasons, all keys cannot be used simultaneously.

When $\alpha_{(i_1,i_2,j_1,j_2)} = 1$, then the secure key between communication pairs is:

$$K = \begin{cases} \{K1, K2, K4\}, & SIZEOF(K2,K4) < (K3,K5) \\ \{K1, K3, K5\}, & SIZEOF(K2,K4) > (K3,K5) \end{cases}$$ (4)

When $\beta_{(i_1,i_2,j_1,j_2)} = 1$ or $\gamma_{(i_1,i_2,j_1,j_2)} = 1$, then the secure key between communication pairs is:

$$K = \begin{cases} \{K1, K2 \oplus K4\}, & SIZEOF(K2 \oplus K4) < (K3 \oplus K5) \\ \{K1, K3 \oplus K5\}, & SIZEOF(K2 \oplus K4) > (K3 \oplus K5) \end{cases}$$ (5)

3　Relay Node Selection Method

We use coalition game method to select relay nodes. In the coalition game, the communication node will select the relay node that makes the key generation rate higher. Therefore, when the alliance is divided, the relay node will be selected according to the key generation rate when generating the security key. The specific algorithm is as follows:

Algorithm 1: game for selecting optimal relay pairs for communication nodes

---

Step # I: Initialization:
1 set the coalition of communication pairs W_i
Step # II: Select relay nodes for every communication pairs:
2 Repeat 2 to 9, untill all communication pairs have been selected
3 Randomly select a communication pair d_{i_1,i_2} from the communication nodes set, and Add d_{i_1,i_2} to W_i
4 Randomly select a potential relay pair r_{i_1,i_2} from the potential relay nodes set
5 Compute R_{relay} of d_{i_1,i_2} with the help of r_{i_1,i_2}, and R_{direct} of d_{i_1,i_2}
6 if $R_{relay} > R_{direct}$
7 Add r_{i_1,i_2} to W_i, and then go to 2
8 else
9 Go to 4
Step # III: Update W_i, select optimal relay pairs:
10 Repeat 10 to 16, untill all communication pairs find their optimal relay pairs
11 Randomly select W_i
12 Randomly select $W_{i'}$
13 if $r_{i_1',i_2'} \in W_{i'}$ cannot make higher R_{relay} of d_{i_1,i_2} compared with $r_{i_1,i_2} \in W_i$
14 Go to 12
15 else
16 Deselect r_{i_1,i_2} from W_i, add $r_{i_1',i_2'}$ to W_i

---

4   Performance Evaluation

As mentioned in the introduction, the fog computing model can reduce the delay by sharing the calculation amount of the cloud computing center to make up for the shortcomings of cloud computing. Therefore, this paper will evaluate the efficiency of relay node selection and dynamic environment selection. In simulation experiments, this article assumes that several fog nodes are randomly distributed in an area with an area of $100 \times 100\,\mathrm{m}^2$ and that several terminal devices are randomly distributed around the fog nodes. The fog nodes and terminal devices can be used as communication nodes and As a relay node. The communication power of the fog node and the terminal equipment is $100\,\mathrm{dBm}$, and the signal frequency is $2.4\,\mathrm{GHz}$. Most of the terminal equipment has a physical connection with the fog node, and the social ties between the terminal equipment are randomly set.

4.1   Efficiency of Selecting Relay Nodes

In the simulation experiment, the total number of devices is equal to the total number of terminal devices plus the total number of fog nodes. Set the minimum total number of devices to 20 and the maximum total number of devices to 400. Add 20 devices at a time. And there are physical connections between the devices. Compare the game theory method used in this paper with the Q-Learning method used in [18] and the Double Q-Learning method used in [19].

The results are shown in the Fig. 3. When the total number of devices is between 20 and 160, Q-Learnings and Double Q-Learning are slightly less efficient in selecting relay nodes than in game theory. When the total number of devices is between 160 and 200, Q-Learnings and Double Q-Learning are slightly more efficient in selecting relay nodes than in game theory. When the total number of devices is greater than 200, Q-Learnings and Double Q-Learning are significantly less efficient in selecting relay nodes than in game theory. The reason analysis is as follows. When there are fewer devices, Q-Learning and Double Q-Learning take more time to train the model than game theory directly compares the benefits. When the total number of devices is between 160–200, Q-Learning and Double Q-Learning trains the model so that the time to select a relay node is lower than game theory. When the total number of devices exceeds 200, as the total number of devices increases, the Q-Learning and Double Q-Learning updates the Q table time continuously. The time required to select a relay node reduced by training the model is not enough to offset the time increased in Q table update. So its efficiency is not as good as game theory.

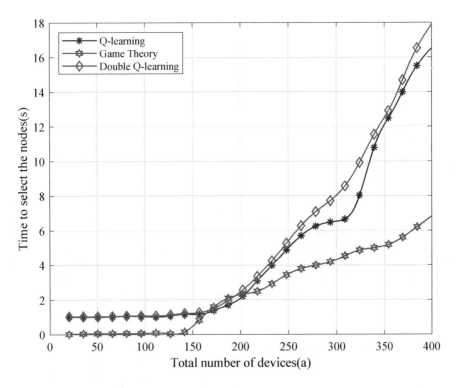

Fig. 3. When the total number of devices is different, the time for each method to select a relay node

In summary, when the total number of devices is small and the total number of devices is large, the efficiency of selecting relay nodes using game theory is higher than Q-Learning and Double Q-Learning.

4.2   Selection Efficiency in a Dynamic Environment

In theory, when the fog node communicates with the terminal device to generate a key to select the relay node, the potential relay node should maintain a physical connection with the fog node and the terminal device, and also have a social relationship with the terminal device. However, in actual situations, potential relay nodes are often moving or transmitting and receiving wireless signals are easily interfered by the environment, and physical connections with fog nodes and terminal equipment cannot be guaranteed. So the potential relay nodes involved in generating the security key are in a dynamic change. In the simulation experiments in this paper, the dynamic changes of potential relay nodes will be simulated, so that the total number of devices in the area is 400, of which 150 devices have physical connections. The physical connection between the devices and the social relationship between the terminal devices are randomly

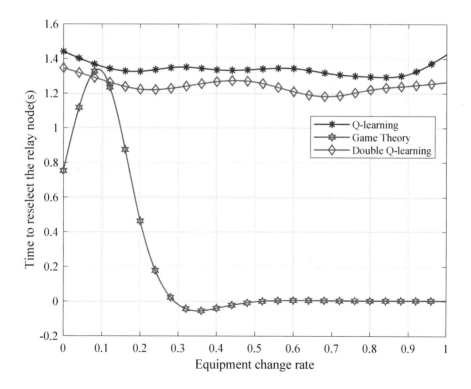

Fig. 4. When the physical connection between the terminal device and the fog node and the terminal device changes, the efficiency of each method to re-select the relay node

generated. The change rate of the terminal device is used to measure the change in the physical connection between the terminal device and the fog node and the terminal device. In experiments, changes in such physical connections are random. The change interval of the terminal equipment change rate is $[0.0, 1.0]$. Whenever the physical connection changes, only the relay node is reselected for the affected communication link, and the unaffected communication link is not changed. The experimental results are shown in the Fig. 4. Because each time the physical connection of the terminal device changes, only the relay node is reselected for the affected link. Therefore, before the physical link is not changed, potential relays have been screened for each group of links through game theory. The node group is arranged in descending order according to the key generation rate. When the physical link changes, only the newly added terminal devices are used as potential relay nodes to calculate its key generation rate, and then it is compared with the original potential relay. The node group comparison can select the potential relay node with the highest key generation rate. For Q-Learning and Double Q-Learning, each time the physical connection changes, the Q table will be updated again, resulting in more time being consumed. Therefore, using game theory to select relay nodes in a dynamically changing environment with physical connections is more efficient than Q-Learning and Double Q-Learning.

5   Conclusion

This article mainly discusses the method of generating the physical layer security key in the fog computing model to solve the problem that the communication between the fog node and the terminal device is easy to be intercepted. This paper builds a fog computing model based on social ties, and uses game theory to select relay nodes that participate in the generation of security keys for communication links. The experimental results show that the method of selecting relay nodes based on game theory has higher efficiency. At the same time, when the communication link is dynamically changing, the method based on game theory can re-select the relay nodes faster and has a stronger Fault tolerance and robustness. In the next work, this article will further explore the use of deep learning to select relay nodes in regions with higher device density.

Acknowledgement. This work was partially supported by the National Natural Science Foundation of China (No. 61801008), National Key R&D Program of China (No. 2018YFB0803600), Beijing Natural Science Foundation National (No. L172049), Scientific Research Common Program of Beijing Municipal Commission of Education (No. KM201910005025).

References

1. Armbrust, M.: A view of cloud computing. Commun. ACM **53**(4), 50–58 (2010)
2. Khan, R., Khan, S.U., Zaheer, R., Khan, S.: Future internet: the internet of things architecture, possible applications and key challenges. In: 10th International Conference on Frontiers of Information Technology, vol. 2012, pp. 257–260. IEEE (2012)

3. Evans, D.: The internet of things: how the next evolution of the internet is changing everything. CISCO white paper, vol. 1, no. 2011, pp. 1–11 (2011)
4. Shi, W., Cao, J., Zhang, Q., Li, Y., Xu, L.: Edge computing: vision and challenges. IEEE IoT J. **3**(5), 637–646 (2016)
5. Alrawais, A., Alhothaily, A., Hu, C., Cheng, X.: Fog computing for the internet of things: security and privacy issues. IEEE Internet Comput. **21**(2), 34–42 (2017)
6. Lakshmanan, S., Tsao, C.-L., Sivakumar, R., Sundaresan, K.: Securing wireless data networks against eavesdropping using smart antennas. In: 2008 The 28th International Conference on Distributed Computing Systems, pp. 19–27. IEEE (2008)
7. Raymond, D.R., Midkiff, S.F.: Denial-of-service in wireless sensor networks: attacks and defenses. IEEE Pervasive Comput. **7**(1), 74–81 (2008)
8. Kannhavong, B., Nakayama, H., Nemoto, Y., Kato, N., Jamalipour, A.: A survey of routing attacks in mobile Ad Hoc networks. IEEE Wirel. Commun. **14**(5), 85–91 (2007)
9. Meyer, U., Wetzel, S.: A man-in-the-middle attack on UMTS. In: Proceedings of the 3rd ACM Workshop on Wireless Security, pp. 90–97 (2004)
10. Wyner, A.D.: The wire-tap channel. Bell Syst. Tech. J. **54**(8), 1355–1387 (1975)
11. Leung-Yan-Cheong, S., Hellman, M.: The gaussian wire-tap channel. IEEE Trans. Inf. Theor. **24**(4), 451–456 (1978)
12. Maurer, U.M.: Secret key agreement by public discussion from common information. IEEE Trans. Inf. Theor. **39**(3), 733–742 (1993)
13. Zeng, K.: Physical layer key generation in wireless networks: challenges and opportunities. IEEE Commun. Mag. **53**(6), 33–39 (2015)
14. Ali, S.T., Sivaraman, V., Ostry, D.: Eliminating reconciliation cost in secret key generation for body-worn health monitoring devices. IEEE Trans. Mob. Comput. **13**(12), 2763–2776 (2013)
15. Thai, C.D.T., Lee, J., Quek, T.Q.: Physical-layer secret key generation with colluding untrusted relays. IEEE Trans. Wirel. Commun. **15**(2), 1517–1530 (2015)
16. Waqas, M., Ahmed, M., Li, Y., Jin, D., Chen, S.: Social-aware secret key generation for secure device-to-device communication via trusted and non-trusted relays. IEEE Trans. Wirel. Commun. **17**(6), 3918–3930 (2018)
17. Waqas, M., Ahmed, M., Zhang, J., Li, Y.: Confidential information ensurance through physical layer security in device-to-device communication. In: IEEE Global Communications Conference (GLOBECOM), vol. 2018, pp. 1–7. IEEE (2018)
18. Tu, S., Yu, J., Meng, Y., Waqas, M., Liu, L.: Secure relay node selection method based on Q-learning for fog computing in 5G network. Telecommun. Sci. **35**(7), 60–68 (2019)
19. Meng, Y., Tu, S., Yu, J.: Detection scheme of impersonation attack based on DQL algorithm in fog computing. Comput. Eng. Appl. **56**(10), 63–68 (2020)

A Research for 2-m Temperature Prediction Based on Machine Learning

Kaiyin Mao, Chen Xue, Changming Zhao, and Jia He[✉]

Chengdu University of Information Technology, Chengdu, China
xiaomaonn@126.com, xuechen714@gmail.com,
{zcm84,hejia}@cuit.edu.cn

Abstract. In this paper, we proposed two improved methods to solve the problems of current machine learning algorithm in the study of weather forecast revision. So this study can be divided into two parts. First of all, contrary to the previous machine learning algorithm with only one machine learning model applied for a vast area, a clustering algorithm is proposed to divide the vast area into multiple areas with different models for forecast correction. Secondly, Considering ECMWF the forecast of ECMWF with multiple initial times, a double-model correction model based on XGBoost machine learning algorithm is proposed. To test the new approach for the daily max and min temperature forecasts, the 2-m surface air temperature in the China area from the ECMWF (European Centre for Medium-Range Weather Forecasts) and the observatory meteorological observation data from china national stations are used. The data ranged from July 2017 to September 2019. We used meal weight interpolation method and sliding time window method to construct sample data. The new approach is compared with the multiple linear regression, random forest and SVM algorithms which are often used in weather forecast correction, it shows better numerical performance. The root mean square error (RMSE) and forecast accuracy are used to evaluate the model, and the forecast is better after the model is revised. The applicability of the model is verified by continuous test data (September 2019), and the experimental results show that the model has good practical value.

Keywords: Forecast correction · 2-m temperature · Machine learning · XGBoost

1 Introduction

With the continuous development of computer technology, the related numerical weather prediction (NWP) model has also been continuously upgraded. At present, NWP model products, as the most important technical support of national meteorological forecasting business institutions, its forecast value will be affected by problems such as errors generated by numerical initialization and errors in the computer model system itself. Therefore, how to post-process the output results of the numerical weather prediction will play a key role in improving the forecast results. In addition, how to improve the accuracy of meteorological forecasting has been the main contents of meteorological

© Springer Nature Switzerland AG 2020
X. Sun et al. (Eds.): ICAIS 2020, LNCS 12239, pp. 184–194, 2020.
https://doi.org/10.1007/978-3-030-57884-8_17

forecasting service, the 2-m surface air temperature as one of the most basic elements of meteorological forecasts, its forecasting accuracy is important for production, life and various security services. This paper builds a correction model for the 2-m surface air temperature based on the XGBoost algorithm in machine learning to provide important data for the meteorological forecasting field.

In the current NWP field, the methods for post-processing errors in temperature numerical forecast mainly include: moving average (MA) [11], Kalman filter [2], statistics [4] and machine learning [13], etc. Among them, the advantage of the MA and Kalman filter is that they do not need to collect a large number of historical sample data, and a small amount of sample data can be used to show a good result. Since the two methods mainly refer to the case where the temperature change is continuous and smooth, there is a good correction for the systematic false positives, but due to the lack of consideration of other elements, when coming to the case of transition weather, the correction is limited. The statistic and machine learning [14] have been highly concerned in the field of meteorological forecasting. Glahn, H.R., D.A. Lowry used multiple linear regression to correct meteorological elements such as wind speed and maximum temperature, and verified the effectiveness of the method [5]. Lemcke C and Kruizinga S use multiple linear regression and logistic regression to correct the 2-m surface highest air temperature, wind speed and rainfall probability of the ECWMF forecast in the Netherlands, which made the numerical model forecast results can be applied to the actual application in the region, but the time-series of meteorological elements was not considered [9]. Keon-tae SOHN, deuk-kyunrha and young-kyung SEO used dynamic linear model to correct the ground temperature forecast for 38 meteorological stations in South Korea, which had better correction effect compared with Kalman filter, however, using the same model in the complex environment in South Korea lacked the consideration related meteorological elements [16]. Sue Ellen Haupt and Branko Kosovic constructed a complete real-time correction forecasting system with multiple numerical model forecast results and probability density functions in data mining, and the system combines big data systems with machine learning algorithms in practical applications, but the time-series of the meteorological forecast and the impact of related meteorological elements are not considered [6]. Rasp S and Lerch S adopted the artificial neural network, and compared with the meteorological observation data from Germany, to correct the 2-m surface air temperature forecast in ECWMF, the study found that based on the 2007–2015 data set, more than 75% of the stations have better forecast results than ensemble forecast and quantile regression forest correction results [15]. Haochen, Jiangjiang et al. used the random forest algorithm to correct the 2-m surface air temperature of the ECWMF in Beijing, and compared with the traditional numerical mode statistical method and the linear regression method in machine learning, but the research only used the data of ECWMF to establish model, which lacks of the correction of systematic errors about meteorological elements in the numerical weather prediction [10].

Because the numerical weather prediction system belongs to a highly nonlinear system and has high requirements for the error of the initial conditions, this paper proposes a forecast correction scheme for reprocessing the corrected results by multi-model weighting. In this paper, two independent machine learning models are used to correct the ECWMF numerical weather prediction data when the initial reporting time is 00 UTC and 12 UTC respectively, and then the model correction results are added according to different weights to obtain the final forecast result. Starting numerical weather prediction at different times of the day will cause differences in the initial conditions of the model, which implies the characteristics of meteorological elements over time. Using a single machine learning model to train meteorological element data in the study area will often ignore it. For this reason, we carried out our research. There are many factors that affecting the meteorological elements, such as the altitude, vegetation, and degree of greening in the area [1]. In order to find the most realistic relationship between the meteorological elements and the prediction model, we clustered the training samples according to meteorological observation stations, and only use one model to train the station data in the same cluster, improving the degree of fitting of the prediction model to the actual meteorological elements. In this paper, the basic model is constructed by XGBoost in machine learning, and the 2-m surface air temperature data forecasted by ECWMF and the observation data of 2552 national meteorological observatories in China are used as training data. Finally, we correct the 2-m surface air temperature results for the next 7 days forecasted by the model.

2   Fundamental and Formula

The post-processing task of numerical weather prediction based on machine learning is a typical data mining task, which includes three stages: data preprocessing, model training and model evaluation. In these stages, some relevant algorithms and evaluation methods will be used. For example, in this paper, inverse distance weight interpolation was used to combine observation data and digital model prediction data, KMeans clustering algorithm was used to build model selector, and XGBoost was used to build prediction model. The main algorithms and models used in this article are described in detail below.

2.1   KMeans

KMeans is a clustering segmentation algorithm for unsupervised learning [7]. The algorithm can be described as: firstly, given a sample set $S = \{s_1, s_2, s_3 \cdots s_n\}$, calculates the Euclidean distance between the sample point and the center point in the sample, and divides the sample into k clusters $C = \{c_1, c_2, c_3 \cdots c_n\}$ according to the calculation results. Then, iteratively calculates and clusters according to the within-cluster sum of squared errors (SSE). Finally, achieves the goal: the distance between the samples in the cluster is as close as possible, and the distance between the clusters is as large as possible, and the calculation of SSE as shown in the formula:

$$\text{SSE} = \sum_{i=1}^{n} \sum_{j=1}^{m} w^{(i,j)} = \left\| x^{(i)} - \mu^{(j)} \right\|_2^2 \tag{1}$$

Where $\mu^{(j)}$ is the mean vector of cluster j, which is the centroid of j. The steps of KMeans algorithm as follows:

1. Select k samples from the data set S as centroid vectors.

2. Calculate the distance from the sample to the centroid, select the closest centroid c_i, and assign the sample to cluster i.
3. Calculate the mean vector of all cluster samples to get a new centroid.
4. Output the clustering result if all centroid vectors have not changed, otherwise skip to step 2.

2.2 XGBoost

XGBoost (Extreme Gradient Boosting) is an open source machine learning project developed by Chen Tianqi team [3]. It efficiently implements the GBDT (Gradient Boosting Decision Tree) algorithm and also makes many improvements to the algorithm and engineering, which can be used for both classification tasks and regression tasks. The algorithm is described as: By establishing K regression trees to make the predicted values as close as possible to the real values. It has strong generalization ability.

XGBoost is widely used in data science competitions and industry, and its main advantages are as follows:

1. Use a number of strategies to prevent over fitting, such as regularization, Shrinkage and Column Subsampling.
2. Objective function optimization only refers to the loss function
3. Support parallelization. Although each tree has a serial relationship with each tree, the same level nodes can use parallel to speed up the calculation. Taking a single node as an example, the selection of the best split point in the node, the calculation of the candidate split point, and the gain can be calculated in parallel, which greatly improves the training speed.
4. Added processing of sparse data.
5. Support cross-validation and early stop to improve training speed.
6. Support custom setting sample weights.

2.3 Temperature Forecast Accuracy

The accuracy rate of meteorological forecast is an important indicator in the forecasting task. Accuracy of temperature Forecast is defined as: when the absolute error value between the forecast result value and the actual observation value is within 2 °C, the forecast result is regarded as correct, otherwise the forecast is wrong [12]. The formula for calculating the accuracy of the forecast is expressed as follows:

$$T = \frac{N_t}{N_f} * 100\% \tag{2}$$

Where N_t is the correct number of forecasts, N_f is the total number of forecasts, and T represents the proportion of forecasted correct temperature values in all forecast results.

3 System Model

In this paper, the forecast data of ECWMF and the observation data of 2552 ground meteorological observation site in China are used as the original data of the model. Then the inverse data weight interpolation method and sliding time window method [8] are used to reconstruct the original data as training samples. The characteristics of the sample data include the actual observation value and the numerical model prediction value of the meteorological element, and the observation value of the 2-m surface air temperature corresponding date is taken as the label value. It should be noted that the data needs to be classified according to the observation site and the forecast time before training the model. The model selector based on KMeans, the regression model based on XGBoost algorithm and the model weighting function constitute the whole model of the experiment. The model building process is mainly divided into the following three stages:

1. Data preprocessing stage: The training data is constructed by IDW interpolation method and sliding time window method.
2. Model training stage: Using the model selector based on KMeans clustering algorithm and the machine learning model based on XGBoost for model training.
3. Result output stage: Weighted and summed the dual mode output results, and outputs the final result.

The modeling process for this paper is shown in the following Fig. 1.

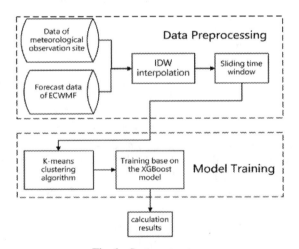

Fig. 1. System structure.

The model selector is built as follows, and the process is as shown in Fig. 2.

In order to reflect the influence of the jitter of meteorological elements on the future forecast results in the numerical forecasting model, this paper uses a dual mode to forecast the temperature of the same day. The ECWMF data with the forecast start time

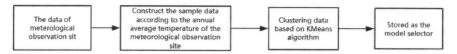

Fig. 2. The process of building model selector

of UTC00 point and UTC12 point is separately trained using the XGBoost model, and finally the two output results are weighted and summed to obtain the final mode forecast result. The weighted summation process can be expressed as the following function:

$$P = \in P_{00} + (1- \in)P_{12} \tag{3}$$

The P is the final forecast result of the model, P_{00} and P_{12} is prediction results of the model trained with the 00 UTC and 12 UTC respectively. The $\in$ is a hyper parameter, which can be adjusted using the grid search method, and the specific process is shown in the Fig. 3

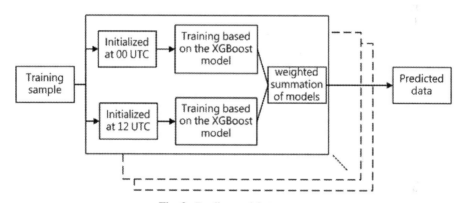

Fig. 3. Predict model structure.

4   Data and Experiments

The experiment uses the data of 2-m surface air temperature, which predicted by ECWMF and the observation data of 2552 meteorological observation sites in China as the initial data. The time range is from July 2017 to September 2019, including 16 meteorological elements such as temperature, air pressure, wind speed, rainfall, and dew point temperature and collected by date. The forecast data of EWMCF is divided in to UTC 00 and UTC 12, and the length of time is 168 h, which is collected every 3 h. The time window is used to reconstruct the sample data set by predicting the maximum temperature and the lowest temperature of the next day. The meteorological data of the previous period is used to construct the time series and the input of the future time, and then use the observation data for the next period of time as the label. The experiment sample is divided the into three parts: the data of September 2019 is used as test data, the data from July 2017 to August 2019 is randomly divided into model training data and test data according to the meteorological observation station.

When reconstructing the input data, we first need to determine the super-parameter K (the number of clusters) of the clustering algorithm. The value of K needs to consider two factors: 1) Evaluation index: SSE, 2) The difference between the amount of experimental data and the meteorological elements forecasted of the sites. The final number of clusters is determined to be 100 (k = 100) in the experiment.

In this paper, three groups of experiments were carried out: 1) In order to verify the advanced nature of the model, three kinds of algorithms commonly used in the correction of meteorological forecasts (Multi Regression, Random Forest, SVM) were selected for comparative experiments, and the main reference indicators are RMSE (Root Mean Square Error); 2) To verify the correction effect of the proposed model, the experiment was carried out after random sampling of the data samples, and the model evaluation criteria are RMSE and prediction accuracy; 3) To verify the practical application ability of the model, The data from September 2019 was used as an input sample for the experiment, and the model evaluation standard was RMSE.

Figure 4 shows the results of the second set of experiments. The maximum daily temperature is corrected using a variety of commonly used predictive models and predictive models proposed by the paper, where the Double XGBoost is the model which the paper proposed, the x-axis represents the RMSE value, and the y-axis represents the forecast duration. By analyzing the experimental results, machine learning algorithms such as Multi Regression can correct the temperature data of ECWMF forecast, and the corrected RMSE is significantly lower than before correction, but the difference of correction effect between the three algorithms is not obvious, the correction effect of the double XGBoost correction algorithm proposed in this paper is significant compared with the other three algorithms. Figure 6 is a comparison diagram of the correction effect of the daily minimum temperature, and the experimental conclusion is similar to that of Fig. 5.

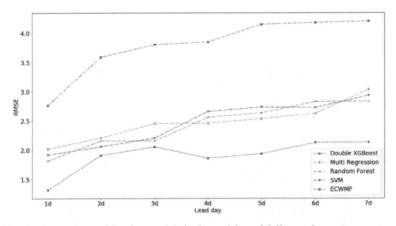

Fig. 4. Comparison with other models in the revision of daily maximum temperature.

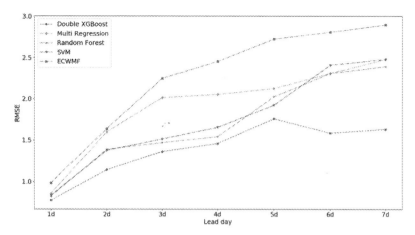

Fig. 5. Comparison with other models in the revision of daily minimum temperature.

Figure 6 is a comparison of the ECWMF model prediction results before and after the use of the model correction. The analysis results show that the RMSE of the ECWMF model for the forecast of the maximum temperature and the lowest temperature in the next 7 days will increase as the forecast duration increases, but the value will eventually stay within a reasonable range, where the RMSE of the highest daily temperature is greater than the RMSE of the lowest daily temperature. After the model is corrected, the RMSE of model forecast still increases with the increase of the forecast duration, but the value of the RMSE is significantly lower than before the correction. The RMSE of the daily minimum temperature after the correction is less than 1 °C, and the RMSE of the lowest temperature forecast in the next 7 days is less than 2 °C. Overall, the RMSE of the daily minimum temperature after corrected is small, and the RMSE of the maximum temperature after corrected is larger than the lowest temperature, but only slightly higher than 2 °C.

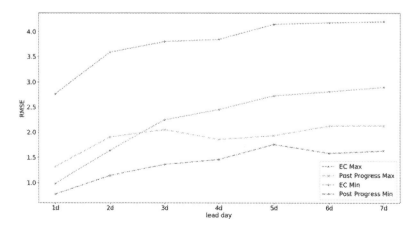

Fig. 6. The maximum and minimum temperatures are revised.

Figure 7 shows the accuracy of the forecast model for the highest and lowest temperature of the next 7 days, where the x-axis presents the forecast accuracy. In general, the forecast values of the daily maximum temperature and the daily minimum temperature gradually decrease with the increase of the forecast duration, and the correct rate of the corrected maximum temperature is significantly lower than the correct rate of the lowest temperature. The correct rate of the lowest temperature forecast in the next 7 days is higher than 69%, and the correct rate of the maximum temperature forecast is higher than 59%.

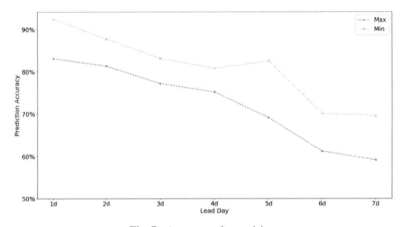

Fig. 7. Accuracy after revision.

In order to verify the practical application ability of the model, the experiment was conducted by taking the samples from September 2019 as input data. Figure 8 is the RMSE of the ECMWF numerical model for the forecast results of the daily maximum temperature and the minimum temperature before and after the correction, where the x-axis represents the RMSE of each meteorological observation site, and the y-axis represents the date value (September 1 to September 30, 2019). The results show that the highest RMSE of ECMWF forecast is significantly higher than the RMSE of the minimum daily temperature, and the RMSE of the maximum temperature after correction is decreased significantly, but is higher than the RMSE of the minimum temperature after correction. In summary, the RMSE of the maximum temperature and the minimum temperature is decreased after the correction, and the floating is within 2 °C, so the correction mode has good practical value. Since the effect of the forecast period from 2 days to 7 days is similar to the effect of the forecast period of 1 day, no further description will be made here.

5   Results and Discussion

This paper constructs a forecast correction model for the 2-m surface air temperature in a wide area, and uses the data of 2552 meteorological observation in China and ECMWF numerical model forecast data from July 2017 to September 2019 correct the

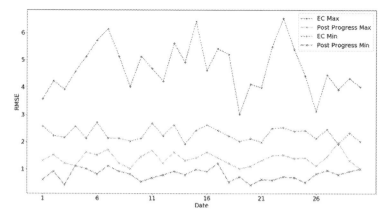

Fig. 8. The RMSE of lead 1 day after revision.

maximum daily temperature and the daily minimum temperature on the 2-m surface. By analyzing the ECWMF forecast, the daily maximum temperature relative error is large, and there is a large correction space. After the model is corrected, the forecasted value of the daily maximum temperature has achieved a good correction effect, but is less than the correction effect of the daily minimum temperature forecast. The best accuracy for the single-day forecast after the correction is over 90%, and the error between the daily maximum temperature and the minimum temperature after correction is still increased with the increasing of duration, but even the forecast accuracy of the maximum temperature of 7 days, which is the worst forecast, can still reach about 60%. Finally, the forecast ability of the model for meteorological elements with a length of one month is verified by experiments. After the correction, the RMSE of the forecast for the maximum daily temperature in each station is below 2.5 °C, and the RMSE of the minimum temperature is below 2 °C. Compared with the ECWMF forecast results, the corrected results of the model have been greatly improved. However, the current work still has defects, the paper does not make a special analysis of the meteorological observation stations with large deviations after the correction, nor does it analyze the correction effect of the model under special weather conditions.

Acknowledgment. This work is supported by National Major Project (No. 2017ZX03001021-005), Sichuan science and Technology Program (2019YFG0212) and Sichuan Science and Technology Program (2018GZ0184).

References

1. Betts, A.K., Ball, J.H., Beljaars, A.C.M., et al.: The land surface-atmosphere interaction: a review based on observational and global modeling perspectives. J. Geophys. Res. **101**(D3), 7209 (1996)
2. Cassola, F., Burlando, M.: Wind speed and wind energy forecast through kalman filtering of numerical weather prediction model output. Appl. Energy **99**, 154–166 (2012)

3. Chen, T., He, T.: Higgs boson discovery with boosted trees. In: Proceedings of the 2014 International Conference on High-Energy Physics and Machine Learning - Volume 42, HEPML 2014, pp. 69–80 (2014). JMLR.org
4. Delle Monache, L., Eckel, F.A., Rife, D.L., et al.: Probabilistic weather prediction with an analog ensemble. Mon. Weather Rev. **141**(10), 3498–3516 (2013)
5. Glahn, H.R., Lowry, D.A.: The use of model output statistics (MOS) in objective weather forecasting. J. Appl. Meteorol. **11**(8), 1203–1211 (1972)
6. Farsiu, S., Robinson, M.D., Elad, M., Milanfar, P.: Fast and robust multiframe super resolution. IEEE Trans. Image Process. **13**(10), 1327–1344 (2004)
7. Haupt, S.E., Kosovic, B.: Big data and machine learning for applied weather forecasts forecasting solar power for utility operations. In: IEEE Symposium Series on Computational Intelligence. IEEE (2015)
8. Peng, K., Leung, V.C.M., Huang, Q.: Clustering approach based on mini batch kmeans for intrusion detection system over big data. IEEE Access **6**, 11897–11906 (2018)
9. Sun, L., Ge, C., Huang, X., Yingjie, W., Gao, Y.: Differentially private real-time streaming data publication based on sliding window under exponential decay. Comput. Mater. Continua **58**(1), 61–78 (2019)
10. Lemcke, C., Kruizinga, S.: Model output statistics forecasts: three years of operational experience in the Netherlands. Mon. Weather Rev. **116**(5), 1077–1090 (1988)
11. Li, H., Yu, C., Xia, J., et al.: A model output machine learning method for grid temperature forecasts in the Beijing area. Adv. Atmos. Sci. **36**(10), 1156–1170 (2019)
12. Liangmin, F., Qiuxue, Z., Lan, K., et al.: Study on 2 m temperature bias correction of EC model in sichuan province. Plateau and Mountain Meteorology Research (2019)
13. Bahrami, K., Shi, F., Rekik, I., Shen, D.: Convolutional neural network for reconstruction of 7T-like Images from 3T MRI using appearance and anatomical features. Springer (2016)
14. Lupo, A.R., Market, P.S.: The application of a simple method for the verification of weather forecasts and seasonal variations in forecast accuracy. Weather Forecast. **17**(4), 891–889 (2002)
15. Ning, M., Guan, J., Liu, P., Zhang, Z., O'Hare, G.M.P.: GA-BP air quality evaluation method based on fuzzy theory. Comput. Mater. Continua **58**(1), 215–227 (2019)
16. Luo, M., Wang, K., Cai, Z., Liu, A., Li, Y., Cheang, C.F.: Using imbalanced triangle synthetic data for machine learning anomaly detection. Comput. Mater. Continua **58**(1), 15–26 (2019)
17. Rasp, S., Lerch, S.: Neural networks for post-processing ensemble weather forecasts. Monthly Weather Rev. (2018)
18. Young-Kyungseo, D.K.: The 3-hour-interval prediction of ground-level temperature in south korea using dynamic linear models **20**(4), 575–582 (2003)

A New Scheme for Essential Proteins Identification in Dynamic Weighted Protein-Protein Interaction Networks

Wei Liu[1,2]($\boxtimes$), Liangyu Ma[1], and Yuliang Tang[1]

[1] College of Information Engineering of Yangzhou University, Yangzhou, China
yzliuwei@126.com
[2] The Laboratory for Internet of Things and Mobile Internet Technology of Jiangsu Province, Huaiyin Institute of Technology, Huaiyin, China

Abstract. Predicting essential protein on the protein-protein interaction network is crucial for understanding the process of cellular organization and development. With the development of high-throughput proteomics technology, essential protein recognition has become a hot topic and focus of research, there are many computational methods for essential proteins detecting. However, these existing methods are mostly predicted in static PPI networks, ignoring the dynamics of the network. Meanwhile, existing methods identify essential proteins in unweighted networks, without considering the tightness and strength of the connections between network nodes, which lead to low accuracy of essential protein identification. Therefore, this paper presents a new essential proteins prediction scheme, called NTMB which integrates a variety of biological information including edge clustering coefficient, common neighbor similarity, Pearson Correlation Coefficient and Subcellular localization score. In order to evaluate the performance of our method NTMB, we conduct a series of experiments on the yeast PPI network and the experimental results shown that the proposed essential protein method NTMB can obtain better results in yeast PPI network than other methods.

Keywords: Essential proteins · Dynamic protein-protein interaction network · Topology feature · Biological information

1 Introduction

Protein is an essential component of all cell and tissue structures [1]. It is the executor of physiological functions and the most important material basis of life activities. Proteins are classified into essential protein and non-essential proteins according to their role in the activities of life. If essential protein in the organism is deleted, it will cause lethality of an organism. So, essential protein is necessary for the survival and reproduction of the organism. Predicting essential protein is important not only for understanding cellular processes but also for designing drug. Initially, identification of essential proteins by traditional experimental methods, such as single gene knockouts [2], RNA interference [3], and conditional knockouts [4]. Although the essential proteins identified by traditional

© Springer Nature Switzerland AG 2020
X. Sun et al. (Eds.): ICAIS 2020, LNCS 12239, pp. 195–206, 2020.
https://doi.org/10.1007/978-3-030-57884-8_18

experimental methods can achieve highly accurate, they are costly and inefficient. As a result, there is an urgent need to propose computational methods to identify essential proteins. These computational methods are not only used to identify essential proteins, but also applied to different fields, such as influence maximization [5], web text mining [6], and big data analysis [7]. In recent years, with the development of high-throughput experimental technology, the available protein interaction data is increasing, and a large amount of sufficient data has been collected into different databases [8]. This creates important conditions for us to identify essential proteins in the protein-protein interaction network.

Existing studies have shown that whether a protein is an essential protein is related to its topological properties in the protein interaction network. Therefore, the centrality-lethality rule has been proposed by Jeong et al. [9] demonstrates that closely related proteins are more likely to be the essential proteins in the protein-protein interaction network. That is to say, a protein interacts with many proteins in the PPI network. If this protein is deleted, the interaction of many proteins will be affected. After that, based on the topological characteristics of the PPI network, the researchers proposed a variety of different centrality methods, such as the simplest degree centrality (DC) [10], eigenvector centrality (EC) [10], betweenness centrality (BC) [11], closeness centrality (CC) [12], subgraph centrality (SC) [13], information centrality (IC) [14]. Wang et al. [15] proposed the use of edge clustering coefficients (NC) to evaluate the strength of interactions between two proteins. Meanwhile, Li et al. proposed an essential protein identification algorithm, named LAC [16] based on local connectivity.

In addition to using network topological features to identify essential proteins, in recent years, some scholars combined the biological significance of proteins in the PPI network and applied biological information to the identification of essential proteins, which significantly improved the accuracy of prediction. For example, Kim et al. [17] proposed a new method called MCGO, which combined protein network module and Go annotation to predict essential proteins. Meanwhile, Pan et al. [18] introduced method PMN, which integrating the weighted interactive network and functional modules. Li et al. proposed a new prediction method PeC [19] based on edge clustering coefficients of protein node and pearson correlation coefficient. Peng et al. proposed an iterative method ION [20] based on the fact that essential proteins are more evolutionarily conservative than non-essential proteins. Tang et al. introduced a modified identification method called WDC [21], which integrates the gene expression data and topological features. Considering the domains information, Peng et al. [22] proposed a method named UDoNC. Based on integrating GO annotations, gene expression data, and PPI network, Lei et al. [23] developed method called HITS and Zhang et al. [24] proposed OGN method. Qin et al. [25] proposed a method called LBCC, which combining the network topological features and protein complexes information.

Although the essential protein identification methods based on network topology and biological information have made good progress, the prediction accuracy is still low. The main reason for this is that existing methods are mostly predicted in static PPI networks, ignoring the dynamics of the network. At the same time, existing methods identify essential proteins in unweighted networks, without considering the tightness and strength of the connections between network nodes. In order to overcome these problems, in this paper, we proposed a new method called NTMB based on network topological feature and Multi-Source biological information on the dynamic weighted protein-protein interaction network. Firstly, we use gene expression data to construct the dynamic PPI network. Then, we employed the Gene Ontology and topological feature to calculate the weighted value on a dynamic PPI network. Finally, we integrated Multi-Source biological data to measure the essentiality of the protein. The experimental results show that the proposed essential protein method NTMB can obtain better results in yeast PPI network than other methods (DC, EC, BC, IC, SC, NC, LAC, PeC, WDC, UDONC, and LBCC).

2 Method

2.1 Construction of the Dynamic PPI Network

We regard the PPI network as an undirected graph $G = (V, E)$, where, V represent the set of protein node and E represent the set of edges which denote the interactions between two proteins. Gene expression data plays an important role in reflecting the dynamic properties of protein-protein interaction network. Therefore, we combine static PPI networks and gene expression data to construct a dynamic PPI network. Suppose $u(p)$ and $\sigma(p)$ denote the arithmetic mean and the standard deviation of gene expression data, respectively. Then $u(p)$ and $\sigma(p)$ can be defined as follows:

$$u(p) = \frac{\sum_{i=1}^{m} Ge_i(p)}{m} \tag{1}$$

$$\sigma(p) = \sqrt{\frac{\sum_{i=1}^{m} (Ge_i(p) - u(p))^2}{m-1}} \tag{2}$$

In order to set an active threshold value, the algorithm calculated it using the three-sigma method [19]. The three-sigma calculation method is as follows:

$$Active\_thresh_k(p) = u(p) + 3 \cdot \sigma(p) \cdot (1 - \frac{1}{1 + \sigma^2(p)}) \tag{3}$$

$$\Pr_i(p) = \begin{cases} 0.99 & if \ T_i'(p) \geq Active\_thresh_3(p) \\ 0.95 & if \ Active\_thresh_3(p) > T_i'(p) \geq Active\_thresh_2(p) \\ 0.68 & if \ Active\_thresh_2(p) > T_i'(p) \geq Active\_thresh_1(p) \\ 0 & if \ T_i'(p) < Active\_thresh_1(p) \end{cases} \tag{4}$$

Since the gene expression data contains three cycles and each cycle has 12 time tamps, we use the average gene expression value of three cycles as the final gene expression value at a certain time point.

$$T_i'(p) = \frac{T(i) + T(i+12) + T(i+24)}{3} (i \in [1, 12]) \tag{5}$$

Where $T(i)$ represents the gene expression data at i time point. For a protein p, it is considered to be active at i time point, when the final gene expression value is greater than the $Active\_thresh_k(p)$. If two proteins, such as p and q, interact with each other in the original network and are active at the same time point, then we believe that the two proteins have an interaction in the dynamic PPI network. Through the above method, the entire PPI network was regarded as 12 sub-networks. The strength of interaction between the two proteins is as follow:

$$weight(p, q) = \frac{\sum_{1 \leq i \leq m} (Pr_i(p) \cdot Pr_i(q))}{|t|} \tag{6}$$

Where $|t|$ represents the number of times that two interacting proteins appear in 12 sub-networks. Finally, we integrate 12 sub-networks into a dynamic PPI network.

2.2 Construct the Dynamic Weighted PPI Network

To maintain topological proximity between two interacting proteins in the original PPI network, we define the loss function [26]:

$$l_1 = \sum_{i \in V} \sum_{j \in V} a_{ij} (\gamma_i - \gamma_j)^2 \tag{7}$$

Where γ_i and γ_j are the vector representation of protein i and protein j, and $a_{ij} = 1$ only when there is interaction between protein i and protein j.

The GO Ontology is the most comprehensive ontology databases which includes the three sub-ontologies: Biological Process (BP), Cellular Component (CC) and Molecular function (MF). The GO ontology plays an important role in the prediction of essential proteins. Therefore, in our method, we select Biological Process (BP) and Molecular function (MF) as protein biological attributes.

According to the protein biological attributes feature on the dynamic PPI network, we construct an attributes matrix $A \in R^{n \times m}$, where n denotes the number of proteins and m denotes the number of GO attributes. We generate a protein attribute affinity matrix $S \in R^{n \times n}$ by the matrix A, each s_{ij} is defined as follow:

$$s_{ij} = \frac{\sum_{k=1}^{m} a_{ik} \times a_{jk}}{\sqrt{\sum_{k=1}^{m} a_{ik}^2} \times \sqrt{\sum_{k=1}^{m} a_{jk}^2}} \tag{8}$$

In order to preserve the protein biological attributes proximity, the loss function is calculated as follow:

$$l_2 = \sum_{i \in V} \sum_{j \in V} (s_{ij} - \gamma_i \gamma_j^T)^2 \tag{9}$$

Since topological properties and biological attributes are very important for the prediction of essential proteins, we combine topological properties and biological attributes to study the representations of protein. Finally, the loss function we defined is shown below:

$$l = \sum_{i \in V} \sum_{j \in V} a_{ij} (\gamma_i - \gamma_j)^2 + \lambda \sum_{i \in V} \sum_{j \in V} (s_{ij} - \gamma_i \gamma_j^T)^2 \tag{10}$$

Where, $\lambda \in (0, 1)$ is a parameter. Each protein is represented as a vector γ_i. Therefore, we obtain a weighted adjacency matrix $W \in R^{n \times n}$.

$$w_{ij} = \begin{cases} \cos\_sim(\gamma_i, \gamma_j) + \sum_{1 \leq i \leq m} (\text{Pr}_i(p) \cdot \text{Pr}_i(q))/|t| & a_{ij} = 1 \\ 0 & a_{ij} = 0 \end{cases} \tag{11}$$

Where, cos_sim represents a function to calculate the cosine similarity between two interacting proteins. Based on the above method, we can obtain the dynamic weighted PPI network.

2.3 Essential Protein Identification on the Dynamic PPI Network

Our proposed algorithm NTMB integrates two biological information and two topological feature. Biological information includes gene expression data, subcellular localization data and topological feature includes edge clustering coefficient, common neighbor similarity.

(1) Edge clustering coefficient

The existing research shows that the edge clustering coefficient is an important index to measure the topological characteristics of the protein-protein interaction network. At the same time, the research proves that the edge clustering coefficient plays an important role in the process of identifying essential proteins. Given an edge (u, v) in an undirected graph G. the edge clustering coefficient of between protein vertices u and v can be defined by the following formula:

$$ECC(u, v) = \frac{|N_u \cap N_v| + 1}{\min\{d_u, d_v\}} \tag{12}$$

Where N_u and N_v represent the set of all the neighboring nodes of the protein node u and v, d_u and d_v represents the degree of node u and v, respectively.

(2) Common neighbor similarity

If there are some common neighbor nodes between two interacting proteins, the interaction between these two proteins becomes strong. The common neighbor similarity (CNS) of edge(u, v) can be defined as followed:

$$CNS(u, v) = \frac{|CNS_u \cap CNS_v|^2}{|CNS_u| \times |CNS_v|} \tag{13}$$

Where, CNS_u and CNS_v represent the in-degree of protein u and protein v, respectively.

(3) Pearson Correlation Coefficient (PCC)

Studies have shown that co-expression genes are tend to encode interacting protein pairs [27]. Therefore, Pearson correlation coefficient can be applied to measure of the strength of their connection between two interacting proteins. The high PCC between the two interacting proteins indicates that the interaction between the two proteins is highly reliable. The PCC between genes X and Y is calculated as follows:

$$PCC(X, Y) = \frac{1}{m} \sum_{i=1}^{m} \left(\frac{g(X, i) - \overline{g}(X)}{\sigma(X)}\right) \cdot \left(\frac{g(Y, i) - \overline{g}(Y)}{\sigma(Y)}\right) \tag{14}$$

Where m is the number of samples in gene expression data. $g(X, i)$ and $g(Y, i)$ represent the expression level of gene X and Y in the sample i, respectively. $\overline{g}(X)$ and $\overline{g}(Y)$ represent the mean expression level of gene X and Y. respectively. $\sigma(X)$ and $\sigma(Y)$ represent the standard deviations of the expression level of gene X and gene Y, respectively. A negative PCC value indicates that a negative correlation based on gene X and gene Y, while a positive PCC value indicates that a positive correlation based on gene X and gene Y.

(4) Subcellular localization score

Subcellular localization consists of different compartments, each of which plays a different role in cellular activities [28]. Studies have shown that the more times a protein is present in the nucleus, the more likely it becomes an essential protein. Let $|n|$ and $|m|$ be the number of protein u and v that appears in the subcellular localization of the Nucleus. C_{max} is the proteins that have the largest number of times appear in the Nucleus. Subcellular localization score of protein u is calculated as follows:

$$SLS(u, v) = \frac{|n| + |m|}{2|C_{\max}|} \tag{15}$$

As we all know, biological information and topological feature play an important role in the identification process of essential proteins. Therefore, we use a method that integrates topological features and biological properties to identify essential proteins. In NTMB, the essentiality of a protein is measured by the importance score. The importance score of a protein consists of the edge clustering coefficient (ECC), common neighbor similarity (CNS), pearson correlation coefficient (PCC) and subcellular localization score (SLS). For a protein u, the importance score (P-score) of protein u is defined as follows:

$$P - score = \alpha w_{ij}(u) + \beta(\sum_{v \in I(u)} ECC(u, v) + \sum_{v \in I(u)} CNS(u, v))$$
$$+ \delta(\sum_{v \in I(u)} PCC(u, v) + \sum_{v \in I(u)} SLS(u, v)) \tag{16}$$

Here, $I(u)$ is the set of all neighbor nodes of protein u, $\alpha, \beta, \delta \in (0, 1)$ is a parameter to adjust the contribution of the weighted value, topological feature and other biological properties similarities.

Based on the above method, the basic framework of the proposed algorithm NTMB (Identification of Essential Proteins by Integrating Network Topology and Multi-Source Biological Information on Dynamic Weighted Protein-Protein Interaction Networks) is as follows:

Algorithm1: NTMB
Input: $G=(V, E)$;
       GO data;
       Gene expression data, attributes matrix $A \in R^{n \times m}$;
       Subcellular localization information;
       k: the number of essential proteins;
       parameter α, β, δ ;
   Output: the set of k identified essential proteins;
1. Construct the dynamic PPI network by gene expression data and three-sigma method
2. Construct the dynamic weighted PPI network by GO ontology and topological feature.
3. Compute the Edge Clustering Coefficient(ECC) by function (12);
4. Compute the Common neighbor similarity(CNS) by function(13);
5. Compute the Pearson Correlation Coefficient (PCC) by function (14);
6. Compute the Subcellular localization score(SLS) by function(15);
7. According to the *weighted value, ECC, CNS, PCC, SLS* calculate the importance score of each node;
8. The importance score *P_score* of protein is obtained by function(16);
9. Sort nodes in descending order by their *P_score* value;
10. Output the top k ranked essential proteins.

3 Experimental Results and Analysis

3.1 Dataset

In order to test the performance of our proposed essential protein prediction algorithm, two yeast protein interaction data were selected to construct a protein interaction network, which was downloaded from two experimental databases: DIP [29], MIPS [30]. The protein details in the two interaction databases are shown in Table 1. The yeast essential proteins dataset was compiled from the following four experimental databases: MIPS [30], SGD [31], DEG [32], and SGDP (http://www.sequence.stanford.edu/group/), which contained 1285 essential proteins identified by biological experiments. Gene expression data were downloaded from the Gene Expression Omnibus (GEO) database [33] (http://www.ncbi.nlm.nih.gov/geo/) with accession number GSE3431. The GO data used in our method is obtained from the GO Consortium [34]. And the subcellular localization information are obtained from the COMPARTMENTS database [35].

Table 1. Information about the two different PPI Datasets

Dataset	Proteins	Interactions	Essential proteins
DIP	5093	24743	1167
MIPS	4546	12319	1016

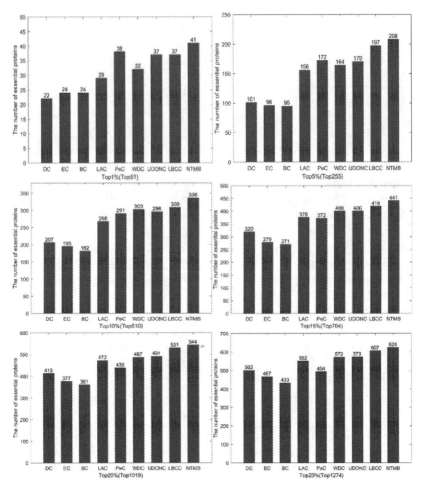

Fig. 1. The total number of true essential proteins predicted by our method NTMB and the other existing methods at six levels on the DIP dataset.

3.2 Compare NTMB with Other Previous Methods

We evaluated the performance of our proposed method NTMB by comparing it with other existed prediction methods: DC [10], EC [10], BC [11], CC [12], SC [13], IC [14], NC [15], LAC [16], PeC [18], WDC [19], UDONC [20], LBCC [23] on two different datasets. It is convenient for us to do the experiment that we only apply UDONC and LBCC on DIP dataset. The importance scores of the proteins can be calculated by using different prediction methods and then we ranked them in descending order according to their importance score of each protein. Six levels (top1, top5, top10, top15, top20, top25 of all proteins) were chosen as candidate essential protein. The last step, the number of truly essential proteins in the candidate essential proteins can be determined as we compared the known essential proteins with candidate essential proteins.

From Fig. 1, we can see the prediction results at six levels on the DIP dataset. The results indicate that our proposed method NTMB is consistently higher than other competing methods DC, EC, BC, LAC, PeC, WDC, UDONC, and LBCC at six levels of ranked proteins. We can see that the total number of essential proteins truly predicted by our method NTMB was 41, 208, 336, 441, 544, and 624 at six levels in Fig. 1. For top 25% identified essential proteins, NTMB achieves more than 20% improvements compared with DC, EC, BC, PeC, and more than 10% improvements compared with LAC.

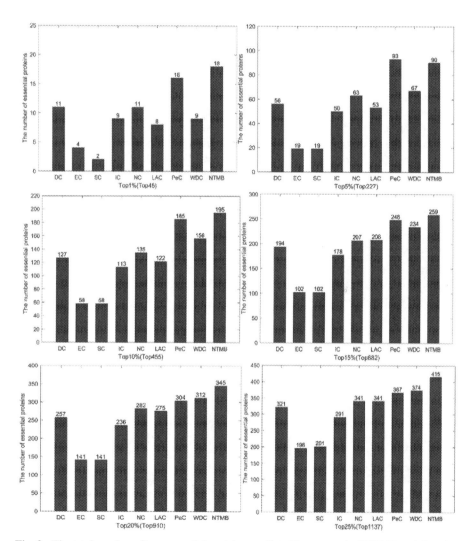

Fig. 2. The total number of true essential proteins predicted by our method NTMB and the other existing methods at six levels on the MIPS dataset.

From Fig. 1, we can see the prediction results at six levels on the DIP dataset. The results indicate that our proposed method NTMB is consistently higher than other competing methods (DC, EC, SC, IC, NC, LAC, PeC and WDC) at six levels of ranked proteins. We can see that the total number of essential proteins truly predicted by our method NTMB was 41, 208, 336, 441, 544, and 624 at six levels in Fig. 2. For top 25% identified essential proteins, NTMB achieves more than 13% improvements compared with PeC.

4 Conclusion

Identifying essential proteins from protein-protein interaction networks is a hot topic in contemporary bioinformatics research. Existing research has proposed a number of methods for identifying essential proteins. In this study, this paper presents a new essential proteins prediction strategy, called NTMB which integrates a variety of biological information including edge clustering coefficient, common neighbor similarity, Pearson Correlation Coefficient and Subcellular localization score. Firstly, we use gene expression data to construct the dynamic PPI network. Then, we employed the Gene Ontology and topological feature to calculate the weighted value on a dynamic PPI network. Finally, we integrated Multi-Source biological data to measure the essentiality of the protein. The experimental results show that the proposed essential protein method NTMB can obtain better results in yeast PPI network than other methods (DC, EC, BC, IC, SC, NC, LAC, PeC, WDC, UDONC, and LBCC). This shows that it is meaningful to identify essential proteins by the NTMB algorithm.

Acknowledgments. This research was supported in part by the Chinese National Natural Science Foundation under Grant Nos. 61702441, 61772454, 61703362, 61602202, Natural Science Foundation of Jiangsu Province under contracts BK20170513, BK20160428, and the Blue Project of Yangzhou University.

References

1. Chen, Y., Xu, D.: Understanding protein dispensability through machine-learning analysis of high-throughput data. Bioinformatics **21**(5), 575–581 (2005)
2. Giaever, G., Chu, A.M., Ni, L., Connelly, C., Riles, L., Véronneau, S.: Functional profiling of the Saccharomyces Cerevisiae genome. Nature **418**(6896), 387–391 (2002)
3. Schulz, G.E., Schirmer, R.H.: Principles of protein structure. Springer Advanced Texts in Chemistry. Springer, Heidelberg, vol. 118(1), pp. 151–152 (1979)
4. Roemer, T., Jiang, B., Davison, J.: Large-scale essential gene identification in Candida albicans and applications to antifungal drug discovery. Mol. Microbiol. **50**(1), 167–181 (2010)
5. Hua, Y., Chen, B., Yuan, Y., Zhu, G., Ma, J.: An influence maximization algorithm based on the mixed importance of nodes. Comput. Mater. Continua **59**(2), 517–531 (2019)
6. Niu, B., Huang, Y.: An improved method for web text affective cognition computing based on knowledge graph. Comput. Mater. Continua **59**(1), 1–14 (2019)
7. Zhou, H., Sun, G., Fu, S., Jiang, W., Xue, J.: A scalable approach for fraud detection in online e-commerce transactions with big data analytics. Comput. Mater. Continua **60**(1), 179–192 (2019)

8. Kim, K.-B.: Web-based computational system for protein-protein interaction inference. J. Inf. Process. Syst. **8**(3), 459–470 (2012)
9. Jeong, H., Oltvai, Z.N., Barabasi, A.-L.: Prediction of protein essentiality based on genomic data. ComPlexUs **1**(1), 19–28 (2003)
10. Vallabhajosyula, R.R., Chakravarti, D., Lutfeali, S.: Identifying hubs in protein interaction networks. PLoS ONE **4**(4), e5344 (2009)
11. Joy, M.P., Brock, A., Ingber, D.E.: High-betweenness proteins in the yeast protein interaction network. J. Biomed. Biotechnol. **2005**(2), 96 (2016)
12. Wuchty, S., Stadler, P.F.: Centers of complex networks. J. Theor. Biol. **223**(1), 45–53 (2003)
13. Estrada, E., Rodríguez-Velázquez, J.A.: Subgraph centrality in complex networks. Phys. Rev. E Stat. Nonlin. Soft Matter Phys. **71**(2), 056103 (2005)
14. Stephenson, K., Zelen, M.: Rethinking centrality: methods and examples. Soc. Netw. **11**(1), 1–37 (1989)
15. Wang, J., Li, M., Wang, H., Pan, Y.: Identification of essential proteins based on edge clustering coefficient. IEEE/ACM Trans. Comput. Biol. Bioinf. **9**(4), 1070–1080 (2012)
16. Li, M., Wang, J., Chen, X., Wang, H., Pan, Y.: A local average connectivity-based method for identifying essential proteins from the network level. Comput. Biol. Chem. **35**(3), 143–150 (2011)
17. Kim, W., Li, M., Wang, J.X.: Essential protein discovery based on network motif and gene ontology. In: IEEE International Conference on Bioinformatics and Biomedicine, Atlanta, pp. 470–475 (2011)
18. Pan, Y., Hu, S., Zhao, B.: Identification of essential protein based on functional modules and weighted protein-protein interaction networks. Int. J. u- and e- Serv. Sci. Technol. **9**(8), 343–350 (2016)
19. Li, M., Zhang, H., Wang, J.X., Pan, Y.: A new essential protein discovery method based on the integration of protein-protein interaction and gene expression data. BMC Syst. Biol. **6**(1), 15 (2012). https://doi.org/10.1186/1752-0509-6-15
20. Peng, W., Wang, J.X., Wang, W.P.: Iteration method for predicting essential proteins based on orthology and protein-protein interaction networks. BMC Syst. Biol. **6**(1), 87 (2012)
21. Tang, X., Wang, J., Zhong, J.: Predicting essential proteins based on weighted degree centrality. IEEE/ACM Trans. Comput. Biol. Bioinf. **11**(2), 407–418 (2014)
22. Peng, W., Wang, J., Cheng, Y., Lu, Y., Wu, F.: UDONC: an algorithm for identifying essential proteins based on protein domains and protein-protein interaction networks. IEEE/ACM Trans. Comput. Biol. Bioinf. **12**(2), 276–288 (2015)
23. Lei, X.J., Wang, S.G., Wu, F.X.: Identification of essential proteins based on improved HITS algorithm. Genes **10**(2), 177 (2019)
24. Zhang, X., Xiao, W., Hu, X.: Predicting essential proteins by integrating orthology, gene expressions, and PPI networks. PLoS ONE **13**(4), e0195410 (2018)
25. Chao, Q., Sun, Y., Dong, Y.: A new method for identifying essential proteins based on network topology properties and protein complexes. PLoS ONE **11**(8), e0161042 (2016)
26. Xu, B., Li, K., Zheng, W.: Protein complexes identification based on go attributed network embedding. BMC Bioinform. **19**(1), 535 (2019). https://doi.org/10.1186/s12859-018-2555-x
27. Von, M.C., Krause, R., Snel, B.: Comparative assessment of large-scale data sets of protein-protein interactions. Nature **417**(6887), 399–403 (2002)
28. Gavin, A.C., Aloy, P., Grandi, P., Krause, R., Boesche, M.: Proteome survey reveals modularity of the yeast cell machinery. Nature **440**(7084), 631–636 (2006)
29. Xenarios, I., Salwínski, L., Duan, X.J.: DIP, the database of interacting proteins: a research tool for studying cellular networks of protein interactions. Nucleic Acids Res. **30**(1), 303 (2002)

30. Mewes, H.W., Frishman, D., Mayer, K.F., Münsterkötter, M., Noubibou, O.: MIPS: analysis and annotation of proteins from whole genomes in 2005. Nucleic Acids Res. **34**, 169–172 (2006)
31. Juvik, T.Y.R., Schroeder, M., Weng, S.: SGD: saccharomyces genome database. Nucleic Acids Res. **26**(1), 73–79 (1998)
32. Zhang, R., Lin, Y.: DEG 5.0, a database of essential genes in both prokaryotes and eukaryotes. Nucleic Acids Res. **37**(Database issue), D455–D458 (2009)
33. Saccharomyces Genome Deletion Project. http://www.sequence.stanford.edu/group/
34. Blake, J., Christie, K., Dolan, M.: Gene ontology consortium: going forward. Nucleic Acids Res. **43**(Database issue), 1049–1056 (2015)
35. Binder, J.X., Pletscherfrankild, S., Tsafou, K.: COMPARTMENTS: unification and visualization of protein subcellular localization evidence. Database **2014** (2014)

On Enterprises' Total Budget Management Based on Big Data Analysis

Hangjun Zhou[1], Jie Li[1(✉)], Jing Wang[1], Jiayi Ren[1], Peng Guo[1,2], and Jianjun Zhang[3]

[1] Hunan University of Finance and Economics, Changsha 410205, China
18274351750@163.com
[2] University Malaysia Sabah, Kota Kinabalu, Malaysia
[3] Hunan Normal University, Changsha, China

Abstract. At present, there is a delay in the transmission and acceptance of information in the total budget management of enterprises, which can only provide a basis for short-term decision-making, but has limitations on long-term decision-making. However, big data analysis can increase the efficiency of capital operations, carry out budget supervision and help companies make long-term decisions. This dissertation attempts to solve the problems above by using the Lasso method and the GM-Model. Based on big data, a series of experiments are carried out on the comprehensive budget management of enterprises to study the methods and application effects of big data analysis and prediction of enterprise income. Finally, through experiments, it is found that the predicted value of the first three years is larger than the actual value, and the deviation is gradually reduced in the following years. However, the actual income in 2001 is almost the same as the predicted value. These results indicate that using this method more accurately requires a large data from different years to support and operate.

Keywords: Big data · Comprehensive budget · Lasso method and GM model

1 Introduction

With the development of information technology such as computers, cloud computing, and the Internet of Things, people's lifestyles are rapidly becoming close to 'data'. Big data has penetrated into all walks of life. This dissertation uses five V features of big data to study and analyze the total budget management and solve the problems in budget preparation, budget execution and budget feedback of enterprises [1]. According to a large number of references, most scholars want to construct a total budget management system based on big data currently. For example, some researchers proposed that based on the centralized, shared and real-time functions of cloud accounting and the frameworks of total budget management, a total budget management can be divided into 5 parts: Infrastructure Layer (IaaS), Hardware Virtualization Layer (HaaS), Data Layer (DaaS), Platform Layer (PaaS) and Software Layer (SaaS) [3].

The total budget of enterprises reflects their financial plan for all production and operation activities in a certain period of time in the future [14]. From the perspective

© Springer Nature Switzerland AG 2020
X. Sun et al. (Eds.): ICAIS 2020, LNCS 12239, pp. 207–218, 2020.
https://doi.org/10.1007/978-3-030-57884-8_19

of the development of total budget management, although a few enterprises have a budget management system already, there are still some problems that the transmission of information in their budget management system is not timely or comprehensively [15]. For example, the existing budget management system has delayed data processing, and it is not sensitive to the changes of external markets. Applying big data to the total budget management of enterprises can make budget preparation, budget execution and budget assessment more accurate, and it can also help to predict the future trends the markets more effectively. In the process of applying big data technology to the total budget management of enterprises, the first step is to determine the original source of data [11]. Basing on the existing data, it uses data seeking technology to conduct deeper research on potential data. After obtaining the data, it is analyzed and processed online using OLAP technology. After the initial analysis, the server cluster and data processing technology are used to collect and analyze the internal and external data of these enterprises [19]. It can be said that the big data technology is the overall analysis and management of internal and external data or surface and potential data of enterprises. In addition, the cloud storage server of big data can guarantee the security and integrity of it, and finally support the total budget management effectively [12].

The specific process of using big data to analyze the total budget is as follows: First, determine the budget targets of the enterprise. Second, select a suitable model to identify the key factors that affect data. The LassoLars variable selection model has *been* chosen in this dissertation. As for the parameters of the evaluation model, this dissertation uses Sum of Squares for Error to determine the residual squared and uses F_oneway of Python SciPy. Stats to determine the F statistic. Third, based on key factors, choose suitable methods to predict future income. In this dissertation, the gray prediction model (GM-Model) is used to identify the degree of deviation between different factors' development trends, that is, carry out a correlation analysis to find the laws of system changes, generate a data sequence with strong regularity, and establish a corresponding equation to predict the future development trends. Fourth, through neural network prediction, the univariate GM (1,1) Model is built according to the selected variables to predict the next year value of the variables sequence, that is, the artificial nerve is constructed according to the selected variable (characteristic variable) and income (class variable, tag variable). The network model (training model) is predicted by the predicted value data of each selected variable obtained by GM (1, 1) and the neural network model, thereby obtaining prediction results of the respective data. Finally, using visualization techniques arrive at corporate budget results.

2 Impacts of Big Data Analysis on Total Budget Management

2.1 Data Integration Model

The traditional budget process is a bottom-up model for submitting budget approvals. It not only consumes a lot of labor force and material resources of the enterprise, but also reduces the timeliness of the budget [14]. In addition, it is prone to insufficient data collection and complicated manual processing, which leads the incomplete preparation plan of enterprises. Big data has ability to quickly process massive amounts of data, laying a technical foundation for improving the levels of budget preparing [18]. For example, the management of the company can set budget targets through the network cloud, and each department and subsidiary will feedback the data in the same way, so that the company can prepare the budget plan in the fastest time. Of course, all the departments of the enterprise can also obtain the budget plan information in the first time through the network cloud, so it helps to realize the integration of data management of the whole enterprise and improve the budget preparation levels. In short, the application of big data to the total budget of enterprises is conducive for them to collect multi-dimensional information, establish a strategic analysis model and optimize the budget management process.

Among traditional budget approaches, a simple incremental (decremental) budge is employed by enterprises. When the current budget is based on the past ones, it will be limited. A lagging budget plan will directly lead to the loss of practical use of the entire budget management. In addition, the data of incremental (decremental) budgets is relatively simple, and the financial department is often very subjective when setting budget targets and budget plans in this approach, resulting in biases of it [10]. By applying big data, establishing a budget data center and absorbing the latest data, the budget plan will be more forward-looking. When unforeseen changes such as economic recessions and inflation have an impact on the future business environment, the data center will play a strategic role at the end of the quarter and the end of the month. It will update its budget targets and alternative budget plans in a timely manner, making it more relevant to the needs of the enterprise.

2.2   Up-and-Down Linkage of Corporate Budget Execution

Budget control includes external control and internal control. External control is the management carried out by superiors for the lower level, and the internal control is the supervision by the employees themselves. In the traditional total budget, enterprises rarely adopt a management model that is comprehensive, full-process, and fully engaged. Due to the lack of communications between departments, various problems such as inefficient management and ineffective real-time control often arise during the execution process. Applying big data to a total budget system is helpful for the enterprises' internal control. It can monitor the budget in real time, optimize the budget execution procedures [4], and make the total budget truly top-down and bottom-up.

Under the development of the big data, enterprises use EPI, OLAP and other technologies to establish a budget data center, which can monitor the expenses, the process and the completion of projects during the budget execution at any time. At the same time, through the analysis of big data, the historical data of the enterprise is compared with the data of the same industry [2]. Through vertical and horizontal analysis, problems in budget execution can be discovered in time, and the information can be fed back to the regulatory and budget departments to enable them to adjust and reassess the budget plans in a timely manner. When major changes occur in the internal and external environment, managers can also adjust the budget targets or strategic objectives in a timely manner through the data center to optimize the allocation of enterprises' resources. Compared with the statistics and analysis of the traditional budget, big data can make a quicker update of them. Managers can understand and monitor the progress of the current total budget management on the Internet in real time, so that the budget control can run through the entire business process.

2.3 Visualization of Total Budget Results

Big data visualization is to present the results of data through images or charts; it provides more accurate results for professional analysis and more reliable decision-making suggestions for information users [6]. The traditional financial budget is manually entered, and the data is rarely processed. Therefore, problems such as untimely information transmissions, inaccurate data analysis and short budget periods are prone to occur. With the help of big data visualization analysis, the total budget can make enterprises to understand and predict its future development more intuitively and identify existing problems and adjust its forecast indicators more time. Therefore, the big data helps the management develop a more suitable budget plan and help the enterprise achieve the profit target at the end.

Big data seeking and analyzing technologies include Cloudera's Flume, Facebook's Scribe, LinkedIn's Kafka, etc., which are mainly used for Python or API for data collection. Data seeking technologies can be based on machine learning, pattern recognition, statistics and data visualization, they help to analyze data of enterprises in a highly automated manner and help information users to adjust market strategies, reduce risks and make correct decisions. The traditional total budget can only analyze and process the internal data of the enterprise. More importantly, as the number of data increases, it cannot process information accurately and quickly. When using the seeking and analyzing technologies of big data, not only can managers understand the dynamic trend of the budgets timely, but they also can make an objective analysis of the changes in the external markets. Thus, they can manage the enterprise effectively and make long-term decisions easily.

3 Total Budget Management Experiments Based on Big Data Analysis

In order to verify the validity and practicability of the model, this dissertation uses the model to simulate and predict. According to the income data of a certain enterprise in Hunan Province, the main processes are: (1) design predictive models, select models, identify key factors affecting fiscal revenue; (2) select forecasting methods and models to predict future income based on key factors. The software analyzes the influencing factors of the enterprise and adopts gray prediction model (GM (1,1)) in the next two years and trains the neural network (NN) with the existing annual sequence. Then, based on the obtained prediction and the predicted value of the annual income in the next two years, the future income can be estimated, the differences between the budgets and the actual value can be judged and the corresponding decisions can be made.

3.1 The Lasso Method and the GM-Model Model

In this dissertation, the Lasso method and the GM-Model model are mainly applied. The L1 regularization of linear regression is usually called Lasso regression. Generally, for high-dimensional feature data, especially the linear relationship is sparse, and Lasso regression is adopted. Or to find the main features in a bunch of features, then Lasso regression is the first choice. There are two commonly used loss function optimization methods for Lasso regression, the coordinate axis descent method and the Least Angle Regression method. The minimum angle regression method is adopted in this paper [8].

GM-Model is the gray prediction model, and it is a method for predicting gray systems. In the gray modeling process, the original sequence of the system is first accumulated, and the cumulative sequence of approximate exponential law is obtained [9]. Then, based on the gray exponential rate of the accumulated generation sequence, the differential equation and the difference equation are established, and the fitted and predicted values of the accumulated generated sequence are obtained, and finally the fitted and predicted values of the original time series are obtained by the subtractive reduction. The gray system is completely known compared to the white and black systems. The white system, that is, the internal characteristics of the system is completely known, and the system information is. The internal information of the black system is ignorant to the outside world, and it can only be observed through its connections with the outside [16]. The grey system is in the middle place, some are known, some are unknown, and there are uncertain relationships among the various factors in the system.

GM (1,1) important concept:

Original sequence: Accumulate (1-AGO, Accumulating Generation Operator)

$$X^{(0)} = \left(x^{(0)}(1), x^{(0)}(2), \ldots, x^{(0)}(n) \right) \tag{1}$$

And the relationship with the original sequence:

$$X^{(1)} = \left(x^{(1)}(1), x^{(1)}(2), \ldots, x^{(1)}(n) \right) \tag{2}$$

$$X^{(1)}(k) = \sum_{i-1}^{k} \left(x^{(0)}(i) \right) \tag{3}$$

Immediately adjacent to mean generation (also weighted generation) and relationship with AGO sequence:

$$Z^{(1)} = \left(z^{(1)}(1), z^{(1)}(2), \ldots, z^{(1)}(n) \right) \tag{4}$$

$$z^{(1)}(k) = a * x^{(1)}(k) + (1 - a) * x^{(1)}(k - 1); a = \frac{1}{2}, k = 2, 3, \ldots, n \tag{5}$$

$$x^{(0)}(1) = x^{(1)}(1) = z^{(1)}(1) \tag{6}$$

GM (1,1) whitening equation:

$$\begin{cases} \frac{dx^{(1)}}{dt} + a * x^{(1)} = b \\ x^{(1)}(t_0) = x^{(1)}(1) \end{cases} \tag{7}$$

The solution and reduction of the whitening equation:

$$\hat{x}^{(1)}(k + 1) = \left(x^{(0)}(1) - \frac{b}{a} \right) e^{-ak} + \frac{b}{a} \tag{8}$$

$$\hat{x}^{(0)}(k+1) = \hat{x}^{(1)}(k+1) - \hat{x}^{(1)}(k) = \left(1 - e^a\right)\left(x^{(0)}(1) - \frac{b}{a}\right)e^{-ak}; k = 1, 2, \ldots, n \tag{9}$$

The model checks the relevant residual sequence:

$$\varepsilon^{(0)} = \left(\varepsilon^{(0)}(1), \varepsilon^{(0)}(2), \ldots, \varepsilon^{(0)}(n)\right)$$
$$= \left(x^{(0)}(1) - \hat{x}^{(0)}(1), x^{(0)}(2) - \hat{x}^{(0)}(2), \ldots, x^{(0)}(n) - \hat{x}^{(0)}(n)\right) \tag{10}$$

p and C in the post-test difference test:

$$\begin{cases} S_1^2 = \frac{1}{n}\sum_{k-1}^{n}\left[x^{(0)}(k) - \bar{x}\right]^2 \\ S_2^2 = \frac{1}{n}\sum_{k-1}^{n}\left[\varepsilon^{(0)}(k) - \bar{\varepsilon}\right]^2 \end{cases}; \bar{x} = \frac{1}{n}\sum_{k-1}^{n}x^{(0)}(k); \bar{\varepsilon} = \frac{1}{n}\sum_{k-1}^{n}\varepsilon^{(0)}(k) \tag{11}$$

$$C = \sqrt{S_2^2/S_1^2} \tag{12}$$

$$p = P\left\{\left|\varepsilon^{(0)}(k) - \bar{\varepsilon}\right| < 0.6745S_1\right\} \tag{13}$$

3.2 Neural Network Prediction

Neural network prediction is a complex network computing system consisting of many large simple highly interconnected neurons [7]. There are two main types of neural network predictions: neural network predictive control based on linearization methods or iterative learning solutions. The linearization method has always been a common method for dealing with nonlinear problems. Through various linear approximations, the solution of the nonlinear control law can be simplified, and the speed of real-time calculation can be improved [17].

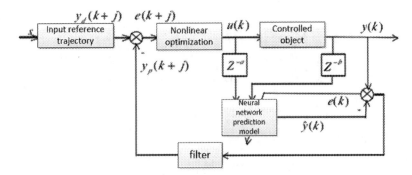

Fig. 1. Neural network prediction structure

In the Fig. 1: S is a set value:

$$a = 1, 2, 3, \cdots, n_u \tag{14}$$

$$b = 1, 2, 3, \cdots, n_y \tag{15}$$

Take the quadratic performance index function:

$$J = \min\left\{\sum_{J=1}^{P} e^2(k+j) + \sum_{j=1}^{L} r_j \Delta u^2(k+j-1)\right\} \tag{16}$$

Where: r_j is the coefficient of control right;

$$\Delta u(k+j-1) = u(k+j-1) - u(k+j-2) \tag{17}$$

$\frac{\partial J}{\partial \Delta u} = 0$ Can be obtained by:

$$\Delta u(k+j-1) = -\frac{e(k+j)}{r_j}\frac{\partial y(k+j)}{\partial \Delta u(k+j-1)} \tag{18}$$

$$e(k+j) = y_d(k+j) - y_p(k+j) \tag{19}$$

Algorithm steps:

It can be seen from the above that the neural network predictive control algorithm steps can be summarized as (j = 1, 2,..., P):

Calculate the desired input reference trajectory:

$$y_d(k+j) \tag{20}$$

Output y*(k) from the neural network prediction model, and generate a predictive output through the filter:

$$y_P(k+j) \tag{21}$$

Calculate the prediction error:

$$e(k+j) = y_d(k+j) - y_p(k+j) \tag{22}$$

Find the quadratic performance function min J (P, L, r), obtain the optimal control law Δ u (k + j − 1), and use u (k) as the first control signal as an input to the controlled object, and then go to step (2).

Evaluate the parameters of the model and the sum of squared residuals. sum of squared residuals. The residual is the difference between the actual tag value and the predicted value for each instance of the data set. There are two kinds of arguments for the sum of squared residuals: SSE (Sum of Squares for Error) is the sum of the squares of the error terms, which reflects the error. The RSS (residual sum of squares) reflects the error term. The formulas of the two are the same F statistic. The test statistic for analysis of variance is the F statistic. The case of analyzing only one factor (variable) is called one-way analysis of variance, also called simple analysis of variance. The analysis that analyzes more than one factor (variable) is called a factorial design.

3.3   Software Structure Introduction

This experiment uses the annual income data of 12 enterprises of the same industry in Hunan Province for nearly 20 years. Since enterprise data does not have problems such as redundancy and lacking, it is not necessary to clean the data, and it can be used directly. Data does not need to be processed by storage technology or deformation work. This dissertation uses pandas and numpy in Python to analyze directly. The operation process is as follows: Firstly, the gray prediction function is defined, $x1$ is defined as a 1-AGO sequence, and after the sequence is generated, it is converted into a matrix, and based on the matrix operation, the minimum value, the maximum value, the mean value, the standard deviation, and the small residual probability are sequentially calculated. After obtaining the parameters, add the predicted column, calculate the row, and predict the predicted value of each row. Forecasts including data modeling, data standardization, feature data processing, obtaining tag data, building models, using the RELU function as an activation function, compiling models, training models, saving model parameters, and finally predicting trends. The operation process includes function determination, calculation of characteristic variables, etc., and finally visualizes the prediction results [13].

```
x1,x2,x3,x4,x5,x6,x7,x8,x9,x10,x11,x12,x13,y
3831732,181.54,448.19,7571,6212.7,6370241,525.71,985.31,60.62,65.66,120,1.029,5321,64.87
3913824,214.63,549.97,9038.16,7601.73,6467115,618.25,1259.2,73.46,95.46,113.5,1.051,6529,99.75
3928907,239.56,686.44,9905.31,8092.82,6560508,638.54,1468.06,81.16,81.16,108.2,1.064,7008,88.11
4282130,261.58,802.59,10444.6,8767.98,6664862,656.58,1678.12,85.72,91.7,102.2,1.092,7694,106.07
4453911,283.14,904.57,11255.7,9422.33,6741400,758.83,1893.52,88.86,114.61,97.7,1.2,8027,137.32
4548852,308.58,1000.69,12018.52,9751.44,6850024,878.26,2139.18,92.85,152.78,98.5,1.198,8549,188.14
4962579,348.09,1121.13,13966.53,11349.47,7006896,923.67,2492.74,94.37,170.62,102.8,1.348,9566,219.91
5029338,387.81,1248.29,14694,11467.35,7125979,978.21,2841.65,97.28,214.53,98.9,1.467,10473,271.91
5070216,453.49,1370.68,13380.47,10671.78,7206229,1009.24,3203.96,103.07,202.18,97.6,1.56,11469,269.1
5210706,533.55,1494.27,15002.59,11570.58,7251868,1175.17,3758.62,109.91,222.51,100.1,1.456,12360,300.55
5407087,598.33,1677.77,16884.16,13120.83,7376720,1348.93,4450.55,117.15,249.01,101.7,1.424,14174,338.45
5744550,665.32,1905.84,18287.24,14468.24,7505322,1519.16,5154.23,130.22,303.41,101.5,1.456,16394,408.86
5994973,738.97,2159.14,19850.66,15444.93,7607220,1696.38,6081.86,128.51,356.99,102.3,1.438,17881,476.72
6236312,877.07,2624.24,22469.22,18951.32,7734787,1863.34,7140.32,149.87,429.36,103.4,1.474,20058,838.99
6529045,1005.37,3187.39,25316.72,20835.95,7841695,2105.54,8287.38,169.19,508.84,105.9,1.515,22114,843.14
6791495,1118.03,3615.77,27609.59,22820.89,7946154,2659.85,9138.21,172.28,557.74,97.5,1.633,24190,1107.67
7110695,1304.48,4476.38,30658.49,25011.61,8061370,3263.57,10748.28,188.57,664.06,103.2,1.638,29549,1399.16
7431755,1700.87,5243.03,34438.08,28209.74,8145797,3412.21,12423.44,204.54,710.66,105.5,1.67,34214,1535.14
7512997,1969.51,5977.27,38053.52,30490.44,8222969,3758.39,13551.21,213.76,760.49,103,1.825,37934,1579.68
7599295,2110.78,6882.85,42049.14,33156.83,8323096,4454.55,15420.14,228.46,852.56,102.6,1.906,41972,2088.14
```

Fig. 2.   The original data.

As can be seen from the above Fig. 2, since the income data of enterprises of the same industry may be similar, we select the income data of eleven companies other than the experimental enterprises for the past 20 years. Among them, $x2$–$x13$ are the specific data of other enterprises in the past 20 years, $x1$ is the 1-AGO sequence, and y is the specific data of the company's income for the past 20 years (Fig. 3).

	Unnamed: 0	x3	x5	x7	x11	y	Development trend forecast
0	1994	448.19	6212.7	525.71	120	64.87	-24.49841309
1	1995	549.97	7601.73	618.25	113.5	99.75	105.7390747
2	1996	686.44	8092.82	638.94	108.2	88.11	190.5442505
3	1997	802.59	8767.98	656.58	102.2	106.07	145.4461365
4	1998	904.57	9422.33	758.83	97.7	137.32	170.2431335
5	1999	1000.69	9751.44	878.26	98.5	188.14	196.1677551
6	2000	1121.13	11349.47	923.67	102.8	219.91	252.8401489
7	2001	1248.29	11467.35	978.21	98.9	271.91	255.7100525
8	2002	1370.68	10671.78	1009.24	97.6	269.1	244.3468628
9	2003	1494.27	11570.58	1175.17	100.1	300.55	289.5619202
10	2004	1677.77	13120.83	1348.93	101.7	338.45	353.4945068
11	2005	1905.84	14468.24	1519.16	101.5	408.86	433.0430908
12	2006	2199.14	15444.93	1696.38	102.3	476.72	561.3566284
13	2007	2624.24	18951.32	1863.34	103.4	838.99	860.8292847
14	2008	3187.39	20835.95	2105.54	105.9	843.14	979.9157104
15	2009	3615.77	22820.89	2659.85	97.5	1107.67	1155.985352
16	2010	4476.38	25011.61	3263.57	103.2	1399.16	1331.444702
17	2011	5243.03	28209.74	3412.21	105.5	1535.14	1484.249023
18	2012	5977.27	30490.44	3758.39	103	1579.68	1700.346069
19	2013	6882.85	33156.83	4454.55	102.6	2088.14	1907.172119
20	2014	7042.31	35046.63	4600.4	101.5	2613.72	2014.585449
21	2015	8166.92	38384.22	5214.78	101.41	3128.77	2263.668213

Fig. 3. The neural network prediction experimental results.

After defining the gray model, the neural network retains x3, x5, x7, and x11 as characteristic data after the analysis, and it uses the data of the enterprises above as a revenue model to predict the development trend of the experimental enterprise.

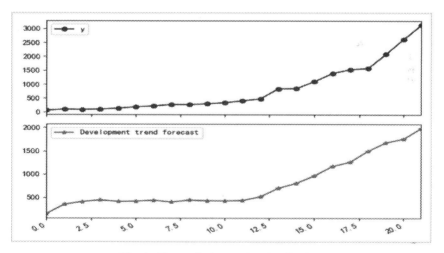

Fig. 4. The prediction results visualization.

After the results are visualized, the upper graph shows the change curve of the real Y value. The figure below shows the change of income value after forecasting. Due to the poor regularity of enterprise income data and the inconsistent selection of neural network structure, the data in the previous three years has some deviation. However, the latter data are more accurate, the prediction results are ideal, and the predicted values are within the error range.

Analysis of experimental results: Through the comparison of factual results and predicted results, using software engineering under big data, it is possible to measure the budget results and development trends of corporate income accurately. Therefore, for budget results and development trends, budget preparation, budget execution and budget assessment can be carried out.

4 Conclusions

This dissertation studies how to apply big data to enterprises' total budget management, and it conducts experiments on sales forecast of X enterprise in Hunan Province. After obtaining valid data on this enterprise' sales revenue from 1994 to 2015, a series of big data technology operations are performed to form a visual prediction curve. As can be seen from Fig. 4, the actual income of X enterprise is roughly the same as the trend of big data forecast revenue, but there are still some deviations. For example, the predicted value in 1994–1996 is slightly larger than the actual value, and the deviation gradually decreases from 1996. The actual income in 2001 is almost the same as the predicted value. But the predicted value of the last two years shows a certain deviation, indicating that the use of this method for big data analysis requires a large amount of effective data support. In general, the model has a more accurate intermediate forecast value, and the predicted values at both ends need to be judged by other experience. The specific reasons for the deviations in this experiment will be carefully analyzed and studied in the next stage.

This dissertation chooses the Lasso method, GM-Model model and neural network prediction, mainly for the following reasons. When the linear relationship of enterprise data is sparse and the main features of enterprise data need to be found, Lasso can solve the problem when establishing the generalized linear model, whether the dependent variable is continuous or discrete. In addition, Lasso has lower data requirements and a wider range of applications, as well as screening variables and reducing the complexity of the model. Neural network prediction has function approximation ability, self-learning ability, complex classification function, associative memory function, and rapid optimization of computing power [20]. It also has the advantages of strong robustness and fault tolerance brought by highly parallel distributed information storage. Similarly, neural network prediction can fully approximate complex nonlinear mapping, and it has the characteristics of learning and adapting the dynamic characteristics of uncertain systems and strong robustness and fault tolerance. Therefore, this dissertation uses neural network method to predict enterprise data. Finally, the GM-Model model is very theoretical, and the gray model can predict both cyclically changing system behaviors and non-periodic changes in system behavior. At the same time, it can be used for both macro and long-term predictions as well as for micro-short-term predictions. The enterprise data just coincides with the non-periodic change system, so this dissertation uses the Lasso method and GM-Model model under big data technology to determine the parameters and calculate the experimental data. After visualizing the results, the calculations can reflect the future revenue trends and revenue values of the company more intuitively.

Compared with traditional budgets, the total budget management under big data technology can feedback the budget information more quickly, observe the changes in the market environment and adjust the budget plan more accurately. It can be seen from experiments that the application of big data technology in total budget management can help enterprises to set accurate budget targets and make effective budget adjustments. Moreover, when using visualization techniques to analyze budget results, budget values and budget trends can be observed more intuitively [5]. In the experimental operation, it is found that big data analysis technology still has shortcomings. When the enterprise budget management objectives are not clear and the budget preparation plan is not comprehensive, the data and results analyzed by big data may have large deviations. Data analysis techniques need to be improved. Many companies don't use or even know big data, so the combination and development of big data and total budget is hindered. Under the development of big data, an enterprise can't operate without data. The application of big data technology to total budget management is an inevitable trend of future development. Therefore, the application of big data is what most companies must face. This dissertation constructs a total budget management framework for enterprises and elaborates the preparation process and compilation methods based on big data technology. It also analyzes the functions and applications of execution, budget and evaluation in detail, and provides theoretical guidance and application reference for the follow-up study of total budget management.

Acknowledgments. This research work is implemented at the 2011 Collaborative Innovation Center for Development and Utilization of Finance and Economics Big Data Property, Universities of Hunan Province; Hunan Provincial Key Laboratory of Big Data Science and Technology, Finance and Economics; Key Laboratory of Information Technology and Security, Hunan Provincial Higher Education. This research is funded by the Open Foundation for the University Innovation Platform in the Hunan Province, grant number 18K103; Open project, grant number 20181901CRP03, 20181901CRP04, 20181901CRP05; Hunan Provincial Education Science 13th Five-Year Plan (Grant No. XJK016BXX001), Social Science Foundation of Hunan Province (Grant No. 17YBA049).

References

1. Tan, H.: Construction of enterprise comprehensive budget management system from the perspective of big data + cloud accounting. Account. Commun. **29**, 88–91 (2018)
2. Mohan, A., Ebrahimi, M., Lu, S., Kotov, A.: Scheduling big data workflows in the cloud under budget constraints. In: 2016 IEEE International Conference on Big Data (Big Data), December 2016, pp. 2775–2784. IEEE (2016)
3. Cheng, P., Wei, F.: Comprehensive budget management of group enterprises based on cloud accounting in the era of big data. Friends Account. **18**, 110–113 (2015)
4. Guo, R.: Research on enterprise's financial management based on big data and cloud computing. In: 2019 4th International Conference on Financial Innovation and Economic Development (ICFIED 2019). Atlantis Press (2019)
5. Luo, S., Wang, Z., Wang, Z.: Big-data analytics: challenges, key technologies and prospects. ZTE Commun. **11**(02), 11–17 (2013)
6. Gandomi, A., Haider, M.: Beyond the hype: big data concepts, methods, and analytics. Int. J. Inf. Manag. **35**(2), 137–144 (2015)

7. Shen, L., Wen, Z.: Network security situation prediction in sciences and engineering. A cloud environment based on grey neural network. J. Comput. Methods 1–15 (2019)
8. Jacob, L., Obozinski, G., Vert, J.P.: Group lasso with overlap and graph lasso. In: Proceedings of the 26th Annual International Conference on Machine Learning, pp. 433–440. ACM, June 2009
9. Wang, Q., Song, X.: Forecasting China's oil consumption: a comparison of novel nonlinear-dynamic grey model (GM), linear GM, nonlinear GM and metabolism GM. Energy **183**, 160–171 (2019)
10. Lv, Y., Duan, Y., Kang, W., Li, Z., Wang, F.Y.: Traffic flow prediction with big data: a deep learning approach. IEEE Trans. Intell. Transp. Syst. **16**(2), 865–873 (2014)
11. Xiaofeng, M., Xiang, C.: Big data management: concepts, techniques and challenges. J. Comput. Res. Dev. **1**(98), 146–169 (2013)
12. Lu, Y., Xu, X.: Cloud-based manufacturing equipment and big data analytics to enable on-demand manufacturing services. Robot. Comput.-Integr. Manuf. **57**, 92–102 (2019)
13. Wei, J., Yin, S., Wang, L.: A big data analytics-based machining optimisation approach. J. Intell. Manuf. **30**(3), 1483–1495 (2019)
14. Shi, X., Zhang, P., Khan, S.U.: Quantitative data analysis in finance. In: Zomaya, A.Y., Sakr, S. (eds.) Handbook of Big Data Technologies, pp. 719–753. Springer, Cham (2017). https://doi.org/10.1007/978-3-319-49340-4_21
15. Ma, S.: A platform of comprehensive budget management system. In: 2016 International Conference on Civil, Transportation and Environment. Atlantis Press, January 2016
16. Jianxin, Z., Zhongzhi, L.: Application of grey neural network combined model in electronic commerce sales forecast. Sens. Lett. **13**(12), 1112–1117 (2015)
17. Lau, H.C.W., Ho, G.T.S., Zhao, Y.: A demand forecast model using a combination of surrogate data analysis and optimal neural network approach. Decis. Support Syst. **54**(3), 1404–1416 (2013)
18. Guo, F., et al.: Research on the relationship between garlic and young garlic shoot based on big data. Comput. Mater. Contin. **58**(2), 363–378 (2019)
19. Wang, B., et al.: Research on hybrid model of garlic short-term price forecasting based on big data. Comput. Mater. Contin. **57**(2), 283–296 (2018)
20. Li, X., et al.: Researching the link between the geometric and Rènyi discord for special canonical initial states based on neural network method. Comput. Mater. Contin. **60**(3), 1087–1095 (2019)

IoT Device Recognition Framework Based on Network Protocol Keyword Query

Zi-Xiao Xu[1], Qin-yun Dai[3], Gang Xu[2,4(✉)], He Huang[1], Xiu-Bo Chen[1,4], and Yi-Xian Yang[1,4]

[1] Information Security Center, State Key Laboratory of Networking and Switching Technology, Beijing University of Posts and Telecommunications, Beijing 100876, China
[2] School of Information Science and Technology, North China University of Technology, Beijing 100144, China
gangxu_bupt@163.com
[3] Shanghai Branch, Coordination Center of China, National Computer Network Emergency Response Technical Team, Shanghai 201315, China
[4] State Key Laboratory of Public Big Data, Guizhou University, Guiyang 550025, Guizhou, China

Abstract. The boost of 5G technology has brought IoT with enhanced feasibility, it is predictable to have a dramatically increased number of the IoT devices with Internet access. However, there will be certain numbers of devices that are extremely vulnerable, particularly with low level of security coding built in. Hackers might have a greater chance to launch DDoS attack. Therefore, identifiable IoT devices linked to the network security vulnerability database might mitigate the risks contributing to a safer cyberspace. This is of great significance to the security of the entire cyberspace. This paper proposes an IoT device identification technology based on network protocol keyword query. Firstly, the traffic packets sent by the IoT device are collected in real time. Then the traffic data carried by different network protocols is extracted by parsing traffic packets. Secondly, the traffic data is filtered and converted into identifiable information related to the identification attribute of the IoT device. Finally, with the search engine of the major Internet of Things e-commerce website, the information about the IoT device is obtained by inputting the protocol keyword.

Keywords: Internet of things security · Products attributes recognition · Network protocols

This work is supported by the NSFC (Grant Nos. 61671087, 61962009, 61572246, 61602232), the Fundamental Research Funds for the Central Universities (Grant No. 2019XD-A02), the Open Foundation of Guizhou Provincial Key Laboratory of Public Big Data (Grant No. 2018BDKFJJ018, 2019BDKFJJ010) and the Scientific Research Foundation of North China University of Technology.

X. Sun et al. (Eds.): ICAIS 2020, LNCS 12239, pp. 219–231, 2020.
https://doi.org/10.1007/978-3-030-57881-8_20

1   Introduction

With the rapid development of the Internet of Things (IoT), various IoT devices, such as network printers, smart speakers and driving recorders, has provided great convenience but also created some security problems. With assigned IP address, IoT devices with low security level could be targeted by hackers for Distributed Denial of Service (DDoS) attacks. For example, in October 2016, hackers used 1.5 million IoT devices to form a botnet network and launched a large-scale DDoS attack, resulting in a large-scale network paralysis. These IoT devices contain a large number of DAHUA camera.

Many scholars [1,2] have begun to study the security of the Internet of Things. This article focuses on how to effectively identify devices to ensure the security of IoT. This paper proposes a property identification framework for IoT devices based on network protocol keyword query. The framework is divided into three parts: network protocol collection framework, protocol information filtering framework and IoT e-commerce automatic query framework.

Firstly, the network traffic packet is collected in real time through the network sniffing tool, and each packet is hierarchically parsed according to the network association layer. Then the packet information carried by different protocols is extracted according to the port characteristics of the protocol. Secondly, the protocol information is effectively segmented after removing non-critical factors such as invisible characters, special symbols, and common English words. Finally, information about the device identification attributes will be obtained after inputting the filtered valid keywords of the network protocol to the search engines of the major e-commerce websites.

2   Related Work

In recent years, more and more security vulnerabilities have gradually emerged with the proliferation of IoT devices. In order to improve the overall security of the Internet of Things, many scholars have invested in the research and development of IoT device identification. In 2010, Cui [4] analyzed the categories and types of weak password devices exposed on the public network, and the identifiable range extended from enterprise devices to consumer electronic devices. In 2015, Radhakrishnan [5] proposed the fingerprint recognition algorithm GTID based on artificial neural network, which can achieve better accuracy. In 2016, Cao [6] extracted the header http field header and status code as the device feature. The device is divided by K-means clustering method based on the cosine similarity between device feature vectors. In the same year, the network space terminal identification framework which uses http slogan information and HTML source code for terminal devices brand recognition is also proposed. In 2017, Ren [7] used machine learning methods to identify whether a device is a video surveillance device based on WEB page information. In the same year, Miettinen [8] proposed a method for identifying types of IoT devices in a specific network. And a device identification framework of GUIDE is proposed by Li [9].

The framework first selects the keyword features of the page through the feature extraction method, and then it constructs the classifier to identify the video monitoring device, which achieves a higher recognition accuracy rate. Meidan [10] applies learning algorithms to network traffic data, which can accurately identify the type of IoT device connected to the network. In 2019, Ya [11] effectively determine 10 IoT devices combined with support vector machine for feature modeling.

Although many research methods for identification of IoT devices have been proposed at home and abroad, there are still many shortcomings. Manual fingerprint extraction can help solve part of the identification, but artificial fingerprints are easy to degrade. Although the automation of device identification is improved when applying machine learning or deep learning methods to analyze network traffic, only the currently existing device type or device brand can be identified. So the recognition granularity is relatively coarse. What's more, the model of the device cannot be distinguished well as page, slogan and traffic characteristic information are similar. Then the existing device identification classifier failed and the features that extracted for different types of devices will be different.

Some scholars [3] have found that the headers of protocols often have rich information related to product attributes, such as device type, brand, model etc. This part of information not only reflects the basic situation of the device, as well as a wealth of semantic information. Known models can help infer the device type and brand. Known types and models can infer the brand. So devices can be efficiently and accurately identified based on the semantic information off the message header. Zou [3] proposed a search-based IoT device identification framework based on this principle. First, it automatically searches for product attributes and other related information from the Internet e-commerce platform to build a product attribute information database. Second, irrelevant information is eliminated according to the natural language processing method. Finally, the information is extracted hierarchically. And the device type, brand, model are extracted respectively. Final, the basic information of the device is inferred based on the semantic information appearing in the slogan information. The experiment proves that the method has higher accuracy and finer granularity than other methods. However, there are the following shortcomings:

1. The process will be intercepted by a firewall and other protection devices when the protocol slogan is crawled. A large number of ports are not open to the outside world, so the slogan information cannot be obtained.
2. The types of protocols are not complete enough, and there are no protocol slogans such as MDNS, DHCP, and DNS. These protocols usually carry information related to device attributes.
3. Some key information may not be obtained because the detected packet information is not perfect and complete.

These shortcomings mainly focus on the aspect of the collection protocol slogan. If more slogan information can be found, more IoT devices can be identified. this paper introduces the message information of the network protocol to increase the information volume of the protocol slogan. The advantages brought by the network protocol are as follows:

1. Some network protocol information cannot be obtained directly by active crawling, but these network protocols can be extracted from traffic packets through port rules. In this way, the types of protocols will be more abundant, and more device-related information can be extracted.
2. All message information can be extracted in the network protocol to avoid missing key information.
3. Network protocol information can be saved offline and used at any time, which can avoid the situation that the slogan information cannot be obtained due to poor network environment.

3   IoT Device Identification Framework

3.1   Overview

In order to realize the accurate and fine-grained identification of IoT devices, this paper designs the IoT device identification framework as shown in Fig. 1. The framework is divided into three parts: network protocol collection framework, protocol information filtering framework, and IoT e-commerce automatic query framework. The network protocol collection framework automatically captures traffic packets from the network space. Then hierarchical analysis on each traffic packet to extract packet information in the protocol is performed. The protocol information filtering framework is used to remove information that is not related to product attributes. The IoT e-commerce automatic query framework inputs the filtered information into the major Internet of Things e-commerce websites for search and query. Then information such as device type, brand and model will be saved.

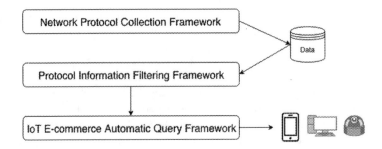

Fig. 1. IoT Device Recognition Framework

3.2   Network Protocol Collection Framework

A large number of traffic packets are generated for every moment in the network. These traffic packets contain rich protocol information. Wireshark is a well-known sniffing tool that displays real-time traffic packets in a network environment. The real-time generated traffic packet in wireshark is shown in Fig. 2.

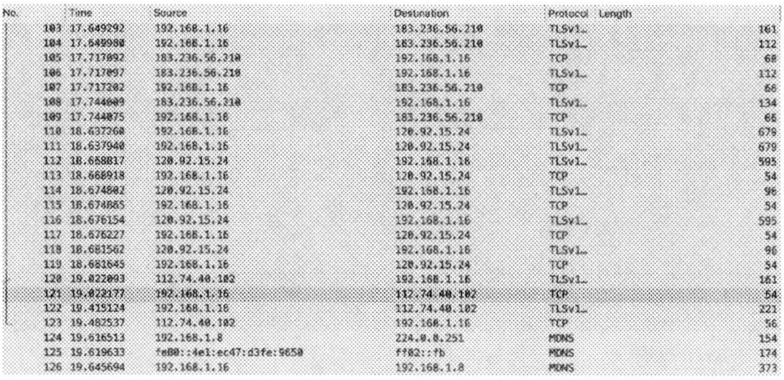

Fig. 2. Traffic packets in the network

Traffic data is transmitted in hexadecimal form. The identifiable information in the red box of Fig. 3 is obtained when encoded and converted. Some of this information is highly relevant to device properties, such as Mac OS X.

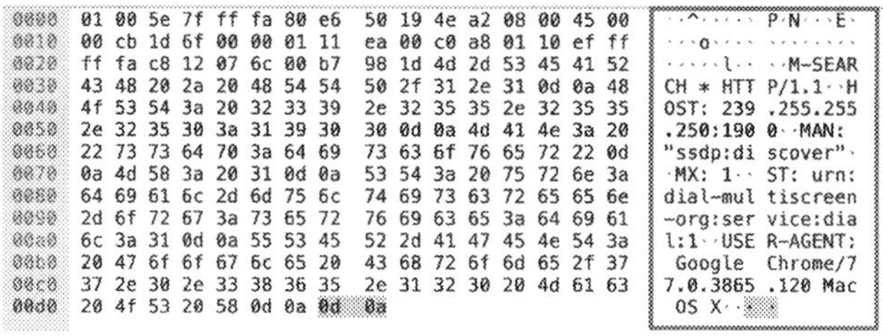

Fig. 3. Data in traffic packets

It should be noted that the network generates a large number of traffic packets every moment. It will greatly reduce the efficiency of IoT device identification if all network protocol information are stored. After observing and analyzing the protocol data, some protocols carry unreadable underlying transmission information. Therefore, the information of this part is not necessary to collect. The protocols that carry a high degree of relevance to the properties of IoT devices are shown in Table 1.

Table 1. Protocols related to IoT device

Protocols	Port	Brief introduciton
FTP	20/21	Transmission between file and server
SSH	22	Connection between client application with server
TELENT	23	Internet remote login
DNS	53	Conversion between domain name and IP
DHCP	67/68	Distribution of IP addresses
HTTP	80	Information returned by the server
RTSP	554	Transmission of multimedia data
UPNP	1900	Discovery and connection between devices
KNET	2345	Communication support for different hardware systems
ONVIF	3702	Provision of network video framework protocol
MDNS	5353	Multicast DNS

In addition to the traffic data of Fig. 3, each traffic packet can also extract source IP, destination IP, source port, destination port and other related information. According to the port characteristics of different protocols in Table 1, the protocol information specified by each device can be obtained by grouping according to IP.

The entire framework is divided into the following three processes:

Step I **Collect traffic packets**

In the Linux environment, the tcpdump command can be used to output network traffic data quickly and easily. The data is output in the pcap file format, which is a commonly used network datagram storage format that can be permanently saved and read repeatedly. In short, network traffic is continuously encapsulated into pcap packages with the tcpdump command.

According to the requirements, it can be divided into two traffic packet collection schemes, one of which is real-time startup, and each traffic packet flowing through the network is stored and downloaded. The second is to start time-sharing. By setting the interval time, the traffic packets are captured every interval. The traffic packet collection process can be completed by starting the timer or starting the tcpdump command in real time.

Step II **Parse traffic packets**

The scapy tool in the programming language python can effectively parse the pcap package. It only needs to input a simple command to quickly get the information of each layer of the network protocol in the traffic packet. At the same time, the source port, destination port, source IP, and destination IP can be obtained. According to the port characteristics of the protocol in Table 1, the port in the pcap packet is compared with the port specified in Table 1. The protocol name and all packet information carried in the protocol will be stored if the ports are equal.

Step III **Remove duplicate data**

There are a large number of duplicate requests and duplicates in the network protocol. Therefore, de-duplication processing is required for the same protocol content of the same IP address before the protocol data is stored in the database.

3.3 Protocol Information Filtering Framework

In the previous chapter, a large amount of network protocol information was obtained by parsing the traffic packets generated in the network. But the information in these protocols is very confusing and complex with a wide variety of characters. So the protocol information must be filtered before the network protocol keyword is queried. The filtered information is mainly divided into two categories: unreadable protocol slogan and readable protocol slogan.

Unreadable protocol slogan contains invisible characters which usually appear as <0x00> or as question marks. Readable protocol slogan obtains the complete protocol information, but the keywords related to IoT devices are mixed with other characters.

To avoid the interference of irrelevant information on the query, it is necessary to gradually eliminate invisible characters and visible characters unrelated to the properties of the Internet of Things product. Four following steps show how the framework works.

1) **Remove Invisible character**

Firstly, all characters are converted into an ASCII code form, and the ASCII code value of each character is judged one by one. If the ASCII value of a character is within 0 to 31 or greater than 127, it will be removed.

2) **Replace punctuation**

Replace all punctuation marks with spaces.

3) **Case replacement**

Convert all letters to lowercase.

4) **Remove unrelated word**

Long characters separated by spaces are converted into a phrase consisting of several characters. Then some words are removed by traversing the phrase.

   a) Common English words collected by web crawlers are removed.

   b) English stop words collected in the NLTK (Natural Language Toolkit) library are removed by using the natural language processing method.

   c) Non-IoT device keywords such as Nginx or Apache are removed. This vocabulary identifies that the device has a high probability of being a WEB server.

3.4 IoT E-Commerce Automatic Query Framework

The IoT device model library was first constructed in the IoT recognition framework proposed Zou [3]. The results is output by matching keywords with IoT device model library. The most similar result is output if there is no matching item in IoT device model library. The framework of this paper replaces this matching model with the IoT e-commerce website query, which has the following two advantages:

1) The IoT device model library needs to be updated regularly. Instead, IoT e-commerce websites are automatic maintained and updated.
2) The query on the e-commerce website provides a correlation search when inputting single or multi-word English letters. The matching process is complex and not accurate enough by just matching with the IoT device model library. At the same time, the query of the IoT e-commerce website can filter the review. The result can be matched even if "Tencent_tingting" is queried. There is a greater difficulty in output if it only matches the IoT device model library.

Fig. 4. search on ZOL

To classify the specific types to which IoT devices belong. The defined product attributes are device type, device brand, device model, and universal unique representation code. Its characteristics are as follows:

– **device type**: the main purpose of the device. Such as mobile phones, stereos, cameras and so on.
– **device brand**: the brand of the device, indicating the ownership of the device. For example, Hikvision, Xiaomi, Baidu.
– **device model**: to the specific model of the device. For example, the iphone11 represents the eleventh generation of iphone.
– **device uuid**: Uniquely identifiable for devices of the same model, this can be distinguished

As shown in Fig. 4, the code label in the webpage can be extracted to obtain the product attributes, including the brand, type and model. On the Amazon website, the name of the item is marked with the model number. The name and model of the item in the title can be extracted together as long as the appropriate

Table 2. Regular expression of device model

Model	Device description	Rules
A6000	Camera	Number and letter
Tichome	Speaker	Pure letter

regular expression is constructed. It is found that all the collected models were found to meet the rules of Table 2.

A single or combined query will be performed after query keywords are obtained. However, one record has multiple keywords, which will constitute multiple query combinations. At the same time, a query will be executed on multiple websites. It's difficult and important to select a final result. The execution flow from multiple returns to the final result is as follows:

Each result item returned from the website marked as $N_0, N_1, \cdots, N_{max}$

1) $\exists n \in N_0, N_1, N_2, \cdots, N_{max}, n = 1$
   Return the results of this e-commerce website and do not compare with other results.
2) $\forall n \in N_0, N_1, N_2, \cdots, N_{max}, n \neq 1$

   The IoT product library is built in advance. Each result item is matched with the IoT product library in turn. Products not in the library will be removed. This process will be executed until the condition of 1) is met.

4   Experiment

The experimental environment consists of 10 common IoT devices as shown in Table 3, including smart speakers, mobile, etc. All devices are in the same LAN. And its IP address is bound to MAC address, which ensure that the original IP address can be restored even after the device is powered off and restarted.

The entire process of obtaining protocol data is described in detail in 3.2, which can be simplified as Algorithm 1. There is a total of 15 records which includes the field IP address, network protocol, and data are displayed in Table 4. Invisible characters are represent by ?.

It can be seen from Table 4 that the protocol data carries keywords related to the Internet of Things properties. Although it can be recognized by human, it is difficult to identify by machine. Irrelevant characters are removed as much as possible while preserving the keywords after the filtering process in Subsect. 3.3.

Table 3. Common IoT devices

IP	Type	Brand	Model
192.168.1.1	Speaker	TmallGenue	FT
192.168.1.2	Mobile	Xiaomi	MI 8 SE
192.168.1.3	TV	XGIMI	4K
192.168.1.4	Speaker	Huawei	Myna
192.168.1.5	Speaker	Tichome	WF62018
192.168.1.6	Speaker	Tencent	tingting9420
192.168.1.7	Speaker	Xiaodu	XDH-06-A3
192.168.1.8	Camera	Dahua	IPC-HFW2125M
192.168.1.9	Speaker	Dingdong	mini
192.168.1.10	Printer	EPSON	L655

Algorithm 1. Collect Traffic Data

Input: traffic data
Output: L< protocol,data>
  while True **do**
    tcpdump → dump.pcap
    data,Port = scrpy(dump.pcap)
    for protocol in Protocols **do**
      if Port == protocol.Port **then**
        L < protocol,data >
      else
        Continue;
      end if
    end for
  end while

A lot of irrelevant data in network protocol are removed while the keywords are reserved. Data in Table 5 is split by spaces, which is easy to remove irrelevant words by matching the dictionary. Besides, a combination of arrays and letters is saved in the data in order to extract potential models.

An ip protocol corresponds to multiple keywords, of which one correspond multiple combined query statements. A finally result that contains the properties of the IoT product is obtained according to the select algorithm in 3.3. The final identified device identification results are presented in a Table 6.

Combined with Tables 6 and 3, it can be concluded that the recognition framework of this paper can correctly identify 7 kinds of other types. The other three devices can correctly identify some attributes.

Table 4. Network protocol data after collecting

	IP	Protocl	Data
1	192.168.1.1	MDNS	?25383401Q0serw_tmallGeniu_"rnetwork audioFec1
2	192.168.1.2	DNS	tracking.miui.com, t4.market.xiaomi.com
3	192.168.1.2	DNS	data.mistat.xiaomi.com,b9.market.xiaomi.com
4	192.168.1.2	HTTP	Dalvik/2.1.0(Linux;U; Android 8.1.0; MI 8 SE MIUI/V10.2.1.0)
5	192.168.1.3	HTTP	Dalvik/2.1.0 (Linux; U; Android 6.0; XGIMI TV Build/MRA58K)
6	192.168.1.4	DHCP	0HUAWEI_Myna_53ce1af2-4a70-49a4-982a-4f11c58231cc0?"0
7	192.168.1.5	MDNS	Tichome−E2EA_http_tcplocal_services_dnssd_udp!x2Android
8	192.168.1.6	HTTP	Linux; U; Android 4.4.2; Tencent_tingting Build/KVT49L
9	192.168.1.7	DHCP	59<*dhcpcd−_:aarch64:AmlogicXiaodu-AudioSpeaker7!3:;
10	192.168.1.7	DNS	xiaodu.a.shifen.com audiolc.n.shifen.com
11	192.168.1.8	UPNP	<a:MessageID>uuid:8f2857db-7932-40df-bc04-6b1092eb24a1
12	192.168.1.8	ONVIF	onvif://www.onvif.org/name/Dahua hardware/IPC-HFW2125M-I1
13	192.168.1.9	DNS	proxy.dingdong.linglongtech.com, aiui-ipv6.openspeech.cn
14	192.168.1.10	MDNS	EPSON7F3058localEPSON L655 Series_printer_tcp_EPSON L655
15	192.168.1.10	UPNP	USN: uuid:335c80b5-c07b-403a-a9b2-72fbdd2dc65c:

Table 5. Network protocol data after filtering

	IP	Protocl	Data
1	192.168.1.1	MDNS	25383401q0serw tmallgeniu rnetwork audiofec1
2	192.168.1.2	DNS	miui t4 xiaomi
3	192.168.1.2	DNS	mistat xiaomi b9 xiaomi
4	192.168.1.2	HTTP	dalvik se miui v10
5	192.168.1.3	HTTP	dalvik XGIMI TV mra58k
6	192.168.1.4	DHCP	0huawei 53ce1af2 49a4 982a 4f11c58231cc0
7	192.168.1.5	MDNS	tichome e2ea http tcplocal dnssd udp x2android
8	192.168.1.6	HTTP	tencent tingting kvt49l
9	192.168.1.7	DHCP	dhcpcd aarch64 amlogicxiaodu audiospeaker7
10	192.168.1.7	DNS	xiaodu shifen audiolc shifen
11	192.168.1.8	UPNP	uuid:8f2857db-7932-40df-bc04-6b1092eb24a1
12	192.168.1.8	ONVIF	onvif www onvif org dahua ipc hfw2125m i1
13	192.168.1.9	ONVIF	dingdong linglongtech aiui ipv6 openspeech cn
14	192.168.1.10	MDNS	l655 series printer tcp epson l655
15	192.168.1.10	UPNP	uuid:335c80b5-c07b-403a-a9b2-72fbdd2dc65c

Table 6. Keyword query result

IP	Keywords	Website	Brand	Type	Model
192.168.1.1	Tamll genie	smzdm	Tmall	Speaker	BOOM
192.168.1.2	miui se	ZOL	Xiaomi	Speaker	8 SE
192.168.1.3	xgimi tv	Joybuy	XGIMI	TV	LUNE 4K
192.168.1.4	myna	ZOL	HuaWei	Speaker	myna
192.168.1.5	tichome	ZOL	Tichome	Speaker	WF62018
192.168.1.6	tencent tingtng	ZOL	Tencent	Speaker	9420
192.168.1.7	audiospeaker7	Joybuy	BH AUDIO	Speaker	Blue
192.168.1.8	hfw2125m	Joybuy	Dahua	Camera	DH-IPC-HFW2125M-AS-I2
192.168.1.9	dingdong	ZOL	Dingdong	Speaker	2G
192.168.1.10	epson l655	ZOL	Epson	Printer	L655

5   Conclusion

This paper proposes a property identification framework for IoT devices based on network protocol keyword query. This framework mainly propose optimization and improvement of the search-based IoT recognition framework proposed by Zou [3]. However, there are some shortcomings:

1) Traffic is generated all the time, which will put a lot of pressure on servers.
2) The traffic packet carries many and miscellaneous irrelevant information. How to effectively propose the keywords related to the IoT identification attribute in the protocol is the focus of future work research.
3) IoT search capabilities are limited. Google and other search engines can be used to find the results of IoT device properties.

References

1. Shi, C.: A novel ensemble learning algorithm based on D-S evidence theory for IoT security. Comput. Mater. Continua **57**(3), 635–652 (2018)
2. Centonze, P.: Security and privacy frameworks for access control big data systems. Comput. Mater. Continua **59**(2), 361–374 (2019)
3. Zou, Y., Liu, S., Yu, N.: IoT Device Recognition Framework based on Web search. J. Cyber Secur. **3**(4), 25–40 (2018). (in Chinese)
4. Cui, A., Stolfo, S.J.: A quantitative analysis of the insecurity of embedded network devices: results of a wide-area scan. In: Proceedings of the 26th Annual Computer Security Applications Conference, pp. 97–106. ACM, December 6, 2010
5. Radhakrishnan, S.V., Uluagac, A.S., Beyah, R.: GTID: a technique for physical deviceanddevice type fingerprinting. IEEE Trans. Depend. Secure Comput. **12**(5), 519–32 (2014)
6. Cao, L.C., Zhao, J.J., Cui, X., Li, K.: Cyberspace device identification based on K-means with cosine distance measure. J. Univ. Chin. Acad. Sci. **33**(4), 562–569 (2016)

7. Ren, R.L., Gu, Y., Cui, J., Liu, S., Zhu, H.S., Sun, L.M.: Web features-based recognition specific-type IoT device in cyber-space. Commun. Technol. **50**(5), 1003–1009 (2017). (in Chinese)
8. Miettinen, M., Marchal, S., Hafeez, I., Asokan, N., Sadeghi, A.R., Tarkoma, S.: IoT sentinel: automated device-type identification for security enforcement in IoT. In: 2017 IEEE 37th International Conference on Distributed Computing Systems (ICDCS), pp. 2177–2184. IEEE, June 5, 2017
9. Li, Q., Feng, X., Wang, H., Sun, L.: Automatically discovering surveillance devices in the cyberspace. In: Proceedings of the 8th ACM on Multimedia Systems Conference, pp. 331–342. ACM, June 20, 2017
10. Meidan, Y., et al.: ProfilIoT: a machine learning approach for IoT device identification based on network traffic analysis. In: Proceedings of the Symposium on Applied Computing, pp. 506–509. ACM, April 3, 2017
11. Tu, Y., Zhang, Z., Li, Y., Wang, C., Xiao, Y.: Research on the internet of things device recognition based on RF-fingerprinting. IEEE Access. **22**(7), 37426–31 (2019)

Application of ARIMA Model in Financial Time Series in Stocks

Jiajia Cheng[1], Huiyun Deng[1], Guang Sun[1(✉)], Peng Guo[1,2], and Jianjun Zhang[3]

[1] Hunan University of Finance and Economics, Changsha 410205, China
simon5115@163.com
[2] University Malaysia Sabah, Kota Kinabalu, Malaysia
[3] Hunan Normal University, Changsha, China

Abstract. In order to study the development of stock exchange between China and the United States during the Sino-U.S. trade war, the stock trends of the two countries were compared and analyzed, combined with the time series prediction, and displayed with the visual result chart. Judging the data's stability from its original time sequence diagram, autocorrelation diagram and p-value, make difference for non-stationary data, then determine the appropriate parameters P and Q in ARIMA model according to autocorrelation diagram and partial autocorrelation diagram, confirm the model for model test, select the model with the lowest AIC, BIC and hqlc values to predict 10% of the total data and visualize. From the visual results, the prediction effect is not very good, there are relatively large errors, and the trend of stock closing price is not consistent. ARIMA model is not very good in the application of stock market, which needs to be improved.

Keywords: ARIMA model · Stock analysis · Time series prediction

1 Introduction

Prediction is a necessity in human life and a mutual question in all disciplines [1]. It is commonly used in financial and economical field. In the stock market, investors aim at predicting the price trend of the stock before determining when to sell or buy the stocks, in order to optimize investment decisions [2, 3]. As the technology and science develops, countries no longer compete with each other by force, but by economic achievements. Contrastive analysis is a way to understand the differences between countries, only after analyzing strength and weaknesses of both sides, can we predict dynamic trends and policy based on existing material [4].

Stock market is a domain of financial play, it can gain more profit compared with bank time deposits. Whether the state economy is stable or inflated, it can be reflected on the trend of the stock market [5]. Therefore, it's important to study Stock Market Volatility. There're many ways to understand the extent of analysis of stock price changes, but we proposed prediction method like Time series Analysis to improve the accuracy of prediction [6–8].

Stock market analyst applied lots of statistical techniques, for example, Autoregressive moving average (ARMA), Autoregressive integral moving average (ARIMA),

© Springer Nature Switzerland AG 2020
X. Sun et al. (Eds.): ICAIS 2020, LNCS 12239, pp. 232–243, 2020.
https://doi.org/10.1007/978-3-030-57884-8_21

ARMA-EGARCH, method of Box and Jenkins' and various soft computing and evolutionary computing [7, 9, 10]. Most of the researchers are proficient in using traditional statistical techniques for predicting precisely. However, traditional statistical techniques are incompatible with non-construction and large database, and it can affect the availability and application of big data prediction. Because data are dynamic and can be produced any time, it's essential to monitor data. The application of rolling window solved this problem, and produced more subsamples which can improve accuracy of data [11]. Through the analysis of the stock data of China and the United States, we can understand the horizontal-time axis to describe the data, ups and downs of the trend. Look at the economic boom and downturn with the emergence of peaks and troughs.

Long term prediction controls the behavior of dynamic economic system, and short-term prediction aims at analyzing the accurate trajectory of economic activities [12, 13]. Many people have used financial time series method to predict stock market, for instance, Zhou, Xingyu (2018), Shah and Habib (2018), Efendi, Riswan, Nureize Arbaiy, and Mustafa Mat Deris (2018), Spiro (2018). However, because of the rapid development of our society and the information explosion, many traditional research methods cannot be used any more. The emerging technology studied in University of Johannesburg can promote the development of collecting data, analyzing data, applying analysis to decision-making and achieving better predictions. The challenge researchers facing is the way to construct time series mode, the way to learn predicting from big data efficiently and how to use learning model to improve prediction when observation is limited. It is also important to determine the impact of constructing predicting system which can process large amount of data. Thus, it is equally important to think about topics like black box of the learning model, the confidence interval of the forecast and uncertainty.

The paper aims at analyzing stock data of the two countries, which can be used in contrastive analysis and prediction. Take the outbreak of stock exchange in 2018 for example, the stock market was shaken and Chinese started to focus on the development of stock market and people's attitude toward stock. In the articles about American's attitude towards stock and everyday stock exchange, it is obvious to find out the difference between us. Therefore, it is necessary to study the price and volatility of stocks and forecast them.

Serve rolling window as a bullet train in the time line, take the average method as internal fuel, use data analysis and visualization to improve engine power and dissect significant information of stock data.

2 Introduction of Model

2.1 Time Series

Time series refers to the sequence of the value of the same statistical index according to the sequence of its occurring time. Time series conceals the relationship between the past and the future, and predict the future by studying the past. Time series analysis is wildly used in many fields like engineering, economics and technology. In the era of big data, time series analysis has become a branch of AI technology, as it can be used in detecting data, prediction and other cases by combining time series analysis and disaggregated model [14].

2.2 Stationarity of Time Series

An essential assumption of classical regression analysis is that the data are stationary. Non-stationary data often lead to 'false regression', as two variables that have no causal relationship are highly correlated. The requirement of stationarity: (1) the mean μ is an irrelevant constant of time t. (2) for random time t and arbitrary time interval k, the autocovariance of time series is only related to k instead of t. From a statistical point of view, the stationarity requires the value of a time series (unknown distribution) must have a certain distribution [15].

2.3 The Principal of ARIMA Model

The full name of ARIMA model is Autoregressive Integrated Moving Average Model, which is also written as ARIMA (p, d, q). It is one of the most commonly used statistics models for predicting time series [16–18].

Parameters of ARIMA model: p—represents the lags of the times series data used in the prediction model. It is also called as AR/Auto-Regressive, which describes the relationship between current value and the historical value. d—the amount of orders of differential differentiation that time series data need to reach a stable state. It is also called as Integrated, assuming y represents the difference of Y at time t. q– represents the lags of the errors used in the prediction model. It is also called as MA/Moving Average, which focus on the accumulation of error terms in the autoregressive model [19].

$$\text{if } d = 0, \ y_t = Y_t$$
$$\text{if } d = 1, \ y_t = Y_t - Y_{t-1}$$
$$\text{if } d = 2, \ y_t = (Y_t - Y_{t-1}) - (Y_{t-1} - Y_{t-2})$$
$$= Y_t - 2Y_{t-1} + Y_{t-2}$$

The mathematical version of ARIMA:

$$\hat{y}_t = \mu + \phi_1 \times y_{t-1} + \cdots + \phi_p \times y_{t-p} + \theta_1 \times e_{t-1} + \theta_q \times e_{t-q}$$

ϕ is the coefficient of AR, θ is the coefficient of MA (Table 1).

Table 1. Model recognition principle.

Model	Autocorrelation coefficient (ACF)	Partial autocorrelation coefficient (PACF)
AR (p)	Trailing	p order truncation
MA (q)	q order truncation	Trailing
ARMA (p, q)	p order trailing	q order trailing

Fig. 1. Experiments are conducted based on the basic procedure of data engineering.

3　Experimental Section

3.1　Experimental Procedure

Data engineering drawing (Fig. 1).

The data engineering process of this article: Get the data: Call the tushare package through pycharm software to get the inventory data of ZTE (000063) and Best Buy (BBY) from March 2018 to October 2019. Clean data: The data used in the experiment is the daily closing price, other data will be deleted. Departmental data: First: the name is unified, and the closing price is changed to "close". Secondly, read a list of dates as the data type with date format and date index. Analysis: When observing the data, we found that ZTE's stock data disappeared for one month in 2018, and the last month had only 9 days. Therefore, influencing factors must be considered when making comparisons. Based on this, the process of the experiment in this paper is formed. This article proposes the results analysis criteria: First, the difference in the amount of data will affect the accuracy of the prediction. Second, compare predictions of the same quantity and time accurately. Third, analyze the similarities and differences of stock trading between the two countries.

3.2　Experimental Operation Process

There are three experimental processes in this section, which are divided into two categories according to the two levels of the sample: small samples and large samples. One experiment for small samples and two experiments for large samples. Experiments are numbered 1, 2, 3 in sequence.

Each experiment follows the following steps: Step 1: Determine whether the original sequence is stable: First. The overall timing of the original timing diagram. Second. Autocorrelation graph to determine whether the correlation is short-term or long-term. Third. The original sequence unit root test table determines whether it is a fixed sequence. Determine whether to proceed to the next step: difference. Step2 is to choose the optimal model when the sequence is stable, in other words, to obtain the values of p and q. Step3: choose the optimal model.

NO. 1 Prediction of Small Sample Data at Best Buy (as BBY) (Figs. 2 and 3) and (Table 2).

Based on the autocorrelogram, it is obvious that autocorrelation coefficients eventually tend to be zero, indicating strong short-term correlation between series. The p value of unit root test statistics is significantly greater than 0.05, and the adf value is greater than any value in the confidence interval. To sum up, this sequence is a non-stationary series (non-stationary series definitely is not a white noise sequence) (Figs. 4 and 5) and (Table 3).

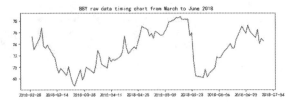

Fig. 2. The original sequence chart of BBY.

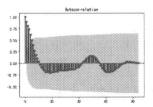

Fig. 3. Autocorrelogram of the original sequence of BBY.

Table 2. Unit root test of the original sequence of BBY.

adf	c value			p
	1%	5%	10%	
−2.0530	−3.5107	−2.8966	−2.5854	0.2638

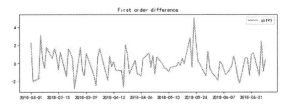

Fig. 4. Sequence diagram after first order difference of BBY.

Two conclusions can be drawn when combine with the graph above: First, the mean and variance of sequence chart after using one order difference method fluctuate smoothly near the 0 axis, the p value after unit root test is less than 0.05, and the value of adf is less than any value in the confidence interval. So, the sequence after using the one order difference method is a stable one and the number of difference d can be set as 1. Second, determine the appropriate p and q values, autocorrelation graph truncates and partial autocorrelation graph appears to be trailing, therefore model MA (1) is chosen.

Although the value of p and q can be determined from the graph, the model is not unique. By testing ARIMA (1, 1, 0), it creates model ARIMA (1, 2, 1), and there is difference between the result of the test. Therefore, it cannot decide which model is

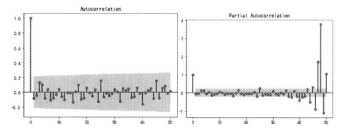

Fig. 5. Autocorrelogram after first order difference of BBY on the left and partial autocorrelogram after first order difference of BBY.

Table 3. Chart of stationarity test after first order difference of BBY.

adf	c value			p
	1%	5%	10%	
−9.6382	−3.5117	−2.8970	−2.5857	0.00000

better. We will choose which model is the best and has minimum error based on the visualizations predicted (Table 4).

Table 4. Examination table of BBY.

	AIC	BIC	HQIC
ARIMA (0, 1, 1)	192.0695	198.5461	194.6209
ARIMA (1, 1, 0)	192.0694	198.5460	194.6208
ARIMA (1, 2, 1)	191.1293	199.7018	194.5009

The best model is ARIMA (0, 1, 1) (Fig. 6).

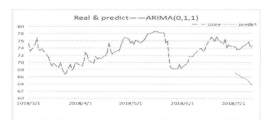

Fig. 6. ARIMA (0, 1, 1) visual result diagram of BBY.

NO. 2 Prediction of Large Sample Data at ZTE (Figs. 7 and 8) and (Table 5).

According to the autocorrelogram, the correlation coefficient eventually tend to be zero, indicating strong short-term correlation between series. The p value of unit root test

Fig. 7. Sequence diagram of the original data of ZTE.

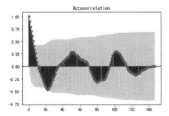

Fig. 8. Autocorrelogram of the original sequence of ZTE.

Table 5. Unit root test of the original sequence of ZTE.

adf	c value			p
	1%	5%	10%	
−2.7357	−3.4756	−2.8814	−2.5773	0.0680

statistics is greater than 0.05 and the value of adf is within 10% of confidence interval. To sum up, this sequence is a non-stationary series (Fig. 9) and (Table 6).

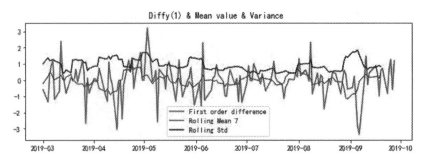

Fig. 9. Sequence diagram after first order difference of ZTE.

The mean and variance of sequence chart after using one order difference method fluctuate smoothly around the 0 axis. The p value of unit root test is less than 0.05 and the

Table 6. Chart of stationarity test after first order difference of ZTE.

adf	c value			p
	1%	5%	10%	
−11.696	−3.4759	−2.8815	−2.5774	0.0000

value of adf is less than any value in the confidence interval. So the time series data after one order difference is smooth and the number of difference d can be set as 1 (Fig. 10).

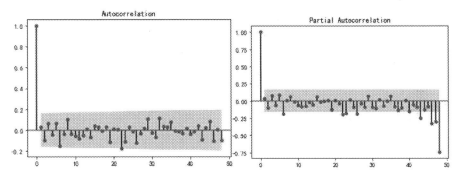

Fig. 10. Autocorrelogram after first order difference of ZTE on the left and partial autocorrelogram after first order difference of ZTE on the right.

Autocorrelogram and partial autocorrelogram are truncate. Choose model ARMA (1, 1) to test model ARIMA (0, 1, 1), ARIMA (1, 1, 0) and ARIMA (1, 2, 1) (Table 7).

Table 7. Examination table of ZTE.

	AIC	BIC	HQIC
ARIMA (0, 1, 1)	421.5435	430.5148	425.1886
ARIMA (1, 1, 0)	421.5435	430.5485	425.2224
ARIMA (1, 2, 1)	426.6630	438.5974	431.5122

The best model is ARIMA (0, 1, 1) (Fig. 11).

NO. 2 Prediction of Large Sample Data at BBY (Figs. 12 and 13) and (Table 8).

Based on the autocorrelogram, the autocorrelation coefficients eventually tend to be zero, indicating strong short-term correlation between series. The p value of unit root test statistics is significantly greater than 0.05, and the adf value is greater than any value in the confidence interval. To sum up, this sequence is a non-stationary series (Fig. 14) and (Table 9).

The mean and variance of sequence chart after using one order difference method are basically stable, the p value is less than 0.05 and the value of adf is less than any

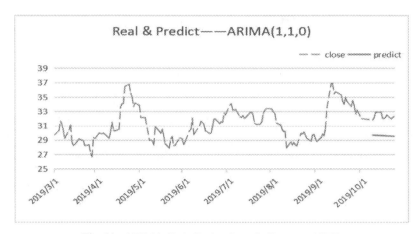

Fig. 11. ARIMA (1, 1, 0) visual result diagram of ZTE.

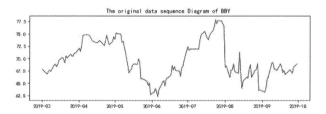

Fig. 12. Sequence diagram of the original data of BBY.

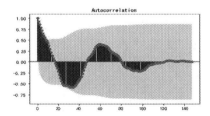

Fig. 13. Autocorrelogram of the original sequence of BBY.

Table 8. Unit root test of the original sequence of BBY.

adf	c value			p
	1%	5%	10%	
−2.3575	−3.4756	2.8814	−2.5773	0.1540

value in the confidence interval. So time series data after one order difference is stable and the number of difference d can be set as 1 (Fig. 15).

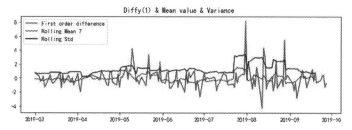

Fig. 14. Sequence diagram after first order difference of BBY.

Table 9. Chart of stationarity test after first order difference of BBY.

adf	c value			p
	1%	5%	10%	
−12.913	−3.4759	−2.8815	−2.5774	0.0000

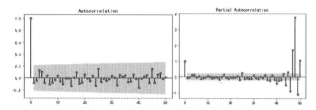

Fig. 15. Autocorrelogram after first order difference of BBY on the left and partial autocorrelogram after first order difference of BBY on the right.

Autocorrelation graph truncates and partial autocorrelation graph appears to be trailing, therefore AR model is chosen and q = 0, the final model is ARIMA (1, 1, 0) (Table 10).

Table 10. Examination table of BBY.

	AIC	BIC	HQIC
ARIMA (0, 1, 1)	530.9071	539.8784	534.5522
ARIMA (1, 1, 0)	530.8892	539.8605	534.5343
ARIMA (1, 1, 1)	532.9485	544.9102	537.8087

ARIMA (1, 1, 0) model is chosen (Fig. 16).

4 Conclusion

In this paper, we have made three experimental time series predictions. Now, answer three target questions, and the answers are as follows.

Fig. 16. ARIMA (1, 1, 0) visual result diagram of BBY.

For the first question: the impact of differences in the amount of data on the accuracy of the prediction. By comparing and analyzing the prediction error of small sample and big sample of BBY, large sample prediction is more quantitative and qualitative. As for the second question: compare the accuracy of prediction with the same amount of data and in the same period of time. I think there are no identical leaves in the world, even imitate, without transcend cannot survive. There is peak value, but it may have happened at the same time and in some cases it is more likely to be coincidental. After reaching the peak value, development may be different. There are many differences between these two stocks, and their trends represent their respective development history. In the trend graph reflecting the whole sample, the first peak is almost synchoronous. But the next climax of BBY directly happened after underestimation. Meanwhile, before the next peak of ZTE, there will be a slow little peak and the climax value is half of the peak value. As for the third question: analyze the similarities and differences of stock trading between the two countries. Similarities: the time period and the sample size of data being chosen are the same. Differences: first of all, in terms of the closing price of the data in the experiment, the closing price of US is significantly higher than the Chinese one, which is about twice as much. It indicates that there is still room for development and improvement in our country. Secondly, in terms of the error in the prediction, in the case of a large sample, the forecast line of the US is closer to the real data, while there appears a noticeable gap in the forecast line of our country. I assume this is because Chinese market has more influence factor than US, especially the political policy. With the development of technology and science, people are able to express their ideas freely. It may because some rumors or private affairs of representatives that affect the stock.

Acknowledgments. This research work is implemented at the 2011 Collaborative Innovation Center for Development and Utilization of Finance and Economics Big Data Property, Universities of Hunan Province; Hunan Provincial Key Laboratory of Big Data Science and Technology, Finance and Economics; Key Laboratory of Information Technology and Security, Hunan Provincial Higher Education. This research is funded by the Open Foundation for the University Innovation Platform in the Hunan Province, grant number 18K103; Open project, grant number 20181901CRP03, 20181901CRP04, 20181901CRP05; Hunan Provincial Education Science 13th Five-Year Plan (Grant No. XJK016BXX001), Social Science Foundation of Hunan Province (Grant No. 17YBA049).

References

1. Dong, X.: Predictive value of routine peripheral blood biomarkers in Alzheimer's disease. Front. Aging Neurosci. **11**, 332 (2019)
2. Zhang, H.: Monetary policy adjustment, corporate investment, and stock liquidity—empirical evidence from chinese stock market. Emerg. Mark. Financ. Trade **13**(55), 3023–3038 (2019)
3. Ma, X., Lv, S.: Financial credit risk prediction in internet finance driven by machine learning. Neural Comput. Appl. **31**(12), 8359–8367 (2019). https://doi.org/10.1007/s00521-018-3963-6
4. Dwumfour, R.A.: Interest rate and exchange rate exposure of portfolio stock returns: does the financial crisis matter? J. Afr. Bus. **20**, 339–357 (2019)
5. Huang, S.-F.: Stock market trend prediction using a functional time series approach. Quant. Financ. **1**(20), 69–79 (2020)
6. Idrees, S.M.: A prediction approach for stock market volatility based on time series data. IEEE Access **7**, 17287–17298 (2019)
7. Fernández, P.: Predicción de variaciones en el precio del petróleo con el modelo de optimización arima, innovando con fuerza bruta operacional. Tec Empresarial **1**(13), 53–70 (2019)
8. Reddy, C.V.: Predicting the stock market index using stochastic time series ARIMA modelling: the sample of BSE and NSE. Indian J. Financ. **8**(13), 7–25 (2019)
9. Fang, S.: Measuring contagion effects between crude oil and Chinese stock market sectors. Q. Rev. Econ. Financ. **68**, 31–38 (2018)
10. Lahmiri, S.: Minute-ahead stock price forecasting based on singular spectrum analysis and support vector regression. Appl. Math. Comput. **320**, 444–451 (2018)
11. Neluka, D.: Is stock return predictability time-varying? J. Int. Financ. Mark. Inst. Money **52**, 152–172 (2018)
12. Feuerriegel, S.: Long-term stock index forecasting based on text mining of regulatory disclosures. Decis. Support Syst. **112**, 88–97 (2018)
13. Weng, B.: Predicting short-term stock prices using ensemble methods and online data sources. Expert Syst. Appl. **112**, 258–273 (2018)
14. Zaji, A.H., Bonakdari, H., Gharabaghi, B.: Developing an AI-based method for river discharge forecasting using satellite signals. Theor. Appl. Climatol. **138**(1), 347–362 (2019). https://doi.org/10.1007/s00704-019-02833-9
15. Chen, G.: Generalized exponential autoregressive models for nonlinear time series: stationarity, estimation and applications. Inf. Sci. **438**, 46–57 (2018)
16. Afeef, M.: Forecasting stock prices through univariate ARIMA modeling. NUML Int. J. Bus. Manag. **2**(13), 130–143 (2018)
17. Khashei, M.: A comparative study of series Arima/Mlp hybrid models for stock price forecasting. Commun. Stat. Simul. Comput. **9**(48), 625–2640 (2019)
18. Nguyen, H.V.: A smart system for short-term price prediction using time series models. Comput. Electr. Eng. **76**, 339–352 (2019)
19. Yang, B.: Stock market forecasting using restricted gene expression programming. Comput. Intell. Neurosci. **2019**, 7198962 (2019)

A Topological Control Algorithm Based on Energy Consumption Analysis and Locomotion of Nodes in Ad Hoc Network

Chaochao Li[1]([⊠]), Jiangxia Tan[2], Hanwen Xu[1], Zichuan Guo[3], and Qun Luo[1]

[1] School of Cyberspace Security, Beijing University of Posts
and Telecommunications, Beijing, China
lichaochaohao@bupt.edu.cn
[2] College of New Media, Beijing Institute of Graphic Communication, Beijing, China
[3] Faculty of Information Technology, Beijing University of Technology, Beijing, China

Abstract. In the wireless ad hoc network, due to the battery energy of the node itself limitation, achieving a balance of the node energy has become a consequential and meaningful research direction of the Wireless Ad hoc Network.

The energy consumption of the node is related to the motion and transmission power of the node. In order to prolong the network lifespan of the wireless Ad hoc, a topological control algorithm based on node energy consumption and locomotion has been proposed. According to network topological actual change, this algorithm evaluates the current statement of stability, such as loss of path, node paralysis, node position variation energy balance and re-optimize network topological structure. The simulation results of NS-3 show that the performance of this algorithm is close to those of higher complexity, which can proportionate the energy of the entire network node, optimize the network throughput rate, and keep the data transmission off the irrelevant disruption. Besides, it can extend the period of wireless Ad hoc network lifetime as well.

Keywords: Ad hoc network · Topological control · Energy balance

1 Introduction

Wireless ad hoc network is a dynamic network with multiple hops and no fixed nodes [1]. Its nodes are always in a mobile state, and its energy, channel, mobility and bandwidth are often limited as well. So, how to extend the life cycle of nodes and networks and enhance communication stability is a popular topic for current researchers. Well, it is better to keep optimizing the network topology and rationally scheduling nodes, in which way, the energy consumption of nodes can be reduced, and the network life cycle can be extended.

In wireless ad hoc network, the nodes are often in a moving state, and the communication between two adjacent nodes will also terminate with the movement of the node, so finally the link will be broken and the path will be invalid between these nodes [2]. As the energy consumption of the node is limited, and considering the sensing range

X. Sun et al. (Eds.): ICAIS 2020, LNCS 12239, pp. 244–256, 2020.
https://doi.org/10.1007/978-3-030-57884-8_22

of the node and the motion path will be affected by various external factors in actual circumstances, it is necessary to study the node energy consumption of the wireless ad hoc network. The main research work of this paper is to predict the stability of the node through power consumption of the node and its position variation, then, optimize the network topology, ensuring the using time of effective path, so that the nodes in the deployment area can work in turn to reduce energy consumption and thus extend network lifetime.

2 Relevant Researches

In wireless ad hoc networks, scheduling problems arise because of sharing wireless channels and the network topology control. Implementing network sharing can have a major impact on the evaluation of network's primary performance criteria, such as end-to-end latency, node power consumption and throughput. The battery provides limited energy to node, so how to save energy becomes a key issue in wireless ad hoc networks. The topology control technology can adjust the transmission power of the wireless node, and reduce the energy consumption of the node by selecting appropriate neighbor nodes, with the premise of ensuring network connectivity, thereby prolonging the network life cycle. Many classical topological control methods have been proposed, in which the topological control algorithm based on the minimum spanning tree (MST) [3] can reduce the node energy consumption by constructing a sparse graph, topological control algorithm based on the shortest path (SP) [4] can reduce routing energy consumption by remaining minimal energy consumption path.

For wireless ad hoc networks, it is possible to predict the lifetime of the network, congestion control of the network, and broadcast boundaries. While, it is a complex task for congestion control prediction of wireless ad hoc networks, due to heterogeneous resource nodes involving the distributed environments and the large fluctuation of the load makes it difficult to predict the load situation after the long-term execution of the Ad hoc network [6]. In the literature [7], a method for calculating the link effective time by receiving the transmission power intensity is proposed. In this paper [8, 9], a mechanism for using a backup node is proposed. A part of the routing information is saved in the backup node, which can be used for recovery when the link is faulty. Researchers all over the world have proposed many topological control algorithms and scheduling algorithms as well, which can be used as a good reference. N. Li [10] proposed the LMTS topology control algorithm, which is based on the LMT of graph theory put in to use with network topology. SINGH. S [11] proposed using the Minimum Drain Rate (MDR) and the Conditional Minimum Drain Rate (CMDR) as network routing evaluation parameters to extend the battery life and path effective time of the node. The literature [12] used a stochastic geometric model to study the optimal scheduling algorithm in ad hoc networks. Personally, a random restricted area around each node in that literature proposes a distributed medium access protocol that maximizes spatial multiplexing at the expense of increasing the amount of information that a certain transmission fails, thus resulting in a lower energy utilization. Another literature [13] implements a multi-level competition protocol for spatial packaging in a distributed manner. The performance of this method can regard nearly as optimal, but the model assumes a fixed transmission distance without power control. While this paper proposes a distributed regional scheduling algorithm NMTCA (Node energy consumption analysis and Mobility Topology Control Algorithm) based on node lifecycle prediction and network topology variation. That algorithm can effectively extend the working time of the network by predicting the working time of nodes and balancing the energy of network nodes.

3 Model Establishment and Analysis on Topological Control

3.1 Model of Energy Consumption

The life cycle of the network is closely related to the remaining energy of the node. The energy is so low that it is difficult to maintain the normal operation of the network nodes, which is called a low-competitive node. Excessive usage of low-competitive nodes will cause them to go dead prematurely and seriously affect the network life cycle. In order to extend the life cycle of the network, establish an energy trust model, understand the residual energy of the node by calculating the trust value of the node energy, we should avoid excessive use of nodes with lower energy when the node communicates and predict the initial network hole, thus can extend the life cycle of the entire network.

The wireless ad hoc network model is set to have N nodes in the network. These N nodes constitute the node set V in the two-dimensional plane, and the network can be simplified into a directed graph $G = (V, E)$. Among them, V is a point collection, and E is an edge collection. An Ad hoc network is a multi-hop network consisting of n nodes. Each node has its own transmit power (assuming that the nodes use omnidirectional antennas). The energy of nodes in the network is mainly used for data transmission and reception. Each node has initial energy and can provide residual energy information at any time. The distribution of nodes is randomly distributed and will not change once deployed. $e = [u, v](u, v \in V)$ is one of the edges in E. E_{uv} indicates the energy consumption of sending a data packet from node u to node v. Considering that data collection adopts one-hop transmission mode, the sensing information and transmission energy consumption will occupy most of the energy consumption of the node. Therefore, the energy consumption in sensing and communication is mainly considered. The energy consumption model used is mainly as follows. For a given node active time t_{active}, the node's energy consumption $E(t_{active})$ can be expressed as the sum of the two parts of energy consumption:

$$E(t_{active}) = E_s(t_{active}) + E_{tx}(t_{active}) = p_s t_{active} + p_{tx} t_{tx} \tag{1}$$

p_s and p_{tx} are node-aware power and node transmission power, respectively.

Considering the power required by the node to send an r-byte packet from the node u in terms of the reception and delivery of a node, which the transmit power [14] is:

$$P_{u\_send}(d) = (\alpha_{11} + \alpha_2 \cdot d^{\mu}) \cdot r \tag{2}$$

While the power required by the node u to receive a data packet of r bytes, that is, the received power [15] is: $p_{u\_receive} = \alpha_{12} \cdot \gamma$, in which $\alpha_{11}, \alpha_{12}, \alpha_2$ is a constant, related to the electronic power consumption of the actual transceiver processing data; $\mu(2 \leq \mu \leq 4)$ is the path loss index.

Then, assuming the network bandwidth is D, the time at which the node sends or receives a byte of data is: $t_u = \frac{8r}{D}$. Assuming that during a certain period of time T of the network operation, node u sends n_u data packets and receives m_u data packet, then the energy consumed by node u in time T is:

$$E_u(d) = n_u(\alpha_{11} + \alpha_2 \cdot d^\mu) \cdot r \cdot t_u + m_u \alpha_{12} \cdot r \cdot t_u \qquad (3)$$

Then the energy consumed by the network in time T is the sum of energy consumed by all nodes in the network, which is: $E_{all} = \sum_{u=1}^{n} E_u$. So based on the above discussion, if the path of node u successfully sending a packet (called a communication) to from k jumping to node v is: $u = u_0, u_1, u_2, \ldots u_{k-1}, u_k,$. The energy cost of this communication can be estimated as:

$$E_{path} = \sum_{j=0}^{k+1} E_u = \sum_{j=1}^{k+1} [(\alpha_{11} + \alpha_2 \cdot d_{u_j}^\mu) \cdot r \cdot t_j] + k\alpha_{12} \cdot r \cdot t_j \qquad (4)$$

According to Eq. (3), it can be concluded that the energy consumption of the node is directly proportional with d, m_u, n_u. And From Eq. (4), it can be concluded that the energy consumption of a communication is directly proportional to the number of route hops k, and the communication radius d of the node is also directly proportional to the transmission power of the node, also the number of packets that the node needs to process, mu, nu is directly proportional to the end-to-end throughput, which indicates that the energy consumption of the Ad hoc network is related to the transmit power of the node, the number of packet forwarding, and end-to-end throughput.

The main purpose of the NMTCA algorithm is to reduce the total energy consumption of nodes in the network and to reduce the loss of transmitted groups of data packets by predicting the link being invalid. If the ideal situation is considered, the node does not distinguish the perceived power and the transmission power, and the power remains the same during the working time. Therefore, the energy consumption of the node in the time period Δt is:

$$E_u(\Delta t) = E_u(t - \Delta t) - E_u(t) \qquad (5)$$

In Eq. (5), $E_u(t - \Delta t)$ and $E_u(t)$ represent the remaining energy of the node the moment before Δt and at the current moment. The current energy trust evaluation index for the node u is:

$$E_{trust}(t) = \frac{E_u(t) - E_u(\Delta t)}{E_u^{\max}} \qquad (6)$$

In Eq. (6), $E_u(\Delta t)$ indicates the energy consumption of the node during the time period Δt before the moment, with $E_u^{\max}$ indicating the initial energy of the node

3.2 Model of Node Position Variation

Nodes in wireless ad hoc network are self-organizing mobile nodes. It is assumed that each node has its own direction and speed. So only within a certain distance, the nodes can communicate directly. When the communication is established, along with the nodes, the distance of the adjacent nodes may change at any time. When the adjacent distance between the nodes exceeds the maximum transmission distance, the link communication will be interrupted and the data passing through it will also be lost. Therefore, if the link failure time can be predicted, topology reconstruction before the link disconnection can effectively improve the data transmission efficiency.

Suppose there are two nodes a and b, the coordinates are (a_x, a_y), (b_x, b_y), and the speeds of the two nodes are x v_1, v_2, and the motion directions are α, β, and then the formula distances of the two nodes after the time are:

$$dist(\Delta t)^2 = [(x_a + v_a \cos \alpha \Delta t) - (x_b + v_b \cos \beta \Delta t)]^2 + [(y_a + v_a \sin \alpha \Delta t) - (y_b + v_b \sin \beta \Delta t)]^2 \tag{7}$$

Assume that the node keeps moving at a constant speed until it touches the boundary of the test area and then changes direction to continue moving. At a moment Δt after the current time, the distance of the node can be calculated according to the above formula. If $dist(\Delta t)^2 \leq d^2$, where d is the effective transmission distance of the node, the link between the nodes is still valid, otherwise the link will be invalid and the route needs to be re-established.

3.3   Analysis on Influence of Topological Control

Implementing topology control technology can reduce the transmit power of the node and reduce the communication radius of the node. Equation (3) shows that the energy consumption of a single node in a communication is mainly proportional to the power of the node communication radius. It can be seen that implementing topology control can reduce the energy consumption of a single node in one communication and prolong the working time of a single node. As shown in Fig. 1, when the node transmits data at the maximum power, the communication radius of the node is 250 m. After the topology control is implemented, it is assumed that the communication radius of the node s is 150 m, which can be obtained by substituting the formula (3). The energy consumption for sending a packet is approximately 0.36 times that of the node transmitting data at the maximum transmit power (using the free-space path loss model, $\mu = 2$). Topology control reduces the transmit power of the node, and also increases the number of times the data packet is forwarded in one communication. According to formula (4), implementing topology control technology will increase the energy consumption of forwarding data in the global network. As shown in Fig. 1, when the nodes send data at the maximum power, the routing path when the node s and the node d communicate is s-1-d, and the routing path after the topology control is selected is s-1-2-5-d. According to formula (4), in one communication between node s and node d, topology control is implemented to increase the energy consumption of forwarding data.

The network that implements topology control can not only reduce the transmit power of the node and reduce the communication radius of the node, but also it can increases the number of times the data packet is forwarded in the network communication and the number of times the bottleneck node forwards the data; and when the network load is high. As the end-to-end throughput of the network implementing the topology control increases, the energy consumption of the data packets and network forwarding data that the node needs to process will also increase. Therefore, whether topology control can reduce network energy consumption and extend network lifetime is mainly related to whether topology control technology can achieve the relationship between the transmit power of the nodes in the network, the number of times of data packet forwarding, and the end-to-end throughput of reaching balanced.

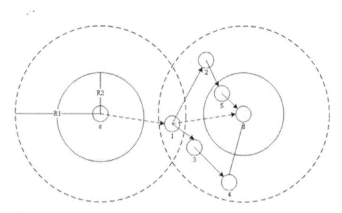

Fig. 1. Status of nodes communication (R1 = 150 m, R2 = 250 m)

4 NMTCA Algorithm

The purpose of the NMTCA algorithm is to predict the validity period of the node by the energy consumption level of the node and the moving path of the node, so that when the topology is constructed, the validity period of the node is taken into consideration, and the path fracture occurring in the node movement can be predicted, thereby enabling. The backup path or the standby node or the reconstruction link is used before the path fails, and the energy balance effect is achieved while ensuring the stability of the data forwarding task, reducing the path failure rate and prolonging the working time and network lifetime of the node.

4.1 Relevant Definition

Definition 1: (link stability index) Let the nodes u, v move at a uniform speed and along their respective fixed directions, and the relative distance between the nodes will change with the movement of the node. The degree of change depends on the node's speed of movement and the direction of motion, when the relative distance between two nodes is too far which is greater than its communication radius, it will be unable to communicate and lead to link instability, so define the link stability index as follows:

$$STAI = \frac{\text{dist}_{uv}(t) + \cos\alpha \cdot dist_{uv}(t + \Delta t) + \cos\beta \cdot dist(t + \Delta t)}{d} \tag{8}$$

The *STAI* index is related to the velocity, direction, motion time, and communication radius of the node. The larger the value of the index, the less likely the link to break.

Definition 2: (link weight) On one hand, the link weight between the nodes depends on the energy of the two nodes, their location and direction in the future. A node with a higher available remaining energy can act as a forwarding node with setting in a smaller link weight. On the other hand, nodes with lower remaining energy should avoid overused as much as possible, which the link weights will be larger for them.

$$weight(uv) = (\frac{1}{E_u} + \frac{1}{E_v}) \cdot [\alpha(\frac{E(u)}{E_0(u)})^x + \beta(\frac{E(v)}{E_0(v)})^y] + \gamma(STAI_{uv})^z \qquad (9)$$

In the formula, E_u and E_v represent the residual energy of the nodes u and v, respectively, x and y are weights, which are inversely proportional to the remaining energy, representing the initial energy, and, α, β, γ are the selection parameters, which sum is 1.

4.2 Algorithm in Topological Optimization

The goal of the NMTCA algorithm is to balance the energy of nodes in the wireless ad hoc network, extend the network life cycle, predict the link stability between the two nodes, and update route to the link will be invalidated in time, thereby reducing the failure in data packet transmission. In the process of constructing the network topology, local network topology information is obtained to ensure the strong connectivity of the topology and minimum cost. The algorithm includes four main stages: node information collection, network topology creation, network topology optimization, and network topology maintenance.

(1) Node information collection

Firstly, the node broadcasts a Hello () data packet, which contains the main information of the node, and nodes position, the current energy status, the direction and speed of the node. When other nodes in the network receive the Hello () data packet, they return an Answer () data packet and add the node information to the neighbor node information table of the current node. After receiving the feedback Answer () data packet, the source node also adds the information to the neighbor node information table. The paths of the two nodes can be established, and the path weights Weight () between the nodes can be calculated according to the acquired node information and the mutual speed and position of the nodes.

(2) Network topology creation

The goal of the algorithm is to construct a network topology that can predict link failure and extend the life cycle of the network. The node first discovers the neighbor nodes using the maximum transmit power. For any node $u \in V$, computes the weights of the node u and neighboring node v to obtain a weighted undirected figure. According to the weight of each edge, a minimum spanning tree is constructed between the clusters G and within the cluster, and the minimum spanning tree of all nodes constitutes a topology construction map $G_{max}(V, E_{max})$. Finally, the distance of the shortest path $path(u, v)$ in the topology G_{max} is calculated as $G'(V, E)$. As a result, the transmission power of each node is determined by the generated topology G'. Since the link weight function changes with the decrease of the remaining energy of the current node, the constructed topology map G' dynamically balances the energy consumption of the nodes in the network. The detailed steps are as follows:

Step 1: Construct the graph G, and set the transmission power of each node to the maximum value P_{max}. If the node pair (u, v) can communicate with each other, add the direct link $\overline{uv}$ to the figure $G(V, E)$. The nodes in the same cluster in the figure G have the same cluster ID, which $C(u)$ is used to represent the cluster ID of the node u. Then copy all the links in the diagram G to the diagram G_{max}.

Step 2: Build the diagram G_{max} and G'. Copy the map G's all the links to the map $G_{max}(V, E_{max})$. Each link weight is determined based on the transmission power of the two-end nodes with its helping node and the current remaining energy level. According to Eq. (6), the link weights in the network are obtained, then the MST algorithm is applied between the clusters and establish a minimum spanning tree, and the links with less energy consumption will be reserved. In each cluster of the graph G_{max}, the MST algorithm is again for optimization and reduction of the number of directly connected links. For each edge of $e(u, v)$ in E_0, calculate the distance of the shortest path $path(u, v)$ in the topology G_{max}. If it is $path(u, v) > t * |uv|$, e will be added into E_{max} and finally construct $G'(V, E)$. Here, the MST weight function takes into account the energy consumption of the nodes at both ends and current remaining energy. The average energy consumption of the nodes in the graph is reduced after the topology optimization in the cluster. The dynamic change of the current residual energy of the node in G' causes the network topology to change, and the overall energy consumption of the network is balanced.

(3) Network topology optimization

The topology optimization includes two aspects: link two-way transmission and transmission power adjustment. Link two-way transmission is achieved by bidirectional processing of unidirectional links occurring in topology construction; transmission power adjustment is based on the farthest transmission link distance of adjacent one-hop neighbor nodes, setting a 120% power according to the standard. The purpose of setting the transmission power by 120% is to ensure that even if the adjacent nodes are far away from each other during the movement of the node, it is still guaranteed that the data can be effectively transmitted on the link in a certain period of time. And Link invalidity will not occur immediately.

(4) Network topology maintenance

In order to avoid that the node cannot communicate with neighboring nodes due to energy exhaustion, or the neighboring node movement excessing the communication distance lead to the break of link, the node is periodically maintained in the topology to calculate the link stability index $STAI$ of the adjacent link and the link weight $weight_{all}$. According to the link weight, the link will be determined whether to be re-created, then set the lowest link weight $STAI_{min}$, and re-establish the link once the link weight is less than the lowest value. Through this process, the link to be invalidated can be predicted, and the standby link is established in advance to ensure normal communication between the nodes in the topology.

5   Experiment Simulation

In order to verify the effectiveness of the algorithm, the performance of the algorithm was investigated by the NS-3 simulation platform simulation and compared with typical routing protocols. NS-3 is an excellent network simulation software, in which the module is so detailed. When simulating the network, each node is just like an empty computer, needy to install the required application, configure the protocol stack, network interface controller, underlying channel and other modules, then join the node to the network can make it starting the simulation. The simulation environment parameter settings are as follows:

Parameter	Value
Network coverage area	2000 m * 2000 m
Maximum transmission radius	250 m
Initial power	50 J
Maximum transmission power	650 mW
Maximum velocity of nodes	10 packets/s
Bytes of data packets	512 Byte
Period of simulation	720 s

The simulation scenario is as follows: In the simulation environment, 10 CBR service flows are randomly generated, and each CBR packet has a size of 512 Bytes. The simulation time is 720 s. The algorithm Hello period is set up to 5 s. In the equation for calculating the link weight, $x = 1$, $y = 10$, $\alpha = 0.34$, $\beta = 0.23$, $\gamma = 0.43$, $W = 0.67$. $n(n \in [50,250])$ nodes are randomly distributed in a two-dimensional area of 2000 m $\times$ 2000 m; the communication radius of each node when transmitting data with maximum transmit power is 250 m; the bandwidth is 2 Mb/s; We use the node energy consumption model in Sect. 3.1, where the parameters are set as: $r = 512$, $\alpha_{11} = 8 \times 10^{-4}$, $\alpha_{12} = 7.2 \times 10^{-4}$, $\alpha_2 = 8 \times 10^{-4}$, $\mu = 2$. In the simulation experiment, the parameters such as node average transmission power, network life cycle, packet delivery rate and routing cost are mainly investigated.

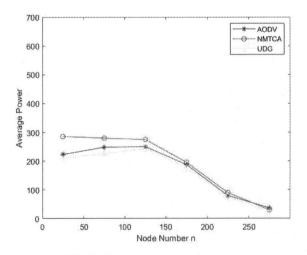

Fig. 2. Transmit power performance

The node average transmission power is the average energy consumed by the network node during the working period, indicating the overall energy consumption of the network. The lower the value is, the better the energy balance of the algorithm and the longer the normal life cycle of the network will be. Figure 2 shows the average transmit power performance of different algorithms at different scales. Compared with the two other algorithms, this study uses the energy-equalization method to effectively predict the link and reduce the energy consumption of the reconstructed link, so the node power is relatively low.

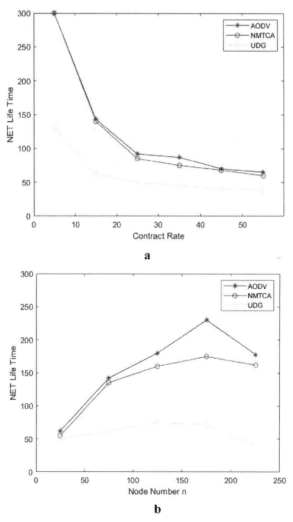

Fig. 3. (a) Relationship of contract rate with net life time, (b) Relationship of node number with net life time

The network lifetime is the period during which the entire topology network works normally. In this algorithm, network life is defined as the time until the first node in the network stops working due to energy exhaustion. Figure 3(a) shows that when the rate of contracting is low (the rate is ≤20 packets) At the same time, the network lifetime of implementing the topology control scheme is twice as much as the UDG scheme. Figure 3(b) shows the trend of network lifetime variation comparing to node density when the network packet rate is 20 packets/s. Figure 3(b) shows that the lifetime of the network increases with the increment of the node density, especially when the node density is large (the number of nodes will be 200), and the lifetime of the network implementing the topology control scheme is 2 times longer than the network lifetime of the UDG scheme. This is because when the network has a low packet transmission rate, the node's transmit power is a key factor affecting the network lifetime, and the node's average communication radius will be as same as the node's transmit power.

It can be seen from Fig. 4 that the average communication radius of the node implementing the topology control scheme is lower than the average communication radius of the UDG scheme, and the average communication radius of the node is further reduced as the node density increases. Therefore, implementing a topology control scheme can reduce the energy consumption of a single node and extend the working time of a single node, thereby prolonging the lifetime of the network.

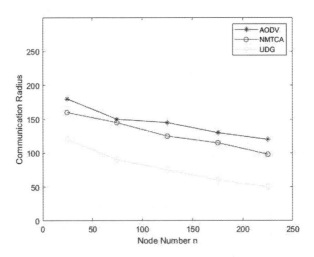

Fig. 4. Relationship of node number with communication radius

Figure 5 shows the relationship between the number of nodes and the cost of routing control. As the number of nodes increases, the link complexity increases sharply, so the routing cost also increases. The algorithm needs to perform motion prediction and energy balance control. Therefore, the required routing cost is more than other algorithms, when comparing with other performance improvements such as average transmission power and network life, and group delivery rate.

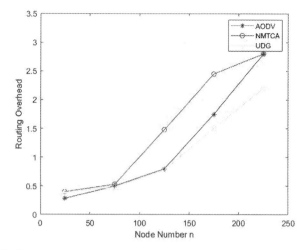

Fig. 5. Relationship of node number with the cost of routing control

6 Conclusion

Saving the energy of the node and prolonging the network working time are the main problems in the design of the wireless Ad hoc network. By evaluating the qualitative result of the link, the NMTCA algorithm chooses the link with higher stability and the active route repair in the routing selection, effectively avoiding the impact of link breaking due to node movement on data transmission, while using energy equalization mechanisms avoids excessive energy consumption of some nodes. Simulation experiments also show that compared with other algorithms, this algorithm can better adapt to the mobile change of Ad hoc network nodes and the energy efficient, also reducing the possibility of link failure, improving packets delivery rate, and extending network life cycle.

References

1. Jiang, X., Liu, M., Yang, C., Liu, Y., Wang, R.: A blockchain-based authentication protocol for WLAN mesh security access. Comput. Mater. Contin. **58**(1), 45–59 (2019)
2. Liu, Y., et al.: QTSAC: an energy-efficient MAC protocol for delay minimization in wireless sensor networks. IEEE Access **6**, 8273–8291 (2018)
3. Liu, Yu., Yang, Z., Yan, X., Liu, G., Hu, B.: A novel multi-hop algorithm for wireless network with unevenly distributed nodes. Comput. Mater. Contin. **58**(1), 79–100 (2019)
4. Cheng, H., Su, Z., Xiong, N., Xiao, Y.: Energy-efficient node scheduling algorithms for wireless sensor networks using Markov Random Field model. Inf. Sci. **329**, 461–477 (2016)
5. Zhang, Y., Xu, W., Huang, J., Wang, Y., Shu, T., Liu, L.: Cross-layer optimal power control and congestion control providing energy saving for ad hoc networks. J. Softw. **24**(4), 900–914 (2013). (in Chinese)
6. Shu, L., Zhang, Y., Yu, Z., Yang, L.T., Hauswirth, M., Xiong, N.: Context-aware cross-layer optimized video streaming in wireless multimedia sensor networks. J. Supercomput. **54**(1), 94–121 (2010)

7. Wan, Z., Xiong, N., Ghani, N., Vasilakos, A.V., Zhou, L.: Adaptive unequal protection for wireless video transmission over IEEE 802.11e networks. Multimedia Tools Appl. **72**(1), 541–571 (2014)
8. Yang, J., et al.: A fingerprint recognition scheme based on assembling invariant moments for cloud computing communications. IEEE Syst. J. **5**(4), 574–583 (2011)
9. Nekrasov, P., Fakhriev, D.: Transmission of real-time traffic in TDMA multi-hop wireless ad-hoc networks. In: IEEE International Conference on Communications, pp. 6469–6474 (2015)
10. Lakani, S., Ghaffarian, H., Fathy, M., et al.: A neighbor-based holdoff reduction scheme for distributed scheduling in wireless mesh networks. In: 6th IEEE International Conference on Wireless and Mobile Computing, Networking and Communications, pp. 733–738 (2010)
11. Agustiningsih, D.R., Teknik, F., Adriansyah, N.M., et al.: Performance analysis of coordinated distributed data scheduling schemes in wireless mesh network. In: International Conference on Computer, Control, Informatics and Its Applications, pp. 43–48 (2013)
12. Khainu, P., Keeratiwintakorn, P.: An adaptive holdoff exponent approach for coordinated distributed scheduling in WiMAX mesh networks. In: 19th IEEE International Symposium on Intelligent Signal Processing and Communication Systems, pp. 1–5 (2011)
13. Karp, B.N.: Geographic routing for wireless networks. Dissertation, Harvard University (2000)
14. Karp, B., Kung, H.T.: GPSR: greedy perimeter state less routing for wireless networks. In: International Conference on Mobile Computing and Networking, pp. 243–254. ACM (2000)
15. Camp, T., Boleng, J., Davies, V.: A survey of mobility models for ad hoc network research. Wirel. Commun. Mob. Comput. **2**(5), 483–502 (2010)

A Simulation Platform for the Brain-Computer Interface (BCI) Based Smart Wheelchair

Xinru Huang, Xianwei Xue, and Zhongyun Yuan$^{(\boxtimes)}$

MicroNano System Research Center, Key Laboratory of Advanced Transducers and Intelligent Control System of Ministry of Education and Shanxi Province, College of Information and Computer Engineering, Taiyuan University of Technology, Taiyuan 030024, China
yuanzhongyun@tyut.edu.cn

Abstract. In this paper, we develop a simulation platform for the Brain-Computer Interface (BCI) based smart wheelchair. The main contribution of our system is to implement an efficient simulation platform for the next generation wheelchair. Meanwhile, due to disabled people training with a real-natural environment that is difficult or costly to access, a means to synthetically recreate a virtual environment is necessary to assist patient training for the wheelchair driving. Emotiv neuroheadset provides an effective and reliable means of performing EEG testing during wheelchair training, which is inexpensive and delivers data of similar quality compared to other EEG recording equipment.

Keywords: Brain-computer interface · Neuronal computing · Smart wheelchair · Human factor · EEG

1 Introduction

Recently, Electric wheelchairs are widely accepted as an efficient tool to help disabled people for the improving of their moving ability. However, some of them encounter critical barriers, and hard to overcome. Especially for patients suffering from neurological injury or disease, like the high paralysis (paralyzed from the neck down), high amputation and Amyotrophic Lateral Sclerosis (ALS), most of them cannot control their wheelchair with a hand-controller. To get around such problems, it is essential to develop a neural-controller technology for a novel smart wheelchair. For example, Tobias Kaufmann and Andreas Herweg et al. proposed a wheelchair based on tactile sensation [1]. Meanwhile, with the advance in the research of brain-computer interface (BCI), many related studies are in progress. In recent years, in the article [2], a human-machine interface established by detecting a user's facial expression and head movement has been proposed. In [3], the author relies on the user's blinking activities to achieve control of wheelchair movement based on BCI technology. And the 3D printing technology is used to create a smart wheelchair, and the deep learning is used to identify the sport mode [4]. These attempts have achieved the desired results.

Development of brain machine interface (BMI) or brain computer interface (BCI) has flourished, and also some significant technological advances have occurred in recent

© Springer Nature Switzerland AG 2020
X. Sun et al. (Eds.): ICAIS 2020, LNCS 12239, pp. 257–266, 2020.
https://doi.org/10.1007/978-3-030-57884-8_23

several years. Since Leigh R. Hochberg et al. [5] and Gopal Santhanam et al. [6] demonstrate that animals or humans can use neural signals from the brain to guide a simple mechanism. The BCI technology has proven to have the potential to enable disabled people to drive intelligent devices through brain activity rather than physical activity [7]. For instance, Electric wheelchair, artificial limb and exoskeleton [9].

Research on BCI systems has mainly involved recording of electroencephalographic (EEG) signals based on surface electrodes, due to invasive techniques are still risky. Briefly, the EEG activity can be used as driver for BCI systems, and to perform certain cognitive tasks. Because of the maturation of EEG sampling and signal processing techniques, EEG based BCI research is become more reliable and accurate. Therefore, many EEG based BCI issue have been proposed. However, among of them, the Emotiv [8], which is the most widely used as a commercial product into intelligent instrument applications. Figure 1 exhibits a smart wheelchair model based on EmoEngine platform (using the Emotiv EPOC [10]). As we mentioned, Emotiv Systems is a technology of developing BCI based on EEG. Generally speaking, Emotiv System in Fig. 1 include the Emotiv EEG neuroheadset, the EPOC (EEG processor on chip) system, wireless function module and a peripheral for programming on OS (Windows, OS X or Linux). The Emotiv system is implemented many applications in simulation and training for human activity on BCI. In these roles, Emotiv system serves as a means to synthetically create the connection in Virtual Reality (VR). As demonstrated in [11].

As we all know, creating a natural environment and making real-training to each patient for the EEG-based smart wheelchair that is difficult and costly access. Even if on the unlimited hardware condition, the time cost of training for a large number of patients that is not be overcome. Therefore, in this paper, we propose a Pre-training system for the EEG-based smart wheelchair in VR. We simulate a simple training in VR, and make a control connection with the Emotiv System. By doing so, patients can make practice for EEG controller on PC before they possess a real smart wheelchair. Finally, they can as soon as better to learn how to utilize the neuro logic (EEG activity) to communicate with the BCI based instrument, and to send commands.

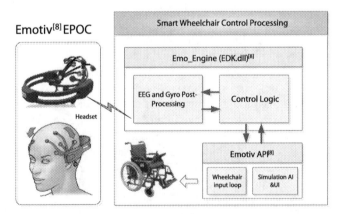

Fig. 1. Integrating the EmoEngine and Emotiv EPOC [8] with a smart wheelchair control processing.

The organization of this article is as follows. After a brief introduction in Sect. 1, Sect. 2 introduces the background and related work for this issue. Section 3 describes the implementation of our scheme. Section 4 presents the experimental results. Finally, Sect. 5 summarizes the article.

2 Background

2.1 BCI System

Since such invasive techniques are still risky, research in human BCI mainly focused on non-invasive methods for monitoring brain activity, such as electroencephalography (EEG), magnetoencephalography (MEG), near-infrared reflectance spectroscopy (NIRS) and functional magnetic resonance imaging (fMRI).

There have been many related studies of different methods based on EEG to help people with disabilities to control the movement of wheelchairs or robotic arms, which proves the feasibility of EEG in detecting human brain activity. For example, in [9], Philips et al. and Millan et al. demonstrated the EEG features of the subjects through training and experimentation between the BCI and the intelligent simulated wheelchair shared control system. The feasibility of non-invasive BCI controlled machinery was finally confirmed. The article [12] showed a 9-DOF robotic arm system that combined a 7-DOF robot arm control and a 2-DOF electric wheelchair control, which also used EEG to detect brain activity.

The position of the electrodes placed on the scalp depicted in Fig. 2 is a very common method used in EEG detection and is referred to as the 10–20 system [9]. The Emotiv EPOC headset is based on this system placing its 14 sensors to detect the wearer's brain activity.

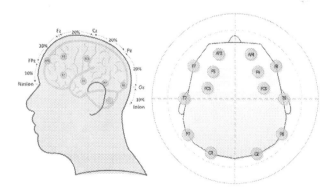

Fig. 2. EEG electrode placement.

2.2 Emotive System

The Emotiv system provides the technology to convert brain waves into digital signals and to control different digital output devices.

In [13], the author used low-cost EEG headset to establish the player's mental state and mobile game interactive environment, reflecting the good potential of EEG headset in the field of mobile applications. The interaction between EEG neuroheadset and mobile phones was also confirmed by the article [14]. The article describes a design for controlling mobile phone applications on iPhone through neural signals using inexpensive EEG headset. In [15], the author uses the Emotiv EPOC neuroheadset to read the data of the user's brain activity in real time on the basis of the brain-computer interface to achieve the purpose of controlling the robot. The experiments conducted in [16] initially explored the EEG evaluation technique using Emotiv EPOC in a virtual reality environment. Therefore, it has been experimentally verified that the Emotiv system can indeed perform different EEG control experiments through various data sources.

The Emotiv system provides four available data streams for experimentation that can be observed and output. The Mental Command can identify specific thought signals such as the movement and rotation of the control cube through the trained system, and the effect is related to the number of trainings.

The Performance Metrics without trainings contains tests for six real-time states in mind [17]: Focus refers to the measurement of the degree of concentration; Engagement is defined as the immersion of the task; Interest accounts for the degree of preference for the current task; Excitement is characterized by an increase in the physiological response; Stress is an indicator for detecting the degree of tension in the current state; Relaxation expresses a measure of the level of recovery after the end of the task.

Sentiment detection can be used to study people's opinions, emotions, assessments, and attitudes from text and reviews [18]. Likewise, Emotiv headsets have the ability to detect facial and eye muscle activity that signals can be used to identify different facial expressions such as smiles and frowns. Independent studies have verified this method is practicable [19]. The Motion Sensors transmit the head motion in the form of an image drawn by the data to the application interface of the Emotiv system. Visual tracking has important applications and very promising prospects in many fields [20]. In recent years, with the development of sensor technology and microelectronics technology, eye tracking technology has been rapid development, and has been widely used in intelligent robotics, human-computer interaction, automatic control and other fields [21].

3   Method

3.1   Signal Acquisition and Processing

The signal is acquired by a neural helmet that houses 16 hydration sensors that can be used for EEG signal acquisition. After being converted to digital form, the brain waves are processed and the results are wirelessly transmitted to the secure cloud storage.

A post-processing software called Emotiv Emo Engine identifies the user's subjective intent by extracting mind-control related EEG features and the results are presented to the application through the Emotiv Application Programming Interface (Emotiv API). One of the kits provided is Cognitiv™ Suite that recognizes the user's intent inside to manipulate the physical actions of real or virtual targets with the measurement of real-time brain messages.

The use of the Emotiv headset to control applications through the mind requires repeated training of mental commands [17]. First, the neutral state is set to a brain-relaxed activity without considering any commands, such as lounging, brain activity in this state will be contrasted with that under the other commands. Next, the training of the mobile control commands such as "forward", "backward", "left" and "right" is performed.

In Emotiv API these commands are associated with the motion of a cube, in the application under the real-time mode, by manipulating the movement of the cube to practice the command and get feedback on whether the command is valid. Alternating the training of the neutral state with other commands and performing repeated exercises is a great help to the effect of the mind control command.

3.2 Virtual Simulation Environment

Virtual environments have become effective feedback medium for studies based on brain-computer interface (BCI). The BCI technology determines and produces control signals in accordance with the user's intent from a variety of different electrophysiological signals, and has a high information transmission rate [22]. An experiment on racing speed and real-time feedback system proves that the feedback effect of neural feedback in the virtual environment is more obvious than the real environment [23].

In this experiment, the operator wears the EEG headset to control the wheelchair for tasks in a virtual simulation environment. The operator faces the simulated path taking the first-person perspective in the simulated world, and advances, draws back, turns left, and turns right through three different mental commands to drive the wheelchair from the starting point to the target according to a predetermined path.

It offers a variety of obstacles requiring the operator to control the wheelchair to avoid on the road in the mission. The user must always be focused on taking appropriate actions. Figure 3 illustrates the reference situation and abovementioned egocentric virtual simulation environment. The virtual screen in front of the operator and the wheelchair will be presented on the output interface to facilitate observation of the movement of the wheelchair under mental orders.

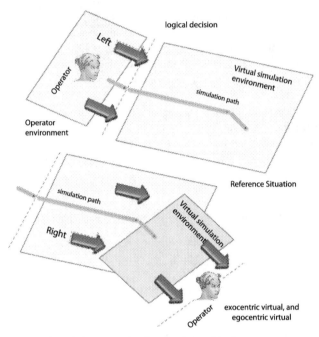

Fig. 3. Reference situation, and egocentric virtual relationships between operator and simulation environment

3.3 Implementation

The procedure of experiment implementing and data handling has shown in Fig. 4. At the beginning, EEG signals captured by the neuroheadset are input and processed to provide the basis of judgement. Before the participant who has finished the training of mental commands starts to control a certain motion of the wheelchair in his mind, the neutral recognition would be determined whether it is detected or not.

When the neutral state is detected, pre-set states will be judged, which refers to brain motion features of the operator who triggers the "forward", "back", "left", and "right" orders. During the process of issuing the command, the brain activity will be presented in the application interface as an image of the brain wave. The next cycle will continue to start from detecting a neutral state.

If the neutral state is not detected, the next decision of the control command cannot be made. The program will go directly to the next cycle to re-detect the neutral state of the input data.

Through the flow described, the entire experimental implementation constitutes a complete logical cycle in order to collect real-time EEG fluctuations in each stage to facilitate analysis of its results. The validity of each command will also be linked to the output signal to verify the feasibility and effectiveness of the mind control.

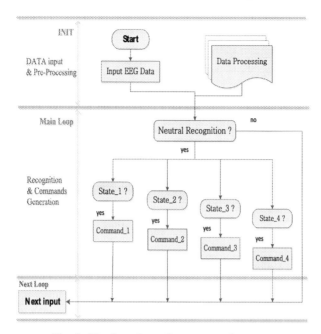

Fig. 4. The flow chart of our proposed strategy.

4 Result

The subject controls the wheelchair to avoid obstacles, and the task of reaching the destination through the preset path is succeeded. Throughout the experiment, participants' thought control was very energy intensive. Therefore, in the pre-training, repeating the training over and over again will be very helpful for the participants to master the command in mind.

Observing the EEG results of the output, we found that the electrodes at various locations on the scalp detected changes. In particular, the activity at the AF3, FC5, T7 and F8 positions are quite intense, indicating that the movement of the frontal, central and anterior temporal part of the subject is greatly active during driving the wheelchair.

Figure 5 shows an image of the observed virtual environment, wheelchair, and its movement process at the application interface. At the same time, the brain wave signals detected by each channel are also output, and it can be clearly seen which parts are most active.

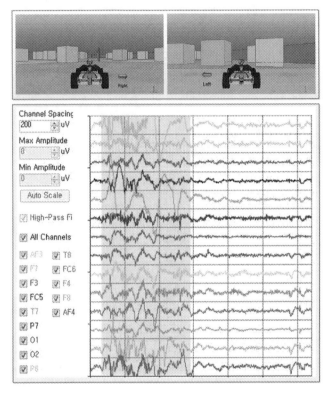

Fig. 5. Examples of the simulator imagery, and samples of all EEG channels.

5 Conclusion

This paper established a simulation platform for the brain-computer interface based smart wheelchair by the neuroheadset. This virtual platform simulates the intelligent wheelchair to achieve independent activities in an indoor environment through the mind control of patients with limited mobility and can avoid various obstacles.

Experiments by operators in virtual environments have demonstrated the feasibility of using mindsets to control smart wheelchairs with moderately priced EEG recording devices and have found a path full of possibilities for people with disabilities who cannot use limbs, facial expressions and other physical movements.

During the trial, eye movements and emotional states were also found to be similar to mind control and can be used to issue instructions. In the future, the sensitivity of controlling the movement of wheelchairs with ideas should be further improved. Subsequent research can extend the capabilities of smart wheelchairs with brain-computer interaction, such as speed control, auto-navigation, and more.

Acknowledgements. The research was partially supported by the National Natural Science Foundation of China (No. 51622507, 61471255, 61474079, 61501316, 51505324), Excellent Talents Technology Innovation Program of Shanxi Province of China (201805D211020), Beijing Natural Science Foundation (7202190).

References

1. Kaufmann, T., Herweg, A., Kübler, A.: Toward brain-computer interface based wheelchair control utilizing tactually-evoked event-related potentials. J. Neuroeng. Rehabil. **11**(1), 7 (2014). https://doi.org/10.1186/1743-0003-11-7
2. Rechy-Ramirez, E.J., Hu, H.: An electric wheelchair controlled by head movements and facial expressions: uni-modal, bi-modal, and fuzzy bi-modal modes. In: Handbook of Research on Investigations in Artificial Life Research and Development, pp. 1–30. IGI Global (2018)
3. Lahane, P., Adavadkar, S.P., Tendulkar, S.V., Shah, B.V., Singhal, S.: Innovative approach to control wheelchair for disabled people using BCI. In: 2018 3rd International Conference for Convergence in Technology (I2CT), pp. 1–5. IEEE (2018)
4. Zgallai, W., et al.: Deep learning AI application to an EEG driven BCI smart wheelchair. In: 2019 Advances in Science and Engineering Technology International Conferences (ASET), pp. 1–5. IEEE (2019)
5. Hochberg, L.R., et al.: Neuronal ensemble control of prosthetic devices by a human with tetraplegia. Nature **442**(7099), 164 (2006)
6. Santhanam, G., Ryu, S.I., Byron, M.Y., Afshar, A., Shenoy, K.V.: A high-performance brain–computer interface. Nature **442**(7099), 195 (2006)
7. Curran, E.A., Stokes, M.J.: Learning to control brain activity: a review of the production and control of EEG components for driving brain–computer interface (BCI) systems. Brain Cogn. **51**(3), 326–336 (2003)
8. Emotiv EPOC. https://emotiv.gitbook.io/epoc-user-manual/. Accessed 08 Nov 2019
9. Frolov, A.A., et al.: Post-stroke rehabilitation training with a motor-imagery-based brain-computer interface (BCI)-controlled hand exoskeleton: a randomized controlled multicenter trial. Front. Neurosci. **11**, 400 (2017)
10. Galán, F., et al.: A brain-actuated wheelchair: asynchronous and non-invasive brain–computer interfaces for continuous control of robots. Clin. Neurophysiol. **119**(9), 2159–2169 (2008)
11. Leeb, R., Friedman, D., Müller-Putz, G.R., Scherer, R., Slater, M., Pfurtscheller, G.: Self-paced (asynchronous) BCI control of a wheelchair in virtual environments: a case study with a tetraplegic. Comput. Intell. Neurosci. **2007**, 79642 (2007)
12. Palankar, M., et al.: Control of a 9-DoF wheelchair-mounted robotic arm system using a P300 brain computer interface: initial experiments. In: 2008 IEEE International Conference on Robotics and Biomimetics, pp. 348–353. IEEE (2009)
13. Coulton, P., Wylie, C.G., Bamford, W.: Brain interaction for mobile games. In: Proceedings of the 15th International Academic MindTrek Conference: Envisioning Future Media Environments, pp. 37–44. ACM (2011)
14. Campbell, A., et al.: NeuroPhone: brain-mobile phone interface using a wireless EEG headset. In: Proceedings of the Second ACM SIGCOMM Workshop on Networking, Systems, and Applications on Mobile Handhelds, pp. 3–8. ACM (2010)
15. Vourvopoulos, A., Liarokapis, F.: Robot navigation using brain-computer interfaces. In: 2012 IEEE 11th International Conference on Trust, Security and Privacy in Computing and Communications, pp. 1785–1792. IEEE (2012)
16. Salisbury, D.B., Dahdah, M., Driver, S., Parsons, T.D., Richter, K.M.: Virtual reality and brain computer interface in neurorehabilitation. In: Baylor University Medical Center Proceedings, vol. 29, no. 2, pp. 124–127. Taylor & Francis (2016)
17. Emotiv BCI. https://emotiv.gitbook.io/emotivbci/. Accessed 08 Nov 2019
18. Zhang, Y., Wang, Q., Li, Y., Wu, X.: Sentiment classification based on piecewise pooling convolutional neural network. Comput. Mater. Contin. **56**(2), 285–297 (2018)
19. Sinyukov, D.A., Li, R., Otero, N.W., Gao, R., Padir, T.: Augmenting a voice and facial expression control of a robotic wheelchair with assistive navigation. In: 2014 IEEE International Conference on Systems, Man, and Cybernetics (SMC), pp. 1088–1094. IEEE (2014)

20. Gao, Z., et al.: Real-time visual tracking with compact shape and color feature. Comput. Mater. Contin. **55**(3), 509–521 (2018)
21. Liu, Z., Wang, X., Sun, C., Lu, K.: Implementation system of human eye tracking algorithm based on FPGA. Comput. Mater. Contin. **58**(3), 653–664 (2019)
22. Wolpaw, J.R., Birbaumer, N., McFarland, D.J., Pfurtscheller, G., Vaughan, T.M.: Brain–computer interfaces for communication and control. Clin. Neurophysiol. **113**(6), 767–791 (2002)
23. Han, D.K., Lee, M.H., Williamson, J., Lee, S.W.: The effect of neurofeedback training in virtual and real environments based on BCI. In: 2019 7th International Winter Conference on Brain-Computer Interface (BCI), pp. 1–4. IEEE (2019)

A Learning Resource Recommendation Model Based on Fusion of Sequential Information

Ruofei Zhu, Zhengzhou Zhu$^{(\boxtimes)}$, and Qun Guo

School of Software and Microelectronics, Peking University,
Beijing 100871, People's Republic of China
zhuzz@pku.edu.cn

Abstract. The accuracy of learning resource recommendation is crucial to realize precise teaching and personalized learning. Recently, deep learning based models have been introduced in recommendations. Due to flexibilities of deep networks, several kinds of architectures have been proposed to tackle different problems recently. With the development of computing power, research on sequence data is increasing, which is one of the most important features for learning user's interests. In order to improve accuracy of learning resource recommendation, we introduce deep learning and propose a deep sequence-fusion network (DSFN) based on fusion of multiple sequential data which is deemed to be more effective in learning resource recommendation than in other field. We take the self-attention mechanism as the core part to design the auxiliary subnet and the prediction subnet for fusing multiple sequential data. The proposed model works with joint training and detached predicting. When training, the two subnets both work by respectively compressing the sequential data into self-attention mechanism. Then the sequences output by self-attention layer flow through multiple feedforward neural networks to produce expected targets. When predicting, there is only prediction subnet working, it performs joint prediction of multiple sequential information by fusing the vector learned on training. The experimental results show that compared with the traditional user-based collaborative filtering algorithm, the proposed model improves the accuracy and recall rate by 20.5% and 13.6% respectively.

Keywords: Learning resource recommendation · Sequential data · Attention mechanism · Deep neural network

1 Introduction

Online learning brings convenience to learners, but the explosive educational resources with uneven quality seriously affects efficiency of online learning. With the development of machine learning, learning resource recommendation technology becomes one of the effective methods to solve the problem of learning resource explosion [4,8]. The most popular methods includes data mining,

X. Sun et al. (Eds.): ICAIS 2020, LNCS 12239, pp. 267–278, 2020.
https://doi.org/10.1007/978-3-030-57884-8_24

content-based and hybird model. Recommendation System actively pushes personalized learning resources to learners [13]. The pushing process is invisible or low-visible to learners, which makes learning process more effective [2]. As the recommendation results is produced on modeling of big data, it also improving learning quality.

Traditional personalized recommendation algorithm technologies include collaborative filtering technology and matrix decomposition technology have been widely used in many areas [1,14]. But the inherent differences between learning resource recommendation and traditional recommendation make it difficult to completely migrate classical technologies to the field of learning resource recommendation. Taking movie recommendation as an example, the audiences' scores to the movies are mostly static because the evaluations do not change over time. But the situation is different in the learning resource recommendation because learning activities are mostly dynamic. The learners' demand for learning resources changes as the courses progress. Despite rate of progress, it is still necessary to consider the dynamic factors like knowledge mastery and learning status [4]. Therefore, we suppose that the dynamic factor which can be learned in sequential data is more important in learning resource recommendation. Thus, we introduce two kinds of sequential data, which are sequential behavior data and sequential resource data, to model dynamic information.

The deep network has strong ability to express sequential information but the related researches is scarce in learning resource recommendation. In this paper, we use the self-attention mechanism of deep network as core part to fuse sequential behaviors data and sequential resources with the help of joint training and detached prediction. The main contributions of this article are:

1. We point out the limitation of traditional recommendation algorithm on processing sequential data, and design model based on multi sequential data. The DSFN uses two kinds of sequential data include sequential resource data and sequential behavior data.
2. We apply the deep neural network to learning resource recommendation , which is proved that it can comprehensively captures the sequential information. In addition, DSFN uses the auxiliary deep subnet to introduce sequential behavior data and uses the prediction deep subnet to introduce sequential resource data.
3. We use joint training to make fusion of multi sequential data more flexible and use detached prediction to make prediction more efficient.

2   Model Architecture

The concept definitions used in this paper are given below:

Definition 1. *Learner is the student who uses the online learning system to study, written as u_i, $0 < i \le q$, q is the total number of learners.*

Definition 2. *Sequential resource data is defined as, a list of online resource viewing records produced by learners and arranged over time, written as* $r^{u_i} = \{r_1^{u_i}, r_2^{u_i}, \cdots, r_l^{u_i}\}$, *l is the length of sequential resource list.*

Definition 3. *Sequential behavior data is defined as, the system interaction behavior generated during the online learning process and arranged over time, written as* $s^{u_i} = \{s_1^{u_i}, s_2^{u_i}, \cdots, s_k^{u_i}\}$, *k is the length of sequential behavior list.*

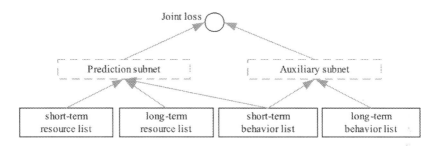

Fig. 1. Sketch view of DSFN model, the two subnets are active while training.

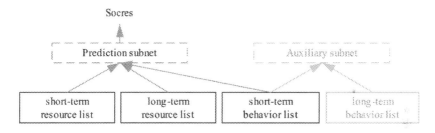

Fig. 2. When predicting, only the prediction subnet is working. Scores produced by softmax function are of corresponding resource items.

The DSFN model consists of two subnets, as shown in Fig. 1. One part is the auxiliary subnet, which is responsible for processing the sequential behavior data. The other part is the prediction subnet, which is responsible for processing the sequential resource data.

2.1   Prediction Subnet

The prediction subnet is responsible for processing the sequential resource data during training. When predicting, it combines the two kinds of sequential data to generate a recommendation list. The detailed structure is shown in the predicting subnetwork section of Fig. 3.

First, the sequential behavior data r is mapped to one-hot vector, then the one-hot vector was compressed to a relatively short vector $e \in R^k$ through a vector compression layer, where k is a hyperparameter. The compression layer is equal to the matrix $E \in R^{n \times k}$, each row corresponding to the resource item r_i. The sequence of vectors is split into long-term and short-term sequences to get fixed length vectors for further processing. We compress the long-term sequence to a single vector $v^{long-term}$. Next, the vector $v^{long-term}$ and the fixed-length short-term sequence are connected to form a vector matrix $G \in R^{(p+1) \times k}$, where p is length of short-term list. Finally, the matrix will feed to self-attention layer [7]. Supposing $Q = GT_q$, $K = GT_k$, $V = GT_v$, where T_v, T_k, T_q is parameter matrix, the formula is shown as (1):

$$Attention(Q, K, V) = softmax(\frac{QK^T}{\sqrt{d_k}}V) \tag{1}$$

Then we splicing each row of the output matrix to a sequence and feed the sequence to multiple fully connected neural networks, it produces the prediction results through the last softmax layer. When predicting, the recommended list is produced according to the probability value. When training, the loss of the predicted subnet is one part of joint loss.

2.2   Auxiliary Subnet

The auxiliary subnet is responsible for processing sequential behavior data. The entire process is similar to the prediction subnet, thus we won't describe it in detail. The joint loss is calculated by simply adding up for efficiency. The main differences between the two subnet are i) The input of the auxiliary training subnet is sequential behavior data. ii) The prediction subnet works in training and predicting but the auxiliary subnet only works during training.

2.3   Sequence Compression

For each learner, the list of the above two types of data will gradually increase. We choose a threshold which divides sequences into long-term and short-term data. We compress long-term data, and keeps short-term data unchanged. Long-term sequence compression can be seen as compressing the matrix $F \in R^{l \times k}$ into a vector $f \in R^k$, where l is the length of long-term sequence. We provide different method as follows.

Max Compression. The max compression capture the most obvious feature by simply selecting the largest item in the matrix. In this paper, the expecting output is a single vector from the input of a matrix. Therefore, the original form of maximum compression cannot be directly applied. We propose an improved max compression as shown in (4).

$$v_i^{max} = \sum_{j=1}^{l} I(v_i^j)v_i^j, 1 < i \leq k \tag{2}$$

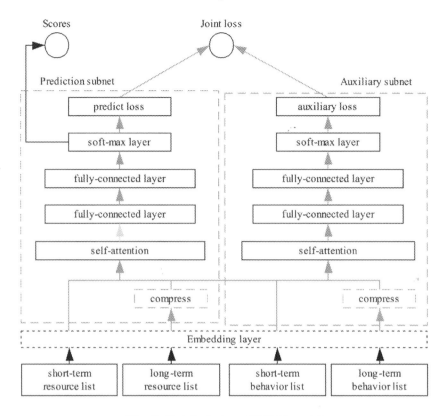

Fig. 3. Detailed view of DSFN model

$$I(v_i^j) = \begin{cases} 1, & v_i = max(v_i^1, v_i^2, ..., v_i^n) \\ 0, & else \end{cases} \tag{3}$$

Average Compression. The average compression allows all the input points to contribute. Compared to the maximum pooled direct discard method, the information of the input feature is more comprehensively captured. Similarly, in order to match the input and output formats, the original average compression method is improved as shown in (6), where n is the input sequence length.

$$v_i^{average} = \frac{\sum_{j=1}^{l} v_i^j}{l}, 1 < i \le k \tag{4}$$

Weighted Compression. Whether it is average or max compression, the input features are rudely combined. The former directly discards non-maximum features, and the latter ignores the differences between features. Taking the background of this article as an example, the two distantly viewed recordings have different effects on the learner's current interests which should have different weight. In order to solve this problem, we combine the features according to the weight of time decay. The formula is shown as.

$$v_i^{weighted} = \sum_{j=1}^{l} w_i v_i^j, 1 < i \leq k \tag{5}$$

$$w_i \propto |time_i - time_{current}| \tag{6}$$

In (4) (6), v_i is the i-th element of target embedding. k is the hyperparameter of embedding size. w_i is the coefficient of time decay.

3   Experiments

Firstly, the experimental part descripts the experimental environment in detail. Then debugs the hyperparameters to optimize the model. Finally, the proposed model is compared with the user-based collaborative filtering algorithm (UCF). And we give necessary discussions are in the corresponding part.

3.1   Experimental Environment

The experiment is based on online learning platform we develop. The service is based on the classic J2EE technology framework and deployed on the Alibaba Cloud server. The video resources are stored on the Alibaba Cloud CDN, which provides smooth access for students. This system is for 332 students from Peking University with a second bachelor's degree. The data is generated by the learner during the online learning process and collected for about 59 weeks. Examples of specific records are shown in Table 1.

Table 1. Examples of records.

UserID	Resource	Behave	Start_time	End_time	Page_time	Happen_time
170**80	g_02_03	8	00:02		03:16	2018-04-21 15:19:13
180**65	b_05_02	1				2018-11-23 12:40:04
180**21	b_05_02	5	17:51	19:52	26:30	2018-11-22 19:50:01

The learning resources are collected from the course of software engineering from three universities. The whole videos are cliped according to the knowledge points. We totally get 1288 different teaching videos. The learning behaviors need to be predefined cause we have to develop corresponding modules in the system to capture it. We totally define 15 behaviors that students generate while watching videos. The specific behavior is defined as shown in Table 2.

Table 2. Learner behavior type

Behavior ID	1	2	3	4	5
Meaning	Open page	Start video	Pause video	Restart video	Fast forward
Behavior ID	6	7	8	9	10
Meaning	Rewind	Leave page	Hide page	Restore page	Full screen
Behavior ID	11	12	13	14	15
Meaning	Exit full screen	Mute	Turn off mute	Adjust volume	Video ends

3.2 Determination of Experimental Parameters

Threshold of Spliting Sequential List. Sequential records need to be divided into long-term and short-term for the purpose we discuss in Sect. 2. The threshold should depend on real distribution to distinguish real-time information from long-term information. We assume that the students are influenced by their behaviors in recent three days. After threshold of time is determined, we find the threshold of learning resources and learning behaviors for each student according to their daily data. Records are randomly selected from fifth students in one month. We calculate the average number of every student in one day as follows.

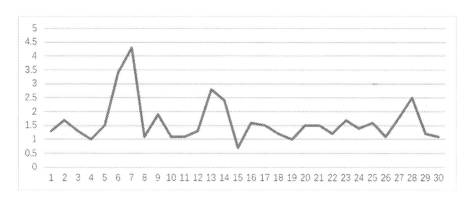

Fig. 4. The average number of learning videos a student watchs per day

As can be seen from the Fig. 4, the number of videos watched by the learners are relatively stable in every week. One student produces an average of two to three records of watching in working day, but the number probably be doubled in non-working day. Considering generalization, we take threshold of learning resources length as eight.

The data we use in Fig. 4 is collected from the same students and time interval as Fig. 3. We can see from Fig. 4 that learners average produce about fifteen records of learning behavior in working day. Thus we select 40 as threshold to split the learning behavior record.

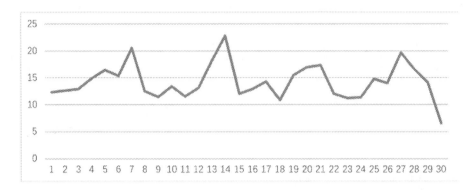

Fig. 5. The average number of behaviors a student produces per day

Comparison of Compression Methods. We compare three methods of compression. We take one hundred students randomly and use ten-fold cross-validation to compare three methods. It can be seen in the Table 3 that weighted compression acts better than others in accuracy rate, but average compression has highest recall rate. This can be explained that average compression makes every feature contributes in a same way, which is focus on generalization. But the weight compression has abilities in seeking the most proper parameter for every feature which focus on memorization.

Table 3. Accuracy and recall rate on different compression method

	Recall rate	Accuracy rate
Max compression	0.513	0.215
Average compression	0.562	0.208
Weight compression	0.497	0.232

Length of Recommend List. Length of recommend list is a crucial parameter of recommend system which influences users' experience and recommendations. We do experiments on one hundred students by using weight attenuation for compressing. It can be seen in Fig. 5 that the accuracy rate shows a downward trend and the recall rate shows an upward trend as the length of the recommendation list increases. In addition to the impact of different calculation methods, it also can be explained that learners would gradually lose the patience as length of recommend list growing. There are obvious changes occurred in length of fifteen of accuracy rate and 10 of the recall rate. In order to balance effect, we choose 10 as the length of the recommend list.

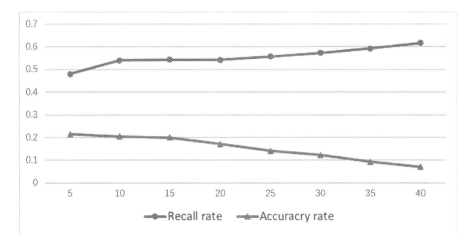

Fig. 6. DSFN accuracy and recall rate varies with recommend list length

3.3   Comparison of DSFN with Traditional UCF Recommendations

As shown in Fig. 7. that an average accuracy increase of 20.5% compare with the UCF algorithm. In addition, the results show that the accuracy of UCF algorithm is fluctuating. The reason is that UCF simply relies on the resources learners have viewed. It is difficult for UCF to model long-term information of each resources, which means it is sensitive to the recent records. While DSFN that relies on fusion of sequential data is relatively stable. But DSFN is limited by its huge amount of parameters, the effect of DSFN is obviously not as good as the traditional UCF algorithm because the training data of the network is limited at the beginning of the semester. To avoid that, we pre-trained the subnet with records that the system collects in last two years.

In Fig. 8, the average recall rate of DSFN increased by 13.6% compared with the UCF algorithm. It can be seen that UCF is as stable as DSFN except a fluctuation in the midterm. This phenomenon is due to the influence of the midterm which makes the learner's online viewing behavior has abruptly changed. Recall rate is focus on generalization which UCF can obtain it with historical records. Because UCF doesn't care the differences between different records, which means they have the same weight for predicting. After sufficient training, DSFN's recall rate is obviously superior to the traditional UCF algorithm in terms of effect and stability.

In summary, we propose a deep sequence-fusion network based on fusion of multi sequential data, we take the self-attention mechanism as the core part and design the auxiliary subnet and the prediction subnet for fusing multi sequential data. The DSFN proposed in this paper is superior to the traditional UCF algorithm in accuracy, recall rate and stability.

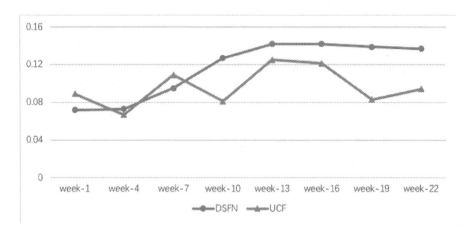

Fig. 7. The accuracy rate between UCF and DSFN

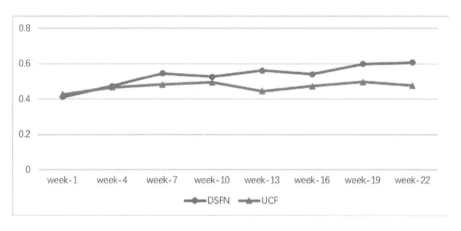

Fig. 8. The recall rate between UCF and DSFN

4 Related Work

The traditional recommendation algorithm has been widely used in learning resource recommendation. For example, A'lvaro Tejeda-Lorentea et al. [5] developed a personalized recommendation system called "AyudasCBI", which uses traditional collaborative filtering algorithm to provide individualized guidance based on students' test scores. Besides, the system provides a warning function which collects the students' negative feedback and reminds them by e-mail after reaching a certain threshold, thus enhancing personalized education. The final experiment showed that students who participated in the study had better academic performance than those who did not. Denley et al. [6] developed an online learning resource recommendation system called "Degree Compass", which uses traditional collaborative filtering algorithms too. The differences are

Denley T et al. have taken students' historical test scores into consideration, which improve the stability of recommendation. The experimental results show that the scores of the students in the control group have improved steadily through the system. All above are based on traditional algorithm with little improvements.

But traditional algorithms have various shortcomings. Wan et al. [11] proposed a hybrid filtering recommendation method for the sparse problem of interpersonal relationships among learners. The method establishes a learner influence model, combines the self-organized recommendation strategy and the sequential mode strategy to recommends learning resources to the learner. Lee et al. [8] improves the FP-Growth algorithm, which compresses the frequent items into a frequent pattern tree, and then divides the compressed data into relevant conditional data for each frequent item respectively, which also improves the efficiency of the algorithm. Wang et al. [9] proposed a hierarchical knowledge structure to model vocabulary, and collected real learning materials from the Internet to introduced a hybrid method to recommend resources. Such works have achieve state-of-the art for improving classical methods. But it is unchanged that they lack the ability in processing the dynamic information such as sequential data.

Dynamic information such as sequential data are gradually introduced in learning resource recommendation. For example, Koren et al. [10] added time vectors to the similarity vector, Wang et al. [12] used sequential information for user clustering to support collaborative filtering. They have made efforts on modeling dynamic information, but restricted by traditional algorithm. Recently, deep learning have been widely applied and gain tremendous achievements, which has inherent advantages for sequential information haven't been fully utilized. We introduce self-attention, one of the most popular method in deep learning, into learning recommendations to process dynamic information. It is verified to be valid in our experiments.

5   Discussion and Future Work

This paper proposes a recommendation model based on deep learning which fuses multiple sequential information. The model has a significant improvement in accuracy and recall rate compared with the traditional collaborative filtering method. However, there are still some shortcomings. For example, some of the hyper-parameters that directly influence the model architecture, such as threshold of splitting list and sequence length, mainly rely on human experience due to the lack of real data, which could be harmful. We are planning to promote our online learning system to other universities to obtain more data from different users. Deeper works such as exploration of better compression methods and optimization of model structure will be supported by enough amount of users.

Acknowledgments. This paper was supported by National Key Research and Development Program of China (Grant No. 2017YFB1402400), Ministry of Education

"Tiancheng Huizhi" Innovation Promotes Education Fund (Grant No. 2018B01004), National Natural Science Foundation of China (Grant No. 61402020, 61573356), and CERNET Innovation Project (Grant No. NGII20170501).

References

1. Liu, G., Meng, K., Ding, J., et al.: An entity-association-based matrix factorization recommendation algorithm. Comput. Mater. Continua **58**(1), 101–120 (2019)
2. Thongsri, N., Shen, L., Bao, Y.: Investigating academic major differences in perception of computer self-efficacy and intention toward e-learning adoption in China. Innov. Educ. Teach. Int., 1–13 (2019)
3. Klašnja-Milićević, A., Vesin, B., Ivanović, M., et al.: Recommender Systems in E-Learning Environments. E-Learning Systems (2017)
4. Cui, L.Z., Guo, F.L., Liang, Y.J.: Research overview of educational recommender systems. In: The 2nd International Conference. ACM (2018)
5. Tejeda-Lorente, Á., Bernabé-Moreno, J., Porcel, C., et al.: A dynamic recommender system as reinforcement for personalized education by a fuzzly linguistic web system. Procedia Comput. Sci. **55**, 1143–1150 (2015)
6. Denley, T.: Degree compass: a course recommendation system. Multimedia (2013)
7. Vaswani, A., Shazeer, N., Parmar, N., et al.: Attention is all you need. In: Advances in Neural Information Processing Systems, pp. 5998–6008 (2017)
8. Lee, M.S., Kim, E., Nam, C.S., et al.: Design of educational big data application using spark. In: 2017 19th International Conference on Advanced Communication Technology (ICACT). IEEE (2017)
9. Wang, S., Wu, H., Kim, J.H., et al.: Adaptive learning material recommendation in online language education (2019)
10. Koren, Y.: Collaborative filtering with temporal dynamics. Commun. ACM **53**(4), 89 (2010)
11. Wan, S., Niu, Z.: A hybrid E-learning recommendation approach based on learners' influence propagation. IEEE Trans. Knowl. Data Eng., 1 (2019)
12. Wang, X., Zhu, Z., Yu, J., et al.: A learning resource recommendation algorithm based on online learning sequential behavior. Int. J. Wavelets Multiresolution Inf. Process. (2018)
13. Hou, M., Wei, R., Wang, T., et al.: Reliable medical recommendation based on privacy-preserving collaborative filtering. Comput. Mater. Continua **56**(1), 137–149 (2018)
14. Wang, G., Liu, M.: Dynamic trust model based on service recommendation in big data. Comput. Mater. Continua **58**, 845–857 (2019)

A Convolutional Neural Network-Based Complexity Reduction Scheme in 3D-HEVC

Chang Liu[1,2,3], Ke-bin Jia[1,2,3], and Peng-yu Liu[1,2,3(✉)]

[1] Beijing University of Technology, Beijing 100124, China
liupengyu@bjut.edu.cn
[2] Beijing Laboratory of Advanced Information Networks, Beijing 100124, China
[3] Beijing Key Laboratory of Computational Intelligence and Intelligent System, Beijing 100124, China

Abstract. View synthesis optimization (VSO) is one of the core techniques for depth map coding in three dimensional high efficiency video coding (3D-HEVC). It improves the quality for synthesized views, while it also introduces heavy computational complexity caused by the calculation of synthesized view distortion change (SVDC) in practice. To reduce the complexity, this paper proposes a convolutional neural network-based VSO scheme in 3D-HEVC. First, the potential factors that may relate to the encoding complexity are explored. Then, based on this, a convolutional neural network (CNN) is embedded into the 3D-HEVC reference software HTM16.0 to predict the depth of coding units (CUs). The complexity of SVDC can be drastically reduced by avoiding the brute-force search for VSO in depth 0 and depth 1. Finally, for depth 2 and depth 3, the zero distortion area (ZDA) is determined based on texture smoothness and the SVDC calculation for that area is skipped. The experimental results show that the proposed scheme can reduce 76.7% encoding time without any significant loss for the 3D video quality.

Keywords: VSO · 3D-HEVC · CNN

1 Introduction

With the development of three dimensional (3D) video services, 3D video coding technology has become a popular research field [1]. Therein, video coding standard is an important foundation for advancing the development of 3D video applications. As the latest two dimensional (2D) video coding standard, High Efficiency Video Coding (HEVC) can achieve a significant bit rates reduction [2]. In view of its superior encoding performance, Video Coding Expert Group (VCEG) of International Telecommunication Union (ITU-T) and Motion Picture Expert Group (MPEG) of International Standard Organization (ISO/IEC), the two major standardization organizations, form a new team called Joint Collaborative Team on 3D Video Coding Extension Development (JCT-3V) [3]. JCT-3V has developed an international 3D video coding standard, namely 3D-HEVC [4].

© Springer Nature Switzerland AG 2020
X. Sun et al. (Eds.): ICAIS 2020, LNCS 12239, pp. 279–290, 2020.
https://doi.org/10.1007/978-3-030-57884-8_25

3D-HEVC standard supports encoding of multi-view video plus depth (MVD) video format [5]. MVD video format includes two or three texture videos and their corresponding depth maps, for joint texture and depth encoding [6]. With the help of depth information, new virtual views between two reference views can be generated by using depth image-based rendering (DIBR) technique [7]. However, the distortion caused by the coded depth map directly leads to the distortion of the virtual view, and the conventional rate distortion optimization (RDO) method does not take into account the distortion of the synthesized virtual views [8]. Therefore, 3D-HEVC adopts a new RDO technology, called view synthesis optimization (VSO) [9]. It takes into account the distortion of the depth map and the distortion of the synthesized view when calculating the depth map coding distortion [10]. In order to establish the relationship between depth map distortion and synthesized view distortion, a synthesized view distortion change (SVDC) model is proposed [11]. Specifically, the 3D-HEVC encoder repeatedly calls the SVDC model to calculate the SVDC value for each coding block (CB), and each calculation process includes the view synthesis process. Therefore, when encoding the depth map, the computational complexity caused by the SVDC model increases as the number of coded views increases, which may affect the efficiency of 3D-HEVC encoder. Thus, how to reduce the complexity of VSO is challenging to improve coding efficiency.

In general, several studies have been conducted to reduce the complexity in 3D-HEVC, and can be mainly classified into two categories: traditional methods, and classification methods. In traditional methods, some heuristic fast decision algorithms have been proposed. L. Yi-Wen et al. [12] proposed to divide depth map into background, middle ground, and foreground by utilizing automatic threshold technique, in order to reduce the coding complexity of the depth map. Yang et al. [13] proposed an efficient view synthesis distortion (VSD) estimation scheme, incorporating into the rate-distortion optimization for depth coding optimization, to measure the effect of depth errors on the VSD for a block given its depth distortion in mean-squared error. Recently, classification methods based on machine learning have flourished for complexity reduction in 3D-HEVC. J. Ruihai et al. [14] proposed a fast coding level decision and mode decision algorithm for 3D depth video using classification and regression tree (CART), where features are extracted to predict the depth level and some decision of coding mode can be skipped early. Saldanha et al. [15] proposed a fast 3D-HEVC depth maps intra-frame prediction based on static Coding Unit (CU) decision trees, which uses data mining to extract the correlation among the encoder context attributes. However, the above methods mainly rely on experts to set thresholds and extract features for classification, and hence have limitations and hinder the practical application of 3D-HEVC. Most recently, considering that convolutional neural network (CNN) is suitable for image and video processing, learning-based methods are introduced to reduce the encoding complexity [16, 17].

In this paper, we propose a CNN-based complexity reduction scheme to reduce the encoding complexity of 3D-HEVC. The main contributions of this paper are listed as follows:

1. We explore the potential factors that may relate to the encoding complexity.
2. We incorporate the CNN model into 3D-HEVC codec to accelerate the process of 3D video coding at depth 0 and depth 1 for both SVDC and VSO.

3. We determine the zero distortion area (ZDA) based on the texture smoothness, which can skip the SVDC calculation at depth 2 and depth 3 for that area.

The remainder of the paper is organized as follows. Section 2 reviews the VSO and the calculation of SVDC in 3D-HEVC. A CNN-based complexity reduction scheme is proposed in Sect. 3. Section 4 and Sect. 5 give the experimental results and the conclusions, respectively.

2 Overview of VSO and SVDC

2.1 View Synthesis Optimization

Previous research has observed that VSO is an indispensably important video coding technology for depth map coding. When encoding the depth map, the coding distortion of the depth map itself directly affects the quality of the synthesized view of the decoding end. However, when calculating the coding distortion, the conventional rate-distortion optimization method only calculates the sum of squared differences (SSD) or the sum of absolute differences (SAD) value between the current block determined in the current coded image frame and the corresponding block in the reference frame. This calculation method cannot reflect the distortion in the synthesis view.

In order to obtain the true virtual view quality at the encoding end, the 3D-HEVC video coding standard introduces the virtual view synthesis process into the encoder. Hence, in the depth map coding process, the virtual view can be generated by the virtual view synthesis technology to obtain the real virtual view distortion. The calculation method of the synthesized view-based RDO method for depth map coding is as follows:

$$J = D_{synth+depth} + \lambda_s \times R \tag{1}$$

where J represents the rate distortion cost currently used for depth map coding, $D_{synth+depth}$ represents the value of both synthesized view distortion and depth map self-coding distortion, R represents the number of coded bits during the encoding process, and λ_s indicates the adjusted Lagrangian multiplier.

In (1), the distortion value definitely does not only consider the distortion of the synthesized virtual view. In order to ensure the fidelity of the depth map coding, the distortion value in the depth map RDO process is obtained by calculating the weighted average of the synthesized distortion and the coded distortion of the depth map itself, as follows:

$$D_{synth+depth} = w_{synth} \times D_{synth} + w_{depth} \times D_{depth} \tag{2}$$

where D_{synth} represents the synthesized view distortion, D_{depth} represents the depth map self-coded distortion calculated by SSD or SAD, and w_{synth} and w_{depth} represent the two weights for D_{synth} and D_{depth}, respectively. VSO introduces the virtual view synthesis process, which can select the optimal candidate according to the virtual view quality. However, it still needs to perform virtual view re-rendering on the reconstructed block, which imposes a heavy computational burden to the encoder.

2.2 Synthesized View Distortion Change

SVDC is proposed to establish the relationship between depth map distortion and synthesized view distortion, thereby improving the VSO performance of depth map coding. However, both dis-occlusion and occlusion can cause depth map distortion in some areas unrelated to virtual view distortion. For example, since the current depth map coding block is occluded by other objects in the synthesized view, the depth map coding block does not affect the synthesized view quality regardless of the distortion. Therefore, for the current depth map coding block, its own distortion cannot be accurately mapped to the distortion in the synthesized view.

In order to solve the above problem, the SVDC model calculates all distortion changes in the synthesized view caused by the change of the depth value in the current depth map encoding block, and simultaneously considers the variation of the depth value of the pixels around the current depth map encoding block. Therefore, the SVDC model is defined as the difference between the distortion values of the two synthesized view, as shown in Fig. 1.

As can be seen from Fig. 1, the calculation process of the SVDC model depends on the difference between the SSD of the three synthesized view, as follows:

$$SVDC = SSD1 - SSD2 = \sum(V1 - V2)^2 - (V2 - V3)^2 \qquad (3)$$

where the virtual view synthesized by the uncoded original texture frame T and the uncoded original depth frame D is denoted as $V2$. $V1$ and $V3$ are two virtual view frames synthesized by the encoded texture frame T1 and the two partially encoded depth frames $D1$ and $D3$, respectively. The difference between $D1$ and $D3$ is that $D3$ also contains the depth information of the distortion in the code block of the current depth map to be determined on the basis of $D1$. $V1$ and $V3$ calculate the sum of the squares of the difference with $V2$, denoted as $SSD1$ and $SSD2$, and finally calculate the difference between $SSD1$ and $SSD2$ to obtain the $SVDC$ value.

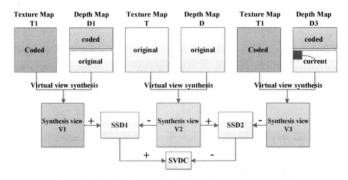

Fig. 1. Synthesized view distortion change model.

3   Proposed CNN-based Complexity Reduction Scheme in 3D-HEVC

3.1 Motivation

As can be seen from Sect. 2, the computational complexity of the VSO is very high. To verify the above conclusion, we test in the 3D-HEVC reference model HTM16.0.

Figure 2 shows the comparison of the encoding time of the VSO method and the RDO method. It can be clearly seen from Fig. 2 that the VSO method has a very high coding complexity compared to the conventional RDO method. For all the test sequences, the depth map encoding time is 2 to 3 times that of the traditional RDO method, and the encoding time also increases as the video resolution increases. It can be seen that the RDO technique based on synthesized view would bring great coding complexity to the depth map coding process. Therefore, an effective depth map fast coding algorithm is needed to reduce the high coding complexity caused by the RDO process based on synthesized view.

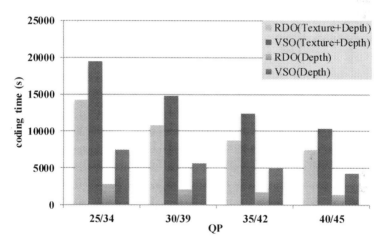

Fig. 2. Coding time comparison of RDO and VSO of 'Kendo: 1024×768'

In order to reduce the computational complexity of VSO, we analyze the moving process of texture map pixels from the original view to the target virtual view. As shown in Fig. 3, it is assumed that the original view numbers are 5 and 3, and the redrawn synthesized view has a number of 4. The pixel values in the reconstructed depth block are used to calculate a disparity vector that is used to help the texture block in the original view move to its position in the synthesized view 4.

As can be seen from Fig. 3, the current depth coding block (red square) is in a texture-flat background area in the same position block (green square) in the corresponding texture map. If the pixel value in the red square is distorted, it would cause its position in the target synthesized view to shift. However, assuming that it shifts from the black square position to the blue square position, since the texture information inside the blue square area is very similar to the texture information inside the black square, in this case, no significant distortion of the synthesized view is caused.

It can be seen that if the position of the current depth coding block in the texture map is relatively smooth, then even if there is distortion in the current depth coding block, the quality of the synthesized view would not be significantly affected. Therefore, if the synthesized view-based VSO process of such pixels is skipped, it would increase the speed of the encoder.

Fig. 3. The moving process of texture map pixels. (Color figure online)

3.2   Zero Distortion Area Decision

The ZDA decision is to analyze the flatness of the current CU in depth map according to the coded information of the previous coded texture frame. As shown in Fig. 4, the coding region in texture video sequence is divided into complex texture region and simple texture region.

The definition of simple texture region is as follows:

$$|P_i - P_{i+1}| \leq T \tag{4}$$

$$FrameLumaDiff = \frac{\sum_{n=0}^{N}\left(\sum_{j=0}^{N_h}\sum_{i=0}^{N_w-1}|l_{i,j} - l_{i+1,j}|\right)}{N(N_w - 1)N_h} \tag{5}$$

$$P\_BlkLumaDiff = |l_{i,j} - l_{i+1,j}| \tag{6}$$

where P_i and P_{i+1} are pixel pairs in a certain horizontal pixel row, and the threshold T is obtained by calculating the average brightness difference of all the adjacent pixels in horizontal pixel rows. For one specific region, *FrameLumaDiff* represents the average brightness difference of the previous texture frame, which is the sum and average of the brightness difference of all the adjacent CUs in the previous texture frame, and

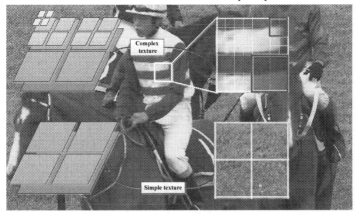

Fig. 4. A diagram of complexity region in texture video

P_BlkLumaDiff represents the brightness difference of the adjacent two CUs in the current coding texture frame. In addition, $N_w \times N_h$ is the CU size, $l_{i,j}$ represents the brightness corresponding to the current pixel position, while (i, j) shows the pixel position. N denotes the number of CUs in the texture frame. The value *FrameLumaDiff* is used as a threshold T to determine whether the VSO process of corresponding CU in depth map is skipped in advance or not.

In summary, the ZDA decision is set as follows:

Step1: For the previous coded video frame, we sum the brightness difference between two adjacent CUs in a frame and record it as *FrameLumaDiff*.
Step2: We use the value *FrameLumaDiff* as a threshold T.
Step3: For the current texture video coding frame, we record the brightness difference between two adjacent CUs as *P_BlkLumaDiff*.
Step4: Compare the values of *FrameLumaDiff* and *P_BlkLumaDiff*, and then decide whether to terminate the VSO process of corresponding CU in depth map. If *FrameLumaDiff* > *P_BlkLumaDiff*, the VSO process is terminated; if *FrameLumaDiff* < *P_BlkLumaDiff* the VSO process is not terminated.

3.3 Implementation of CNN-based Complexity Reduction Scheme

To determine the depth range of CU partition, we adopt a well-trained CNN architecture [18] to predict the structured output of the CU partition in the form of hierarchical CU partition map (HCPM). On one hand, for CUs with output depths of 0 and 1, the VSO process can be terminated directly. On the other hand, the CUs with output depths of 2 and 3 are further determined based on the ZDA decision. The CU partition at different depth is listed in Table 1.

The original CNN architecture in [18] is utilized for HEVC reference HM. In our problem, in order to test its compatibility with 3D-HEVC reference HTM, we embed it into HTM, as shown in Fig. 5. The output of CNN can be regarded as an optional mode to make the corresponding decisions. The switch after the prediction is determined by the CU depth.

Table 1. CU partition at different depth.

Size	Depth 3	Depth 2	Depth 1	Depth 0
64 × 64	Split	Split	Split	Non-split
32 × 32	Split	Split	Non-split	Non-split
16 × 16	Split	Non-split	Non-split	Non-split
8 × 8	Non-split	Non-split	Non-split	Non-split

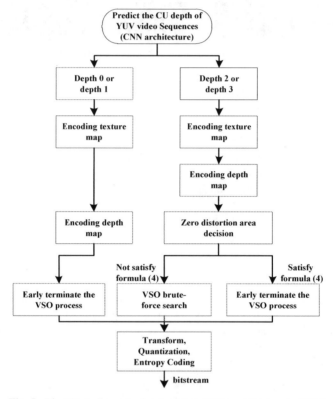

Fig. 5. The block diagram of embedding CNN architecture in HTM.

4　Experimental Results

4.1　Configuration and Settings

In the experiment, we select various types, different resolutions and frame rates of HTM standard video test sequences, including "Balloons", "Kendo", "Newspaper", "Poznan_Hall2", "PoznanStreet", and "Undo-Dancer" to evaluate our proposed scheme. We set QP as (25, 34), (30, 39), (35, 42), and (40, 45). The number of frames in the test sequence is 50. The remaining parameters follow the prescribed conventional test conditions. And it is tested on HTM16.0. Table 2 lists the specific parameter settings. All the experiments are conducted on Intel(R) Xeon(R) CPU E31230 @ 3.20 GHz 8.00 GB RAM.

Table 2. Configuration parameters for different test groups

Sequences	Resolution	Frame rate	Camera numbers
Balloons	1024 × 768	30	3, 1, 5
Kendo	1024 × 768	30	3, 1, 5
Newspaper	1024 × 768	50	4, 2, 6
Poznan_Hall2	1920 × 1088	25	6, 7, 5
PoznanStreet	1920 × 1088	25	4, 5, 3
Undo-dancer	1920 × 1088	25	5, 1, 9

4.2 Comparison of Complexity Reduction

In order to prove that the proposed scheme can significantly reduce the coding complexity of 3D-HEVC, we compare it with the original coding method and the method in [19]. Table 3 reports the results of complexity reduction in ΔT, defined as follows:

$$\Delta T = \frac{T_{reference} - T_{proposed}}{T_{reference}} \times 100\% \tag{7}$$

where $T_{reference}$ and $T_{proposed}$ represent the encoding time of reference method and proposed method, respectively. From Table 3, we can see that the proposed scheme saves 76.7% and 72.8% encoding time compared with the original method and the method in [19], respectively. Among them, the encoding time of "Undo-Dancer" sequence has the largest savings. The reason is that the background of the "Undo-Dancer" sequence does not change and the object only moves in the foreground. The encoding time of "Newspaper" sequence has the least savings, since the background of the "Newspaper" sequence always changes. In addition, it can be seen that when encoding time increases, the proposed scheme time saving increases. This is because as QP increase, the probability of skipping texture-flat region has increased.

4.3 Comparison of RD Performance

In order to evaluate the RD performance of different algorithms, the Bjontegaard Delta Bitrate (BDBR) and Bjongaard Delta Peak Signal-to noise Rate (BDPSNR) are measured. Here BDBR indicates the rate saving of the two methods under the same coding quality. BD-PSNR indicates the PSNR-Y comparison of the two methods at the same bitrate.

We analyze the RD performance of two typical test sequences. Figure 6 show the RD curves of the proposed scheme compared to original encoder on "Newspaper (1024 × 768 resolution with complex motion)" and "Poznan_Hall2 (1920 × 1088 resolution with simple motion)". From Fig. 6, we can see that under the same coding quality, the bitrate of the proposed method increases slightly. But there is no obvious loss in subjective quality. It can be seen from Table 4 that compared with the original coding algorithm, average 0.0% and 0.05% BD-Rate loss for coded views. In summary, the analysis above indicates that the proposed scheme can further improve the performance of complexity reduction for 3D-HEVC, with acceptable loss of RD performance.

Table 3. Comparison of coding time between different methods

Sequences	QP	Coding time (s)			ΔT_1 (%)	ΔT_2 (%)
		Original	[19]	Proposed method		
Balloons	(25, 34)	5096.3630	4385.9294	1452.1530	71.5	66.8
	(30, 39)	5386.8340	4648.3352	1392.9740	74.1	70.0
	(35, 42)	5321.2400	4660.7300	1365.7880	74.3	70.7
	(40, 45)	5404.1850	4936.0736	1250.8080	76.8	74.6
Kendo	(25, 34)	4655.7810	3692.0445	1179.6150	74.6	68.1
	(30, 39)	4951.6000	4319.9220	1118.0890	77.4	74.1
	(35, 42)	4937.5680	4156.4658	1131.0530	77.1	72.8
	(40, 45)	5140.6310	4197.8418	1029.9790	79.9	75.5
Newspaper	(25, 34)	6428.8730	5490.3541	1837.3760	71.4	66.5
	(30, 39)	6593.4230	5321.1196	1758.2190	73.3	66.9
	(35, 42)	6340.0810	5202.4427	1721.9170	72.9	66.9
	(40, 45)	6284.5220	4877.9338	1515.3240	75.8	68.9
Ponan_Hall2	(25, 34)	6811.7400	5903.2704	1593.0350	76.6	73.0
	(30, 39)	7278.9000	6546.4022	1567.1210	78.4	76.1
	(35, 42)	7474.3020	6843.2654	1504.0450	79.8	78.0
	(40, 45)	7975.1090	7043.3938	1406.9050	82.3	80.0
PonanStreet	(25, 34)	11413.2880	11051.3383	3208.1880	71.8	70.9
	(30, 39)	11274.2160	10276.8663	2892.3920	74.3	71.8
	(35, 42)	9812.6720	8511.8603	2598.6920	73.5	69.5
	(40, 45)	9523.6030	7509.1390	2228.6300	76.5	70.3
UndoDancer	(25, 34)	13881.2600	11615.3617	2500.5900	81.9	78.5
	(30, 39)	13331.1020	11168.8935	2412.9550	81.9	78.4
	(35, 42)	13566.1650	11272.9206	2335.1330	82.8	79.3
	(40, 45)	13699.9390	11223.0818	2255.6630	83.5	79.9
Average		8024.3081	6868.9580	1802.3600	76.7	72.8

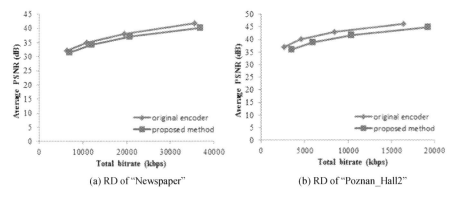

(a) RD of "Newspaper"   (b) RD of "Poznan_Hall2"

Fig. 6. RD performance curves and subjective quality.

Table 4. Coding performance of the proposed method with 3D-HEVC

Sequences	Video 0	Video 1	Video 2	Video PSNR/video bitrate	Video PSNR/total bitrate
Balloons	0.0%	0.0%	0.0%	0.0%	1.5%
Kendo	0.0%	0.0%	0.0%	0.0%	0.9%
Newspaper	0.0%	0.0%	0.0%	0.0%	−1.6%
Poznan_Hall2	0.0%	0.0%	0.0%	0.0%	−0.6%
PoznanStreet	0.0%	0.0%	0.0%	0.0%	0.1%
Undo-Dancer	0.0%	0.0%	0.0%	0.0%	−0.1%
1024 × 768	0.0%	0.0%	0.0%	0.0%	0.3%
1920 × 1088	0.0%	0.0%	0.0%	0.0%	−0.2%
Average	0.0%	0.0%	0.0%	0.0%	0.05%

5 Conclusions

In this paper, we propose a CNN-based complexity reduction scheme in 3D-HEVC, which consists of three main steps. The specific steps of this scheme are to explore the potential factors that may relate to the encoding complexity, avoid the brute-force search for VSO by embedding a CNN into the 3D-HEVC reference software HTM16.0, and utilize the ZDA decision to further skip the VSO process. The experimental results show that compared with the original HTM16.0 encoder, the proposed scheme can saves 76.7% encoding time, without any significant loss for the 3D video quality. In the future, we can further improve the coding performance by modifying the original CNN architecture.

Acknowledgement. This work was supported in part by the Project for the National Natural Science Foundation of China under Grants No. 61672064, Beijing Laboratory of Advanced Information Networks under Grants No. PXM2019_014204_500029, and the Beijing Natural Science Foundation under Grant No. 4172001.

References

1. Gerhard, T., Ying, C., Karsten, M., Jens-Rainer, O., Anthony, V., et al.: Overview of the multiview and 3D extensions of high efficiency video coding. IEEE Trans. Circ. Syst. Video Technol. **26**(1), 35–49 (2016)
2. Li, Z., Meng, L., Xu, S., Li, Z., Shi, Y., Liang, Y.: A HEVC video steganalysis algorithm based on PU partition modes. Comput. Mater. Con. **59**(2), 563–574 (2019)
3. Gerhard, T., Krzysztof, W., Ying, C., Sehoon, Y.: 3D-HEVC Draft Text 7. JCT-3V, doc. JCT3V-K1001, Geneva, CH (2015)
4. Sullivan, G.J., Boyce, J.M., Chen, Y., Ohm, J., Segall, C.A., et al.: Standardized extensions of high efficiency video coding (HEVC). IEEE J. Sel. Top. Sig. Process. **7**(6), 1001–1016 (2013)
5. Zhang, Q., Wu, Q., Wang, X.: Early SKIP mode decision for three-dimensional high efficiency video coding using spatial and interview correlations. J. Electron. Imaging **23**, 053017 (2014)
6. Zheng, Z., Huo, J., Li, B., Yuan, H.: Fine virtual view distortion estimation method for depth map coding. IEEE Sig. Process. Lett. **25**(3), 417–421 (2018)
7. Guo, L., Zhou, L., Tian, X., Chen, Y.: Hole-filling map-based coding unit size decision for dependent views in three-dimensional high-efficiency video coding. J. Electron. Imaging **25**(3), 033020 (2016)
8. Zhang, Y., Kwong, S., Hu, S., Kuo, C.-C.J.: Efficient multiview depth coding optimization based on allowable depth distortion in view synthesis. IEEE Trans. Image Process. **23**(11), 4879–4892 (2014)
9. Ma, S., Wang, S., Gao, W.: Low complexity adaptive view synthesis optimization in HEVC based 3D video coding. IEEE Trans. Multimedia **16**(1), 266–271 (2014)
10. Yang, M., Zheng, N., Zhu, C., Wang, F.: A novel method of minimizing view synthesis distortion based on its non-monotonicity in 3D video. IEEE Trans. Image Process. **26**(11), 5122–5137 (2017)
11. Gerhard, T., Heiko, S., Karsten, M., Thomas, W.: 3D video coding using the synthesized view distortion change. In: 2012 Picture Coding Symposium, Krakow, Poland, May 2012
12. Liao, Y.-W., Chen, M.-J., Yeh, C.-H., Lin, J.-R., Chen, C.-W.: Efficient inter-prediction depth coding algorithm based on depth map segmentation for 3D-HEVC. Multimed. Tools Appl. **78**(8), 10181–10205 (2018). https://doi.org/10.1007/s11042-018-6547-7
13. Yang, M., Zhu, C., Lan, X., Zheng, N.: Efficient estimation of view synthesis distortion for depth coding optimization. IEEE Trans. Multimedia **21**(4), 863–874 (2019)
14. Jing, R., Zhang, Q., Wang, B., Cui, P., Yan, T., Huang, J.: CART-based fast CU size decision and mode decision algorithm for 3D-HEVC. SIViP **13**(2), 209–216 (2018). https://doi.org/10.1007/s11760-018-1347-0
15. Mario, S., Gustavo, S., Cesar, M., Luciano, A.: Fast 3D-HEVC depth maps intra-frame prediction using data mining. In: 2018 IEEE International Conference on Acoustics, Speech and Signal Processing (ICASSP), Calgary, AB, Canada, April 2018
16. Cui, Q., McIntosh, S., Sun, H.: Identifying materials of photographic images and photorealistic computer generated graphics based on deep CNNs. Comput. Mater. Con. **55**(2), 229–241 (2018)
17. Fang, W., Zhang, F., Sheng, V.S., Ding, Y.: A method for improving CNN-based image recognition using DCGAN. Comput. Mater. Con. **57**(1), 167–178 (2018)
18. Xu, M., Li, T., Wang, Z., Deng, X., Yang, R., et al.: Reducing complexity of HEVC: a deep learning approach. IEEE Trans. Image Process. **27**(10), 5044–5059 (2018)
19. Dou, H., Chan, Y.-L., Jia, K.-B., Siu, W.-C.: Segment–based view synthesis optimization scheme in 3D-HEVC. J. Vis. Commun. Image Represent. **42**, 104–111 (2016)

Air Quality Index Prediction Based on Deep Recurrent Neural Network

Zhiyu Chen, Yuzhu Zhang, Gang Liu[(⊠)], and Jianwei Guo

Changchun University of Techology, Changchun 130012, Jilin, China
lg@ccut.edu.cn

Abstract. Aiming at the problem of missing values in air pollutant data and the single structure of prediction model, it is also need to consider that air pollutants will be affected by meteorological data and change quickly. Therefore, this paper mainly studies short-term (hourly) air quality index prediction. First, original pollutant concentration data are converted into individual air quality index of each pollutant item through calculation formula of AQI. Then, according to the missing data attributes and length of missing time, a combined missing processing method is proposed. After correlation analysis and feature selection, the air quality index prediction model using deep recurrent neural network based on gated recurrent unit (GRU-BPNN) is constructed to obtain the final predicted value, it is then classified to obtain its corresponding AQI level. Based on the real data sources obtained in Changchun, a large number of experiments have been carried out to prove that our model can improve the performance of air pollution prediction.

Keywords: Data conversion · Missing processing · Gated recurrent unit

1 Introduction

With the rapid development of industrialization and urbanization, all kinds of harmful substances are discharged into the air, causing air pollution and seriously affecting people's health and daily life [1]. Air quality monitoring stations have been set up in many cities, and people are paying more and more attention to urban air quality. In addition to monitoring, there is a growing need to predict future air quality. Because accurate prediction can help people reduce the harm caused by air pollution, and air quality prediction can provide reference for urban residents' life and government decision-making, it is of great social significance to predict air quality [2].

However, this work faces many challenges, one is data source incompleteness. The reason for incomplete data is related to the air quality monitoring equipment. The situation that monitoring station has no output generally occurs when the equipment is under maintenance, there is a fault or other problems, so that there is a large amount of data missing in a period of time. In addition, air quality is affected by humidity, wind direction, wind speed and other meteorological factors, which makes the prediction of air quality uncertain.

© Springer Nature Switzerland AG 2020
X. Sun et al. (Eds.): ICAIS 2020, LNCS 12239, pp. 291–304, 2020.
https://doi.org/10.1007/978-3-030-57884-8_26

Traditional air quality prediction models can be divided into numerical prediction models and statistical prediction. Numerical prediction is a mathematical model to simulate the diffusion process of air pollutants in the air, such as urban airshed model (UAM) [3], a nested grid air quality prediction model system developed by the institute of atmospheric physics, Chinese Academy of Sciences [4]. However, the information of pollution sources is difficult to obtain in practical application, and usually assumed, so the accuracy of its forecast is not high. Statistical forecasting is the prediction of future air quality by analyzing the statistical rules between input and output related to air pollution [5, 6]. Donnelly used a nuclear regression model combined with meteorological data to predict the hourly concentration and daily maximum concentration of NO2 [7]. Zheng used the combination of ANN and linear regression to predict the future air quality, and the forecast could be updated hourly and refined to the site level [8]. Ma combined fuzzy theory and genetic algorithm to optimize the standard BP neural network, and made the fuzzy hybrid algorithm for air quality evaluation GA-BP model [9]. Yang proposed a model for prediction of urban AQI based on KNN algorithm, and added wind factors into KNN prediction results to improve prediction accuracy [10]. Cheng used the combined model of ARIMA and WNN to predict the water level of river runoff [11]. This takes advantage of different models to improve the overall model. Wan used a two-stage neural network model to predict the hourly pollutant concentration, so as to predict AQI value more accurately [12]. In recent years, as a model that considers the time sequence, recurrent neural network (RNN) has been widely used in traffic congestion prediction, medical treatment and other fields, as well as in air quality prediction [13–17].

Based on the above theoretical research and challenges, this paper established a deep RNN prediction model for pollutant sub-index time series with missing values. The main work includes:

1. The data used in this paper is no longer the concentration value of air pollutants, but concentration value of each pollutant is converted into air quality sub-index of each pollutant item through environmental standards, which facilitates the processing of missing values;
2. For the missing data, a combined missing processing algorithm is designed by analyzing the attributes of missing data and the missing time;
3. On the basis of missing processing algorithm and considering influence of meteorological factors on air quality, the establishment of deep RNN based on GRU and the prediction of air quality grade are realized.

2   Basic Conception

2.1   Definitions

Definition 1: Air quality index (AQI) and AQI level. The AQI is a digital label used by government agencies to communicate current levels of air pollution to the public. Air quality reflects the degree of air pollution, which is determined by the concentration of pollutants in the air.

The concentration of air pollutants is converted into an air quality index, which varies depending on the pollutant. AQI is divided into a number of ranges, each range corresponds to a level. As shown in Table 1, a higher air quality index indicates a higher level, this means that the more serious the air pollution, the greater the harm to human health. Therefore, the classification of AQI makes people have a more intuitive understanding of the degree of air pollution.

Table 1. AQI values, levels and categories.

AQI values	AQI levels	Categories
0–50	I	Excellent
51–100	II	Good
101–150	III	Light pollution
151–200	IV	Middle level pollution
201–300	V	Heavy pollution
>300	VI	Severe pollution

Definition 2: Time series samples. Data from each monitoring station can be represented by a collection, $S = \{X1, X2, \ldots, XN\}$, N stand for timestep, each timestep is a d-dimensional vector, which represents collection of concentration observations of 6 pollutants at a certain time in a monitor station, $X_T = \{C1, C2, \ldots, C6\}$, where C1-C6 are the concentrations of SO2, NO2, PM10, CO, O3 and PM2.5.

Thus, data dimension of each monitor station is (N, D), and time series samples generated by each monitor station are three-dimensional data with dimension (N, T, D), where N is the number of samples, T is the number of timesteps contained in a sample, and D is the number of characteristics of input data. X^m, X^{m+1}, $\ldots$, X^{m+T+1} represent the feature vector measured in the past T hours, is the observed value of pollutant item D in the past k hour.

2.2 Problem Statement

In this paper, air quality time series data of the past 24 h and meteorological parameters are used to predict the air quality index of each station in the next 1 h, and then the predicted value is judged to obtain the AQI level of each monitor station. This study predicted the AQI level by predicting the value of AQI, rather than directly predicting the AQI level through the model.

3 Research Method

3.1 Calculation of AQI

In 2012, China formulated environmental air quality standards (GB3095-2012) in accordance with the air pollution prevention and control law of the People's Republic of China, as shown in Table 2. This standard stipulates the standard classification of ambient air quality and the concentration limits of various pollutant items at various levels, which is the scientific basis for evaluating air quality [18].

Table 2. Analysis of ambient air quality pollutant concentration.

IAQI	Limit of pollutant item concentration (Average)									
	SO2 (24 h)	SO2 (1 h)	NO2 (24 h)	NO2 (1 h)	PM10 (24 h)	CO (24 h)	CO (1 h)	O3 (1 h)	O3 (8 h)	PM2.5 (24 h)
0	0	0	0	0	0	0	0	0	0	0
50	50	150	40	100	50	2	5	160	100	35
100	150	500	80	200	150	4	10	200	160	75
150	475	650	180	700	250	14	35	300	215	115
200	800	800	280	1200	350	24	60	400	265	150
300	1600	–	565	2340	420	36	90	800	800	250
400	2100	–	750	3090	500	48	120	1000	–	350
500	2620	–	940	3840	600	60	150	1200	–	500

AQI Calculation Process

1. Compare the limits of graded concentrations of each pollutant, individual air quality index (IAQI) is calculated by the actual measured concentration values of PM2.5, PM10, SO2, NO2, O3, CO (where PM2.5 and PM10 are 24-h average concentrations), the calculation formula is as follows:

$$IAQI_p = \frac{IAQI_{Hi} - IAQI_{Lo}}{BP_{Hi} - BP_{Lo}}(C_p - BP_{Lo}) + IAQI_{Lo} \tag{1}$$

Where $IAQI_P$ is air quality sub-index of pollutant item P, C_P is actual concentration of Pollutant item P; BP_{Hi} is the high value of the limit value of pollutant concentration close to C_P in the table; BP_{Lo} is the low value of the limit value of pollutant concentration close to C_P in the table; $IAQI_{Hi}$ is the air quality sub-index corresponding to BP_{Hi} in the table; $IAQI_{Lo}$ is the air quality sub-index corresponding to BP_{Lo} in the table.

2. The maximum value of each pollutant's IAQI is selected and determined as AQI.

$$AQI = \max\{IAQI_1, \ IAQI_2, \ldots, IAQI_n\} \tag{2}$$

Where IAQI is the sub-index of air quality, and n in the pollutant item.

3. Determine the air quality grade and category according to the classification standard of AQI.

3.2 Missing Processing Algorithm

One of the most common problems with data quality is that some records do not have attribute values. The historical data of some monitoring stations in a certain period of time will be missing to varying degrees. In this study, missing processing algorithm mainly considers two aspects: missing data attributes and length of missing period.

Missing Data Attribute. By analyzing the converted data, it can be found that the whole data set includes two types of data: IAQI of each pollutant and AQI (target label). Different methods should be used to fill the gaps in these two categories:

1. For the missing of IAQI, the filling method should be further determined according to its missing time.
2. For the absence of AQI, according to the formula calculated by AQI, it can be known that AQI is obtained from the maximum value of IAQI of various pollutants. Therefore, IAQI of each pollutant was first analyzed to determine whether there was any absence. If there is a deficiency in the sub-index, it should be processed first. When all sub-indexes exist, the deficiency of AQI should be filled according to its definition. The maximum value of sub-index should be assigned to AQI.

Length of Missing Time. In the past, only a single method was used to deal with data missing without specific analysis of the missing time. Based on neural network structure with the same structure, this paper proposes a combined missing value processing method for IAQI according to length of missing time, as shown in Table 3.

Table 3. Combined missing value processing method.

		Missing data attribute	
		IAQI of each pollutant	AQI
Length of missing time	1 h	Forward fill method/Mean replacement	Definition
	1 h < t < 24 h	Moving average method	Definition

Forward Fill Method: Replace the missing value with the nearest valid data from the current missing time (i.e., the data from previous time), as shown below:

$$X_t^{d'} = X_{t-1}^d \tag{3}$$

Where X_{t-1}^d represents the effective observation value of d-dimension component data at the previous moment.

Mean Replacement: Average value of two effective data before and after the current missing moment (i.e., the data at previous moment and the data at next moment) is used to replace the missing value, as shown below:

$$X_t^{d'} = \frac{\left(X_{t-1}^d + X_{t+1}^d\right)}{2} \tag{4}$$

Where X_{t+1}^d represents effective observation value of d-dimension component data at next moment of the current missing moment.

Moving Average Method: It is necessary to specify a window as k(this paper use k = 3), which limits the number of data that can be seen each time. Moving average method is average value obtained by sliding the time window along timeline, when calculating the average, using the first $t - 1, t - 2, ..., t - k$ time step data. The following formula is shown:

$$X_t^{d'} = \left(X_{t-1}^d + X_{t-2}^d + ... + X_{t-k}^d\right)\Big/ k \tag{5}$$

Where X_{t-k}^d represents d-dimension component data of k time steps away from the current missing moment.

3.3 Correlation Analysis

Pearson correlation coefficient is used in this paper, which is a statistcal indicator of the degree of closeness between two variables. For x and y variables, Pearson's correlation coefficient is calculated as follows:

$$r = \frac{\sum_i (x_i - \bar{x})(y_i - \bar{y})}{\sqrt{\sum_i (x_i - \bar{x})^2}\sqrt{\sum_i (y_i - \bar{y})^2}} \tag{6}$$

Where r ranges from -1 to 1. It approaching 1 means that x is positively correlated with y, Simlarly, r approaching -1 means that x is negatively correlated with y, and r equals 0 means that x is not correlated with y.

In this paper, IAQI data and various meteorological data are used for correlation analysis of AQI data. First, the correlation analysis results of AQI and various IAQI are obtained, as shown in Fig. 1.

Table 4 shows the influence of meteorological factors on air quality index. In Fig. 1, it can be seen that correlation coefficient is at least 0.147. For selection of meteorological features, the correlation coefficient between meteorological factors and AQI should be greater than 0.15. Therefore, the first 6 meteorological factors that have a high impact on AQI are adopted as meteorological parameters, They are station pressure, sea level pressure, humidity, 2 min wind direction, 10 min wind direction, instantaneous wind direction.

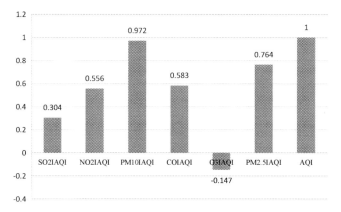

Fig. 1. AQI and various IAQI correlation analysis.

Table 4. AQI and meteorological factors correlation coefficient.

Attribute	Correlation coefficient
Station pressure	0.193
Sea level pressure	0.194
Temperature	0.11
Dew point temperature	0.049
Humidity	0.154
2 min wind direction	0.198
2 min wind speed	0.078
10 min wind direction	0.188
10 min wind speed	0.06
Instantaneous wind direction	0.189
Instantaneous wind speed	0.064

4   Main Model

4.1   Framework

The framework established in this paper consists of data conversion, data preprocessing, data generation and prediction model training. In this framework, specific steps are described as follows:

Data Conversion. The collected pollutant concentration information need to convert into IAQI of each pollutant item.

Missing Value Processing. Deletion processing method is determined according to missing data attributes and missing length, which makes the incomplete time series data with missing values complete.

Correlation Analysis. The correlation analysis between AQI and influence factors is carried out to reveal the main characteristics of air quality change.

Standardization. In this paper, the normalization of min-max is used, which is a linear transformation of data, and its conversion function is as follows:

$$x^* = \frac{x - x_{min}}{x_{max} - x_{min}} \tag{7}$$

Where x_{max} is the maximum value that this attribute appears, x_{min} is the minimum value that this attribute appears, x^* is the result of data X being normalized.

Data Generation. Time sequence samples were generated by setting the number of timestep is 24, and then time sequence data for training, verification and testing models were generated according to a certain proportion.

Deep RNN Model. Deep RNN is the first stage network model, which is used to process the time series data of pollutant IAQI. The second stage network model is constructed by BP network, the purpose is to optimize prediction error and verify the impact of meteorological data on air quality.

Model Evaluation. Predicted values need to anti standardization, AQI level was divided according to classification standard. If predicted AQI level was consistent with real level, the prediction was correct, so as to calculate accuracy of the model.

4.2 Prediction Model

The prediction model can be divided into two stages. The first stage is a recurrent neural network model based on GRU, input data are time series samples of various pollutant IAQI data of air quality monitoring stations in the past 24 h, the model is composed of GRU layer and full connection layer. In order to avoid overfitting, Droupout method was adopted [19]. Model of the second stage uses BP network. The input includes the output of first stage and the sequence of meteorological parameters extracted after correlation analysis. Output data is predicted value of air quality index in the next hour. For convenience of presentation, the prediction model proposed in this paper is called GRU-BPNN, as shown in Fig. 3.

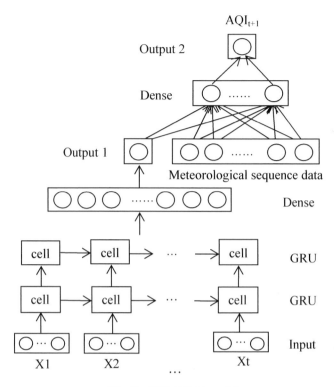

Fig. 3. GRU-BPNN model.

5 Experiment

5.1 Dataset

The city targeted by this research is Changchun, Jilin province, and the data sources used are divided into three categories: 1) Air quality monitoring data, update every hour, including 7 data: PM2.5 concentration, PM10 concentration, O3 concentration, SO2 concentration, NO2 concentration, CO concentration and AQI value; 2) Meteorological monitoring data shall be updated every hour, including 11 aspects: pressure of the station, sea level pressure, air temperature, dew point temperature, humidity, 2-min average wind direction, 2-min average wind speed, 10-min average wind direction, 10-min average wind speed, instantaneous wind direction and instantaneous wind speed; 3) Station information, including station name, latitude and longitude information.

The data span from September 1, 2018 to September 30, 2018, a total of one month, and use 7,200 pieces of information from 10 air quality monitoring stations in Changchun. Data were divided into training set, verification set and test set in a ratio of 6:2:2.

5.2   Parameter Setting

In this paper, three deep circulation neural networks were designed according to different deletion processing methods in the first stage: 1) Deep RNN based on forward fill approach; 2) Deep recurrent neural network using the combination of forward fill and moving average method; 3) Deep RNN is composed of means replacement and moving average method; Where Deep RNN is divided into two types: LSTM-BPNN and GRU-BPNN which is the main model designed in this study. The structural parameters of neural network are shown in Table 5.

Table 5. Structural parameters of neural network.

Missing value handing	Network name	Network layer	
		First stage	Second stage
Forward Fill	F-LSTM-BPNN	2 LSTM + 2 Dense	2 Dense
	F-GRU-BPNN	2 GRU + 2 Dense	2 Dense
Forward fill + Moving average	FA-LSTM-BPNN	2 LSTM + 2 Dense	2 Dense
	FA-GRU-BPNN	2 GRU + 2 Dense	2 Dense
Mean replacement + Moving average	MA-LSTM-BPNN	2 LSTM + 2 Dense	2 Dense
	MA-GRU-BPNN	2 GRU + 2 Dense	2 Dense

Python language and deep learning library Keras are used to build and train the above neural network structure, and the training parameters of each stage model are shown in Table 6.

Table 6. Training parameters of neural network model.

Parameter	Describe	Value	
		First stage	Second stage
Timesteps	Number of time steps	24	3
Inputs	Number of neurons in input layer	144	19
Outputs	Number of neurons in output layer	1	1
HLayers	Number of Layers	4	2
Hunits	Number of neurons per layer	{50, 100, 10, 1}	{13, 1}
Activations	Activation functions used by each layer	{relu, relu, relu, linear}	{relu, linear}
Dropout	Random concealment of a proportionate number of units and their associated weights	=0.1	
Loss	Loss function	Mse	
Optimizer	Optimization method	Adam	
Batchsize	Number of samples taken from each training set	417	
Numepoch	Number of training	100	

5.3 Evaluation Standard

MAE (mean absolute error) and RMSE (root mean squared error) are used to evaluate our models in this paper, calculation formula are as follow:

$$MAE = \frac{1}{m} \sum_{i=1}^{m} \left| (y - \hat{y}_i) \right| \tag{8}$$

$$RMSE = \sqrt{\frac{1}{m} \sum_{i=1}^{m} (y - \hat{y}_i)^2} \tag{9}$$

We use another evaluation standard is prediction accuary (ACC), it is ratio of the number of AQI levels judged to be same as the real ones in all test samples, which is defined as follow:

$$ACC = \frac{n_T}{n_T + n_F} \tag{10}$$

Where n_T and n_F are the number of correct and wrong predictions respectively, n_T + n_F is the total number of test samples.

6 Result

After training, prediction results of these models on test set are shown in Table 7, and the following conclusions are obtained: 1) Among three missing value processing methods proposed, the single forward fill method has the lowest prediction accuracy, and processing method of MA method has the best effect. FA method is in between. 2) With the same missing processing method and the same input data, prediction accuracy of deep RNN based on GRU is higher than that based on LSTM.

Table 7. Prediction accuracy of deep RNN.

Network name	Evaluation methodology		
	MAE	RMSE	ACC
F-LSTM-BPNN	14.165	22.027	0.782
F-GRU-BPNN	12.758	20.448	0.801
FA-LSTM-BPNN	10.955	18.436	0.838
FA-GRU-BPNN	9.361	15.109	0.869
MA-LSTM-BPNN	10.854	16.981	0.848
MA-GRU-BPNN	9.344	15.660	0.883

In addition, in order to more intuitively verify the effects of different models, this paper selected the data from 1 am, September 26, 2018 to 0 am, September 27, 2018, comparison between predicted result and actual value is shown in Fig. 4. Due to a long time of data loss between 12 am and 6 pm, it can be found that:After data completion with forward fill method is presented as a straight line, which makes effect of neural network learning greatly different from the real value; Using combination missing value precessing method, difference with the real value is obviously reduced. However, it is worth noting that the number of hours between GRU-BPNN model and the real value is higher than that of LSTM-BPNN through comparison of AQI classification. It can be seen that Deep RNN based on GRU is more consistent with real value, and the MA-GRU-BPNN model has the best effect.

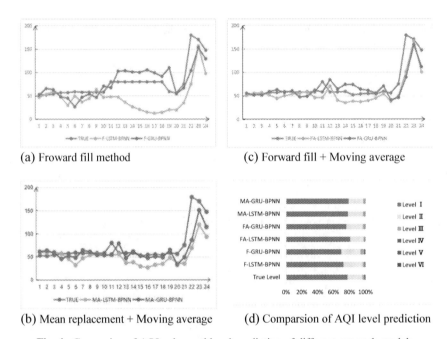

(a) Froward fill method

(c) Forward fill + Moving average

(b) Mean replacement + Moving average

(d) Comparsion of AQI level prediction

Fig. 4. Comparion of AQI value and level prediction of different network models.

The result of air quality level prediction is shown in Fig. 4-(d), although MA-GRU-BPNN has the highest prediction accuracy,f or a few cases of severe pollution (Level IV, V, VI), it may be difficult to learn the relationship between data perhaps because of small proportion of emergencies in the data set, so the prediction accuracy of severe pollution is not very ideal.

7   Conclusion

This paper mainly studies hourly air quality prediction. First, according to the calculation formula of AQI, concentration data in 10 monitoring stations in Changchun city are converted into air quality sub-index data that is conducive to missing processing; Aiming at the situation of missing data, a combination method of missing processing is proposed. Deep RNN was constructed to predict AQI, then BP network with meteorological factors was introduced to optimize the prediction results. By constructing 6 networks and conducting training to evaluate effect of model on the test data set, it was confirmed that GRU-BPNN based on mean replacement and moving average combined processing method had the best accuracy in AQI prediction.

Due to limited data set used in this study and AQI is affected by a variety of factors, it is necessary to consider other factors besides meteorological factors in future research, and improve the prediction accuracy of sudden severe weather conditions.

References

1. Kampa, M., Canstanas, E.: Human health effects of air pollution. Environ. Pollut. **151**(2), 1–367 (2008)
2. Zheng, Y., Capra, L., Wolfson, O., et al.: Urban computing: concepts, methodologies, and applications. ACM Trans. Intell. Syst. Technol. **5**(3), 1–55 (2014)
3. Sistla, G., Zhou, N., Hao, W., et al.: Effects of uncertainties in meteorological inputs on urban airshed model predictions and ozone control strategies. Atmos. Environ. **30**(12), 1–2025 (1996)
4. Wang, Z., Xie, F., Wang, X., et al.: Development and application of nested air quality prediction modeling system. Atmos. Sci. **5**, 52–64 (2006)
5. Dong, T., Zhao, J., Hu, Y.: AQI levels prediction based on deep neural network with spatial and temporal optimizations. Comput. Eng. Appl. **53**(21), 17–23 (2017)
6. Kumar, U., Jain, V.K.: ARIMA forecasting of ambient air pollutants (O3, NO, NO2 and CO). Stoch. Env. Res. Risk Assess. **24**(5), 751–760 (2010)
7. Donnelly, A., Misstear, B., Broderick, B.: Real time air quality forecasting using integrated parametric and non-parametric regression techniques. Atmos. Environ. **103**, 53–65 (2015)
8. Zheng, Y., Liu, F., Hsieh, H. P.: U-air: when urban air quality inference meets big data. In: Proceedings of the 19th ACM SIGKDD International Conference on Knowledge Discovery and Data Mining, pp. 1436–1444. ACM, Chicago (2013)
9. Ma, N., Guan, J., Liu, P., et al.: GA-BP air quality evaluation method based on fuzzy theory. Comput. Mater. Continua **58**(1), 215–227 (2019)
10. Yang, F., Wang, B., Chen, Y., et al.: K-nearest neighbor urban forecasting algorithm considering wind factors. Appl. Res. Comput. **36**(06), 1679–1682 (2019)
11. Cheng, Y., Lou, Y., Ye, F., Li, L.: Research on hydrological time series prediction based on combined model. In: Zou, B., Li, M., Wang, H., Song, X., Xie, W., Lu, Z. (eds.) ICPCSEE 2017. CCIS, vol. 727, pp. 558–572. Springer, Singapore (2017). https://doi.org/10.1007/978-981-10-6385-5_47
12. Wan, Y., Xu, F., Yan, C., et al.: An air quality prediction method integrating meteorological parameters and pollutant concentrations. Comput. Appl. Softw. **35**(8), 113–117 (2018)
13. Zhu, H., Meng, F., Rho, S., et al.: Long short term memory networks based anomaly detection for KPIs. Comput. Mater. Continua **60**(1), 147–161 (2019)
14. Wang, D. Cao, W., Li, J., et al.: DeepSD: supply-demand prediction for online car-hailing services using deep neural networks. In: ICDE, pp. 243–254. IEEE, San Diego (2017)
15. Maamar, A., Benahmed, K.: A Hybrid model for anomalies detection in AMI system combining K-means clustering and deep neural network. Comput. Mater. Continua **60**(1), 15–39 (2019)
16. Fan, J., Li, Q., Zhu, Y., et al.: Aspatio-temporal prediction framework for air pollution based on deep RNN. Sci. Surv. Mapp. **42**(7), 76–83 + 120 (2017)

17. Hou, J., Li, Q., Lin, S., et al.: PM2.5 concentration real-time forecasting method based on GRU model. Sci. Surv. Mapp. **43**(7), 79–86 (2018)
18. Liu, P.: Understanding the Individual Air Quality Index (IAQI) calculation formula and calculating quickly. Heilongjiang Environ. J. **38**(2), 25–27 (2014)
19. Srivastava, N., Hinton, G., Krizhevsky, A., et al.: Dropout: a simple way to prevent neural networks from overfitting. J. Mach. Learn. Res. **15**(1), 1929–1958 (2014)

Sparse Representation for Face Recognition Based on Projective Dictionary Pair Learning

Yang Liu, Jianming Zhang$^{(\boxtimes)}$, and Yangchun Liu

School of Computer and Communication Engineering,
Changsha University of Science and Technology, Changsha 410114, Hunan Province, China
jmzhang@csust.edu.cn

Abstract. In order to improve the recognition rate and speed of face recognition, the paper proposes a sparse representation for face recognition algorithm based on Gabor feature and projective dictionary pair learning. Firstly, we extract the local feature of the face image at multiple scales and orientations by using Gabor transform, in that Gabor feature shows significant robustness to the variations in expression, illumination and angle. And the augmented Gabor feature reduced dimension by principle component analysis can be obtained to form a new training data instead of original training sample. Secondly, a synthesis dictionary with reconstructing capability and an analysis dictionary with the capability of quickly obtaining representation coefficients are learned jointly during the training phase. Finally, the face is identified by the reconstructed error. The experimental results on ORL, extended YaleB, AR and CMU-PIE face database show that our proposed algorithm not only get a high recognition rate but also improve recognition speed efficiently.

Keywords: Face recognition · Gabor feature · Synthesis dictionary · Analysis dictionary

1 Introduction

Face recognition [1] is a biological recognition technology based on facial feature. Compared with other biological recognition technologies (such as voice recognition, fingerprint recognition, etc.), face recognition has advantage of convenience. So face recognition is used widely in video surveillance, e-commerce, security defense, and it is also an important research topic of computer vision and pattern recognition.

Sparse representation [2] based on a linear model is an emerging signal processing method in recent years, which represents a signal as the linear combination of some atoms in the over-complete dictionary, and has achieved big success in image classification [3], image restoration [4], and image denoising [5]. In the sparse representation, the over-complete dictionary plays a very important role. So far, the research of over-complete dictionary can be divided into two models: sparse representation based on analysis dictionary and sparse representation based on synthesis dictionary. In analysis dictionary based sparse representation [6], the sparse representation coefficients can be

© Springer Nature Switzerland AG 2020
X. Sun et al. (Eds.): ICAIS 2020, LNCS 12239, pp. 305–316, 2020.
https://doi.org/10.1007/978-3-030-57884-8_27

calculated quickly by using the inner product between the predefined analysis dictionary and samples. Such an explicit and fast coding algorithm makes analysis dictionary attract great attention in sparse representation. But it cannot model the complex local structures of images effectively. In synthesis dictionary based sparse representation [7–12], a desired synthesis dictionary can be learned from the training samples by using some dictionary learning algorithms, such as KSVD (K Singular Value Decomposition [13]). The learned synthesis dictionary can model the complex local structures of images effectively. However, its sparse representation coefficients are usually obtained by L_0-norm or L_1-norm sparse coding process, which has more expensive computational cost than analysis dictionary based sparse representation.

As the learned synthesis dictionary can better model the complex image local structures, it thus leads to a much better image reconstruction than analysis dictionary. In order to apply synthesis dictionary to the classification task, we need to learn a desired dictionary with discrimination ability by exploiting the class label during the dictionary learning. In [7], the sparse representation based classification algorithm (SRC) framework based face recognition has great robustness for noise and disguise. The SRC directly uses the training samples as the over-dictionary, and the sparse coefficients of testing samples is solved via the minimum of L_1-norm. Finally, the classification result is obtained by reconstruction errors. Because of redundancy and noise in face images, Yang et al. [8] proposed the metaface learning (MFL) of face images under the framework of SRC. MFL aims to learn a more compact and robust dictionary for each class of the training samples, so it gets better recognition results than SRC. During dictionary learning process, the problem that how to make the learned dictionary more adapted for classification tasks has attracted great attention. To make each learned sub-dictionary as independent as possible, Ramirez et al. [9] added incoherence term to the objective function of dictionary learning. Zhang et al. [10] proposed discriminative K-SVD (D-KSVD) algorithm, which could learn a linear classifier and a dictionary with discrimination capability by incorporating the classification error term into the objective function. In [11], label consistent K-SVD (LC-KSVD) algorithm incorporated discriminative sparse-code error term into objective function and get better recognition result than D-KSVD. Yang et al. [12] proposed Fisher discrimination dictionary learning algorithm (FDDL), which imposed Fisher discrimination criterion into the coding coefficients. Both the reconstruction error and sparse coding coefficients were used for classification.

In [7–12], the dictionary is learned from the original training samples. Although these methods in [7–12] can achieve good recognition results, they only use the global information of images, and not make full use of image local feature information. Gabor wavelet can simulate the outline of single-cell receptive field in the human cerebral cortex, and capture the local structure information corresponding to the spatial frequency, spatial position and all directions. On one hand, Gabor feature [14] can extract the image local features effectively, which shows significant robustness to the variations in expression, illumination and angle. In this paper, we apply the Gabor local feature of images instead of the original training samples to learn dictionary. On the other hand, sparse representation coefficients are usually obtained by L_0-norm or L_1-norm sparse coding process which has larger computational burden in most discriminative dictionary learning algorithms [8–12]. So a simple linear projection is used to replace the nonlinear sparse coding. That is, we can use the inner product between analysis dictionary and samples instead of L_0-norm or L_1-norm sparse coding process to obtain the sparse coefficients. Combining the advantages of two algorithms, we propose a face recognition algorithm based on Gabor feature and projective dictionary pair learning (GDPL). Firstly, the algorithm extracts the local feature of the image at multiple scales and orientations by using Gabor transformation. And the augmented Gabor feature reduced dimension by principle component analysis (PCA) can be obtained to form a new training data instead of original training sample. Secondly, a synthesis dictionary with reconstructing capability and an analysis dictionary with the capability of quickly obtaining representation coefficients are learned jointly from the new training data. Finally, the face can be identified by the reconstructed error. The experimental results on ORL, extended YaleB, AR and CMU-PIE face database show that our proposed algorithm not only get a high recognition rate but also improve recognition speed efficiently.

2 Discriminative Dictionary Learning Framework

Let $X = [X_1, X_2, \ldots X_K] \in \mathbb{R}^{p \times n}$ be the set of original training samples from K classes, where $X_i = [x_{i,1}, x_{i,2}, \ldots, x_{i,n_i}] \in \mathbb{R}^{p \times n_i}$ represents the ith class of the training samples and $x_{i,j}$ is a p-dimensional column vector. Through exploiting the class label information of training samples, discriminative dictionary learning algorithms can learn a synthesis dictionary with discrimination capability from X for classification tasks. Most discriminative dictionary learning frameworks are defined as following:

$$\min_{D,A} ||X - DA||_F^2 + \lambda ||A||_P + \Psi(D, A, Y) \tag{1}$$

where, $\lambda \geq 0$ is a regularization parameter, Y is the class label matrix of the training samples X, and A is the sparse coefficient matrix of X on the learned synthesis dictionary D. In Eq. (1), the fidelity term $||X - DA||_F^2$ guarantees the reconstructing capability of dictionary D; $||A||_P$ represents the sparse coefficient constraint term; $\Psi(D, A, Y)$ represents some discrimination function, which guarantees the discriminative capability of D and A. In most discriminative dictionary learning algorithms, the sparse coefficients of samples are usually obtained by L_0-norm or L_1-norm sparse coding process with expensive computational cost.

3 Face Recognition Algorithm Based on Gabor Feature and Projective Dictionary Pair Learning

3.1 Gabor Feature Extraction

Gabor feature can be extracted through the convolution of image with multiple scales and orientations Gabor filters. And the Gabor filter (kernel) is defined as:

$$\psi_{\mu,v}(z) = \frac{||k_{\mu,v}||^2}{\sigma^2} e^{-\frac{||k_{\mu,v}||^2 ||z||^2}{2\sigma^2}} (e^{ik_{\mu,v}z} - e^{-\sigma^2/2}) \tag{2}$$

where $z = (x, y)$ denotes the pixel; $k_{\mu,v} = k_v e^{i\phi_\mu}, k_v = k_{max}/f^v, \phi_\mu = \pi\mu/8, k_{max}$ is the maximum frequency; f is the spacing factor between kernels in the frequency domain;

the parameter σ represents the ratio of the Gaussian window width to wavelength; μ and ν stand for the orientation and scale of filter respectively.

Let $I(x, y)$ represents a face image, the convolution of $I(x, y)$ with a Gabor filter outputs $G_{\mu,\nu}(z) = I(x, y) * \psi_{\mu,\nu}(z)$. The Gabor filtering coefficient is a complex that can be rewritten as $G_{\mu,\nu}(z) = M_{\mu,\nu}(z) \cdot \exp(i\theta_{\mu,\nu}(z))$, where $M_{\mu,\nu}(z)$ denotes the magnitude and $\theta_{\mu,\nu}(z)$ denotes the phase. Because the magnitude information contains the variation of local energy in the image, we use the magnitude as the description of the image feature. As a multi-scale and multi-orientation feature extraction method, Gabor filter will produce a highly redundant feature, so it is necessary to reduce the Gabor feature dimension by using PCA.

In [15], the augmented Gabor feature vector ω is defined by uniform down-sampling, normalization and concatenation of the Gabor filtering coefficients:

$$\omega = [\alpha_{0,0}^{(\rho)^T} \, \alpha_{0,1}^{(\rho)^T} \, \ldots \, \alpha_{4,7}^{(\rho)^T}]^T \tag{3}$$

Where $\alpha_{\mu,\nu}^{(\rho)}$ is the concatenated column vector of magnitude matrix $M_{\mu,\nu}^{\rho}(z)$ down-sampled by a factor of ρ. Augmented Gabor feature vector is a local feature descriptor. Before dictionary learning, we will extract the local Gabor feature of the training images, and then the training data become $G = \omega(X) = [\omega(X_1), \omega(X_2), \ldots, \omega(X_K)]$, where $\omega(X_i) = [\omega(x_{i,1}), \omega(x_{i,2}), \ldots, \omega(x_{i,n_i})]$.

3.2   Face Recognition Algorithm Based on Gabor Feature and Projective Dictionary Pair Learning

Given an analysis dictionary P that can make the coefficient matrix be formulated by $A = PX$, thus it will largely reduce the time complexity in testing. In this paper, a synthesis dictionary D and an analysis dictionary P are learned jointly during dictionary learning phase, namely projective dictionary pair learning framework (DPL):

$$\{P^*, D^*\} = \underset{P,D}{\arg\min} \, ||X - DPX||_F^2 + \Psi(D, P, X, Y) \tag{4}$$

Where $\Psi(D, P, X, Y)$ standing for some discrimination function determines the discrimination ability of the DPL algorithm. Matrix D and P construct a dictionary pair, where the analysis dictionary P can quickly obtain the representation coefficients from training samples X, and the synthesis dictionary D can effectively reconstruct X. A synthesis dictionary $D = [D_1, D_2, \ldots, D_K] \in \mathbb{R}^{p \times mK}$ and a structured analysis dictionary $P = [P_1; P_2; \ldots; P_K] \in \mathbb{R}^{mK \times p}$ are learned in dictionary learning phase, where $\{D_k \in \mathbb{R}^{p \times m}, P_k \in \mathbb{R}^{m \times p}\}$ constructs a sub-dictionary pair from class k and m is the number of the atoms in sub-dictionary D_k. For the analysis dictionary P, the sub-dictionary P_k almost project the training sample X_i ($i \neq k$) to a nearly null space:

$$P_k X_i \approx 0, \, \forall i \neq k \tag{5}$$

Under the constraint of Eq. (5), the coefficient matrix PX is nearly a diagonal matrix. In addition, for the synthesis dictionary D, the sub-dictionary D_k can better reconstruct the training samples X_k via the projective coefficient matrix $P_k X_k$. That is, the kth sub-dictionary pair can minimize the reconstruction error from the kth class of the samples:

$$\min_{P,D} \sum\nolimits_{k=1}^{K} ||X_k - D_k P_k X_k||_F^2 \tag{6}$$

With Eq. (5) and Eq. (6), the DPL model becomes the following problem:

$$\{P^*, D^*\} = \underset{P,D}{\arg\min} \sum\nolimits_{k=1}^{K} ||X_k - D_k P_k X_k||_F^2 + \lambda ||P_k \bar{X}_k||_F^2, s.t ||d_i||_2^2 \le 1 \tag{7}$$

Where $\bar{X}_k$ is the complementary samples of X_k in the all training samples set X, and $\lambda > 0$ is a regularization parameter. Although the DPL model in Eq. (7) enforces group sparsity [16] on the coefficient matrix PX.

The objective function in Eq. (7) is usually non-convex, so we introduce a variable matrix A and convert Eq. (7) to Eq. (8):

$$\{P^*, A^*, D^*\} = \underset{P,A,D}{\arg\min} \sum\nolimits_{k=1}^{K} ||X_k - D_k A_k||_F^2 + \tau ||P_k X_k - A_k||_F^2 + \lambda ||P_k \bar{X}_k||_F^2, s.t ||d_i||_2^2 \le 1 \tag{8}$$

The objective function in Eq. (8) can be solved by alternatively updating matrix A and dictionary pair $\{D, P\}$. The minimization problem can be alternated between the following two sub-problems:

1) Fix synthesis dictionary D and analysis dictionary P, update A:

$$A^* = \underset{A}{\arg\min} \sum\nolimits_{k=1}^{K} ||X_k - D_k A_k||_F^2 + \tau ||P_k X_k - A_k||_F^2 \tag{9}$$

The objective function in Eq. (9) is a standard least squares problem [17], and its closed-form solution is obtained as follows:

$$A_k^* = (D_k^T D_k + \tau I)^{-1} (\tau P_k X_k + D_k^T X_k) \tag{10}$$

2) Fix coefficient matrix A, update dictionary D and P respectively:

$$\begin{cases} P^* = \underset{P}{\arg\min} \sum\nolimits_{k=1}^{K} \tau ||P_k X_k - A_k||_F^2 + \lambda ||P_k \bar{X}_k||_F^2 \\ D^* = \underset{D}{\arg\min} \sum\nolimits_{k=1}^{K} ||X_k - D_k A_k||_F^2, \ s.t ||d_i||_2^2 \le 1 \end{cases} \tag{11}$$

The closed-form solution of analysis dictionary P can be solved by the following formula:

$$P_k^* = \tau A_k X_k^T (\tau X_k X_k^T + \lambda \bar{X}_k \bar{X}_K^T + \gamma I)^{-1} \tag{12}$$

The optimal solution of synthesis dictionary D can be solved by introducing a variable matrix S:

$$\min_{D,S} \sum_{k=1}^{K} ||X_k - D_k A_k||_F^2, \ s.t. D = S, \ ||s_i||_2^2 \le 1 \tag{13}$$

The ADMM algorithm [18] is a very efficient solution to Eq. (13):

$$\begin{cases} D^{(r+1)} = \mathrm{argmin}_D \sum_{k=1}^{K} ||X_k - D_k A_k||_F^2 + \rho ||D_k - S_k^{(r)} + T_k^{(r)}||_F^2 \\ S^{(r+1)} = \mathrm{argmin}_S \sum_{k=1}^{K} \rho ||D_k^{(r+1)} - S_k + T_k^{(r)}||_F^2, \ s.t. ||s_i||_2^2 \le 1 \\ T^{(r+1)} = T^{(r)} + D_k^{(r+1)} - S_k^{(r+1)} \end{cases} \tag{14}$$

In each iterative process, the closed-form solution of the variable matrix A and analysis dictionary P can be solved easily. In addition, the ADMM based optimization of synthesis dictionary D converges very fast. So the training time of DPL algorithm is much less than most discriminative dictionary learning algorithms [8–12].

Let $X = [X_1, X_2, \ldots X_K]$ be the set of original training samples from K classes. The local Gabor feature set $G = \omega(X) = [\omega(X_1), \omega(X_2), \ldots, \omega(X_K)] = [G_1, G_2, \ldots, G_K]$ of the training images X can be extracted, where $G_i = [\omega(x_{i,1}), \omega(x_{i,2}), \ldots, \omega(x_{i,n_i})]$ denotes Gabor feature for the ith class of the training samples. Before dictionary learning, the extracted Gabor feature set is used to form new training data instead of original training sample. If the extracted Gabor feature set is applied to DPL model, the Gabor feature projective dictionary pair learning (GDPL) model will be obtained as follows:

$$\{P^*, D^*\} = \mathrm{argmin}_{P,D} \sum_{k=1}^{K} ||G_k - D_k P_k G_k||_F^2 + \lambda ||P_k \bar{G}_k||_F^2, \ s.t. ||d_i||_2^2 \le 1 \tag{15}$$

During the training process of GDPL model, the analysis sub-dictionary P_k^* is trained to project the training data from class k to significant coding coefficients. For the data from classes other than k, the learned P_k^* will project them to very small coefficients. On the other hand, the synthesis sub-dictionary D_k^* is trained to well reconstruct the data from class k via its projective coefficients $P_k^* G_k$, so the reconstruction error $||G_k - D_k^* P_k^* G_k||_F^2$ will be smaller than $||G_i - D_k^* P_k^* G_i||_F^2 (i \ne k)$. That is to say, the learned sub-dictionary pair from class k can minimize the reconstruction error of the data from class k. If the testing image y is from class k and its Gabor feature vector is $g = \omega(y)$, the reconstruction error $||g - D_k^* P_k^* g||_2^2$ will be smaller than $||g - D_i^* P_i^* g||_2^2, \ i \ne k$. So we can use the class-specific reconstruction error to identify the class label of the testing image y, and the classification function of GDPL algorithm is formulated as follows:

$$\mathrm{identity}(y) = \mathrm{argmin}_i ||g - D_i P_i g||_2 \tag{16}$$

According to the above analysis, the implementation of face recognition algorithm based on Gabor feature and projective dictionary pair learning can be summarized in Table 1.

Table 1. Face recognition algorithm based on GDPL

Gabor feature projective dictionary pair learning phase:

Input: The Gabor feature set $G=[G_1, G_2, ..., G_K]$ of the samples from K classes, parameters λ, τ, γ, m

Output: a synthesis dictionary D, an analysis dictionary P

Process:

1) initialization: $D^{(0)}$, $P^{(0)}$, iteration $t=1$;

2) **for** $k=1:K$ **do**

3) $A_k^{(t)} = (D_k^T D_k + \tau I)^{-1}(\tau P_k G_k + D_k^T G_k)$

4) $P_k^{(t)} = \tau A_k G_k^T (\tau G_k G_k^T + \lambda \bar{G}_k \bar{G}_k^T + \gamma I)^{-1}$

5) update synthesis sub-dictionary $D_k^{(t)}$ based on (14)

6) **end for**

7) if the difference between the energy in two adjacent iterations is less than the threshold, the iteration stops; else, $t=t+1$, and return to step 2)

Classification phase:

Input: the Gabor feature g of testing sample y, a synthesis dictionary D, an analysis dictionary P

Output: the class label of testing sample: $\text{identity}(g)=\text{argmin}_i \|g - D_i P_i g\|_2$

---

4   Experimental Results and Analysis

To evaluate the performance of the proposed algorithm, we compare our method with other six related works on the ORL, extended YaleB, AR, and CMU-PIE face database respectively, including SRC [4], GSRC [11], K-SVD [7], LC-KSVD1 [8], LC-KSVD2 [8], and FDDL [9]. All experiments are done in the computer with 2.71 GHz and RAM of 2.00 GB. ORL database consists of 400 face images from 40 individuals, each providing 10 images under different expressions and angles (as shown in Fig. 1). The extended YaleB database consists of 21888 face images from 38 individuals, each providing 576 images under different poses and illumination conditions. In the paper, the simulation performed in the subset, which contains 2432 face images of 38 individuals under 64 different illumination conditions (as shown in Fig. 2). The AR face database contains 4000 face images from 126 individuals under different illumination conditions, expressions and disguise. We use 1400 face images from 100 individuals under different illumination conditions and expressions (as shown in Fig. 3). The CMU-PIE face database contains 41368 images of 68 individuals including the variations of pose, expression and illumination (as shown in Fig. 4). During the training phase, GDPL converges very fast and can obtain quickly synthesis dictionary D and analysis dictionary P. So the training time of GDPL is far less than other discriminative dictionary learning methods. During testing phase, these comparative algorithms usually use L0-norm or L1-norm sparse coding process to obtain sparse representation coefficients, which is computationally expensive. So their recognition times are generally greater than GDPL.

Fig. 1. Face images under different expressions and angles in ORL database

Fig. 2. Face images under different illumination conditions in extended Yale B face database

Fig. 3. Face images under different expressions and illumination in AR face database

Fig. 4. Face images under different expressions in CMU-PIE face database

4.1 Experiments on ORL Database

Various amounts of samples are selected randomly for training in each class, with the remaining samples for testing. The size of original images in ORL database is 112*92. Each image is down-sampled to 14*12 and form a 168-dimensional column vector. In order to analyze the influence of the number of samples on the recognition rate, we carried out experiments in four cases, and the number of samples in each class was 5, 6, 7 and 8 respectively. In GSRC and GDPL algorithm, we extract the local Gabor feature of face images, then use PCA to reduce dimension and form a 168-dimensional column vector. The parameters are $\lambda = 0.003$, $\tau = 0.05$, $\gamma = 0.001$ in GDPL, and the number of atoms in synthesis sub-dictionary is equal to the amount of training samples per class. Table 2 shows the recognition rate of various algorithms under different amount of training images. Table 3 shows training time and testing time under different amount of training images. As shown in Table 2 and Table 3, with the number of training samples increasing in each class, both recognition rate and training time increase gradually. Compared with other algorithms, the proposed algorithm gets the highest recognition rate. On the whole, the performance of GDPL is the best.

Table 2. Recognition rate (%) of different algorithms on ORL database

Algorithms	5 samples	6 samples	7 samples	8 samples
SRC	88.0	95.0	95.8	96.3
GSRC	96.0	97.5	97.5	98.7
K-SVD	88.0	93.8	95.0	96.3
LC-KSVD1	88.0	94.4	95.0	97.5
LC-KSVD2	89.0	94.4	95.8	97.5
FDDL	88.5	93.2	95.8	98.7
GDPL	98.5	99.4	100	100

Table 3. Training Time and Testing Time of different algorithms on ORL database

		5 samples	6 samples	7 samples	8 samples
Training Time (s)	SRC	0	0	0	0
	GSRC	0	0	0	0
	K-SVD	3.934	5.698	7.483	9.704
	LC-SVD1	11.766	16.786	22.165	28.476
	LC-SVD2	12.263	17.505	24.658	30.540
	FDDL	129.346	197.193	257.099	311.877
	GDPL	3.757	3.913	4.284	4.384
Testing Time (s)	SRC	0.316	0.269	0.240	0.205
	GSRC	0.217	0.210	0.203	0.197
	K-SVD	0.033	0.035	0.033	0.032
	LC-SVD1	0.033	0.034	0.034	0.032
	LC-SVD2	0.033	0.034	0.037	0.033
	FDDL	62.204	57.882	52.495	39.566
	GDPL	0.036	0.032	0.024	0.015

4.2 Experiments on Extended YaleB Database

We randomly select 32 images for training in each class with the remaining images for testing. The size of original images in extended YaleB database is 192*168. Each image is down-sampled to 24*21 and form a 504-dimensional column vector. In GSRC and GDPL algorithm, we extract the local Gabor feature of face images, then use PCA to reduce dimension and form a 504-dimensional column vector. The parameters are $\lambda = 0.003, \tau = 0.05, \gamma = 0.001$ in GDPL, and the number of atoms in synthesis sub-dictionary is 15. Table 4 shows the experimental results for various algorithms. Since the number of training samples is big, all algorithms can obtain a higher recognition rate, but the GDPL algorithm achieves the highest recognition rate of 98.1%.

Table 4. The experimental results on extended YaleB database

	Recognition rate (%)	Training time (s)	Testing time (s)
SRC	96.5	0	5.90
GSRC	97.1	0	5.96
K-SVD	93.7	93.09	1.27
LC-LSVD1	94.0	268.23	1.55
LC-KSVD2	95.0	273.64	1.54
FDDL	96.7	31950	5642.8
GDPL	98.1	27.26	0.80

4.3 Experiments on AR Face Database

We randomly select 7 images for training in each class with the remaining images for testing. The size of original images in extended AR database is 165*120. Each images is down-sampled to 28*20 and form a 560-dimensional column vector. In GSRC and GDPL algorithm, we extract the local Gabor feature of face images, then use PCA method to reduce dimension and form a 560-dimensional column vector. The parameters are $\lambda = 0.003, \tau = 0.05, \gamma = 0.001$ in GDPL. Table 5 shows the experimental results of various algorithms. Since the number of training samples is big, all algorithms can obtain a higher recognition rate, but the GDPL algorithm achieves the highest recognition rate of 96.0% and has at least 3.0% improvement over other methods.

4.4 Experiments on Multi-PIE Face Database

In the Multi-PIE face database, we only use five frontal poses (C05, C07, C09, C27 and C29) for experiment. Hence, there are 170 images for each class. We randomly select 85 images for training in each class with the remaining images for testing. All the images were cropped to 32*32 and form a 1024-dimensional column vector. In GSRC and GDPL algorithm, we extract the local Gabor feature of face images, then use PCA to reduce dimension and form a 1024-dimensional column vector. The parameters are $\lambda = 0.003, \tau = 0.05, \gamma = 0.001$ in GDPL. Table 6 shows the experimental results of various algorithms. Since the number of training samples is big, all algorithms can obtain a higher recognition rate, but the GDPL algorithm achieves the highest recognition rate of 99.2% and has at least 1.2% improvement over other methods.

Table 5. The experimental results on AR database

	Recognition rate (%)	Training time (s)	Testing time (s)
SRC	87.9	0	3.916
GSRC	93.0	0	4.023
K-SVD	89.3	131.68	1.949
LC-LSVD1	89.6	272.37	1.987
LC-KSVD2	90.5	271.45	1.993
FDDL	92.0	23450	3724.5
GDPL	96.0	73.77	1.276

Table 6. The experimental results on CMU-PIE face database

	Recognition rate (%)	Training time (s)	Testing time (s)
SRC	97.2	0	20.50
GSRC	98.0	0	21.15
K-SVD	95.3	286.47	7.53
LC-LSVD1	96.1	593.19	8.24
LC-KSVD2	96.7	586.28	8.21
FDDL	97.9	64264	9564.8
GDPL	99.2	87.63	3.45

5   Conclusions

This paper proposes a Gabor feature and projective dictionary pair learning algorithm for face recognition. Because Gabor feature shows significant robustness to the variations in expression, illumination and angle, the local Gabor features of images are extracted instead of original training sample to learn dictionary. In addition, the representation coefficients are usually obtained by L_0-norm or L_1-norm minimum process with expensive computational cost in most discriminative dictionary learning frameworks. This paper learns jointly a synthesis dictionary with reconstructing capability and an analysis dictionary with the capability of quickly obtaining representation coefficients during dictionary learning. The experimental results show that our algorithm can get a high recognition rate and improve recognition speed efficiently. Our future work is to make the learned synthesis dictionary more adapt for face recognition by adding more effective regularization term.

Acknowledgements. The "Double First-class" International Cooperation and Development Scientific Research Project of Changsha University of Science and Technology under Grant No. 2019IC34.

References

1. Li, S., Liu, F., Liang, J., et al.: Optimization of face recognition system based on Azure IoT edge. Comput. Mater. Continua **61**(3), 1377–1389 (2019)
2. Xu, Y.F., Zhu, Q.S.: A simple and fast representation-based face recognition method. Neural Comput. Appl. **22**(7–8), 1543–1549 (2013). https://doi.org/10.1007/s00521-012-0833-5
3. Wu, Q.F., Li, Y.S., Lin, Y.T., et al.: Weighted sparse image classification based on low rank representation. Comput. Mater. Continua **56**(1), 91–105 (2018)
4. Dong, W.F., Zhang, D.S., Shi G.T.: Centralized sparse representation for image restoration. In: IEEE International Conference on Computer Vision (ICCV), pp. 1259–1266. IEEE, Piscataway (2011)
5. Dong, W.F., Li, X.S., Zhang, D.T., et al.: Sparsity-based image denoising via dictionary learning and structural clustering. In: IEEE Conference on Computer Vision and Pattern Recognition (CVPR), pp. 457–464. IEEE, Piscataway (2011)
6. Chen, Y.F., Pock, T.S., Bischof, H.T.: Learning l1-based analysis and synthesis sparsity priors using bi-level optimization. In: NIPS Workshop, pp. 1–5, NIPS foundation, New York (2012)
7. Wright, J.F., Yang, A.Y.S., Ganesh, A.T., et al.: Robust face recognition via sparse representation. IEEE Trans. Pattern Anal. Mach. Intell. **31**(2), 210–227 (2009)
8. Yang, M.F., Zhang, L.S., Yang, J.T., et al.: Metaface learning for sparse representation based face recognition. In: The 17th IEEE International Conference on Image Processing (ICIP), pp. 1601–1604. IEEE, Piscataway (2009)
9. Ramirez, I.F., Sprechmann, P.S., Sapiro, G.T.: Classification and clustering via dictionary learning with structured incoherence and shared features. In: IEEE Conference on Computer Vision and Pattern Recognition (CVPR), pp. 3501–3508. IEEE, Piscataway (2010)
10. Zhang, Q.F., Li, B.S.: Discriminative K-SVD for dictionary learning in face recognition. In: IEEE Conference on Computer Vision and Pattern Recognition (CVPR), pp. 2691–2698. IEEE, Piscataway (2010)
11. Jiang, Z.F., Lin, Z.S., Davis, L.S.T.: Learning a discriminative dictionary for sparse coding via label consistent K-SVD. In: IEEE Conference on Computer Vision and Pattern Recognition (CVPR), pp. 1697–1704. IEEE, Piscataway (2011)
12. Yang, M.F., Zhang, D.S., Feng, X.T.: Fisher discrimination dictionary learning for sparse representation. In: IEEE International Conference on Computer Vision (ICCV), pp. 543–550. IEEE, Piscataway (2011)
13. Aharon, M.F., Elad, M.S., Bruckstein, A.T.: K-SVD: an algorithm for designing overcomplete dictionaries for sparse representation. IEEE Trans. Sig. Process. **54**(11), 4311–4322 (2006)
14. Yang, M.F., Zhang, L.S.: Gabor feature based sparse representation for face recognition with gabor occlusion dictionary. In: Daniilidis, K., Maragos, P., Paragios, N. (eds.) ECCV 2010. LNCS, vol. 6316, pp. 448–461. Springer, Heidelberg (2010). https://doi.org/10.1007/978-3-642-15567-3_33
15. Liu, C.F., Wechsler, H.S.: Gabor feature based classification using the enhanced Fisher linear discriminant model for face recognition. IEEE Trans. Image Process. **11**(4), 467–476 (2002)
16. Sun, Y.F., Liu, Q.S., Tang, J.T., et al.: Learning discriminative dictionary for group sparse representation. IEEE Trans. Image Process. **23**(9), 3816–3828 (2014)
17. Nocedal, J.F., Wright, S.J.S.: Least-Squares Problems. Springer, New York (2006). https://doi.org/10.1007/978-0-387-40065-5
18. Chartrand, R.F., Wohlberg, B.S.: A nonconvex ADMM algorithm for group sparsity with sparse groups. In: IEEE International Conference on Acoustics, Speech, and Signal Processing (ICASSP), Piscataway, pp. 6009–6013. IEEE (2013)

An Improved HF Channel Wideband Detection Method Based on Scattering Function

Lantu Guo[1], Yanan Liu[2,3(✉)], and Wenxin Li[2]

[1] Beijing Institute of Technology, Beijing, China
[2] Harbin Engineering University, Harbin, Heilongjiang, China
lc19851225@126.com
[3] China Research Institute of Radiowave Propagation, Qinggdao, Shandong, China

Abstract. The ionospheric HF communications channel is a typical dispersive channel. Its rapid change, time-domain broadening, frequency-domain dispersion and deep fading caused by multipath propagation seriously affect the transmission quality. In this paper, the m sequence is used for wideband detection of HF link, and the link delay information and Doppler spread information are obtained. At the same time, aiming at the problem of poor adaptability of Doppler spread estimation algorithm under different SNR, an improved Doppler spread estimation method based on dynamic threshold is proposed, which achieved good results under different SNR.

Keywords: Channel detection · Delay estimation · Doppler spread estimation

1 Introduction

In the face of the increasingly complex communication environment and the need of long-distance communication transmission, it is of great significance to ensure the communication feasibility of long-distance link [1, 2]. How to make full use of channel detection technology, fully understand the characteristics of link channel, establish low power consumption, low interference communication system link is the key problem of communication link building [3, 4].

Many attempts have been made to detect channel link quality [5, 6]. Among them, multipath effect and Doppler effect are the main factors that affect the quality of signal transmission. If the receiver does not perform effective channel estimation and equalization, the performance of the system will seriously deteriorate. Therefore, the accuracy of channel estimation will directly affect the result of equalization.

Many channel analysis work is based on the study of multipath and Doppler problems.

In 1995, Wagner L S carried out a channel detection observation along a high latitude link (1249 km) [7]. He supplemented the propagation of HF across auroral channels before and after solar maximum by using the signal amplitude, delay and Doppler spread characteristics.

Considering the presence of the mid-latitude channel on the HF radio system, Warrington E M studied a series of channel features on a link between Sweden and

© Springer Nature Switzerland AG 2020
X. Sun et al. (Eds.): ICAIS 2020, LNCS 12239, pp. 317–326, 2020.
https://doi.org/10.1007/978-3-030-57884-8_28

the United Kingdom 1,400 km along the mid-latitude channel. The results reveal the strong correlation between channel multipath effect, Doppler effect and time and season [8]. In 2006, the comparative analysis of several measurement paths was given [9]. In 2011, In order to obtain the maximum value of sunspot, Warrington E M et al. measured the multipath and Doppler parameters on the 1440 km long mid-latitude link [10].

Vilella et al. carried out cross-latitude link quality detection from Antarctica to Spain, which can be used for preliminary analysis of channel availability and multipath measurements [11]. In 2008, Vilella et al. improved the detection equipment. A longer time and band range of channel detection is achieve [12]. In 2013, Vilella et al. have given all-day channel detection results covering the whole shortwave link and added confidence intervals to the complex multipath delay to improve the reliability of the evaluation [13]. By 2015, he summarized more than 10 years of exploration work on the channel from Antarctica to Spain, including the hardware facilities of the system, narrowband detection method and broadband detection method [14].

In this paper, the multipath delay and Doppler spread of the channel are obtained by correlation detection algorithm. Then, the related work on channel detection is given. In the following part, an algorithm to estimate the channel parameters is introduced by using correlation detection method. And the fourth section analyzes the simulation result. At last, we give the conclusion of this paper.

2 Basic Theory

2.1 Channel Scattering Function

This paper uses the channel estimation algorithm of cyclic correlation of multi-segment m sequence to send N segments m sequence continuously as training sequence $r[i]$, and the receiver uses a segment m sequence $s[i]$ to do cyclic correlation with it:

$$\phi[n] = \sum_{i=1}^{N} r[n+i]s[i] \tag{1}$$

Where n is the number of points sampled from the m sequence. Then we transform the obtained correlation function to obtain the impulse response of the channel:

$$h[n, \tau] = \phi[nlN_s + \tau], n \in \left[0, \left[\frac{\Delta t F_m}{lN_s}\right] - l\right], \tau \in [0, lN_s - l] \tag{2}$$

Where, l represents the number of m sequences, and N_s represents the number of samples of each point. Δt is the interval between sending m sequences in a loop. The frequency when sampling is set to F_m. The autocorrelation function of channel impulse response is:

$$R_h[\sigma, \tau] = \sum_{n} h^*[n, \tau]h[n+\sigma, \tau]x+y = z \tag{3}$$

Where, σ represent the time variable. Then, the scattering function can be obtained by Fourier transform of the above formula:

$$R_s[\tau, v] = \sum_{\sigma} R_h[\sigma, \tau] e^{-j2\pi\sigma v} \tag{4}$$

Where v represents Doppler shift.

2.2 Channel Parameter Acquisition

The estimation of multipath delay and Doppler spread by using the scattering function is mainly done by adding observation window to the scattering function. The delay window is $[\tau_1, \tau_2]$, and the frequency spread window is $[v_1, v_2]$. The power delay curve is $\phi[\tau]$ represented by thinning $R_s[\tau, v]$ along the Doppler window $[v_1, v_2]$, as shown below:

$$\phi[\tau] = \sum_{v=v_1}^{v_2} R_s[\tau, v] \tag{5}$$

Further, we de-noise it as:

$$\tilde{\phi}[\tau] = \phi[\tau] - \frac{1}{T - (\tau_2 - \tau_1)} \sum_{m \notin [\tau_1, \tau_2]} \phi[m] \tag{6}$$

Finally, we get the delay value, which above the maximum of 70%:

$$\tau_{min} = \overset{min}{\tau} \ \tilde{\phi}[\tau] > 0.7 \max\left(\tilde{\phi}[\tau]\right) \tag{7}$$

$$\tau_{max} = \overset{max}{\tau} \ \tilde{\phi}[\tau] < 0.7 \max\left(\tilde{\phi}[\tau]\right) \tag{8}$$

Therefore, the composite delay is:

$$\tau_c = \tau_{max} - \tau_{min} \tag{9}$$

Similarly, The Doppler spread distribution curve can also be obtained according to the obtained scattering function as:

$$\phi[v] = \sum_{v=v_1}^{v_2} R_s[\tau, v] \tag{10}$$

Then, through a series of operations such as denoising and adding threshold, we can obtain the composite Doppler spread as:

$$v_c = v_{max} - v_{min} \tag{11}$$

Table 1. ITU-R HF channel model

Channel model	Describe	Delay (ms)	Doppler spread (Hz)
iturHFLQ	Low latitude, quiet conditions	[0 0.5]	0.5
iturHFMM	Mid-latitude, medium conditions	[0 1]	0.5
iturHFMD	Mid-latitude, interference conditions	[0 2]	1
iturHFHQ	High latitude, quiet conditions	[0 1]	0.5

3 Simulation Analysis

The short-wave channel model used in the simulation process is four typical HF channels under ITU-R standard, as shown in Table 1.

In the simulation, m sequence was used as the detection signal. SNR range:−5 db ~ 10 dB. At each SNR, 200 experiments were performed. The signal and related parameters are set as shown in Table 2. Where, the sampling rate of the signal is 10000 Hz, and the multipath resolution is the reciprocal of the sampling frequency, that is, 0.1 ms. The detection time of the signal is 10 s, the extended resolution of Doppler is the reciprocal of the detection time, that is, 0.1 Hz.

Table 2. m Sequence and related parameter setting

Parameter	
m Sequence length	127
Sampling rate (Hz)	10000
Detection time (s)	10
Doppler observation window (Hz)	[−5, 5]
Delay observation window (ms)	[−3, 3]

The following is an example of itu-r HFMD channel model to illustrate the channel parameter estimation process.

The channel impulse response obtained by the receiver through correlation operation and matrix transformation of the received signal is shown in Fig. 1. As can be seen from the figure, a total of 800 probes were completed in 10 s. Each probe contains two responses, indicating that there are two propagation paths in the short-wave channel. The normalized pulse amplitudes of different channels are not the same, and the signals will fluctuate irregularly in the course of passing through the channel. It is proved that the ionospheric environment is random and changeable during the propagation of sky wave in short-wave communication, and it is necessary to estimate the parameters of the ionospheric reference channel.

Furthermore, the obtained channel impulse response is transformed to obtain the scattering function of the short-wave channel, as shown in Fig. 2.

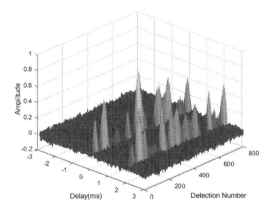

Fig. 1. Channel impulse response

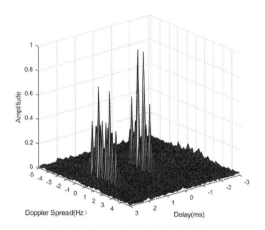

Fig. 2. Channel scattering function

Since the range of the doppler observation window set in Table 2 is $[-5, 5]$ Hz, and the range of the multipath observation window is $[-3, 3]$ ms. Therefore, only the scattering function matrix within the limit of the observation window can be captured in the simulation process for analysis. In addition, because the spread resolution of doppler is 0.1 Hz, that is, the frequency interval between two adjacent coordinates in the intermediate frequency domain of the scattering function is 0.1 Hz. Therefore, the selection range of the coordinate box in the frequency domain is $[-50, 50]$. Similarly, because the multipath time-delay resolution is 0.1 ms, that is, the time interval between the two adjacent coordinates in the time domain in the scattering function is 0.1 ms. Therefore, the time domain coordinate box is selected as $[-30, 30]$. Abundant channel information can be obtained by analyzing the scattering function in different profiles.

By adding threshold to the obtained delay profile curve, the multipath delay estimation of the channel can be extracted. The results of time delay estimation for the four channel models are shown in Fig. 4. As can be seen from the figure, within the range of

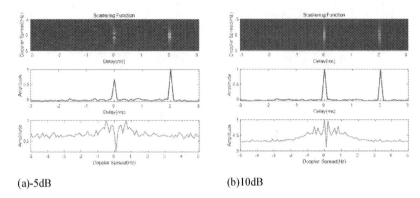

(a)-5dB   (b)10dB

Fig. 3. Scattering function and its corresponding delay and Doppler propagation profile

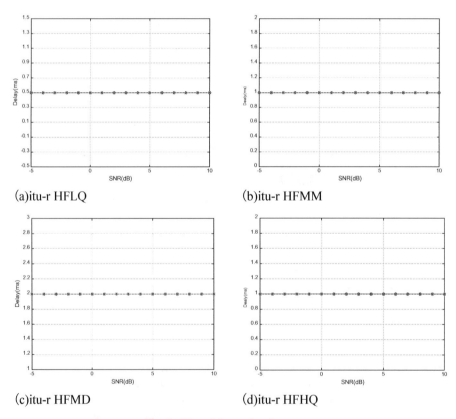

(a)itu-r HFLQ   (b)itu-r HFMM

(c)itu-r HFMD   (d)itu-r HFHQ

Fig. 4. Time-delay estimation curve

−5–10 dB SNR, the delay estimation results obtained based on the m sequence correlation detection method are accurate. Even under the itu-r HFLQ model, when the delay is estimated at 0.5 ms, less than 1 ms, the delay value can be accurately estimated.

Similarly, the Doppler spread estimate of the channel can be extracted by adding threshold to the obtained Doppler spread profile. The results of Doppler spread estimation for the four channel models are shown in Fig. 5.

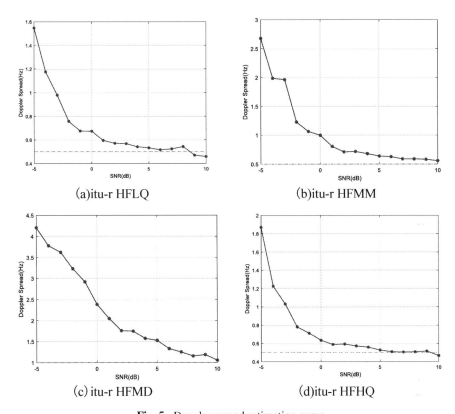

(a)itu-r HFLQ

(b)itu-r HFMM

(c) itu-r HFMD

(d)itu-r HFHQ

Fig. 5. Doppler spread estimation curve

As can be seen from the figure, at −5 dB, the Doppler spread values estimated by the correlation detection algorithm of m sequence have large errors, which are all higher than the real values. This is because when the signal noise is relatively low, the Doppler spread profile curve is greatly affected by the noise, so that the part of the Doppler spread profile curve below the extraction threshold also exceeds the threshold under the influence of the noise, which is considered as the signal component in the Doppler spread. With the increase of SNR, the estimated value of Doppler spread gradually approaches the real value. Starting from 5 dB, the curve of Doppler spread estimation gradually flattens out. In addition, for the two channel models, itu-r HFLQ and itu-r HFHQ, when the SNR is around 6 dB, the error between the estimated value and the real value is the smallest. When the SNR continues to rise to 10 dB, the estimated result is lower than the real value, and the error between the real value and the estimated value increases slightly. This is because in the parameter estimation of Doppler spread profile curve, the selected threshold is fixed, and is usually selected around 6 dB to take into account different noise environments. Therefore, when the SNR is around 6 dB, the algorithm has the highest estimation accuracy for Doppler spread. When the SNR is off 6 dB, a certain estimation error will be introduced due to the mismatch of threshold setting.

Therefore, we consider to improve the original Doppler spread calculation method, and propose a dynamic threshold calculation method that can be adjusted according to different noise environment as follows:

$$\theta = \frac{1}{v_2 - v_1} \sum_{v=v_1}^{v_2} \phi(v) - \xi \tag{12}$$

Where, θ is the variable that determines the dynamic range of threshold, and θ is a regulating factor with a range of [0, 1]. According to the analysis of multiple experimental results, here we set it as 0.5.

Therefore, the Doppler spread is expressed as:

$$v_{\min} = \overset{\min}{v} \tilde{\phi}[\tau] > (0.7 + \theta)\max\left(\tilde{\phi}[v]\right) \tag{13}$$

$$v_{\max} = \overset{\max}{v} \tilde{\phi}[\tau] < (0.7 + \theta)\max\left(\tilde{\phi}[v]\right) \tag{14}$$

$$v_c = v_{\max} - v_{\min} \tag{15}$$

In order to better compare the estimation errors before and after the improvement of noise resistance of Doppler spread estimation algorithm. The mean normalized mean square error is introduced as the evaluation criterion. The mean value of normalized mean square error of each parameter is taken as the standard of parameter estimation and evaluation. Normalized mean square error is expressed as:

$$NMSE_r = E[\frac{(\hat{x} - x_0)^2}{x_0^2}] \tag{16}$$

Where, x_0 represents the true value of the estimated parameter, and $\hat{x}$ represents the estimated value of the estimated parameter.

Then the mean normalized mean square error of channel parameter estimation is:

$$\overline{NMSE}_{r,\hat{\theta}} = \left(NMSE_{r,\hat{\theta}_1} + NMSE_{r,\hat{\theta}_2} + \cdots NMSE_{r,\hat{\theta}_N}\right)/N \tag{17}$$

Where, θ represents the parameters of channel estimation and N represents the total number of paths.

The comparison between the original Doppler spread estimation algorithm and the improved Doppler spread estimation algorithm is shown in Fig. 6.

Can be seen from the diagram, the four kinds of shortwave channel, within the scope of the signal-to-noise ratio, dynamic threshold is worth to Doppler spread NMSE are lower than the original fixed threshold method: when the fixed threshold to 5 dB four shortwave channel model of Doppler estimate NMSE in between, with the increase of signal-to-noise ratio, Doppler spread NMSE gradually decreases, and basic on; When

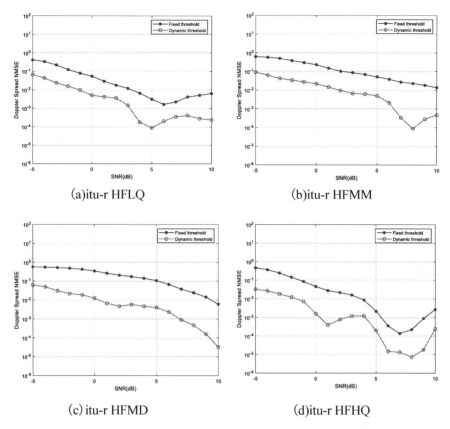

(a)itu-r HFLQ

(b)itu-r HFMM

(c) itu-r HFMD

(d)itu-r HFHQ

Fig. 6. Doppler extended NMSE contrast

the dynamic threshold is adopted, the Doppler estimation NMSE of the four short-wave channel models of −5 dB are all in between. With the increase of SNR, the Doppler spread NMSE gradually decreases, and finally all are lower than, and the minimum NMSE reaches below. In addition, it can be seen from the simulation results that for the short-wave models itu-r HFLQ, itu-r HFMM and itu-r HFHQ, when the SNR is around 6 dB, the Doppler spread NMSE reaches the minimum value. This is because the dynamic threshold algorithm in this paper is expanded on the basis of the original fixed threshold, so the estimated result accuracy trend is still the same as before.

Therefore, the noise resistance of Doppler spread estimation is significantly improved by using the correlation detection algorithm after dynamic threshold, and the mean normalized mean square error of Doppler spread estimation is reduced by at least 1 order of magnitude at any SNR.

4   Conclusion

Because of the complexity of channel environment, the requirement of channel quality has been improved. How to use channel detection technology to obtain effective channel information is a key problem to enhance the performance of communication system. In this paper, four typical HF channels were estimated by multipath and Doppler spread using wideband detection method. At the same time, an adaptive Doppler estimation method based on the original scattering function estimation algorithm is proposed, which greatly improves the anti-noise performance of the Doppler spread estimation algorithm.

References

1. Bergadà, P., Deumal, M., Vilella, C., et al.: Remote sensing and skywave digital communication from Antarctica. Sensors **9**(12), 10136–10157 (2009)
2. Feng, X., Zhang, X., Xin, Z., Yang, A.: Investigation on the Chinese text sentiment analysis based on convolutional neural networks in deep learning. Comput. Mater. Continua **58**(3), 697–709 (2019)
3. Wang, J., Yu, G., Lei, W., Wu, W., Lim, S.-J.: An asynchronous clustering and mobile data gathering schema based on timer mechanism in wireless sensor networks. Comput. Mater. Continua **58**(3), 711–725 (2019)
4. Long, S., Zhao, M., He, X.: Yield stress prediction model of RAFM steel based on the improved GDM-SA-SVR algorithm. Comput. Mater. Continua **58**(3), 727–760 (2019)
5. Badia, D., Bergadà, P., Salvador, M., et al.: Spread spectrum high performance techniques for a long haul high frequency link. IET Commun. **9**(8), 1048–1053 (2015)
6. Joan, P., David, A., Joan, T., et al.: Remote Geophysical observatory in Antarctica with hf data transmission: a review. Remote Sens. **6**(8), 7233–7259 (2014)
7. Wagner, L.S., Goldstein, J.A., Rupar, M.A., et al.: Delay, doppler, and amplitude characteristics of hf signals received over a 1300-km transauroral sky wave channel. Radio Sci. **30**(3), 659–676 (1995)
8. Warrington, E.M., Stocker, A.J.: Measurements of the Doppler and multipath spread of HF signals received over a path oriented along the mid-latitude trough. Radio Sci. **38**(5), 1 (2003)
9. Warrington, E.M., Stocker, A.J., Siddle, D.R.: Measurement and modeling of HF channel directional spread characteristics for northerly paths. Radio Sci. **41**(2), 1–13 (2006)
10. Warrington, E.M., Stocker, A.J.: The effect of solar activity on the Doppler and multipath spread of HF signals received over paths oriented along the midlatitude trough. Radio Sci. **46**(2), 1–11 (2011)
11. Vilella, C., Bergadà, P., Deumal, M., Pijoan, J.L., Aquilué, R.: Transceiver architecture and Digital Down Converter design for long distance, low power HF ionospheric links. In: Proceedings of the Ionospheric Radio Systems and Techniques, London, UK, pp. 95–99, 18–21 July 2006
12. Vilella, C., Miralles, D., Pijoan, J.L.: An Antarctica-to-Spain HF ionospheric radio link: sounding results. Radio Sci. **43**(4), 1–17 (2008)
13. Ads, A.G., Bergadà, P., Vilella, C., et al.: A comprehensive sounding of the ionospheric HF radio link from Antarctica to Spain. Radio Sci. **48**(1), 1–12 (2013)
14. Hervas, M., Alsina-Pages, R.M., Orga, F., et al.: Narrowband and wideband channel sounding of an Antarctica to Spain ionospheric radio link. Remote Sens. **7**(9), 11712–11730 (2015)

Research on Quality Control of Marine Monitoring Data Based on Extreme Learning Machine

Yuanshu Li[1], Feng Liu[2], Kai Wang[1], Hai Huang[1], Fenglin Wei[1], and Hong Qi[1(✉)]

[1] College of Computer Science and Technology, Jilin University, Changchun, China
qihong@jlu.edu.cn
[2] Department of Computer Science and Technology, College of Information and Controlling Engineering, Jilin Institute of Chemical Technology, Jilin, China

Abstract. The collection and research of marine monitoring data are the main work of marine science. How to obtain the observation data of un-known sea area is of great significance to marine scientific research. Using the data provided by the Argo project, in this work we aim to effectively predict the marine monitoring data with the extreme learning machine which is simple in structure with short training time, in order to achieve the accuracy of marine information. According to the size of the sample for adjusting the corresponding model parameters, the pre-diction of marine data can be achieved. By comparing the forecast data of the unknown sea area and the data originally provided by the Argo plan, the accuracy of the ocean data is higher and the trend of the reflected data is more detailed.

Keywords: Marine monitoring data · Extreme learning machine · Prediction · Accuracy

1 Introduction

The changes of the ocean constantly affect the survival and development of mankind. In order to make more friendly and efficient use of the marine resource environment to develop human activities in various fields, we must pay close attention to and understand the dynamics and changes of the ocean at all times. This paper mainly aims to study and predict ocean observation information and makes up for the shortcomings that it is hard to observe ocean data on the ground in reality. This makes it particularly important to collect and process effectively marine data. This paper firstly makes a general explanation of the collection method and processing method used in marine data research. Then it explains that neural network algorithm is applied to the processing of ocean observation data, which is completely feasible according to the existing technical conditions. But there are some obvious disadvantages of neural network algorithm in the past, that is, using neural network algorithm to train an algorithm model, which takes a lot of time and is unfriendly in this era of information that pursues efficiency. So how to use more efficient and fast algorithms to study ocean observation data becomes difficult.

© Springer Nature Switzerland AG 2020
X. Sun et al. (Eds.): ICAIS 2020, LNCS 12239, pp. 327–338, 2020.
https://doi.org/10.1007/978-3-030-57884-8_29

The emergence of extreme learning machines (ELMs) makes up for this deficiency [1]. ELM proposes a theory that hidden neurons are important for a wide range of networks, but do not need to be adjusted. ELM is a feed-forward neural network for classification or regression with a single layer of hidden layer nodes where the weights that connect input to the hidden layer node are randomly assigned and never updated. Compared with other neural network algorithms, it is faster and does not require artificial adjustment of training parameters.

Argo is an acronym for "Array for Real-time Geostrophic Oceanography" [2]. Data are collected 24 h a day, transmitted via satellite to the Internet database, and then combined data sharing in various locations to form an observation array that provides a global sample of data. This paper begins with the optimization of the Argo dataset to save it in a reasonable format, dimension, and precision. Then 3/4 of the data samples are trained through the extreme learning machine, and the results are compared with the other 1/4 of the samples to see the training results. Finally, according to the training results, adjust the data sample or the extreme learning machine model and repeat the above process until the expected reasonable results.

The structure of this paper is as follows: in section two, we give the research of the extreme learning machine (ELM) and the Argo data set; in section three, we introduce the theory of the extreme learning machine model; the experimental results are presented in section four with a brief analysis of the experimental results; the final part puts forward relevant conclusions.

2 Preliminary

2.1 ELM

In 2012, Mohammed proposed a method based on curvelet image decomposition of human faces and a subband that exhibits a maximum standard deviation which is dimensionally reduced using an improved dimensionality reduction technique [3]. In 2014, Wan et al. proposed a new method for prediction intervals formulation based on the ELM and the pairs bootstrap, which is effective for probabilistic forecasting of wind power generation with a high potential for practical applications in power systems [4]. Bai et al. developed an efficient training algorithm based on iterative computation, which scales quadratically with regard to the number of training samples. Sparse ELM outperforms support vector regression (SVR) in the aspects of generalization performance, training speed and testing speed. In 2017, Losifidis et al. [5] proposed graph embedded ELM (GEELM) algorithm, which is able to naturally exploit both intrinsic and penalty SL criteria that have been (or will be) designed under the graph embedding framework. In addition, they extended the proposed GEELM algorithm in order to be able to exploit SL criteria in arbitrary (even infinite) dimensional ELM spaces. In 2016, Huang et al. [6] proposed a computationally efficient method for traffic sign recognition (TSR). This proposed method consists of two modules: (1) extraction of histogram of oriented gradient variant (HOGV) feature and (2) a single classifier trained by extreme learning machine (ELM) algorithm.

2.2 Argo Dataset

In response to traditional ocean data and surveys, countries around the world cooperate and propose Argo programs. Argo is an acronym for "Array for Real-time Geostrophic Oceanography" and the Chinese meaning is "Geo-To-Oceanography Real-Time Observation Array". Real-time refers to a 24-h, non-stop collection of data, and the data collected are transmitted by satellite to the Internet database Then, through the combination of data sharing in various places, Argo forms an observation array, providing a global sample of data [7]. This is the role of the Ocean Observation Network. So how to achieve this huge data collection work requires a device called a profile buoy. The device carries a variety of instruments that detect ocean data, and it has a unique advantage - easy to launch, either by sailing ships or by flying ordinary aircraft. It makes up for the shortcomings of traditional ocean survey ships in terms of safety, rapidity and even traveling to the rare sea area. It can also collect a wide range of data, available in a variety of models, and can serve as a project for any marine scientific research. And it is able to work at a depth of 1000 m or deeper, so it won't drift elsewhere for a long time, thus meeting the steady requirements of data collection [8–10].

3 Extreme Learning Machine Model

ELM is considered as a special type of FNN in the study, or an improvement of FNN and its reverse propagation algorithm, characterized by the random or artificial weight of the hidden layer node and no need to update. Its learning process only calculates the output weight.

3.1 SLFN

The SLFN function with L hidden layer nodes can be mathematically described as:

$$f_L(x) = \sum_{i=1}^{L} \beta_i G(a_i, b_i, x), x \in R^n, a_i \in R^n \tag{1}$$

Here a_i and b_i represent the learning parameters of the hidden layer nodes. β_i is the weight that connects the ith hidden layer node to the output node. The output value of the ith hidden layer node is represented by $G(a_i, b_i, x)$ based on input x. Implicit nodes with g(x) activation functions: R-R (e.g., S-shaped and threshold) [11].

$$G(a_i, b_i, x) = g(a_i x + b_j), b_i \in R \tag{2}$$

Here a_i indicates the weight vector of the input layer connected to the ith hidden layer node. In addition, b_i represents the deviation of the ith hidden layer node R_n. x is the inner product of the vector ai and x in R_n. RBF hidden layer nodes can be found by activating the function g(x).

$$G(a_i, b_i, x) : R \rightarrow R$$

$$G(a_i, b_i, x) = g(//x - a_i//), b_i \in R^+ \tag{3}$$

a_i and b_i represent the center and influence of the first RBF node. A collection of all positive real values is represented by R^+. The RBF network is a special case of SLFN with RBF nodes in its hidden layer. For N any different sample $(x_i, t_i) \in R^n \times R^m$, x_i is $n \times 1$ input vector, and t_i is $m \times 1$ target vector [12].

$$f_L(x_j) = \sum_{i=1}^{L} \beta_i G(a_i, b_i, x_j), j = 1, \ldots, N. \tag{4}$$

$$H(\tilde{a}, \tilde{b}, \tilde{x}) = \begin{pmatrix} G(a_1, b_1, x_1) & \ldots & G(a_L, b_L, x_1) \\ & \ldots & \\ G(a_1, b_1, x_N) & \ldots & G(a_L, b_L, x_N) \end{pmatrix}_{N \times L}$$

$$\tilde{a} = a_1, \ldots, a_L$$

$$\tilde{b} = b_1, \ldots, b_L$$

$$\tilde{x} = x_1, \ldots, x_N \tag{5}$$

$$\beta^T = [\beta_1^T, \beta_2^T, \ldots, \beta_L^T]_{L \times m}^T$$

$$T^T = [t_1^T, t_2^T, \ldots, t_L^T]_{N \times m}^T \tag{6}$$

H is the hidden output matrix of SLFN, where the nth column of H is the output of the ith implicit node based on the input $x_1, \ldots, x_N$.

3.2 ELM Principle

ELM designed with L hidden neurons can learn L different samples with zero error SLFN. Even if the number of hidden layer neurons (L) is less than the number of samples in the input layer, EML can still assign random parameters to the hidden layer node and calculate the output weight by false inversion of H, giving only a small error of 0. The hidden layer node parameters of ELM a_i and b_i do not need to be tuned throughout the training process and can be easily assigned with random values [13].

Theorem 1: Give SLFN with L RBF hidden layer nodes and the activation function $g(x)$ of infinite differentials in the random intervals of R. Then for any L different input vector $\{x_i | x_i \in R^n, i = 1, \ldots, L\}$ and any uninterrupted probability distribution which is randomly produced, $\{(a_i, b_i)\}_{i=1}^{L}$, the hidden output matrix is reversible and the probability is 1. The hidden output matrix H of the SLFN is reversible, and the $//H\beta - T// = 0$ [14].

Theorem 2: Given any small positive value of 0 and activation function $g(x): R\text{-}R$, which is infinitely differential in any interval, there is $L \leq N$, so that for any different input vector, $\{x_i | x_i \in R^n, i = 1, \ldots, L\}$, $\{(a_i, b_i)\}_{i=1}^{L}$ will randomly generate the probability 1 distribution based on any continuous $\|H_{N \times L}\beta_{L \times M} - T_{N \times M}\| < \varepsilon$.

ELM's hidden layer node parameters should not be adjusted throughout the training process because they can easily be given random values to become linear systems. The output weights can be estimated from this [15]

$$\beta = H^+ L \tag{7}$$

Where H^+ represents H's Moore-Penrose broad inversion, the hidden layer output matrix H can be calculated in several ways. The implementation of ELM uses SVD (singular value decomposition) 24 to calculate H's Moore-Penrose broad inversion because it can be used in all cases. Therefore, ELM is considered as a method of bulk learning.

Algorithm 1. ELM:

Given a training set $N = \{(x_i, t_i) \mid x_i \in R^n, t_i \in R^n, i = 1, \cdots, N\}$, the number of hidden layer nodes is L:

1. Specify the input collection in a random way for the hidden layer, i-1,...,L, and a data

   element corresponds to a weight a_i and a bias b_i

2. Calculate the hidden layer to get the output weight H .

3. Calculate the output weight vector $\beta : \beta = H^+ L$.

---

4 Experiments

The test platforms for this experiment are Windows 10, Python 2.7.3, numpy 1.6.1, scipy 0.10.1, and scikit-learn 0.13.1. This experiment is to train the extreme learning machine model. In this paper, the research focuses on selecting the appropriate number of hidden layer nodes, and discussing the effect of random factors on the extreme learning machine, and finally comparing the accuracy of each activation function to arrive at the model of the extreme learning machine.

4.1 Training Extreme Learning Machine Model

Taking the activation function "sine" as an example, this paper explores the experimental results of the number of hidden layer nodes (Table 1), and uses the training set for the model training of the extreme learning machine. Take the temperature of 12 months in 2013 as an example.

From Fig. 1, it can be seen that as long as the choice of the number of hidden layer nodes is within the reasonable range, the extreme learning machine model can achieve an accuracy of up to 99%. But in the data prediction, the fitting phenomenon will appear sometimes, resulting in a large deviation of the forecast data. With the increase of the number of nodes, the random factor has less and less influence on the accuracy of the model. When the number of hidden layer nodes reaches more than 20, the accuracy curve tends to be stable and close to 1. Here is the effect of the adjustment activation function on the accuracy of the model (Fig. 2).

Table 1. Adjusting parameter results record.

activation_func	n_hidden	random_state	Accuracy rate
Sine	5	0	0.269987138091
	5	5	0.52200507827
	5	10	0.470357003321
	10	0	0.779702617159
	10	5	0.566649036734
	10	10	0.455618232351
	15	0	0.585897391339
	15	5	0.899130586651
	15	10	0.879947442724
	20	0	0.850895136351
	20	5	0.953468211885
	20	10	0.889518222815
	25	0	0.931417655963
	25	5	0.959695839157
	25	10	0.952606485821
	30	0	0.962499068949
	30	5	0.952049686282
	30	10	0.962499286485

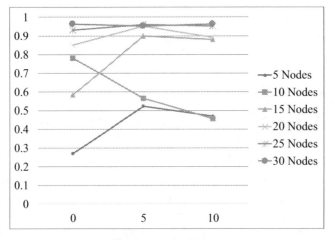

Fig. 1. Data in Table 1.

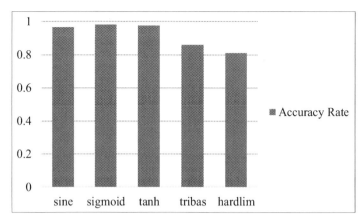

Fig. 2. Effect of different activation functions on model accuracy.

In Fig. 2, it is found that sigmoid, sine and tanh are more accurate as activation functions. Comparing the characteristics of these activation functions, we can find that the value domain of sigmoid is 0,1, and the output of the function is not centered on 0. The characteristic range of the data in this study is near 0, so the use of tanh function is in line with the requirements of this experiment. The scope of the mapping is near 0, so now the tanh function is more widely used as an activation function.

4.2 Using Extreme Learning Machines to Predict Unknown Sea Information

This experiment uses the global ocean monitoring data set provided by the Argo program with an accuracy of $1° \times 1°$degree. The purpose of this summary is to predict ocean information with more detailed precision. Location information contains longitude, dimension, and depth. Latitude range is from $14°$ N to $10°$ N, and longitude range is from $110°$ E to $115°$ E. And the original data set is spaced at $1°$. The depth range is 0–1000 m and the interval is 10 m. However, for the problem in this paper, the depth information obtained is limited from 0 to 300 m, the interval is 5 m, the latitude and longitude are set to latitude 10.5 N to 13.6 N and longitude 110.5 E to 114.6 E, and the interval is $0.25°$.

The next step is to standardize this location information. Once the standardization process is complete, it can be applied directly to the model's predictions. The predicted data set is standardized data, and it is necessary to reversely standardize the data and make a restore of the data. The finely refined location information and predicted data information are combined output and saved in Table 2.

Using extreme learning machine to simulate predicted sea level information is significantly more granular than images originally drawn by the Argo dataset. But the shape of the contours of the two images and the areas corresponding to the colors are very inconsistent. Although the same point is compared, such as the red dot in the lower left corner, it is found that the original data is $26.9205°$ and the prediction data is $26.9952°$. The difference between the two is only about $0.07°$. However, then look at the data reports of both (Table 3, Table 4) and find that the predicted data minimum is $25°$, the

Table 2. Temperature data with a position accuracy of 0.25°.

Lat.	Lon.	Dep.	Temp.
10.5	110.5	285	13.13
10.5	110.5	290	13.04
10.5	110.5	295	12.95
10.5	110.5	300	12.87
10.5	110.75	0	26.51
10.5	110.75	5	26.12
10.5	110.75	10	25.73
10.5	110.75	15	25.35
10.5	110.75	20	24.97

maximum value is 28°, and the average value is 27.33°. The original data minimum is 26°, the maximum value is 27°, and the average is 26.87°.

Table 3. Prediction of ocean surface temperature data sample report (a).

50%-tile	75%-tile	90%-tile	95%-tile	99%-tile	Minimum	Maximum
27.41	27.82	28.09	28.17	28.28	25.35	28.32

Table 4. Raw ocean surface temperature data sample report (a).

1%-tile	5%-tile	10%-tile	25%-tile	50%-tile	75%-tile	90%-tile	95%-tile	99%-tile	Mean
26.19	26.32	26.43	26.68	26.93	27.08	27.23	27.29	27.35	26.87

Although the predicted model accuracy is 96, for a small range of ocean surface temperature data, the average difference is nearly 0.5°, which means a low accuracy. The data sample is too small to prove the validity of the model. Therefore, the data for high-precision locations is predicted based on small sample data. In the selection of samples, we consider that temperature is related to the month, so we select the January 2013 data set to filter the ocean surface temperature, i.e. longitude, dimension, and temperature data at a depth of 0. The total number of data is 24 and they are shown in Table 5.

After that, according to the previous experimental steps, the data experiences test set, training set, x matrix, y matrix separation, standardized processing, and finally goes into the extreme learning machine model processing. In addition, the position information of 0.25° × 0.25° needs to be regenerated, for a total of 289.

Then we have to use the same standardizer to standardize the data for the location information. Pre-processed data is trained by using the extreme learning machine

Table 5. Ocean surface temperature data for January 2013.

Lat.	Lon.	Temp.	Lat.	Lon.	Temp.	Lat.	Lon.	Temp.
10.5	110.5	26.92	11.5	113.5	27.23	13.5	111.5	26.74
10.5	111.5	26.99	11.5	114.5	27.36	13.5	112.5	26.62
10.5	112.5	27.07	12.5	110.5	26.78	13.5	113.5	26.72
10.5	113.5	27.21	12.5	111.5	26.95	13.5	114.5	26.83
10.5	114.5	27.37	12.5	112.5	26.98	14.5	110.5	26.26
11.5	110.5	26.81	12.5	113.5	27.11	14.5	111.5	26.43
11.5	111.5	26.99	12.5	114.5	27.23	14.5	112.5	26.2
11.5	112.5	27.08	13.5	110.5	26.61	14.5	113.5	26.13

model. In the earlier data prediction, in order to prevent the occurrence of overfitting phenomenon, the number of nodes in the hidden layer is generally about 30 for getting reasonable prediction results. But now there is very little sample data, which will lead to the phenomenon of underfitting, that is, the model does not learn enough of the sample data, resulting in the correct rate of negative value and so on. So we can significantly increase the number of hidden layer nodes. The model used in this study is: the hidden layer node is set to 2000, the random factor is 4, and the activation function is sigmoid. The accuracy of the obtained model is 96.6%. The resulting forecast data is counter-standardized, and the resulting data are as shown in Table 6.

Table 6. Forecast ocean surface temperature data.

Lat.	Lon.	Temp.
10.5	110.5	26.92
10.5	110.75	26.94
10.5	111	26.95
10.5	111.25	26.97
10.5	111.5	26.99
10.5	111.75	27
10.5	112	27.02
10.5	112.25	27.04
10.5	112.5	27.07

By comparing Tables 5 and 6, there is little difference between the forecast data and the raw data. The following is a visualization of these two datasets, as shown in Figs. 3 and 4.

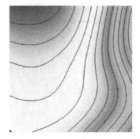

Fig. 3. Higher precision region sea level temperature profile.

Fig. 4. Original regional sea level temperature profile.

By comparing Fig. 3 with Fig. 4, it can be found that the contour shape and temperature area of Fig. 3 are basically the same. The contour comparison of Fig. 3 is sleeker, and the temperature area is more delicate. The original data sample is only 25, and after the extreme learning machine forecast, the data sample increases to 289, more than 10 times the amount of data. Comparing the reports of the two images, we can see in more detail that the difference between the two is very small, seeing Tables 7 and 8.

Table 7. Forecasting ocean surface temperature data sample report.

1%-tile	5%-tile	10%-tile	25%-tile	50%-tile	75%-tile	90%-tile	95%-tile	99%-tile
26.21	26.37	26.47	26.71	26.94	27.08	27.23	27.30	27.35

As shown in the table, whether the maximum value and minimum value or, average, median, root Square and other parameters, the difference slot sits around 0.02° combining with visual images, we can verify that the extreme learning machine model can be used for the prediction of ocean data. The extreme learning machine model in this paper is used for predicting ocean observation data. This experiment visualizes the data and compares it with the raw data by simulating the data results. Through reasonable adjustment, the extreme learning machine model can realize the prediction of ocean data, so it can be concluded that the extreme learning machine can be applied to the quality control of marine monitoring data.

Table 8. Raw ocean surface temperature data sample report.

Median	Geometric mean	Harmonic mean	Root mean square	Trim mean (10%)
26.94	26.89	26.89	26.89	26.90
Interquartile mean	Midrange	Winsorized mean	Tri mean	
26.93	26.75	26.90	26.92	

5 Conclusions

When using the ELM model to predict the data, increasing the number of hidden layer nodes to improve the accuracy of the model is a common mistake. Overfitting will appear, so that the predicted results become abnormal. The improvement for these problems is to find the right number of nodes, avoiding model deviations and overfitting. Bayesian ELM allows the introduction of prior knowledge and obtains confidence intervals without the need to apply computationally intensive methods. The results of this paper show that the method can reduce the overfitting probability of the model. In the processing of small sample data, we can increase the hidden layer nodes, otherwise it will lead to insufficient model training and the phenomenon of underfitting. This experiment is feasible for the quality control of marine data based on the extreme learning machine, makes good use of the advantages of the fast training speed of the extreme learning machine, can have a good prediction of the monitoring data in the unknown sea area, and provide more high-precision marine data for future scientific researches.

Acknowledgments. This work was supported in part by the National Natural Science Foundation of China (61872160, 51679105, 51809112).

References

1. Huang, G., Song, S., Gupta, J.N., Wu, C.: Semi-supervised and unsupervised extreme learning machines. IEEE Trans. Cybern. **44**(12), 2405–2417 (2014)
2. Zheng, A., Labrinidis, A., Chrysanthis, P. K., Lange, J.: Argo: architecture-aware graph partitioning. In: 2016 IEEE International Conference on Big Data (Big Data), pp. 284–293. IEEE (2016)
3. Mohammed, A.A., Minhas, R., Wu, Q.J., Sid-Ahmed, M.A.: Human face recognition based on multidimensional PCA and extreme learning machine. Pattern Recogn. **44**(10–11), 2588–2597 (2011)
4. Wan, C., Xu, Z., Pinson, P., Dong, Z.Y., Wong, K.P.: Probabilistic forecasting of wind power generation using extreme learning machine. IEEE Trans. Power Syst. **29**(3), 1033–1044 (2013)
5. Jiang, Y., Gou, Y., Zhang, T., Wang, K., Hu, C.Q.: A machine learning approach to argo data analysis in a thermocline. Sensors **17**(10), 2225 (2017)

6. Qin, H.D., Wang, C., Jiang, Y., Deng, Z.C., Zhang, W.: Trend prediction of the 3D thermo-cline's lateral boundary based on the SVR method. EURASIP J. Wirel. Commun. Networking **2018**(1), 252 (2018)

7. Jiang, Y., Zhang, T., Gou, Y., He, L.L., Bai, H.T., Hu, C.Q.: High-resolution temperature and salinity model analysis using support vector regression. J. Am. Intell. Humanized Comput. (2018). https://doi.org/10.1007/s12652-018-0896-y

8. Jiang, Y., Zhao, M.H., Hu, C.Q., He, L.L., Bai, H.T., Wang, J.: A parallel fp-growth algorithm on world Ocean Atlas data with multi-core CPU. J. Supercomput. **75**(2), 732–745 (2019)

9. Zhao, M.H., Hu, C.Q., Wei, F.L., Wang, K., Wang, C., Jiang, Y.: Real-time underwater image recognition with FPGA embedded system for convolutional neural network. Sensors **19**(2), 350 (2019)

10. Qin, H.D., Chen, H., Sun, Y.C.: Distributed finite-time fault-tolerant containment control for multiple ocean bottom flying nodes. J. Franklin Inst. (2019). https://doi.org/10.1016/j.jfrank lin.2019.05.034

11. Liu, Y.Q., Wang, X.X., Zhai, Z.G., Chen, R., Zhang, B., Jiang, Y.: Timely daily activity recognition from headmost sensor events. ISA Trans. **94**, 379–390 (2019)

12. Qin, H., Chen, H., Sun, Y., Chen, L.: Distributed finite-time fault-tolerant containment control for multiple ocean bottom flying node systems with error constraints. Ocean Eng. **189**, 106341 (2019)

13. He, L.L., Bai, H.T., Ouyang, D.T., Wang, C.S., Wang, C., Jiang, Y.: Satellite cloud-derived wind inversion algorithm using GPU. Comput. Mater. Continua **60**(2), 599–613 (2019)

14. Li, X., Liang, Y.C., Zhao, M.H., Wang, C., Jiang, Y.: Few-shot learning with generative adversarial networks based on WOA13 data. Comput. Mater. Continua **60**(3), 1073–1085 (2019)

15. He, L.L., et al.: A method of identifying thunderstorm clouds in satellite cloud image based on clustering. Comput. Mater. Continua **57**(3), 549–570 (2018)

Study on the Extraction of Target Contours of Underwater Images

Luyan Tong[1], Fenglin Wei[2], Yupeng Pan[2], and Kai Wang[2(✉)]

[1] Changchun Architecture and Civil Engineering College, Changchun, China
[2] College of Computer Science and Technology, Jilin University, Changchun, China
wangkai87@jlu.edu.cn

Abstract. Image is an important medium for human to obtain information, and the study of underwater image information plays an important role in marine scientific research. This paper studies the target contour extraction in the complex underwater environment. In this paper, we focus on the external contour calibration, and the study aims to segment the contours of many images with prominent texture, chromatic aberration, and shape differences. We also improve underwater image denoise and enhancement technology, restore target area features as much as possible, and process color channel selection with K-means algorithm. At the same time, the external contour extraction and the filling of the target area are partially optimized. Multi-scale watershed algorithm is used to improve the performance of the algorithm. Finally, the Canny and Grabcut algorithms are selected to detect the edges of the image. The experimental results verify that our method is feasible to extract the contours of fish in complex environments.

Keywords: Underwater image pre-processing · Target contour extraction · Image segmentation · Target classification and recognition

1 Introduction

The ocean covers most of the surface of the earth. In recent years, the monitoring of the marine environment has always been a hot topic. Due to the complexity of the underwater environment, human beings could hardly enter it without preparation, resulting in the appearance of more emerging technologies to track and extract subsea material. With the development of computer vision and image processing technology, as well as the use of camera and image data, more research topics and complex algorithms have been proposed, such as Remote-controlled vehicles (ROVs) and Autonomous underwater voyager (AUV) [1–3].

Because of the low underwater visibility, the visual range is only a few meters away, and water has a strong weakening effect on light, leading to a reduction of visibility. So the accurate detection of the target has long been the focus of researches around the world [4]. As an important medium of information collection, various related technologies have been widely used in various fields of life, such as industry, daily traffic, clinical research, intelligent networking, high-precision identification, etc. Applying

© Springer Nature Switzerland AG 2020
X. Sun et al. (Eds.): ICAIS 2020, LNCS 12239, pp. 339–349, 2020.
https://doi.org/10.1007/978-3-030-57884-8_30

image processing algorithm to underwater image processing is helpful to underwater target tracking, underwater resource exploration and other fields. This paper studies the extraction of the contours of the target image under the underwater natural light. By using the image enhancement algorithm, the "fish" in the processing image is pre-processed by computer vision, the comparison information of the target and the background of the picture is extracted, and then the two-valued image is processed at a threshold. Based on these, the Canny and Grabcut algorithms are used to extract the contours.

The structure of this paper is as follows: in section two, we give a study of underwater image recognition; in section three, we introduce the underwater environmental information, the edge detection Canny algorithm and the Grabcut algorithm theory; the experimental results are shown in section four with a brief analysis; the last part presents relevant conclusions.

2 Related Work

In 2007, Kashif et al. proposed an approach based on slide stretching. The objective of this approach is twofold. Firstly, the contrast stretching of RGB algorithm is applied to equalize the color contrast in images. Secondly, the saturation and intensity stretching of HSI is used to increase the true color and solve the problem of lighting. Interactive software has been developed for underwater image enhancement. In 2010, Raimondo Schettini and Silvia Corchs reviewed some of the methods that have been specifically developed for the underwater environment. The conditions for which each of them has been originally developed are highlighted and they two compared the performance. In 2011, Chiang and Chen proposed a novel systematic approach to enhance underwater images through a dehazing algorithm, to compensate the attenuation discrepancy along the propagation path [5], and to take the influence of the possible presence of an artificial light source into consideration. The performance of the proposed algorithm for wavelength compensation and image dehazing (WCID) is evaluated objectively and subjectively by utilizing ground-truth color patches and videos downloaded from the YouTube website. Both results demonstrate that images with significantly enhanced visibility and superior color fidelity are obtained by the proposed WCID. In 2013, Hitam et al. [6] presented a new method called mixture Contrast Limited Adaptive Histogram Equalization (CLAHE) color models and it was specifically developed for underwater image enhancement. The method operates CLAHE on RGB and HSV color models and both results are combined together using Euclidean norm. The results show that the proposed approach significantly improves the visual quality of underwater images by enhancing contrast, as well as reducing noise and artifacts. In 2019, Park et al. presented R-CNN, Fast R-CNN and Faster R-CNN methods to automatically detect and recognize the predators in underwater videos. They devised a data model with three classes (seal, dolphin, background). Based on the results, the best model of predator detection using visual deep learning models is Faster R-CNN with 2000 proposals.

3 Methodology

Due to the low underwater visibility, we first analyzed the underwater imaging environment and made an enhanced treatment. Then we used the traditional Canny algorithm and Grabcut algorithm for picture segmentation. Finally, a hybrid Gaussian model was established to optimize Grubcut algorithm.

3.1 Analysis of Underwater Image Features

Due to the unusually significant effect of the water body itself on the reception and scattering of light, and the presence of dissolved organic matters and visible tiny particles in the underwater environment, the images taken underwater often have technical problems such as low visibility, loud noise and uneven display [6, 7]. Underwater light transmission has a certain trajectory to find, which is usually exponential attenuation, even though experiments show that processed pure water has a significant effect on the attenuation of the image. According to the Lambert-Beer theory, the light L is as follows when the transmission distance is D:

$$L = L_0 e^{-cD} \tag{1}$$

In (1), c is a volume attenuation function, in units of m^{-1} that represent a positive change between the distance of light travel in water and the attenuation of energy. Normal attenuation coefficients of deep-sea water, coastal water and bay water are 0.05, 0.2, 0.33 respectively.

As the depth increases, the light into the water begins to decrease. The deeper you go, the darker the underwater image is. When at a certain depth, not only does the amount of light decrease, but the color also fades. The red color starts to disappear at a depth of about 3 m, the orange is lost at a depth of 5 m, and the yellow will eventually disappear at 10 m. Blue travels the farthest in the water, which makes the underwater image dominated only by blue. Besides excessive blue, problems including low brightness and low contrast exist in blurry images.

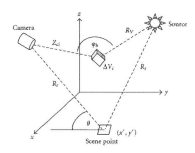

Fig. 1. Coordinate system of the Jaffe-McGlamery model.

We will follow the image imaging model of Jaffe-McGlamery, as shown in Fig. 1. According to the model, underwater images can be made with three components in different directions, namely direct components, forward and background components. The underwater image experiment should conduct tracking from light source to camera, and camera-to-light studies, where direct components represent light that is not directly reflected by objects scattered in the water and forward component is a kind of light reflected by an object. So, the total radiation illumination can be expressed as:

$$E_T = E_d + E_f + E_b \tag{2}$$

E_d, E_f and E_b represent three reflective components. This study mainly uses the color correction method of the ACE model which belongs to a type of unsupervised color equalization algorithm [8]. ACE is applied to videos shot in underwater environments. With the influence of the depth of water and artificial illumination, these videos show a strong and uneven color cast. The internal parameters of the ACE algorithm are appropriately adjusted to meet the requirements of the naturalness of image and histogram shapes and to recover this type of aquatic images. Figure 2 shows the two original sample images and the ACE versions after recovery.

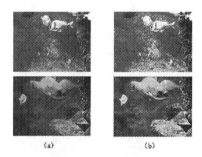

(a)　　　　　(b)

Fig. 2. Original images compared with ACE correction versions.

This paper uses an underwater image of integrated color model for enhancement and the primary application is the use of sliding stretching method. First, use the RGB algorithm to balance the change of color contrast in the image. Second, apply HSI saturation and strength tensile to enhance real color and address lighting issues for calibration of the range from light blue to dark blue. For relatively stable models, spectrometer measurements of different color plates are made. Then, color the spectral data using the Bill's Law:

$$I(z') = I(z) \exp[c(z)z - c(z')z'] \tag{3}$$

I is the pixel intensity used for depth in the image, and z and $c(z)$ are the corresponding attenuation coefficients calculated from spectral data, in which the resulting image is more similar to a shot in shallower depth than in reality. All highlight images are "lifted" to a depth of 1.8 m, and the data can eventually be returned to the original RGB space, as shown in Fig. 3.

3.2 Edge Detection

The theory of edge detection is divided into two parts, one of which is to detect the intensity changes that occur in a wide range of natural images on different scales [9].

Fig. 3. Image comparison before and after RGB algorithm enhancement.

The second is that the intensity change of the image comes from the surface or from the reflection or lighting boundary, and has the characteristic of spatial localization. Common image edge extraction algorithms are divided into Sobel operators, Laplace operators, and the most widely used Canny edge detection.

Canny Edge Detection. Canny Edge Detection is an edge detection algorithm based on image gradient. Canny algorithm first uses the Gaussian filters to smooth images. The original image is represented by $L(x, y)$. Smoothed images can be represented by $R(x, y)$:

$$R(x, y) = G(x, y) \cdot L(x, y) \tag{4}$$

$G(x, y) = \frac{1}{2\pi\sigma^2} e^{\frac{x^2+y^2}{2\sigma^2}}$ is a convolution nucleus that represents a Gaussian filter function, and σ is standard deviation. Smoothing effect has a linear positive correlation with σ. Then the first-order partial derivative is calculated by the Sobel algorithm, and the gradient direction of each pixel is obtained:

$$G(x, y) = \sqrt{G_x^2(x, y) + G_y^2(x, y)} \tag{5}$$

$$\theta(x, y) = \tan^{-1}[\frac{G_y(x, y)}{G_x(x, y)}] \tag{6}$$

Then non-maximum suppression is conducted with the gradient value or along gradient direction. Leave the maximum pixel value in the uniform gradient direction among the surrounding pixels, otherwise suppressed. The gradient amplitude can be determined:

$$M(i, j) = \sqrt{P_x(i, j)^2 + P_y(i, j)^2} \tag{7}$$

M's gradient direction is:

$$\theta(i, j) = \tan^{-1}[\frac{P_y(i, j)}{P_x(i, j)}] \tag{8}$$

$P_x(i, j)$ and $P_y(i, j)$ are the first-order derivatives in the x and y directions, respectively. The Canny algorithm uses high and low thresholds to resolve edge pixels, which are set step by step as T_H and T_L, when the edge pixel gradient value is higher than T_H. Mark gradient values as weak edge points when they are below a high threshold but above a low threshold. Pixels less than the low threshold are suppressed. Canny algorithm is a method of finding the best edge measurement. Each edge is marked only once. For all edge points, iteratively search for points with pixel values greater than the low threshold in 8 neighborhoods and mark them as edge points.

Grabcut Image Segmentation Algorithm. The split model diagram in the figure below shows that the energy equation is based on the max flow algorithm, the minimum energy cutting edge is solved globally at a time, and the energy equation is:

$$E(L) = \alpha R(L) + B(L) \tag{9}$$

In (9), the dotted line between S and each pixel is $R(L)$, the solid line between the pixels is $B(L)$, and the coefficient a is a weight factor which determines which factor is more important. $R(L)$ is calculated as shown in Eqs. 10 and 11, indicating the probability that the pixel l_p belongs to the foreground obj, or to the background bkg probability. If the more likely the pixel point is to the foreground, the smaller the $R_p(1)$ energy is, and the more applicable the theory of the max flow algorithm is. The Graph cut algorithm calculates the foreground and background probabilities based on the respective proportion of the pixel l_p grayscale value in the foreground and background grayscale histogram, and the Grab cut is based on the Gaussian hybrid model:

$$R_p(1) = -\ln P_r(l_p|obj) \tag{10}$$

$$R_p(0) = -\ln P_r(l_p|bkg) \tag{11}$$

Calculate boundary item $B(L)$:

$$B(L) = \sum_{\{p,q\}\in N} B_{<p,q>} \cdot \delta(l_p, l_q) \tag{12}$$

$$\delta(l_p, l_q) = \begin{Bmatrix} 0, & if \ l_p = l_q \\ 1, & if \ l_p = l_q \end{Bmatrix} \tag{13}$$

$$B_{<p,q>} \in \exp(-\frac{(I_p - I_q)^2}{2\sigma^2}) \tag{14}$$

B is to calculate the difference between a pixel l_p and adjacent pixel l_q. The greater the value of $B_{<p,q>}$, the greater the probability that the two pixels should be split, and the smaller the value of $B_{<p,q>}$. In Grabcut, the difference between two pixels is calculated not on grayscale, but by BGR three channels.

3.3 Edge Detection Optimization

Watershed algorithm can better preserve the edge portion of the image and ensure that the space between each area is small enough [10]. Due to the lack of over-segmentation caused by this algorithm, the image is pre-processed using a multi-scale Sobel operator, as shown in the Eq. (15):

$$G(f) = (f \oplus B) - (f \odot B) \tag{15}$$

$\oplus$ and $\odot$ represent expansion and corrosion operations respectively, and B is the structural element. $G(f)$ relies primarily on the value of B. For multiscale morphological gradient operators, definition is as follows:

$$MG(f) = \frac{1}{n} \times \sum_{i=1}^{n} \left\{ \left[(f \oplus B_i) - (f \odot B_i) \right] \odot B_{i-1} \right\} \tag{16}$$

i is any natural number in (16), and B_i is a collection of $i * i$-sized pixels. Because the algorithm is more noise-resistant and has the best retention of edge information, this paper uses this algorithm for pre-image processing [11].

The background GMM and foreground GMM are initialized by the K-means algorithm, where the multi-dimensional Gaussian hybrid model is:

$$D(x) = \sum_{i=1}^{k} \pi_i g_i(x, \mu_i, \sum), \sum_{i=1}^{K} \pi_i = 1, 0 \ll \pi_i \ll 1 \tag{17}$$

x in (17) represents the channel vector of BGR, π_i represents the mixed scale coefficient, μ_i is the mean for each Gaussian probability distribution, and g_i is the probability equation for the ith Gaussian model:

$$P_k(x_i, \theta) = \frac{1}{\sqrt{(2\pi)^d \left| \sum i \right|}} \exp\left[\frac{1}{2} (x - \mu)^T \sum{}^{-1} (x - \mu) \right] \tag{18}$$

For the input image, all pixels are represented by the mixed characteristics of the K GMM, representing the density function under the Class K Gaussian distribution, and Σ_i is the covariance [12]. The maximum log-likelihood can be estimated through the above two equations as follows:

$$L(x) == - \sum_{-1}^{N} \log\left[\sum_{K=1}^{K} \pi_k P_K(x_i, \theta) \right] \tag{19}$$

For the entire image, Gibbs energy is:

$$E(\alpha, m, \theta, x) = U(\alpha, m, \theta, x) + V(\alpha, x)$$
$$= \sum_{n} D(\alpha_n, k_n, \theta, x_n) + V(\alpha, x) \tag{20}$$

$U(\alpha, m, \theta, x)$ represents the data item, and $V(\alpha, x)$ is the smoothed data, which further obtains:

$$D(\alpha_n, k_n, \theta, x_n) = -\log \pi_k + \frac{1}{2} [x_n - u(\alpha_n, k_n)]^T \sum (\alpha_n, k_n)^{-1} [x_n - u(\alpha_n, k_n)] \tag{21}$$

Finally, iterate on the Gaussian hybrid parameter model and get the Eq. (22):

$$V(\alpha, x) = \gamma \sum_{(m,n)\in c} [\alpha_m \neq \alpha_n] \exp\left(-\beta \|x_m - x_n\|^2\right) \qquad (22)$$

When and only when $\alpha_m \neq \alpha_n$, $[\alpha_m \neq \alpha_n] = 1$; when and only $\alpha_m = \alpha_n$, $[\alpha_m \neq \alpha_n] = 0$ [13]. The optimized energy function can eliminate the excess GMM component, which can improve the division accuracy to some degrees. If the difference between pixels in two neighborhoods is small, there is an increased likelihood that both belong to the same goal or background. Conversely, if the differences are great, it means that two pixels may be on the edge of the image and are more likely to be split.

4 Experiment

The experimental test platform used in this paper is Windows10, python 3.6, and opencv 4.0. In this paper, the results of Canny image segmentation algorithm and Grabcut image segmentation method are compared and analyzed. In order to ensure that the results are comparable, the remaining parameters are selected consistently. The evaluation of the final result of the image can be explained from the subjective and objective aspects. Subjective evaluation mainly depends on the first intuitive feeling of human eyes about the segmentation results as the evaluation criterion. But the results of subjective evaluation are usually influenced by individual factors, which causes non-universality. So expected conclusions cannot be drawn from the mathematical models. The objective evaluation method is a relatively reliable quantitative analysis based on the combination of logical thinking and scientific calculation, which can avoid the influence of subjective factors.

4.1 Two-Value Threshold Segmentation

Image segmentation is to divide the image space into areas of interest. In this paper, the contour detection of fish in underwater images is achieved for obtaining black and white images through graphic segmentation. The black-and-white two-value threshold segmentation method takes the target and background in the graph as two components, takes the threshold as the demarcation point, and makes contour extraction of the image with strong contrast between the target and the background grayscale to obtain the boundary between the closed area and the connected area. Results with two-value processing are shown in Figs. 4 and 5.

Figure 4 represents processed color images, with a, b, c, d, e, and f representing two-value graphs, inverted graphs, multi-pixel value graphs, high thresholds, and low threshold plots.

4.2 Edge Detection

The Canny algorithm's determination of the edges of underwater images is shown in Fig. 6.

The Grabcut method's results on the division of the underwater image's foreground and target are shown in Figs. 7, 8, 9 and 10.

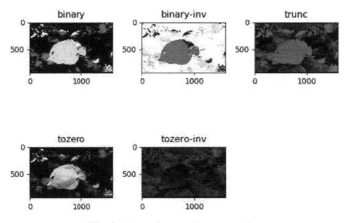

Fig. 4. Normal two-value processing.

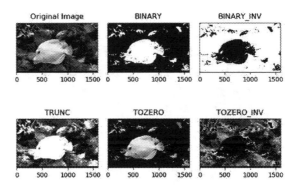

Fig. 5. Black and white flip two-valued results graph.

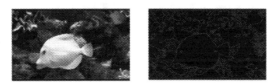

Fig. 6. Canny method image segmentation results.

According to Fig. 6, the Canny algorithm is able to extract the outline of the entire picture. According to the Grabcut method used in this paper, the target fish can be extracted from the picture, which basically removes the complex background behind the fish. In Fig. 7 the color differences between fish and environment is larger, while in Fig. 8 and 9, the difference between fish and environment is smaller. In Fig. 10, the color difference between the two fish is smaller. The method in this paper can extract different targets.

Fig. 7. Grabcut method image segmentation result graph (1).

Fig. 8. Grabcut method image segmentation result graph (2)

Fig. 9. Grabcut method image segmentation result graph (3)

Fig. 10. Grabcut method image segmentation result graph (4)

5 Conclusion

In this paper, the graphics processing technology of the complex underwater environment is briefly summarized. In view of the serious color cast problem of the image, we use the color cast correction algorithm to make color correction of the spectrum on the basis of Bill's law, restores the original features of the image as much as possible. Then, we use the Median filter ingesting algorithm to smooth the target area, retains more image grayscale information, and further improves the contour quality. In the edge detection stage, we use Canny algorithm to detect the inner and outer contours, and use Gaussian filter function to calculate the relationship between image smoothness and σ, and obtain the two-valued image of the target profile. The optimization of target areas with K-means clustering analysis is achieved. The Gaussian mixed parameter model is established, and the image is pre-segmented by multi-scale watershed algorithm. Finally, Canny image segmentation algorithm and CV segmentation algorithm are selected for comparative analysis, so as to obtain a more ideal segmentation effect. In the future work, the study of underwater image imaging law and the further optimization of the existing noise reduction and enhancement technologies will be carried out. According to the characteristics of fish tail fins, dorsal fins and fish eyes, we will design a classification system to achieve the detection and identification of fish targets.

Acknowledgments. This work was supported in part by the National Natural Science Foundation of China (61872160, 51679105, 51809112).

References

1. Jiang, Y., Gou, Y., Zhang, T., Wang, K., Hu, C.Q.: A machine learning approach to argo data analysis in a thermocline. Sensors **17**(10), 2225 (2017)
2. Qin, H.D., Wang, C., Jiang, Y., Deng, Z.C., Zhang, W.: Trend prediction of the 3D Thermocline's lateral boundary based on the SVR method. EURASIP J. Wirel. Commun. Networking **2018**(1), 252 (2018)
3. Jiang, Y., Zhang, T., Gou, Y., He, L.L., Bai, H.T., Hu, C.Q.: High-resolution temperature and salinity model analysis using support vector regression. J. Ambient Intell. Humanized Comput., 1–9 (2018)
4. Jiang, Y., Zhao, M.H., Hu, C.Q., He, L.L., Bai, H.T., Wang, J.: A parallel FP-growth algorithm on world Ocean Atlas data with multi-core CPU. J. Supercomputing **75**(2), 732–745 (2019)
5. Chiang, J.Y., Chen, Y.C., Chen, Y.F.: Underwater image enhancement: using Wavelength Compensation and Image Dehazing (WCID). In: International Conference on Advanced Concepts for Intelligent Vision Systems, pp. 372–383. Springer, Heidelberg (2011)
6. Hitam, M.S., Awalludin, E.A., Yussof, W.N.J.H.W., Bachok, Z.: Mixture contrast limited adaptive histogram equalization for underwater image enhancement. In: 2013 International Conference on Computer Applications Technology (ICCAT), pp. 1–5. IEEE (2013)
7. Zhao, M.H., Hu, C.Q., Wei, F.L., Wang, K., Wang, C., Jiang, Y.: Real-time underwater image recognition with FPGA embedded system for convolutional neural network. Sensors **19**(2), 350 (2019)
8. Qin, H.D., Chen, H., Sun, Y.C.: Distributed finite-time fault-tolerant containment control for multiple Ocean bottom flying nodes. J. Franklin Inst. 1–23 (2019)
9. Liu, Y.Q., Wang, X.X., Zhai, Z.G., Chen, R., Zhang, B., Jiang, Y.: Timely daily activity recognition from headmost sensor events. ISA Trans. **94**, 379–390 (2019)
10. Qin, H., Chen, H., Sun, Y., Chen, L.: Distributed finite-time fault-tolerant containment control for multiple ocean bottom flying node systems with error constraints. Ocean Eng. **189**, 106341 (2019)
11. He, L.L., Bai, H.T., Ouyang, D.T., Wang, C.S., Wang, C., Jiang, Y.: Satellite cloud-derived wind inversion algorithm using GPU. Comput. Mater. Continua **60**(2), 599–613 (2019)
12. Li, X., Liang, Y.C., Zhao, M.H., Wang, C., Jiang, Y.: Few-shot learning with generative adversarial networks based on WOA13 data. Comput. Mater. Continua **60**(3), 1073–1085 (2019)
13. He, L.L., et al.: A method of identifying thunderstorm clouds in satellite cloud image based on clustering. Comput. Mater. Continua **57**(3), 549–570 (2018)

Short-Term Demand Forecasting of Shared Bicycles Based on Long Short-Term Memory Neural Network Model

Ming Du[1], Dandan Cao[2], Xi Chen[2], Shurui Fan[2(✉)], and Zirui Li[2]

[1] The 54th Research Institute of China Electronics Technology Group Corporation,
Shijiazhuang 050081, China
[2] Tianjin Key Laboratory of Electronic Materials Devices, School of Electronics
and Information Engineering, Hebei University of Technology, Tianjin 300401, China
fansr@hebut.edu.cn

Abstract. Shared bicycles have strong liquidity and high randomness. In order to more accurately predict the short term demand for shared bicycles, the long short-term memory (LSTM) neural network model was used as the tool to predict, on the basis of crawling the weather characteristics data of bicycles shared by Citi Bike in New York City, and analyzing the influence of time factor and meteorological factors on the demand for bicycles. On the purpose of verify our method, the traditional RNN and back propagation (BP) neural network were compared with LSTM neural network. The experimental results show that the main factors affecting the demand for shared bicycles including temperature, holidays, seasons and morning and evening peak time periods. Compared with traditional BP neural network and cyclic neural network RNN algorithm, LSTM has high robustness and strong generalization ability. The prediction result curve is consistent with the real result curve, the prediction accuracy is the highest with 0.860 and the root mean square error is the smallest with 0.090. It can be seen that the LSTM model can be used to predict the short-term demand for shared bicycles.

Keywords: Shared bicycles · LSTM neural network · Demand forecast

1 Introduction

With the arrival of the "Internet +" era and the emergence of sharing economy, innovation has gradually emerged in the field of trip mode in China. The scale of shared bicycles system has also grown [1]. Though the sharing economy is developing rapidly, how to accurately predict the short-term demand for shared bicycles in a certain area to achieve reasonable dispatch is an urgent problem to be solved [1]. Now, there are already some machine learning methods, including the ensemble learning methods [2], which have been widely used in the study of shared bicycles. The traditional linear OLS model, the two-category and the multi-category Logit model were used to predict the demand for bicycles in reference [3], but required a large amount of observation data. These models have obvious limitation and the regression relationship obtained does not match the

© Springer Nature Switzerland AG 2020
X. Sun et al. (Eds.): ICAIS 2020, LNCS 12239, pp. 350–361, 2020.
https://doi.org/10.1007/978-3-030-57884-8_31

actual situation. Bacciu et al. (2017) [4] used support vector machine and random forest models for prediction, but did not elaborate on how to predict the short-term demand for shared bicycles.

In order to improve the calculation accuracy and accurately predict the daily demand for shared bicycles in New York City, the prediction method has been transformed from the machine learning algorithms, such as random forest, SVM and ANN [5], to the deep learning algorithm [6, 7]. BP neural network has good organization and adaptability, and its data samples can solve nonlinear problems through learning [8], however, the shortcoming is that it is more likely to form local maximum and minimum during the training process [9]. Feng, Hillston et al. used the Markov chain model with time-dependent rate to predict the future availability of bicycles between different stations and verified the validity of the model [10]. The cyclic neural network RNN is good at processing continuous time series data, but the problem is that vanishing gradient and exploding gradient are prone to occur in the operation. The improved model for the RNN model is the LSTM model, which has been widely used in many timing researches and has achieved good results [11, 12].

The data of demand for bicycles is non-linear time series. Learning historical data by using the RNN with the LSTM, that is, the long and short-term memory unit, can predict the demand of a certain area in a short time [13]. The factors of influencing the demand for shared bicycles include time, weather, holidays, gender, location and other factors [14]. Kim studied the impact on the demand for shared bicycles from natural climate and temperature changes such as different weather and time periods [15, 16]. Zhang et al. [17] also found that the location, size, and service range of the station have an impact on demand for bicycles. El-Assi, Wafic et al. [18] analyzed the impact of surrounding geographical environment and weather factors on bicycle rental. References [19, 20] also describe various methods for feature extraction and feature selection. Based on the data of shared bicycles in New York, this paper analyzed the characteristics of the data and main factors which influences the demand for shared bicycles, and thus proposed the cyclic neural network algorithm LSTM to predict the short-term demand for bicycles. Then, compared with the prediction results of traditional RNN and BP neural network models. The results show that the LSTM model is expected to have a better predictive effect on the data set.

2   Network Structure and Principle of LSTM Model

LSTM neural network, improve the cyclic neural network RNN [21], was originally proposed by Hochreiter and Schmidhuber [22]. Compared with RNN, the advantages of LSTM neural network are that it not only increases the memory storage unit which stores past information, but also solves the vanishing gradient problem and the lack of long-term dependencies by training the data through the back-propagation algorithm. LSTM is widely used in various aspects, such as natural language translation and speech recognition [23, 24], and predicting time series, which has good effects.

The network structure of LSTM works with the gate mechanism, including a memory cell and three control gates, which are input gate, output gate and forget gate [25]. The structure is shown in Fig. 1. The three boxes represent the state of the cells at

352   M. Du et al.

different timings, the small boxes in the middle box are the feedforward network layers with activation function sigmoid, and the two small boxes with built-in tanh are the feedforward network layers with the activation function tanh. In Fig. 1, X_t represents the input at time t, and h_t represents the status value of the cell at time t. Each gate of LSTM works as follows:

First, calculate the value of the input gate and the candidate state value $\tilde{C}_t$ of the input cell at time t, expressed as Eqs. (1) and (2):

$$i_t = \delta(W_i * (X_t, h_{t-1}) + b_i) \tag{1}$$

$$\tilde{C}_t = \tan h(W_c * (X_t, h_{t-1}) + b_c) \tag{2}$$

Secondly, calculate the activation value of the forget gate at time t, expressed as Eq. (3):

$$f_t = \delta\left(W_f * (X_t, h_{t-1}) + b_f\right) \tag{3}$$

Through the above two steps, the updated status value of the cell at time t can be calculated, expressed as Eq. (4):

$$C_t = i_t * \tilde{C} + f_t * C_{t-1} \tag{4}$$

Calculate the value of the output gate, expressed as Eqs. (5) and (6):

$$O_t = \delta(W_o * (X_t, h_{t-1}) + b_o) \tag{5}$$

$$h_t = O_t * \tan h(C_t) \tag{6}$$

After the above four steps, the LSTM can effectively use the input to make it have a long-term memory.

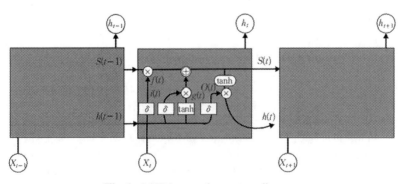

Fig. 1. LSTM network structure diagram.

3   Establishment of LSTM Network Model

First, build the tensorflow framework of the CPU version in the Windows operating system environment. Tensorflow can not only implement deep learning algorithms, but also implement algorithms such as regression forecasting and cluster analysis. The LSTM model is built by the LSTMCell module provided by Tensorflow, which encapsulates the hidden layers of the LSTM in tensorflow and contains three gates, the forget gate, the input gate, and the output gate. The number of hidden layers setted in this text is 10, which is determined according to the actual situation. In order to avoid over-fitting, this paper adopted the dropout mechanism to enhance the generalization of the model.

In recent years, the process of building a network was based on network nodes, and the process of constructing the LSTM framework using tensorflow was constructed by the number of network layers.

In the process of building and training the model, the most important part is the initialization of the parameters. Different types of feature parameters require different types of initialization, which has a very significant impact on the training effect. In this paper, the initialization method for the continuous parameter variables was:

Dummies = pd.get_dummies(rides[each], prefix = each, drop_first = False)

The initialization method for discrete parameter variables is: Mean, std = data[each]. mean (), data[each].std(). The batch random gradient descent method was used to train the model in this paper.

4   Experimental Process and the Results Analysis

4.1   Experimental Environment

This experiment was carried out in Windows7 system, using Anaconda Navigator3 (Jupyter Notebook), Python3.6 as the experimental platform for simulation experiment, and the neural network models such as LSTM provided by tensorflow were used in this simulation experiment.

4.2   Data Acquisition

This experiment used the method of web crawling to crawl the historical weather data of New York from the meteorological service center of the National Oceanic and Atmospheric Administration's official website (NOAA). The data range from January 2015 to December 2018, with a total of 48 months and a total of 35064 data. Some of the data are missing, and their data fields are shown in Table 1.

4.3   Impact Factors Analysis

Impact of Meteorological Factors. Figure 2 shows the thermodynamic diagram of the correlation between the total number of shared bicycles rentals and the five meteorological factors in the New York from 2015 to 2018.

According to Fig. 2, there are correlations between the four meteorological factors and the demand for shared bicycles. Temperature is positively related to the rental

354   M. Du et al.

Table 1. Data set used in the experiment.

Attribute	Implication	Attribute	Implication
dteday	Date	Weather Type	Weather Type
season	Season	tem-max	Maximum Temperature
yr	Year	tem_min	Minimum Temperature
mnth	Month	temp	Average Temperature
hr	Hour	hum	Humidity
holidy	Holiday	snow_fall	Snowfall
weekday	Week	snow_depth	Snow Depth
workingday	Workday	windSpeed	Wind Speed
TLC	Precipitation	subscriber	Registered User
customer	Unregistered User	cnt	Loan Amount

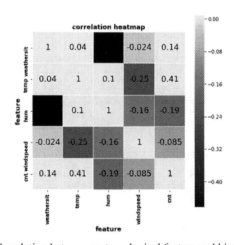

Fig. 2. Correlation between meteorological factors and bicycle usage

number- cold weather causes a decrease in demand, and humidity is negatively correlated to the rental number - rain and snow weather cause a decrease in demand. Clearly, the demand for bicycles has the highest correlation with temperature and humidity, which are 0.41 and −0.19, respectively.

Impact of Time Factors. The usage of shared bicycles is also related to the time factor. By analyzing and fitting the data, the following results were obtained.

Overall Variation. The usage of shared bicycles is affected by time factor. By analyzing the data of shared bicycle projects in New York from 2015 to 2018, the timing change law is shown in Fig. 3.

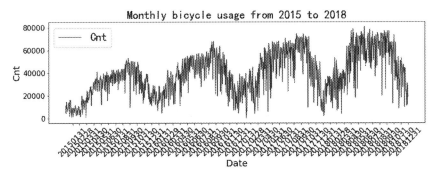

Fig. 3. Line chart of date and shared bicycles demand.

During the period from January 2015 to December 2018, the total usage of shared bicycles increased year by year. Every year from January, the number of bicycle users was increasing rapidly. Until June, the number reached a peak and then slowly decreased in October. After October, it was drastically reduced. In the colder seasons like January, February or December, the number of bicycles used in working days is higher than in non-working days. However, in the warm or cool season that makes people feel comfortable, it is the opposite. Thus, it can be concluded that time factors such as month and year will have a significant impact on the amount of rent. As the impact of the month and season on the number of leases is consistent, and the month is more detailed, we removed seasonal features and remained month features.

Specific Influencing Factors. Figures 4 and 5 further analyzed the influence of the average hourly and weekly usage of shared bicycles from 2015 to 2018, and plotted the column and line charts as follows.

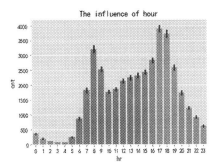
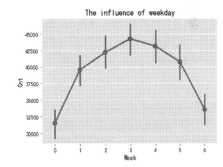

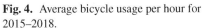

Fig. 4. Average bicycle usage per hour for 2015–2018.

Fig. 5. Average bicycle usage per week in 2015–2018.

The Fig. 4 and 5 show that there is a relatively large number of bicycle hirers on weekdays, with two peak periods per day, one is around 7–8 o'clock in the morning and another is around 5–6 o'clock in the afternoon, which is exactly at rush hour. In addition, the using rate is higher from 12 noon to 4 pm, which further reflects that the time period is an important factor affecting the demand for bicycles, especially during peak hours.

4.4   Data Preprocessing

Supplement to Missing Factors. Data acquired by the above method is incomplete, and the missing data needs to be filled in order to be used in the next prediction.

Virtual Variables. Create binary virtual variables for discrete variables such as season, month, and weather by using the get_dummies() function in the Pandas library.

Adjustment of Target Variables. Adjust the target variable: In order to train the model more easily, continuous variables such as temperature, humidity, and wind speed needed to be normalized to have a mean value of 0 and a standard deviation of 1. Meanwhile, saved the conversion factor, and the data can be restored in the subsequent prediction. As shown in Eq. (7):

$$X_{ij} = \frac{x_{ij} - \min_{1 \leq i \leq N} x_{ij}}{\max_{1 \leq i \leq N} x_{ij} - \min_{1 \leq i \leq N} x_{ij}} \tag{7}$$

In Eq. (7), *max* represents the maximum value of the feature in the selected data, and *min* represents the minimum value of the data feature. x_{ij} represents original data, and X_{ij} represents normalized data.

4.5   LSTM Network Structure Setting

Evaluation Indicator. In order to evaluate the performance of the LSTM prediction model, this paper used RMSE (Eq. (8)) to evaluate the prediction performance of the model. RMSE is the physical quantity used to measure the deviation between the predicted value and the true value. The smaller the value is, the better the model performs.

$$RMSE = \sqrt[1/2]{\frac{1}{n} \sum_{i=1}^{n} (\widehat{y_i} - y_i)^2} \tag{8}$$

In Eq. (8), y_i represents the true value of the sample, $\tilde{y}_i$ represents the predicted value. The predicted value needed to be reverse converted between the raw data interval before the RMSEwas calculated.

Preparation of Network Model. The study used a single-layer LSTM model to predict the demand for shared bicycles in New York. The experiment used the LTSMCell model provided in tensorflow in the deep learning library to predict the hourly demand for bicycles, and used the add function to linearly stack multiple network layers. After continuous debugging and searching, the process of determining network structure parameters is as follows:

Layer Setting. This paper built a single-layer LSTM model structure. The total dimension of the input shared bicycle data was 60, with the values of hidden layers, input layers and output layer were 10, 12 and 1, respectively.

Parameter Setting. The activation function was ReLU function. The number of training samples per batch was 10. The time-step length was setted to 10, the learning rate was setted to 0.0005, the truncation length of the training data was 50, and the size of the vector was 256. In order to prevent over-fitting during training, we setted the discard rate Dropout for each layer of network nodes to 0.75.

Dimensional Transformation. While inputting the features, the tensor needs to be converted into two-dimensional data for calculation, the result was used as the input of the hidden layer, and finally convert the tensor into three-dimensional data as the input of the LSTMCell. Processed the data in batches by using get_batches.

4.6   Contrast Experiment and Result Analysis

Network Training. The experimental data was obtained from the New York City from 2015 to 2018, the total number of shared bicycles usage per hour was 35064, of which 35043 were training data, and the remaining data was used for testing. The number of samples for each training was setted to 10, epoch is the number of training rounds, which was setted to 100. Each round specifies the corresponding test data of the model, and outputs each training record. While training the model, the number of iterations can be changed. The more the number of iterations, the more accurate the prediction is. However, the disadvantage is that it takes a long time. The number of iterations in this experiment was 6600, the loss function was RMSE. Selected adaptive moment estimation Adam as the optimizer. Figure 6 shows the decline of the loss function of the model during the training process. It can be seen from Fig. 6 that the loss function of the model on the training set is decreasing and approaches to 0. Figure 7 shows the change in prediction accuracy of the model on the verification set. It can be seen from Fig. 7 that the accuracy of the verification set gradually increases and approaches to 0.8, which shows that the parameter we selected is the optimal combination of parameters of the model based on the data set.

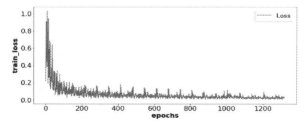

Fig. 6. Training set loss curve.

Prediction Results. The trained model is the optimal model. It was used to test the remaining 21 days of data, and inversely normalized the predicted and actual values of the model. The predicted model has the RMSE value of 0.090, and then compared with the actual value. The result is shown in Fig. 8. It can be seen from Fig. 8 that the LSTM

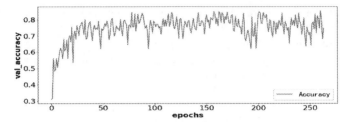

Fig. 7. Verification set accuracy curve.

model can well predict the data, except for the last 10 days. This is because the 10 days are holidays, the demand for bicycles is not the same as usual. The predicted hourly usage curve is in accordance with the actual usage curve, satisfying the empirical error requirement in the regression prediction, which proves that the model has a good fitting effect. Therefore, the LSTM prediction model is feasible in shared bicycles demand forecasting.

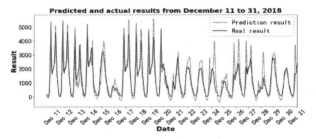

Fig. 8. LSTM neural network prediction result graph.

Performance Analysis. Next, the model was evaluated from two perspectives, which were deviation and performance index.

Comparison of Prediction Deviation. The BP neural network model and the RNN cyclic neural network were used to predict the same shared bicycles dataset, inversely normalized the prediction results and the real results, and then calculated the difference between the predicted value and the actual demand value. The difference is the prediction error, and its curve is shown in Fig. 9. It can be seen after making a comparison among them that the LSTM neural network model offers the best performance, which not only well predict the trend of the data, but also has the smallest prediction error. However, the prediction results and prediction errors of the shared bicycles data by the RNN cyclic neural network and the BP neural network are slightly worse than the LSTM model. Therefore, the LSTM model is more reasonable in terms of shared bicycles demand forecasting.

Comparison of Evaluation Indicators. After the above three models be fitted with all the data, the RMSE and prediction accuracy are shown in Table 2. According to the

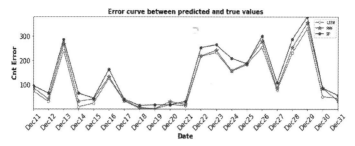

Fig. 9. Three network models prediction error comparison chart

performance index results of each model in Table 2, the LSTM model has a minimum RMSE value with 0.090, and the prediction accuracy is as high as 0.860. The model has the best fitting effect and the variable has the strongest ability to interpret the predicted value. RNN is relatively weaker than LSTM, with the RMSE of 0.131 and the prediction accuracy of 0.780. This is due to the exploding gradient and vanishing gradient problem of RNN during model training, and LSTM can solve this problem. BP neural network has the worst performance. The RMSE value is the largest and the prediction accuracy is the lowest. Overall, the LSTM model has the best performance.

Table 2. Comparison of evaluation indicators of different models.

Model	LSTM	RNN	BP
RMSE	0.090	0.131	0.153
Accuracy	0.860	0.780	0.712

5   Conclusions

In this paper, we focused on how to accurately predict the hourly level of shared bicycles demand in an area. By crawling the shared bicycle dataset of New York and analyzing the impact of various characteristic variables on the total bicycle rental, we finally chose the LSTM neural network to predict the hourly demand for shared bicycles in New York.

Experimental results have shown that factors affecting the demand for bicycles include temperature, holidays, seasons, and rush hour, among which the main reason are temperature and rush hour (7–8 am and 5–6 pm).

Compared with the traditional BP neural network algorithm and the cyclic neural network RNN algorithm, the LSTM model can achieve the highest accuracy of 0.860 in demand prediction than other models. With fairly small error, the prediction result curve of LSTM model is consistent with the real result curve, which can be used to predict the short-term demand of shared bicycles.

This article studies the prediction of the demand for shared bicycles in New York City per hour. In order to help the dispatch center system to make reasonable dispatch, the next work will use the grid division method to predict each area regular amount of cycling between regions.

References

1. Schuijbroek, J., Hampshire, R.C., Hoeve, W.J.V.: Inventory rebalancing and vehicle routing in bike sharing systems. Eur. J. Oper. Res. **257**(3), 992–1004 (2017)
2. Michael, Z., Majid, S.Z.: Understanding ride splitting behavior of on-demand ride services: an ensemble learning approach. Transp. Res. Part C-Emerg. Technol. **76**, 51–70 (2017)
3. Campbell, A., Cherry, C., Ryerson, M., et al.: Factors influencing the choice of shared bicycles and shared electric bikes in Beijing. Transp. Res. Part C **67**(6), 399–414 (2016)
4. Bacciu, D., Carta, A., Gnesi, S., et al.: An experience in using machine learning for short-term predictions in smart transportation systems. J. Logical Algebraic Methods Program. **87**(2), 52–66 (2017)
5. Chen, Y., Xiong, J., Weihong, X., et al.: A novel online incremental and decremental learning algorithm based on variable support vector machine. Cluster Comput. **22**(8), S7435–S7445 (2019)
6. Chen, Y., Wang, J., Chen, X., et al.: Single-image super-resolution algorithm based on structural self-similarity and deformation block features. IEEE Access **7**, 58791–58801 (2019)
7. Chen, Y., Wang, J., Xia, R., et al.: The visual object tracking algorithm research based on adaptive combination kernel. J. Ambient Intell. Humaniz. Comput. **10**(12), 4855–4867 (2019). https://doi.org/10.1007/s12652-018-01171-4
8. Lin, Y., Dou, W.: Research on urban public bicycle demand forecast based on network model. Appl. Res. Comput. **34**(09), 2692–2695 (2017)
9. Liu, M., Shi, J.: A cellular automata traffic flow model combined with a BP neural network based microscopic lane changing decision model. J. Intell. Transp. Syst. **23**(4), 309–318 (2019)
10. Feng, C., Hillston, J., Reijsbergen, D.: Moment-based availability prediction for bike-sharing systems. Perform. Eval. **117**, 58–74 (2017)
11. Fischer, T., Krauss, C.: Deep learning with long short-term memory networks for financial market predictions. Eur. J Oper. Res. **270**(2), 654–669 (2018)
12. Zhang, J., Jin, X., Sun, J., et al.: Spatial and semantic convolutional features for robust visual object tracking. Multimedia Tools Appl. **79**(21), 15095–15115 (2019)
13. Ai, Y., Li, Z., Gan, M., et al.: A deep learning approach on short-term spatiotemporal distribution forecasting of dockless bike-sharing system. Neural Comput. Appl. **31**(5), 1665–1677 (2019)
14. Cai, S., Long, X., Li, L., et al.: Determinants of intention and behavior of low carbon commuting through bicycle-sharing in China. J. Cleaner Prod. **212**, 602–609 (2019)
15. Kim, K.: Investigation on the effects of weather and calendar events on bikesharing according to the trip patterns of bike rentals of stations. J. Transp. Geogr. **66**, 309–320 (2018)
16. Ran, L., Li, F.: Analysis of bike-sharing travel characteristics and travel behavior. J. Transp. Inf. Saf. **35**(06), 93–110 + 114 (2017)
17. Zhang, Y., Thomas, T., Brussel, M., et al.: Exploring the impact of built environment factors on the use of public bikes at bike stations: case study in Zhongshan, China. J. Transp. Geogr. **58**, 59–70 (2017)
18. El-Assi, W., Mahmoud, M.S., Habib, K.N.: Effects of built environment and weather on bike sharing demand: a station level analysis of commercial bike sharing in Toronto. Transportation **44**(3), 589–613 (2017)
19. Chen, Y., Wang, J., Liu, S., et al.: The multiscale fast correlation filtering tracking algorithm based on a features fusion model. Concurrency Comput.-Pract. Experience (2019). https://doi.org/10.1002/cpe.5533
20. Zhang, J., Jin, X., Sun, J., et al.: Dual model learning combined with multiple feature selection for accurate visual tracking. IEEE Access **7**(1), 43956–43969 (2019)

21. Hochreiter, S., Jurgen, S.: Long short-term memory. Neural Comput. **9**(8), 1375–1780 (1997)
22. Polson, N.G., Sokolov, V.O.: Deep learning for short-term traffic flow prediction. Transp. Res. Part C-Emerg. Technol. **79**, 1–17 (2017)
23. Qu, Z., Cao, B., Wang, X., et al.: Feedback LSTM network based on attention for image description generator. Comput. Mater. Continua **59**(2), 575–589 (2019)
24. Shen, Y., Li, Y., Li, Y., Sun, J., et al.: Hashtag recommendation using LSTM networks with self-attention. Comput. Mater. Continua **61**(3), 1261–1269 (2019)
25. Deng, X., Wan, L., Ding, H., et al.: Traffic flow prediction based on deep neural networks. Comput. Eng. Appl. **55**(02), 228–235 (2019)

An Effective Bacterial Foraging Optimization Based on Conjugation and Novel Step-Size Strategies

Ming Chen[1], Yikun Ou[2], Xiaojun Qiu[2], and Hong Wang[2(✉)]

[1] College of Economics, Shenzhen University, Shenzhen, China
[2] College of Management, Shenzhen University, Shenzhen, China
ms.hongwang@gmail.com

Abstract. Bacterial Foraging Optimization (BFO) is a high-efficient meta-heuristic algorithm that has been widely applied to the real world. Despite outstanding computing ability, BFO algorithms can barely avoid premature convergence in computing difficult problems, which usually leads to inaccurate solutions. To improve the computing efficiency of BFO algorithms, the paper presents an improved BFO algorithm: Conjugated Novel Step-size BFO algorithm (CNS-BFO). It employs a novel step-size evolution strategy to address limitations brought by fixed step size in many BFOs. Also, the improved BFO algorithm adopts Lévy flight strategy proposed in LPBFO and the conjugation strategy proposed in BFO-CC to enhance its computing ability. Furthermore, Experiment on 24 benchmark functions are conducted to demonstrate the efficiency of the proposed CNS-BFO algorithm. The experiment results suggest that the proposed algorithm can deliver results with better quality and smaller volatility than other meta-heuristics, and hence sufficiently mitigate the limitation of premature convergence facing many meta-heuristics.

Keywords: BFO · Lévy flight · Conjugation · Adaptive step size

1 Introduction

The so-called nature-inspired algorithm has drawn considerable attention since its very beginning. Because of its excellent ability to tackle many optimization problems (especially multi-minimum and multi-constraints problems) [17], meta-heuristic algorithm is considered one of the most effective measures to address difficult real-world problems for researchers. Due to its advantages in solving those problems, meta-heuristic algorithm remains popular in varying fields.

The most popular nature-inspired meta-heuristics include particle swarm optimization algorithm (PSOA) [8], bacterial foraging optimization algorithm (BFOA) [19], ant colony optimization algorithm (ACOA) [5], genetic algorithm (GA) [7], and so on. Nature-inspired algorithms are mostly derived from observations of natural phenomena or biological behaviors. For example, particle swarm

© Springer Nature Switzerland AG 2020
X. Sun et al. (Eds.): ICAIS 2020, LNCS 12239, pp. 362–374, 2020.
https://doi.org/10.1007/978-3-030-57884-8_32

optimization algorithm [8] is inspired by swarm characteristics of birds seeking foods, and genetic algorithm [7] simulates the process of natural selection and genetic construction. Up until now, researchers are keen on proposing novel meta-heuristics for different optimizations, such as Novel Multi-Hop algorithm for wireless network [11], Novel Improved Bat algorithm for path planning [10], Genetic Frog Leaping for text document clustering [1].

Performance of nature-inspired meta-heuristics is mostly determined by the process of exploitation and exploration [24], which refer to the convergence to best candidate solutions and the diversion to broader search space separately. Exploitation is introduced to perform local search so as to speed up convergence, usually leading to premature convergence in which the algorithms get stuck in local optima. In contrast to exploitation, exploration centers on seeking various solutions by broadening search area to find sufficient candidate solutions for the problem, usually slowing down the convergence process and wasting computational efforts [20]. Lopsided emphasis on either search process can lead to low-efficiency in computation. Hence researchers have exerted enormous efforts in balancing the exploration and exploitation for meta-heuristics [25].

Bacterial foraging optimization algorithm is proposed by Passino in 2002 [19]. Under observation, researchers found the way that Escherichia coli approaches to nutrients and be eliminated under certain adverse conditions. Based on these observation, Passino built a basic four-stage process in the original BFO algorithm. Since the introduction of BFO algorithm, it has been broadly employed in all sorts of fields, including robotic cells [12], power generation [6], sensor network [4], and so on.

In order to boost the computation efficiency of BFO, numerous improved BFO algorithms have been proposed in recent years. Hybridization of BFO algorithm with other meta-heuristics attracts much attention from researchers. Researchers have proposed hybrid BFO algorithms combined with all sorts of algorithms, such as GABFO [9] and DEBFO [2]. Benefiting from advantages inherent in other meta-heuristics, hybrid BFOs dramatically improve their ability to find solutions with decent accuracy.

Aside from hybridizing BFO with other algorithms, introduction of communication mechanisms to BFOs also proved quite efficient in improving BFOs' computing efficiency. The introduction of communication mechanisms serves a purpose of taking better advantage of solutions found by different bacteria in BFOs, increasing the rate of convergence [16]. In this way, BFOs' ability to find solutions with good quality can be enhanced. Hence some researchers have been studying on introducing effective communication mechanisms to BFOs. For instance, social learning mechanism is incorporated into classical BFO algorithm by Yan et al. in 2012 [22]; Adaptive Comprehensive Learning BFO (ALCBFO) [21] is introduced by L. Tan et al. in 2015 to improve balance between exploitation and exploration in BFO; novel chemotaxis and conjugation strategies (BFO-CC) are employed by Yang et al. [23] to propel the development of BFO.

Besides, improvements on chemotaxis step are frequently studied by researchers. Original BFO is featured with fixed step size, which is unhelpful for BFO to fine-tune solutions. Therefore, by improving the step size of BFOs, accurate solutions are more likely to be found by BFOs. Some of the BFOs with improved step size include Adaptive Bacterial Foraging Optimization (ABFO) proposed by Dasgupta et al. in 2009 [3]. Linear decreasing BFO (BFO-LDC) and Non-linear Decreasing BFO (BFO-NDC) are modeled by Niu et al. in 2011 [14]. Gravitation search strategy is combined with chemotactic step by Zhao et al. in the proposed Effective BFO (EBFO) [26]. Recently, Pang B et al. proposed LPBFO [18], which incorporates Lévy flight to the reconstruction of BFO's step-size, proving quite efficient in mitigating premature convergence in BFOs.

Despite huge efforts exerted by researchers, the problem of premature convergence remain rather tricky. In order to propel the development of BFO algorithm, the paper proposes a Conjugated Novel Step-size BFO algorithm(CNS-BFO) to shed some light on the solution of premature convergence. The proposed algorithm employs a novel chemotaxis step-size evolution strategy that enable bacteria to approach best solutions efficiently. Moreover, the proposed CNS-BFO algorithm employs conjugation strategy proposed in BFO-CC algorithm and Lévy flight strategy proposed in LPBFO to enhance the computation ability of BFOs. The total of the three strategies are combined to mitigate the premature convergence facing BFOs, and hence improve their computational efficiency.

The rest of the paper is organized as follows. Classical BFO algorithm is explained in section two. In section three, the paper introduces the proposed CNS-BFO algorithm. Extensive experiments in both unimodal and multi-modal cases are conducted to test the performance of CNS-BFO algorithm in section four. Finally, section five draw conclusions on the proposed CNS-BFO.

2   Classical BFO Algorithm

Bacterial foraging optimization proposed by Passino in 2002 [19] is inspired by biological behaviors of E. coli bacteria. To summarize, classical bacterial foraging optimizations consist of three basic stages: chemotaxis, reproduction, and elimination-dispersal. Subsections below show how the three behaviors work.

2.1   Chemotaxis

E. coli bacteria move towards places of abundant nutrients by rotating flagella. The rotation of flagella can be in either clockwise direction or counter-clockwise direction. Counter-clockwise direction rotation pushes the bacteria to move towards nutrients, while clockwise direction leads to bacteria' s tumbling away from the current position to search more nutrients. Taking both actions, bacteria conduct chemotaxis to approach nutrients gradually.

Suppose bacterium i is denoted as $\theta^i(j,k,l)$, in which j denotes the j-th chemotaxis, k denotes the k-th reproduction, and l denotes the l-th elimination& dispersal, the behaviors of the i-th bacterium can be represented as follows:

$$\theta^i(j+1,k,l) = \theta^i(j,k,l) + C(i)\frac{\Delta(i)}{\sqrt{\Delta^T(i)\Delta(i)}} \tag{1}$$

where $C(i)$ denotes the step size of chemotaxis and $\Delta(i)$ denotes random direction vector in which elements are within $[-1, 1]$. If the health status gets better after tumbling, bacteria will keep on swimming for several steps, in order to further approach optimal solutions.

2.2 Reproduction

After the stage of chemotaxis, fitness values of bacteria over the whole stage of chemotaxis are summarized. The smaller the summation of fitness values, the healthier the bacterium. The computation of health status J_{health} can be represented by the following equation:

$$J_{health} = \sum_{i=1}^{S} \sum_{j=1}^{N_c} J(i, j, k, l) \tag{2}$$

where N_c denotes the number of chemotaxis and J is the evaluated value of the health status of the i bacterium at the k reproduction and the l elimination. In classical BFOs, only the first half bacteria ranked by health status can survive, with their offsprings take the place of the other half of bacteria. Reproduction is designed to strengthen local search ability of BFO while maintaining stable population size of bacteria. The process of reproduction can be represented as follows:

$$\theta^{i+Sr}(j, k, l) = \theta^{i}(j, k, l) \tag{3}$$

where Sr is half of the population size.

2.3 Elimination & Dispersal

To improve global search ability, classical BFOs introduce the stage of elimination & dispersal. After stages of chemotaxis and reproduction, some bacteria face elimination triggered by adverse situations by a possibility P_{ed}. Bacteria will be dispersed to random places within the search range. By introducing stage of elimination & dispersal, the diversity of bacteria is improved and results in stronger global search ability.

3 Conjugated Novel Step-Size BFO Algorithm

In this Section, features of the proposed CNS-BFO are introduced. Pseudocode for the proposed BFO algorithm is shown in Algorithm 1.

3.1 Lévy Flight Step Length Strategy

Levy distribution, proposed by Lévy Paul [13], is an inverse-gamma distribution, whose formula can be expressed as $P_s = s^{-\lambda}$, in which s is a step size and λ is between 1 and 3. Subject to Lévy distribution, Lévy flight is a sort of random moving which can generate frequent small step sizes and occasional big step sizes for bacteria. The characteristics of Lévy flight enable bacteria to move with ununiformed step size, strengthening BFO's capability of exploitation and exploration at the same time: on the one hand, frequent small step sizes fortify the local search ability of bacteria, whereby the exploitation ability of the algorithm can be improved. On the other hand, occasional big step sizes prevent bacteria from over concentrating on exploiting the current position, guiding bacteria to a broader search area, whereby the exploration ability of BFO can be strengthened.

The paper adopts the Lévy flight strategy modified by Bao Pang et al. in 2018 [18]. The strategy can be represented by the equation below:

$$C^{'}(i) = \frac{\alpha}{t(i)} |\frac{u}{|v|^{\frac{1}{\beta}}}| \tag{4}$$

where α is a parameter for controlling the strength of change in step-size, $t = j + (k-1)N_c + (l-1) \cdot rand$. β is a random number between 0.3 and 1.99. u and v are two normal random variables with means equal to zero and with variances of σ_u and σ_v respectively. σ_u and σ_v are calculated as follows:

$$\sigma_u = \frac{\Gamma(1+\beta)sin(\frac{\pi\beta}{2})}{\Gamma[\frac{(1+\beta)}{2}]2^{\frac{\beta-1}{2}}\beta} \quad \sigma_v = 1 \tag{5}$$

Apparently, with bacteria go through more chemotaxis, the step sizes will dwindle as $t(i)$ gets greater, and hence convergence can be accelerated at later stages.

3.2   Novel Step Evolution Strategy

In the original BFO algorithm, chemotaxis step size is set fixed for each bacterium. Fixed step size saps bacterium's strength to exploit, leading to premature convergence easily. To address the limitation, a novel step-size evolution strategy is applied here. After bacteria being ranked and reproduced, step lengths of bacteria will be updated by the following equation:

$$C^{''}(i, k+1) = C(i, k+1)\boldsymbol{b} \tag{6}$$

$$C(i, k+1) = [C^{''}(i, k+1) - C(i, k+1)] \cdot (1 - \frac{1}{lk+1}) \tag{7}$$

where $\boldsymbol{b}$ denotes a $p \times 1$ vector whose elements are all 1, p is the dimension of bacteria. This novel step-size evolution strategy is introduced to make bacteria move adaptively and efficiently. As moving step sizes significantly impact the ability of bacteria to search glocal solution, careful moving can dramatically improve the odds of finding the best solution. The novel step-size evolution strategy allows bacteria to move with bigger step sizes at the beginning, and allow them to fine-tune solutions at later stage with smaller step size. Combined with Levy-flight step length strategy, the adaptive moving strategy can efficiently mitigate the problem of premature convergence.

3.3   Conjugation Strategy

Conjugation strategy proposed in BFO-CC algorithm by Cuicui Yang et al. in 2016 [23] is a novel strategy simulating the process of exchanging genetic makeup among creatures. The process of conjugation is illustrated in Fig. 1. The proposed BFO-CC algorithm introduces conjugation mechanism to bacteria, providing a channel for bacteria to communicate among each other, so that the diversity of bacteria can be improved. In this way, conjugation strategy enhances bacteria's capability of searching for global best. In BFO-CC algorithm, each bacterium chooses a bacterium $\theta^{i'}$ and a conjugation point pt randomly to perform conjugation. The equation for conjugation mechanism is given as follows:

$$\theta^i_{new} = \theta^i(j,k,l) + \boldsymbol{a} \cdot [\theta^{i'}(j,k,l) - \theta^i(j,k,l)] \tag{8}$$

where $\boldsymbol{a}$ is a p-dimension vector in which the values of the elements from pt to $pt + L - 1$ are random numbers uniformly distributed from 0 to 1, with the rest of elements being 0. L indicates the length of exchange of dimensions. In this way, learning process is made more flexible and adaptable, for bacterium can not only learn from other bacterium, but also learn with varying strength on varying dimensions. By taking the conjugation strategy, the bacterium can take advantage of solutions found by each other, whereby avoid premature convergence.

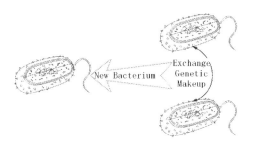

Fig. 1. Conjugation of bacteria

4   Experiments and Analysis

4.1   Experiment Setting

To prove the efficiency of the proposed CNS-BFO, it is to be tested along with other sorts of meta-heuristics: GA [7] and PSO [8]. Besides, as a variant of BFO, CNS-BFO is also to be compared with standard BFO [19] and several popular BFO variants: BFO-LDC [14], BFO-NDC [14], and SRBFO [15] algorithms.

To make sure that the comparisons are fair, algorithms are set to initialize at the same initial positions. The population sizes for the algorithms are set to 50.

Algorithm 1. Pseudocode for CNS-BFO Algorithm

1: Initialize parameters
2: **for** $l = 1,2,...,N_{ed}$ **do**
3:   **for** $k = 1,2,...,N_{re}$ **do**
4:     **for** $j = 1,2,...,N_c$ **do**
5:       **for** $i = 1,2,...,S$ **do**
6:         Compute J^i, and set $J_{last} = J^i$
7:         Tumble by equation (4), then compute J^i
8:         **while** $m < N_s$ **do**
9:           **if** $J^i < J_{last}$ **then**
10:             Update θ^i
11:           **else**
12:             Stop
13:           **end if**
14:         **end while**
15:         **if** $rand < p_{con}$ **then**
16:           Compute θ_{new} by equation (8)
17:           **if** $J^i < J_{last}$ **then**
18:             $\theta^i = \theta_{new}$
19:           **end if**
20:         **end if**
21:       **end for**
22:     **end for**
23:     Compute the health status J^i and sort bacteria by J^i
24:     **for** $j = 1$ to Sr **do**
25:       split
26:       **for** $e = 1$ to N_n **do**
27:         Update $C_{i,k+1}$ by equation (6) and (7)
28:       **end for**
29:     **end for**
30:   **end for**
31:   **for** m $= 1,2,...,$S **do**
32:     Eliminate each bacterium if $rand < p_{ed}$
33:   **end for**
34: **end for**

The total number of FEs is set to 10,000. All the algorithms are tested for 30 times independently on all the test functions. Furthermore, parameters of PSO, GA, BFO, BFO-LDC, BFO-NDC and SRBFO are set as recommended at their source papers, as referred to [7,8,14,14,15,19] respectively. Parameters of CNS-BFO algorithms are set as follows: $N_c = 1,000$, $N_{re} = 10$, $N_{ed} = 2$, $P_{ed} = 0.25$, $Sr = \frac{S}{2}$, $n = 30$, $s = 2$, and $g = 2$, $\alpha = 1,000$, $\beta = 1.5$, $p_{con} = 0.2$, $N_n = 30$, $L = 2$. To comprehensively test the efficiency of CNS-BFO, 24 popular test functions are used in the experiment. Test functions consist of 10 unimodal function (f_1-f_{10}), 10 multi-modal function ($f_{11}-f_{20}$), and 4 2-dimensional multi-modal function ($f_{21}-f_{24}$). Generally multi-modal test functions are considered more difficult than unimodal functions to solve, since algorithms are easier to be trapped in multiple local optima in the case of multi-modal functions. Benchmark functions included in the experiment are shown in Table 1.

Table 1. Test functions: name, search range, and global optimum

Fun	Name	Search range	f_{opt}	Fun	Name	Search Range	f_{opt}
f_1	Sphere	$[-100, 100]^D$	0	f_{13}	Alpine N. 1	$[0, 10]^D$	0
f_2	Sum Squares	$[-10, 10]^D$	0	f_{14}	Xin-She Yang	$[-5, 5]^D$	0
f_3	Powell Sum	$[-1, 1]^D$	0	f_{15}	Xin-She Yang N. 2	$[-2\pi, 2\pi]^D$	0
f_4	Schwefel 2.20	$[-100, 100]^D$	0	f_{16}	Levy	$[-10, 10]^D$	0
f_5	Schwefel 2.21	$[-100, 100]^D$	0	f_{17}	Periodic	$[-10, 10]^D$	0.9
f_6	Schwefel 2.22	$[-100, 100]^D$	0	f_{18}	Quartic	$[-1.28, 1.28]^D$	0
f_7	Schwefel 2.23	$[-10, 10]^D$	0	f_{19}	Salomon	$[-100, 100]^D$	0
f_8	Griewank	$[-600, 600]^D$	0	f_{20}	Powell	$[-4, 5]^D$	0
f_9	Zakharov	$[-5, 10]^D$	0	f_{21}	Schaffer N. 1	$[-100, 100]^D$	0
f_{10}	Dixon-Price	$[-10, 10]^D$	0	f_{22}	Bohachevsky N. 1	$[-100, 100]^D$	0
f_{11}	Ackley	$[-32, 32]^D$	0	f_{23}	Bohachevsky N. 2	$[-100, 100]^D$	0
f_{12}	Rastrigin	$[-5.12, 5.12]^D$	0	f_{24}	Eggcrate	$[-5, 5]^D$	0

4.2   Experiment Results and Analysis

Means and standard deviations of the optima attained by algorithms are showcased in Table 2. Note that the best results have been marked with boldface, and that all of the actual fitness values have been taken the log of to contrast performance of different algorithms. Hence when certain algorithm reaches the optimal value 0 before maximum FEs, its convergence curve seems to "disappear" for the remaining iterations. Also, Fig. 2 reveals convergence graphs on a part of test functions. Only convergence graphs for f_1, f_8, f_{10}, f_{16} are displayed, for the convergence graphs for remaining functions are considered quite similar to the four selected graphs: graph for f_1 represents those for function 1, 2, 3, 4, 5, 6, 7, 11, 13, 14, 18, 19, 20, 24; graph for f_8 represents those for function 12, 21, 22, 23; graph for f_{10} represents those for function 15, 17; graph for f_{16} represents that for function 19.

Clearly, the proposed CNS-BFO outperforms other meta-heuristics in most cases. Table 2 reveals dramatic advantages of CNS-BFO as compared to other algorithms. To begin with, in 23 out of 24 cases, means of best results obtained by CNS-BFO are superior to those by other algorithms. Take f_1 for example, the mean of best result obtained by CNS-BFO is $1.24E^{-50}$, while the best results of other algorithms at most reach $3.56E^{-02}$, a level far subordinate than the result attained by CNS-BFO. Even in cases where CNS-BFO do not attain sufficient advantage (such as f_{17}), it still has the edge over the its counterparts. Besides, it is noticeable that CNS-BFO are exceptionally good at low-dimension functions

Table 2. Means and standard deviations on $f_1 - f_{24}$ over 30 runs

Func		PSO	GA	BFO	BFO-LDC	BFO-NDC	SRBFO	CNS-BFO
f_1	Mean	3.56E−02	3.08E+01	8.42E−02	7.48E+02	6.31E−01	7.51E−02	**1.24E−50**
	Std.	6.17E−02	2.12E+01	1.33E−02	1.70E+02	1.22E−01	1.42E−02	**3.71E−50**
f_2	Mean	2.82E+00	2.94E+00	7.32E−01	7.69E−03	5.27E+00	5.34E−01	**2.69E−47**
	Std.	6.17E+00	1.99E+00	1.30E−01	1.47E−03	9.01E−01	8.18E−02	**1.45E−46**
f_3	Mean	2.84E−03	9.05E−07	2.47E−04	8.44E−07	3.49E−01	1.17E−04	**2.67E−54**
	Std.	4.50E−03	1.34E−06	8.86E−05	3.68E−07	2.72E−01	5.11E−05	**1.46E−53**
f_4	Mean	3.45E+01	9.80E+00	1.12E+00	1.34E+02	3.00E+00	1.01E+00	**9.46E−25**
	Std.	8.72E+01	4.42E+00	9.86E−02	1.93E+01	2.99E−01	8.95E−02	**4.42E−24**
f_5	Mean	8.42E−01	9.73E+00	1.64E+00	3.07E+01	6.84E−01	2.04E−01	**8.90E−20**
	Std.	4.35E−01	3.44E+00	1.11E+00	1.34E+00	1.05E−01	2.57E−02	**2.48E−19**
f_6	Mean	1.55E+12	2.33E+02	3.33E+08	3.67E+09	8.07E+04	1.37E+04	**1.20E−24**
	Std.	2.23E+12	7.06E+01	8.82E+08	1.16E+10	1.40E+05	2.55E+04	**6.49E−24**
f_7	Mean	3.54E−05	5.61E+00	7.39E−08	1.37E−17	3.00E−03	2.27E−08	**2.21E−218**
	Std.	1.15E−04	1.55E+01	5.08E−08	1.02E−17	3.91E−03	2.10E−08	**0**
f_8	Mean	6.95E+00	9.19E−01	1.48E−02	1.31E+00	7.83E−02	2.80E+00	**0**
	Std.	4.52E−15	2.33E−01	7.95E−03	1.08E+00	1.36E−02	7.83E−01	**0**
f_9	Mean	3.45E+00	1.33E+01	1.53E+00	2.51E+02	8.06E+01	8.02E−01	**1.26E−04**
	Std.	7.33E+00	5.76E+00	3.59E−01	8.51E+02	2.67E+01	1.86E−01	**4.19E−04**
f_{10}	Mean	1.66E+01	1.09E+05	1.40E+00	**6.52E−01**	1.11E+01	1.22E+00	6.67E−01
	Std.	3.70E+01	2.82E+05	1.51E−01	**1.12E−01**	3.42E+00	9.64E−02	2.38E−16
f_{11}	Mean	2.62E+00	1.03E+00	7.92E−01	9.37E−01	2.83E+00	5.56E−01	**2.66E−15**
	Std.	6.32E−01	6.49E−01	1.23E−01	9.46E−01	2.40E−01	7.52E−02	**0**
f_{12}	Mean	4.54E+01	2.42E+01	5.37E+01	6.18E+01	1.29E+02	2.45E+01	**0**
	Std.	1.56E+01	8.39E+00	7.31E+00	1.16E+01	1.20E+01	2.38E+00	**0**
f_{13}	Mean	3.67E+00	9.70E−01	3.44E+00	2.78E+00	9.60E+00	1.69E+00	**4.76E−27**
	Std.	1.86E+00	6.25E−01	6.08E−01	1.98E+00	1.92E+00	4.14E−01	**1.81E−26**
f_{14}	Mean	3.81E+02	1.25E−01	8.49E+00	2.55E+02	1.32E+02	4.88E−03	**2.17E−25**
	Std.	2.06E+03	2.51E−01	7.04E+00	2.71E+02	9.36E+01	2.02E−03	**8.93E−25**
f_{15}	Mean	8.30E−04	2.82E−05	6.26E−05	1.17E−05	3.53E−03	8.93E−06	**5.73E−06**
	Std.	6.01E−04	7.51E−07	1.66E−05	1.22E−06	2.20E−03	8.82E−07	**2.65E−09**
f_{16}	Mean	6.55E+00	5.81E+01	1.68E+01	1.65E+01	2.92E+00	1.49E+01	**4.37E−06**
	Std.	4.73E+00	9.53E+01	3.81E+00	3.46E+00	1.21E+00	3.29E+00	**1.40E−05**
f_{17}	Mean	1.00E+00	1.01E+00	1.08E+00	1.00E+00	1.50E+00	1.07E+00	**9.13E−01**
	Std.	4.13E−03	1.02E−02	1.15E−02	1.40E−04	8.80E−02	1.62E−02	**3.46E−02**
f_{18}	Mean	2.08E−02	1.76E+00	1.23E−02	1.48E−06	5.85E−01	7.19E−03	**5.81E−89**
	Std.	5.02E−02	1.74E+00	4.32E−03	4.29E−07	2.10E−01	2.23E−03	**2.28E−88**
f_{19}	Mean	9.59E−01	1.21E+00	1.54E+01	1.52E+01	1.15E+01	1.37E+01	**9.99E−02**
	Std.	2.01E−01	3.46E−01	4.66E−02	5.24E−01	7.04E−01	1.23E+00	**2.62E−02**

(continued)

Table 2. (*continued*)

Func		PSO	GA	BFO	BFO-LDC	BFO-NDC	SRBFO	CNS-BFO
f_{20}	Mean	1.83E−02	1.89E+03	5.37E−01	5.82E−03	8.84E+00	3.34E−01	**8.73E−27**
	Std.	3.28E−02	2.07E+03	1.33E−01	1.38E−03	2.82E+00	8.04E−02	**4.78E−26**
f_{21}	Mean	2.26E−04	2.34E−01	1.25E−01	9.00E−02	9.03E−02	6.69E−03	**0**
	Std.	4.90E−04	2.02E−01	4.33E−02	4.53E−02	4.05E−02	2.02E−02	**0**
f_{22}	Mean	3.56E−01	2.07E+03	6.04E−06	8.33E−08	1.92E−04	3.75E−06	**0**
	Std.	2.12E−01	2.78E+03	6.64E−06	7.33E−08	2.20E−04	3.15E−06	**0**
f_{23}	Mean	2.48E−01	1.41E+03	5.72E−06	4.83E−08	1.98E−04	3.25E−06	**0**
	Std.	8.89E−02	2.40E+03	5.10E−06	4.17E−08	2.85E−04	2.60E−06	**0**
f_{24}	Mean	1.50E−03	2.14E+01	3.16E+00	2.21E+00	1.44E−04	5.08E−06	**8.02E−54**
	Std.	1.19E−03	1.91E+01	4.55E+00	4.08E+00	1.61E−04	5.49E−06	**3.37E−53**

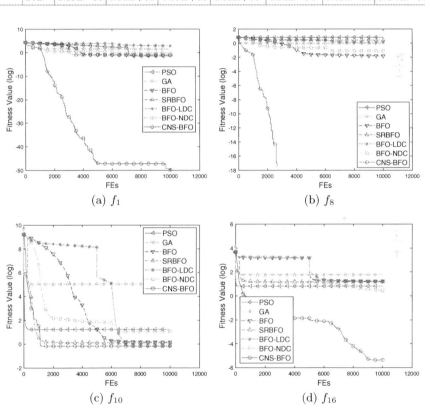

(a) f_1

(b) f_8

(c) f_{10}

(d) f_{16}

Fig. 2. Convergence graphs on f_1, f_8, f_{10}, f_{16} over 30 runs

(from f_1 to f_{24}), in which CNS-BFO obtained the optimal solution in three out of four functions. The same conclusion can be drawn by looking at Fig. 2. Convergence graph for f_8 suggests that CNS-BFO can rapidly converge to best solution in some cases. In graph for f_1 and f_{16}, even though CNS-BFO fails to achieve best results over the 10000 iterations, it can effectively approach far better solutions than its counterparts. Both the figure and table show that CNS-BFO can effectively obtain good solutions in both unimodal and multi-modal problems.

Furthermore, CNS-BFO also achieves satisfying results in the aspect of stability. Table 2 suggests that performance of CNS-BFO comes with very small volatility in most cases, which can be measured by the standard deviations of results over 30 runs. Over the 24 test functions, CNS-BFO has sufficiently smaller standard deviations than other meta-heuristics. Thus, it proves that the proposed CNS-BFO has strong stability in solving complicated problems.

Overall, the proposed CNS-BFO is proved sufficient enough to address both multidimensional problems and unimodal problems. It is also proved to be quite competitive in low-dimension problems. Also, low volatility of the results computed by CNS-BFO shows that it is capable of delivering results with strong stability.

5   Conclusion

Conjugated Novel Step-size BFO algorithm (CNS-BFO) is proposed to improve the performance of BFO algorithm. Improvements lie in the way of constructing a novel step evolution strategy and incorporating both Lévy flight step-size strategy and conjugation strategy. By modifying the step-size strategy and introducing a learning mechanism, CNS-BFO strikes a good balance between exploitation and exploration, whereby significantly mitigates the problem of premature convergence in BFO algorithms. Benchmark functions tests are conducted to compare the performance of the CNS-BFO algorithm with that of other meta-heuristics. The experiment results prove that the proposed CNS-BFO algorithm significantly can deliver solutions with good quality and stability in both unimodal and multi-modal cases, revealing its outstanding capability of solving difficult problems. Future work might focus on the the improvement of the proposed algorithm or its applications in some areas, such as feature selection.

Acknowledgment. This work is supported by the National Natural Science Foundation of China (Grant No. 71901152), Natural Science Foundation of Guangdong Province (2018A 030310575), Natural Science Foundation of Shenzhen University (85303/00000155), Project supported by Innovation and Entrepreneurship Research Center of Guangdong University Student (2018A073825), and Research Cultivation Project from Shenzhen Institute of Information Technology (ZY201717).

References

1. Alhenak, L., Hosny, M.: Genetic-frog-leaping algorithm for text document clustering. Comput. Mater. Contin. **61**, 1045–1074 (2019)

2. Biswas, A., Dasgupta, S., Das, S., Abraham, A.: A synergy of differential evolution and bacterial foraging optimization for global optimization. Neural Netw. World **17**(6), 607 (2007)
3. Dasgupta, S., Das, S., Abraham, A., Biswas, A.: Adaptive computational chemotaxis in bacterial foraging optimization: an analysis. IEEE Trans. Evol. Comput. **13**(4), 919–941 (2009)
4. Deepa, S.R., Rekha, D.: Bacterial foraging optimization-based clustering in wireless sensor network by preventing left-out nodes. In: Mandal, J.K., Sinha, D. (eds.) Intelligent Computing Paradigm: Recent Trends. SCI, vol. 784, pp. 43–58. Springer, Singapore (2020). https://doi.org/10.1007/978-981-13-7334-3_4
5. Dorigo, M., Di Caro, G.: Ant colony optimization: a new meta-heuristic. In: Proceedings of the 1999 Congress on Evolutionary Computation-CEC99 (Cat. No. 99TH8406), vol. 2, pp. 1470–1477. IEEE (1999)
6. Hernández-Ocaña, B., Hernández-Torruco, J., Chávez-Bosquez, O., Calva-Yáñez, M.B., Portilla-Flores, E.A.: Bacterial foraging-based algorithm for optimizing the power generation of an isolated microgrid. Appl. Sci. **9**(6), 1261 (2019)
7. Holland, J.H., et al.: Adaptation in Natural and Artificial Systems: An Introductory Analysis with Applications to Biology, Control, and Artificial Intelligence. MIT Press, Cambridge (1992)
8. Kennedy, J.: Particle swarm optimization. In: Sammut, C., Webb, G.I. (eds.) Encyclopedia of Machine Learning, pp. 760–766. Springer, Boston (2010). https://doi.org/10.1007/978-0-387-30164-8_630
9. Kim, D.H., Abraham, A., Cho, J.H.: A hybrid genetic algorithm and bacterial foraging approach for global optimization. Inf. Sci. **177**(18), 3918–3937 (2007)
10. Lin, N., Tang, J., Li, X., Zhao, L.: A novel improved bat algorithm in UAV path planning. J. Comput. Mater. Contin. **61**, 323–344 (2019)
11. Liu, Y., Yang, Z., Yan, X., Liu, G., Hu, B.: A novel multi-hop algorithm for wireless network with unevenly distributed nodes. Comput. Mater. Contin. **58**(1), 79–100 (2019)
12. Majumder, A., Laha, D., Suganthan, P.N.: Bacterial foraging optimization algorithm in robotic cells with sequence-dependent setup times. Knowl.-Based Syst. **172**, 104–122 (2019)
13. Mantegna, R.N.: Fast, accurate algorithm for numerical simulation of Levy stable stochastic processes. Phys. Rev. E **49**(5), 4677 (1994)
14. Niu, B., Fan, Y., Wang, H., Li, L., Wang, X.: Novel bacterial foraging optimization with time-varying chemotaxis step. Int. J. Artif. Intell. **7**(A11), 257–273 (2011)
15. Niu, B., Bi, Y., Xie, T.: Structure-redesign-based bacterial foraging optimization for portfolio selection. In: Huang, D.-S., Han, K., Gromiha, M. (eds.) ICIC 2014. LNCS, vol. 8590, pp. 424–430. Springer, Cham (2014). https://doi.org/10.1007/978-3-319-09330-7_49
16. Niu, B., Liu, J., Bi, Y., Xie, T., Tan, L.: Improved bacterial foraging optimization algorithm with information communication mechanism. In: 2014 Tenth International Conference on Computational Intelligence and Security, pp. 47–51. IEEE (2014)
17. Niu, B., Liu, J., Wu, T., Chu, X., Wang, Z., Liu, Y.: Coevolutionary structure-redesigned-based bacterial foraging optimization. IEEE/ACM Trans. Comput. Biol. Bioinform. **15**(6), 1865–1876 (2017)
18. Pang, B., Song, Y., Zhang, C., Wang, H., Yang, R.: Bacterial foraging optimization based on improved chemotaxis process and novel swarming strategy. Appl. Intell. **49**(4), 1283–1305 (2018). https://doi.org/10.1007/s10489-018-1317-9

19. Passino, K.M.: Biomimicry of bacterial foraging for distributed optimization and control. IEEE Control Syst. Mag. **22**(3), 52–67 (2002)
20. Sahib, M.A., Abdulnabi, A.R., Mohammed, M.A.: Improving bacterial foraging algorithm using non-uniform elimination-dispersal probability distribution. Alexandria Eng. J. **57**(4), 3341–3349 (2018)
21. Tan, L., Lin, F., Wang, H.: Adaptive comprehensive learning bacterial foraging optimization and its application on vehicle routing problem with time windows. Neurocomputing **151**, 1208–1215 (2015)
22. Yan, X., Zhu, Y., Zhang, H., Chen, H., Niu, B.: An adaptive bacterial foraging optimization algorithm with lifecycle and social learning. Discrete Dyn. Nat. Soc. **2012** (2012)
23. Yang, C., Ji, J., Liu, J., Yin, B.: Bacterial foraging optimization using novel chemotaxis and conjugation strategies. Inf. Sci. **363**, 72–95 (2016)
24. Yang, X.S.: Nature-Inspired Optimization Algorithms. Elsevier, Amsterdam (2014)
25. Yang, X.S., Deb, S., Fong, S.: Metaheuristic algorithms: optimal balance of intensification and diversification. Appl. Math. Inf. Sci. **8**(3), 977 (2014)
26. Zhao, W., Wang, L.: An effective bacterial foraging optimizer for global optimization. Inf. Sci. **329**, 719–735 (2016)

A Coding Scheme Design for the Shape Code of Standardized Yi Characters in Liangshan

Qiyan Hu[1,2], Xiaobing Zhao[2,3(✉)], and Liang Zhang[2,3]

[1] School of Chinese Ethnic Minority Languages and Literatures, Minzu University of China,
Beijing 100081, China
977918599@qq.com
[2] Minority Languages Branch, National Language Resource and Monitoring Research Center,
Beijing 100081, China
[3] School of Information Engineer, Minzu University of China, Beijing 100081, China

Abstract. Liangshan standardized Yi language is a phonographic syllabic language which takes Shengzha dialect of the northern Yi language as the basic dialect and Xide pronunciation as the standard pronunciation. Most standardized Yi characters in Liangshan, Sichuan province are single-character forms composed by basic strokes. According to the principles of resemblance, rotatability, consistency and assimilation, this paper disassembles Yi characters into eight basic strokes including horizontal, vertical, semi-enclosed, full-enclosed, and the broken line, curve, oblique and hook, and then accordingly maps them to the keyboard codes resembling English letters. In addition, the 9-key form coding is adopted to classify and encode the basic strokes so as to make the coding process feature the accessibility of learning, moderate code length, low code repetition, fast input and strong extensibility. The last but not least, the coding form is applicable to convert the strokes to numeric keys and to code the traditional Yi character.

Keywords: Standardized Yi character · Coding rules · Performance analysis

1 Introduction

The Yi is one of the ethnic minorities with a long history in China, and its people mainly live in Sichuan, Guizhou, Yunnan and Guangxi provinces [1]. The Yi language belongs to the Yi Branch of the Tibeto-Burman Group of the Sino-Tibetan Language Family, and is subdivided into six major dialects including the northern, central, western, southern, eastern, and southeastern dialects including five sub-dialects and more than 40 local dialects. In addition, the northern, eastern, southern and southeastern dialects have their own characters except the central and western dialects [2]. The Yi character is an ancient writing which is called Cuanwen, Weishu, Yizi, or Luowen (different forms in Chinese characters) by Han History Records. According to archaeological data, Yi Character has gone through the development of pictures-pictograms-ideograms-phonetic writings [3]. In December 1999, The Yi Coding Character set for Information Exchange was included in The International Information Standards Collections (the edition in 2000). The set was

X. Sun et al. (Eds.): ICAIS 2020, LNCS 12239, pp. 375–383, 2020.
https://doi.org/10.1007/978-3-030-57884-8_33

established according to the standardized implementation of the Yi Language Standard Program approved by the State Council in 1980, in which a total of 1,165 characters including 819 commonly used characters, 345 sub-high tone characters and 1 substitute tone character are included [4]. Since the founding of the People's Republic of China, the Yi language experts have contributed their wisdom in various aspects and designed various Yi language input methods and yielded fruitful success to meet the requirements of the development of the Yi language.

At present, the relatively advanced Yi language input methods include Shamalayi input method, Acai Yi input method, Sogou and Vista system's input method and Chuxiong Yi language input method. Shamalayi input method includes Yi simplified Pinyin code, whole Pinyin code and stroke code. The whole Pinyin code is encoded with Yi phonetic symbols, and the number of keystrokes is up to 5 without repeated codes. The simplified Pinyin code focuses on the pronunciation of the Yi language, and uses its Pinyin or Pinyin code as the input code, which easily performs the input with 4 keystrokes at most. The stroke code disassembles the strokes of characters according to the order in which they are written in the Yi language, and then maps the disassembled ones to the English letters or number keys on the keyboard. Its rules are intuitive and easy to learn. However, lots of coding elements need to be memorized when using the input method [5]. Acai input method and Vista system's input method adopt the whole Pinyin coding method, which can realize the free mixed input of Chinese characters, Yi characters, numbers and symbols [6]. Similarly, whole Pinyin code input also works in performing of the Sogou Yi input method where Yi and Chinese mixed inputs can be realized. The stroke input rule of Chuxiong Yi language takes the roots of the Yi character as the encoding principle, maps the roots to the corresponding English letters on the computer keyboard in the form of "shape-orientation", and realizes the input with 4 keys [7]. Nevertheless, the current input methods come with their own disadvantages such as complex and unmemorable rules, low input speed and high repeated code rate.

With the development of mobile terminals and natural language processing technology, applications based on natural language processing technology have emerged [8]. It is necessary to use modern technology to improve the processing of Yi language. As a powerful scientific project, the design of input methods requires constant verification to get practical results. Therefore, on the basis of previous researches and the standardized Yi Coding Character set in Liangshan, the current study proposes a coding scheme based on segmented glyph mode, which uses variable length coding. The scheme has simple rules, short code, low repetition rate, and fast input speed.

2 Scheme Design

Liangshan standardized Yi character is a phonographic syllabic language which takes Shengzha dialect of the northern Yi language as the basic dialect and Xide pronunciation as the standard pronunciation [9]. Most of its characters are single characters forming by the method of structural translocation, stroke addition, etc. The word construction system includes radical, stroke, stroke sequence and writing structure. Stroke is the smallest unit in the form structure [2]. In view of this, based on the characteristics of its character strokes, this paper splits all characters into 8 kinds of basic strokes: horizontal, vertical, semi-enclosed, full-enclosed, and the broken line, curve, oblique and hook, and then maps them to eight shapes resembling the capitalized English letters, while map the sub-high notes and substitute note to the P key. Furthermore, this program does not stipulate the basic typing sequence of 8 stroke types, but allows the users to combinatorically input them randomly to realize automatic matching by the program itself.

2.1 Coding Classification

The orientation, size and sequence of the basic strokes are not distinguished when coding them. In other words, the shapes of the stroke or the stroke codes are the same. Following the principles of resemblance, rotatability, assimilation and consistency, the eight basic strokes are mapped to the English letters in the right-hand keyboard one to one correspondence.

Resemblance. A certain English letter defines a resemblant stroke. For example, the strokes "丨, ı 丨" are similar to the capital letter I, "∪,U" to a capital "U", "0" to capitalized "O", and "Ɨ,Ɨ" to the capitalized "N" in English, so the similar capitalized letters are adopted to code the corresponding strokes.

Rotatability. The basic strokes with different size, position and orientation share the same shape, which can be seen as the same strokes rotating randomly in a plane [10]. The stroke pairs such as "∪" and "⌒,⊂,⊃", "U" and "∩", " ⌐ " and "⊿, >, <", "Ɛ" and "Ɛ,ᵚ", left oblique "\, \, ╲, ╲" and right oblique "╱ , ╱, ╱" sharing the same shape but different from each other in size and orientation, can be regarded as the same kind of stroke and coded by the same letters.

Assimilation. Basic strokes are classified into 8 categories according to the principle of their shape assimilation. Among them, the strokes ⌐,⌐,Ʒ,C ,⌐ ,Ʋ,Ʋ similar to the "semi-enclosed" in the strokes of Chinese character, are presented by "U"; the strokes Ɛ, S,ᵚ,ᵚ,ᶘ,ʃ,ɹ,ʃ,ᵚᵚ,ɕ,ʳ,ᶘ,ʆ , ʆ,ɕ resembling the lowercase English letter "m" with curved shapes, are classified as curve strokes and coded by "M".

Consistency. The basic strokes are mapped to the English letters on the keyboard in a case-insensitive manner [11]. The strokes "ᵚ,Ɛ,Ʒ,Ɛ,ᵚ" are similar to English lowercase letter "m", which are coded uniformly with the letter M. The classification and coding of the basic strokes are shown in Table 1.

2.2 Coding Rules

The users can combine the strokes randomly for the free input sequence of 8 types of basic strokes. And, the repeated strokes are encoded by "stroke letter + repeated times", while the encoding letter P is added into the commonly used character encoding for the sub-high tone character.

Table 1. Basic strokes classification and coding

No.	Stroke Classification	Basic Strokes	No. of Strokes	Corresponding Code
1	Horizontal	─、 -	2	L
2	Vertical	∣、 ∣ 、 ∣	3	I
3	Semi-enclosed	C、 ⌣、 ⌒、 ⟩、 U、 ⌐、 ↲、 ⊃、 ⊏、 C、 ⊐、 Ʋ、 ∪	13	U
4	Full-enclosed	◌、 O、 ○、 △	4	O
5	Broken Line	∫、 ⌐、 ⌐、 ⌐、 ⌐、 ⌐、 ⌐、 ⌐、 ⌐、 ↘、 ⟩、 ⟨、 ⌐、 ⌐	15	N
6	Curve	Ɛ、 Ʒ、 Ɛ、 S、 ш、 ш、 ȣ、 ȣ、 ﻝ、 ℃、 ﻢﻢ、 ♂、 ʳ、 ⟨、 ⅋、 ⟨、 ♂	17	M
7	Oblique	\、 ⟩、 \、 /、 /、 ⟨	7	K
8	Hook	↲、 ∟、 ⌐	3	J
9	Sub-high/Substitute tone	⌐ ⱳ		P

Description: "horizontal" strokes include long horizontal and short horizontal ones, "vertical" strokes include long vertical and short vertical ones. "semi-enclosed" stroke includes unclosed curve structure, while "full-enclosed" stroke is closed structure. "broken line" refers to the one with a folding stroke, "curve" means a curve stroke completely written down by one stroke, "oblique line" strokes in-clude the left oblique line stroke and the right oblique line stroke.

Sequence. This scheme does not stipulate the sequence of stroke input. That's to say, when the user inputs strokes, the sequence of stroke input in the basic character is not limited, which allows the users to input stroke combinations randomly, and finally the program can realize automatic matching. The character encoding process is shown in Fig. 1.

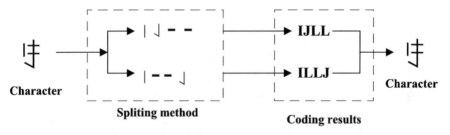

Fig. 1. Diagram of basic stroke coding of glyph

Figure 1 illustrates only two coding forms of character "ꀭ". The character can also be coded by JIL2, JL2I, IJL2, IL2J, L2JI, L2IJ, among them, the code inputs of JILL and JIL2 are equivalent to each other. The program can realize automatic detection and matching according to the coding sequence the user performs.

Repeated Strokes. In order to simplify encoding process and shorten encoding length, this scheme adopts the coding form of "stroke letter + repeated times", and the strokes repeated 3 times or less can also be encoded according to the strokes. For example, the character "ꇆ" is coded by MI4 or I4M, rather than MIIII and IIIIM; "ꑟ" coded by U4N or NU4, rather than UUUUN and NUUUU.

Sub-high Tone Character. Among the 1165 characters in Liangshan, 345 are the sub-high tone characters. "*Yi Language Specification Scheme*" regulates: no specific key is set for sub-high tone which is coded by middle-leveled tone characters with an additional symbol " ˘ " [12]. In this scheme, the letter "P" is added into the commonly used encoding characters of the sub-high tone characters. For example, the character "ꇆ" is coded by MP or PM, "ꀈ" by I5P or PI5, "ꐎ" by UIP, UPI, IPU, IUP, PUI, PIU.

3   Performance Analysis

The national standard regards the coding level and software level as a unified keyboard input system for performance evaluation. In GB18031 *General Requirements for Chinese Character Input of Numeric Keyboard*, it is mentioned that the system performance index includes qualitative index (learnability) required at the coding level, quantitative index (average code length and the selection rate of repeated codes) [13]. In this paper, the efficiency and feasibility of the input method are analyzed by qualitative and quantitative index statistics. Moreover, the sub-high tone characters and the substitute characters are excluded. In previous studies, professor Wang Jiamei's team designed a coding scheme of numeric keys and letter keys based on features of the standardized Yi characters in Sichuan and Yunnan province, as well as the ancient Yi characters. Professor Wu Xie designed the shape code input method with the components of Yi characters in Guizhou province as meta-codes [14]. This paper refers to the "scheme and design of Yi language input method based on free split mode" designed by Feng Hao and Wang Jiamei, whose character set and encoding forms are similar. It is found that the encoding rules of this scheme are simple, easy to remember, the average code length is moderate, the repeated code rate is low, and the input efficiency is high.

3.1   Learnability and Memorability

Learnability, that is, "the time of character input system should be as short as possible, and fit the user's habits of thinking" [13]. The scheme is not completely restricted by the character structure and stroke writing sequence. It adopts the shape encoding method to classify the character strokes reasonably, which fits the thinking habits of users and it is easier to learn and remember. In addition, users are not required to be familiar with the pronunciation of the Yi characters. Beginners can learn the rules and type them in about five minutes.

3.2 Quantitative Index

The quantitative index of the scheme presents moderate average code length, low maximum repetition rate of a single character, high accuracy of character matching. Detailed statistical results are shown in Table 2.

Average Code Length. The scheme is a variable-length coding, and the average code length can be used as a reference index of input efficiency. According to GB18031 *General Requirements for Input of Chinese Characters on Numeric Keyboard*, the average code length of Chinese characters should be <6 with word-by-word field input method [15]. As can be seen from Table 2, the average code length of the two schemes is less than 6 keys which shows slight difference with 3 keys. The characters with the length of 3 and 4 meta-keys show the highest distribution rate, so when characters are input, they present the characteristics of less keystrokes and higher input speed.

Code Repetition Situation. The code repetition is identified based on the code repetition rate, which is applied to measure the input efficiency of the input method [15]. As can be seen from Table 2, the characters under 10 times of code repetition in the target scheme account for 93.89% of the characters, that is, more than 90% of the characters do not need to turn the page for search. The maximum repetition of a single character is 15 times, and a page turn can achieve matching. The maximum repetition times of a single character in the "standardized Yi language form code scheme" set by National University of Yunnan is 24 times, and the corresponding character can be found on 3 pages at most. Therefore, the target scheme is more accurate in character matching, less character needing to be searched on the page, and faster in input.

4 Function Extension

The scheme adopts 9-key encoding form which can be transplanted to the traditional Yi character set, converted into digital keys, transplanted to the mobile phone keyboard, mobile phone and other small handy devices, which shows its higher applicability.

4.1 Convertibility to Numeric Keys

After being converted to numeric keys, the Numbers 1–9 are separately encoded. The inputter is not required to be familiar with the Pinyin and writing structure of the characters, which facilitates the inputting of traditional characters, variant characters and rare characters, saves keyboard resources and is more widely used.

Coding Rules. In encoding, the 8 basic strokes are mapped to Number 1 to 8 respectively, and the sub-high tone and substitute characters are mapped to Number 9, where 1 corresponds to "horizontal", 2 to "vertical", and 3 to "semi-enclosed", 4 to "full-enclosed", 5 to "broken line", 6 to "curve", 7 to "hook", 8 to "oblique", and sub-high tone and substitute characters correspond to the number 9. When encoding, repeated strokes are directly encoded with repeated numbers. Strokes repeated 3 times or more are retained to be encoded 3 times, and the excessive repetition are neglected. For example "卅" character is coded as 52221 rather than 5222221.

Table 2. Performance statistical results of performance

Scheme source	Yunnan Minzu University			The target scheme		
Character set	standardized Yi character in Sichuan province			standardized Yi character in Liangshan		
No. of Characters	3755			819		
Encoding methods	Form code			Form code		
Code key	26 key			9 key		
Easy to learn	Relatively difficult			Easy to learn & remember		
Extensibility	Difficult			Easy		
Code length distribution	Code length	No. of characters	Percent	Code length	No. of characters	Percent
	1	122	3.26%	1	10	1.22%
	2	1159	31.01%	2	127	15.51%
	3	1579	42.25%	3	330	40.29%
	4	704	18.84%	4	256	31.26%
	5	154	4.12%	5	89	10.87%
	6	19	0.51%	6	7	0.85%
Average code length	2.9			3.4		
Repeated code rate	Number of repeated code	No. of characters	Percent in character set	Number of repeated code	No. of characters	Percent in character Set
	0	1313	39.74%	0	167	20.39%
	2 – 3	934	28.27%	2 – 3	261	31.87%
	4 – 5	516	15.62%	4 – 5	172	21.00%
	6 – 10	513	15.53%	6 – 10	169	20.63%
	11 – 15	20	0.61%	11 – 15	50	6.11%
	16 – 24	8	0.24%	/	/	/
	$\in[1-10]$	3276	99.16%	$\in[1-10]$	769	93.89%
	$\in > 10$	28	0.84%	$\in > 10$	50	6.11%

Performance Analysis. After the conversion to numeric keys, the length of character code and the repetition number of a single character are similar to the encoding of letter keys, so the number of characters that need to be searched through page-turning is small, which contributes to high input efficiency. The statistical results are shown in Table 3.

Table 3. Performance statistics of the numeric key scheme

Length of code	No. of characters	Repetition of codes	No. of characters	Percent
1	10	0	133	16.24%
2	112	2 – 3	250	30.53%
3	281	4 – 5	190	23.20%
4	253	6 – 10	185	22.59%
5	130	11 – 15	61	7.45%
6	29	/	/	/
7	3	/	/	/
8	1	$\in[1-10]$	758	92.55%
Average length of Code: 3.6		$\in > 10$	61	7.45%

As it is shown in Table 3, the moderate average code length of the characters is 3.6. The number of characters without code repetition is 133, accounting for 16.24%. The characters with 2–5 keys of code repetition shows the highest distribution rate. The number of characters with less than 10 repetitions accounts for 92.55%. Only about 7.5% characters needed to be searched through page turning. The scheme also shows its advantages in character matching accuracy.

4.2 Traditional Yi Character

The shape coding method is suitable for traditional Yi characters coding. The traditional Yi characters, commonly known as "old Yi characters", is a kind of symbolic square characters whose strokes are the smallest structural units. In addition, traditional characters set features large in quantity, lots of variant and rare characters [16]. Therefore, it is suitable to adopt the stroke split mode to classify and encode the character set. In this way, it is not restricted by the pronunciation of characters, but exemplifies simple rules, learnability and memorability.

5 Conclusion

In conclusion, the structures and stroke writing sequences in this scheme design is not completely consistent with the original ones of Yi characters considering that users may not be familiar with them. According to the structure of Chinese characters and the stroke shape assimilation of Yi characters, this scheme adopts the 9-key shape encoding method based on character splitting to classify and encode Yi characters. Furthermore, both the character stroke split and classification are reasonable and clear. When inputting stroke codes, the sequence of stroke input of the basic character is not limited, which allows the users to input stroke combinations randomly. According to the statistical analysis of its performance, the scheme features accessibility of learning, moderate code length, low code repetition, fast input and strong extensibility, as well as extensive application such as converting the strokes to numeric keys.

On the basis of this paper, the next research will try to combine the handwriting of yi with image recognition and input method based on Arabic character handwriting recognition technology [17]. At the same time, based on this input method, we can use relevant methods to evaluate the readability of yi textbooks [18].

Acknowledgment. This work is supported by The Collection and Database Creation of the Current Minority Ethnic Characters (0610-1041BJNF2328/19).

References

1. Sun, W.: Study on the Use of Normative Yi Language in Sichuan. Minzu University of China, Beijing (2010)
2. Ma, J.: Study on the Origin and Development of Yi Language. The Ethnic Publishing House, Beijing (2011)
3. Yunnan, S., Guizhou, G., Yi Language Cooperation Group.: Yunnan, Sichuan, Guizhou, Guangxi, Yi Language Collection. The Nationalities Publishing House of Yunnan, Kunming (2000)
4. La, D.: Chinese Yi language information processing joining the international information standard –a brief introduction to the development of Yi language information processing international standard. J. Southwest Univ. Nationalities (Philos. Soc. Sci.), 159–160 (2000)
5. Shamalayi: Computer Information Processing in Yi Language. Sichuan Minorities Press, Chengdu (2000)
6. Shamalayi: Research and discussion on computer information processing in Yi language. J. Southwest Univ. Nationalities (Philos. Soc. Sci.) **23**(04), 6–10 (2002)
7. Zhang, X., Zheng, Y.: A dynamic structure stroke of Chinese character coding input method. J. Chin. Inf. Process. **17**(03), 59–64 (2003)
8. Wang, S., et al.: Natural language semantic construction based on cloud database. Comput. Mater. Continua **57**(3), 603–619 (2018)
9. Huang, J.: Yi Language and Philology. The Ethnic Publishing House, Beijing (2003)
10. Hao, F., Hui, W., Jiamei, W.: Design and implementation of Yi language input method based on free split model. J. Comput. Appl. **30**(S1), 307 (2010)
11. Hao, F., Ying, Z., Jiamei, W.: Promotion and application of free coding scheme of Yi language input method. Mod. Comput. **09**, 3–4 (2011)
12. Shamalayi: A review of the practical effect of standardizing Yi language program for years. J. Southwest Univ. Nationalities (Philos. Soc. Sci.) **31**(08), 28–29 (2010)
13. Kelan, Z.: Study on Evaluation System of Chinese Character Digital Input Method. Suzhou University, Suzhou (2005)
14. Wu, X.: Design and implementation of computer coding input method for Guizhou Yi language. J. Yunnan Nationalities Univ. (Nat. Sci. Ed.) **23**(05), 387–390 (2014)
15. Hu, G., Wang, J., Zhang, J., Sun, S., Tang, X., Zhao, H.: Implementation of Yi language input method in yunnan based on windows platform. Comput. Syst. Appl. **24**(12), 40–41 (2015)
16. Ding, C.: Modern Yi Language. The China Minzu University Press, Beijing (1991)
17. El Mamoun, M., Mahmoud, Z., Kaddour, S.: SVM model selection using PSO for learning handwritten arabic characters. Comput. Mater. Continua **61**(03), 995–1008 (2019)
18. Wang, Z., Zhao, X., Song, W., Wang, A.: Readability assessment of textbooks in low resource languages. Comput. Mater. Continua **61**(1), 213–225 (2019)

Police: An Effective Truth Discovery Method in Intelligent Crowd Sensing

Ming Zhao and Jia Jiao[✉]

School of Computer Science and Engineering, Central South University, Changsha, China
Jiajiao@csu.edu.cn

Abstract. With the progressively increasing number of smart mobile devices, mobile crowdsensing (MCS) becomes prevalent and pervasive in real life. People use devices as sensors to report claims about entities. Therefore, how to find the true information from the data uploaded by people is a key issue. Iterative updates, optimization or probabilistic models are three main aspects that most truth discovery focused. There is no denying the fact that these methods show their advantages and some limitations. They ignore the connection between the entities and focus on the data only in a single time node, without considering the trend of the data over a while. In this paper, we propose a new Probabilistic mOdel for real-vaLued sensIng data on Correlated Entities named police. This model using time series analysis to predict the entity's probabilistic time distribution over a period of time. In this way, the efficiency of truth discovery can be improved. Moreover, this proposed model can be applied to correlated entities. If there are not have enough reliable users to observe entities, it is impossible to get accurate information, so we take the correlation among entities into consideration to ensure accuracy. Entities' association will increase the difficulty of solving the problem. However, we have proposed a timing grouping method, by dividing the entities into related groups and iterating through the block coordinate descent method. The experiments on real-world demonstrate that the proposed methods satisfy properties better than the existing truth discovery frame from conflicting information reported on correlated entities.

Keywords: Truth discovery · Probabilistic time series · Correlated entity

1 Introduction

Crowdsensing becomes prevalent and pervasive in data collection due to the reason that participants could proactively report their observations. Human in this special paradigm acts as the sensors or use the mobile devices (e.g., iWatch, smartphones) equipped with sensors have given rise to intelligent crowdsensing. This paradigm outsources sensory data collection to users. Hence, it becomes convenient for people to information sharing and information sense. The potential of the sensory data is tremendous. Because the crowd wisdom can be harnessed for wide-ranging applications such as air quality monitoring, traffic conditions monitoring, gas price predicting. On the cloud server, useful information can be aggerated.

© Springer Nature Switzerland AG 2020
X. Sun et al. (Eds.): ICAIS 2020, LNCS 12239, pp. 384–398, 2020.
https://doi.org/10.1007/978-3-030-57884-8_34

In real practice, because of various factors such as lack of effort, poor sensor quality, background noise, users' sensory data on the same entity may not always be true. Thus, discover true information among these conflicting sensory data is the key issue. In recent years, several methods [1–6] have been studied to manage uncertain data. Since the data changes with time are highly irregular, we adopt the GARCH (Generalized Auto Regressive Conditional Heteroskedasticity) model which is an effective dynamical model.

When using the GARCH model to generate probabilistic time series, data quality is another challenge cannot be ignored. On the same entity, different people may report conflicting sensory data. Intuitively, the information reported by reliable users we should pay attention to. However, users' reliability is usually unknown prior and should be calculated according to their data. To crackdown this issue, a family of algorithms [7–10] has been proposed and widely studied. In [11] users whose uploading data are closer to the ground truth, the more reliable grade they are, and in the next iteration, the data will be counted. In our paper, we jointly consider truth discovery and the real-world applications' correlation. Noticed that adjacent areas have the same temperature in a short period of time, we encode this correlation into the truth discovery model.

To propose an effective truth discovery model for crowdsensing with object correlations, it is important to consider the following questions: (1) How to efficiently predict the truth distribution over the next period of time? (2) How to use entities' correlation to improve the accuracy of prediction? (3) How to group entities to improve optimization efficiency?

Aiming to solve the above problems, we proposed an effective truth discovery method. The main contributions of this paper are as follows:

(1) We adopt GARCH to generate probabilistic time series for real-valued sensing data on correlated entities. This method models the readings of each sensor through probability distribution. We use this method to deal with evolutionary probability distributions.
(2) To minimize the difference between the inferred truth and sensory data, ensure the users' weight, we defined the energy minimization problem $min\ f\left(D^{(*)}, W\right)$. The similarity of related entities is taken as the constraint condition for calculation, so the inferred truth and users' weight can be searched.
(3) We use kNN(k-Nearest Neighbor) algorithm to partition entities and combine the iteration process with the block coordinate descent with convergence guarantee.

The remaining sections are organized as follows. The problem definition in Sect. 2. We describe the details about the proposed framework and solutions in Sect. 3. Sections 4 and 5 demonstrate experimental results. The related work in Sect. 6 and we conclude the paper in Sect. 7.

2 Problem Definition

In this section, we introduce the system overview, describe the symbol notation. Secondly, we state the framework of the probabilistic model for real-valued sensing data correlated entities, and the optimization solution is given to the problem.

2.1 System Overview

Figure 1 shows this system's framework. Observers make an observation on several entities and then generating probabilistic time series. After this, uploading observation to the cloud-based platform server via mobile phones. Based on the GARCH model, we generate a probabilistic time series from raw data, then upload them to the server. On the server-side, the proposed truth discovery frame will be conducted. Each observer's reliability and the inferred truths for entities are calculated at the same time. As mentioned above, the correlation between the entities is important and can be measured, the detail of the proposing algorithm will be discussed in Sect. 3.

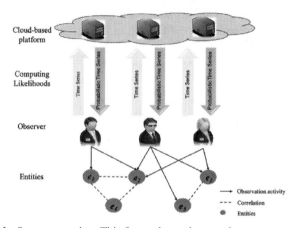

Fig. 1. System overview. This figure shows the paper's system overview.

2.2 Proposed Framework

We formally define the elements of our problem as follows. Indexed each entity with i, and e_i denotes the i-th entity. We use k to index each observer. A time series is represented as $S_i^{(k)}$, which indicates that it is observed by observer k on entity e_i. Here $D = \{S_1, S_2, \cdots, S_n\}$ is a set of time series. $S_i =< r_1^k, r_2^k, \cdots r_m^k >$ is a time series, made up of timestamped values. A reading collected by observer k at time j is marked as $r_j^k \in S_i$. Since each observers' reliability is not known before, the raw data r_j^k it provides may not be correct. The inferred truth should close to the observations given by reliable observers, and entities which have correlation with should have similar true values. We use $s_i^{(*)}$ to denote the entity e_i truths, and $D^{(*)}$ to denote all entities' truths. We use w_k denote the weight of user k. We use W to record all user weight. The most important symbols used in this paper are summarized in Table 1. We can formally state the problem with these notations as follows.

Problem Statement. We set K users and N entities in this paper. Time series $\{S_1, S_2, \ldots, S_N\}$ collected by all the K users as input. The set of entities that have correlations with entity e_i denoted as $C(i)$. The expected outputs are the truths $D^{(*)}$ and

Table 1. Notations

Symbol	Definition
k	observers' index
w_k	observer weight of observer k
W	all observer weights set
i	entities' index
e_i	the i-th entity
I	all entities sets
I_m	the m-th sets
$s_i^{(*)}$	the true value of time series of entity i
r_j^k	observation provided by observer k at timestamp j
$s_i^{(k)}$	time series provided by observer k on entity i
$D^{(*)}$	the set of true time series of entities
$C(i)$	entities which correlate with entity i
$C(o, o')$	the degree of similarity

the observer weights W. We formulate the problem according to the intuitions that the reliable users' observations should be small to the deviation, the problem is defined as follows:

$$min f\left(D^{(*)}, W\right) = \sum_{i=1}^{N} \left\{ \sum_{K=1}^{K} w_k \left\| s_i^{(*)} - s_i^{(k)} \right\|^2 + \alpha \sum_{i' \in C(i)} C(o, o') \left\| s_{i'}^{(*)} - s_i^{(*)} \right\|^2 \right\}$$

$$s.t \sum_{k=1}^{K} exp(-w_k) = 1 \tag{1}$$

Here α is a hyperparameter that balances the effect of correlated object truths, and the value of α can be set in different datasets. $C(o, o')$ is a similarity function that captures the correlation degree of two entities. We use the Gaussian kernel (Eq. (2)) to define two entities' temporal correlation coefficient. In Eq. (2), if two entities temporal distance $d(o, o')$ is within a threshold δ, they are correlated. The distance becomes larger, the faster $C(o, o')$ approaching 0.

$$C(o, o') = \begin{cases} exp\left(-\dfrac{d(o, o')^2}{\sigma^2}\right), & if\ d(o, o') \leq \delta \\ 0 & , otherwise \end{cases} \tag{2}$$

Equation (1) the objective function $f(D^{(*)}, W)$, shows our purpose. The first term $\sum_{i=1}^{N} \sum_{k=1}^{K} w_k \left\| s_i^{(*)} - s_i^{(k)} \right\|^2$ aims at minimizing the difference on entities from the observers with higher reliability degrees are weighted higher. The intention behind the term is that the more deviation from the corresponding truths, the reliable observers' penalties are higher. The second term $\sum_{i=1}^{N} \sum_{i' \in C(i)} C(o, o') \left\| s_{i'}^{(*)} - s_i^{(*)} \right\|^2$ models similarity.

In order to avoid the weight go to negative infinity, we use constraints in Eq. (1). With the proposed framework in Eq. (1), we search for two sets of variables, one is truths $D^{(*)}$ and the other is observer weights W. Facing two variables, block coordinate descent approach [12] can solve the problem. We use the kNN algorithm to divide entities into independent sets. For example, as we can see that in Fig. 1, e_1, e_2, e_4 are correlated with each other, and e_3, e_5 are correlated, if a new entity e_6 need to be observed and partition, which set is e_6 belong to? We will discuss the method of constructing disjoint sets in Sect. 3.2.

In this part, assuming we have used kNN algorithm, the independent sets are obtained. We denote m as the index for the independent set. The set containing all the entities are denoted as $I = \bigcup_{m=1}^{M} I_m$, and I_m denotes the m-th subset. Naturally, Eq. (1) can be rewritten as follows:

$$\min f\left(D^{(*)}, W\right) = \sum_{I_m \subset I} \sum_{i \subset I_m} \left\{ \sum_{k=1}^{K} w_k \left\| s_i^{(*)} - s_i^{(k)} \right\|^2 + \alpha \sum_{i' \in C(i)} C(o, o') \left\| s_{i'}^{(*)} - s_i^{(*)} \right\|^2 \right\}$$

$$s.t \sum_{k=1}^{K} exp(-w_k) = 1 \tag{3}$$

3 POLICE: Proposed Probabilistic Model

3.1 Proposed Solution

Computing likelihoods is the first step of our framework: we use GARCH model to process raw data. The GARCH method uses Gaussian probability density function $N\left(\widehat{r_i}, \widehat{\sigma}^2\right)$ to model $p_i(R_i)$. The two parameters' details are as follows.

Predicting Expected True Values. Given a time series S, a reading at timestamp j can be modeled by the ARMA [13] model as $r_j^k = \widehat{r_j^k} + a_i$. Noted that $\widehat{r_j^k}$ is the expected true value and a_i follows a normal distribution $N\left(0, \sigma_a^2\right)$. From its past values, at time j the expected true value $\widehat{r_j^k}$ can be calculated as follows:

$$\widehat{r_j^k} = \delta_0 + \sum_{b=1}^{p} \delta_b r_{j-b}^k + \sum_{b=1}^{q} \gamma_j^k a_{j-b} \tag{4}$$

Where p, q are both nonnegative number, two parameters represent respectively the autoregressive and moving average orders. Here δ_0 is a constant, and $j > \max(p, q)$.

Inferring Variances. According to the reading r_j^k at time j as $r_j^k = \widehat{r_j^k} + a_i$, we can then define the conditional variance given all the information available until time $j - 1$ as

$\sigma_j^2 = \mathbb{E}(a_j^2 | F_{j-1})$. Where $\mathbb{E}(a_j^2 | F_{j-1})$ is the variance of a_j. Then, based on the GARCH model, we can measure the variance as a linear function of a_j^2 as:

$$a_j = \sigma_j \in_j, \quad \sigma_j^2 = \theta_0 + \sum_{c=1}^{h} \theta_c a_{j-b}^2 + \sum_{c=1}^{s} \lambda_c \sigma_{j-b}^2 \tag{5}$$

Where (h, s) specifies the model orders, $\in_j$ is a distributed random variables, $\theta_0 > 0, \theta_c \geq 0, \lambda_c \geq 0, \sum_{c=1}^{max(h,s)} (\theta_c + \lambda_c) < 1$. In this paper, we set h = 1, s = 1 in the following common practice to reduce the difficulty. The GARCH model has the ability to calculate the variances.

Algorithm 1 describes the detail of the approach. Step 3 is to compute the expected true value $\hat{r}_t$ by ARMA, step 4 is to calculate the $\widehat{\sigma_t^2}$ by GARCH. The complexity of the algorithms is mainly affected by the step1 and 2.

Algorithm 1   Inferring $\hat{r}_t$ and $\widehat{\sigma_t^2}$

Input: Observers' time series S_i, parameters(p,q) of ARMA model and scaling factor f.

Output: Inferred $\hat{r}_t$, inferred variance $\widehat{\sigma_t^2}$

1: Estimated an ARMA(p,q) model on S_i and obtain a_i.

2: Estimated a GARCH(1,1) model using $a_i's$

3: Using ARMA(p,q) model to infer $\hat{r}_t$

4: Using GARCH(1,1) model to infer $\widehat{\sigma_t^2}$

Return: $\hat{r}_t$, $\widehat{\sigma_t^2}$

In our paper, there are N entities and K observers, if we take into account the time series from each observer separately the level of complexity cannot be ignored. To tackle this challenge, we propose a method which combines observer weight in generating probabilistic time series.

In Eq. (3), observer weights W and truths $S_1^{(*)}, S_2^{(*)}, \ldots, S_M^{(*)}$ should be simultaneously minimized. Block coordinate descent approach [12] is a useful method to achieve this goal. The details are as follows:

Update W while fixing $\{S_1^{(*)}, S_2^{(*)}, \ldots, S_M^{(*)}\}$: All sets of truths are fixed, we compute the observer weights W according to the difference of truths and the observations made by the observer:

$$W \leftarrow argmin f\left(D^{(*)}, W\right)|. \ s.t. \ \sum_{k=1}^{K} exp(-w_k) = 1 \tag{6}$$

Lagrange multiplier method can be used to solve optimization problems under constraints. The Lagrangian of Eq. (3) is given as:

$$L\left(w_{k=1}^{K}, \lambda\right) = \sum_{l_m \subset l} \sum_{i \subset l_m} \left\{ \sum_{k=1}^{K} w_k \left\| s_i^{(*)} - s_i^{(k)} \right\|^2 + \alpha \sum_{i' \in C(i)} C(o, o') \left\| s_i^{(*)} - s_{i'}^{(*)} \right\|^2 \right\} + \lambda \left(\sum_{k=1}^{K} exp(-w_k) = 1 \right),$$

where λ is the parameter before the constraint in Lagrange multiplier method. Let the partial derivative of Lagrangian with respect to w_k be 0, we get:

$$\sum_{I_m \subset I} \sum_{i \subset I_m} \left\| s_i^{(*)} - s_i^{(k)} \right\|^2 = \lambda exp(-w_k) \tag{7}$$

From the constraint that $\sum_{k=1}^{K} exp(-w_k) = 1$, we can derive that

$$\lambda = \sum_{k=1}^{K} \sum_{I_m \subset I} \sum_{i \subset I_m} \left\| s_i^{(*)} - s_i^{(k)} \right\|^2 \tag{8}$$

We can then derive the update rule for each observer's weight by plugging Eq. (8) into Eq. (7):

$$w_k = -\log\left(\frac{\sum_{I_m \subset I} \sum_{i \subset I_m} \left\| s_i^{(*)} - s_i^{(k)} \right\|^2}{\sum_{I_m \subset I} \sum_{i \subset I_m} \sum_{k'=1}^{K} \left\| s_i^{(*)} - s_i^{(k')} \right\|^2} \right) \tag{9}$$

where k' denotes the index of an observer. The rule shows that the observation reported by user is more close to the ground truths, the higher weight the user has.

Update $S_m^{(*)}$ while fixing $\{W, S_1^{(*)}, \ldots, S_{m-1}^{(*)}, S_{m+1}^{(*)}, \ldots, S_M^{(*)}\}$: Observer weights W are fixed in this case, and except $S_m^{(*)}$ the sets of truths are also fixed. We update the truth for each entity in $S_m^{(*)}$ by minimizing the objective function:

$$S_m^{(*)} \leftarrow argmin f\left(D^{(*)}, W\right) \tag{10}$$

Let the derivative of Eq. (3) be 0 with respect to $s_i^{(*)}$, then for each $i \in I_m$, or equivalently for each $s_i^{(*)} \in S_m^{(*)}$, we have the following update rule:

$$s_i^{(*)} = \frac{\sum_{k=1}^{K} w_k s_i^{(k)} + \alpha \sum_{i' \in C(i)} C(o, o') s_{i'}^{(*)}}{\sum_{k=1}^{K} w_k + \alpha \sum_{i' \in C(i)} C(o, o')} \tag{11}$$

As the correlated variables are divided into different sets, we can use Eq. (11) to solve $s_i^{(*)}$. According to Eq. (11), $S_m^{(*)}$ is updated by averaging over the observed values weighted by observer weights $\sum_{k=1}^{K} w_k s_i^{(k)}$ and the values of its correlated entities $\sum_{i' \in C(i)} s_{i'}^{(*)}$.

Algorithm 2 summarizes the method until the convergence criterion is satisfied, the iteration updates truths and observer weight finished.

Algorithm 2 Computing observer weight w_k and truths $s_i^{(\bullet)}$

Input: raw time series from K observers: $\{S^{(1)}, \ldots, S^{(K)}\}$

Output: true time series sets $D^{(\bullet)} = \{s_i^{(\bullet)}\}_{i=1}^{N}$, observer weights $W = \{w_k\}_{k=1}^{K}$.

1: Initializing the truths $D^{(\bullet)}$; Dividing entities into different independent sets $\{I_1, \ldots, I_M\}$;

2: **repeat**

3:     Updating observer weights W according to Eq(10);

4:   **for** m=1 to M **do**

5:     **for** $i \in I_m$ **do**

6:       Updating the truth of the i_{th} entity $s_i^{(\bullet)}$ from the set I_m according to Eq(11);

7:     **end for**

8:   **end for**

9: **until** Convergence criterion is satisfied;

10: **return** $D^{(\bullet)}$ and W.

3.2 Constructing Independent Sets

In Sect. 2.2, we discussed that only the entities are divided into disjoint sets, the Eq. (1) can combine iterative solution. In this section, we discuss the method to form disjoint independent sets with temporal correlation.

Due to data arrives sequentially in the application, temporal correlation among the truths of an entity exists. For example, nearby segments of the air quality may have the same condition, the values of entities are similar in a short period of time. The temporal correlation can also be incorporated into our mode to infer truths. First, we find the most similar sets in the existing entity categories, and then decide which group the entity to be classified belongs according to the categories of the kNN. The classification effectiveness depends largely on the k value (the good or bad choice of nearest neighbor number). In this paper, according to experience, we take k as the largest integer not exceeding the square root of the training sample number. An example is shown in Fig. 2 which we have eleven samples which be separated into two sets, one is triangle, the other is square. And there is a circle sample need to be classified. When k = 3, the nearest samples are two triangle entities and one square entity, so the circle one belongs to the triangle sets; when k = 5, however, the circle one belongs to the square sets. As we can see, the different k we choose, the different result is. According to the experience, the k we decide is the largest integer number that not exceeding the square root of the training sample number. In Fig. 2, there are eleven samples, 3 is the largest integer number that not exceeding the square root of eleven. So the circle one is classified into the triangle sets.

4 Theoretical Analysis

In this section, we analyze the proposed method's time complexity. We also prove that by using block coordinate descent method, the energy cost in Eq. (3) is minimized.

We can use $\mathcal{Y}$ to denote the set of blocks of variables where $\mathcal{Y} = \left\{W, S_1^{(*)}, S_2^{(*)}, \ldots, S_M^{(*)}\right\}$

$$minimize \, f(y) \quad s.t. \quad y \in \mathcal{Y} \tag{12}$$

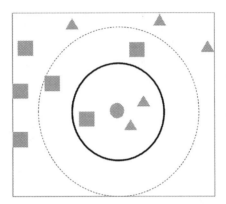

Fig. 2. Separating entities based on k-Nearest Neighbor.

Suppose for each $y_i \in \mathcal{Y}_i$,

$$f(y_1, \ldots, y_{i-1}, \xi, y_{i+1}, \ldots, y_m)$$

viewed as a function of ξ while the other blocks of variables are fixed, attains a unique minimum $\bar{\xi}$ over $\mathcal{Y}_i$, and is monotonically non-increasing in the interval from y_i to $\bar{\xi}$. Let $\{y^k\}$ be the sequence generated by the following updating rule

$$y_i^{k+1} \leftarrow argminf\left(y_1^{k+1}, \ldots, y_{i-1}^{k+1}, \xi, y_{i+1}^k, \ldots, y_m^k\right)$$

In [12], we know that every limit point of $\{y^k\}$ is a stationary point. The one last thing we have to do is that Algorithm 2 meets the conditions.

Case 1. When fixing $\{S_1^{(*)}, S_2^{(*)}, \ldots, S_M^{(*)}\}$ we update W. In this case, as $\alpha \geq 1, f(W)$ involves linear combination negative logarithm functions and constants, and thus it is convex. Besides, the constraint is linear in w_k, which is affine. Thus, f satisfies the above conditions when updating weight W in Eq. (9).

Case 2. When fixing $\{W, S_1^{(*)}, \ldots, S_{m-1}^{(*)}, S_{m+1}^{(*)}, \ldots, S_M^{(*)}\}$ we update $s_i^{(*)}$. By fixing observer weights W, Eq. (3) is a summation of quadratic functions concerning $s_i^{(*)}$. It is convex since these quadratic functions are convex and summation operation preserves convexity. Thus, a unique minimum for $s_i^{(*)}$ can be achieved with the update rule in Eq. (11).

Therefore, using block coordinate descent to minimize energy function f is valid.

Time Complexity Analysis: Assuming there are N entities and K observers. We calculate the square error of each observers' observations and truths, the step of updating the observers' weight costs O(KN) time. For each entity, the step of updating the truth, we calculate the weighted sum of observations from K observers, the sum of correlated values and the sum of K observers' weights. It also needs O(KN) time. Overall, one iteration requires O(KN) time. Therefore, the number of observations is linearly related to the time complexity.

5 Experiments

5.1 Experiment Setup

The datasets, baseline methods, and performance evaluated metrics are described in this part. Both of them are used to evaluate the proposed POLICE method.

Evaluated Performance Metrics. Based on the continuous data in the datasets, we supposed to find the difference between inferred truth and ground truth. The difference can be measured by mean absolute error(MAE) and root mean square error (RMSE). No matter MAE or RMSE are measuring the difference the predicted truth and the ground truth. The lower number is, the predicted truth and the ground truth is more close.

Baseline Methods. We denoted our proposed method as TD-police in the experiment. Compared it with the following truth discovery methods. CRH is a method that resolves conflicts among multiple sources of heterogeneous data types. GTM is a Bayesian probabilistic model for resolving conflicts on continuous data. Besides these two methods, the mean of the values of observations(Mean) and median of values of observations (Median) are also compared.

Datasets. Weather Dataset: this dataset contains temperature readings collected from three weather forecast platforms (*Darksky*[1], *WorldWeatherOnline*[2], *Wunderground*[3]). From these three platforms, we collect 40320 claims in 42 different locations in Washington. The ground truth of the weather condition also be crawled. We define an entity is a temperature of location in a time node. For each platform, two different observers record the temperature forecasts, thus six observers are formed in total. Two entities are correlated if the suburbs they belong to are near to each other. In this situation, Euclidean distance can be applied to install correlation.

Gas Price Dataset: in intelligent crowdsensing application, collect gas price from crowd users and aggerate their records to get correct gas price is the key issue. In this experiment, we collect 3460 gas stations gas prices of 22 major cities for one day in the US from *Gasbuddy*[4] website as the ground truth. Gas prices are reported by 30 users who have different reliability degrees. In the same city, the correlation exists among gas stations. We find that if the distance between is within 6 km, the gas price is similar. Therefore, 6 km is the threshold to determine if two stations' gas prices are correlated, the correlation can be installed by the Gaussian kernel.

Table 2 summarized two datasets' statistics. The average number of correlated entities per entity in the datasets shown in the last column.

5.2 Evaluations on Computing Likelihoods

In this section, the running time of the computing likelihood component (GARCH) is analyzed. To figure out the GARCH model's performance, two real-world datasets (weather datasets and gas price datasets) are used. We set the time series length from 20 to 160, in order to reflect the GARCH model performance. The statistics are shown in Table 3. As we can see, with the increase in length, the running time increases. But the

394   M. Zhao and J. Jiao

Table 2. Datasets statistics

Dataset	Entities	Observers	Claims in total	Time-stamps
Weather dataset	42	6	40320	160
Gas price dataset	3460	30	103800	1

speed of increasing running time is much slower than increasing time series length. In Table 3, the consuming time of the GARCH model increase 2 times, but the time length from 20 to 160 increases 8 times. In a word, the GARCH model is an efficient model for real-world computation.

Table 3. Running time

Time series length	Weather dataset	Gas price dataset
20	0.1218	0.1109
40	0.1463	0.1389
80	0.1837	0.1642
120	0.2102	0.1934
160	0.2533	0.2316

5.3 Weather Condition Experiments Results

Table 4 and Fig. 3 show the performance of TD-police and baseline methods. From Fig. 3, we can see that our proposed method is superior to the baseline approach, both in MAE and RMSE. This is because our proposed method not only consider the correlation information on weather condition but also measures the reliability of observers. All of these making the difference between our predictions and the actual values smaller.

5.4 Gas Price Experiments Results

Table 5 summarized 30 users' sensory data results. Figure 4 clearly shows in different metrics such as MAE and RMSE, the experiment results are shown that with the increase of coverage, the baseline methods' error is significantly reduced. While the proposed method has been in a stable state. Especially when the coverage rate is low, we can see that the performance of our method better than baseline methods.

6 Related Work

Probabilistic Data Streams. The research of probabilistic data streams [15–17], which has drawn much concern. Most existing research efforts related to probabilistic data

Table 4. Weather condition estimation

Metric	MAE				RMSE			
Method	Coverage							
	40%	60	80%	100%	40%	60%	80%	100%
TD-police	1.23	1.131	1.022	0.96	2.478	1.767	1.483	1.372
CRH	1.332	1.149	1.032	0.989	2.864	1.978	1.705	1.662
GTM	1.342	1.201	1.047	0.962	2.763	2.234	1.798	1.555
Mean	1.672	1.435	1.376	1.221	3.653	2.632	2.464	2.314
Median	1.518	1.396	1.231	1.198	3.627	2.925	2.463	2.076

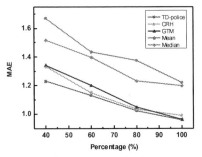 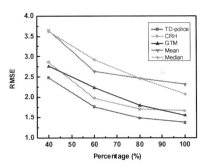

Fig. 3. Effects on MAE and RMSE. In the weather dataset, the left figure shows the MAE situation, and the right figure shows the RMSE situation. We can see that our TD-police is superior than other methods.

Table 5. Gas price estimation

Metric	MAE					RMSE				
Method	Coverage									
	20%	40%	60%	80%	100%	20%	40%	60%	80%	100%
TD-police	0.194	0.182	0.172	0.168	0.159	0.07	0.047	0.042	0.038	0.029
CRH	0.43	0.327	0.287	0.265	0.249	0.288	0.167	0.129	0.10	0.095
GTM	0.407	0.301	0.26	0.231	0.212	0.263	0.141	0.102	0.085	0.070
Mean	0.468	0.376	0.335	0.289	0.271	0.343	0.222	0.174	0.13	0.112
Median	0.490	0.364	0.307	0.275	0.250	0.381	0.211	0.151	0.121	0.101

streams focus on processing queries such as projection, selection, aggregation [5, 18]. In [18], T.T.L. Tran et al. employs a unique data model that is flexible and allows efficient computation. Moreover, some researchers focus on clustering [19] for probabilistic data

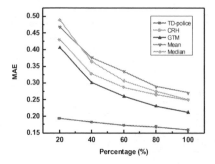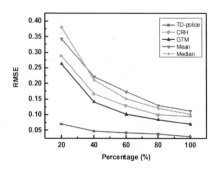

Fig. 4. Effects on MAE and RMSE. In the gas price dataset, the left figure shows the MAE situation, the right figure shows the RMSE situation. When coverage rate is low, our proposed method TD-police more stable than other existing methods.

streams. The approach mentioned above differs from our dynamic probabilistic data streams because it does not apply kNN to time series forecasting.

Crowd Sensing. The research of crowdsensing [20–24], due to the installation of sensors on mobile devices has attracted much attention. K. Ota et al. [20] propose a new incentive mechanism that ensures the quality and the usability of information for application requirements. Wenyan Pan et al. [24] develop a diagnosis system for citrus diseases based on mobile services computing. These papers focus on the design of the system rather than the intelligent crowd sensing aspect of truth discovery.

Truth Discovery. Most existing research about truth discovery solution is majority voting. H. Jin et al. [25] propose a payment mechanism incentive users to try their best report sensory data. In this paper, workers' data quality and the underlying truths are estimated. X. Yin et al. [26] noticed that the information on the web are not always correct and no one can guarantee the correctness. They figured out a new solution, finding true facts from conflicting information on many subjects. In [27] Y. Zheng et al. adopted a method for multi-valued attributes, this method incorporating the quality, which iteratively derives the accuracy. R. W. Ouyang et al. [28] proposed two unsupervised models, each of them are applied in spatial events. These models are capable of effectively handling various types of uncertainties and automatically discovering truths. However, these methods overlook entities' temporal correlations. In this paper, the proposed framework can be applied in correlated entities.

7 Conclusion

To sum up, in this paper, we proposed a new probabilistic model for real-valued sensing data on correlated entities named police. In our model, dynamic uncertain sensory data uploaded by users can be captured. First, we use a GARCH model to obtain the probabilistic time series distribution from the raw data. Since the credibility of the users is known a prior, the data they provide cannot be treated equally. Therefore, in the second step, we transformed this problem into the truth discovery problem of the correlated entity. Noticing that the correlated entities should have similar truth values, we used the kNN algorithm to group the entities into disjoint sets, and then by conducting block coordinate descent to update truths and user weight iteratively. Theoretical analysis and experimental results prove that the desirable properties are validated.

References

1. Cheng, R., Kalashnikov, D.V., Prabhakar, S.: Evaluating probabilistic queries over imprecise data. In: SIGMOD, pp. 551–562 (2003)
2. Hua, M., Pei, J., Zhang, W., Lin, X.: Ranking queries on uncertain data: a probabilistic threshold approach. In: SIGMOD, pp. 673–686 (2008)
3. Dalvi, N., Suciu, D.: Efficient query evaluation on probabilistic databases. In: JVLDB, pp. 523–544 (2007)
4. Olteanu, D., Huang, J., Koch, C.: Sprout: lazy vs. eager query plans for tuple-independent probabilistic databases. In: ICDE, pp. 640–651 (2009)
5. Cormode, G., Garofalakis, M.: Sketching probabilistic data streams. In: SIGMOD, pp. 281–292 (2007)
6. R´e, C., Letchner, J., Balazinksa, M., Suciu, D.: Event queries on correlated probabilistic streams. In: SIGMOD, pp. 715–728 (2008)
7. Li, Q., Li, Y., Gao, J., Su, L., Zhao, B., Demirbas, M.: A confidence aware approach for truth discovery on long-tail data. In: PVLDB (2014)
8. Li, Q., Li, Y., Gao, J., Zhao, B., Fan, W., Han, J.: Resolving conflicts in heterogeneous data by truth discovery and source reliability estimation. In: SIGMOD, pp. 1187–1198 (2014)
9. Li, Y., Li, Q., Gao, J., Su, L., Zhao, B.: On the discovery of evolving truth. In: SIGKDD, pp. 675–684 (2015)
10. Yin, X., Han, J., Yu, P.S.: Truth discovery with multiple conflicting information providers on the web. IEEE Trans. Knowl. Data Eng. **20**(6), 796–808 (2008)
11. Yin, X., Han, J., Philip, S.Y.: Truth discovery with multiple conflicting information providers on the web. IEEE Trans. Knowl. Data Eng. **20**(6), 796–808 (2008)
12. Bertsekas, D.P.: Nonlinear programming. Athena scientific Belmont (1999)
13. Shumway, R.H., Stoffer, D.S.: Time series Analysis and its Applications. Springer, New York (2000). https://doi.org/10.1007/978-1-4757-3261-0
14. Brelaz, D.: New methods to color the vertices of a graph. Commun. ACM **22**(4), 251–256 (1979)
15. Liu, L., Li, W., Jia, H.: Method of time series similarity measurement based on dynamic time warping. Comput. Mater. Continua **57**(1), 97–106 (2018)
16. Li, H., Zhang, N., Zhu, J., Wang, Y., Cao, H..: Probabilistic frequent itemset mining over uncertain data streams. Expert Syst. Appl. **112**, 274–287 (2018)
17. Cheng, R., et al.: Managing uncertainty in spatial and spatiotemporaldata. In: ICDE, pp. 1302–1305 (2014)
18. Tran, T.T.L., Peng, L., Diao, Y., McGregor, A., Liu, A.: CLARO: modeling and processing uncertain data streams. VLDB J. **21**(5), 651–676 (2012)
19. Aggarwal, C.C., Yu, P.S..: A framework for clustering uncertain data streams. In: IEEE 24th International Conference Data Engineering, pp. 150–159 (2008)
20. Ota, K., Dong, M., Gui, J.: QUOIN: incentive mechanisms for crowd sensing networks. IEEE Netw. **32**(2), 114–119 (2018)
21. Xiao, Z., Yang, B., Tjahjadi, D.: An efficient crossing-line crowd counting algorithm with two-stage detection. Comput. Mater. Continua **60**(3), 1141–1154 (2019)

22. Cao, G., Campbell, A.: Semantics analytics of origin-destination flows from crowd sensed big data. Comput. Mater. Continua **61**(1), 227–241 (2019)
23. Qin, J., Luo, Y., Xiang, X., Tan, Y., Huang, H.: Coverless image steganography: a survey. IEEE Access **7**, 171372–171394 (2019). https://doi.org/10.1109/ACCESS
24. Pan, W., Qin, J., Xiang, X., Wu, Y., Tan, Y., Xiang, L.: A smart mobile diagnosis system for citrus diseases based on densely connected convolutional networks. IEEE Access **7**(1), 87534–87542 (2019). https://doi.org/10.1109/ACCESS
25. Jin, H., Su, L., Nahrstedt, K.: Theseus: incentivizing truth discovery. In: ACM on Mobile Ad Hoc Networking and Computing, pp. 1–10 (2017)
26. Yin, X., Han, J., Yu, P.S.: Truth discovery with multiple conflicting information providers on the web. IEEE Trans. Knowl. Data Eng. **20**(6), 796–808 (2008)
27. Zheng, Y., Luo, J., Yin, M.: Truth discovery on categorical multi-valued attributes with source relations. In: IEEE International Conference on Computer and Communications, pp. 1820–1824 (2018)
28. Ouyang, R.W., Srivastava, M., Toniolo, A.: truth discovery in crowdsourced detection of spatial events. IEEE Trans. Knowl. Data Eng. **28**(4), 1047–1060 (2016)

Improve Data Freshness in Mobile Crowdsensing by Task Assignment

Ming Zhao and Shanshan Yang[✉]

School of Computer Science and Engineering, Central South University, Changsha, China
shanshanyang@csu.edu.cn

Abstract. In real-time crowdsensing, the key factor is the freshness of information. However, due to the randomness and uncertainty of mobile users, some tasks will not get status updates, and leaving a lot of outdated information. To solve this problem, we use efficient task assignment strategy to ensure the freshness of the data at the receiver's side. We take space and time into consideration. First, we develop a matrix-based location matching mechanism for the service provider to achieve accurate location-based task allocation without disclosing the location of mobile users and the sensing area of tasks. Then we use "age of information" (AoI) as a metric to represent freshness of information and we consider a simple linear AoI-aware allocation strategy algorithm to reduce the average AoI across the system and analyze its AoI performances. Extensive simulations are performed to validate our analytical results. Both analysis and simulation results verify the effectiveness of the proposed scheduling strategy.

Keywords: Crowdsensing · AoI · Task assignment

1 Introduction

Fueled by the rapid development of smart mobile devices (e.g., smartphones, smartwatches, etc.) installed a series of sensors (e.g., compass, GPS, accelerometer, etc.), newly emerged crowdsensing systems have been developed [1, 2]. Although mobile crowdsensing holds a great potential to fundamentally change our modern society, a key factor affecting its future large-scale adoption is the freshness of the crowdsensing information, which can be measured by a fundamental metric termed "Age-of-Information" (AoI). The AoI is proposed to capture the freshness of information updates in a system, which has received great attention because of rapid deployment of real-time applications [3–5].

To get fresh information quickly, task allocation is one of the most important part in mobile crowdsensing. If tasks can be accurately assigned to users in the same region, it will improve the speed of information collection. Previous studies also have concluded that the distance between a worker and the task location negatively affect the quality of the result [6, 7]. For example, if a task is assigned to a remote user, the user will spend some time on the journey to complete the task, which will cause the data requester can't get the latest information quickly, and may cause the user to lose interest in the

© Springer Nature Switzerland AG 2020
X. Sun et al. (Eds.): ICAIS 2020, LNCS 12239, pp. 399–410, 2020.
https://doi.org/10.1007/978-3-030-57884-8_35

crowdsensing task because of the long distance. Without user's participation, which also in turn degrades the information freshness. While assigning tasks based on location, a series of privacy exposure problems will arise without any protection.

Furthermore, as there will be constant task requests, users will often complete the current task, resulting in the backlog of previous tasks and unable to update. It could result in two undesirable consequences: (i) too many users flock to current task, which leads to redundant sampling and severe queueing congestion; and (ii) all other tasks suffer from large AoI because of under-sampling. The service provider's goal is to achieve minimum average AoI. And we need a reasonable arrangement to ensure information freshness in crowdsensing.

Existing works on task allocation in mobile crowdsensing mainly focus on security and designing incentive mechanisms to improve user participation, while information freshness has not been extensively explored. Therefore, how to effectively allocate crowdsensing tasks to ensure a low AoI for each task remains a challenge. In this paper, we overcome the above challenges and propose the allocation strategy to guarantee the freshness of location-based mobile crowdsensing information. This paper has two main contributions as shown below.

- We propose a matrix-based location matching method to allow the service provider to allocate the crowdsensing tasks based on the locations of mobile users and the sensing areas of tasks. Specifically, the mobile user and the data requester utilize a customized random matrix to hide the location and the sensing area, respectively. Thus, the service provider can assign crowdsensing tasks to users in the same region without learning the location of the mobile user and the sensing area of data requester.
- We propose an AoI-aware allocation scheduling algorithm to ensure the whole system holds low AoI. Specifically, the service provider will assign tasks to the mobile user based on the last update time of the task and the queue length. Therefore, the following two problems are solved: first, the previous tasks will be unable to be updated because only the new task is concerned by the user; Secondly, AoI is improved due to due to the under-sampling.

The remainder of this paper is organized as follows. We present a high-level overview of our framework and formulate the problems in Sect. 2. The proposed intimate location-based allocation policy and AoI-aware allocation scheduling algorithm are presented in Sects. 3. We evaluate the performance of the proposed allocation policy in Sect. 4. We review related work in Sect. 5 and we conclude the paper in Sect. 6.

2   Related System Overview and Problem Statement

2.1   System Model

We use Fig. 1 to assist our description of a location-based mobile crowdsensing system. The system consists of a mobile sensing platform, which consists of multiple data requesters, and multiple mobile users, which are connected to the platform via the service provider. Mobile users arrive at random and data requesters post tasks at random.

The tasks require mobile users to continuously upload and update data. Multiple tasks can be published in one area and users can only accept one task at a time.

Next, we will formally define the system model. At first of all, the data requesters and the mobile users should register at the trusted authority. And then the data requester sends its sensing task st which need mobile users to collect data in a specific area to the service provider and the mobile user send its current location to the service provider as well. We assume that there is a set $U = \{u_1, u_2, \ldots, u_N\}$ of mobile users willing to participate in crowdsensing and use $ST = \{st_1, st_2, \ldots, st_M\}$ to denote the set of tasks, where N and M are the numbers of users and tasks, respectively. Upon receiving the message from user, the service provider firstly checks whether user's location is in the sensing area. If mobile user u_i in sensing task st_j's sensing area, the service provider then assigns the sensing task st_j to the mobile use $u_i (i \in \{1, 2 \ldots, N\}, j \in \{1, 2, \ldots, M\})$. After the mobile user u_i accept the sensing task st_j, it will sense the event or phenomena and collect data, and then upload the sensing report to data requester via the service provider. After receive the sensing report, the data requesters will distribute appropriate rewards to the mobile users according to their contributions on the task.

We assume that the time is slotted and in each time slot t, each sensing task st has some state information $ST_j[t]$ that is time-varying and to be sampled by mobile users. The service provider relies on randomly arriving users to sample and report the states of the tasks. And the service provider maintains a record for each task, whose value in time slot t is denoted as $V_j[t]$. We use $U_j[t]$ be the most recent update time up to time slot t for task st's record. So, the age (freshness) of record $V_j[t]$ in time slot t can be represented as $T_j[t] = t - U_j[t]$.

Let $A_j[t]$ be the number of users who select the sensing task st_j in time slot t, which are independently and identically distributed (i.i.d.) over time and $A_j[t] \leq A_{max}$ for some $A_{max} < \infty$. On the other hand, we let $S_j[t]$ denote the number of users that can be served by sensing task st_j in time slot t. We assume that $S_j[t]$ are independently and identically distributed over time with $S_j[t] \leq S_{max}$, for some $S_{max} < \infty$. We use C denote channel state. For any arrival process that lies strictly within the maximal satisfiable region $\wedge(0, C)$, it stabilizes all virtual queues in a sense. And we use $W_j[t]$ to

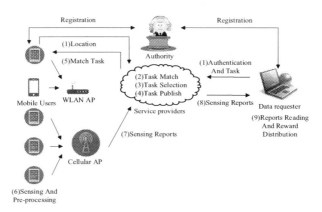

Fig. 1. System model of mobile crowdsensing

denote the number of users who wait for service in sensing task st in time slot t. And for each sensing task guarantees a minimum timely throughput is λ_j. Note that we do not consider packet loss in this work.

We eager to get the total weighted sum of the average steady-state AoI over all sensing task in the crowdsensing, i.e., $\sum_{j=1}^{M} \theta_j \mathbb{E}[T_j[t]]$, where $\theta_j \geq 0$ is a weighting parameter associated with sensing task st_j and gives the preference of sensing task st_j towards the information freshness.

2.2   Problem Statement

In the location-based mobile crowdsensing system, the AoI of the whole system put up a poor show due to the randomness and uncertainty of mobile users. And we use task allocation to solve this problem. The following we will describe in detail.

a) **Privacy Exposure Problem in Location Matching:** Previous studies have concluded that the distance between a user and the task location negatively affect the quality and freshness of the result [6, 7]. In addition, assigning location-based tasks will involve the location privacy of data requesters and mobile users, it may cause a series of privacy exposure problems.

In order to solve the above problems, we decide each task is associated with a geographical sensing region and only those mobile users within the region can receive the task information and can apply for it. Moreover, the sensing area of tasks and the location of mobile users should not be exposed to others. The sensing area of a task can't be visible to the service provider and the mobile users who are not in the sensing area. And the mobile users are only aware of whether their geographical positions are in the sensing area or not.

b) **AoI Instability Problem:** In a multi-tasking mobile crowdsensing system, there will be multiple sensing tasks in an area. However, due to the random selected by users in crowdsensing, the information of some sensing task st_j's record could be old and hence $V_j[t]$ may be outdated and inaccurate. In general, mobile users will participate in newly released tasks and ignore the previous ones. That causes the data requester having unnecessarily outdated status information because of a lack of updates. And this will yield AoI instability in mobile crowdsensing. For example, in Fig. 2, task st_a releases first and task st_b releases later. Clearly, we can see that the AoI of task st_b is relatively low while task st_a's AoI is large and grows linearly with respect to time. Due to the stochastic arrivals of participating users, some tasks may have too many participants at a time. There will be channel congestion and data transmission delays. So we should also take queueing congestion into consideration. And we also need to ensure the minimum timely throughput for each task, that is, the length of the waiting queue.

To solve the above problem, we propose a task allocation strategy algorithm to improve the information freshness by reducing the AoI. In other studies, researchers use incentive mechanisms to motivate users to complete corresponding tasks. Note that in

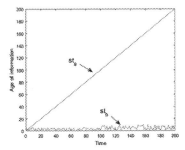

Fig. 2. AoI instability in mobile crowdsensing

our study, each task gave appropriate and adequate rewards to the participants who were willing to participate in the task. To ensure information freshness in crowdsensing, we will decide a linear AoI-aware allocation scheduling algorithm to reduce the average AoI across the system while keeping queueing congestion at each task low.

3 Our Construction

In this section, we use allocation strategy to solve the information freshness problem at the receiver's side from two perspectives of space and time. First we consider a privacy-protecting matrix-based location matching mechanism to assign tasks to users in the same area precisely and privately. And then we propose a linear AoI-aware allocation scheduling algorithm to reduce average AoI across the system and improve the freshness of task status information. And we also analyze the AoI performance of our proposed algorithm.

3.1 Matrix-Based Location Matching Mechanism with Privacy Protection

This phase is run by the authority and the service provider to bootstrap the mobile crowdsensing service. Let $(\mathbb{G}, \mathbb{G}_T)$ be two cyclic groups with a prime order p, where p is λ bits, and $\hat{e} : \mathbb{G} \times \mathbb{G} \to \mathbb{G}_T$ be a bilinear map. The authority chooses a random $\mathcal{G} \in \mathbb{G}_T$, and defines a cryptographic hash function $\mathcal{H} : \{0, 1\}^* \to \mathbb{Z}_p^*$ and a pseudo-random function $\mathcal{F} : \mathbb{Z}_p \times \{0, 1\}^* \to \mathbb{Z}_p^*$. The service provider employs a matrix $L_{m \times n}$ to represent the geographical area (i.e., longitude and latitude) that the crowdsensing service can cover. Each entry in the matrix represents a small grid in the sensing region. Given the longitude of Hunan province is from 108.47°E to 114.15°E and the latitude is from 24.38°N to 30.08°N, we have a 57 × 57 matrix or a 568 × 570 matrix more precisely to represent it.

When a data requester has something to request the sensing data from mobile users, it will first generate a sensing task. The statement of the sensing task is denoted as st = (task, area, reward), indicating that the task what and where to sense, and also what award can offer. Other attributes like acceptance conditions and task deadlines can be added as well.

To protect the location privacy in the crowdsensing, the data requester generates a matrix $\hat{L}_{m \times n}$ to denote the sensing area. As depicted in Fig. 3, the entry in $\hat{L}_{m \times n}$

corresponding to each position in the sensing area is a random value chosen from $\mathbb{Z}_p^*$; otherwise, the value for a location outside is set to be zero. In order to mask the sensing area in $\hat{L}_{m \times n}$, the service provider picks $m \times n$ random values from $\mathbb{Z}_p^*$ to generate an invertible matrix $\hat{M}_{m \times n}$ and computes $\hat{N}_{n \times n} = \hat{L}_{m \times n}^T \times \hat{M}_{m \times n}$, where $\hat{L}_{m \times n}^T$ is the transpose of the matrix $\hat{L}_{m \times n}$. Note that all non-zero entries in $\hat{L}_{m \times n}$ should be distinct. Then, the data requester sends (v, c1, c2, $\hat{N}_{n \times n}$, reward) to the service provider, along with its ID. Anonymity or pseudonym techniques can be used for ID to protect the Identity information of each mobile user and data requester. Note that identity protection is not the focus of this paper, any suitable identity protection techniques can be adopted into our framework.

After the service provider received the task, it assigns the task number num, and stores (num, $V_j[t]$, c1, c2, $\hat{N}_{n \times n}$, reward) in its database.

When a mobile user u_i wants to participate in the crowdsensing task, u_i firstly generates a matrix $\tilde{L}_{m \times n}$ according to its current location. The corresponding entry of the users' location in $\tilde{L}_{m \times n}$ is a random value chosen from $\mathbb{Z}_p^*$ and all the other entries are zero. The non-zero entries in $\tilde{L}_{m \times n}$ should be different. To protect this location information, it also generates a random invertible matrix $\tilde{M}_{m \times n}$ by picking $m \times n$ random values from $\mathbb{Z}_p^*$, and calculates $\tilde{N}_{n \times n} = \tilde{M}_{m \times n}^T \times \tilde{L}_{m \times n}$, where $\tilde{M}_{m \times n}^T$ is the transpose of the matrix $\tilde{M}_{m \times n}$. Finally, u_i sends ($\tilde{N}_{n \times n}$) to the service provider, along with its ID.

When the service provider received the encrypted location, it will check which sensing area matches the user's location. It calculates $\bar{N}_{n \times n} = \tilde{N}_{n \times n} \times \hat{N}_{n \times n}$ and checks whether $\bar{N}_{n \times n}$ is zero matrix or not. If $\bar{N}_{n \times n}$ is non-zero matrix, u_i successfully matches st. Then, the service provider releases (num, $V_j[t]$, c1, c2, $\hat{N}_{n \times n} \widehat{N}_{n \times n}$, reward) for u_i. If there is no task matching u_i, the service provider responds failure.

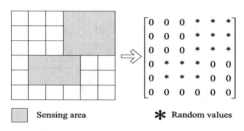

Fig. 3. Sensing area and alternative matrix

3.2 A Linear AoI-Aware Allocation Scheduling Algorithm

The proposed age-based and real-time allocation scheduling algorithm is given in Algorithm1.

Algorithm1: AoI-aware Algorithm

In time slot t, the AoI-aware algorithm selects a schedule $st_j^*[t]$ such that

$$st_j^*[t] = \arg\max\big(\xi \alpha_j T_j[t] + (1-\xi)Q_j[t]\big), \forall t.$$
$$j \in \{1,2,\ldots,m\}$$

where $0 \le \xi \le 1$ and $\alpha_j \ge 0, \forall j$, are some control parameters.

The scheduling decision is made on a per-time basis. The idea is that, at the beginning of time slot t, the service provider selects a maximum weighted sensing task $st_j^*[t]$ to the arriving users by treating $(\xi \alpha_j T_j[t] + (1-\xi)Q_j[t])$ in Algorithm1 over the time slot. Here, the virtual-queue length $Q_j[t]$ and the AoI $T_j[t]$ are the two driving factors in making scheduling decisions.

We use $\mathcal{G}_j^*[t]$ to denote the event that at least a use take part in the sensing task st_j at time slot t. And let $\mathbb{I}_{\mathcal{G}_j^*[t]}$ denote the indicator variable such that $\mathbb{I}_{\mathcal{G}_j^*[t]} = 1$ if event $\mathcal{G}_j^*[t]$ happens, and $\mathbb{I}_{\mathcal{G}_j^*[t]} = 0$ otherwise.

The dynamics of queue-length and age of sensing task st_j can be described as follows:

$$Q_j[t+1] = \max\{Q_j[t] + A_j[t] - S_j[t], 0\}, \forall j \tag{1}$$

$$T_j[t+1] = \begin{cases} T_j[t] + 1, & if \ \mathbb{I}_{\mathcal{G}_j^*[t]} = 0, \\ 0, & otherwise, \end{cases} \tag{2}$$

Note that ξ is a parameter to control the data freshness performance of the algorithm and α_j is a parameter related to the sensing task preference towards the data freshness. We will show that our policy can provide guarantees on AoI with $0 < \xi < 1$.

3.3 Data Freshness Guarantee

In this section, we analyze the AoI performance of our proposed algorithm and derive an upper bound of data freshness under the AoI-aware algorithm, which indicates that the data freshness under our strategy can be guaranteed.

Then, we have the following theorem regarding the AoI performance.

Theorem 1. *The data freshness under the AoI-aware algorithm by choosing* $\alpha_j = \frac{\theta_j}{\lambda_j}, \forall j$ *is upper bounded by the following:*

$$\sum_{j=1}^{M} \theta_j \mathbb{E}\big[T_j[t]\big] \le \frac{G}{1+\epsilon} \sum_{j=1}^{M} \alpha_j + \frac{1}{2(1+\epsilon)}\left(\frac{1}{\xi}-1\right)\sum_{j=1}^{M} \mathbb{E}\big[A_j[t]\big] \tag{3}$$

where $G \triangleq \min\{A_{max}, tS_{max}\}$, ϵ *is some positive constant satisfying that* $\lambda + \epsilon 1$ *still lies within the maximal satisfiable region, and* 1 *is an* $L - dimensional$ *vector of all ones and* $\epsilon > 0$ *satisfies* $\lambda + \epsilon 1 \in \wedge(0, C)$.

Proof. We consider the following quadratic Lyapunov function: $F(\mathbf{Q}, \mathbf{T}) \triangleq \frac{1-\xi}{2} \sum_{j=1}^{M} (Q_j[t])^2$. Then we have:

$$\triangle F(\mathbf{Q}, \mathbf{T}) = \mathbb{E}\left[F(Q_j[t+1], T_j[t+1]) - F(Q_j[t], T_j[t]) | \mathbf{Q}, \mathbf{T} \right]$$

$$= \mathbb{E}\left[\frac{1-\xi}{2} \sum_{j=1}^{M} (Q_j[t+1])^2 - \frac{1-\xi}{2} \sum_{j=1}^{M} Q_j[t] | \mathbf{Q}, \mathbf{T} \right]$$

$$\leq \frac{1-\xi}{2} \sum_{j=1}^{M} \mathbb{E}\left[(Q_j[t] + A_j[t] + S_j[t])^2 - (Q_j[t])^2 | \mathbf{Q}, \mathbf{T} \right]$$

$$\leq (1-\xi) \sum_{j=1}^{M} \mathbb{E}\left[Q_j[t] A_j[t] | \mathbf{Q}, \mathbf{T} \right] + \frac{1-\xi}{2} \sum_{j=1}^{M} \mathbb{E}\left[(S_j[t])^2 | \mathbf{Q}, \mathbf{T} \right]$$

$$- \sum_{j=1}^{M} \mathbb{E}\left[(1-\xi) Q_j[t] \min\left\{ A_j[t], S_j[t] \mathbb{I}_{\mathcal{G}_j^*[t]} \right\} | \mathbf{Q}, \mathbf{T} \right] \tag{4}$$

Take the expectation of both sides of the steady state distribution of $(\mathbf{Q}, \mathbf{T})$, i.e., $(\bar{\mathbf{Q}}, \bar{\mathbf{T}})$. We use the fact that $\mathbb{E} \triangle F(\bar{\mathbf{Q}}, \bar{\mathbf{T}}) = 0$, so we have:

$$0 \leq (1-\xi) \sum_{j=1}^{M} \lambda_j \mathbb{E}[Q_j[t]] + \frac{1-\xi}{2} \sum_{j=1}^{M} \mathbb{E}\left[(A_j[t])^2 - (S_j[t])^2 \right]$$

$$- \sum_{j=1}^{M} \mathbb{E}\left[(1-\xi) Q_j[t] \min\left\{ A_j[t], S_j[t] \mathbb{I}_{\mathcal{G}_j^*[t]} \right\} \right] \tag{5}$$

which equivalent of

$$\sum_{j=1}^{M} \mathbb{E}\left[(1-\xi) Q_j[t] \min\left\{ A_j[t], S_j[t] \mathbb{I}_{\mathcal{G}_j^*[t]} \right\} \right]$$

$$\leq (1-\xi) \sum_{j=1}^{M} \lambda_j \mathbb{E}[Q_j[t]] + \frac{1-\xi}{2} \sum_{j=1}^{M} \mathbb{E}\left[(A_j[t])^2 - (S_j[t])^2 \right] \tag{6}$$

Adding $\xi \sum_{j=1}^{M} \alpha_j \mathbb{E}\left[T_j \min\left\{ A_j[t], S_j[t] \mathbb{I}_{\mathcal{G}_j^*[t]} \right\} \right]$ to both sides, we have:

$$\sum_{j=1}^{M} \mathbb{E}\left[\left((1-\xi) Q_j[t] + \xi \alpha_j T_j \right) \min\left\{ A_j[t], S_j[t] \mathbb{I}_{\mathcal{G}_j^*[t]} \right\} \right]$$

$$\leq (1-\xi) \sum_{j=1}^{M} \lambda_j \mathbb{E}[Q_j[t]] + \frac{1-\xi}{2} \sum_{j=1}^{M} \mathbb{E}\left[(A_j[t])^2 - (S_j[t])^2 \right]$$

$$+ \xi \sum_{j=1}^{M} \alpha_j \mathbb{E}\left[T_j \min\left\{ A_j[t], S_j[t] \mathbb{I}_{\mathcal{G}_j^*[t]} \right\} \right] \tag{7}$$

Since the arrival process is strictly within the maximal satisfiable region $\wedge(0, \mathbf{C})$ and the optimal strategy in linear programming always occurs at extreme points, under proposed our AoI-aware algorithm, we have:

$$\sum_{j=1}^{M}(1+\epsilon)\lambda_j\big((1-\xi)Q_j[t]+\xi\alpha_j T_j[t]\big)$$
$$\leq \sum_{j=1}^{M}\mathbb{E}\Big[\big((1-\xi)Q_j[t]+\xi\alpha_j T_j[t]\big)\min\big\{A_j[t],\,S_j[t]\mathbb{I}_{\mathcal{G}_j^*[t]}\big\}\big|\mathbf{Q},\mathbf{T}\Big] \tag{8}$$

Then, for the steady state distribution of $(\mathbf{Q},\mathbf{T})$, take the expectation of both sides of the inequality, and we have:

$$\sum_{j=1}^{M}(1+\epsilon)\lambda_j\mathbb{E}\big((1-\xi)Q_j[t]+\xi\alpha_j T_j[t]\big)$$
$$\leq \sum_{j=1}^{M}\mathbb{E}\Big[\big((1-\xi)Q_j[t]+\xi\alpha_j T_j[t]\big)\min\big\{A_j[t],\,S_j[t]\mathbb{I}_{\mathcal{G}_j^*[t]}\big\}\Big] \tag{9}$$

By selecting $\theta_j=\alpha_j\lambda_j$, $\forall j$ and plugging (9) into (8), we have:

$$\sum_{j=1}^{M}\theta_j\mathbb{E}\big[T_j[t]\big]=\sum_{j=1}^{M}\alpha_j\lambda_j\mathbb{E}\big[T_j[t]\big]$$
$$\leq \tfrac{1}{1+\epsilon}\sum_{j=1}^{M}\alpha_j\mathbb{E}\Big[T_j[t]\min\big\{A_j[t],\,S_j[t]\mathbb{I}_{\mathcal{G}_j^*[t]}\big\}\Big]$$
$$+\tfrac{1-\xi}{2\xi(1+\epsilon)}\sum_{j=1}^{M}\mathbb{E}\Big[\big(A_j[t]\big)^2-\big(S_j[t]\big)^2\Big] \tag{10}$$
$$\leq \tfrac{G}{1+\epsilon}\sum_{j=1}^{M}\alpha_j\mathbb{E}\Big[T_j[t]\mathbb{I}_{\mathcal{G}_j^*[t]}\Big]+\tfrac{1-\xi}{2\xi(1+\epsilon)}\sum_{j=1}^{M}\mathbb{E}\Big[\big(A_j[t]\big)^2-\big(S_j[t]\big)^2\Big]$$

where G denotes the maximum number of packets that can be transmitted in one time slot.

The evolution of (1) can be rewritten as follows:

$$T_j[t+1]=T_j[t]\big(1-\mathbb{I}_{\mathcal{G}_j^*[t]}\big)+1=T_j[t]-T_j[t]\mathbb{I}_{\mathcal{G}_j^*[t]}+1 \tag{11}$$

Multiplying α_j on both sides of (11), and summing over all sensing tasks st, we have:

$$\sum_{j=1}^{M}\alpha_j T_j[t+1]=\sum_{j=1}^{M}\alpha_j T_j[t]-\sum_{j=1}^{M}\alpha_j T_j[t]\mathbb{I}_{\mathcal{G}_j^*[t]}+\sum_{j=1}^{M}\alpha_j \tag{12}$$

We take expectation on both sides of (12) and then we have:

$$\sum_{j=1}^{M}\alpha_j\mathbb{E}\Big[T_j[t]\mathbb{I}_{\mathcal{G}_j^*[t]}\Big]=\sum_{j=1}^{M}\alpha_j \tag{13}$$

We plug (13) into (10), we can get formula (3).

Note that when θ_j and λ_j are given for all sensing task, we can always find α_j for the AoI-aware algorithm such that $\theta_j=\alpha_j\lambda_j$ holds. From theorem 1, it can be seen that the considered data freshness is upper bounded. Despite all the factors above are not controllable, the AoI-aware algorithm provides a way of tuning data freshness performance via a controllable parameter ζ. We can see that the performance of data freshness improves with the increase of ζ. It is worth noting that when $\zeta=1$, the algorithm yields the best AoI performance.

4   Performance Evaluation

4.1   Location Privacy Preservation

The location of the sensing region is represented as a matrix $\hat{L}_{m \times n}$, which is randomized by a random matrix $\hat{M}_{m \times n}$ to generate $\hat{N}_{n \times n}$. The location of the mobile user is represented as a matrix $\tilde{L}_{m \times n}$, which is randomized by a random matrix $\tilde{M}_{m \times n}$ to generate $\tilde{N}_{n \times n}$. After receiving these two matrices, the service provider cannot learn any information about the mobile user's location and the data requester's sensing area from $\hat{N}_{n \times n}$ and $\tilde{N}_{n \times n}$, respectively. The service provider multiplies the above two matrices to get $\bar{N}_{n \times n}$. If there is no overlapping in the location of users and crowdsensing area, $\bar{N}_{n \times n}$ must be a zero matrix. If there has overlapping grid (where the corresponding entry is $\hat{L}_{ab}$ in $\hat{L}_{m \times n}$ and is $\tilde{L}_{ab}$ in $\tilde{L}_{m \times n}$, respectively), the entries in b-row of $\hat{N}_{n \times n}$ are nonzero, as well as the entries in b-column of $\hat{N}_{n \times n}$. Then, the service provider only knows that there are some overlapping locations in the b-column of the sensing area. But it is unable to distinguish which location is overlapped from m locations. Furthermore, $\tilde{N}_{n \times n} \times \hat{N}_{n \times n}$ and $\hat{N}_{n \times n} \times \tilde{N}_{n \times n}$ cannot provide any information to the service provider. On the whole, matrix matching strategies accurately match tasks to users in the same region while the location of mobile user and the sensing area is not exposed to service providers and other entities.

4.2   AoI of System

In this section, we present a simulation to demonstrate the performance of the proposed AoI-aware task allocation strategy algorithm.

In this set of simulations, we consider a multi-task mobile crowdsensing system including 200 sensing task and 100 random arrival users. The tasks were released randomly and sequentially. After the task was released, AoI of each task was counted after 100 time slots. Figure 4 gives histograms of AoI over all the simulated time slots. We first look at the AoI performance of the system without an allocation strategy. As show in Fig. 4(a), we can see that the AoI of the system is polarized without any strategy, a small part of AoI is relatively low, but a large part of AoI is high. This indicates that a large number of tasks have not been updated for a long time. Next we will show our policy with various ξ. In Fig. 4(b), $\xi = 0$, and we can see that AoI of some tasks is still on the high side. It didn't take into account the AoI of the overall system. Figure 4(c)-€ show the AoI performance with $\xi = 0.1$, $\xi = 0.5$ and $\xi = 1$, respectively. With the increase of ξ, the number of large values of AoI (such as > 50 time slots) decreased, indicating a better performance of data freshness and in Fig. 4€ we can see that the AoI performance best and it's going up dramatically.

In general, the simulation results verify the ability of the proposed AoI-aware allocation strategy algorithm to improve the freshness of data. However, if the data freshness is important to applications, we could trade for it by increasing ξ.

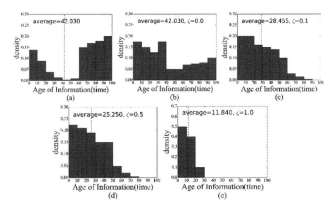

Fig. 4. AoI performance with various ξ

5 Related Work

With the rapid increase of portable mobile devices carrying sensors, a novel sensing paradigm – Mobile Crowdsensing (MCS) [8] has become an effective way to sensing and collect environmental, human society and individuals. The service provider plays a major role on task publishing and data collection, but optimized multi-task allocation has not received attention. One of the key issues for MCS platforms is how to find users to complete tasks, while taking into account the different initial locations and movement costs of users. Some recent results have focused on the centralized task assignment in MCS, which aims to improve energy efficiency or maximize social surplus [9]. They set up a task schedule with the goal of maximizing the completion of the task. Different from the above literature, which focuses on the collection of location-dependent data without any time constraints, we consider the case where the service provider aims to collect time-sensitive and location-dependent information for its data requester through distributed decisions of mobile users while protect location privacy.

6 Conclusion

In this paper, we let the server provider assign tasks to ensure the freshness of the collected information in mobile crowdsensing. To solve such a problem, we think in terms of space and time. First of all, we consider the task assignment of regional location and design a location-based assignment strategy to support accurate task allocation, which also guarantees the location privacy. Then we use AoI to be a metric of information freshness, minimum throughput is also considered, and design a linear AoI-aware allocation scheduling algorithm to achieve minimum average AoI in a multi-task mobile crowdsourcing system. And we show that the proposed algorithm is capable of improving the information freshness by reducing the AoI in the simulation experiments. In general, these results serve as an exciting first step toward optimizing information freshness in mobile crowdsensing systems.

References

1. Cao, N., et al.: Semantics analytics of origin-destination flows from crowd sensed big data. Cmc-computers Mater. Continua **61**(1), 227–241 (2019)
2. Wang, T., Wu, T., Ashrafzadeh, A.H., He, J.: Crowdsourcing-based framework for teaching quality evaluation and feedback using linguistic 2-tuple. Cmc-computers Mater. Continua **57**(1), 81–96 (2018)
3. Luo, Y., Qin, J., Xiang, X., Tan, Y., Liu, Q., Xiang, L.: Coverless real-time image information hiding based on image block matching and Dense Convolutional Network. J. Real-Time Image Process., 1–11 (2019)
4. Sun, L., Ge, C., Huang, X., Yingjie, W., Gao, Y.: Differentially private real-time streaming data publication based on sliding window under exponential decay. Cmc-computers Mater. Continua **58**(1), 61–78 (2019)
5. Dong, G., Gao, J., Huang, L., Shi, C.: Online burst events detection oriented real-time microblog message stream. Cmc-computers Mater. Continua **60**(1), 213–225 (2019)
6. Alt, F., Shirazi, A.S., Schmidt, A., Kramer, U., Nawaz, Z.: Location-based crowdsourcing: extending crowdsourcing to the real world. In: 6th Nordic Conference on Human-Computer Interaction: Extending Boundaries, pp. 13–22. AMS, Copenhagen (2010)
7. Kazemi, L., Shahabi, C., Chen, L.: Trustworthy query answering with spatial crowdsourcing. In: 21st ACM Sigspatial International Conference on Advances in Geographic Information Systems, pp. 314–323. GeoTruCrowd, America (2013)
8. Ganti, R.K., Ye, F., Lei, H.: Mobile crowdsensing: current state and future challenges. IEEE Commun. Mag. **49**(11), 32–39 (2011)
9. He, S., Shin, D. H., Zhang, J., Chen, J.: Towards optimal allocation of location dependent tasks in crowdsensing. In: Proceedings of IEEE INFOCOM, Toronto, Canada (2014)

Access Control Based on Proxy Re-encryption Technology for Personal Health Record Systems

Baolu Liu and Jianbo Xu[✉]

School of Computer and Engineering, Hunan University of Science and Technology,
Xiangtan 411201, Hunan, China
baoluliu1220@163.com, jbxu@hnust.edu.cn

Abstract. As an emerging electronic medical technology, PHR (Personal Health Record) systems play a key role in disease monitoring and improves medical efficiency. Because medical data carries patient privacy data, if it is falsified or illegally obtained, it will cause medical accidents and endanger the lives of patients. Therefore, strict medical data access control is required. Ciphertext-based attribute encryption (CP-ABE) supports access control of encrypted data, which is a promising technology for PHR systems. However, the current CP-ABE scheme consumes a large amount of computing resources during the encryption and decryption process. In addition, it is difficult to implement dynamically changing and a large variety of user private key updates. In this paper, a non-dual linear proxy re-encryption method is proposed, which reduces the computational overhead on both the encryption and decryption ends. At the same time, the concept of attribute group is used to realize the dynamic update of ciphertext.

Keywords: Privacy protection · Attribute encryption · Proxy re-encryption · Attribute revocation

1 Introduction

With the development of smart healthcare, PHR is a cloud-based system architecture [1, 2, 26] with powerful storage, sharing and analysis of user health, disease history and more. With the development of smart healthcare, the traditional electronic health care system has gradually transformed into a cloud-based health cloud service [3]. At present, PHR system is widely used in telemedicine consultation, disease prevention, etc. [4]. However, due to the sensitivity of digital medical data in the real-world PHR system, if it is illegally traded or tampered with, it will cause medical disputes or insurance problems [5–7]. Therefore, the confidentiality of the patient's medical data must be guaranteed. In recent years, the ciphertext-based attribute encryption (CP-ABE) [8–10] proposed by research scholars is a promising technology that implements fine-grained access control for PHR data. The access tree formed by the attribute is used as the public key to encrypt the PHR data, and the user private key is described by the attribute set. The user can decrypt the ciphertext only if the user attribute set satisfies the access tree. However, in some specific cases, some problems have hindered the

© Springer Nature Switzerland AG 2020
X. Sun et al. (Eds.): ICAIS 2020, LNCS 12239, pp. 411–421, 2020.
https://doi.org/10.1007/978-3-030-57884-8_36

application of CP-ABE in PHR systems. First, CP-ABE is closely related to bilinear pairs and therefore requires a large amount of computational resources to perform. To enhance security and functionality, most CP-ABE schemes use more bilinear pairing calculations [12–14]. On the other hand, the large and diverse user types with dynamic changes make it difficult to implement effective client management. Flexible access control with fine-grained revocation is one of the difficulties in solving this problem. At present, various scholars have proposed various revocation schemes [15–17], but it is important that the computational overhead is still huge. Therefore, in order to solve the computational overhead, the attribute revocation is not flexible, and the computational efficiency, security, and flexibility are improved, and the CP-ABE scheme needs to be improved [18].

1.1 Related Work

Sahai and Waters [8] first proposed a fuzzy identity based encryption (FIBE). The following year, based on FIBE, Goyal et al. [17] proposed a key strategy ABE (KP-ABE) in which each private key is associated with an access policy and each ciphertext is associated with a set of attributes. Bethencourt et al. [9] proposed the ciphertext strategy ABE (CP-ABE) and pointed out that CP-ABE is more suitable for establishing access control than KP-ABE, because data owners can directly define access policies by themselves. Ostrovsky et al. [12] constructed a CP-ABE that can perform an AND gate. Lewko et al. [14] and Waters both proposed a CP-ABE of the type of linear secret share scheme (LSSS). In order to reduce the computational overhead in the encryption process, [17] proposed a multi-authorization scheme based on CP-ABE, which can achieve efficient decryption, and the user private key has multiple authorization centers to distribute, ensuring that users are insecure due to insecure single authorization centers. The private key is dangerous to leak. [23] proposed a PHR sharing scheme based on multi-authority ABE, which implements on-demand revocation and dynamic policy update. Xhafa et al. [24] established a PHR access control based on multi-access ABE. It not only supports hidden access policies, but also supports user responsibility. Hong et al. [19] proposed an access control for a CP-ABE-based PHR system. Without pairing operations, it significantly reduces computational overhead, but does not provide fine-grained revocation. H. Hong, D. Chen et al. [15, 16] a secure and flexible EHR sharing method based on proxy re-encryption technology, but they still have some disadvantages, such as high computational cost on the patient side when updating policies. Therefore, inspired by the attribute-based conditional proxy re-encryption (ABCPRE) technique [22], we propose a privacy-protected PHR scheme.

Aiming at the problems of high computational cost and unrecognized inflexibility in the current ciphertext encryption and decryption process, this paper proposes a re-encrypted and unpaired ciphertext encryption scheme based on re-encryption (PRP-CP-ABE, proxy revocable and pairing-free CP-ABE). Reduce computational overhead from the encryption side and the decryption side. The remainder of the paper is organized as follows: Sect. 2 shows some preliminary knowledge, Sect. 3 introduces the proposed system model, Sect. 4 gives a proposed detail algorithm design, Sect. 5 performs a security and performance analysis. Finally, we draw conclusion and prospect future work.

2 Preliminaries

Definition 1 (Access Tree). Let Γ be a tree representing an access policy. Each non-leaf node of the tree represents a threshold gate, described by its children and a threshold value. If num_x is the number of children of a node x and k_x is its threshold value, then $0 < k_x \leq num_x$. When $k_x = 1$, the threshold gate is an OR gate and when $k_x = num_x$, it is an AND gate. Each leaf node x of the tree is described by an attribute and a threshold value $k_x = 1$. To facilitate working with the access tree, we define a few functions. We denote the parent of the node x in the tree by $parent(x)$. The function $att(x)$ is defined only if x is a leaf node and denotes the attribute associated with the leaf node x in the tree. The access tree Γ also defines an ordering among the children of every node, that is, the children of a node x are numbered from 1 to num_x. The function $index(x)$ returns such a number associated with the node x. For each access tree, the index values are uniquely assigned to nodes in an arbitrary manner.

Definition 2 (Attribute Group). Let $Att = \{\theta_1, \theta_2, \cdots, \theta_n\}$ denote the universe of all attributes and $\mathcal{U} = \{u_1, u_2, \cdots, u_n\}$ denote the universe of all users that described by attribute sets. Let G_t be an attribute group, which contains all users that hold an attribute θ_t. All users in each attribute group G_t share a key $K_i \in Z_p^*$, we call it attribute group key. Let $\mathcal{G} = \{G_1, G_2, \cdots, G_n\}$ denote the universe of all attribute groups and the $\mathcal{K} = \{K_1, K_2, \cdots, K_n\}$ denote the universe of all attribute group keys.

3 Proposed System Model

3.1 PRP-CP-ABE for PHR System

Figure 1 shows the PHR-based system entity model in the medical environment. The first layer is the data acquisition layer, which mainly consists of body sensors. The second layer is the middle layer, which is mainly used for identity authentication and data forwarding. Layer 3 and cloud services. The layer is the focus of this paper. The third layer is responsible for data encryption, decryption and access control. The cloud service layer is mainly responsible for data storage. The scheme model proposed in this paper is mainly the third layer and the cloud service layer, which mainly consists of 6 entities.

When the DO wants to upload data, use the patient-defined access policy to construct the access tree to encrypt the medical data to obtain the ciphertext CT_{init}, and then according to the attribute group key of the different attributes in the access tree, the ciphertext CT_{init} Re-encryption, and finally get the complete ciphertext CT, uploaded to the CSP through the secure channel. When there is a DU requesting medical data, the CSP sends the CT to the PDS for partial decryption, and then sends it to the DU through the secure channel, and the DU decrypts the plaintext by using the private key.

Trusted Center (TC): Local deployment, mainly publishing system public parameters, system public key, user private key, this article assumes that TC is trusted.

Cloud Service Provider (CSP): Cloud deployment, CSP is responsible for data storage, data re-encryption, this paper believes that CSP is not credible but honest, such as: Alibaba Cloud.

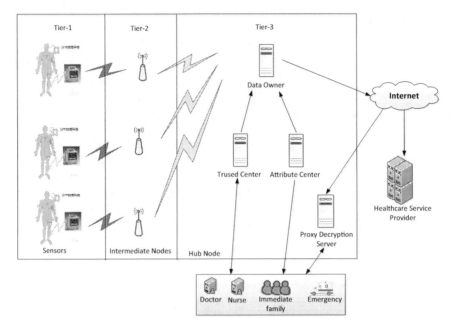

Fig. 1. Medical data system model based on PHR

Attribute Center (AC): Local deployment, responsible for publishing and updating in-system attribute collections, attribute group collections, and user attributes.

Data Owner (DO): Locally deployed to encrypt and upload patient medical data.

Proxy Decryption Server (PDS): Local deployment, PDS is responsible for most of the decryption work, reducing user decryption calculation overhead. This paper considers PDS to be honest and trustworthy.

Data User (DU): DU has different attributes, who can decrypt some ciphertext data locally, and obtain plaintext, such as doctors, nurses, ambulances, family members of patients, etc.

$(\mathcal{G}, PK, MSK) \leftarrow Setup(U, Att, \lambda)$: First, the AC generates a system attribute set, distributes the user attributes, and then generates an attribute group set $\mathcal{G}$ according to the user attributes. The TC selects λ as a security parameter to generate a system public key PK and a system master key MSK.

$(SK, PK_{session,k}) \leftarrow KeyGen(PK, MSK, S, id)$: The TC selects the system public key PK, the system master key MSK, user attribute S and id as input, generates the private key SK of the user u_k, and the public session key $PK_{session,k}$.

$CT_{init} \leftarrow Encryption(PK, \Gamma, M)$: The DO selects the access tree Γ, the system public key PK, and the plaintext M as input, and outputs the initial ciphertext CT_{init}.

$(Hdr, CT) \leftarrow ReEncryption(\mathcal{G}, CT_{init})$: Then DO selects the attribute group set $\mathcal{G}$, the initial ciphertext CT_{init} as input, the output message header Hdr and the re-encrypted ciphertext CT.

$Mor \perp \leftarrow Decryption(Hdr, CT, SK, id)$: The PSD selects the re-encrypted ciphertext CT, the message header Hdr, and the user private key SK as input. If the user attribute satisfies the access policy, the *CT* can be partially decrypted, and part of the ciphertext is output to the user, and the user decrypts locally, and if satisfied, outputs *M*, Otherwise output ⊥.

$\left(Hdr', CT'\right) \leftarrow Update\left(G_t', CT\right)$: When the attribute group G_t in the AC is updated to G_t' and Hdr', the DO selects the encrypted ciphertext *CT* with the attribute group key updated by the attribute θ_t in the access tree, and is updated to CT'.

3.2 Complexity Assumption

Our PRP-CP-ABE is based on the following complexity assumption:

Definition 4 (Hashed Diffie-Hellmen Assumption (HDH)). Define a cyclic group $\mathbb{G}$ with a prime order p. Let g be a generator of $\mathbb{G}$ and $(\cdot)$ be a random oracle. There are three random elements $a, b, c \in Z_p^*$. Given g^a and g^b, the Hashed Diffie-Hellman problem is to distinguish $H\left(g^{ab}\right)$ from $H(g^c)$. We define the advantage of $\mathcal{B}$ as $Adv_{HDH}(\mathcal{B}) = \mathcal{E}$, if there is a probabilistic polynomial-time algorithm $\mathcal{B}$ that can solve this problem with a probability that satisfies:

$$\left| \begin{array}{c} Pr[\mathcal{B}(g^a, g^b, H(g^{ab})) = 1] \\ -Pr[\mathcal{B}(g^a, g^b, H(g^c)) = 1] \end{array} \right| \geq \mathcal{E} \qquad (1)$$

The HDH assumption is given by: There is no probabilistic polynomial-time algorithm can distinguish $H\left(g^{ab}\right)$ from $H(g^c)$ with a non-negligible advantage $\mathcal{E}$.

4 Proposed PRP-CP-ABE Detail Scheme

4.1 Setup Algorithm

The AC publishes a global attribute set, and the TC publishes global public parameters, which are shared among all entities.

Step 1. The AC issues an authorization attribute set $Att = \{\theta_1, \theta_2, \cdots, \theta_n\}$, distributes the user attribute and the identity *id*, and constructs the attribute group set $\mathcal{G}$ according to the user attribute.

Step 2. TC take a security parameter λ as an input. Choose a large prime p with λ-bit length and generate a cyclic group $\mathbb{G}$ with order p. Let g denote a generator of $\mathbb{G}$. Select $\alpha, \beta \in Z_p^*$ and $t_1, t_2, \cdots, t_n \in Z_p^*$ for all attributes in *Att*. Then define two hash functions $H_1 : (\{0, 1\}^{l_1}, \mathbb{G}) \to \{0, 1\}^{l_2}$ and $H_2 : \mathbb{G} \to \{0, 1\}^{l_1+l_2}$, where l_1 is the length of a plaintext and l_2 is the length of a bit string.

Step 3. The TC output master key is:

$$MSK = \{\alpha, \beta\} \qquad (2)$$

TC publish system public key is:

$$PK = \{A, B\}, A = g^\alpha, B = g^\beta \qquad (3)$$

TC publish system public parameter is:

$$PP = \langle PK, p, \mathbb{G}, g, Att : \{\theta_i \in Att : T_i\}, \mathcal{G}, A, B, H_1, H_2\rangle, T_i = g^i \qquad (4)$$

4.2   User Private Key Generation Algorithm

The user private key generation algorithm completes the user private key generation by the TC using the public parameter PP, the system master key MSK, the user u_k having the attribute set S, and the identity id.

Step 1. TC selects two random numbers $r, \mu \in Z_p^*$, then calculates $D_1 = (id \cdot \alpha)\beta^{-1}$, and calculates $D_i = t_i - r$ for each attribute $\theta_i \in S$ of the user.
Step 2. The TC generates the user u_k private session key $SK_{session,k} = \mu$ and the public session key $PK_{session,k} = g^\mu$. User private key is:

$$SK = \langle D_1, \{\forall \theta_i \in S : D_i\}, SK_{session,k} \rangle \tag{5}$$

Step 3. TC sends SK to user u_k, $PK_{session,k}$ to DO.

4.3   Encryption Algorithm

The encryption algorithm is defined by the plaintext M, the access policy, and the DO constructs the access tree Γ according to the access policy formulated by the patient, and encrypts the plaintext M.

Step 1. For each node in the access tree, DO selects $s \in Z_p^*$ as the shared secret value, constructing a random polynomial according to the following rules: starting from the root node, selecting one for each non-leaf node in the access tree. The polynomial q_x, whose order is d, should satisfy $d_x = k_x - 1$, k_x is the threshold of the node. For the root node, let $q_x(0) = s$, and q_x be randomly selected at other d_x nodes. For non-leaf nodes x, $parent(x)$ represents the parent node of x in the access tree, which satisfies $q_x(0) = q_{parent(x)}(index(x))$, and the values of other d_x nodes are randomly defined.
Step 2. DO encrypts plaintext M with symmetric key s and calculates:

$$\sigma = H_1(M, A^s), \ C_0 = H_2(A^s) \oplus (M \| \sigma), \ C_1 = B^s \tag{6}$$

Step 3. The DO encrypts the symmetric key s with the access tree Γ, and for the access tree Γ, $\forall \theta_i = att(x)$ calculates:

$$C_{2,x} = g^{q_x(0)}, \ C_{3,x} = T_i^{q_x(0)} \tag{7}$$

$$CT_{init} = \langle C_0, C_1, \{\forall att(x) \in \Gamma : C_{2,x}, C_{3,x}\} \rangle \tag{8}$$

4.4   Re-encryption Algorithm

The proxy re-encryption algorithm takes as input the initialization ciphertext CT_{init} and the attribute group set $\mathcal{G}$, and the TC generates the attribute group key set $\mathcal{K} = \{K_1, K_2, \cdots, K_n\}$ through the attribute group set $\mathcal{G}$, and the DO encrypts CT_{init} with the attribute group key K_i.

Step 1. TC selects the random number $\gamma \in Z_p^*$ to generate a public session key $PK_{session} = g^\gamma$. For each user $u_k \in \mathcal{U}$, calculate the factor $X_k = PK_{session,k}^\gamma$. For each $G_i \in \mathcal{G}$ construct a polynomial:

$$f_i(X) = \prod_{k=1}^{v}(X - X_k) = \sum_{j=0}^{v} a_{i,j}X^j \tag{9}$$

Where v represents the number of users in G_i, constructing the vector $P_i = \{P_{i,0}, P_{i,1}, \cdots, P_{i,v}\}$, where $P_{i,0} = K_i + a_{i,0}$, $\forall j \in \{2, 3, \cdots v\} : P_{i,j} = a_{i,j}$, TC sends K_i to DO.

Step 2. DO encrypts CT_{init} with symmetric key K_i:

$$CT = \left\langle C_0, C_1, \left\{\forall att(x) \in \Gamma : C_{2,x}^{K_i}, C_{3,x}^{K_i}\right\}\right\rangle \tag{10}$$

Step 3. DO generated message header sent to PDS:

$$Hdr = \langle PK_{session}, \forall G_i \in \mathcal{G} : P_i \rangle \tag{11}$$

4.5 Decryption Algorithm

Logically, decryption includes attribute group key decryption and access tree decryption. From the entity model, decryption includes proxy server decryption and user local decryption.

Step 1. When the DU requests medical data from the PDS, it first sends $SK_{session,k}$ to PDS, and the PDS downloads the corresponding ciphertext from the CSP, and then PDS calculates $PK_{session}^{SK_{session,k}} = (g^\gamma)^\mu = X^k$, PDS extraction attribute group key:

$$P_{i,0} + \sum_{j=1}^{v} P_{i,j} \cdot X_k^j = K_i + \sum_{j=0}^{v} a_{i,j} \cdot X_k^j = K_i \tag{12}$$

Step 2. For each node in the access tree Γ, if the user attribute $\theta_i \in \Gamma$, then the PDS calculate:

$$DecryotNode(x) = \left(C_{2,x}^{D_i}/C_{3,x}^{D_i}\right)^{1/K_i} = g^{-r \cdot q_x(0)} \tag{13}$$

Step 3. PDS sends C_0, C_1 and (13) results to DU, DU calculate:

$$T_0 = C_1^{D_1} \cdot g^{-r \cdot q_x(0)} = g^{id \cdot \alpha \cdot s}, \ C_0 \oplus H_2\left(T_0^{id^{-1}}\right) = M \| \sigma \tag{14}$$

Step 4. DU calculate:

$$H_1\left(M, T_0^{id^{-1}}\right) = \sigma \tag{15}$$

If the user id meets the requirements, then M can be calculated. Conversely, then output $\perp$.

4.6   Attribute Update Algorithm

When the attribute θ_v is revoked in the attribute set Att, the AC changes the attribute group $G_v = \{u_1, u_2, \cdots, u_v\}$ to $G'_v = \{u_1, u_2, \cdots, u_{v'}\}$.

Step 1. TC generates a new attribute group key $K'_t \in Z^*_p$, then construct a new polynomial $f_t(X) = \prod_{k=1k=1}^{v'} (X - X_k) = \sum_{j=0}^{v'} a_{t,j} X^j$, Update vector $P'_t = \left(K'_t + a'_{t,0}, a'_{t,1}, \cdots, a'_{t,v'} \right)$, Update header $Hdr' = \left\langle PK_{session}, \{\forall G_i \in \mathcal{G} \backslash G_t : P_i\}, P'_t \right\rangle$, TC sends K'_t to DO.

Step 2. DO downloads ciphertext from CSP and encrypts ciphertext CT with K'_t, uploads it to CSP again.

$$CT' = \left\langle C_0, C_1, \left\{ \forall \theta_t \in Att : C_{2,t}^{K'_t/K_i}, C_{3,t}^{K'_t/K_i} \right\}, \{\forall att(x) \right.$$
$$\left. \in \Gamma \backslash att(\theta_t) : C_{2,x}, C_{3,x}\} \right\rangle \qquad (16)$$

5   Security and Performance Analysis

5.1   Security Analysis

Confidentiality. As stated in the proof of confidentiality, it is guaranteed that no illegal users can access the plaintext of the health record. The parameter $g^{\alpha \cdot s}$ cannot be extracted, so the decryption result cannot be derived. We note that the confidentiality of PRP-CP-ABE can be guaranteed based on the HDH assumption, and we give the following theorem as its official statement: If the HDH assumption is difficult to crack, our scheme is safe under the ciphertext attack selected in the random prediction model. Prove that, based on the PRP-CP-ABE security model, we designed a challenge game and provided a reduction from the PRP-CP-ABE to HDH assumptions. We say that if there is no probability polynomial time algorithm that can break the HDH hypothesis with a non-negligible advantage, then there is no probability polynomial time adversary that can attack PRP-CP-ABE.

Collusion Attack. For public key encryption, it is highly possible that collusion attack takes place if a new authorized private key can be generated with illegal collusion of some unauthorized users. Considering our PRP-CP-ABE, each private key is randomized by inserting a unique pair of a random element $r \in Z^*_p$ and an identity id. Even if PHR receivers collude and combine their private keys, they still cannot recover A^s. Therefore, the proposed PRP-CP-ABE is inherently secure against collusion attack.

Forward and Backward Secrecy. We recommend efficient immediate revocation through re-encryption algorithms and attribute group-based update algorithms [20]. For each initial ciphertext CT_{init}, encrypted with the access policy Γ, the CSP will re-encrypt it with the attribute group key associated with the attribute in Γ and generate a header Hdr for decryption. Once some DUs are revoked attribute θ_t, the TC will immediately update the corresponding attribute group G_t. Therefore, the TC updates the corresponding attribute group key K_t to K'_t, so that the header and all ciphertexts encrypted by the access policy with the attribute θ_t are updated. Those who no longer have θ_t can never get K'_t.

Therefore, forward secrecy can be guaranteed. Similarly, once some DUs obtain a new attribute θ_t, the message header encrypted with the attribute θ_t is updated and all ciphertexts are updated. Even if these users hold the previous ciphertext, they still cannot extract the correct health record. Therefore, backward secrecy can also be guaranteed.

5.2 Performance Analysis

In terms of efficiency, traditional CP-ABE requires a large amount of computing resources to perform encryption, revocation, and decryption. For encryption and undo, the computation of exponentiation and multiplication takes up more computational resources, and pairing operations result in more computational overhead in decryption. First, we compare the encryption efficiency in these schemes with the exponentiation and multiplication costs of the encryption algorithm and the re-encryption algorithm. Obviously, PRP-CP-ABE has the least computational overhead in the encryption algorithm.

XFZCL15 [24] and HCS16 [15, 16] do not support undo, so no undo overhead is incurred. PTMW06 [10] suggests adding an extension time for each attribute, but it does not provide a detailed construct of the undo. LHS17 [20] introduced attribute groups to achieve immediate revocation, but it consumes a lot of computing resources. Finally, we conclude that the computational overhead of PRP-CP-ABE implementation revocation is minimal. In terms of decryption overhead, as shown in Fig. 2, we analyzed the number of pairing calculations in different scenarios. PTMW06 [10] uses SW05's large attribute set system, and YWRL10 [18] supports non-monotonic access strategies, so a large number of Pairing calculation. We note that $\mathbb{A}^+$ and $\mathbb{A}^-$ represent a collection of all positive attributes and a set of all negative attributes in A, respectively.

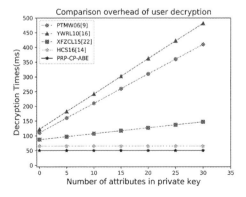

Fig. 2. Decryption time cost comparison

The PRP-CP-ABE scheme and HCS16 [15, 16] are much smaller than the PTMW06 [10] and XFZCL15 [24] schemes in terms of user decryption overhead. This is because the PRP-CP-ABE scheme and HCS16 [15, 16] have a large number of agents to re-decrypt. Completed by the server, the amount of user decryption calculations is fixed and does not increase as the attribute set increases.

6 Conclusion

In this paper, we propose a revocable and unpaired ciphertext policy attribute-based encryption for medical health data management. Based on the attribute group, we built an efficient, fine-grained and instant property update. In addition, the proposed scheme does not require pairing calculation, which is more effective than similar schemes. At the same time, the proxy decryption server is added to decrypt the partial ciphertext at the decryption end, and the local user only needs to decrypt symmetrically to obtain plaintext, reducing local users. In future work, if the TC becomes malicious, can we guarantee the security of the health record and the security of the user's private key in terms of the diverse roles in the personal health record system? On the other hand, it is necessary to design a more efficient access strategy that can reduce the size of ciphertext.

Acknowledgments. This work was supported by the National Natural Science Foundation of China (Grant Nos. 61872138).

References

1. Lanranjo, L., Neves, A.L., Vilanueva, T., Cruz, J., Brito de Sa, A., Sakellarides, C.: Patients' access to their medical records. Acta Med. Port. **26**, 265–270 (2013)
2. Pearce, C., Bainbridge, M.: A personally controlled eletronic health record for Austrilia. J. Am. Med. Inform. Assoc. **21**(4), 707–713 (2014)
3. Yoo, H., Chung, K.: PHR based diabetes index service model using life behavior analysis. Wirel. Personal Commun. **93**(1), 161–174 (2017)
4. Roehrs, A., da Costa, C.A., de Oliveira, K.S.F.: Personal health records: a systematic literature review. J. Med. Internet Res. **19**(1), 100–120 (2017)
5. Li, M., Yu, S., Zheng, Y., Ren, K., Lou, W.: Scalable and secure sharing of personal health record in cloud computing using attribute-based encryption. IEEE Trans. Parallel Distrib. Syst. **24**(1), 131–143 (2013)
6. Chen, T.S., Liu, C.H., Chen, C.S., Bau, J.G., Lin, T.C.: Secure dynamic access control scheme of PHR in cloud computing. J. Med. Syst. **26**(6), 4005–4020 (2012)
7. Sahai, A., Waters, B.: Fuzzy identity-based encryption. In: Proceedings of EoroEncrypt, pp. 457–473 (2005)
8. Bethencout, J., Sahai, A., Waters, B.: Ciphertext-policy attribute-based encryption. In: Proceedings of IEEE Symposium Security Privacy, pp. 321–334 (2007)
9. Pirretti, M., Traynor, P., McDaniel, P., Waters, B.: Secure attribute-based systems. In: Proceedings of ACM CCS, pp. 99–112 (2006)
10. Ahmad, M.S., Musa, H.E., Nadarajah, R., Hassan, R., Othman, N.E.: Comparison between Android and iOS operating system in terms of security. In: Proceedings of CITA, pp. 1–4 (2013)

11. Barak, O., Cohen, G., Toch, E.: An anonymizing mobility data using cloaking. Pervasive Mob. Comput. **28**, 102–112 (2016)
12. Zhang, Y., Zheng, D., Chen, X., Li, H.: Efficient attribute based data sharing in mobile clouds. Pervasive Mob. Comput. **28**, 35–149 (2016)
13. Rahimi, M.R., Ren, J., Liu, C.H., Vasilakos, A.V., Venkatasubramanian, N.: Mobile cloud computing: a survey, state of art and future directions. Mob. Netw. Appl. **19**(2), 133–143 (2014)
14. Hong, H., Chen, D., Sun, Z.: A practical application of CPABE for mobile PHR system: a study on user accountability. SpringerPlus **5**(1), 1320 (2016)
15. Goyal, V., Pandey, O., Sahai, A., Waters, B.: Attribute based encryption for fine-grained access control of encrypted data In: Proceedings of ACM CCS, pp. 89–98 (2006)
16. Yu, S., Wang, C., Ren, K., Lou, W.: Attribute based data sharing with attribute revocation. In: Proceedings of the ASIACCS, pp. 261–270 (2010)
17. Hong, H., Sun, Z.: High efficient key-insulated attribute based encryption scheme without bilinear pairing operations. SpringerPlus **5**(1), 131 (2016)
18. Lin, G., Hong, H., Sun, Z.: A collaborative key management protocol in ciphertext policy attribute-based encryption for cloud data sharing. IEEE Access **5**, 9464–9475 (2017)
19. Win, K.T.: A review of security of electronic health records. HIM J. **34**(1), 13–18 (2005)
20. Wungpornpaiboon, G., Vasupongayya, S.: Two-layer ciphertext-policy attribute-based proxy re-encryption for supporting PHR delegation. In: Proceedings of ICSEC, pp. 1–6 (2015)
21. Qian, H.L., Li, J.G., Zhang, Y.C., Han, J.G.: Privacy preserving personal health record using multi-authority attribute based encryption with revocation. Int. J. Inf. Secur. **14**(6), 487–497 (2015)
22. Xhafa, F., Feng, J., Zhang, Y., Chen, X., Li, J.: Privacy-aware attribute-based PHR sharing with user accountability in cloud computing. J. Supercomput. **71**(5), 1607–1619 (2014). https://doi.org/10.1007/s11227-014-1253-3
23. Xie, X., Ma, H., Li, J., Chen, X.: New Ciphertext-Policy Attribute-Based Access
24. Sukhodolskiy, I.A., Zapechnikov, S.V.: An access control model for cloud storage using attribute-based encryption. In: Young Researchers in Electrical and Electronic Engineering, pp. 578–581. IEEE (2017)
25. Sandhia, G.K., Kasmir Raja, S.V., Jansi, K.R.: Multi-authority-based file hierarchy hidden CP-ABE scheme for cloud security. SOCA **12**, 295–308 (2018)
26. Wang, S., Zhou, J., Liu, J.K., et al.: An efficient file hierarchy attribute-based encryption scheme in cloud computing. IEEE Trans. Inf. Forensics Secur. **11**(6), 1265–1277 (2016)
27. Centonze, P.: Security and privacy frameworks for access control big data systems. Comput. Mater. Continua **59**(2), 361–374 (2019)
28. Tang, Y., Lian, H., Zhao, Z., Yan, X.: A proxy re-encryption with keyword search scheme in cloud computing. Comput. Mater. Continua **56**(2), 339–352 (2018)
29. Sun, Y., Yuan, Y., Wang, Q., Wang, L., Li, E., Qiao, L.: Research on the signal reconstruction of the phased array structural health monitoring based using the basis pursuit algorithm. Comput. Mater. Continua **58**(2), 409–420 (2019)

Generating More Effective and Imperceptible Adversarial Text Examples for Sentiment Classification

Xiaohu Du[✉], Zibo Yi, Shasha Li, Jun Ma, Jie Yu, Yusong Tan,
and Qinbo Wu

College of Computer Science and Technology, National University of Defense
Technology, Changsha, Hunan, China
{duxiaohu18,yizibo14,shashali,majun,yj,yusong.tan,qinbo.wu}@nudt.edu.cn

Abstract. In this paper, we propose a novel white-box attack against word-level CNN text classifier. On the one hand, we use an Euclidean distance and cosine distance combined metric to find the most semantically similar substitution when generating perturbations, which can effectively increase the attack success rate. We've increased global search success rate from 75.8% to 85.8%. On the other hand, we can control the dispersion of the location of the modified words in the adversarial examples by introducing the coefficient of variation(CV) factor, because greedy search sometimes has poor readability for the modified positions in adversarial examples are close. More dispersed modifications can increase human imperceptibility and text readability. We use the attack success rate to evaluate the validity of the attack method, and use CV value to measure the dispersion degree of the modified words in the generated adversarial examples. Finally, we use the combination of these two methods, which can increase the attack success rate and make modification positions in generated examples more dispersed.

1 Introduction

Adversarial example is a small modification of an input that can change the judgment of a classifier but is not easily detected by humans. These adversarial examples can expose some weaknesses of the classifier, which can be used to improve the model, such as adding them to the training set for adversarial training [17,19]. Liang et al. [9] point out that two problems need to be solved if we want to generate a very effective adversarial example. The first one is to make the adversarial examples maintain their original meaning. The second difficulty is to make it difficult for humans to detect that the adversarial examples are forged or modified. They propose three strategies: insert, delete and modify. These three strategies can be applied to words or characters. If the attacker can get classifier internal information, we call it white-box attack, which produces some gradient-based methods. Tsai et al. [12] propose a Global Search attack method against white-box classifier. This attack aims at word level classifier and

© Springer Nature Switzerland AG 2020
X. Sun et al. (Eds.): ICAIS 2020, LNCS 12239, pp. 422–433, 2020.
https://doi.org/10.1007/978-3-030-57884-8_37

contrasts with misspelling noise and greedy search approach. When generating candidate words, the global search only uses the Euclidean distance to find the nearest word. We consider combining Euclidean distance and cosine distance to find similar words, and find that some of the original attack failed examples can be re-attack successfully when we use the cosine distance. In the process of generating the adversarial examples, The existing methods modify the word without considering the dispersion of the modified position in the whole sentence. Especially the word replacement position of greedy attack is often close in the sentence, which greatly reduces the readability (Fig. 1).

Original text: The Pallbearer is a disappointment and at times extremely boring with a love story that just doesn't work partly with the casting of Gwyneth Paltrow (Julie). Gwyneth Paltrow walks through the entire film with a confused look on her face and its hard to tell what David Schwimmer even sees in her.⟨br /⟩⟨br /⟩However The Pallbearer at times is funny particularly the church scene and the group scenes with his friends are a laugh but that's basically it. Watch The Pallbearer for those scenes only and fast forward the rest. Trust me you aren't missing much.

Greedy text:on despite has given tempered well outside well surprisingly boring with a love story that just doesn't work partly with the casting of Gwyneth Paltrow (Julie). Gwyneth Paltrow walks through the entire film with a confused look on her face and its hard to tell what David Schwimmer even sees in her.⟨br /⟩⟨br /⟩However The Pallbearer at times is funny particularly the church scene and the group scenes with his friends are a laugh but that's basically it. Watch The Pallbearer for those scenes only and fast forward the rest. Trust me you aren't missing much.

Global text: The Pallbearer is a artist and at times extremely boring with affection love story that just doesn't work partly with the american of Paltrow Paltrow (Julie). Paltrow Paltrow walks through the entire film with a confused look on her face and its hard to tell what David Schwimmer even sees in her.⟨br /⟩⟨br /⟩However The Pallbearer at times is funny particularly the church scene and the group scenes with his friends are a laugh but that's basically it. Watch The Pallbearer for those scenes only and fast forward the rest. Trust me you aren't missing much.

Global text with CV: The Pallbearer is affection artist and at times extremely boring with a love story that just doesn't work partly with the casting of Gwyneth Paltrow (Julie). Paltrow Paltrow walks through the entire film with a confused look on her face and its hard to tell what David Schwimmer even sees in her.⟨br /⟩⟨br /⟩However The Pallbearer at times is funny particularly the church scene and the group scenes with his friends are affection laugh but that's basically it. Watch The Pallbearer for those scenes only and fast forward the rest. Trust me you aren't missing much.

Fig. 1. An original text and three adversarial examples of successful attack

---

Our method is essentially to use different metric to measure embedding distance to find similar words in generating candidate words and introduce the CV in the Generate Adversary Function in [12]. We compare results of different CV weight on the success rate of the attack, and finally get the best CV weight for generating adversarial examples.

In summary, our contributions in this work are as follows: (1) We use different metric of Euclidean distance and cosine distance to find the similar words when generating perturbations and candidate words, the result show that the combining metric can increase attack successful rate. (2) We propose a method to generate adversarial examples in which the modified words have high semantic relevance and their positions are more dispersed. (3) Finally, we prove that global search attack with coefficient of variation is more similar to the original text and more imperceptible for human, which is verified by human evaluation.

2   Related Work

An important problem is that the generated adversarial examples must not only fool the target model, but also keep the perturbation undetected. A good adversarial example should convey the same semantics as the original text, so we need some metrics to quantify the similarity. There are three metrics that used on vectors and documents [13, 15].

Euclidean Distance. In text, Euclidean Distance measures the linear distance between two vectors in Euclidean space.

Cosine Similarity. Cosine similarity represents the semantic similarity of two words by calculating the cosine of the angle between two vectors. Cosine distance is more concerned with the direction difference between two vectors. The smaller the angle between two vectors (the larger the cosine value), the greater the similarity.

Word Movers Distance (WMD). WMD [6] mainly reflects the distance between documents, so it is not suitable for finding similar words. Its semantic representation can be based on the embedding vector obtained by word2vec. Of course, it can also be based on other word embedding methods. This algorithm constructs the document distance into a combination of the semantic distances of the words in the two documents. For example, the Euclidean distance is obtained from the word vectors corresponding to any two words in the two documents and then weight and sum. The WMD distance between two documents A and B is:

$$WMD(A, B) = \sum_{i,j=1}^{n} T_{ij} \cdot D(\overrightarrow{i}, \overrightarrow{j}) \tag{1}$$

Where $D(\overrightarrow{i}, \overrightarrow{j})$ is the Euclidean distance of the word vectors corresponding to the two words i and j. The Bag Of Words model is used to get the word frequency of the word in the document (as the weight of a word in the document), and the problem then becomes how to "carry" all the word units of document A to the corresponding word units of document B with the minimum cost, and finally get the weight matrix T. WMD algorithm is a special case of EMD [10] algorithm, and some improved algorithms based on WMD, such as WCD and RWMD.

The first adversarial attack originates in 2014, Szegedy et al. [11] find that the deep neural network used for image classification could be deceived by tiny pixel perturbations, which are added to the input image [8,16,18]. They find that there is a high rate of deception in the image classifier, but humans did not detect the change. Jia et al. [3] is the first to consider adversarial examples generation on text-based deep neural networks. Since then, people begin to pay attention to generating adversarial examples for text. Ebrahimi et al. [2] propose a gradient-based method to generate adversarial examples. This attack aims at character level RNN classifier and achieves good results. Cheng et al. [1] propose a white-box method to generate adversarial examples against NMT and improve the robustness of NMT. Greedy attack proves to be a very effective method of attack [5,12,14], Yang et al. [14] even show that greedy attack achieves state-of-the-art attack rates across various kinds of models. But it also has some problems, Tsai et al. [12] propose it may not guarantee to produce the optimal results and sometimes spends too much time because the algorithm needs to search the candidate words for every iteration, at the same time, due to the nature of greedy, the algorithm can replace sub-optimal words that do not contribute much to the final goal in an earlier position. Another limitation is that replaced words tend to be in a close area of the sentence, especially in the front. This greatly reduces the readability of the sentence and even destroys the semantic meaning of the original text. They propose a "Global Search" algorithm, which obtains candidate words by calculating a perturbation, and then replace the words in the position where the perturbation is larger. The larger the perturbation, the more sensitive the classifier is to the change of the word. Results of global search prove to generate better examples than the greedy attack and higher attack success rate, which is the baseline used in our experiments. In addition, Lei et al. [7] propose a greedy method can be very time consuming when the space of attacks is large and give the optimization scheme.

3   Methods

3.1   Greedy Search Attack

Greedy search as our comparative experiment. We follow the greedy search algorithm in [5,12], which starts from the first word and then selects the word that has the greatest influence on the success of the attack according to different labels in the k nearest of each word. To find the word with the closest Euclidean distance or cosine distance in word space E. If the modification of the word causes the classifier logit output to be larger than the original and the label is pos or the logit output is smaller than the original and the label is neg, the word is replaced, otherwise the word is not modified. Finally, The attack is successful until the label of the classifier output is different from the original, and the total number of replacement words is lower than our threshold.

3.2   Global Search Attack with Coefficient of Variation

We follow the global search algorithm in [12] and add the coefficient of variation factor in the process of generating example. The algorithm defines an objective function $\mathcal{J}(\delta)$ that maximizes the perturbation and the original input to learn a perturbation δ, and calculates the gradient of the δ of the objective function $\mathcal{J}(\delta)$ every iteration, then updates the δ via back propagation, finally the perturbed embedding and perturbation are obtained. Then we can find the candidate words for each word in the original text by constantly looping the perturbed embedding. Finally we replace the words selectively. The greater the perturbation magnitude, the word confusion classifier is more efficient, so global search prefers to replace the high perturbation words. But we combine the perturbation and CV values to determine the modification order (Fig. 2).

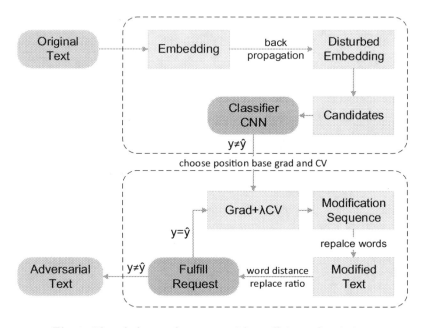

Fig. 2. The whole attack process with coefficient of variation

The whole algorithm is described in Algorithm 1. The algorithm is divided into two parts including generating candidate words and modifying the candidate words which have the most influence on the classifier. We use the whole text as the research object. E_x represents word embedding of the original input text x. y and $\hat{y}$ represent the original and adversarial label. In the first loop, we obtain a perturbed embedding E_x' by gradient-based method, and then traverse every word in E_x', to find the word with the closest Euclidean distance or cosine distance in word space E, and then generate a list of candidate words. We attack the classifier with this list of candidate words. If the label is not the same as the original input, the loop ends. Otherwise, algorithm continues to calculate the gradient to get a new E_x'. After obtaining a list of valid candidate words, we will make a modified order selection. First, we will take the gradient of the first loop. We set a full 0 tensor with the same dimensions as the original text, with *selected* for recording the position of the selected word, *list* for the sequence of selected words, and *positions* for the last modified order. Then start the loop selection, First we select the index number *selectedX* which value is 0 from the *selected*, the number of words in the text is the traversal range, and the variable is represented by x, when $selected[x] \neq 1$, calculating the coefficient of variation CV_1 of x added to the *list* and the CV_2 of the *selectedX* added to the *list*, if the weighted sum of gradient and CV_1 is greater than the sum of gradient and CV_2, we then replace *selectedX* with x, add x to *positions*, update *selected* and *list*, let *selected[selectedX]*=1, add *selectedX* to *list*. Finally, a modification sequence is generated. Then we modify them in order until the output of the classifier is different from the original text. In the process we can set the word distance threshold, word replacement rate and other parameters, and compare different results under different parameters.

Algorithm 1: Global Search with coefficient of variation

1: $y \leftarrow f_{sigmoid}(E_x)$
2: Initialize $\delta \leftarrow 0, success = False,$candidates $\leftarrow \varnothing$
3: **while** *not success* **do**
4:     $E'_x \leftarrow E_x + \delta$ ◁ *back propagation*
5:     **for** e *in* E'_x **do**
6:         **if** *Euclidean distance* **then** $idx = argmin\|e - E\|_2$
7:         **if** *Cosine distance* **then** $idx = argmax(cosine\_similarity(e, E))$
8:         candidates $\leftarrow$ *candidates append idx*
9:     $\widehat{y} \leftarrow f_{sigmoid}($candidates$)$
10:    **if** $\widehat{y} \neq y$ **then** *break*
11: $grad \leftarrow \|\delta\|_2$
12: $selected \leftarrow 0, list \leftarrow \varnothing, \Omega \leftarrow \varnothing$
13: **while** *not success* **do**
14:    $selectedX = selected.index(0)$
15:    **for** x *in total(grad) and* $selected[x] \neq 1$ **do**
16:    $CV_1 = CV(list\ append\ x), CV_2 = CV(list\ append\ selectedX)$
17:    **if**$grad[x] + \lambda \cdot CV_1 > grad[selectedX] + \lambda \cdot CV_2$ **then**
18:        $selectedX \leftarrow x$
19:    $\Omega \leftarrow \Omega\ append\ x$
20:    $selected[selectedX] = 1, list \leftarrow list\ append\ selectedX$
21:    **for** i *in* Ω **do**
22:        $w \leftarrow i-th\ word\ in\ x$
23:        **if** *distance* < *threshold and* $\frac{i}{len(x)} < r$ **then** $x'_i \leftarrow w$
24:        **if** $\widehat{y} \neq y$ **then**
25:            $success = True$
26:            **return** x'
27:        **else**
28:            **return** None

We can control the degree of dispersion of the word's position by the weight value λ of the CV. When λ is too large, the modified words will be more dispersed. The order of modification is closer to the global search algorithm when λ is smaller, when λ=0, the algorithm becomes the global search.

4   Experiments

4.1   Dataset And Target Model

The dataset contains 25,000 IMDB film reviews specifically for sentiment analysis. The mood of the comment is binary. Labels are pos(positive) and neg(negative). We randomly segment 25,000 data into 20,000 examples as training sets and 5,000 examples as test sets.

We use the CNN model and some of its hyperparameters [4] as our target model. In this model, filter windows of 3, 4, 5 with 100 feature maps each, dropout rate of 0.5, and a max pooling layer, batch size has been modified to 64. We use 20,000 examples train the model and test the model with 5,000 examples. The result of this model is that the accuracy of training sets is 1 and the accuracy of testing sets is 0.89.

4.2   Evaluation

4.2.1   Different Metric of Embedding Distance

Our experimental goal is to generate adversarial examples to confuse the classifier. As long as the prediction result of the adversarial example is different from the original comment result, the attack is considered successful. At the same time, we need to exclude the influence of classifier accuracy on the experimental results, so we select 500 correctly classified examples. Our experiments have the same word distance threshold and word replacement rate, the word Euclidean distance is set 50 and the word replacement rate is set 0.1, k is set 35 in the greedy attack, finally, to do the greedy attack and global attack and add the cosine distance experiment (Table 1).

Table 1. The total success rate of combining metric and only using cosine distance of greedy attack and global attack

Vector angle	10	40	50	60	70	80	90
Combining greedy attack	0.658	0.66	0.662	0.694	0.72	0.72	0.72
Combining global attack	0.774	0.774	0.78	0.788	0.808	0.833	0.858
Greedy using cosine	0	0.316	0.498	0.66	0.68	0.682	0.682
Global using cosine	0.046	0.052	0.098	0.2	0.33	0.518	0.644

When we generate the candidate words, the perturbed embedding E_x' cannot be accurately mapped to the word in the embedding space, we need to use the E'_x looking for all the nearest words as the candidate words, there are two ways to measure the distance between two words in the embedding space, Euclidean distance and cosine distance. Our experiments use the cosine distance to generate candidate words after the failure of candidate words generated by Euclidean distance. Cosine distance may find a successful candidate words and attack successfully when the example which candidate words generated by Euclidean distance fails to attack.

We calculate the attack success rate corresponding to several different cosine distance thresholds. We use the cosine distance after the failure of Euclidean distance attack and calculate the final success rate. The experiments show that the cosine distance can increase the number of successful attacks in the comments on the failure of the Euclidean distance attack. The smaller the cosine distance is, the more successful rate the attack will be, because the word angle is larger and their similarity is smaller. Greedy attack and global attack without cosine distance had original success rates of 65.8% and 75.8%. In the global attack added cosine distance, even if the angle is set to 10, the final success rate will increase to 77.4%, while the greedy attack has no new successful examples. There is no significant change in the number of increases between the angles of 10 and 50. An angle of more than 50 will have a significant increase in the number of successful attacks. In addition, we do an experiment to find candidate words only by cosine distance as a contrast, we find that in the case of high cosine distance threshold, the attack success rate is very low, but finally when the cosine distance threshold is 0, it can also achieve a high attack success rate, which shows that in terms of word similarity, the effect of cosine distance is worse than that of European distance. We can use the cosine distance as a supplement of European distance to improve the overall attack success rate. The evaluation of adversarial examples shows that greedy attacks still appear that the replacement locations are sometimes close, or even in the front of the text, which readability is not good. While the replacement position of global attack is relatively random. The following human evaluation also shows this conclusion (Fig. 3).

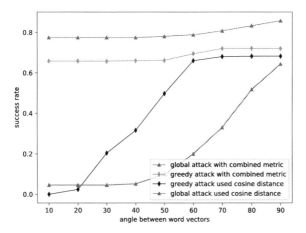

Fig. 3. The total success rate of greedy attack and global attack added cosine distance

4.2.2   Global Attack with Combined Metric and CV

Because greedy attack may cause word modification positions close and result in poor readability, global search improves this situation to some extent, but it does not have the ability to effectively control the dispersion of word modification positions. Because the modified word is selected from big to small according to the gradient magnitude. We propose a method to control the dispersion of the modified position by adding the coefficient of variation, and control the dispersion degree of the modified position by the weight of CV. We introduced the CV factor into the global search with combining metric. Finally, we use global attack with the combined metric and set the cosine distance threshold to 0.9848 to maintain the word's high similarity, and compare the attack success rate under different CV weight. We select 500 comments used by greedy attacks and calculate the attack success rate under different CV weight (λ).

In order to prove the validity of adding the CV factor when determining the modification order, we select 50 comments that are correctly classified by the classifier and the algorithm attack is successful, and calculate the CV value of the adversarial examples under different λ, When the CV weight value is 0, it is the global search attack.

Experiments show that when λ is larger, the degree of dispersion of the modified positions of the adversarial example is larger (CV value is larger), and the final attack success rate has a slight decrease overall, which indicates that the word with larger gradient magnitude has more influence on the classifier. We analysis the adversarial examples and find that some examples of the original attack failure will be re-attacked successfully after joining CV, and some examples of the original successful attack will fail the attack after joining CV, as a whole, the attack success rate basically decreases as λ increases, but this change is very small. The experiments also show that even if we set the λ to 30 after we add the cosine distance, the attack success rate still reaches 76.2%, which still exceeds the original global attack success rate of 75.8%. So our methods can improve the attack success rate and make the modification positions in adversarial example more dispersed (Table 2).

Table 2. The attack rate and CV value under different λ

λ	0	1	5	10	20	30
Success rate	0.774	0.776	0.766	0.768	0.77	0.762
CV value	0.487	0.589	0.70	0.73	0.827	0.886

4.2.3   Human Evaluation

In order to illustrate that adversarial text with more dispersion words modification could lead more similar to the original text and more imperceptible for human, we select 5 volunteers to score the similarity between the adversarial examples and the original text, from 1 to 5. We randomly chose 10 comments with positive and negative examples and their adversarial examples generated by greedy search, global search and the global search with CV. This is used to prove the similarity between the original text and adversarial examples. On the other hand, we select 15 comments including the original text, greedy text, global text and global text with CV, and then ask volunteers to identify which of them are modified adversarial examples, this is used to prove the imperceptibility for human. After counting all the scoring results, in terms of similarity, we calculate the average score, the global text with cv similarity score is 3.642, the global text score is 3.612, and the greedy text score is 2.91, because the greedy text often modifies the front words in a large amount, it can obviously find the difference with the original text. In terms of imperceptibility, the global text with CV has a detectable ratio of 0.37, the global text has a detectable ratio of 0.4, and the greedy text has a detectable ratio of 0.46. We find that the score of global attack and global attack with CV is close, and they are better than greedy attack.

5   Conclusion and Future Work

In this paper, we use different metric for finding candidate words in perturbation-based adversarial examples attack, and find that candidate words found by different similarity metric in word space may compensate for the failure of the other method. At the same time, we propose the factor of adding CV to the modified positions of the word in the algorithm for the problem of greedy search to modify the positions of the words may be too close. Our algorithm proves that the position of the word modification in adversarial examples can be more dispersed after adding the CV factor. It compensates for the poor readability of the greedy attack examples. Our methods can maintain a high attack success rate and make the modification positions in adversarial example more dispersed by a combined metric and CV factor. We must also consider the quality of the adversarial examples when pursuing a high attack success rate.

In our experiments, we use Glove as the word vector, it can't match all the words of IMDB, especially the names of people and places, which results in some UNK cases when converting vector and text. It will greatly improve this situation to train a good word vector for a specific dataset. A word vector that better reflects the similarity of words can also enhance the effect of our work. At the same time, Adversarial examples can reveal the vulnerability in the NLP model, and we can use it to improve the robustness of the model, which will be the future work.

References

1. Cheng, Y., Jiang, L., Macherey, W.: Robust neural machine translation with doubly adversarial inputs. In: Proceedings of the 57th Annual Meeting of the Association for Computational Linguistics, pp. 4324–4333. Association for Computational Linguistics, Florence, July 2019. https://doi.org/10.18653/v1/P19-1425, https://www.aclweb.org/anthology/P19-1425

2. Ebrahimi, J., Rao, A., Lowd, D., Dou, D.: HotFlip: white-box adversarial examples for text classification. In: Proceedings of the 56th Annual Meeting of the Association for Computational Linguistics (Volume 2: Short Papers), pp. 31–36. Association for Computational Linguistics, Melbourne, July 2018. https://doi.org/10.18653/v1/P18-2006, https://www.aclweb.org/anthology/P18-2006

3. Jia, R., Liang, P.: Adversarial examples for evaluating reading comprehension systems (2017)

4. Kim, Y.: Convolutional neural networks for sentence classification. CoRR abs/1408.5882 (2014). http://arxiv.org/abs/1408.5882

5. Kuleshov, V., Thakoor, S., Lau, T., Ermon, S.: Adversarial examples for natural language classification problems (2018). https://openreview.net/forum?id=r1QZ3zbAZ

6. Kusner, M., Sun, Y., Kolkin, N., Weinberger, K.: From word embeddings to document distances. In: International Conference on Machine Learning, pp. 957–966 (2015)

7. Lei, Q., Wu, L., Chen, P.Y., Dimakis, A.G., Dhillon, I.S., Witbrock, M.: Discrete adversarial attacks and submodular optimization with applications to text classification (2019)

8. Li, M., Sun, Y., Su, S., Tian, Z., Wang, Y., Wang, X.: DPIF: a framework for distinguishing unintentional quality problems from potential shilling attacks. CMC-Comput. Mater. Continua **59**(1), 331–344 (2019)

9. Liang, B., Li, H., Su, M., Pan, B., Shi, W.: Deep text classification can be fooled. In: IJCAI (2018)

10. Rubner, Y., Tomasi, C., Guibas, L.J.: A metric for distributions with applications to image databases (1998)

11. Szegedy, C., et al.: Intriguing properties of neural networks. arXiv preprint arXiv:1312.6199 (2013)

12. Tsai, Y.T., Yang, M.C., Chen, H.Y.: Adversarial attack on sentiment classification. In: Proceedings of the 2019 ACL Workshop BlackboxNLP: Analyzing and Interpreting Neural Networks for NLP, pp. 233–240. Association for Computational Linguistics, Florence, August 2019. https://doi.org/10.18653/v1/W19-4824, https://www.aclweb.org/anthology/W19-4824

13. Wang, W., Tang, B., Wang, R., Wang, L., Ye, A.: A survey on adversarial attacks and defenses in text. CoRR abs/1902.07285 (2019). http://arxiv.org/abs/1902.07285

14. Yang, P., Chen, J., Hsieh, C.J., Wang, J.L., Jordan, M.I.: Greedy attack and gumbel attack: generating adversarial examples for discrete data. arXiv preprint arXiv:1805.12316 (2018)

15. Zhang, W.E., Sheng, Q.Z., Alhazmi, A., Li, C.: Adversarial attacks on deep learning models in natural language processing: a survey. arXiv preprint arXiv:1901.06796 (2019)

16. Zhang, Y., Cheng, Q.: An image steganography algorithm based on quantization index modulation resisting scaling attacks and statistical detection. Comput. Mater. Continua **56**(1), 151–167 (2018)

17. Zhao, S., Cai, Z., Chen, H., Wang, Y., Liu, F., Liu, A.: Adversarial training based lattice lstm for chinese clinical named entity recognition. J. Biomed. Inform. **99**, 103290 (2019)

18. Zhao, W., Li, P., Zhu, C., Liu, D., Liu, X.: Defense against poisoning attack via evaluating training samples using multiple spectral clustering aggregation method. CMC-Comput. Mater. Continua **59**(3), 817–832 (2019)
19. Zhao, W., Long, J., Yin, J., Cai, Z., Xia, G.: Sampling attack against active learning in adversarial environment. In: Torra, V., Narukawa, Y., López, B., Villaret, M. (eds.) MDAI 2012. LNCS (LNAI), vol. 7647, pp. 222–233. Springer, Heidelberg (2012). https://doi.org/10.1007/978-3-642-34620-0_21

QR Code Detection with Faster-RCNN Based on FPN

Jinbo Peng[1,2], Song Yuan[1,2(✉)], and Xin Yuan[1,2]

[1] School of Computer Science and Technology, Wuhan University of Science and Technology, Wuhan 430065, China
yuansong_2002@163.com
[2] Hubei Province Key Laboratory of Intelligent Information Processing and Real-time Industrial System, Wuhan University of Science and Technology, Wuhan 430065, China

Abstract. Nowadays, the QR code is often used in many popular fields, such as payment and social networking. Therefore, it is particularly important to quickly and accurately detect the position of QR code in real complex scenes. Traditional QR code detection methods mainly use hand-engineered features for detection. However, the QR code photos we take may be blurred due to pixel, distance, and other problems, and may even produce some rotations and deformations because of the complex scenes. Under such circumstances, the traditional QR code detection methods may not be so applicable. Faster-RCNN was originally used for multiple object detection, but we adjusted it slightly and applied it to the detection of QR code. At the same time, we made a small dataset under complex scenes for training Faster-RCNN networks. However, in complex scenes, the size of the QR code vary greatly due to the distance of shooting, so we add an FPN module to the Faster-RCNN to improve the detection performance for small and multi-scale QR code. Experimental results show that our method has achieved good performances in the detection of QR code in complex scenes.

Keywords: QR code · Object detetction · Convolutional neural network

1 Introduction

QR code was proposed by DensoWave in 1994. Due to its High-density coding, large information capacity, low cost, easy production, and other advantages. It has been widely popularized in many fields and has also produced many QR detection applications, such as ZXing, Zbar, Boofcv, etc. These applications all use some traditional Haar-like Features, Histograms of Oriented Gradients [5] (HOG) and other hand-engineered features for detection, so the pixel requirements for photos are relatively high. In the case of complex scenes, when QR code is blurred, deformed or the area of QR code in the photos taken by the

© Springer Nature Switzerland AG 2020
X. Sun et al. (Eds.): ICAIS 2020, LNCS 12239, pp. 434–443, 2020.
https://doi.org/10.1007/978-3-030-57884-8_38

(a) curve   (b) brightness   (c) glare

(d) outdoor   (e) lots   (f) nominal

Fig. 1. QR code pictures in complex scenes: (a) Twisted QR code picture. (b) QR code in a dark environment. (c) QR code under strong light. (d) QR code in outdoor. (e) A picture containing many QR codes. (f) Picture of small size QR code.

photographer is relatively small (as shown in Fig. 1), the traditional detection methods are no longer applicable.

In this paper, we make some adjustments to the Faster-RCNN algorithm for multiple object detection and add an FPN [15] module for QR code detection of a single object. Faster-RCNN is an object detection algorithm using convolutional neural network, and it uses a convolution neural networks (ConNets) [13,14] module to extract feature map, and then obtains the region proposals [23] through RPN layer. After that, the region proposals are sent to the Fast R-CNN [7] module to output the bounding box of the detection object and the corresponding category. For convolutional neural networks, it plays an important role in many fieldssuch as Text Detection and Recognition [25], Binaural Sound Source Localization [27] and Food Recognition [17]. Different depths correspond to different levels of semantic features. Shallow network resolution is high, learn more detail features, deep network resolution is low, learn more semantic features. In Faster-RCNN, whether the RPN network or Fast-RCNN network, detection is based on a single high-level feature. An obvious defect of this method is that it is unfriendly to small objects. Therefore, we add an FPN module to the Faster-RCNN module. FPN module improves the detection performance of small objects by constructing a unique feature pyramid to deal with multi-scale changes in object detection.

To realize QR code single-object detection, we have created a small dataset of QR code for complex scenes. In the dataset, the total number of photos we collected is 1,000. The labeling format of photos is in accordance with VOC2007 [6]. In the aspect of the model, we make some adjustments to the Faster-RCNN code and use our dataset to train the network. In the convolution neural network module for extracting feature map, we use three different networks for experiments respectively. They are Vgg16 [20], Resnet50 [10], Resnet101 [10], and the mAP achieved 86.95% on Vgg16, then we put the dataset on Faster-RCNN network with FPN module added for training, the feature extraction network uses Resnet101, finally, our mAP achieved 90.77%.

The remainder of our paper is organized as follows: Sect. 2 introduces the related work of QR code detection, Faster-RCNN object detection and FPN. In Sect. 3, we introduce our network structure. Section 4 demonstrates our dataset and shows our experimental results. Finally, our work is concluded in Sect. 5.

2   Related Work

Tradition Method for QR Detection. In the past, there were many traditional methods are used for QR code detection, such as [3,22,24]. These methods all use some hand-engineered features such as Haar-like features, histograms of oriented gradients (HOG). At the same time, there are many open-source softwares on the market, such as ZXing, Zbar, Boofcv, and others also use hand-engineered features for QR code detection, which provide convenience for some embedded devices to detect and identify QR code.

Faster-RCNN for Objection Detection. Faster-RCNN is improved by classic RCNN [8], which is a milestone method for using CNN to object detection problems. With CNN's good classification and feature extraction performance, it uses Region Proposal [23] to solve some object detection problems. Fast-RCNN adds a SPP Net [9] to RCNN, which improves its detection performance. Faster-RCNN adds an RPN module based on Fast-RCNN to obtain Region Proposals instead of original Selective Search method, which makes the speed and precision of object detection well. After Faster-RCNN was put forward, many people have used it in specific object detection fields, such as Pedestrian detection [26], Face detection [12,21] and so on.

Feature Pyramid Networks for Object Detection. In computer vision, it is a fundamental challenge to recognize objects at vastly different scales. Featured image pyramids [2] (Fig. 2(a)) is a standard solution about the problem, but this method is too computationally intensive. The Faster-RCNN detection algorithm uses the method of extracting the last layer of features (Fig. 2(b)), besides, the image feature pyramid can be used to process multi-scale objects, the multi-level structure of the depth network itself can also be used to extract multi-scale features (Fig. 2(c)), such as SSD [16] detection algorithm. FPN adopts the method that top-level features pass through the top sampling and low-level feature fusion, and each layer is independently predicted (Fig. 2(d)), YOLOv3 [18] object detection algorithm also draws on this idea of FPN and has achieved good results.

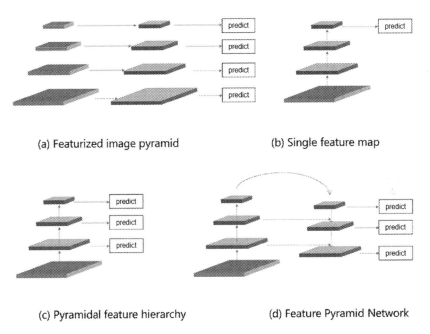

(a) Featurized image pyramid   (b) Single feature map

(c) Pyramidal feature hierarchy   (d) Feature Pyramid Network

Fig. 2. This figure refers to some figures in FPN's paper. (a) Building a feature pyramid by using an image pyramid. Features are computed independently on each image scale (b) Some detection models choose to use features of a single scale to achieve faster detection. (c) Pyramidal feature hierarchy computed by convolution neural network is used as prediction. (d)Feature Pyramid Network is as fast as (b) and (c), but more accurate.

3 Faster-RCNN Based on FPN

3.1 Faster-RCNN

Faster-RCNN is an object detection method based on CNN, and uses some basic conv, relu and pooling layers to compose the feature extraction network to extract feature maps of the input image. Meanwhile, the conv layers and relu layers don't change the size of the image, conv layers generally adopt convolution kernel with the size of 3*3, pad and stride are both 1, and each pooling layer reduces the picture to 1/2 of the original size. Therefore, there is a multiple relationship between the size of the original input image and the feature map. Commonly used feature extraction networks include Vgg, Resnet, Moblenet [11], etc.

The feature map extracted by the feature extraction network will be passed to the RPN module, which is used to extract Region Proposals, which may contain the objects. Before entering the RPN network, the location of each pixel of the feature map will generate 9 different anchor boxes [19]. Each pixel corresponds to an area on the input image, because of the certain multiple relationships between input image and feature map, in the RPN module, two operations are carried out for each Anchor box: classification and regression. Classification is used to judge whether the anchor is an object to be detected and belongs to a Bi-categorization task, regression is aimed to fine-tune anchor boxes, and the number of bounding boxes will be very large. To eliminate some overlapping boxes, we perform NMS processing on these bounding boxes and extract the first N bounding boxes (for example 300) to obtain the final region proposals.

The region proposals obtained by the RPN pass through the Fast-RCNN detection module, and the ROIPooling in the Fast-RCNN detection module can map each region proposal to the corresponding position of the feature map, then pass through the full connection layer, finally softmax is used to judge whether it is a QR code and the probability that it is a QR code, and regression will obtain the position offset of each region proposal to get a more accurate bounding box.

3.2   FPN

FPN module is used in RPN and Fast-RCNN in Faster-RCNN. Taking Resnet101 as the feature extraction network as an example, the output of conv2, conv3, conv4, conv5 are extracted as a feature list and marked as Clist$\{C2, C3, C4, C5\}$, and the step size compared with the original image is $\{4, 8, 16, 32\}$. Next, we will obtain Plist$\{P2, P3, P4, P5\}$ corresponding to Clist by Top-down paths and lateral connections [15], P6 is obtained by down-sampling C5. After obtaining the final feature list $\{P2, P3, P4, P5, P6\}$, each feature in the feature list is followed by an RPN module, and the final output is synthesized. Region proposals are obtained as input of ROIPooling in the Fast-RCNN detection module.

However, the region proposals obtained in the RPN module are from feature maps of different scales, so different feature layers P_k should be used as the input of ROIPooling for region proposals of different scales. We use the calculation formula proposed by FPN's authors to calculate the feature layers. P_k corresponding to different region proposals:

$$k = \left\lfloor k_0 + \log_2 \sqrt{wh/244} \right\rfloor \tag{1}$$

w refers to the corresponding width of the input image, h refers to the corresponding height of input image, k_0 is the initial value, and the initial value is 5. Connect the outputs of multiple ROIPooling modules for classification and regression. Figure 3 shows the whole process.

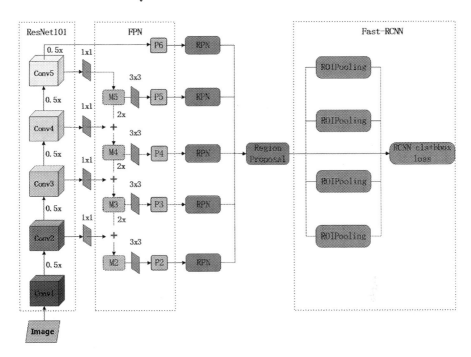

Fig. 3. Faster-RCNN based on FPN. ResNet101 module: Extracting feature list Clist{$C2, C3, C4, C5$} from the input image by Resnet101 network. FPN module: Obtaining feature list Plist{$P2, P3, P4, P5, P6$} by Top-down paths and lateral connections. Every feature in Plist gets Region Proposals by an RPN module. Fast-RCNN module: Using the specified formula to divide the Region Proposals obtained by different depth features as the input of different ROIPooling, finally, using the output for regression of bounding box and classification of categories.

4   Experiments

4.1   Datasets

We have made a small QR code dataset. The QR code images in the train set are a total of 1,000 pieces that we collected on the Internet, and test set are a total of 150 images which we used from the QR code related experiment paper by boofcv's author [1], and the test set of images contains large-scale, small-scale and various complex scenes of QR code images. We labeled the ground truth of the QR code in train set and the test set according to the format specified by VOC2007.

4.2   Experiments on Datasets

We used the self-made QR code dataset to train and test the model, and mAP was used as our evaluation index. The trained models include standard Faster-RCNN model and Faster-RCNN model with the FPN module added. We use

Vgg16, Resnet50, Resnet101 networks as feature extraction networks of the
Faster-RCNN model for training and testing respectively. The code we used
is implemented by chen [4]. We use the Resnet101 network as the feature extraction network of the Faster-RCNN model with an FPN module for training and
testing. GPU is TelsaP4, an 8G video memory.

Table 1. Detection results on my own test set. The detectors are FPN+Faster-RCNN
and Faster-RCNN, but using different feature extraction networks for training and
testing.

Model	Net	mAP(%)
FPN+Faster-RCNN	Resnet101	**90.77**
Faster-RCNN	Resnet101	84.18
Faster-RCNN	Resnet50	83.87
Faster-RCNN	Vgg16	**86.95**

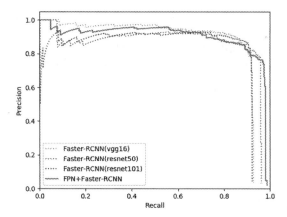

Fig. 4. Precision-Recall. Detection on my own test set using FPN+Faster-RCNN model
and Faster-RCNN model with different feature extraction networks.

Finally, our results can be seen from Table 1. Training the Faster-RCNN
model, and use the Vgg16 network as a feature extraction network, and the
highest value of mAP reaches 86.95%. Training the Faster-RCNN model with
the FPN module added, the mAP can reach 90.77%, which is nearly 4% higher
than the highest mAP of Faster-RCNN without the FPN module. Figure 4 shows
the PR curve corresponding to each model, and the Fig. 5 shows the detection
results of QR codes.

(a) curve   (b) brightness   (c) glare

(d) outdoor   (e) lots   (f) nominal

Fig. 5. The examples of detection results using the FPN+Faster-RCNN model on my test set. The areas of QR code are labeled by red bounding boxes, and each bounding box represents that there may be a QR code, and a softmax score between 0 and 1 represents the probability of QR code. The score threshold of these images is 0.6.

5   Conclusion

We apply Faster-RCNN, a classic object detection algorithm based on CNN, to detect the QR code, and add an FPN module to improve the detection performance of Faster-RCNN for multi-scale QR code of small objects in complex scenes. Finally, good results are achieved.

References

1. Abeles, P.: Study of QR code scanning performance in different environments. v3. https://boofcv.org/index.php?title=Performance:QrCode
2. Adelson, E.H., Anderson, C.H., Bergen, J.R., Burt, P.J., Ogden, J.M.: Pyramid methods in image processing. RCA Eng. **29**(6), 33–41 (1984)
3. Belussi, L., Hirata, N.: Fast QR code detection in arbitrarily acquired images. In: 2011 24th SIBGRAPI Conference on Graphics, Patterns and Images, pp. 281–288. IEEE, Maceio, Alagoas (2011)

4. Chen, X., Gupta, A.: An implementation of faster RCNN with study for region sampling. arXiv preprint arXiv:1702.02138 (2017)
5. Dalal, N., Triggs, B.: Histograms of oriented gradients for human detection. In: Schmid, C., Soatto, S., Tomasi, C. (eds.) International Conference on Computer Vision & Pattern Recognition, vol. 1, pp. 886–893. IEEE Computer Society, San Diego (2005). https://hal.inria.fr/inria-00548512
6. Everingham, M., Van Gool, L., Williams, C.K., Winn, J., Zisserman, A.: The pascal visual object classes (VOC) challenge. Int. J. Comput. Vis. **88**(2), 303–338 (2010)
7. Girshick, R.: Fast R-CNN. In: Proceedings of the IEEE International Conference on Computer Vision, pp. 1440–1448. IEEE, Santiago, Chile (2015)
8. Girshick, R., Donahue, J., Darrell, T., Malik, J.: Rich feature hierarchies for accurate object detection and semantic segmentation. In: Proceedings of the IEEE Conference on Computer Vision and Pattern Recognition, pp. 580–587. IEEE, Columbus (2014)
9. He, K., Zhang, X., Ren, S., Sun, J.: Spatial pyramid pooling in deep convolutional networks for visual recognition. IEEE Trans. Pattern Anal. Mach. Intell. **37**(9), 1904–1916 (2015)
10. He, K., Zhang, X., Ren, S., Sun, J.: Deep residual learning for image recognition. In: Proceedings of the IEEE Conference on Computer Vision and Pattern Recognition, pp. 770–778. IEEE, Las Vegas (2016)
11. Howard, A.G., Zhu, M., Chen, B., Kalenichenko, D., Wang, W., Weyand, T., Andreetto, M., Adam, H.: Mobilenets: efficient convolutional neural networks for mobile vision applications. arXiv preprint arXiv:1704.04861 (2017)
12. Jiang, H., Learned-Miller, E.: Face detection with the faster R-CNN. In: 2017 12th IEEE International Conference on Automatic Face & Gesture Recognition (FG 2017), pp. 650–657. IEEE, Washington, DC (2017)
13. Krizhevsky, A., Sutskever, I., Hinton, G.E.: Imagenet classification with deep convolutional neural networks. In: Advances in Neural Information Processing Systems, pp. 1097–1105 (2012)
14. LeCun, Y., Boser, B., Denker, J.S., Henderson, D., Howard, R.E., Hubbard, W., Jackel, L.D.: Backpropagation applied to handwritten zip code recognition. Neural Comput. **1**(4), 541–551 (1989)
15. Lin, T.Y., Dollár, P., Girshick, R., He, K., Hariharan, B., Belongie, S.: Feature pyramid networks for object detection. In: Proceedings of the IEEE Conference on Computer Vision and Pattern Recognition, pp. 2117–2125. IEEE, Honolulu (2017)
16. Liu, W., et al.: SSD: single shot multibox detector. In: Leibe, B., Matas, J., Sebe, N., Welling, M. (eds.) ECCV 2016. LNCS, vol. 9905, pp. 21–37. Springer, Cham (2016). https://doi.org/10.1007/978-3-319-46448-0_2
17. Pan, L., Qin, J., Chen, H., Xiang, X., Li, C., Chen, R.: Image augmentation-based food recognition with convolutional neural networks. CMC Comput. Mater. Continua **59**(1), 297–313 (2019)
18. Redmon, J., Farhadi, A.: Yolov3: an incremental improvement. arXiv preprint arXiv:1804.02767 (2018)
19. Ren, S., He, K., Girshick, R., Sun, J.: Faster R-CNN: towards real-time object detection with region proposal networks. In: Advances in Neural Information Processing Systems, pp. 91–99 (2015)
20. Simonyan, K., Zisserman, A.: Very deep convolutional networks for large-scale image recognition. arXiv preprint arXiv:1409.1556 (2014)
21. Sun, X., Wu, P., Hoi, S.C.: Face detection using deep learning: an improved faster rcnn approach. Neurocomputing **299**, 42–50 (2018)

22. Szentandrási, I., Herout, A., Dubská, M.: Fast detection and recognition of QR codes in high-resolution images. In: Proceedings of the 28th Spring Conference on Computer Graphics, pp. 129–136. ACM, Budmerice (2013)
23. Uijlings, J.R., Van De Sande, K.E., Gevers, T., Smeulders, A.W.: Selective search for object recognition. Int. J. Comput. Vis. **104**(2), 154–171 (2013)
24. Viola, P., Jones, M.: Rapid object detection using a boosted cascade of simple features. In: Proceedings of the 2001 IEEE Computer Society Conference on Computer Vision and Pattern Recognition, p. I. IEEE, Kauai, HI (2001)
25. Wu, X., Luo, C., Zhang, Q., Zhou, J., Yang, H., Li, Y.: Text detection and recognition for natural scene images using deep convolutional neural networks. CMC Comput. Mater. Continua **61**(1), 289–300 (2019)
26. Zhang, L., Lin, L., Liang, X., He, K.: Is faster R-CNN doing well for pedestrian detection? In: Leibe, B., Matas, J., Sebe, N., Welling, M. (eds.) ECCV 2016. LNCS, vol. 9906, pp. 443–457. Springer, Cham (2016). https://doi.org/10.1007/978-3-319-46475-6_28
27. Zhou, L., Wang, L., Chen, Y., Tang, Y.: Binaural sound source localization based on convolutional neural network. CMC Comput. Mater. Continua **60**(2), 545–557 (2019)

General Virtual Images Construction Using Pixel Scrambling for Face Recognition

Yingnan Zhao[(✉)] and Jie Wu

School of Computer and Software, Nanjing University of Information Science
and Technology, Nanjing 210044, China
zh_yingnan@126.com, 919930583@qq.com

Abstract. Face recognition encounters the problem that multiple samples of the same object may be very different owing to the deformation of appearances. To synthesize reasonable virtual images of objects is a good way to solve it, which has been successfully used in face recognition. However, the schemes proposed are usually not suitable for other applications. In this paper, we propose a novel and general scheme to generate virtual images for deformable faces. We introduce the idea of pixel scrambling to our scheme and allow the pixels within an image block to be randomly rearranged to reflect possible variations of the appearance of objects. We demonstrate that virtual images obtained using pixel scrambling and original images are complementary in representing the appearance of objects. Extensive classification experiments on face databases show that the proposed virtual image scheme is very competent and can be combined with a number of classifiers to achieve surprising accuracy improvement.

Keywords: Face recognition · Virtual image · Sparse representation

1 Introduction

Many latest technologies have recently emerged in the field of face recognition [1,9], including the well-known deep learning and its variations [3]. However, most of them need plenty of samples to guarantee the final performance. Virtual sample construction becomes a complementary strategy. Previous schemes of generating virtual images have some drawbacks.

First, these schemes have no wide applications. Image classification on deformable objects encounters the following difficulty. Due to the appearance deformation, the same position of different images of the same object may have varying pixel intensities. This thereby increases the difficulty to accurately recognize the object [2]. One typical example of the deformable object is the face, whose images (i.e. face images) always change with the variations of illuminations, facial expressions and poses [8]. As a consequence, the data uncertainty of

© Springer Nature Switzerland AG 2020
X. Sun et al. (Eds.): ICAIS 2020, LNCS 12239, pp. 444–454, 2020.
https://doi.org/10.1007/978-3-030-57884-8_39

limited face images seem to be high in representing the faces, which is disadvantageous for correct recognition of faces [12]. In particular, for a face recognition problem, when a test sample is compared with the training samples of the same subject for the classification purpose, the test sample may not be correctly recognized owing to the possible great deviation between the test sample and training samples. One feasible way to decrease the data uncertainty of face images is to obtain virtual face images that convey possible appearances of the faces. In other words, because the original and virtual face images simultaneously provide proper appearances of the faces, the use of them is able to reduce the uncertainty of the representation of face images. This implies that the faces can be more easily recognized. A few schemes of generating virtual face images have achieved promising accuracy improvement of face recognition. For example, reference [14] obtained surprising improvement in face recognition by properly using virtual face images. Face synthesis images such as face sketch images can also be applied to law enforcement and digital entertainment [6]. These studies offer good solutions to produce more available samples by exploiting the characteristics of face images. The use of face synthesis images can be very beneficial to pose-robust face recognition [5]. The face synthesis algorithm proposed in [17] can remove the illumination effects of face images and lead to better face recognition performance. 3D face synthesis also plays an important role in computer vision and virtual reality [4].

Second, the shortcomings of previous schemes decrease their availability. For example, the symmetrical face scheme proposed in [16] may produce unnatural face images, though it can obtain attractive accuracy improvement in face recognition. The mirror face image used in [13] is very easy to obtain, but it is mainly suitable for the face, a kind of special objects with the geometrical symmetry. In other words, the mirror image of a general original image is usually not a proper representation of this image. The face synthesis algorithm proposed in [11] is able to synthesize face images from sparse samples, but the important procedure to determine suitable transformations for given samples is still empirical and almost impracticable for real-world applications.

It seems that it is crucial to design a widely applicable virtual image scheme. It had better be mathematically tractable and its rationale should be easy to interpret. If the scheme is not dependent on the special nature of images, then it is suitable for general images. We aim at generating competent virtual images of objects and achieving a high classification accuracy. With this paper, we propose a novel scheme to generate virtual images for deformable objects. Our scheme first divides an original image into non-overlapping blocks. Then for each block, we randomly exchange intensities of all pixels. After all blocks are dealt with, we treat the new image as a virtual image. The rationale of pixel scrambling in our scheme is based on the following fact. The deformation of the appearance of objects makes pixel intensities within the same block changeable. For example, different facial expressions of a person will lead to varying image block sat the same position of different face images. For example, a slight smile will cause the pixel shift of some blocks in comparison with the original image.

The experiments show that the proposed virtual images provide important information of deformable objects and can win very satisfactory accuracy for image classification and face recognition. This paper has the following contributes. 1) It proposes a kind of competent virtual images for deformable objects, which is very beneficial to reduce the data uncertainty.2) The proposed scheme is simple, easy to understand and implement. 3) The scheme has wide applications and is suitable for general deformable objects and even other objects whose images are affected by illuminations.

The other parts of this paper are organized as follows. Section 2 presents the proposed novel scheme to generate virtual images from the original images and interprets the reasonability of the proposed virtual images. Section 3 shows the experimental results. Section 4 provides the conclusions.

2   The Proposed Scheme and Algorithm

2.1   Scheme to Produce Virtual Images Using Pixel Scrambling

The proposed scheme includes the following steps.

1) Each original training sample is divided into non-overlapping image blocks each having the size of a by b pixels. For all pixels within each block, they are respectively numbered by $1,2,3,\ldots k(k = a * b)$ in the order of from the left to right and from the top to bottom.
2) For each block of an original image, an-dimensional array whose entries are randomly indexed by 1 to k is generated. We denote this array by $s[i]$, $i = 1,2,\ldots,k$.
3) For each block of the virtual image, the pixels within it are identically numbered as those of the corresponding block of the original image. The values of the first to k-th pixels of a block of the virtual image are respectively set to the values of the $s[i]$-th to $s[k]$-th pixels of the same block of the original image.
4) Every virtual image and original image is converted into vectors.

Figure 1 shows illustrations of an original image and our random pixel scrambling scheme. In order to visually present the difference of the way to convert the original image into a column vector and the way to convert the virtual image into a column vector, we also provide Fig. 2. Figure 2 (a) shows how the original image is converted into a column vector by concatenating the rows in sequence. Figure 2 (b) shows how the possible virtual image is converted into a column vector. The blue parts depict the conversion for the first row in Fig. 2 (a) and the first block in Fig. 2 (b). We use the left matrix in the form of table in Fig. 2 (b) to denote a virtual image. For simplicity of presentation, the difference between the original and virtual images is just the red image block (only pixels of this image block is scrambled), but in real algorithm the pixels of all image blocks of an original image will be scrambled to produce a virtual image.

40	35	50	70	80	60
36	48	60	75	86	78
65	72	80	88	111	100
70	80	83	110	125	99
83	105	89	140	136	87
77	99	102	130	128	103
90	116	125	166	142	105
73	106	120	170	155	92

(a)

40	35	50	70	80	60
36	48	60	75	86	78
65	72	110	80	111	100
70	80	88	83	125	99
83	105	89	140	136	87
77	99	102	130	128	103
90	116	125	166	142	105
73	106	120	170	155	92

(b)

40	35	50	70	80	60
36	48	60	75	86	78
65	72	110	80	111	100
70	80	83	88	125	99
83	105	89	140	136	87
77	99	102	130	128	103
90	116	125	166	142	105
73	106	120	170	155	92

(c)

40	35	50	70	80	60
36	48	60	75	86	78
65	72	88	83	111	100
70	80	110	80	125	99
83	105	89	140	136	87
77	99	102	130	128	103
90	116	125	166	142	105
73	106	120	170	155	92

(d)

Fig. 1. Illustrations of an original image and the proposed random pixel scrambling scheme. (a) represents an original image. If this image is divided into blocks each having the size of 2 by 2 pixels, there will be 12 $(4 \times 3 = 12)$ blocks. (b), (c) and (d) show three possible variations (other possible variations are not shown) of the red block in (a), obtained using our random pixel scrambling scheme. After each block is processed by our random pixel scrambling scheme, a virtual image is obtained.

Figure 3 shows some examples of original images and the corresponding virtual images. We see that the virtual image is still a natural image. Because it is a modification of the corresponding original image, it has obvious deviation from the original image and represents some possible variations of the object. As a result, the simultaneous use of the original images and virtual images allows more information of the object to be provided, and better recognition accuracy can be achieved. An intuitive explanation of the benefit of more available training samples is as follows. Because the object is deformable and the number of samples in the form of images is always smaller than the dimension of samples, the difference between images of the same object is usually great [34]. Consequently, the test sample may be very different from the limited training samples of the same object. However, when virtual images are also used as training samples, the possibility that the training sample with the minimum distance to the test sample is from the same object as the test sample will be increased. As we know, if the training sample with the minimum distance to the test sample is from the same object as the test sample, we can obtain the correct recognition result of the test sample via the nearest neighbor classifier. Thus, more accurate recognition of objects can be obtained under the condition that the original

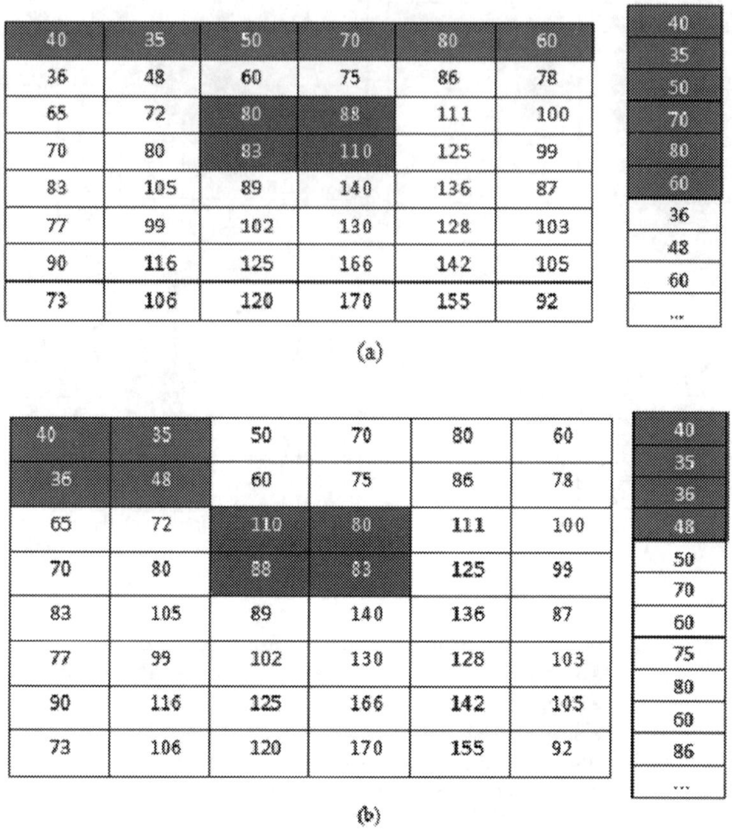

40	35	50	70	80	60
36	48	60	75	86	78
65	72	80	88	111	100
70	80	83	110	125	99
83	105	89	140	136	87
77	99	102	130	128	103
90	116	125	166	142	105
73	106	120	170	155	92

Column vector (a): 40, 35, 50, 70, 80, 60, 36, 48, 60, ...

(a)

40	35	50	70	80	60
36	48	60	75	86	78
65	72	110	80	111	100
70	80	88	83	125	99
83	105	89	140	136	87
77	99	102	130	128	103
90	116	125	166	142	105
73	106	120	170	155	92

Column vector (b): 40, 35, 36, 48, 50, 70, 60, 75, 80, 60, 86, ...

(b)

Fig. 2. The way to convert the original image into a column vector and the way to convert the virtual image into a column vector.

images and virtual images are simultaneously used as training samples and the nearest neighbor classifier is exploited. We also say that the simultaneous use of the original images and virtual images enables the data uncertainty to be reduced.

Figure 4 shows correlation coefficients of the class-specific scores, derived from the original face images and virtual images, of the first to 500-th test samples in the GT face database using collaborative representation [25]. The concrete procedure can be found in [11]. We see that the correlation coefficients corresponding to different test samples are all not very high. This partially means that the original face images and virtual images of the same subject are somewhat complementary, so the combination of them is able to obtain better recognition result.

It is interesting that pixel scrambling can be integrated with dedicated algorithms to achieve the image encryption [35]. We demonstrate that slight pixel

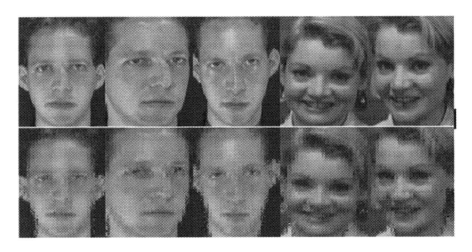

Fig. 3. Examples of original images (the top row) and the corresponding virtual images (the bottom row).

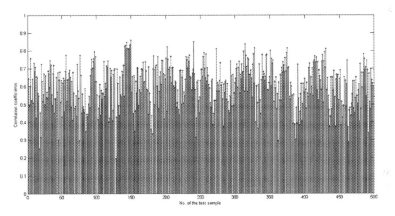

Fig. 4. Correlation coefficients of the class-specific scores, derived from the original face images and virtual images, of the first to 500-th test samples obtained using collaborative representation. The Georgia Tech (GT) face database is used and the first five face images per subject are used as training samples and the other ten images are used as test samples.

scrambling can obtain reasonable additional representation of deformable objects which is useful for improving the accuracy of object classification.

2.2 Algorithm to Integrate Original Images and Virtual Images

First of all, it should be pointed out that our algorithm exploits three kinds of virtual training samples. The first kind of virtual training samples is the mirror face image presented in [14]. The second kind of virtual training samples is obtained by applying our pixel scrambling scheme to the original face image. The third kind of virtual training samples is obtained by applying our pixel scrambling scheme to the mirror face image.

For a test sample, an arbitrary representation based classification algorithm is respectively applied to the original training samples and virtual training samples. Suppose that there are C classes. Let $d_1, ..., d_C$ denote the class residuals of the test sample with respect to the original training samples of the first to the C-th classes. Let $e_1, ..., e_C$ denote the class residuals of the test sample with respect to the first kind of virtual training samples of the first to the C-th classes. Let $f_1, ..., f_C$ and $g_1, ..., g_C$ denote the class residuals of the test sample with respect to the second and third kinds of virtual training samples.

The class residuals of the test sample are integrated using

$$q_j = t_1 d_j + t_2 e_j + t_3 f_j + t_4 g_j, j = 1, ..., C \qquad (1)$$

Let $r = argmin_j q_j$. We assign the test sample to the r-th class.

As for the weight, we use an automatic procedure to determine it. The automatic procedure is as follows. Let $d_1, ..., d_C$ stand for the sorted results of $d_1, ..., d_C$ and suppose that $d_1 \leq ... \leq d_C$. Accordingly, $e_1, ..., e_C$, $f_1, ..., f_C$, and $g_1, ..., g_C$ have the similar meanings. We define $t_{10} = d_2 - d1$ and $t_{20} = e_2 - e1$. t_{30} and t_{40} are defined as $t_{10} = f_2 - f1$ and $t_{10} = g_2 - g1$. We respectively define

$$t_1 = \frac{t_{10}}{t_{10} + t_{20} + t_{30} + t_{40}}, \qquad (2)$$

$$t_2 = \frac{t_{20}}{t_{10} + t_{20} + t_{30} + t_{40}}, \qquad (3)$$

$$t_3 = \frac{t_{30}}{t_{10} + t_{20} + t_{30} + t_{40}}, \qquad (4)$$

$$t_4 = \frac{t_{40}}{t_{10} + t_{20} + t_{30} + t_{40}} \qquad (5)$$

We would like to point out that a similar weight setting algorithm was proposed in [7], which is the first completely adaptive weighted fusion algorithm and obtained an almost perfect fusion result. However, the algorithm in [15] can fuse only two kinds of scores.

3   Experimental Results

In experiments, SRC [10] and CR [18] are respectively used as the classifier. Because our scheme randomly scrambles the pixels of an original image, so we run it five times and take the average of the rates of classification errors as the final result. In the context "Our method integrated with CR" is used to present the result of applying our method to improve CR. "Our method integrated with SRC" is used to present the result of applying our method to improve SRC.

3.1 Experiment on the FERET Database

We first use a subset of the FERET face dataset to perform an experiment. The subset contains 1400 face images from 200 subjects and each subject has seven different face images. This subset was composed of images in the original FERET face dataset whose names are marked with two-character strings: 'ba', 'bj', 'bk', 'be', 'bf', 'bd', and 'bg'. Every face image was resized to a 40 by 40 image. We respectively select the first 2, 3 and 4 face images of each subject as original training samples and use the remaining face images as testing samples. Figure 5 presents some cropped face images from the FERET face database. The size of the image block in our method is set to $a = 2$ and $b = 2$. Table 1 shows the rates of classification errors (%) of different methods on the FERET database. We see that our method outperforms naïve CR and SRC.

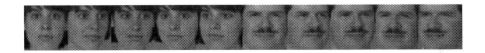

Fig. 5. Some face images from the FERET face database.

Table 1. Experimental results on the FERET database.

Number of training samples per person	2	3	4
CR	41.60	55.63	44.67
Our method +CR	38.84	46.43	42.53
SRC	41.90	49.25	36.67
Our method +SRC	35.88	37.72	32.00

3.2 Experiment on the GT Database

In the GT face database there are 750 face images from 50 persons and every person provided 15 face images. The original images are color images with clutter background taken at the resolution of 640480 pixels and show frontal and tilted faces with different facial expressions, lighting conditions and scales. We used the manually cropped and labeled face images to conduct the experiment. Figure 6 shows some of these images. We further resized them to obtain images with the resolution of 4030 pixels. They were then converted into gray images. In our experiment the first 3–10 face images of every person are used as training samples and the other images are employed as test samples. The size of the image block in our method is set to $a = 2$ and $b = 4$. Table 2 shows the rates of classification errors (%) of different methods on the GT database. We see that our method also performs better than original CR and SRC.

Fig. 6. Some face images from the FERET face database.

Table 2. Experimental results on the GT database.

Number of training samples per person	3	4	5	6	7	8	9	10
CR	54.67	52.91	51.20	44.44	41.50	40.57	39.00	36.00
Our method +CR	51.77	48.76	46.52	41.91	36.55	34.40	32.40	29.36
SRC	55.33	53.82	51.20	44.44	41.75	41.43	37.33	4.40
Our method +SRC	53.20	52.29	49.52	42.09	38.95	37.54	34.80	32.72

3.3   Experiment on the GT Database

The ORL face database includes 400 face images taken from 40 subjects each having 10 face images. In this database some images were taken at different times and with various lightings, facial expressions (open/closed eyes, smiling/not smiling), and facial details (glasses/no glasses). All images were taken against a dark homogeneous background with the subjects in an upright, frontal position (with tolerance for some side movement). We resized every image to form a 56 by 46 image matrix. Figure 7 shows some images from this ORL dataset. The first 1, 2, 3, 4, 5 and 6 face images of each subject were used as training samples and the others were exploited as test samples, respectively. The size of the image block in our method is set to $a = 4$ and $= 8$. Table 3 shows rates of classification errors (%). It again tells us that our method is better than naïve CR and SRC.

Fig. 7. Some images from this ORL dataset.

Table 3. Experimental results on the ORL database.

Number of training samples per person	1	2	3	4	5	6
CR	31.94	16.56	13.93	10.83	11.50	8.13
Our method +CR	30.78	16.38	13.71	11.33	11.40	9.50
SRC	32.78	17.81	16.07	12.50	12.00	10.00
Our method +SRC	30.83	16.69	14.50	9.42	10.30	7.75

4   Conclusions

In this paper, the proposed novel scheme not only can efficiently generate natural virtual images for deformable objects but also can lead to notable accuracy improvement. Moreover, the rationale of the proposed scheme is easy to understand. As shown earlier, pixel scrambling of an original image can reflect possible change of the appearance of a deformable object, so the obtained virtual image is a proper representation of the object. Besides the proposed scheme is applicable for deformable objects, it can also provide useful representation for a general object because of the following factors. Within a small enough region of a general image, in most cases the difference of the values of pixels is usually little. As a result, pixel scrambling of a small region of an image will obtain reasonable and new representation of this region. It should be pointed out that a non-deformable object also usually has varying images owing to changeable illuminations, imaging distances and views. Thus, after we use the proposed scheme to produce virtual images for general objects, we can also integrate the virtual images and original images to obtain better classification performance.

Acknowledgments. This work is supported in part by the Priority Academic Program Development of Jiangsu Higher Education Institutions, Natural Science Foundation of China (No. 61103141, No. 61105007 and No. 51405241).

References

1. Li, S., Liu, F., Liang, J., Cai, Z., Liang, Z.: Optimization of face recognition system based on Azure IoT edge. CMC Comput. Mater. Continua **109**(3), 1377–1389 (2019)
2. Li, Y., Chen, C., Allen, P.K.: Recognition of deformable object category and pose, pp. 5558–5564 (2014)
3. Ma, B., Dui, G., Yang, S., Xin, L.: A method for improving CNN-based image recognition using DCGAN. CMC Comput. Mater. Continua **109**(6), 537–554 (2018)
4. Nguyen, H.T., Ong, E.P., Niswar, A., Huang, Z., Rahardja, S.: Automatic and real-time 3D face synthesis, pp. 103–106 (2009)
5. Sharma, A., Jacobs, D.W.: Bypassing synthesis: PLS for face recognition with pose, low-resolution and sketch, pp. 593–600 (2011)
6. Song, Y., Bao, L., Yang, Q., Yang, M.-H.: Real-time exemplar-based face sketch synthesis. In: Fleet, D., Pajdla, T., Schiele, B., Tuytelaars, T. (eds.) ECCV 2014. LNCS, vol. 8694, pp. 800–813. Springer, Cham (2014). https://doi.org/10.1007/978-3-319-10599-4_51
7. Wang, S., Chen, H., Peng, X., Zhou, C.: Exponential locality preserving projections for small sample size problem. Neurocomputing **74**(17), 3654–3662 (2011)
8. Wang, S., Yang, J., Zhang, N., Zhou, C.: Tensor discriminant color space for face recognition. IEEE Trans. Image Process. **20**(9), 2490–2501 (2011)
9. Wang, X., Xiong, C., Pei, Q., Qu, Y.: Expression preserved face privacy protection based on multi-mode discriminant analysis. CMC Comput. Mater. Continua **57**(1), 107–121 (2018)

10. Wright, J., Yang, A.Y., Ganesh, A., Sastry, S., Ma, Y.: Robust face recognition via sparse representation. IEEE Trans. Pattern Anal. Mach. Intell. **31**(2), 210–227 (2009)
11. Xu, H., Zha, H.: Manifold based face synthesis from sparse samples, pp. 2208–2215 (2013)
12. Xu, Y., et al.: Data uncertainty in face recognition. IEEE Trans. Syst. Man Cybern. **44**(10), 1950–1961 (2014)
13. Xu, Y., Li, X., Yang, J., Lai, Z., Zhang, D.: Integrating conventional and inverse representation for face recognition. IEEE Trans. Syst. Man Cybern. **44**(10), 1738–1746 (2014)
14. Xu, Y., Li, X., Yang, J., Zhang, D.: Integrate the original face image and its mirror image for face recognition. Neurocomputing **131**, 191–199 (2014)
15. Xu, Y., Lu, Y.: Adaptive weighted fusion: a novel fusion approach for image classification. Neurocomputing **168**, 566–574 (2015)
16. Xu, Y., Zhu, X., Li, Z., Liu, G., Lu, Y., Liu, H.: Using the original and 'symmetrical face' training samples to perform representation based two-step face recognition. Pattern Recogn. **46**(4), 1151–1158 (2013)
17. Zhang, L., Wang, S., Samaras, D.: Face synthesis and recognition from a single image under arbitrary unknown lighting using a spherical harmonic basis morphable model. In: CVPR 2005, vol. 2, pp. 209–216 (2005)
18. Zhang, L., Yang, M., Feng, X.: Sparse representation or collaborative representation: which helps face recognition? pp. 471–478 (2011)

Knowledge Graph Construction
of Personal Relationships

Yong Jin[✉], Qiao Jin, and Xusheng Yang

Wuhan Fiberhome Putian Information and Technology Ltd., Wuhan 430074, China
{yongjin,qjin,xsyang}@fiberhome.com

Abstract. Knowledge graph has attracted much attention in recent years. It is a high-level natural language processing (NLP) problem, which includes many NLP tasks such as named entity recognition, relation extraction, entity alignment, etc. In this paper, we focus on the entity of persons in the large amount of text data, and then construct the graph of personal relationships. Firstly we investigate how to recognize person names from Chinese text. Secondly, we propose a comprehensive approach including Improved BiGated Recurrent Unit and syntactic analysis to extract the relations between different persons. Then, we align the person entities through entity alignment techniques and some manual proofreading work. Finally, we apply this graph construction process in text records for experimentation. This process performs effectively and efficiently to construct the graph of personal relationships from unstructured Chinese text, and this graph can provide significant relationship insights in texts.

Keywords: Knowledge graph · Personal relationships · Entity recognition · Relation extraction · Entity alignment.

1 Introduction

1.1 Background

Knowledge graph, is to represent entities and their relationships, which are extracted from large amount of unstructured texts. It is originally introduced by Google to enhance the power of search engine [1]. Until the end of 2017, Google knowledge graph has tripled its size up to 570 million entities and 18 billion facts [2], which has greatly improved the efficiency and accuracy of Google searches. In recent years, knowledge graph technique has been paid much attention in both academic and engineering domains.

Knowledge graph is a high-level NLP problem, which includes a series of fundamental NLP tasks such as named entity recognition(NER)–extract entities from text [3,4], relation extraction–extract relationships from text [12–14], entity alignment–match entities [6], data fusion–integrate all data together into graph database [7]. However, Chinese knowledge graph especially includes a

© Springer Nature Switzerland AG 2020
X. Sun et al. (Eds.): ICAIS 2020, LNCS 12239, pp. 455–466, 2020.
https://doi.org/10.1007/978-3-030-57884-8_40

step called word segmentation, different from English. For English text, it can be automatically segmented by space because each word normally indicates individual meaning; while for Chinese text, a single Chinese character usually can not represent full meaning, e.g. "们" goes widely with "我", "你" or "他" to form full words "我们", "你们" or "他们". Chinese word segmentation is an initial step for other Chinese NLP tasks, hence the other Chinese NLP models need special processing as well.

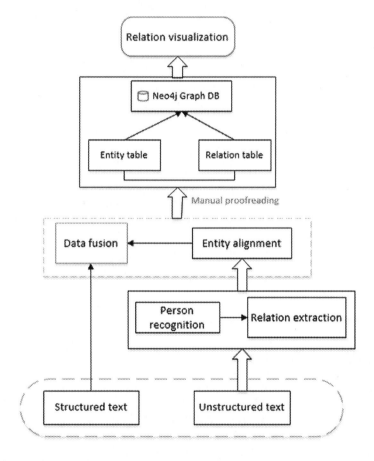

Fig. 1. The united process to construct the graph of personal relationships

1.2   United Process for Graph Construction

In this paper, we study the graph construction of personal relationships based on Chinese text in public security domain because personal relations play an important role in case investigation.

Hence, we propose a united process to construct a graph of personal relationships w.r.t Chinese text according to the following process in Fig. 1. This figure illustrates that Chinese personal relationships begins from word segmentation, followed by person entity recognition, relation extraction, person entity alignment, and data fusion with manual proofreading. Lastly, all data are integrated into graph database that is used for further application such as graph visualization, search engine, or other relation analysis.

In this paper, we firstly introduce some background and propose a united construction process for personal relationships. Then we come up with specific ideas in detail for extractions of personal relationships. After that we demonstrate this united process with case study in crime records text. Finally, it is followed by conclusion and some future work.

2   Extraction of Personal Relationships

2.1   Person Recognition

The first step to construct the graph of personal relationships is to extract person entities from text. Person recognition is a subtype of named entity recognition, which is also fundamental task to construct the graph of personal relationships. Especially in Chinese, it is difficult to recognize person names because (1) Chinese names can be composed of almost any single word, e.g. "力齐", "姬从良", which leads to a fact that the name is very hard to be indexed into Chinese vocabulary; (2) Chinese family names can be one single character such as "张", "李", 王" or double character such as "司马", "欧阳", 上官"; and meanwhile, given names are usually comprised of one or two Chinese characters. Hence, we propose a method to recognize Chinese names based on open-source projects jieba [8] and pangu [9]. This method not only incorporates the segmentation results from jieba, but also uses the dictionaries such as single-character family names, double-character family names, and also finally use Floyd algorithm to semantically disambiguate Chinese names.

In detail, Chinese person recognition process is depicted in Fig. 2, which mainly contains the following steps:

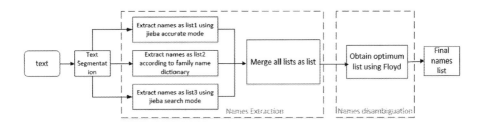

Fig. 2. The process for Chinese names recognition

(1) Split text using jieba segmentation with default mode (exact segmentation mode), and then extract the words with POS tag "nr" (person name) as *list1*;

(2) Candidate names generation, generating all candidate Chinese names as *list2* according to the family name and given name dictionaries;

(3) Segment the text using jieba with the search mode, and obtain all the segmentation words as *list3*;

(4) Combine the above three lists together, called *list*, and rank this list with the position of each word in ascending order and the length of each word in descending order;

(5) Representation of semantic matrix, building an adjacent matrix M based on each *list* from (4), the element value of each cell in this matrix equals the minimum gap from the starting position to current position. Specifically, the length of each row and column equals to the number of words for each *list*. Denote the ith word occurs at the starting character position of $p(i)$, with length $l(i)$, and the jth word occurs at the starting character position of $p(j)$. Then the adjacent matrix is defined as follows:

$$|M(i,j)| = \begin{cases} p(j) - p(i) - l(i), & \text{the } i^{th} \text{ word occurs before the } j^{th} \text{ word; (1)} \\ 999, & \text{the } i^{th} \text{ word occurs after the } j^{th} \text{ word. (2)} \end{cases}$$

where the value 999 indicates the ith word is too far away from the jth word.

(6) Disambiguation of Chinese names (keep the most probable names and remove the least ones), using Floyd algorithm (minimum distance between two points) [15] based on adjacent matrix M, and keep the best path with minimum distance of 0.

Here are some evaluation measures in Table 1 for the method of Chinese names recognition based on the news data of 2014 People's Daily. It contains 329794 names in total. It can be drawn from the table that our method performs better than jieba and pangu because it has the highest F1 score of 0.905.

Table 1. Measures for Chinese names recognition.

Measures	jieba	pangu	This method
Precision	70.73%	70.22%	86.62%
Recall	97.09%	94.15%	94.65%
F1 score	0.818	0.848	0.905

Meanwhile, in order to further distinguish these persons from each other, four fixed attributes (gender, height, birth place, and birth date) are also extracted using syntactic analysis with related keywords.

2.2 Relation Extraction

Relation extraction plays a key role in the construction of knowledge graph. After the person entities are extracted, they are isolated but not linked together, so we need to extract the relations among these person entities. Due to the various types of relations among people in the large amount of data, relation extraction can be considered as supervised or unsupervised NLP task, or even both. In this paper, we combine supervised relation classification model and unsupervised extraction together. In general, for the eleven relation types including "父母", "夫妻", "祖孙", "师生", "兄弟姐妹", "合作", "情侣", "好友", "亲戚", "同门" and "上下级", except "unknown", we employ "Chinese Relation Extraction by BiGRU with Character and Sentence Attentions" [10], which is especially modified for Chinese based on the Chinese character vectors from the ideas in [13] and [14]. For other relations under the type of "unknown", such as "敲诈", "绯闻女友", we mainly accommodate syntactic analysis to extract personal relations.

For the task of relation extraction using Bi-directional Gated Recurrent Units and dual attention model (BiGRU+2ATT), its experimental accuracy is around 70% [10], it has been exposed to three points of challenges as below:

(1) Challenge of the length of sentences. It means, if the length of text (referring to two person entities) increases to a certain extent, the relation may change. E.g. the relation of "赵兰坤" and "连震东" in one text "赵兰坤在西安 生下连震东" and another text "中华人民共和国张遭遇最危难时刻,赵兰坤在西 安生下连震东" is not consistent. But it should be the same actually.

(2) Challenge of the change of person entities given the same surrounding context. It indicates that if two person entities has changed in the same context, their relations may change, which is not reasonable. E.g. the relation of "赵幂" and "陈小威" in this sentence "赵幂与陈小威的细节显示两人已结婚" should be consistent with the relation of "刘花" and "李晓阳" in another sentence "刘花与李晓阳的细节显示两人已结婚", while the original method produces different relations.

(3) Challenge of position turnover of two person entities. It means if two person entities in one sentence are swapped with each other, the relation may change. E.g. the relation of "赵兰坤" and "连震东" in one sentence "赵兰坤在西安生下连震东" (relation of 父母) is not similar as that in "连震东在西安生下 赵兰坤" (relation of "夫妻"). But actually they should the same relation type (parents 父母) only with the relation's direction swapped.

In this paper, we deeply investigate these three challenges w.r.t supervised relation extraction using Improved BiGRU attention model. On one hand, long text should be split and the text including person entities should be paid more attention. On the other hand, the word vectors of Chinese characters do not make any sense if these characters occur in a Chinese name, in other words, the vector of a name "赵兰坤" is not simply a combination of three character vectors of "赵", "兰", and "坤". Hence, the model can be improved correspondingly through more rational text representations.

We deal with these challenges as follows:

(i) Split text based on punctuation marks such as , . ; ! ?, e.g., one text is split to $\{S1, S2\}$ and another text is split to to $\{S1', S2', S3'\}$;

(ii) Referring to those parts including specific person entities, e.g., two person entities exist in $\{S1, S2\}$ and $\{S1', S3'\}$ respectively, and only these parts are used for relation extraction. –w.r.t the first challenge, the relevant text is selected while the irrelevant text is ignored.

(iii) Perform person recognition in those relevant text, and then for those text including two persons or more, each character of Chinese names is replaced by single mark '#'. For example, "赵幂与陈小威的细节显示两人已结婚" is replaced by "##与###的细节显示两人已结婚". –w.r.t the second and third challenges, the characters in Chinese names do not make sense, so they are handled as the same character.

Through these specific improvements regarding the three challenges of Chinese BiGRU+2ATT model, based on the same data set with totally around 80000 examples, we split 80% as training set and the reamining 20% as test set, we perform the Improved BiGRU model and the experimental has increased to about 86%, which is about 15% higher than previous baseline.

For unsupervised relation extraction task under the type of "unknown", we mainly reply on syntactic analysis method, like the regular expression of "name–SBV–word–VOB/POB–name" followed by syntactic parsing of each text. Through this method, we manually choose some keywords related with relations such as "打伤", "故意伤害", "放风", "通风报信", "诈骗", "骗钱" etc. Then we summarize some typical sentence syntactic structures, e.g. "personA—打伤 —personB" (- represents other words). Finally the syntactic analysis is represented using keywords and expression of sentence parsing structures.

The relation extraction process in this paper is illustrated in Fig. 3. It incorporates supervised (the eleven relations labeled in BiGRU+2ATT model) and unsupervised approaches to extract personal relations as many as possible.

2.3   Person Alignment

Even if the person entities and their relations are extracted, in order to produce the positive effect as much as possible, we need to not only recall all the entities and their relations from text, but also make the entities in the graph as globally unique. One person may have several different names/nicknames, or different names probably refer to the same person. For example, "张三" and "小张" probably means the same one; "李四" extracted from this article may be the same as "李四" extracted from another article, or may not. Hence, person alignment (also called entity matching) is necessary to construct the graph, otherwise, the knowledge in this graph will provide confused information for users.

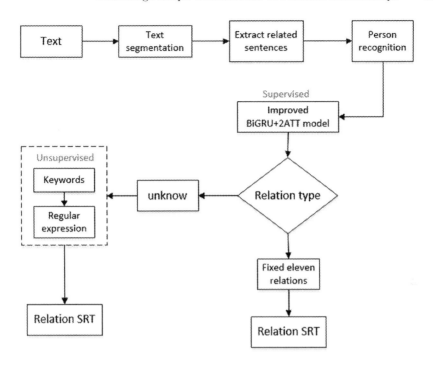

Fig. 3. The process for personal relation extraction

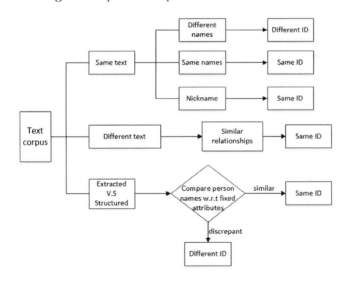

Fig. 4. The process for person alignment

Graphically, the task process of person alignment is depicted in Fig. 4. Specifically, the task in this paper conforms to the following principles:

(1) Different person names in one text are assigned to different person entities (different ID).

(2) The same person names in one text are assigned to the same person entity (same ID).

(3) Compare the nicknames and similar full names using word window (close in several words) and related words such as "外号", "小名", "也叫", e.g. for this text "张三, 外号大锤,他在北京电子厂上班...大锤也参与到这次诈骗中来了", "张三" and "大锤" refer to the same person.

(4) Compare extracted person names with those in personal table, the same person names will be assigned to the same person entity if the same names correspond to similar gender, birth date, birth place or height.

(5) The same person names in different text who have similar relationships are assigned to the same person entity (same ID). E.g. in one text "张三" has relationships with {"张小三", "王小花", "王五"}; while in another text "张三" has relationships with {"王小花", "赵六"}, then "张三" in these two text refers to the same person as well as "王小花".

2.4 Data Fusion

Even NLP techniques are employed in person recognition, relation extraction, and person alignment, they can not make one hundred percent correct. So some manual Proofreading work is inevitable here to make sure the knowledge extracted from text is true. This proofreading includes but not limited to the following contents:

(1) Check whether the attributes of person is reasonable, e.g., DOB of one person is 190206, this maybe unreasonable because the birth year 1902 is too old;

(2) Check whether nicknames are all extracted and whether they can match any full name, because nicknames sometimes are difficult to extracted such as "大锤", "老黑".

(3) Check whether the relation is correct or omitted, especially the relations extracted with syntactic analysis, e.g., person A is a friend of person B, but also a robber of B;

(4) No duplicate IDs for the person entity to make sure ID is globally unique.

After processing unstructured text and entity matching, we need perform data fusion work into a graph database [11], to display personal relationships directly in a graph and to prepare for the semantic computation based on knowledge such as knowledge inference, minimum path.

Theoretically, the knowledge graph is stored in graph database [11], and the fundamental data storing format is semantic relation triples (SRT). A SRT is normally denoted as (P, R, Q), here P and Q both represent different entities, R indicates the relationship between P and Q, and this relation can be undirected, unidirectional or bidirectional, which is up to the specific applications. For example, the person entity "张三" and its attribute "DOB" value 198306 is

actually stored as a SRT (张三, DOB, 198306); the relation of personn entities
"张三" and "王小花" is represented as a SRT (张三 夫妻 王小花).

While in practice, due to the easiness of graph database neo4j [11], we only
need to integrate all the data into two primary forms of tables: one is entity
table, and the other one is relation table. The entity table includes a global
ID that distinguishes this entity from all other entities, and other attributes
such as date of birth(DOB), gender, nickname, education, birth place, current
address, nation. The relation table includes person ID, relations and attributes
of relations e.g. relation type, relation source.

The process of data fusion is illustrated in Fig. 5. In this process, we mainly
integrate all entities and attributes, and relations respectively. Then, data normalization is performed to keep all words expressions and formats consistent,
which needs much manual proofreading work.

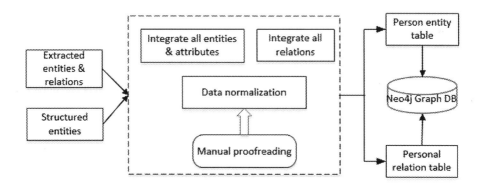

Fig. 5. The process for data fusion and manual proofreading

3 Case Study

3.1 Data Collection

In this paper, we apply these techniques mentioned above into a specific domain
in public security. We choose several desensitized data sets provided by one
Public Security Bureau (PSB) of a prefecture-level city in China. These data sets
consists of structured and unstructured text. Specifically, there are structured
data including case table (with case ID, case time, case address, and other basic
case information) and personal table (with person ID, date of birth, birth place,
current address, and other basic personal information). The other unstructured
data is arrest records table that are normally long text. In these arrest records,
there are much valuable information such as motivation, crime marks. Here we
focus on personal relationships because people investigation is one significant
role of PSB and personal relationships can help crack a criminal case to some
extent.

The record table consists of several fields such as case ID, record time, and one field named record text. The total number of records is about 210000, average number of lines for each record roughly equals 18, average number of each record is approximately 768 Chinese words. The following person entity recognition and relation extraction mainly refer to this dataset.

For the person alignment and data fusion work, both the case table and personal table will be analyzed together with the extracted relationships from record text.

3.2 SRT Construction

The main goal is to construct SRTs for all person entities. For records text, we apply the tasks of person recognition, relation extraction, person alignment, and data fusion.

In the data fusion task, if the person extracted from text can not match any one in the personal table, he/she should be assigned another globally unique ID auto-generated by the system.

Lastly, we integrate all the extracted data and structured data and form two types of data which are listed in Table 2 and Table 3 (in Chinese format). Due to confidentiality reasons, all the data examples in this paper are desensitized from real data. Entity table includes global ID and other attributes such as name, pinyin (Chinese spelling of names), nickname, DOB, gender, education, nation, height(cm), native address, current residential address, career, contact number. Relation table includes the first entity ID and the second entity ID, relation, relation type, and relation source.

Table 2. Examples of person entity table.

globalID	name	pinyin	nickname	DOB	gender	education	nation	height(cm)	...
900101	张三	zhangsan	大锤	198306	男	初中	汉	171	...
900201	王小花	wangxiaohua	豆腐西施	198904	女	高中	汉	160	...
900301	阿克木	akemu	–	197103	男	小学	维吾尔	175	...

Table 3. Examples of relation table.

globalID1	globalID2	relation	relation type	relation source
900101	900201	夫妻	亲属	record
900101	900301	盗窃	社会	record

When the person entity table and relation table are ready, we use the *import* statement of neo4j database [11] to convert the relational data into graph data format. In total, we have extracted around 0.3 million person entities including suspects, reporters and witnesses; and around 30 thousands of SRTs w.r.t personal relations including 102 personal relationships with family relations and social relations.

Then the graph of personal relationships could be generated using the query language *Cypher* of neo4j. For example, a subgraph of personal relationships is depicted in Fig. 6. It is not only intuitive but also reflects some relationships hidden in large unstructured text.

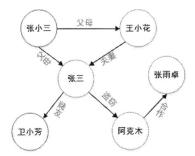

Fig. 6. The process for data fusion and manual proofreading

4 Conclusions

In this paper, we propose a united approach to generate knowledge graph of personal relationships from large amount of Chinese text. In detail, we firstly propose a new method for Chinese name recognition incorporating single-character and double-character family name dictionaries and Floyd semantic disambiguation algorithm. And then we improve the Chinese BiGRU+2ATT model through three challenges, coupled with syntactic analysis to extract the relations among different person entities. Thirdly, person entities are aligned consistently w.r.t structured and extracted entities and relations. Finally, we make some manual proofreading work to check some inevitable errors and integrate all data together into entity table and relation table for graph database.

In practice, we apply this approach in records mining of one PSB in China. It performs effectively and efficiently with people knowledge graph. In this personal relation graph, police officers can easily observe the personal relations given a person or several persons, which is useful to track one person in case analysis.

However, in the task of relation extraction, syntactic analysis is partly used because of some unknown types of relations and insufficient labeled examples. While in the future, more labeled examples with diversified relations could be collected for supervised relation extraction because it is normally performs better and costs less manual work in relation prediction as well. Besides, for person alignment, some considerations can be further taken into account. Take attributes of person entities for example, if the attributes of one entity are extracted correctly as many as possible, it could be matched more accurately.

References

1. Singhal, A.: Introducing the Knowledge Graph: Things, Not Strings. Google Official Blog (2012). Accessed 10 Dec 2017
2. Casey Newton: Google's Knowledge Graph tripled in size in seven months. CNET. CBS Interactive (2012). Accessed 10 Dec 2017
3. Nadeau, D., Sekine, S.: A survey of named entity recognition and classification. Lingvisticae Investigationes (2007)
4. Sang, T.K., Erik, F., De Meulder, F.: Introduction to the CoNLL-2003 shared task: language-independent named entity recognition, CoNLL (2003)
5. Ritter, A., Clark, S., Mausam, Etzioni, O.: Named entity recognition in tweets: an experimental study. In: Proceeding of Empirical Methods in Natural Language Processing (2011)
6. Zhuang, Y., Li, G., Zhong, Z., et al.: Hike: a hybrid human-machine method for entity alignment in large-scale knowledge bases. ACM (2017)
7. Liggins, M.E., Hall, D.L., Llinas, J.: Handbook of Multisensor Data Fusion: Theory and Practice, 2nd edn. CRC, Boca Raton (2008). (Multisensor Data Fusion)(Multisensor Data Fusion)(Multisensor Data Fusion)
8. jieba. https://github.com/fxsjy/jieba
9. pangu. http://pangusegment.codeplex.com/
10. ChineseRE. https://github.com/crownpku/Information-Extraction-Chinese/tree/master/RE_BGRU_2ATT
11. Robinson, I., Webber, J., Eifrem, E.: Graph Databases, 2nd edn. (Lu Liu and Yue Liang translated). Posts & Telecom Press, Beijing (2015)
12. Zeng, D., Liu, K., Chen, Y., Zhao, J.: Distant supervision for relation extraction via piecewise convolutional neural networks. In: Proceedings of EMNLP (2015)
13. Zhou, P., Shi, W., Tian, J., et al.: Attention-based bidirectional long short-term memory networks for relation classification. In: Proceedings of the Association for Computational Linguistics, Berlin (2016)
14. Lin, Y., Shen, S., Liu, Z., Luan, H., Sun, M.: Neural relation extraction with selective attention over instances. In: Proceedings of the Association for Computational Linguistics, Berlin (2016)
15. Cormen, T.H., Leiserson, C.E., Rivest, R.L., Sten, C.: Introduction to Algorithm, 3rd edn. The MIT Press, Cambridge (2009)

Blind Spectrum Sensing Based on the Statistical Covariance Matrix and K-Median Clustering Algorithm

Jiawei Zhuang[1], Yonghua Wang[1,2(✉)], Pin Wan[1,3], Shunchao Zhang[1], Yongwei Zhang[1], and Yi Li[1]

[1] School of Automation, Guangdong University of Technology, Guangzhou 510006, China
[2] State Key Laboratory of Management and Control for Complex Systems, Institute of Automation, Chinese Academy of Sciences, Beijing 100190, China
sjzwyh@163.com
[3] Hubei Key Laboratory of Intelligent Wireless Communications, South-Central University for Nationalities, Wuhan 430074, China

Abstract. Spectrum sensing is a fundamental function for cognitive radio systems, which can improve spectrum utilization. In this article, a blind spectrum sensing method based on the sample covariance matrix and K-median clustering algorithm is proposed to further improve the sensing performance. Specifically, to obtain a two-dimensional signal feature vector, the received signal matrix is rebuilt into two sub-matrices by the decomposition and reorganization (DAR) method. Moreover, two statistical covariance matrices are constructed by the sub-matrices, respectively. The ratios between the sum of some elements from the sample covariance matrices and the sum of diagonal elements from those matrices are used as a signal feature vector. It is demonstrated that the new signal feature vector and the feature vector based on the improved covariance absolute value method are equivalent. Furthermore, K-median clustering algorithm is trained by signal feature vectors to obtain a classifier. Indeed, this classifier can directly be used to detect whether the PU signal is absent or not. Simulation results report that the proposed algorithm has better sensing performance than some popular sensing algorithms based on random matrix theory or information geometry.

This work was supported in part by the National Natural Science Foundation of China under Grant 61971147, in part by the special funds from the central finance to support the development of local universities under Grant 400170044 and Grant 400180004, in part by the project supported by the State Key Laboratory of Management and Control for Complex Systems, Institute of Automation, Chinese Academy of Sciences under Grant 20180106, in part by the Foundation of National and Local Joint Engineering Research Center of Intelligent Manufacturing Cyber-Physical Systems and Guangdong Provincial Key Laboratory of Cyber-Physical Systems under Grant 008, and in part by the higher education quality projects of Guangdong Province and Guangdong University of Technology.

X. Sun et al. (Eds.): ICAIS 2020, LNCS 12239, pp. 467–478, 2020.
https://doi.org/10.1007/978-3-030-57884-8_41

Keywords: Spectrum sensing · Covariance matrix · Decomposition and reorganization · K-median clustering algorithm

1   Introduction

1.1   Previous Solutions

Cognitive radio technology is considered to be an effective method to improve the imminent problem of the low spectrum utilization for wireless communication systems [1,2]. Spectrum sensing is one of the key technologies among cognitive radio technologies, which can discovery the spectrum holes. There are some traditional spectrum sensing algorithms, such as energy detection, cyclostationary detection and matched filter detection. It is well known that these traditional algorithms have disadvantages. For example, the energy detection has poor sensing performance under the noise uncertainty. Besides, the cyclostationary detection does not detect whether the PU signal is existing or not when the cyclic frequencies of the PU signal are unknown. Likewise, the matched filter detection does not work when perfect information of the channel responses from the primary user (PU) to the receiver is not obtained. To resolve the shortcomings of these traditional algorithms, some spectrum sensing schemes based on random matrix theory have been proposed. For instance, [3] proposed a sensing scheme based on maximum-minimum eigenvalue (MME), which had good sensing performance at low SNR. [4] proposed a sensing method that used the ratio of maximum eigenvalue to average eigenvalue as signal feature to detect the spectrum holes. [5] proposed a sensing method, which used the ratio of some elements from covariance absolute value (CAV) as signal feature to perform spectrum sensing. This method is called as the improved CAV-based method. [6] proposed a spectrum sensing method based on the ratio of maximum eigenvalue to the trace (RMET), which had better sensing performance in some cases. The above sensing scheme based on random matrix theory improve the sensing performance, which do not require the prior information and work well under the noise uncertainty. However, all the above sensing method have to calculate the decision threshold that is difficult to obtain an accurate threshold in the practical applications.

Recently, it is trend that machine learning algorithms are applied to cognitive radio systems [7]. To further improve the sensing performance, some popular sensing schemes based on the clustering algorithms are proposed to avoid obtaining the decision threshold. For example, [8] proposed a sensing method based on K-means clustering algorithm. This method classified the signal energy features by K-means clustering algorithm, which is different from the traditional energy detection method. [9] proposed a sensing scheme based on the eigenvalues of covariance matrix and K-means clustering algorithm. This scheme worked better than the conventional sensing schemes based on random matrix theory under the same conditions. [10] proposed a sensing algorithm based on information geometry and Fuzzy-c-means (FCM) clustering algorithm. This algorithm obtained signal feature in the statistical manifold and used FCM clustering algorithm to make a decision. Apparently, the mentioned sensing methods based on the clustering algorithms have better sensing performance than the sensing schemes based on the decision threshold.

1.2   Contributions

Motivated by the random matrix theory and machine learning algorithms, aiming at improving the sensing performance, a new blind spectrum sensing scheme based the sample covariance matrix and K-median clustering algorithm is proposed in this article. The basic steps of the proposed scheme are shown in Fig. 1. Secondary users (SUs) collect the sensing signal and these signals are sent to the fusion center (FC). Moreover, the received signal matrix is regrouped into two sub-matrices by the decomposition and reorganization (DAR) method [10]. Furthermore, two different statistical covariance matrices are obtained by two sub-matrices, respectively. A new two-dimensional signal feature vector is established by two different statistical covariance matrices. Besides, K-median clustering algorithm is trained by the signal feature vectors to obtain a classifier. Indeed, this classifier can make a global decision via the signal feature vector. Especially, this classifier does not need to be trained again, which can directly perform spectrum sensing next time. Given what has been described above, the main contributions of this paper are as follows:

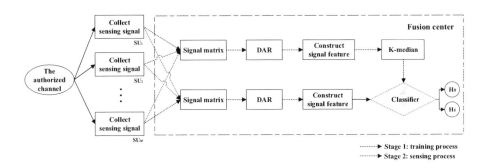

Fig. 1. The basic steps of the proposed scheme.

1. A new two-dimensional signal feature vector is proposed. The received signal matrix is rebuilt into two sub-matrices by the DAR method. Furthermore, two different sample covariance matrices are constructed by two sub-matrices, respectively. Indeed, the ratios between the sum of some elements from the sample covariance matrices and the sum of diagonal element from these matrices are used as a signal feature vector.
2. A new classifier based on K-median clustering algorithm is proposed. K-median clustering algorithm is trained by a feature vector set. After this algorithm is completed, two cluster centers are obtained. Moreover, a new

classifier is given by two cluster centers. Indeed, this classifier can make a
global decision without calculating the decision threshold.

2   Scenario and Signal Model

2.1   Cooperative Spectrum Sensing Scenario

The cooperative spectrum sensing scenario is apparently shown in Fig. 2. Assume
that there are M SUs and one PU in the cognitive radio network. Each SU sends
the sensing signal to the FC and does not makes a decision. The FC makes a
global decision to detect whether the PU exists or not. SUs are allowed to use
the authorized channel when the PU signal is absent. However, SUs will not be
allowed to access the authorized channel when PU is using the channel.

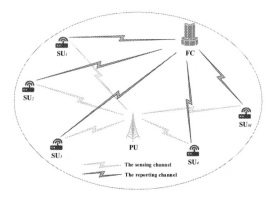

Fig. 2. Cooperative spectrum sensing scenario.

2.2   Signal Model

Assume that each SU has a sensing antenna and PU has a sensing antenna. It
is well known that the spectrum sensing problem can be expressed as a binary
hypothesis, i.e., the PU signal is present ($\mathcal{H}_1$) or absent ($\mathcal{H}_0$). In other words,
$\mathcal{H}_0$ represents that a SU can access the authorized channel and $\mathcal{H}_1$ represents
that SUs cannot access this channel. Based on the above two assumptions, the
mathematical model of the sensing signal can be expressed by

$$\begin{cases} \mathcal{H}_0 : x_i(n) = w_i(n), \\ \mathcal{H}_1 : x_i(n) = s_i(n) + w_i(n), \end{cases} \tag{1}$$

where $i(i = 1, 2, \ldots, M)$ denotes the number of SUs and $n(n = 1, 2, \ldots, N)$
indicates the number of sampling points. $x_i(n)$ represents the sensing signal of
ith SU. $s_i(n)$ represents the PU signal. $w_i(n)$ represents the Gaussian white noise
signal with a mean of 0 and a variance of σ^2. Suppose that $s_i(n)$ and $w_i(n)$ are
independent of each other.

3   Construct Signal Feature Vector

The received signal matrix $\mathbf{X}$ is obtained by the received signals, which can be expressed as

$$\mathbf{X} = \begin{bmatrix} x_1 \\ x_2 \\ \vdots \\ x_M \end{bmatrix} = \begin{bmatrix} x_1(1) & x_1(2) & \dots & x_1(N) \\ x_2(1) & x_2(2) & \dots & x_2(N) \\ \vdots & \vdots & \ddots & \vdots \\ x_M(1) & x_M(2) & \dots & x_M(N) \end{bmatrix}. \tag{2}$$

To construct a two-dimensional signal feature vector, the DAR method is used to obtain two different signal sub-matrices. Hence, based on (2), $\mathbf{X}$ can be rebuild into $\mathbf{X}^1$ and $\mathbf{X}^2$, which are be given by

$$\mathbf{X}^1 = \begin{bmatrix} x_1(1) & x_1(2) & \cdots & x_1(p) \\ \vdots & \vdots & \vdots & \vdots \\ x_1((g-1)p+1) & x_1((g-1)p+2) & \cdots & x_1(gp) \\ \vdots & \vdots & \vdots & \vdots \\ x_M(1) & x_M(2) & \cdots & x_M(p) \\ \vdots & \vdots & \vdots & \vdots \\ x_M((g-1)p+1) & x_M((g-1)p+2) & \cdots & x_M(gp) \end{bmatrix} \tag{3}$$

and

$$\mathbf{X}^2 = \begin{bmatrix} x_1(1) & x_1(g+1) & \cdots & x_1((p-1)g+1) \\ \vdots & \vdots & \vdots & \vdots \\ x_1(g) & x_1(g+g) & \cdots & x_1((p-1)g+g) \\ \vdots & \vdots & \vdots & \vdots \\ x_M(1) & x_M(g+1) & \cdots & x_M((p-1)g+1) \\ \vdots & \vdots & \vdots & \vdots \\ x_M(g) & x_M(g+g) & \cdots & x_M((p-1)g+g) \end{bmatrix}, \tag{4}$$

respectively, where $N = p \times g$. Thus, two different statistical covariance matrices are calculated by

$$\mathbf{R}^1 = E[(\mathbf{X}^1)(\mathbf{X}^1)^{\mathsf{T}}] \tag{5}$$

and

$$\mathbf{R}^2 = E[(\mathbf{X}^2)(\mathbf{X}^2)^{\mathsf{T}}], \tag{6}$$

respectively. Undoubtedly, $\mathbf{R}^1$ and $\mathbf{R}^2$ can be expressed as

$$\mathbf{R}^1 = \begin{bmatrix} r_{11}^1 & r_{12}^1 & \cdots & r_{1M}^1 \\ r_{21}^1 & r_{22}^1 & \cdots & r_{2M}^1 \\ \vdots & \vdots & \ddots & \vdots \\ r_{M1}^1 & r_{M2}^1 & \cdots & r_{MM}^1 \end{bmatrix} \tag{7}$$

and

$$\mathbf{R}^2 = \begin{bmatrix} r_{11}^2 & r_{12}^2 & \cdots & r_{1M}^2 \\ r_{21}^2 & r_{22}^2 & \cdots & r_{2M}^2 \\ \vdots & \vdots & \ddots & \vdots \\ r_{M1}^2 & r_{M2}^2 & \cdots & r_{MM}^2 \end{bmatrix}, \tag{8}$$

respectively. According to (7) and (8), the signal features can be given by

$$t^1 = \frac{\sum_{1 \leq i < j \leq M} |r_{ij}^1|}{\sum_{1 \leq i \leq M} |r_{ii}^1|} \tag{9}$$

and

$$t^2 = \frac{\sum_{1 \leq i < j \leq M} |r_{ij}^2|}{\sum_{1 \leq i \leq M} |r_{ii}^2|}, \tag{10}$$

respectively. When the PU signal is absent, $t^1 = 0$ and $t^2 = 0$. However, $t^1 > 0$ and $t^2 > 0$ when the PU signal is present. According to this difference, t^1 and t^2 can be used to detect the status of the authorized channel. Therefore, the signal feature vector $\mathbf{T} = [t^1, t^2]$ is also used to detect the status of the authorized channel.

Theorem 1. *The new signal feature vector and the improved CAV-based feature vector based are equivalent.*

Proof. See Appendix.   ∎

4   Decision Based on K-Median Clustering Algorithm

In this section, we will use K-median clustering algorithm to make a global decision for spectrum sensing. The framework of K-medians clustering algorithm is similar to K-means clustering algorithm, but K-means clustering algorithm is affected by the extreme values easily [11]. This disadvantage will lead to poor sensing performance. However, K-median clustering algorithm can overcome this disadvantage well. Different from K-means clustering algorithm, K-median clustering algorithm uses the Manhattan distance to analyze the difference between points in the two-dimensional plane. The Manhattan distance can be calculated by

$$d(\xi_1, \xi_2) = |X_1 - X_2| + |Y_1 - Y_2|, \tag{11}$$

where $|\cdot|$ denotes the absolute value operation, $\xi_1 = (X_1, Y_1)$ and $\xi_2 = (X_2, Y_2)$ indicate the point in the two-dimensional plane, respectively. Moreover, this algorithm uses the median value among the Manhattan distance values to update the cluster center in the same cluster, rather than the average value.

First of all, a training set $\mathbf{D}$ including enough signal feature vectors is needed. Secondly, the number of classes are set to 2, and the training set is used as the input of K-median clustering algorithm. Next, two cluster centers are used as the output after K-median clustering algorithm is successfully trained. Finally, a classifier based on two cluster centers is constructed. The mathematical model of this classifier can be represented as

$$\Theta(\mathbf{T}_d) = \frac{|\mathbf{T}_d - \mathbf{\Psi}_1|}{|\mathbf{T}_d - \mathbf{\Psi}_2|} \geq \gamma, \qquad (12)$$

where $\mathbf{\Psi}_1$ and $\mathbf{\Psi}_2$ mean the cluster center that belonged to $\mathcal{H}_0$ and the cluster center that belonged to $\mathcal{H}_1$, respectively, $\mathbf{T}_d$ stands for the dth signal feature vector in the set $\mathbf{D}$, and γ denotes the parameter to adjust the probability of false alarm (P_f). If $\Theta(\mathbf{T}_d) \geq \gamma$ is satisfied, it means that the authorized channel is available, otherwise, this channel is busy. The specific process for obtaining the classifier based on K-median clustering algorithm is shown in Table 1.

Table 1. The specific process for obtaining the classifier.

Algorithm 1:	The specific process for obtaining the classifier
Begin:	
	Input $\mathbf{D} = [\mathbf{T}_1, \mathbf{T}_2, \ldots, \mathbf{T}_d]$.
End:	
	A classifier based on K-median clustering algorithm.
K-median:	
	(1) Randomly select 2 samples in the set $\mathbf{D}$ as the cluster centers.
	repeat:
	(2) Calculate the Manhattan distance between each feature vector in the set $\mathbf{D}$ and the cluster centers $\mathbf{\Psi}_k (k \in \{1,2\})$ in the same cluster.
	(3) Update $\mathbf{\Psi}_k$ by the the median value among the Manhattan distance values.
	(4) If $\mathbf{\Psi}_k$ no long changes, the algorithm will stop, otherwise it continues.
	(5) Output: Two optimal cluster centers.
	end
	(6) The classifier based on K-median clustering algorithm is obtained by two optimal cluster centers. The mathematical model of this classifier can be formulated as (12).

Once the classifier is obtained, we can directly input the signal feature vector into the classifier to detect the status of the PU. Moreover, the classifier not to be trained next time.

5   Simulations

In this section, the sensing performance of the proposed algorithm will be analyzed. To ensure the accuracy and reliability of the simulation results, the PU signal is uniformly set to AM signal. Moreover, the set **D** is obtained by 2000 signal feature vectors to train K-median clustering algorithm, and the test set **S** is constructed by 4000 the signal feature vectors to analyze the sensing performance. In the simulations, we compare the proposed algorithm with some popular spectrum sensing methods, such as the sensing method based on the maximum-minimum eigenvalue (MME) and K-mesns clustering algorithm, the sensing scheme based on the difference of maximum and minimum eigenvalues (DMME) and K-mesns clustering algorithm, the sensing algorithm based on the ratio of maximum eigenvalue to the trace (RMET) and K-medoids clustering algorithm, and the sensing method based on information geometry (IG) and FCM clustering algorithm. Furthermore, P_f and the probability of detection P_d are used to measured the sensing performance [2].

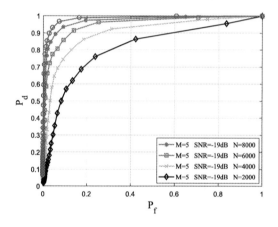

Fig. 3. The sensing performance of the proposed algorithm with different N.

In Fig. 3, the sensing performances of the proposed algorithm are displayed when $M = 5$, SNR $= -19$ dB and different N, where M indicates the number of SUs and N is the number of sampling points. The proposed algorithm has the best sensing performance when $N = 8000$, however, it has the worst sensing performance when $N = 2000$. It is obviously demonstrated that the proposed algorithm can work better with N increasing, since more information about the PU signal can be obtained as N grows.

In Fig. 4, the sensing performances of the proposed algorithm are described when $N = 2000$, SNR $= -20$ dB and different M. As can be apparently seen in it, the proposed algorithm has the best sensing performance when $M = 10$, but it has the worst sensing performance when $M = 2$. The FC can receive more the sensing signals when more SUs participate in spectrum sensing together, thus, the FC can make a more accurate decision.

In Fig. 5, the sensing performances of different sensing methods are introduced when $M = 5$, SNR $= -18$ dB and $N = 2000$. Likewise, In Fig. 6, the sensing performances of different sensing schemes are presented when $M = 7$, SNR $= -20$ dB and $N = 4000$. Obviously, the proposed algorithm has better

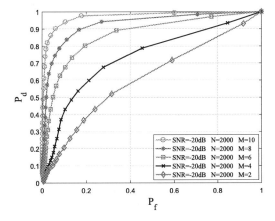

Fig. 4. The sensing performance of the proposed algorithm with different M.

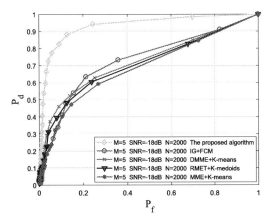

Fig. 5. The sensing performance for different sensing algorithms when SNR $= -18$ dB.

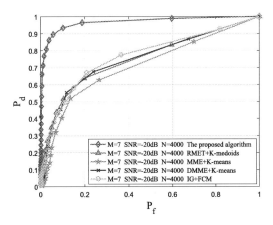

Fig. 6. The sensing performance for different sensing algorithms when SNR $= -20$ dB.

sensing performances than some popular sensing methods under different simulation conditions. Some popular sensing methods use the eigenvalues of covariance matrix to perform spectrum sensing. Different from some popular sensing schemes, the proposed method use more the knowledge of the received signals. Consequently, the proposed method can makes a accurate global decision.

6   Conclusion

To further improve spectrum sensing, a new blind sensing method based on the statistical covariance matrix and K-median clustering algorithm is proposed in this article. Different from some popular sensing schemes based on the eigenvalues of covariance matrix, the proposed algorithm uses some elements from covariance matrix and the DAR method to construct a new signal feature vector. Moreover, this algorithm uses K-median clustering algorithm to obtain the classifier. Since K-median clustering algorithm works batter than K-means clustering algorithm, the classifier based on K-median clustering algorithm has batter classification performance than the classifier based on K-means clustering algorithm. In simulation, we analyze the impact of number of SUs and number of sampling points on the sensing performance, respectively. Compared with some popular sensing algorithms, the proposed algorithm can improve the sensing performance for spectrum sensing well.

Appendix

In this section, it will be demonstrated that the new signal feature vector and the improved CAV-based feature vector are equivalent. Assume that a 3×3 statistical covariance matrix is given by

$$\mathbf{C} = \begin{bmatrix} c_{11} & c_{12} & c_{13} \\ c_{21} & c_{22} & c_{23} \\ c_{31} & c_{32} & c_{33} \end{bmatrix}. \tag{13}$$

The matrix $\mathbf{C}$ is rebuilt into two sub-matrices by the DAR method. Two sub-matrices can be written as

$$\mathbf{C}^1 = \begin{bmatrix} c_{11}^1 & c_{12}^1 & c_{13}^1 \\ c_{21}^1 & c_{22}^1 & c_{23}^1 \\ c_{31}^1 & c_{32}^1 & c_{33}^1 \end{bmatrix} \tag{14}$$

and

$$\mathbf{C}^2 = \begin{bmatrix} c_{11}^2 & c_{12}^2 & c_{13}^2 \\ c_{21}^2 & c_{22}^2 & c_{23}^2 \\ c_{31}^2 & c_{32}^2 & c_{33}^2 \end{bmatrix}, \tag{15}$$

respectively.

According to [5], the improved CAV-based feature vector can be obtain by

$$\mathbf{U} = [t^1_{\text{ICAV}}, t^2_{\text{ICAV}}], \tag{16}$$

where t^1_{ICAV} and t^2_{ICAV} can be formulated as

$$t^1_{\text{ICAV}} = \frac{\sum_{1 \le i \le j \le 3} |c^1_{ij}|}{|c^1_{11}| + |c^1_{22}| + |c^1_{33}|} \tag{17}$$

and

$$t^2_{\text{ICAV}} = \frac{\sum_{1 \le i \le j \le 3} |c^2_{ij}|}{|c^2_{11}| + |c^2_{22}| + |c^2_{33}|}, \tag{18}$$

respectively. The Manhattan distance is used to analyze the similarity between two points in this article. Note that $\mathbf{U} = [1,1]$ when the PU signal is absent or $\mathbf{U} = [t^1_{\text{ICAV}}, t^2_{\text{ICAV}}]$ when the PU signal is present, where $t^1_{\text{ICAV}} > 1$ and $t^2_{\text{ICAV}} > 1$. Therefore, the Manhattan distance between points which belong to $\mathcal{H}_0$ and $\mathcal{H}_1$, respectively, can be written as

$$
\begin{aligned}
d(\mathbf{U}_{\mathcal{H}_1}, \mathbf{U}_{\mathcal{H}_0}) &= |t^1_{\text{ICAV}} - 1| + |t^2_{\text{ICAV}} - 1| \\
&= \left| \frac{\sum_{1 \le i \le j \le 3} |c^1_{ij}|}{|c^1_{11}| + |c^1_{22}| + |c^1_{33}|} - 1 \right| + \left| \frac{\sum_{1 \le i \le j \le 3} |c^2_{ij}|}{|c^2_{11}| + |c^2_{22}| + |c^2_{33}|} - 1 \right| \\
&= \frac{|c^1_{12}| + |c^1_{13}| + |c^1_{23}|}{|c^1_{11}| + |c^1_{22}| + |c^1_{33}|} + \frac{|c^2_{12}| + |c^2_{13}| + |c^2_{23}|}{|c^2_{11}| + |c^2_{22}| + |c^2_{33}|}.
\end{aligned} \tag{19}
$$

Moreover, the new signal feature vector can be obtained by

$$\mathbf{T} = [t^1, t^2], \tag{20}$$

where t^1 and t^2 are written as

$$t^1 = \frac{|c^1_{12}| + |c^1_{13}| + |c^1_{23}|}{|c^1_{11}| + |c^1_{22}| + |c^1_{33}|} \tag{21}$$

and

$$t^2 = \frac{|c^2_{12}| + |c^2_{13}| + |c^2_{23}|}{|c^2_{11}| + |c^2_{22}| + |c^2_{33}|}, \tag{22}$$

respectively. Note that $\mathbf{T} = [0,0]$ when the authorized channel is idle or $\mathbf{T} = [t^1, t^2]$ when the authorized channel is busy, where $t^1 > 0$ and $t^2 > 0$. Hence, the Manhattan distance between points which belong to $\mathcal{H}_0$ and $\mathcal{H}_1$, respectively, can be calculated by

$$
\begin{aligned}
d(\mathbf{T}_{\mathcal{H}_1}, \mathbf{T}_{\mathcal{H}_0}) &= |t^1 - 0| + |t^2 - 0| \\
&= \left| \frac{|c^1_{12}| + |c^1_{13}| + |c^1_{23}|}{|c^1_{11}| + |c^1_{22}| + |c^1_{33}|} - 0 \right| + \left| \frac{|c^2_{12}| + |c^2_{13}| + |c^2_{23}|}{|c^2_{11}| + |c^2_{22}| + |c^2_{33}|} - 0 \right| \\
&= \frac{|c^1_{12}| + |c^1_{13}| + |c^1_{23}|}{|c^1_{11}| + |c^1_{22}| + |c^1_{33}|} + \frac{|c^2_{12}| + |c^2_{13}| + |c^2_{23}|}{|c^2_{11}| + |c^2_{22}| + |c^2_{33}|}.
\end{aligned} \tag{23}
$$

Based on (19) and (23), it is easy job for us to find that $d(\mathbf{U}_{\mathcal{H}_1}, \mathbf{U}_{\mathcal{H}_0}) = d(\mathbf{T}_{\mathcal{H}_1}, \mathbf{T}_{\mathcal{H}_0})$. Consequently, it has been demonstrated that the new signal feature vector and the improved CAV-based feature vector are equivalent.

References

1. Hou, Y., Huang, H., An, D., Zhou, Z.: Research status and prospect of InSAR performance evaluation methods. In: 2nd International Conference on Electronic Information and Communication Technology (ICEICT), pp. 486–490. IEEE, Harbin (2019)
2. Zhuang, J., Wang, Y., Zhang, S., Wan, P., Sun, C.: A multi-antenna spectrum sensing scheme based on main information extraction and genetic algorithm clustering. IEEE Access **7**, 119620–119630 (2019)
3. Wael, C., Armi, N., Rohman, B.: Spectrum sensing for low SNR environment using maximum-minimum eigenvalue (MME) detection. In: International Seminar on Intelligent Technology and Its Applications, pp. 435–438. IEEE, Lombok (2016)
4. Zhang, S., Yang, J., Guo, L.: Eigenvalue-based cooperative spectrum sensing algorithm. In: Second International Conference on Instrumentation, Measurement, Computer, Communication and Control, pp. 375–378. IEEE, Harbin (2012)
5. Upadhya, V., Jalihal, D.: Almost exact threshold calculations for covariance absolute value detection algorithm. In: National Conference on Communications (NCC), pp. 1–5. IEEE, Kharagpur (2012)
6. Ahmed, A., Hu, Y., Noras, J., Pillai, P., Abd-Alhameed, R., Smith, A.: Random matrix theory based spectrum sensing for cognitive radio networks. In: Internet Technologies and Applications, pp. 479–483. IEEE, Wrexham (2015)
7. Huang, X., Zhai, H., Fang, Y.: Robust cooperative routing protocol in mobile wireless sensor networks. IEEE Trans. Wirel. Commun. **12**(7), 5278–5285 (2008)
8. Kumar, V., Kandpal, D., Jain, M., Gangopadhyay, R., Debnath, S.: K-mean clustering based cooperative spectrum sensing in generalized $\kappa - \mu$ fading channels. In: Twenty Second National Conference on Communication, pp. 1–5. IEEE, Guwahati (2016)
9. Zhang, Y., Wan, P., Zhang, S., Wang, Y., Li, N.: A spectrum sensing method based on signal feature and clustering algorithm in cognitive wireless multimedia sensor networks. Adv. Multimed. **2017**(4), 1–10 (2017)
10. Zhang, S., Wang, Y., Li, J., Wan, P., Zhang, Y., Li, N.: A cooperative spectrum sensing method based on information geometry and fuzzy c-means clustering algorithm. EURASIP J. Wirel. Commun. Network. **2019**(1), 1–12 (2019). https://doi.org/10.1186/s13638-019-1338-z
11. Zhu, H., Shi, Y.: Brain storm optimization algorithms with k-medians clustering algorithms. In: Seventh International Conference on Advanced Computational Intelligence (ICACI), pp. 107–110. IEEE, Wuyi (2015)

An Efficient Web Caching Replacement Algorithm

Fan You[1], Tong Liu[1], Xiaoyu Peng[1], Jiahao Liang[1], Baili Zhang[1,2(✉)], and Yinian Zhou[3]

[1] Southeast University, Nanjing 211189, China
zhangbl@seu.edu.cn
[2] Research Center for Judicial Big Data, Supreme Court of China, Nanjing 211189, China
[3] School of Computer and Software, Nanjing University of Information Science and Technology, Nanjing 210044, China

Abstract. One of the keys to improve the performance of Content Delivery Network (CDN) is the efficiency of caching. To get the goal of improving caching efficiency, the main aim of current CDN caching algorithms is to obtain a higher caching hit ratio, while the validation and freshness of outdated pages have not received due consideration in these replacement models. In this paper, a new improved cache profit model is clearly defined, and the validation and freshness factors of Web pages have been taken into adequate account. Based on the profit model, a new caching replacement algorithm-EWCR (Efficient Web Caching Replacement) is proposed, and it is proved to be optimal based on reasonable assumptions. To conclude, a series of comparative experiments verifies the efficiency of EWCR in Web caching replacement.

Keywords: Web cache replacement · Profit model · Proxy server

1 Introduction

Network congestion is becoming one of the main factors restricting the performance of Web services, due to the rapid growth of network traffic [1–3]. To cope with the situation, CDN is proposed to improve network performance [4, 5]. CDN proxy server and Web caching mechanism play a significant role in reducing the whole Internet bandwidth consumption, guaranteeing network load balancing and speeding up user visiting speed [6–9]. Due to the constraints of storage space and the vast amount of network traffic, the CDN proxy server cannot save all the Web pages and objects unconditionally. Thus, there has been a lot of researches on the strategies of Web caching replacement [4, 10]. Among them, the Least Recently Used algorithm (LRU) [7, 11] and the Least Frequently Used algorithm (LFU) are of the most representative. To solve the problem of cache pollution, various improved algorithms have been proposed, such as LRU-K [12], LCN-R [13], GDSF [14], BP Neural Network [15], Hybrid [16] and TSP [17]. However, all the algorithms mentioned above normally aim at maximizing cache hit ratio, but the expense of validating and updating pages in cache are failed to be equally treated. This

© Springer Nature Switzerland AG 2020
X. Sun et al. (Eds.): ICAIS 2020, LNCS 12239, pp. 479–488, 2020.
https://doi.org/10.1007/978-3-030-57884-8_42

paper argues for the validation and freshness of outdated pages in the choice of cache replacement. In view of this, from the perspective of optimal cache efficiency, this paper fully considers various factors affecting cache performances including update efficiency. The paper firstly constructs a new improved cache profit model of Web page and object caching, then converts the problem of cache replacement to a function optimality problem. After that, a new replacement algorithm-EWCR is recommended. Finally, based on the rational assumption, the optimality of algorithm is proved. Compared with the previous research, the proposed model takes fuller account of various factors that may affect caching performance.

2 Problem Description

In this section, some relevant definitions for caching mechanism are provided based on Web page user query and cache profit model.

Definition 1. Suppose $C_q(p_i, Web)$ is the cost of generating the Web page p_i by querying data though Web server, and f_{q_i} is the probability that page p_i is accessed by the user. Without considering the proxy server cache, the query generation cost for querying the entire Web page set $P' = \{p_1, p_2, \ldots, p_m\}$ is:

$$C_Q(P) = \sum_{p_i \in P} f_{q_i} \times C_q(p_i, Web) \tag{1}$$

Considering the existence of cache in proxy server, given a set of cached Web pages $P' = \{p_1, p_2, \ldots, p_m\}$, user query request on those pages can be directly responded by cache matching without access to the Web server. Therefore:

Definition 2. Define $C_q(p_i, Cache)$ as the cost that generates Web page $C_q(p_i, Cache)$ directly through cache, under the conditions of considering the proxy server cache and ignoring the page updates, the query generation cost for users to query the entire Web page set P is:

$$C'_Q(P) = \sum_{p_i \in P'} f_{q_i} \times C_q(p_i, Cache) + \sum_{p_i \in P-P'} f_{q_i} \times C_q(p_i, Web) \tag{2}$$

Definition 3. If the cache page is out of date, a timely update is needed to keep data freshness. Suppose f_{u_i} is the probability of updating Web page p_i in the cache, $C_u(p_i, Web)$ is the corresponding update cost, then the update cost of the entire pages in the cache is:

$$C_U(P') = \sum_{p_i \in P'} f_{u_i} \times C_u(p_i, Web) \tag{3}$$

The page update process is equivalent to the query cost that retrieve a page p_i from the Web server (only considering the HTML page, and ignoring the generative relations between Web objects), therefore, $C_u(p_i, Web) \approx C_q(p_i, Web)$. Furthermore, the update cost of the entire pages in the cache is:

$$C_U(P') = \sum_{p_i \in P'} f_{u_i} \times C_u(p_i, Web) \approx \sum_{p_i \in P'} f_{u_i} \times C_q(p_i, Web) \tag{4}$$

Due to the existence of the cache, a part of Web pages can be accessed directly by matching back-up copy in the cache server instead of the Web server. Therefore, the generation cost of user query for the entire page set P will change as the following:

$$
\begin{aligned}
\Delta C = {} & C_Q(P) - C'_Q(P) - C_U(P') \\
= {} & \sum_{p_i \in P'} f_{q_i} \times C_q(p_i, \text{Web}) + \sum_{q_i \in P - P'} f_{q_i} \times C_q(p_i, \text{Web}) - \sum_{p_i \in P'} f_{q_i} \times C_q(p_i, \text{Cache}) \\
& - \sum_{p_i \in P - P'} f_{q_i} \times C_q(p_i, \text{Web}) - \sum_{p_i \in P'} f_{u_i} \times C_q(p_i, \text{Web}) \\
= {} & \sum_{p_i \in P'} (f_{q_i} \times C_q(p_i, \text{Web}) - f_{q_i} \times C_q(p_i, \text{Cache}) \qquad (5)
\end{aligned}
$$

Definition 4. Suppose $C_q(p_i, \text{Web})$ is the cost that we access Web page p_i by Web server. When p_i is cached and updated in time, the cost will be $C_q(p_i, \text{Cache})$. Meanwhile, we define f_{q_i} to be the probability that page p_i is accessed by users, and f_{u_i} is the probability that the update operation happens, and then the expected profit that page p_i can be obtained by caching is:

$$
B(p_i) = f_{q_i} \times C_q(p_i, \text{Web}) - f_{q_i} \times C_q(p_i, \text{Cache}) - f_{u_i} \times C_q(p_i, \text{Web}) \qquad (6)
$$

Furthermore, the profit that the entire query set P can be obtained by caching is:

$$
\begin{aligned}
B(P) = {} & \sum_{p_i \in P'} \left(f_{q_i} \times C_q(p_i, \text{Web}) - f_{q_i} \times C_q(p_i, \text{Cache}) - f_{u_i} \times C_q(p_i, \text{Web}) \right) \\
= {} & \sum_{p_i \in P'} f_{q_i} \times C_q(p_i, \text{Web}) \times \left(1 - \frac{C_q(p_i, \text{Cache})}{C_q(p_i, \text{Web})} - \frac{f_{u_i}}{f_{q_i}} \right) = \sum_{p_i \in P'} B(p_i) \\
& \qquad\qquad\qquad\qquad\qquad\qquad\qquad\qquad\qquad\qquad\qquad\qquad\qquad\qquad (7)
\end{aligned}
$$

Definition 5. Given a Web page access set P, the problem of selecting a page to be cached is equivalent to the following optimization problem: selecting a caching page set p' under certain constraints to maximize the profit.

$$
B(P) = \sum_{p_i \in P'} B(p_i). \qquad (8)
$$

Usually, calculating the maximum $B(P)$ can be restrained by space, update cost and other conditions. In this paper, the most common space constraints are discussed in detail, and the selection problem under the constraint of update cost can be discussed similarly.

3   Web Page Replacement Algorithm-EWCR

Proxy server is always faced with massive Web pages, and these pages need to be selectively cached. Therefore, an EWCR algorithm is proposed for cache set selection and replacement, based on the constraint of cache space SC.

3.1   The Implementation of Algorithm-EWCR

Algorithm 1. EWCR Algorithm
Input: p_x, P', S_{free} // $p_x \epsilon P$ is an arbitrary page
Output: P', S_{free}

```
1.   s_x = sizeof(p_x)
2.   // s_x is the size of page p_x waiting to enter cache
3.   B(p_x) = EstimateBenifit(p_x)
4.   ϕ(p_x) = B(p_x)/s_x
5.   if ( s_x ≤ S_free      )
6.      // S_free is the remaining space for the current cache
7.      insert p_x into cache
8.      sort by ϕ({p_x},{∅},SignA)
9.      // the identifier id, the parameter ϕ, and the size
10.     // of all the page in P' are stored by the
11.     // non-decreasing order of ϕ in the array SignA,
12.     // and the new page p_x is inserted into the
13.     // corresponding position in SignA according to the
14.     // size of ϕ(p_x)
15.     P' = P' ∪ {p_x}
16.     S_free = S_free − s_x
17.     Return (1)
18.end if
19.pre_evit = {∅} // pre-deleted view set
20.count = 0
21.while (s_x > S_free )
22.     // the available space is not enough to save the
23.     // view
24.     p_j = SignA[count].id
25.     pre_evit = pre_evit ∪ {p_j}
26.     // according to the order of ϕ, the pages in P' are
27.     // set to be pre-deleted one by one until the space
28.     // requirement is met
29.     S_free = S_free + SignA[count].s
30.     count = count + 1
31.end while
32.ϕ(p_max) = max{{ϕ(p_j)|p_j ∈ pre_evit}}
33.// obtain the view with maximum ϕ value in pre_evit set
34.if (ϕ(p_max) < ϕ(p_x))
35.     // make sure p_x is the optimal
36.     delete pre_evit cache
37.     insert p_x into cache
38.     sort by ϕ({p_x},pre_evit,SignA)
39.     // sort by value ϕ, and at the same time, calculate ϕ
40.     // for all elements periodically
41.     P' = P' − pre_evit
42.     P' = P' ∪ {p_x}
43.     S_free = S_free − s_x
44.     Return (0)
45.else
46.     Return (-1)
47.end if
```

The basic of EWCR is that, whenever a new Web page appears, if the remaining space of the cache is enough, then it will be cached directly. However, if the cache space is full or insufficient to store the page, then a selection replacement is required. The principle of replacement is to ensure that total profit of cache $B(P)$ keeps increasing, so that every selection replacement decision the algorithm made is an effective approach to the maximum of cache profit is guaranteed. Because of the space constraints, the principle has been transformed into eliminating pages according to ϕ value.

3.2   Discussion on the Replacement Measure Value ϕ

Practically, the EWCR algorithm mainly focused on calculation and comparison of cache replacement measure value ϕ, which is the common feature of all cache algorithms. To be exact, the main difference of cache algorithms is the specific definition of replacement measure value ϕ. In the classic LRU algorithm and the improved algorithm LRU-K both are the representation of the memory cache algorithms, aiming at maximizing cache page hit ratio and minimizing replacement frequency of memory pages, in order to reduce the expense of page-disk replacement. To summarize, this type of algorithm is appropriate for the memory page replacement with equal page size the disk page replacement cost is similar. But it is unsuitable for Web objects with various parameters (such as the probability the query is referred to, the update cost of query and the update probability of query). To solve this problem, a series of replacement measure models are proposed [1]. Among them, GreedyDual-Size [18], LCN-R [13], Hybrid [19] and TSP [17] are typical and fundamental. These algorithms no longer simply pursue the goal of cache hits, they tend to reduce average response time ant the total query cost meanwhile. More than that various factors are also taken into consideration in their replacement measure models. However, the updating problem has not be considered explicitly. Based on the above studies, in this paper, an improved measure value ϕ is defined by:

$$\varphi(p_i) = \frac{B(p_i)}{s_i} = \frac{f_{q_i} \times C_q(p_i, \text{Web})}{s_i} \times \left(1 - \frac{C_q(p_i, \text{Cache})}{C_q(p_i, \text{Web})} - \frac{f_{u_i}}{f_{q_i}}\right) \qquad (9)$$

Where $C_q(p_i, \text{Web})$ is the cost to generate the page, equal to the cost to calculate p_i anew. There are two methods to calculate $C_q(p_i, \text{Web})$, one is based on the previous calculation cost or the accumulated average cost, the other makes use of c_s, b_s for calculation, just like the Hybrid algorithm. The main difference lies in the utilization of $C_q(p_i, \text{Cache})$ and f_{u_i}. $C_q(p_i, \text{Cache})$ refers to the cost of retrieving p_i from the cache and fetching the result. In general, $C_q(p_i, \text{Web}) << C_q(p_i, \text{Web})$, thus the measure value can be simplified as:

$$\varphi(p_i) = \frac{B(p_i)}{s_i} = \frac{f_{q_i} \times C_q(p_i, \text{Web})}{s_i} \times \left(1 - \frac{f_{u_i}}{f_{q_i}}\right) \qquad (10)$$

From the above equation, it can be concluded that it is more worthwhile to cache a page with higher access probability and a lower updating probability. On the contrary, if f_{q_i} and f_{u_i} are analogous, the value of $\phi(p_i)$ closes to zero, at this point, caching is not needed for the page. Furthermore, for a static page with very small f_{u_i}, caching strategies can certainly bring a significant effect. However, not all the above can concluded in the previous models.

Compared with other parameters, the estimation of access probability and update probability in the model is complicated. O'Neil [12] use the LRU-K mechanism to estimate the frequency. By adjusting the value K, it overcomes the deficiency of LRU. Therefore, it is a relatively good method to express instantaneous frequency. However, it ignores the characteristics of a long-term distribution of the query. For this reason, Scheuermann [13] adds the total number of the query occurrences, is added to the LUR-k, which can be considered as a consideration for long-term cumulative value. The disadvantage of his work is that, in order to offset the excessive effort of a long-term accumulation, a longer accumulative period is set and the accumulated value of the query that has not occurred in this period is cleared. Apparently, this approach is not sufficiently reliable and may cause abrupt changes in frequency values. To solve the problem, EWCR algorithm adopt AIF (attenuator of instantaneous frequency) algorithm [20] which use an attenuation factor for gradually returning to zero. In AIF algorithm, if no query happens during a short period, the cumulative number of queries may decline according to a certain attenuation coefficient, to avoid a large shaking in the materialized set caused by jumping of frequency values.

3.3   Discussion of the Optimality of the Algorithm

Let's discuss how EWCR algorithm guarantees that $B(P)$ keeps increasing after each replacement, that is the optimality of the algorithm. Under the constraints of the proxy server's cache space, maximizing cache profit can be described by the following optimization functions.

$$\text{Max}\left(\sum_{p_i \in P'} B(p_i)\right) \tag{11}$$

$$\sum_{p_i \in P'} s_i \leq S_{cache} \tag{12}$$

The problem defined by (11) and (12) belongs to NP-complete knapsack problem. No optimal algorithm is applicable to the problem, unless a hypothesis is made that cache size S_{Cache} is big enough for a single page s_i. In this situation, s_{Cache} can be filled with small s_i by approximate evaluation (this approximate evaluation is rational, because a single page is small enough, compared with disk cache space which often several hundred M or even G in size). Under the hypothesis, (12) can be revised as:

$$\sum_{p_i \in P'} s_{i_i} = S_{cache} \tag{13}$$

For the problem defines in (11) and (13), it can be proved that algorithm EWCR can obtain an optimal solution.

Theorem 1. The Web page cache set obtained by EWCR algorithm is the optimal solution to the problem defined in (11) and (13).

Proof by Contradiction

Suppose M is the page by the selected by EWCR algorithm, and $\sum_{p_i \in M} B(p_i)$ is the total profit obtained by the cache space $\sum s_i \approx S_{Cache}$. Suppose there is another set N, and total profit of the page cache under the same space constraint is bigger than $M's$. That is,

$$\exists \sum_{p_i \in N} B(p_i) > \sum_{p_i \in M} B(p_i), \sum_{p_i \in N} s_i = \sum_{p_i \in M} s_i$$

$$\sum_{p_i \in N-I} B(p_i) > \sum_{p_i \in M-I} B(p_i), \sum_{p_i \in N-I} s_i = \sum_{p_i \in M-I} s_i, I = N \cap M$$

$$let \; \phi(p_n) = max(\{\phi(p_i)|p_i \in N - I\})$$

$$then, \; \sum_{p_i \in N-I} B(p_i) = \sum_{p_i \in N-I} \phi(p_i) \cdot s_i \le \sum_{p_i \in N-I} \phi(p_n) \cdot s_i = \phi(p_n) \cdot \sum_{p_i \in N-I} s_i$$

$$let \; \phi(p_m) = min(\{\phi(p_i)|p_i \in M - I\})$$

$$then, \; \sum_{p_i \in M-I} B(p_i) = \sum_{p_i \in M-I} \phi(p_i) \cdot s_i \ge \sum_{p_i \in M-I} \phi(p_m) \cdot s_i = \phi(p_m) \cdot \sum_{p_i \in M-I} s_i$$

$$thus, we \; have : \; \phi(p_n) > \phi(p_m)$$

From the above derivation, it can be concluded that there is a page p_m in the set $M - I$, the value $\phi(p_m)$ is less than the value $\phi(p_n)$ of the page p_n in the set $N - I$. However, in the EWCR algorithm, page p_i is stored in reverse order according to $B(p_i)/s_i$ ($\phi(p_i)$), until space S_{Cache} is packed. When a new page p_x comes, pages in cache with a small ϕ value than $\Phi(p_x)$ is eliminated gradually, until the remaining space is big enough to hold page p_x. The final page set M the algorithm selected (p') has the following features: in the page set P accessed by the user, all the pages with a large $\Phi(p_x)$ value is contained in set M, while no other sets contain a higher $\Phi(p_x)$ value than in M, thus occurs contradiction. To sum up, there exists no other sets owning a larger profit value than M. Therefore, the proposed EWCR algorithm has a maximum total profit $\sum_{p_i \in M} B(p_i)$ on set M with the space constraints.

4 Experiments

In order to verify the effectiveness of the profit model and algorithm proposed in this paper, a series of comparative experiments are conducted on EWCR algorithm, the representative LRU-K and Hybrid algorithm.

The first experiment was simulation experiment based on the query generator. The experimental environment includes a Web/DB server (Dell PowerEdge 6666, Windows Server 200, Oracle 8 database), (IBM eServer xSeries 440, Windows Server 2003, Squid Web Proxy Cache), and three clients (Windows XP). In the experiment, each client randomly generates 5000 queries according to parameters such as page type, data type, hot spot area, update probability and so on by running a simulated query generator, 10% of the 5000 queries are used for buffered training set, which are not used as the basis for statistics and analysis. In total, five groups of experiments are conducted, the parameters of each group are different and the proportion of page updates is gradually increased. According to previous experimental conditions, the cache space of the proxy server is taken as 200M. The specific experimental results are shown in Fig. 1.

Figure 1(left) is the comparison on query response time based on EWCR, LRU-K (K = 2) and Hybrid. According to the figure, EWCR lacks advantages in the case of low update probability, and as the probability of updating increases, the advantages of its profit model gradually manifest. Figure 1(right) is a comparison of the cache hit ratio of three algorithms in different query sets. The results show that there is no significant difference between EWCR, LRU-K and Hybrid algorithm.

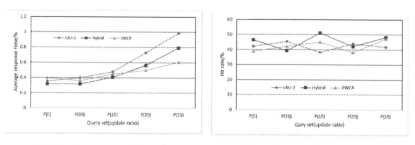

Fig. 1. Comparison of query response time (left) and cache hit ratio (right)

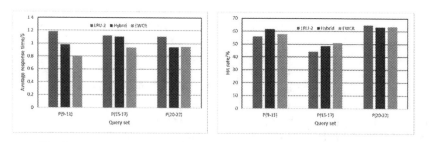

Fig. 2. Comparison of query response time (left) and cache hit ratio (right)

Next, an export proxy server of a college education network is selected, and a comparison experiment of the algorithms is conducted in the actual environment. In the experiment, three periods 9:00–11:00, 15:00–17:00 and 20:00–22:00 are selected, and 5000 access experiments were performed respectively in three groups to compare the response time and cache hit ratio of different proxy servers by adopting different algorithms. The results are shown in Fig. 2. Figure 2 indicates that, in the actual network environment, the performance of EWCR algorithm is not as superior as the simulated environment. The result is related to the non-repeatability of the user query and the complexity of the query types in the actual environment, especially the existence of digital video stream. In our profit model, no special treatment is conducted about digital video stream, thus further research in this aspect will be one of our future work.

5 Conclusion

Cache replacement strategy of Web pages in CDN proxy servers mainly focuses on improving cache hit ratio previously, and it has not considered the validation and freshness of outdated pages. This results in a long response time and expensive system cost. In this paper, an improved cache profit model is constructed, in which the factors of Web page update are fully explored. A proxy server cache replacement algorithm EWCR is proposed based on the model, and its theoretical optimality has been verified statistically. Finally, a series of comparative experiments are conducted, verifying that the algorithm can significantly improve the response time of Web query, especially when data update frequently. Compared with LRU-K and Hybrid, the EWCR algorithm presents a superior performance.

References

1. Vakali, A.: Proxy cache replacement algorithms: a history-based approach. World Wide Web **4**(4), 277–297 (2001). https://doi.org/10.1023/A:1015133818512
2. Li, Y., Li, J., Chen, J., Lu, M., Li, C.: Seed selection for data offloading based on social and interest graphs. Comput. Mater. Con. **57**(3), 571–587 (2018)
3. Liu, Y., Liu, L., Yan, Y., Feng, F., Ding, S.: Analyzing dynamic change in social network based on distribution-free multivariate process control method. Comput. Mater. Con. **60**(3), 1123–1139 (2019)
4. Pathan, A., Buyya, R.: A Taxonomy and Survey of Content Delivery Network. Springer, Heidelberg (2008)
5. Dafre, N., Shrawankar, U., Kapgate, D.: Pattern based cache management policies. Int. J. Comput. Sci. Eng. **2**, 28–35 (2014)
6. Jo, B., Piran, Md.J., Lee, D., Suh, D.Y.: Efficient computation offloading in mobile cloud computing for video streaming over 5G. Comput. Mater. Con. **61**(2), 439–463 (2019)
7. Wang, J.: A survey of web caching schemes for the internet. In: Proceedings of SIGCOMM, Cambridge, Massachustts, USA, pp. 36–46 (1999)
8. Anitha, R., Mukherjee, S.: Bloom filter-based framework for cache management in large cloud metadata databases. Int. J. High Perform. Comput. Networking **10**(1–2), 148–155 (2017)
9. Sadashiv, N., Dilip Kumar, S.M.: Broker-based resource management in dynamic multi-cloud environment. Int. J. High Perform. Comput. Networking **12**(1), 94–109 (2018)
10. Hasslinger, G., Ntougias, K., Hasslinger, F., et al.: Performance evaluation for new web caching strategies combining LRU with score based object selection. Comput. Netw. **125**, 172–186 (2017)
11. Podlipnig, S., Boszormenyi, L.: A survey of web cache replacement strategies. ACM Comput. Surv. **35**(4), 374–398 (2003)
12. O'Neil, E.J., O'Neil, P.E., Weikum, G.: The LRU-K page, replacement algorithm for database disk buffering. In: Proceedings of the 1993 ACM SIGMOD International Conference on Management of Data, pp. 297–306. ACM Press, New York (1993)
13. Scheuermann, P., Shim, J., Vingralek, R.: A case for delay- conscious caching of web documents. In: Proceedings of the 6th International WWW Conference, Santa Clara, April 1997, pp. 997–1005 (1997)

14. Arlitt, M.F., Cherkasova, L., Dilley, J.: Evaluating content management techniques for web proxy caches. ACM SIGMETRICS Perform. **27**(4), 3–11 (2000)
15. Cobb, J., ElAarag, H.: Web proxy cache replacement scheme based on backpropagation neural network. J. Syst. Softw. **81**(9), 1539–1558 (2008)
16. Aimtongkham, P., So-In, C., Sanguanpong, S.: A novel web caching scheme using hybrid least frequently used and support vector machine. In: International Joint Conference on Computer Science & Software Engineering. IEEE (2016)
17. Yang, Q., Zhang, H.H., Zhang, H.: Taylor series prediction: a cache replacement policy based on second-order trend analysis. In: Proceedings of the 34th Hawaii International Conference on System Science. IEEE Computer Society, Piscataway (2001)
18. Cao, P., Irani, S.: Cost-aware WWW proxy caching algorithm. In: Proceedings of the USENIX Symposium on Internet Technologies and Systems, Monterey, CA, pp. 193–206 (1997)
19. Wooster, R., Abrams, M.: Proxy caching that estimates pages load delays. In: Proceedings of the 6th International World Wide Web Conference, Santa Clara, CA, pp. 977–986 (1997)
20. Zhang, B.L., Sun, Z.H.: Dynamic cache optimization algorithm for static materialized views. J. Softw. **17**(5), 1213–1221 (2006)

Carrier-Phase Based Ranging Algorithm with Multipath Suppression in RFID System

Xiaohui Fu, Liangbo Xie$^{(\boxtimes)}$, and Mu Zhou

School of Communication and Information Engineering,
Chongqing University of Posts and Telecommunications, Chongqing, China
xielb@cqupt.edu.cn

Abstract. Due to the effect of multipath in indoor environment, the accuracy of the phase difference-based method in RFID localization system is around meter-level. In order to increase the localization accuracy, a multipath suppression algorithm is proposed. By using the property that indoor channel frequency response remains unchanged in a coherence time, the system can emulate large bandwidth through frequency hopping technique and obtain channel frequency response (CFR) corresponding to multiple frequency points. The multipath effect, which has significant impact on the accuracy of ranging, is suppressed by multipath suppression algorithm using CFR and the large bandwidth. Utilizing the phase information after the multipath suppression, a phase tolerance method is employed to solve the phase cycle ambiguity to achieve the distance estimation. Simulation results show that the proposed multipath suppression algorithm can effectively suppress the multipath effect, and the accuracy of distance estimation is less than 1.4 cm with a probability of 87% under the multipath channel model.

Keywords: Phase cycle ambiguity · Multipath suppress algorithm · Phase error tolerance · Ranging

1 Introduction

Passive RFID tags have attracted wide attention due to their small size, easy deployment and low cost. Tag-based indoor positioning technology has become a research hotspot [1, 2]. Passive RFID positioning methods are similar to traditional wireless positioning system methods, which can be divided into two categories: range-based and non-range-based [3]. In the non-ranging method, a large number of reference tags should be arranged in the undetermined area in advance, reference tags whose RSS are the most similar to that of target tag are chosen for estimating the target's position, However, the placement of reference tags greatly limits their application scenarios. The location method based on ranging usually obtains the distance information by measuring the signal strength, transmission delay, phase and other parameters [4–6]. The signal strength fluctuates with time and is greatly affected by the direction of the target tag, which leads to a large error in the ranging method based on signal strength. The distance measurement based on transceiver delay requires strict time synchronization between the transmitter and the

© Springer Nature Switzerland AG 2020
X. Sun et al. (Eds.): ICAIS 2020, LNCS 12239, pp. 489–500, 2020.
https://doi.org/10.1007/978-3-030-57884-8_43

receiver, which makes the system more complex and costly. RFID ranging based on phase has the potential of high-precision ranging. In this paper, the phase information is used as the index to achieve ranging.

The phase based ranging method needs to solve the phase cycle ambiguity, which is usually solved by the phase difference-based method, mainly divided into two types: Phase difference in frequency domain (FD-PDOA) and Phase difference in space domain (SD-PDOA) [7, 8]. FD-PDOA calculates the distance between the tag and the reader by measuring the phase at two different frequencies when the tag is at the same position. SD-PDOA estimates the direction of the tag relative to the antenna array by measuring the phase of the antenna array at the same frequency, and the intersection point of multiple directions is the tag position. These two methods make use of the slight difference of frequency or distance to keep the phase cycle number consistent and avoid solving the phase cycle number. However, the method based on phase difference has a high requirement for phase accuracy, and its positioning accuracy is in meter level in the case of multipath environment.

Literature [9] proposed a positioning algorithm based on phase difference using multi-frequency carrier ranging, which used the Chinese remainder theorem to select frequency combination and obtained a larger non-fuzzy distance. In reference [10], the multi hypothesis Kalman filter is used to fuse the phase information of the tag signal and the mileage information of the mobile robot for positioning. The experimental results show that the positioning error can reach 4 cm when there are few tags arranged. This method can't locate the static tag and needs to combine the mileage information, so the application scenarios of the tag are limited. In reference [11], a method of combining multi frequency points to solve the cycle ambiguity is proposed. The self-interference phenomenon is eliminated by custom-designed tag, and the high-precision positioning of the moving car is realized by frequency hopping. However, the existing commercial tag can't be used in the system, and the range of application is limited due to the low precision of distance estimation in the multi-path environment.

Therefore, a multipath suppression algorithm against multipath interference is proposed in this paper. Through frequency hopping, the system can expand the bandwidth, improve the resolution of multipath, and suppress the multipath of the acquired data to get the value close to the direct path phase. Then the phase cycle ambiguity is solved by the phase tolerance method. Compared with the direct phase cycle ambiguity solution, the accuracy of phase cycle ambiguity solution is higher after multipath suppression. Simulation results show that the accuracy of distance estimation is less than 1.4 cm with a probability of 87% under the multipath channel model.

The remainder of the paper is structured as follows. The technical background on RFID system is introduced in Sect. 2. The multipath suppression algorithm is detailed in Sect. 3. The explanation and evaluation of solving phase cycle ambiguity method is given in in Sect. 4. Finally, Sect. 5 concludes the paper.

2   System Overview

The RFID system built in this paper consists of a transmitter, a receiver, a tag and a computer, the clock of transmitter and receiver is synchronous. The transmitter sends a pure carrier signal, and the passive tag is activated after receiving the signal and modulates its own digital coding information to the signal. After receiving the reflection signal of the tag, the receiver demodulates it to obtain the encoding information of the tag, and finally sends it to the computer for processing to obtain the tag identity information and location information [12, 13]. The bandwidth of a single carrier signal is only tens to hundreds of kHz. The system uses frequency hopping technology to send 730 MHz–940 MHz carrier signal with 5 MHz interval in a short time to meet the needs of multipath suppression algorithm. When the carrier signal angular frequency is w, the signal at the receiver can be expressed as

$$V_B(t) = A_B S(t) \cos(w(t + \tau) + \varphi_B) \tag{1}$$

where A_B is the signal amplitude, $S(t)$ is the digital encoding information stored in the tag, φ_B is the initial phase of the signal, and τ is the Time of Flight (TOF) [14].

The receiver and transmitter keep strict clock synchronization, and the local oscillator signal is divided into two orthogonal signals after passing through the phase shifter, and they are

$$V_R(t) = A_R \cos(wt + \varphi_B) \tag{2}$$

$$V_R'(t) = -A_R \sin(wt + \varphi_B) \tag{3}$$

where A_R is the amplitude of the orthogonal signal.

The signal obtained by the receiver is multiplied by two orthogonal signals, and then I and Q signals are obtained through low-pass filtering, the expressions are as follows

$$V_I(t) = \frac{1}{2} A_B A_R S(t) \cos(w\tau) + I_{DC} \tag{4}$$

$$V_Q(t) = -\frac{1}{2} A_B A_R S(t) \sin(w\tau) + Q_{DC} \tag{5}$$

where I_{DC} and Q_{DC} are the DC components of the signal, which are mainly generated by the leakage signal, and can be directly filtered out at the receiving block. After filtering out the DC components, the sum of I and Q signals can be expressed as

$$V(t) = V_I(t) + j \times V_Q(t)$$
$$= \frac{1}{2} A_B A_R S(t) e^{-jw\tau} \tag{6}$$

However, in indoor multipath environment, the receiver signal includes a large number of multipath transmission signals in addition to line of sight transmission signals and leakage signals. The signal transmission diagram is shown in Fig. 1. The signals in these paths are superimposed on the receiver to form a mixed signal, among which only the phase information extracted from the sight distance signal is useful, while the multipath signal will affect the accurate phase extraction [15].

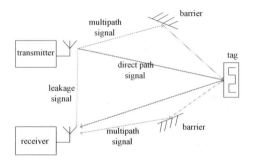

Fig. 1. Multipath signal propagation diagram of RFID system

Assuming the carrier frequency is f and there are k paths from transmitter to receiver, then the Channel Frequency Response (CFR) obtained by receiver can be expressed as

$$h_k = a_0 e^{-j2\pi f t_0} + \sum_{l=1}^{L} a_l e^{-j2\pi f t_l} \tag{7}$$

where a_0 and t_0 are the amplitude and time delay corresponding to the direct path, respectively, a_l and b_l are the amplitude and time delay corresponding to the reflection path, respectively.

After obtaining CFR information, if the phase is directly obtained from Eq. (7), the phase will contain multipath interference, and if it is directly used to solve the phase cycle ambiguity, the accuracy will be low. In this paper, CFR information was first substituted into the multipath suppression algorithm to suppress the multi-path energy and obtain the phase value close to the direct path, and then this phase was used to solve the phase cycle ambiguity to obtain the estimation result, the corresponding distance estimation algorithm flow of this system is shown in Fig. 2.

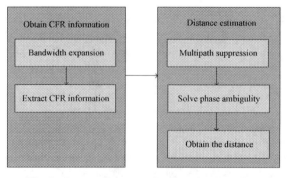

Fig. 2. System distance estimation algorithm flow

3 Multipath Suppression Algorithm

The transmitter signal will cause reflection, diffraction and scattering due to the presence of indoor shielding, resulting in multipath phenomenon. In order to reduce the interference of multipath effect on the phase of direct path and improve the accuracy of solving the phase cycle ambiguity, a multipath suppression algorithm is proposed in this section. It can improve the ratio of direct path and reflected path energy, the rough estimated distance d_0^c is obtained from CFR information through inverse Fourier transform. The distance difference between $\tilde{d}_0^c$ and the direct path is small, but the difference between $\tilde{d}_0^c$ and the reflection path is large. Using this feature, we can suppress multipath energy and give the phase after multipath suppression

$$\theta_k = \angle \sum_{i=m}^{n} h_i e^{j \frac{2\pi}{c}(f_i - f_k)\tilde{d}_0^c} \tag{8}$$

where $h_i = a_0 e^{-j\frac{2\pi}{c}f_i d_0} + \sum_{l=1}^{L} a_l e^{-j\frac{2\pi}{c}f_i d_l}$, as described in Sect. 2, the system sends a frequency hopping signal of 730 MHz–940 MHz at 5 MHz interval in a short time, we can get $1 \leq i \leq 43$ and $1 \leq m \leq n \leq 43$. Then Eq. (8) can be rewritten as

$$\theta_k = \angle a_0 e^{-j\frac{2\pi}{c}f_k d_0} \sum_{i=m}^{n} e^{j\frac{2\pi}{c}(i-k)\Delta f\left(\tilde{d}_0^c - d_0\right)} + \angle \sum_{l=1}^{L} a_l e^{-j\frac{2\pi}{c}f_k d_l} \sum_{i=m}^{n} e^{j\frac{2\pi}{c}(i-k)\Delta f\left(\tilde{d}_0^c - d_l\right)} \tag{9}$$

where $a_0 e^{-j\frac{2\pi}{c}f_k d_0}$ and $\sum_{l=1}^{L} a_l e^{-j\frac{2\pi}{c}f_k d_l}$ are the direct diameter and reflection diameter of CFR, respectively.

In order to avoid introducing a new phase in the direct path, the value of k should be $(m+n)/2$, the difference between $\tilde{d}_0^c$ and d_0 is small, and the part multiplied by the direct path is shown in Fig. 3, the accumulated items in Fig. 3 are dense and concentrated in the first and fourth quadrants, and the sum of the components projected to the x-axis is large, we can get $\sum_{i=m}^{n} e^{j\frac{2\pi}{c}(i-k)\Delta f\left(\tilde{d}_0^c - d0\right)} \approx n - m$. The difference between $\tilde{d}_0^c$ and d_l is large, and the part multiplied by the reflection diameter is shown in Fig. 4, the accumulated items in Fig. 4 are scattered and distributed in each quadrant, and the components projected to the x-axis cancel each other, so their sum is much smaller than that in Fig. 3, we can get $\sum_{i=m}^{n} e^{j\frac{2\pi}{c}(i-k)\Delta f\left(\tilde{d}_0^c - d0\right)} \ll n - m$. Therefore, we have achieved the suppression of multipath interference, and can obtain the radiation angle to get the value close to the direct path phase.

On the premise of the system frequency hopping described in Sect. 2, the larger the $n - m$ value in Eq. (8), the better the multipath suppression effect. When calculating the direct path phase corresponding to some frequencies, we always guarantee $n - m \geq 23$ to have a good multipath suppression effect.

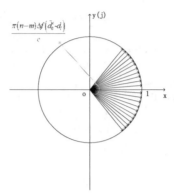

Fig. 3. Direct path coefficient

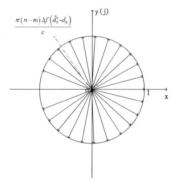

Fig. 4. Reflection path coefficient

The above only suppresses the multipath interference near the middle part of the frequency hopping signal, in order to improve the accuracy of solving the phase cycle ambiguity, this paper proposes a method to suppress the multipath interference of the frequency hopping signal at both ends, substituting the CFR information into the formula

$$\theta_m = \angle \sum_{i=m}^{n} h_i e^{j\frac{2\pi}{c}(f_i - f_m)\tilde{d}_0^c}$$

$$= \angle a_0 e^{-j\frac{2\pi}{c} f_m d_0} \sum_{i=m}^{n} e^{j\frac{2\pi}{c}(i-m)\Delta f\left(\tilde{d}_0^c - d_0\right)} + \angle \sum_{l=1}^{L} a_l e^{-j\frac{2\pi}{c} f_m d_l} \sum_{i=m}^{n} e^{j\frac{2\pi}{c}(i-m)\Delta f\left(\tilde{d}_0^c - d_l\right)}$$

(10)

Compared with Eq. (9), Eq. (10) has lost symmetry, new phase errors will be introduced into the direct path, but it can still enhance the direct path and suppress the reflection path, if the number of superpositions is the same, the phase error introduced will be the same. In the same way, we guarantee $n - m \geq 23$ to ensure a good multipath suppression effect, using Eq. (9) and Eq. (10) to calculate the phase corresponding to the same frequency, and then the introduced phase error can be calculated by making a difference between the two phases. By subtracting the phase error, the phases after multipath suppression can be obtained, which are near both ends of the frequency hopping signal. Using a larger bandwidth will improve the accuracy of solving phase cycle ambiguity.

In order to verify the performance of the multipath suppression algorithm, a multipath channel model is established to simulate the received multipath signals, we compare the phase errors before and after the multipath suppression algorithm.

The transmission channel can be regarded as time invariant in a short time, the channel impulse response can be expressed as a sequence with constant amplitude and delay

$$h(t) = \sum_{i=1}^{N} a_i \delta(t - t_i) \tag{11}$$

where a_i and t_i are respectively the amplitude and delay corresponding to the i-th path, n represents the number of paths, and the statistical characteristics of these parameters obey a certain probability distribution. The amplitude of each path is generated from Rice distribution, the amplitude of direct path is set as 2.5 times of the average value of multipath amplitude, and the arrival time interval of each path is generated from exponential distribution to get the delay of each path, and the average arrival time interval of each path is set as 10 ns [16].

500 channel models are randomly generated, the cumulative probability distribution of phase error caused by multipath and phase error after multipath suppression algorithm is shown in Fig. 5. It can be seen that the phase error after multipath suppression is smaller, the phase error corresponding to 90% probability before and after multipath suppression algorithm is 0.58 rad and 0.37 rad respectively, which indicates that multipath suppression algorithm has obvious inhibitory effect of multipath interference.

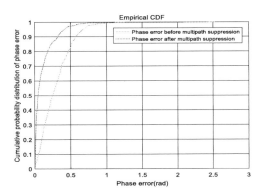

Fig. 5. Comparison of phase errors before and after the multipath suppression algorithm

The phase cycle ambiguity still exists after multipath suppression algorithm, the next section will introduce the algorithm of solving phase cycle ambiguity and compare the accuracy of solving phase cycle ambiguity before and after the multipath suppression algorithm.

4   Solving the Phase Cycle Ambiguity

4.1   Principle of Phase Tolerance Method

Solving phase ambiguity is the key step of the phase-based distance estimation method. This paper introduces the phase tolerance method [11], which is more robust than the traditional least variance method [17], and gives the relationship between the phase error and the phase ambiguity solution.

If the phase of corresponding frequency f_k obtained by the receiver is θ_k, where $\theta_k \in [0, 2\pi)$, the estimated distance can be expressed as

$$d^{(k)} = n_k \times \lambda_k + \frac{\theta_k}{2\pi} \times \lambda_k \tag{12}$$

where n_k is the number of phase cycles to be calculated corresponding to the frequency f_k, assuming that the receiver obtains four phases at different frequencies, we get the distance corresponding to the fractional part of the wavelength by $\frac{\theta_k}{2\pi} \times \lambda_k$. On the basis of this, the multiple possible distances corresponding to the frequency f_k can be obtained by adding the integral times of the wavelength, the principle of solving phase cycle ambiguity is shown in Fig. 6. The distances close to each other at four frequencies are classified into one category, the least variance method selects the category with the minimum variance and calculate the average as the result of distance estimation. The specific steps and theoretical analysis of the error tolerance method to solve phase cycle ambiguity are introduced in detail below.

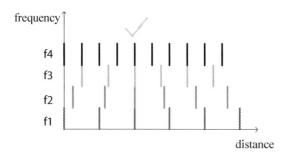

Fig. 6. Principle of solving phase cycle ambiguity

The error tolerance method takes the distance under the maximum frequency as a reference, finds the number of phase cycles under other frequencies to lead to closest to the distance, and we will get a sequence containing the number of phase cycles under all frequencies. By traversing the ranging range, we will get a number of integer sequences, in which a correct integer sequence is included.

In order to introduce parameter phase error $\Delta\theta_k$, Eq. (12) is rewritten as

$$d^{(k)} = d + \frac{\Delta\theta_k}{2\pi} \times \lambda_k \tag{13}$$

where d represents the actual distance, $\Delta\theta_k$ represents the phase error corresponding to frequency f_k. Assuming that the absolute value of phase error at all frequencies is less than $\Delta\theta_{max}$, subtract the corresponding distance at any two frequencies and substitute it into Eq. (12) and Eq. (13), we can obtain

$$
\left| d^{(k)} - d^{(l)} \right| = \left| n_k \times \lambda_k + \frac{\theta_k}{2\pi} \times \lambda_k - n_l \times \lambda_l - \frac{\theta_l}{2\pi} \times \lambda_l \right|
$$

$$
= \left| \frac{\Delta\theta_k}{2\pi} \times \lambda_k - \frac{\Delta\theta_l}{2\pi} \times \lambda_l \right| < \frac{\Delta\theta_{max}}{2\pi} \times (\lambda_k + \lambda_l) \ \forall k, l \qquad (14)
$$

It is known that the correct integer sequence satisfies Eq. (14), but the wrong integer sequence will also satisfy Eq. (14) when $\Delta\theta_{max}$ is set too large. In order to distinguish the correct cycle sequence from multiple integer sequences and make $\Delta\theta_{max}$ as large as possible, the value of $\Delta\theta_{max}$ is derived below.

If another integer sequence satisfies Eq. (14), then

$$
-\Delta\theta_{max} < \frac{2\pi \times \left((n_k + b_k) \times \lambda_k + \frac{\theta_k}{2\pi} \times \lambda_k - (n_l + b_l) \times \lambda_l - \frac{\theta_l}{2\pi} \times \lambda_l \right)}{\lambda_k + \lambda_l}
$$

$$
< \Delta\theta_{max} \quad \forall k, l, \sum_k b_k^2 > 0 \qquad (15)
$$

By making a difference between Eq. (15) and Eq. (14), we have

$$
\frac{\pi \times |b_k \times \lambda_k - b_l \times \lambda_l|}{\lambda_k + \lambda_l} < \Delta\theta_{max} \quad \forall k, l, \sum_k b_k^2 > 0 \qquad (16)
$$

If Eq. (16) deduces contradiction, then Eq. (15) does not exist, and Eq. (14) will only contain correct integer sequence. Because the distance difference between the sequences adjacent to the correct sequence is smaller than that of other error sequences, that is to say, the unique solution can be guaranteed when the adjacent sequence does not exist (enough frequency is needed to ensure that only one correct integer sequence is included in the maximum range), so we can get $b_k = 1$, $b_l = 1$, substituting it into Eq. (16) leads to

$$
\Delta\theta_{max} = \frac{\pi \times (f_{max} - f_{min})}{f_{max} + f_{min}} \qquad (17)
$$

we call it phase error tolerance.

When the phase error of all frequencies is less than $\Delta\theta_{max}$, we can get the correct integer sequence with 100% probability, it also gives the relationship between the phase error and the bandwidth. However, the phase error in the actual measurement may exceed $\Delta\theta_{max}$. In this case, we will enlarge $\Delta\theta_{max}$ until an integer sequence passes. When the correct integer sequence is solved, due to the phase error is small, it can achieve the centimeter level estimation distance. When the phase error is large, the solution of phase cycles may be wrong, and the result is mostly the sequence adjacent to the correct integer sequence.

4.2 Performance Comparison

When the same value exceeding the tolerance of phase error is added randomly, the cumulative probability distribution of distance estimation error is shown in Fig. 7. The accuracy of the error tolerance method for ambiguity resolution is 92%, while the least variance method is 78%. The error tolerance method is better than the least variance method when the phase error exceeds the phase error tolerance. The main reason is that the class with the smallest variance is not necessarily the optimal solution, while the error tolerance method provides the same error tolerance for phase at each frequency, which is more reasonable.

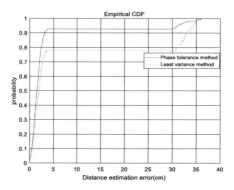

Fig. 7. Performance comparison with large phase error

The multipath suppression algorithm and the phase error tolerance method were introduced above, and we combined them to evaluate the distance estimation performance of the system in the multipath environment. The channel model proposed in Sect. 3 is used to simulate the multipath environment, and the distance estimation algorithm of the system is compared with the commonly used phase difference-based method.

The bandwidth used by the phase difference-based method is generally about 20 MHz [7], while the algorithm in this paper is based on the premise of a large bandwidth (210 MHz), so it's necessary to unwrap the phase firstly and then use the phase difference-based method to estimate the distance, the performance comparison is shown in Fig. 8. It can be seen that the traditional phase difference method has a uniform error distribution within 40 cm, the algorithm in this paper achieves centimeter accuracy with 87% probability. In other cases, it differs by one wavelength. The accuracy of distance estimation in this paper is significantly higher than that of the traditional method. The main reason is that this paper obtains the phase close to the direct path through the multipath suppression algorithm and estimates the distance using the unwrapped phase. The accuracy of solving phase cycle ambiguity is 87% after the suppression algorithm, which is 19% higher than that without multipath suppression algorithm, which indicates that the multi-path suppression algorithm can suppress multipath energy.

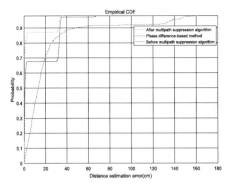

Fig. 8. Distance estimation error in multipath environment

5 Conclusion

In this paper, a multipath suppression algorithm is proposed for the indoor multipath interference. The algorithm is used to suppress the multipath energy and obtain the value close to the direct path phase. Then a phase error tolerance method with stronger robustness is used to solve phase cycle ambiguity and obtain the estimated distance. Simulation results show that the proposed multipath suppression algorithm can effectively suppress the multipath effect, and the accuracy of distance estimation is less than 1.4 cm with a probability of 87% under the multipath channel model. We believe it provides a new idea for high precision positioning.

Acknowledgements. This work was supported partly by the National Natural Science Foundation of China (No. 61704015), and General program of Chongqing Natural Science Foundation (special program for the fundamental and frontier research) (No. cstc2019jcyj-msxmX0108).

References

1. Zhang, J.: A review of passive RFID tag antenna-based sensors and systems for structural health monitoring applications. Sensors **17**(2), 265–270 (2017)
2. Chai, J.: Reference tag supported RFID tracking using robust support vector regression and Kalman filter. Adv. Eng. Inform. **32**, 1–10 (2017)
3. Ma, H.: The optimization for hyperbolic positioning of UHF passive RFID tags. IEEE Trans. Autom. Sci. Eng. **14**, 1590–1600 (2017)
4. Jiang, Y., Zhong, X., Guo, Y., Duan, M.: Communication mechanism and algorithm of composite location analysis of emergency communication based on rail. Comput. Mater. Contin. **57**(2), 321–340 (2018)
5. Wang, Y., et al.: Location privacy in device-dependent location-based services: challenges and solution. Comput. Mater. Contin. **59**(3), 983–993 (2019)
6. Khan, U.H.: Localization of compact circularly polarized RFID tag using TOA technique. Radioengineering **26**(1), 147–153 (2017)
7. Chen, H., Liu, K., Ma, C., Han, Y., Su, J.: A novel time-aware frame adjustment strategy for RFID anti-collision. Comput. Mater. Contin. **57**(2), 195–204 (2018)

8. Povala, A., Jiri Sebesta, J.: Phase difference of arrival distance estimation for RFID tags in frequency domain. In: IEEE International Conference on RFID-Technologies and Applications, RFID-TA, Sitges, Spain (2011)
9. Li, X., Zhang, Y., Amin, M.G.: Multifrequency-based range estimation of RFID tags. In: IEEE International Conference on RFID, pp. 147–154 (2009)
10. Digiampaolo, E.: Mobile robot localization using the phase of passive UHF RFID signals. IEEE Trans. Industr. Electron. **61**(1), 365–376 (2014)
11. Ma, Y., Hui, X., Kan, E.C.: 3D real-time indoor localization via broadband nonlinear backscatter in passive devices with centimeter precision. In: International Conference on Mobile Computing & Networking. ACM (2016)
12. Hricova, R., Balog, M.: Introduction of RFID system into transport and defining its model of return on investment. Int. J. Eng. Res. Afr. **18**, 130–135 (2015)
13. Kabachinski, J.: An introduction to RFID. Biomed. Instrum. Technol. **39**(2), 131 (2005)
14. Pichler, M., Schimback, E.: Indoor localization of passive UHF RFID tags based on phase-of-arrival evaluation. IEEE Trans. Microw. Theory Tech. **61**(12), 4724–4744 (2013)
15. Wang, J.: Where's my card? RFID positioning that works with multipath and non-line of sight. Comput. Commun. Rev. **43**(4), 51–62 (2013)
16. Scherhaufl, M.: UHF RFID localization based on phase evaluation of passive tag arrays. IEEE Trans. Instrum. Meas. **64**(4), 913–922 (2015)
17. Ma, Y., Selby, N., Adib, F.: Minding the billions: ultra-wideband localization for deployed RFID tags. In: Proceedings of the 23rd Annual International Conference on Mobile Computing and Networking, pp. 248–260 (2017)

Research on Vehicle Routing Problem Based on Tabu Search Algorithm

Jinhui Ge[1] and Xiaoliang Liu[2(✉)]

[1] Department of Mathematics, Tonghua Normal University, Tonghua 134000, China
[2] School of Computer Science and Technology, Hunan University of Technology and Business, Changsha 410205, China
v1zone@163.com

Abstract. Based on the description of the vehicle routing problem, an improved tabu algorithm is proposed. In the solution process, a double-layer operation is used to change the neighborhood structure; a dynamic tabu table is constructed, so that when the tabu object enters the tabu table, it is based on where The tabu length varies during the search phase; set some variable parameters, verify the effect of the parameters on the solution through simulation, and control the degree of convergence of the parameters by the parameters. Through the above improvements, the stability of the solution and the global search ability are improved.

Keywords: Double-layer operation · Dynamic tabu list · Vehicle routing problem · Tabu algorithm

1 Background of the Problem

Driven by economic globalization and informatization, the modern logistics industry has expanded from providing traditional transportation services to society to a comprehensive logistics system with modern technology, management, and information technology as the backbone, which can be attributed to social computing issues. Logistics distribution is in the logistics industry An important link directly connected to consumers, and Vehicle Routing Problem (VRP) is a key step in logistics system optimization. It was first proposed by Dantzig and Ramsen in 1959 [1], and its model was named "Truck Transportation Assignment Problem". Later known as "vehicle routing problem", has become the forefront and research hotspot in the field of operations research and combinatorial optimization. In the past ten years, experts and scholars at home and abroad have conducted a lot of theoretical research and experimental analysis on this problem and its solution Great progress has been made and some fruitful results obtained [2–10]. As far as research issues are concerned, there are theoretical research and applied research, of which the theoretical research mainly includes VRP with loading constrains, VRPLC; VRP with simultaneous pickup and delivery, VRPSPD; (dynamic vehicle routing problem, DVRP; open vehicle routing problem, HO VRP; heterogeneous multi-type fleet vehicle loading problem in finished vehicle logistics, HVLP–FVL; (multi-objective vehicle routing problem, MOVRP; in terms of solving intelligent algorithms, including

X. Sun et al. (Eds.): ICAIS 2020, LNCS 12239, pp. 501–509, 2020.
https://doi.org/10.1007/978-3-030-57884-8_44

ant colony algorithm, genetic algorithm, simulated annealing algorithm, predator search algorithm, tabu search algorithm, Particle swarm algorithm, quantum algorithm, decentralized search algorithm, etc. At present, the vehicle routing problem has been closely related to the actual situation of real life logistics distribution from theoretical research, and it is widely used in automobile transportation, maritime transportation, air transportation, pipelines Transportation and communications, power, industrial management, computer applications, etc.

In order to further study the generality of the solution algorithm, the tabu search algorithm will be improved, and compared with ant colony algorithm and genetic algorithm, more extensive research and applications will be obtained.

2   VRPS Descriptions and Mathematical Model

2.1   Problem Description

Vehicle Routing Problem, referred to as VRP, is generally described as: for a given series of customers (delivery points or pickup points), determine the appropriate vehicle driving route so that it starts from the center point, completes tasks in an orderly manner, and finally returns to the center Point, and to meet certain constraints (such as vehicle volume, customer demand, delivery restrictions, etc.) to minimize the total transportation cost (number of vehicles, driving distance).

2.2   Mathematical Model

$$x_{ijk} = \begin{cases} 1 \text{ if i can be completed by vehicle} \\ 0 \quad \text{otherwise} \end{cases} \tag{1}$$

$$\min Z = \sum_{i \in N} \sum_{j \in N} \sum_{k \in V} c_{ij} x_{ijk} \tag{2}$$

$$\sum_{k \in V} \sum_{i \in N} x_{ijk} = 1, \ \forall j \in C \tag{3}$$

$$\sum_{i \in C} g_i \sum_{j \in N} x_{ijk} \leq q, \quad \forall k \in V \tag{4}$$

$$\sum_{j \in N} x_{0jk} = 1, \quad \forall k \in V \tag{5}$$

$$\sum_{i \in N} x_{ihk} - \sum_{j \in N} x_{hjk} = 0, \quad \forall h \in C, \ k \in V \tag{6}$$

$$\sum_{i \in N} x_{iok} = 1, \quad \forall k \in V \tag{7}$$

$$x_{ijk} = 0, \quad \text{or} \quad 1, i, j \in N, k \in V \tag{8}$$

In the mathematical model c_{ij} Express from point i to the point j total traveled cost can be distance, cost, time, etc.

The objective function (1) seeks to minimize total sum of the distance or time costs. Constraints (2) ensures that each customer is visited exactly one vehicle; Constraints (3) is the vehicle capacity limit; Constraints (4)–(6) to ensure that the vehicle returns to the depot from the depot; Constraints (7) is the decision variable.

3 Tabu Search Algorithm for Vehicle Routing Problem

3.1 Basic Procedures of the Tabu Search

Tabu search [10] (Tabu search, TS) algorithm is a search algorithm that expands the neighborhood and optimizes gradually globally. It simulates the best features of human memory. Glover first proposed the idea of tabu search in 1986. It is a deterministic iterative optimization algorithm. Due to its fast search speed and strong local "climbing" ability, its main steps are:

(1) Given the algorithm parameters, an initial solution x is generated randomly, and the tabu table is left blank.

(2) Determine whether the termination conditions are met? If satisfied, end the algorithm and output the optimal result. otherwise:

(3) Use the neighborhood function of the current solution x to generate its neighborhood solution, and choose the solution y from it.

(4) Does y satisfy the contempt rule? If it meets, then use this candidate solution y instead of x to become the current solution, and put y into the tabu list, and replace the "best so far" state, go to step (2); otherwise:

(5) Determine the tabu attributes of the candidate solution, select the best state corresponding to the non-tabu object in the candidate solution set as the new current solution, and add its corresponding tabu object to the tabu table.

(6) Go to step (2).

3.2 Algorithm Design

The application of the tabu algorithm to solve the vehicle routing problem mainly includes the representation of the solution in the algorithm, the formation of the initial solution, the neighborhood structure, the tabu object, the tabu length, the candidate set, the termination criterion, and the contempt criterion, etc. The method of equal distribution forms the initial solution, the double-layer operation constructs the neighborhood, and the design of dynamic tabu lists and other work to improve the overall optimization ability.

Solution Representation

For a vehicle routing problem with L customers, the representation of this solution is to directly generate L permutations of natural numbers that are not repeated between L and 1, indicating the order of customers. Assuming n customers, 2n points are generated., 1 ~ n indicates that each customer is a real point, n + 1 ~ 2n indicates a virtual distribution center is a virtual point, and they are all considered as distribution centers. A permutation of these 2n points is a solution. For our solution It is required that the first number must be greater than n, that is, the first point must be a virtual distribution center, because the vehicle must depart from the distribution center. For 5 customers, the initial solution can be expressed as 7—> 4—>5—>6—>10—>2—>8—>9—>3—>1, the initial solution is expressed as three paths: distribution center—> customer 4—> customer 5—> distribution center; distribution center—> customer 2 —>Distribution Center; > Distribution Center—> Customer 3—> Customer 1—> Distribution Center. And meet certain restrictions, such as vehicle load limit, if the total demand of all customers is greater than the vehicle's load capacity, a car It is impossible to complete the task in one distribution. In this way, these solutions can be divided into two categories according to whether the obtained solutions meet the requirements of these factors. What is satisfied is a feasible solution, what is not satisfied is an infeasible solution, and then the distance is calculated for the feasible solution to find the shortest delivery method. The following is a specific method to determine whether it is a feasible solution and a distance calculation based on the structure of the solution.

(1) The number of vehicles is constrained. What needs to be explained here is why n virtual points are added. According to the worst case, it is intended to equip each customer with a car, so n virtual points are added, which results in the upper limit of inserted virtual points. This is the case where there is no constraint on the number of vehicles. If the number of vehicles is limited, for example, when there are n customers and m vehicles, m virtual points are inserted, so that at most m paths are generated. How many paths are there, it is necessary to use The number of vehicles, so that the number of vehicles will not be greater than the number of existing vehicles, which results in a lower limit for the inserted virtual points. The closer the number of inserted virtual points is to the upper limit, the more likely it is that the vehicle will be insufficient and the closer it will be. The lower limit, because the constraint is too tight, the more difficult it is to generate an initial solution, which needs to be determined according to the specific situation.

(2) Load capacity constraints. After the previous step, you can calculate which customers are on a car, and then calculate the total according to the customer's demand scale to see if it exceeds the vehicle's load capacity. If it exceeds, it means that this solution is not possible Solution.

(3) For distance calculation, according to the structure of the proposed solution, the distance between two adjacent points can be divided into four categories.

Dummy point-> Real point: The data can be found in the question, marking the beginning of a path.

Imaginary point-> Imaginary point: The specified distance is 0.

Real Point-> Real Point: Data can be found in the question.

Real point-> virtual point: The data can be found in the question, marking the end of a path.

According to the above rules, the distance between any two points in a solution can be calculated, and then the sum is the distance of the entire transportation plan.

Formation of Initial Solution

A good initial solution can determine the tabu search space, and it can also speed up the search speed, but the constraints on the load capacity are rigid, which means that some infeasible solutions will be generated, and the fewer the number of vehicles, the fewer feasible solutions will be. The quality depends on this feasible initial solution. Therefore, when constructing the initial solution, we must ensure the feasibility of the solution. This article uses the method of evenly distributing customers to each car under the price adjustment that meets vehicle constraints to construct the initial solution.

Neighborhood Structure

The tabu search algorithm is an algorithm based on neighborhood search technology. Determining the neighborhood operation method is an important step in constructing the algorithm, according to the needs, a double-layer operation method is designed, as shown in Fig. 1. In the inner layer, the two points in the initial solution are exchanged according to the rules, traversing all possible, collecting some better feasible solutions as the candidate set for the next neighborhood operation and tabuing the optimal solution. The 2-opt exchange is mainly used here. Two points, the outer layer, randomly select a solution in the candidate set obtained in the inner layer operation as the initial solution for the next neighborhood operation for the inner layer operation.

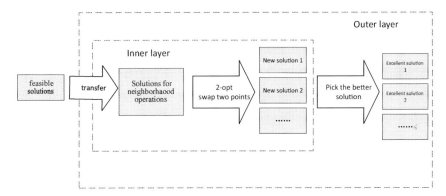

Fig. 1. Double-layer structure

The two points exchanged in the inner operation may be on the same path, or they may be on different paths, so a strategy of simultaneous optimization between and within paths is implemented. Because traversal is all possible, the solution can be made to converge as much as possible The outer layer operation is to randomly select a solution among the better solutions to enter the next cycle, which allows the solution to diverge and avoid falling into a local optimal solution. The double-layer operation will not diverge too much and make the solution unstable, It will not be too convergent and fall into a local optimum, and take overall planning into consideration.

Identification of Contraindicated Objects
The tabu list is a collection of local optimal solutions. One of the main purposes of the tabu list is to prevent loops during the search process and avoid falling into the local optimum. The second is to adjust the size of the search to diverge or converge. The tabu list is a tabu search algorithm. The core, its size greatly affects the speed of the search and the quality of the settlement. The tabu object refers to the elements in the tabu table. In this paper, the best solution obtained in each iteration is put into the tabu table as a tabu object.

Determination of Tabu Length
Tabu length refers to the number of times a tabu object stays in the tabu table during the search process. In order to better connect with reality, optimize the structure of the tabu table, and better control the convergence and divergence of the solution, this paper uses a dynamic tabu table to make each When a contraindicated object enters the contraindication list, the length of the contraindication also changes according to the stage it is in. Its size is expressed as:

$$L = L_{min} + \left[\frac{L_{max} - I}{n} \right] \tag{7}$$

L_{min} Minimum value of tabu length, L_{max} The maximum number of times the optimal solution is maintained, I Represents the number of times the current optimal solution remains unchanged, n Is an arbitrary integer, intended to adjust the magnitude of the tabu length change, such as n = 10 The time indicates that if the optimal solution remains unchanged, the tabu list length is reduced every 10 searches.

Determining the Candidate Set
The set of better solutions obtained after each iteration is used as a candidate set for the next iteration, and a random one is selected for the next iteration. Here it is necessary to explain "better" and arrange the obtained solutions from good to bad. The first few can be taken as the better solution. The more solutions taken, the more divergent, the less solutions taken, the more convergent. The specific amount can be determined according to the size of the data.

Determination of Termination Criteria
When the total number of iterations reaches a given value, or the current best solution does not change within a given number of consecutive iteration steps, the algorithm terminates. The number of steps given in this paper is L_{max}.

Contempt Criterion
In the tabu search, it may happen that all candidate solutions are tabu. The contempt criterion can lift a state to achieve more efficient and optimized performance. This article uses the contempt criterion based on the adaptation value. If a candidate solution is contraindicated, but its target value is better than "best so far", the candidate solution is unbanned and the "best so far" is updated.

4 Experimental Calculation and Result Analysis

4.1 The Simulation Results

In order to compare and use the data in [6] for experiments, the vehicle routing problem of the logistics distribution system with 8 stores, 1 distribution center, and 2 vehicles (each with a load capacity of 8 tons), between the store and the store The specific data of the distance are shown in Table 1.

Table 1. Characteristics of tasks

Shop	Distance								
	0	1	2	3	4	5	6	7	8
0	0	4	6	7.5	9	20	10	16	8
1	4	0	6.5	4	10	5	7.5	11	10
2	6	6.5	0	7.5	10	10	7..5	7.5	7.5
3	7.5	4	7.5	0	10	5	7	9	15
4	9	10	10	10	0	10	7.5	7.5	10
5	20	5	10	5	10	0	7	9	7.5
6	10	7.5	7.5	9	7.5	7	0	7	10
7	16	11	7.5	9	7.5	9	7	0	10
8	8	10	7.5	15	10	7.5	10	10	0
9	1	2	1	2		1	4	2	2

4.2 Results and Discussions

For comparison, the parameters of the simulation algorithm are set as $L_{min} = 3, L_{max} = 50, n = 10$ and use the mathematical software mathematica to program and solve. The minimum value obtained by 10 experiments is 69, and the corresponding paths are $0 \rightarrow 8 \rightarrow 4 \rightarrow 7 \rightarrow 2 \rightarrow 0$ and $0 \rightarrow 6 \rightarrow 5 \rightarrow 3 \rightarrow 1 \rightarrow 0$.

Table 2. Comparison of optimization results

Algorithm	DPGA [6]	PGA [7]	ICOGA [8]	IPSO [9]	ITSA
The average	69.575	69.375	69.6	68.425	69

- DPGA: Double population genetic algorithm
- PGA: Partheno-genetic algorithm
- ICOGA: Improved crossover operator genetic algorithm
- IPSO: Improved PSO
- ITSA: Improved tabu search algorithm

From the results of the simulation experiments in Table 2, it can be seen that the improved tabu search method in this paper is very close to the hybrid pso algorithm, and is superior to the two-population genetic algorithm, the one-parent genetic algorithm, and the improved cross-operator algorithm.

Using the tabu search algorithm designed in this paper to solve the vehicle routing problem, not only can obtain a high-quality solution, but also the algorithm has a faster convergence speed, higher calculation efficiency, and more stable calculation results, showing a good optimization performance, indicating the algorithm If it is possible to design a more reasonable dynamic structure and apply this problem to logistics practice\solve the last mile problem of logistics distribution, just like city distribution and cold fresh distribution, the quality of the solution will be further improved. The combinatorial optimization problems and social computing problems have certain theoretical significance and application value.

References

1. Dantzig, G., Ramser, J.: The truck dispatching problem. Manag. Sci. **6**, 80–91 (1959)
2. Chao, I.M.: A tabu search method for the truck and trailer routing problem. J. Comput. Oper. Res. **29**(1), 33–51 (2002)
3. Scheuerer, S.: A tabu search heuristic for the truck and trailer routing problem. Comput. Oper. Res. **33**(4), 894–909 (2006)
4. Brandao, J.: A tabu search algorithm for the open vehicle routing problem. Eur. J. Oper. **157**(3), 552–564 (2004)
5. Li, J., Binglei, X., Guo, Y.: Genetic algorithm for vehicle scheduling problem with non-full load. Syst. Eng. Theory Method. Appl. **93**, 235–239 (2000)
6. Zhao, Y., Wu, B., Jiang, L., et al.: Double populations genetic algorithm for vehicle routing problem. Comput. Integr. Manuf. Syst. **103**, 303–306 (2004). (in Chinese)
7. Xiao, P., Li, M., Zhang, J.: Partheno-genetic algorithmfor vehicle routing problem. Comput. Technol. Autom. **19**(1), 26–30 (2000). (in Chinese)
8. Zhang, L., Chai, Y.: Improved genetic algorithm for vehicle routing problem. Syst. Eng. Theory Pract. **22**(8), 79–84 (2002). (in Chinese)
9. Luo, X., Shi, H.-B.: Improved particle swarm optimization for vehicle routing problem with non-full load. J. East China Univ. Sci. Technol. Nat. Sci. Ed. **32**(7), 767 (2006)
10. Baker, B.M., Ayechew, M.A.: A genetic algorithm for the vehicle routing problem. Comput. Oper. Res. **30**(5), 787–8001 (2003)
11. Jiang, D., Yang, S., Du, W.: A study on the genetic algorithm for vehicle routing problem. Syst. Eng. Theory Pract. **196**, 44–45 (1999). in Chinese
12. Xian-sheng, L., Hua, Z., Fei, L., Naixiu, G., Lu, Y.: City delivery vehicle dispatching model and its algorithm. J. Jilin Univ. (Eng. Technol. Ed.) **36**(4), 618–621 (2006)
13. Zhang, X.-N., Fan, H.-M.: Hybrid scatter search algorithm for capacitated vehicle routing problem. Control Decis. **13**, 1937–1944 (2015)
14. Pang, Y., Luo, H., Xing, L., Ren, T.: A survey of vehicle routing optimization problems and solution methods. Control Theory Appl. **36**(10), 1574–1582 (2019)
15. Zhang, C., Zhao, Y., Zhang, J., et al.: Location and routing problem with minimizing carbon. Comput. Integr. Manuf. Syst. **23**(12), 2768–2777 (2017)
16. Chen, Y., Shan, M., Wang, Q.: Research on heterogeneous fixed fleet vehicle routing problem with pick-up and delivering. J. Cent. S. Univ. (Sci. Technol.) **46**(5), 1938–1945 (2015)

17. Sun, L., Ge, C., Huang, X., Wu, Y., Gao, Y.: Differentially private real-time streaming data publication based on sliding window under exponential decay. Comput. Mater. Continua **58**(1), 61–78 (2019)
18. Jiang, W., et al.: A new time-aware collaborative filtering intelligent recommendation system. Comput. Mater. Continua **61**(2), 849–859 (2019)
19. Liu, Y., Yang, Z., Yan, X., Liu, G., Hu, B.: A novel multi-hop algorithm for wireless network with unevenly distributed nodes. Comput. Mater. Continua **58**(1), 79–100 (2019)

Unsupervised Situational Assessment for Power Grid Voltage Stability Monitoring Based on Siamese Autoencoder and k-Means Clustering

Xiwei Bai[1,2]([⊠]) and Jie Tan[2]

[1] School of Artificial Intelligence, University of Chinese Academy of Sciences, Beijing, China
baixiwei2015@ia.ac.cn
[2] Institute of Automation, Chinese Academy of Sciences, Beijing, China

Abstract. Accurate situational assessment and severity rating are of great importance to the voltage stability of power grid. Traditional approaches depend heavily on the network parameters and component models, which restrict their applications. In this paper, an unsupervised situational assessment scheme is proposed to achieve a voltage stability margin-based, three-class situation categorization via the knowledge-aided siamese autoencoder and k-Means clustering. The distribution characteristic of voltage stability margin is utilized to provide support for searching optimal feature subspace that enables k-Means to minimize intra-class and maximize inter-class differences through the siamese architecture. Experiments on IEEE-39 system prove that the proposed scheme outperforms classical approaches in multiple indicators, which proves it a useful situational assessment tool for power grid voltage stability monitoring.

Keywords: Situational assessment · Voltage stability · Siamese autoencoder · k-Means

1 Introduction

A smart grid should be capable of judging its overall status and forecast potential risks via the real-time comprehensive situational assessment (SA) ability. The top level task for modern power grid is the safety operation and prevention of voltage instability, which could be caused by equipment failures, incorrect operation and load increment, etc. Therefore, the so-called overall status can be determined as voltage stability margin (VSM), which is represented as the active power distance between the current operating point (OP) and the voltage collapse point (VCP) [1]. VSM is a critical indicator for voltage stability monitoring of power grid. Under normal circumstances, it should be positive and steady. Any disturbance that leads to negative VSM may incur large-scale blackouts with heavy economic losses.

To obtain VSM, the power flow distribution of VCP should be determined [2]. Traditional methods are model-based, which require the precise dynamic characteristics

X. Sun et al. (Eds.): ICAIS 2020, LNCS 12239, pp. 510–521, 2020.
https://doi.org/10.1007/978-3-030-57884-8_45

of components like generators and loads. The most frequently-used continuous power flow (CPF) method [3] approaches VCP by increasing load demands gradually under a given increment direction. The computational complexity of CPF is high due to the iteration operation. To solve this problem, supervised data-driven methods [4, 5] build machine learning models to estimate VSM based on the simulation results of traditional methods. However, both two types neglect that the accurate modelling of components and parameter identification of buses and branches are often hard to realize.

In recent years, with the development of wide-area measurement system (WAMS), synchronous data acquisition can be achieved by phasor measurement units (PMU) deployed on the major pivot points of power grid [6]. The massive volume of PMU data contains valuable information that reflects the variation of VSM. Without system modelling, unsupervised situational assessment (uSA) method utilizes the historical and real-time PMU data to achieve correlation mining and knowledge discovery for voltage stability monitoring. Generally speaking, the two major implementation types of uSA are dimensionality reduction [7] and clustering [8]. The former achieves feature extraction through unsupervised metric learning techniques including linear methods typified by principal component analysis (PCA), locality preserving projection (LPP) and non-linear methods such as kernel principal component analysis (KPCA) and Laplacian eigenmaps (LE), etc. The latter achieves data categorization through clustering techniques [9] including partitioning, hierarchical and density-based methods. In short, it is expected that uSA can distill relevant information about VSM and classify the stability situation into several stages according to its severity. However, the obtained information may not be acceptable due to the lack of instruction, thus the intervention of external knowledge is needed. The accumulated experiences and rules of expert and operators can bridge the gap between the historical data and their corresponding VSM. Although the direct connections may not be established, the distribution characteristics of VSM are embedded in these knowledge. In this paper, an initial label that reflects the severity of situation will be extracted and given to each acquired data samples according to the VSM distribution characteristics.

To achieve effective feature extraction with consideration of the initial labels, namely minimize the distance among samples with same labels and maximize the distance among samples with different labels, the autoencoder (AE) [10] is introduced. AE is a special neural network that learns a compressed representation to reconstruct the original data. Different from above methods, the flexible structure of AE enables multiple input/output and the learned feature can be controlled by customized loss function. Unfortunately, traditional AE uses only reconstruction loss, thus the training process is totally autonomous. In this paper, a siamese autoencoder (SAE) is proposed to solve this problem. SAE integrates the structure of AE and siamese network (SN) [11] and utilizes a weighted reconstruction-contrastive loss to learn a compressed representation. Next, the partitioning-based k-Means clustering algorithm [12, 13] is selected to achieve data classification for assessment of the severity of voltage stability situation.

The rest part of this paper is organized as follow. Section 2 introduces the content of situational assessment for power grid voltage stability. Section 3 sketches the concepts of AE and SN. Section 4 illustrates the proposed SAE and uSA scheme in details. In

Sect. 5, the IEEE-39 system is used to evaluate the performance of proposed method. Section 6 concludes the whole paper.

2 Problem Statement

The uSA aims to monitor the variation of VSM for power grid using only PMU data and experimental knowledge. It comes down to a clustering problem which minimizes the deviation of VSM in each cluster. For a historical data set $X = [x_1, x_2, \ldots, x_n] \in \mathbb{R}^{N \times n}$ and the corresponding set of VSM $Y = [y] \in \mathbb{R}^{N \times 1}$, uSA divides X into m subsets that obtains the minimal intra-class deviation, where N is the sum of available historical PMU data and n is the number of monitored variables. Let $X = \{X_1, X_2, \ldots, X_m\}$ and $Y = \{Y_1, Y_2, \ldots, Y_m\}$ be the union of variable and VSM subsets, the objectives of uSA are:

$$\min J_{\text{intra}} = \frac{1}{m} \sum_{i=1}^{m} \sigma(Y_i) \tag{1}$$

$$\max J_{\text{inter}} = \frac{2}{(m-1)m} \sum_{i=1}^{m-1} \sum_{j=i+1}^{m} \left| \bar{Y}_i - \bar{Y}_j \right| \tag{2}$$

where $\bar{Y}_i$ and $\sigma(Y_i)$ refer to the mean and standard deviation of the ith and j^{th} VSM subset. J_{intra} and J_{inter} are designed to evaluate the intra/inter-class differences. The main objective is minimizing J_{intra} to increase clustering accuracy. On this basis, larger J_{inter} signifies more representative clusters. New sample can be classified into m categories, each represents a type of severity. It is worth noting that Y is actually unknown, thus it is only used to evaluate the performance of uSA. The VSM is defined as the active power difference ΔP between OP and VCP:

$$\Delta P = P_{\text{max}} - P_0 \tag{3}$$

where P_{max} and P_0 are the maximal and current total active power. To achieve the aforementioned objective, a feature subspace $S = [s_1, s_2, \ldots, s_p] \in \mathbb{R}^{N \times p}$ is learned to remove useless information and enhance the performance of clustering.

In this paper, the SAE is proposed to obtain the best S and the k-Means algorithm is employed to achieve m subset division, namely $S = \{S_1, S_2, \ldots, S_m\}$.

3 Preliminaries

3.1 Autoencoder

AE (Fig. 1) is a neural network-based compression model which aims to learn a latent representation of data. Network input x is reconstructed at output $\hat{x}$ through a hidden layer with compressed data h. AE is mainly comprised of two symmetric-structured encoder and decoder which implement transformation $x \rightarrow h$ and $h \rightarrow \hat{x}$ respectively. Both

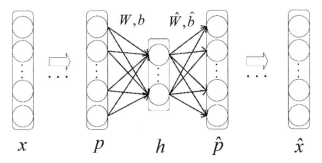

Fig. 1. The architecture of autoencoder

encoder and decoder can have single or multiple layers. For single-layered structure, namely $x = p$ and $\hat{x} = \hat{p}$ in Fig. 1, the encoding and decoding process is described as:

$$h = \sigma_{encoder}(Wp + b) \tag{4}$$

$$\hat{p} = \sigma_{decoder}\left(\hat{W}h + \hat{b}\right) \tag{5}$$

where W, b, $\hat{W}$, and $\hat{b}$ are the weights and biases, $\sigma_{encoder}(\bullet)$ and $\sigma_{decoder}(\bullet)$ are the corresponding activation functions. AE aims to minimize the reconstruction loss:

$$\min L_R\left(W, b, \hat{W}, \hat{b}\right) = \|p - \hat{p}\|^2 \tag{6}$$

For multiple-layered structure, x is transformed to p through several fully-connected layers. Additional weights and biases are added to the loss function and trained together. In this situation, the architecture is called deep autoencoder (DAE).

3.2 Siamese Network

SN learns a metric to judge whether two samples are similar. The architecture is shown in Fig. 2. Given a sample pair x_1, x_2 and their binary labels c, a SN model M calculates the corresponding output tensors $M(x_1)$ and $M(x_2)$ simultaneously. A distance measurement d (usually Euclidian distance) is employed to measure the difference between two output tensors. If the two samples are from same category ($c = 1$), d should approach zero, otherwise ($c = 0$) should approach a predetermined non-zero value t. For binary labels, $t = 1$. To meet the above requirement, the following contrastive loss is used to obtain the optimal d:

$$\min L_C(W, c, d) = \frac{1}{2}\left[cd^2 + (1 - c)\max(0, t - d)^2\right] \tag{7}$$

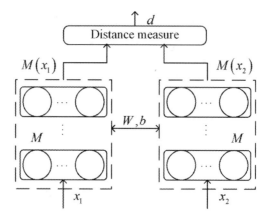

Fig. 2. The architecture of siamese network

4   Unsupervised Situational Assessment

The framework of the proposed uSA scheme is shown in Fig. 3. It consists of three main parts: density-based labelling, feature learning via SAE and subset division via k-Means.

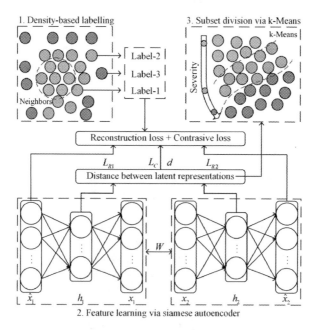

Fig. 3. The framework of uSA

4.1 Density-Based Labelling

The first part aims to create an initial label for each sample according to the experiences of operator. In reality, the power grid is stable in most cases and the number of cases decreases with the deterioration of the situational severity. On the basis of this principle, the local density of samples is relevant with VSM, namely the severity indicator. In this paper, the local density is defined as the number of neighbors within a determined radius r and the density-based labelling process divides the whole sample set into m subsets according to the local densities and a set of partition value B. The whole labelling process can be finished in three steps:

1. $D \leftarrow$ Euclidian distance between each sample pair in X
2. $N_c \leftarrow$ number of neighbors within r for each row in D
3. Divide X into m subsets according to B and N_c

The symbol D refers the distance matrix and N_c refers to the vector of neighbor number. The input contains historical dataset X, radius r, number of subset m, boundary value set B and the output is the union of subset $\{X_1, X_2, \ldots, X_m\}$.

4.2 Feature Learning via Siamese Autoencoder

The second part aims to learn the latent representations of data guided by their initial labels. In Fig. 3, the SAE is a combination of two AEs with shared weights and different inputs x_1 and x_2. Naturally, two AEs give two latent representations h_1 and h_2 in their hidden layers. Next, same as SN, the Euclidian distance d between h_1 and h_2 is calculated and compared with the binary label c, which is determined by the initial labels of the input sample pair. The loss function of SAE consists of three parts: two reconstruction losses L_{R1}, L_{R2} and a contrastive loss L_C.

So what is the difference between traditional AE and SAE? Actually, SAE is a supervised learning approach with self-labelled data. The result of AE is uncontrollable, depending on the network architecture, yet SAE utilizes the initial labels to restrict the latent representations. Those with identical labels would have shorter distance and vice versa. Due to the inaccuracy of initial labels, SAE gives three loss functions different weights, namely:

$$\min L_{SAE} = \alpha L_{R1} + \alpha L_{R2} + \beta L_C \tag{8}$$

Input reconstruction is generally the primary task thus L_{R1} and L_{R2} will be given identical large weight α. The contrastive loss is given relative small weight β to provide appropriate corrections for the original representations.

4.3 Subset Division via k-Means

The third part aims to achieve subset division using clustering approach. After feature learning, the feature subspace S is determined and samples can be easily partitioned using k-Means. The k-Means algorithm is widely applicable owing to its efficiency and accuracy. In this paper, S is divided into m subsets (clusters), which corresponds with the labelling procedure. The cluster centers will be stored for online applications. Newly acquired samples will be assigned to the cluster which has the shortest distance between the cluster center and themselves.

5 Experiments

The IEEE-39 power system (Fig. 4) is used to evaluate the performance of uSA. There are total 39 buses (nodes) and 46 branches in this system. The symbol "G" stands for generator and the triangle symbol stands for the electrical load. IEEE-39 contains 10 valid generator node and 17 load nodes in total except node 31 and 39.

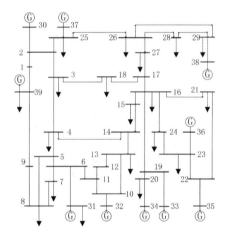

Fig. 4. The IEEE-39 system

5.1 Experimental Data

To simulate real conditions, the active/reactive power of all loads and the active power of all generators varies between ±50% of their base values. Let P_{L-i}, Q_{L-i} and P_{G-j} be the random power consumption of ith load nodes and random power generation of jth generator nodes, P^b_{L-i}, Q^b_{L-i} and P^b_{G-j} be their base value, the following equations hold:

$$P_{L-i} = P^b_{L-i}(0.5 + \varepsilon_{P_{L-i}}) \tag{9}$$

$$Q_{L-i} = Q^b_{L-i}(0.5 + \varepsilon_{Q_{L-i}}) \tag{10}$$

$$P_{G-j} = P^b_{G-j}(0.5 + \varepsilon_{P_{G-j}}) \tag{11}$$

where $\varepsilon_{P_{L-i}}$, $\varepsilon_{Q_{L-i}}$ and $\varepsilon_{P_{G-j}}$ are uniform distributed random variables in the range of 0 and 1.

For each case, the real and imaginary parts of voltage phasor of all nodes forms the input vector. The VSM is calculated using MATPOWER CPF tool [14]. Total 20000 cases are generated and divided into two parts. One with 15000 cases for training, the other 5000 cases are for testing.

5.2 Experimental Settings

For voltage stability monitoring, it is reasonable to qualitatively divide the situation into three level: "normal", "alert" and "serious" according the value of VSM. Therefore in the experiment, a three-subset division problem ($m = 3$) is considered. The partition value set $B = \{0.1, 0.4, 1\}$, namely 10% samples are "serious", 30% are "alert" and the rest 60% are "normal". The radius r is set to 0.5 after all samples are normalized to the range between 0 and 1. The number of subsets and B can be adjusted as needed.

For precision and visualization requirements, the architecture of basic AE in SAE contains five fully-connected layers, which have 40-20-3-20-40 neurons respectively. The activation function and optimizer of network are set to "tanh" and "Adam". The ratio between α and β is initially set to 10:3.

Before training, a random pairwise matching procedure should be implemented for samples in X because the proposed SAE only accepts sample pairs as input.

5.3 Experimental Results and Comparative Analysis

In this section, the PCA and DAE is introduced for performance comparison. The structure of DAE is identical with the basic AEs in SAE. Results of subset division through k-Means is shown in Fig. 5.

The level of severity is depicted with different color: "red" for "serious", "green" for "alert" and "blue" for normal. The left half represents the ground truth (GT), which is quantified by VSM according to B. The right half represents the clustering results. The 3D scatter plots of PCA and DAE reveal a rule that the value of VSM reduces from inside to outside. It is not beneficial for clustering algorithm because the boundaries between clusters are indistinct, thus k-Means cut the whole dataset into three parts directly in Fig. 5(a, b). The results are impractical because there is no way to know that samples on the left and right edges are from the same category. Therefore, the obtained low-dimensional features cannot well describe the actual VSM-based distance among samples. The above problem is solved by SAE. The initial label given to each sample contains empirical knowledge that indicates their VSM-based distance. For example, the "serious" samples localize mostly in the marginal area with low local density, hence they are given same labels regardless of their Euclidian distance. SAE uses these knowledge, reduces the distance between samples of identical category and increase the distance between samples from different category via the SN architecture. In Fig. 5(c), SAE bends the shape of PCA/DAE plot and make it possible to cluster similar samples with large distance.

The performance comparison is shown in Table 1. The internal indices including the mean value and standard deviation of the corresponding VSM is employed to evaluate the model performance due to the subjectivity of GT. DAE and SAE are trained under same condition.

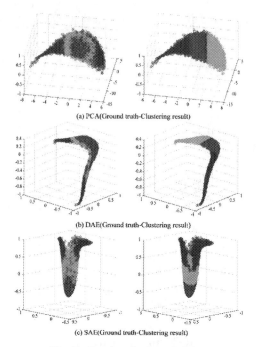

(a) PCA(Ground truth-Clustering result)

(b) DAE(Ground truth-Clustering result)

(c) SAE(Ground truth-Clustering result)

Fig. 5. Results of subset division

Table 1. Performance comparison on training set

Index	Standard deviation				Mean value			
Subset	S1	S2	S3	J_{intra}	S1	S2	S3	J_{inter}
PCA	9.99	9.37	7.41	8.92	30.57	40.99	51.58	14.00
DAE	12.40	9.69	7.38	9.82	39.51	42.46	53.56	9.37
SAE	**8.80**	**5.91**	**6.83**	**7.19**	**28.15**	**41.44**	**52.30**	**16.09**
GT	6.40	3.94	6.08	5.47	21.36	37.22	52.35	20.66

As aforementioned, the main objective of uSA is to minimize the J_{intra} and maximize the J_{inter} of VSM. Due to the lack of external knowledge, the results of PCA and DAE are far apart from GT. DAE cannot well separate S1 and S2 because of the similar mean value, which explains the reason of their low J_{inter}. PCA obtains relative better performance in this respect. The proposed SAE surpasses the above two approaches in both two overall indices and a large proportion of local indices. The reductions of J_{intra} are 19.39% and 26.78% compared with PCA and DAE.

The dimension of latent representation influences the reconstruction accuracy of AE directly. Therefore, it is meaningful to explore the relationship between it and the VSM-based performance indices. In Fig. 6, SAEs with 3 to 15 neurons in their hidden layers are tested on IEEE-39 system under same condition. The plot describes the variations of

J_{intra} and J_{inter}. Unfortunately, the performance of SAE does not rise with the increase of hidden neurons. The improvement of one index often comes at the expense of the other. By comparison, the latent representation with four and nine dimensions shows relatively better effect but not prominent, thus three dimension is appropriate for both accuracy and visualization concerns.

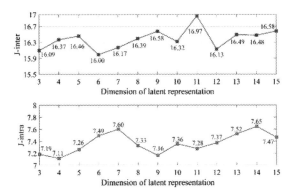

Fig. 6. Influence of the latent representation

The ratio between contrastive loss and reconstruction loss also influences the result of categorization. Experiments are conducted on SAE with loss ratio β/α from 0.1 to 1. Due to the randomness in training neural networks and the precision of calculation, results here are slight different from the above. In Fig. 7, the ratio 0.3 is selected because it reaches the minimal value of the main objective. After it, J_{intra} increases continuously while J_{inter} remains basically invariant.

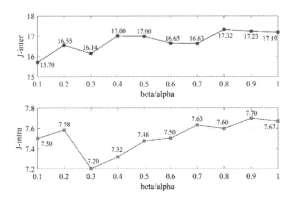

Fig. 7. Influence of the loss ratio

At last, the performance comparison on testing set among three approaches are shown in Table 2. Samples are assigned to the nearest cluster center. To clarify, the number of samples are unevenly distributed in the three clusters of PCA, which causes large J_{inter}. For the main objective J_{intra}, the proposed SAE obtains the best performance. The reductions of J_{intra} are 4.97% and 28.05% compared with PCA and DAE.

Table 2. Performance comparison on testing set

Index	Standard deviation				Mean value			
Subset	S1	S2	S3	J_{intra}	S1	S2	S3	J_{inter}
PCA	4.71	7.75	9.89	7.45	14.58	33.40	47.11	21.69
DAE	12.18	9.86	7.48	9.84	39.83	42.60	53.34	9.00
SAE	**8.51**	**5.93**	**6.80**	**7.08**	**28.44**	**41.28**	**52.32**	**15.92**

6   Conclusions

In summary, a uSA scheme is proposed in this paper to meet the requirements of voltage stability monitoring. Without system modelling and parameter identification, uSA solves a multi-level severity categorization problem through a knowledge-aided SAE model and k-Means clustering algorithm. The relevance between local density and VSM is utilized to learn a VSM-based distance metric for clustering. Experiments are conducted on the IEEE-39 system. The most appropriate loss ratio and number of neurons in the hidden layer are explored and compared. Results show that the proposed SAE reduces 19.39% and 26.78% of J_{intra} for the training set and 4.97% and 28.05% for the testing set compared with PCA and DAE, which proves that the uSA scheme can achieve effective situational assessment by severity categorization. For practical applications, uSA can monitor the variation of voltage stability and assist operators in the decision-making process.

Acknowledgements. This work was supported by the National Natural Science Foundation of China under Grant U1801263, U1701262 and the National Key Research and Development Program under Grant 2018YFB1703400.

References

1. Zhou, D.Q., Annakkage, U.D., Rajapakse, A.D.: Online monitoring of voltage stability margin using an artificial neural network. IEEE Trans. Power Syst. **25**(3), 1566–1574 (2010)
2. Semlyen, A., Gao, B., Janischewskyj, W.: Calculation of the extreme loading condition of a power system for the assessment of voltage stability. IEEE Trans. Power Syst. **6**(2), 307–315 (1991)
3. Ajjarapu, V., Christy, C.: The continuation power flow: a tool for steady state voltage stability analysis. IEEE Trans. Power Syst. **7**(1), 416–423 (1992)
4. Suganyadevi, M.V., Babulal, C.K.: Support vector regression model for the prediction of loadability margin of a power system. Appl. Soft Comput. J. **24**, 304–315 (2014)

5. Shakerighadi, B., Aminifar, F., Afsharnia, S.: Power systems wide-area voltage stability assessment considering dissimilar load variations and credible contingencies. J. Modern Power Syst. Clean Energy **7**(1), 78–87 (2019)

6. De La Ree, J., Centeno, V., Thorp, J.S., Phadke, A.G.: Synchronized phasor measurement applications in power systems. IEEE Trans. Smart Grid **1**(1), 20–27 (2010)

7. Mohammadi, H., Dehghani, M.: PMU based voltage security assessment of power systems exploiting principal component analysis and decision trees. Int. J. Electr. Power Energy Syst. **64**, 655–663 (2015)

8. Wang, X.Z., Yan, Z., Ruan, Q.-T., Wang, W.: A kernel-based clustering approach to finding communities in multi-machine power systems. Eur. Trans. Electr. Power **19**(8), 1131–1139 (2009)

9. Yang, K., Tan, T., Zhang, W.: An evidence combination method based on DBSCAN clustering. Comput. Mater. Continua **57**(1), 269–281 (2018)

10. Hinton, G.E., Salakhutdinov, R.R.: Reducing the dimensionality of data with neural networks. Science **313**(5786), 504–507 (2006)

11. Chopra, S., Hadsell, R., LeCun, Y.: Learning a similarity metric discriminatively, with application to face verification. In: 2005 IEEE Computer Society Conference on Computer Vision and Pattern Recognition, CVPR 2005, San Diego, CA, United states, 20 June 2005–25 June 2005, vol. I, pp. 539–546: IEEE Computer Society

12. Maamar, A, Benahmed, K.: A hybrid model for anomalies detection in AMI system combining K-means clustering and deep neural network. Comput. Mater. Continua **60**(1), 15–39 (2019)

13. Ling, T., Chong, L., Jingming, X., Jun, C.: Application of self-organizing feature map neural network based on K-means clustering in network intrusion detection. Comput. Mater. Continua **61**(1), 275–288 (2019)

14. Zimmerman, R.D., Murllo-Sanchez, C.E., Thomas, R.J.: MATPOWER: steady-state operations, planning, and analysis tools for power systems research and education. IEEE Trans. Power Syst. **26**(1), 12–19 (2011)

The Mathematical Applications of Majorization Inequalities to Quantum Mechanics

Mengke Xu[1], Zhihao Liu[1,2], Hanwu Chen[1,2(✉)], and Sihao Zheng[2]

[1] School of Cyber Science and Engineering, Southeast University, Nanjing 211189, China
xmk0000@126.com, hw_chen@seu.edu.cn
[2] School of Computer Science and Engineering, Southeast University, Nanjing 211189, China

Abstract. We supplement some foundational theorems which connect majorization inequalities with quantum mechanics. To begin, we give the sufficient condition for bi-partite pure states entanglement transformation from the control to the controlled states which are impossible to be transformed even by catalysts. Furthermore, a sufficient and necessary condition about the existence of an entangled assisted state is discussed. Also, the sufficient and necessary condition to give rise to the same given ensemble for a group of vectors is demonstrated. At last, owing to the unique distinguishability of orthogonal states, it is natural to establish local distinguishability of two bi-partite orthogonal quantum states by using ensembles of majorization in LOCC.

Keywords: Majorization inequalities · Entanglement transformation · Quantum mechanics · Local distinguishability in LOCC

1 Introduction

Majorization inequality is a significant and flexible mathematical tool to measure the disorder of two vectors. The study of various properties of majorization inequalities can be used to derive a wide variety of geometric inequalities, analysis matrix theory and numerical analysis [11]. Nowadays the applications of majorization inequalities are towards in inter-discipline such as quantum mechanics and von Neumann entropy [13].

Especially, quantum entanglement plays an important role and it is an essential resource in quantum mechanics [2, 7, 9, 16]. Owing to the contributions which quantum entanglement made to quantum teleportation and quantum cryptography [1, 5, 18, 20], how to quantify entanglement transformation is necessary to consider. Unfortunately, in practice, it becomes very difficult to preserve the bi-partite pure entanglement as the two systems may interact with other systems. Consequently, a very natural question is, how many bi-partite pure entangled states can be distilled from multipartite states by means of local operations and classical communication (LOCC)? 2018 Li gave a sufficient and necessary condition to determine whether a given tripartite pure state could be transformed to the bi-partite entangled state under stochastic local operations classical communication (SLOCC) [10]. Hence, we only focus on the bi-partite pure states entanglement transformation. Nielsen found an interesting connection which was so concise

© Springer Nature Switzerland AG 2020
X. Sun et al. (Eds.): ICAIS 2020, LNCS 12239, pp. 522–531, 2020.
https://doi.org/10.1007/978-3-030-57884-8_46

yet rather surprising between entanglement transformation and majorization inequalities for two bi-partite entangled pure states in LOCC in 1999 [12]. However there exist some entangled states which cannot transform one controlled state into another control state when they are not satisfied with the Nielsen theorem. On the one hand, to make the impossible entanglement transformation to possible, Jonathan and Plenio discovered that the entanglement catalyst had the ability to achieve entanglement transformation [8]. Furthermore, a necessary and sufficient condition about existence of a 2×2 catalyst was found by Sun xiaoming and Duan runyao [17]. On the other hand, Feng yuan proposed the catalyst-assisted probabilistic entanglement transformation [6].

According to the proposition 12.11 and theorem 12.13 [14], majorization inequalities closely contacted with doubly stochastic matrices and unitary matrices. Therefore some profound theorems in quantum mechanics were described by majorization inequalities. For example, Nielsen derived that matrix majorization could be characterized in terms of quantum operations [13]. In addition, Partovi discussed the majorization formulation of uncertainty [15].

Besides quantum uncertainty principle, it is crucial to analyze the distinguishability of states. Jonathan provided that any two orthogonal quantum states shared between bi-partite were perfectly distinguished by LOCC [19]. Although two nonorthogonal quantum states were not perfectly distinguishable whenever a finite number of copies were availabel, Duan runyao showed that two different unitary operations, no matter orthogonal or not, could always be perfectly distinguishable [3, 4].

The purpose of this paper is to show that majorization inequalities can be utilized in quantum mechanics. First of all, we induct and derive the application in two bi-partite pure states in LOCC. Specially, a sufficient condition for bi-partite pure states entanglement transformation from the control to the controlled state is discussed. Also we give a sufficient and necessary condition about the existence of an entangled assisted state. Then, a unitary freedom in the ensemble for density matrices is studied under majorization inequalities. At last, the local distinguishability of orthogonal quantum states by majorization inequalities is discovered.

This paper is organized as follows. In Sect. 2, we introduce the basic definitions about majorization inequalities. In Sect. 3.1, we apply the theory of majorization inequalities to analyze the bi-partite pure states entanglement transformation. In Sect. 3.2, we deal with unitary freedom in the ensemble for density matrices by using majorization inequalities. Section 3.3 gives a sufficient and necessary condition to local distinguish orthogonal states by majorization inequality in LOCC. Conclusion is sketched in Sect. 4.

2 Majorization Inequalities and Notation

To begin, we give some significant definitions and lemmas of majorization inequalities which will be widely used.

Definition 1. *Let x be a vector from R^d and $x^\downarrow = (x_1, \cdots, x_d)$ denotes a vector consisting of coordinates of x in decreasing order $x_1 \geq x_2 \geq \cdots \geq x_d$. We say that y majorizes x, which can write as $x \prec y$, if the conditions*

$$\sum_{i=1}^{d} x_i \leq \sum_{i=1}^{d} y_i \tag{1}$$

That is to say

$$
\begin{aligned}
x_1 &\leq y_1 \\
x_1 + x_2 &\leq y_1 + y_2 \\
&\vdots \\
x_1 + x_2 + \cdots + x_d &= y_1 + y_2 + \cdots + y_d.
\end{aligned} \tag{2}
$$

Definition 2. *A $n \times n$ matrix $D = (d_{ij})$ is double stochastic if $d_{ij} \geq 0$, for $i,j = 1, \cdots, n$, and*

$$\sum_{i=1}^{n} d_{ij} = 1, \ \sum_{j=1}^{n} d_{ij} = 1. \tag{3}$$

Lemma 1. *The following statements are equivalent,*

1. $x \prec y$.
2. $x = \sum_i p_i P_i y$, *for some set of probabilities p_i, and permutation matrices P_i.*
3. $x = Dy$ *for doubly stochastic matrix D.*

Lemma 2. *Suppose A and B are Hermitian operators. Then $A \prec B$ if and only if there is a probability distribution p_i and unitary matrices U_i such that*

$$A = \sum_i p_i U_i B U_i^+. \tag{4}$$

Lemma 3. *Suppose $A, B \in H^{n \times n}$, so*

$$\lambda(A \otimes B) = \lambda(A) \otimes \lambda(B) \tag{5}$$

Lemma 4. *If $A \in H^{n \times n}$ and it is a Hermitian matrix with diagonal elements $a = (a_{11}, a_{22}, \cdots, a_{nn})$ and eigenvalues $\lambda = (\lambda_1, \lambda_2, \cdots, \lambda_n)$, then*

$$a \prec \lambda \tag{6}$$

3 The Mathematical Application of Majorization Inequalities to Quantum Mechanics

Quantum mechanics is a mathematical framework for physical quantum theories. The rules of quantum mechanics may be counter-intuitive. For example, quantum entanglement, the no-cloning theorem and the non-orthogonal states cannot be reliably distinguished, which are not coincided with the classical physics. It is essential to understand these unique and surprising phenomena in terms of mathematical tool–majorization inequalities. Firstly, the connection between majorization inequalities and bi-partite pure states entanglement transformation in LOCC is taken into account in Sect. 3.1.

3.1 The Mathematical Application of Majorization Inequalities to Bi-partite Pure States Entanglement Transformation in LOCC

Li studied the problem that transformed a tripartite pure state to a bi-partite one using SLOCC. So we pay attention to the bi-partite pure states entanglement transformation.

Suppose Alice and Bob share an entangled pure state $|\psi\rangle$. Given that they only can perform local operations and classical communication, the question is what other types of entanglement $|\varphi\rangle$ can they transform $|\psi\rangle$ into? Nielsen gave a sufficient and necessary theorem which made use of majorization inequalities to describe bi-partite pure state entanglement transformation.

Lemma 5 [14]. *A bi-partite pure entanglement state $|\psi\rangle$ may be transformed to another entanglement state $|\varphi\rangle$ by LOCC if and only if $\lambda_{|\psi\rangle} \prec \lambda_{|\varphi\rangle}$.*

Where $\lambda_{|\psi\rangle}$ and $\lambda_{|\varphi\rangle}$ are consisted of the square of Schmidt coefficients of $|\psi\rangle$ and $|\varphi\rangle$ in decreasing order, so, $|\psi\rangle = \sum_i \sqrt{\lambda_i} |i^A\rangle |i^B\rangle$.

However, let us consider an interesting problem. Suppose that there exist two entangled pure states $|\psi_1\rangle$, $|\psi_2\rangle$, which have $|\psi_1\rangle \to |\psi_2\rangle$, i.e. $\lambda_{|\psi\rangle} \prec \lambda_{|\varphi\rangle}$. So the converse transformation of $|\psi_2\rangle \to |\psi_1\rangle$ is impossible according to Lemma 5. In order to solve the problem, one entangled assisted state $|\psi_3\rangle$ can be consumed.

Theorem 1. *There exists an entangled assisted state $n \times n$ $(n \geq 2)$ $|\psi_3\rangle$ and a separable pure state $|P\rangle$ for two entangled pure states $n \times n$ $(n \geq 4)$ $|\psi_1\rangle$, $|\psi_2\rangle$ with $\lambda_{|\psi_1\rangle} \prec \lambda_{|\psi_2\rangle}$, $\lambda_{|\psi_2\rangle} \nprec \lambda_{|\psi_1\rangle}$. If $\lambda_{|\psi_3\rangle} \prec \lambda_{|\psi_1\rangle}$, then $|\psi_2\rangle|\psi_3\rangle$ can be transformed into $|\psi_1\rangle|P\rangle$, where $|P\rangle$ is orthogonal to $|\psi_1\rangle$.*

Proof. For the case $\lambda_{|\psi_1\rangle} \prec \lambda_{|\psi_2\rangle}$, $\lambda_{|\psi_2\rangle} \nprec \lambda_{|\psi_1\rangle}$, it is impossible to transform $|\psi_2\rangle$ into $|\psi_1\rangle$ only by means of entangled catalysts. This needs to consume an entangled state. Suppose $\lambda_{|\psi_3\rangle} \prec \lambda_{|\psi_1\rangle}$, and for a separable pure state $|P\rangle$, its eigenvalues are 1 or 0. Thus

$$\lambda_{|\psi_3\rangle} \otimes \lambda_{|\psi_2\rangle} \prec \lambda_{|\psi_3\rangle} \otimes \lambda_{|\psi_P\rangle} \prec \lambda_{|\psi_1\rangle} \otimes \lambda_{|\psi_P\rangle} \tag{7}$$

So $\lambda_{|\psi_2\rangle|\psi_3\rangle} \prec \lambda_{|\psi_1\rangle|P\rangle}$

From Lemma 5 we can achieve the transformation from $|\psi_2\rangle|\psi_3\rangle$ into $|\psi_1\rangle|P\rangle$. Then by measuring the $|P\rangle$, $|\psi_1\rangle$ can be obtained.

This completes the proof.

Especially, analogous to the existence of a 2×2 quantum entanglement catalysts [17] from one 4×4 state to another, the condition for the existence of an entangled assisted state is given here.

Theorem 2. *There exists an entangled assisted state 2×2 $|\psi_3\rangle$ with the eigenvalue vector of $\lambda_{|\psi_3\rangle}^\downarrow = (c, 1-c)$, $(c \geq 0.5)$ and a separable pure state $|P\rangle$ for two entangled pure state $4 \times 4 |\psi_1\rangle$, $|\psi_2\rangle$ with $\lambda_{|\psi_1\rangle} \prec \lambda_{|\psi_2\rangle}$, $\lambda_{|\psi_2\rangle} \nprec \lambda_{|\psi_1\rangle}$. If and only if $\beta_1 \leq \alpha_1 + \alpha_2$, and $c \leq \left\{ \frac{\alpha_1}{\beta_1}, \frac{\beta_1}{\beta_1 + \beta_2}, \frac{\alpha_1 + \alpha_2 + \alpha_3 - \beta_1}{\beta_2} \right\}$*

or

$$\frac{\beta_1}{\beta_1 + \beta_2} \leq c \leq \min\left\{\frac{\alpha_1}{\beta_1}, \frac{\alpha_1 + \alpha_2}{\beta_1 + \beta_2}, \frac{\alpha_1 + \alpha_2 + \alpha_3 - \beta_1}{\beta_2}\right\},$$

then $|\psi_2\rangle|\psi_3\rangle$ can be transformed into $|\psi_1\rangle|P\rangle$, where $|P\rangle$ is orthogonal to $|\psi_1\rangle$.

Proof. Suppose $\lambda_{|\psi_1\rangle}^{\downarrow} = (\alpha_1, \alpha_2, \alpha_3, \alpha_4)$, $\lambda_{|\psi_2\rangle}^{\downarrow} = (\beta_1, \beta_2, \beta_3, \beta_4)$, $\lambda_{|\psi_3\rangle}^{\downarrow} = (c, 1 - c), (c \geq 0.5)$.

Let us first consider the necessary. The first case, assume

$$\lambda_{|\psi_2\rangle|\psi_3\rangle}^{\downarrow} = (\beta_1 c, \beta_1(1 - c), \beta_2 c, \beta_2(1 - c), \beta_3 c, \beta_3(1 - c), \beta_4 c, \beta_4(1 - c)),$$

where we only consider the former three elements because of the $\alpha_1 + \alpha_2 + \alpha_3 + \alpha_4 = 1$, and regardless of the rest part. So Lemma 5 tells us that

$$\begin{cases} \beta_1 c \leq \alpha_1, \\ \beta_1(1 - c) \geq \beta_2 c, \\ \beta_1 c + \beta_1(1 - c) \leq \alpha_1 + \alpha_2, \\ \beta_1 c + \beta_1(1 - c) + \beta_2 c \leq \alpha_1 + \alpha_2 + \alpha_3, \end{cases} \quad (8)$$

By calculation, which implies

$$\begin{cases} \beta_1 \leq \alpha_1 + \alpha_2, \\ c \leq \left\{\frac{\alpha_1}{\beta_1}, \frac{\beta_1}{\beta_1 + \beta_2}, \frac{\alpha_1 + \alpha_2 + \alpha_3 - \beta_1}{\beta_2}\right\} \end{cases} \quad (9)$$

Now we consider the second case.

$$\lambda_{|\psi_2\rangle|\psi_3\rangle}^{\downarrow} = (\beta_1 c, \beta_2 c, \beta_1(1 - c), \beta_2(1 - c), \beta_3 c, \beta_3(1 - c), \beta_4 c, \beta_4(1 - c))$$

Similarly, by definition, we obtain

$$\begin{cases} \beta_1 c \leq \alpha_1, \\ \beta_1(1 - c) \leq \beta_2 c, \\ \beta_1 c + \beta_2 c \leq \alpha_1 + \alpha_2, \\ \beta_1 c + \beta_1(1 - c) + \beta_2 c \leq \alpha_1 + \alpha_2 + \alpha_3, \end{cases} \quad (10)$$

Therefore, it holds that

$$\begin{cases} \beta_1 \leq \alpha_1 + \alpha_2, \\ \frac{\beta_1}{\beta_1 + \beta_2} \leq c \leq \min\left\{\frac{\alpha_1}{\beta_1}, \frac{\alpha_1 + \alpha_2}{\beta_1 + \beta_2}, \frac{\alpha_1 + \alpha_2 + \alpha_3 - \beta_1}{\beta_2}\right\} \end{cases} \quad (11)$$

Of course, the sufficient condition is apparently. Which completes the proof.

To be more specific, two examples are listed.

Example 1. Let $\lambda_{|\psi_1\rangle}^{\downarrow} = (\alpha_1, \alpha_2, \alpha_3, \alpha_4) = (0.25, 0.25, 0.25, 0.25)$,

$$\lambda_{|\psi_2\rangle}^{\downarrow} = (\beta_1, \beta_2, \beta_3, \beta_4) = (0.4, 0.4, 0.1, 0.1), \lambda_{|\psi_3\rangle}^{\downarrow} = (c, 1 - c)(c \geq 0.5).$$

Obviously, $\beta_1 \leq \alpha_1 + \alpha_2$, i.e.$0.4 \leq 0.25 + 0.25$. Then, due of the condition

$$\frac{\beta_1}{\beta_1 + \beta_2} \le c \le \min\left\{\frac{\alpha_1}{\beta_1}, \frac{\alpha_1 + \alpha_2}{\beta_1 + \beta_2}, \frac{\alpha_1 + \alpha_2 + \alpha_3 - \beta_1}{\beta_2}\right\},$$

which holds $0.5 \le c \le 0.625$. Thence, when $0.5 \le c \le 0.625$, there exists an entangled assisted state $|\psi_3\rangle^\downarrow = (c, 1 - c)$ to make $|\psi_2\rangle|\psi_3\rangle \to |\psi_1\rangle|P\rangle$.

Example 2. Let $|\psi_1\rangle^\downarrow = (\alpha_1, \alpha_2, \alpha_3, \alpha_4) = (0.25, 0.25, 0.25, 0.25)$,

$$|\psi_2\rangle^\downarrow = (\beta_1, \beta_2, \beta_3, \beta_4) = (0.6, 0.2, 0.1, 0.1), \lambda_{|\psi_3\rangle^\downarrow} = (c, 1 - c)(c \ge 0.5).$$

Obviously, $\beta_1 \not< \alpha_1 + \alpha_2$, i.e. $0.6 \not< 0.25 + 0.25$. Thence, there exists not an entangled assisted state to perform $|\psi_2\rangle|\psi_3\rangle \to |\psi_1\rangle|P\rangle$.

3.2 The Mathematical Application of Majorization Inequalities to Unitary Freedom in the Ensemble for Density Matrices

In order to precisely describe an incompletely known quantum system, the density operators are provided. More evidently, suppose that a number of states $|\psi_i\rangle$ constitute a quantum system, where p_i is respective probabilities. Then the set of $\{p_i, |\psi_i\rangle\}$ is called an ensemble of pure states.

Naturally, Nielsen proved unitary freedom and discovered what classes of ensembles had identical density matrix. Combined with the fact that majorization inequalities were closely relevant to unitary matrices, Nielsen also discovered how a density matrix ρ could be decomposed into ensemble of the given probabilities vectors $\{p_i\}$.

Here, we premeditate what classes of a set of R^d vectors have identical given density matrix as far as majorization inequalities.

Theorem 3. *Suppose ρ is a given density matrix. For a set of vectors $\{\widetilde{|\varphi_i\rangle}\}$, where $d = (d_1, d_2, \cdots d_i, \cdots), d_i = \widetilde{\langle\varphi_i|\varphi_i\rangle}, \sum_i d_i = 1$, such that $\rho = \sum_i \widetilde{|\varphi_i\rangle}\widetilde{\langle\varphi_i|}$ if and only if $d \prec \lambda(\rho)$. If ρ and d have different dimensions, some zeroes are added to smaller vector.*

Proof. Let us first consider the necessary. Suppose the system ρ as a new system A. Then we introduce an auxiliary quantum system B and assume that the system B has a set of normal orthonormal bases $\{|i\rangle\}$. Defining a pure state of the joint system AB as $|\Phi\rangle$, so the bi-partite pure state $|\Phi\rangle$ can be written

$$|\Phi\rangle = \sum_i \widetilde{|\varphi_i\rangle}|i\rangle \tag{12}$$

Let ρ^A and ρ^B be the corresponding reduced density matrices, while

$$\rho^B = tr_A(|\Phi\rangle\langle\Phi|) = \sum_{i,j} |i\rangle\langle j|\widetilde{\langle\varphi_i \mid \varphi_j\rangle} \tag{13}$$

So the diagonal elements of ρ^B are $\langle \widetilde{\varphi_i \mid \varphi_j} \rangle$, when $i = j$. According to Lemma 4, $\mathrm{diag}(\rho^B) = \mathrm{d} \prec \lambda(\rho^B)$

Based on the Schmidt decomposition, $\lambda(\rho^B) = \lambda(\rho^A) = \lambda(\rho)$. Thus we can write down $d \prec \lambda(\rho)$.

Sufficient part is as follows.

Assume $d \prec \lambda(\rho)$, where $d = (d_i)$ is a set of probabilities vectors. Hence there exists a corresponding set of normalized state vectors $\{|\varphi_i\rangle\}$, such that

$$\rho = \sum_i p_i |\varphi_i\rangle \langle \varphi_i| \tag{14}$$

Where defined by $\widetilde{|\varphi_i\rangle} = \sqrt{p_i}|\varphi_i\rangle$.

Which completes the proof.

The interesting question, which classes of ensembles do give rise to a particular density matrix, has many applications in quantum computation and information, notably in the quantum error correction.

3.3 The Mathematical Application of Majorization Inequalities to Local Distinguishability of Two Bi-partite Orthogonal Quantum States

Only two states are orthogonal can we perfectly distinguish them in quantum mechanics. Jonathan proved that two states which were orthogonal remained distinguishable by LOCC [19]. Suppose that Alice and Bob share a bi-partite orthogonal pure state $|\psi_1\rangle$ or $|\psi_2\rangle$. They know the concrete form about $|\psi_1\rangle$ or $|\psi_2\rangle$, but have no idea which one they possess. Additional, they cannot implement quantum communications. The strategy that Alice and Bob adopt is simple. They can always find a set of bases in which the two orthogonal states can be represented as

$$|\psi_1\rangle = \sum_i \lambda_i \left| v_i^A \right\rangle \left| v_i^B \right\rangle \tag{15}$$

$$|\psi_2\rangle = \sum_i \sigma_i \left| v_i^A \right\rangle \left| v_i^{B\perp} \right\rangle \tag{16}$$

Where $\{|v_i^A\rangle\}$ are orthogonal basis sets for Alice, $\{|v_i^B\rangle\}$ and $\{|v_i^{B\perp}\rangle\}$ form orthogonal basis sets for Bob, and $|v_i^B\rangle \perp |v_i^{B\perp}\rangle$. So Alice only simply measures her part of the system in $\{|v_i^A\rangle\}$, and communicates the result i to Bob. For Bob, he can distinguish locally between $\{|v_i^B\rangle\}$ and $\{|v_i^{B\perp}\rangle\}$.

We are now ready to present the characterization of majorization inequalities of locally distinguishability for two orthogonal states in LOCC.

Theorem 4. *Let $|\psi_1\rangle$ and $|\psi_2\rangle$ be two bi-partite orthogonal pure states on Hermitian space. Suppose that Alice and Bob share an unknown $|\psi_m\rangle$ $(m = 1, 2)$. Then Alice and Bob can distinguish the unknown $|\psi_m\rangle$ is $|\psi_1\rangle$ or $|\psi_2\rangle$ in LOCC. If and only if $|\psi_m\rangle$ can be transformed into an ensemble $\left\{p_i, \left|v_i^A\right\rangle\left|v_i^B\right\rangle\right\}$ or $\left\{p_i, \left|v_i^A\right\rangle\left|v_i^{B\perp}\right\rangle\right\}$, i.e. $\lambda_{\psi_1} \prec \sum_{i=1}^{n} p_i \lambda_{\psi_i}$, or $\lambda_{\psi_2} \prec \sum_{i=1}^{n} p_i \lambda_{\psi_i \perp}$, where $|\psi_i\rangle = \left|v_i^A\right\rangle\left|v_i^B\right\rangle$.*

Proof. It is convenient to only consider $|\psi_m\rangle = |\psi_1\rangle$. Let us first consider the necessary. For $\rho = |\psi_1\rangle\langle\psi_1|$, $\rho^A = tr_B(|\psi_1\rangle\langle\psi_1|)$ is the initial state of Alice's system, and suppose Alice performs a quantum measurement described by measurement matrices M_i and communicates the result i to Bob. For Bob performs unitary transformations based on the results. We get

$$(M_i \otimes U_i)\rho(M_i^+ \otimes U_i^+) = p_i\sigma_i, \tag{17}$$

$$\sigma = \sum_{i=1}^n p_i\sigma_i = \sum_{i=1}^n p_i|\psi_i\rangle\langle\psi_i| \tag{18}$$

where $|\psi_i\rangle = |v_i^A\rangle|v_i^B\rangle$.

Considering Alice's system alone, Alice's states from ρ^A to σ_i^A. No matter the measurements, we get $M_i\rho^A M_i^+ = p_i\sigma_i^A$, where $\sigma_i^A = tr_B(|\psi_i\rangle\langle\psi_i|) = tr_B(\sigma_i)$. We deduce from the relationship of majorization and measurement that $\lambda_{\rho^A} \prec \sum_{i=1}^n p_i\lambda_{\sigma_i^A}$, which is equal to $\lambda_{\psi_1} \prec \sum_{i=1}^n p_i\lambda_{\psi_i}$.

Sufficiency. Now suppose $\lambda_{\psi_1} \prec \sum_{i=1}^n p_i\lambda_{\psi_i}$, so $\lambda_{\psi_1} = \sum_{i,j=1}^n p_iq_jP_j\lambda_{\psi_i}$, then there exists a quantum measurement described by measurements matrices M_{ij} such that

$$M_{ij}\rho^A M_{ij}^+ = p_iq_j\sigma_i^A, \ p_{ij} = p_iq_j, \ \sum_j p_{ij} = p_i \tag{19}$$

The procedure to get ensemble for Alice and Bob is that Alice performs the measurement sets $\{M_{ij}\}$. Here we introduce an ancillary system which effectively stores the value of i. Suppose the post-measurement state is $|\psi_{ij}\rangle = \sum_{i,j}\sqrt{p_{ij}}|\psi_i\rangle|i^R\rangle$, $tr_B(|\psi_{ij}\rangle\langle\psi_{ij}|) = \sigma_{ij}^A$.
Because of

$$tr_R(|\psi_{ij}\rangle\langle\psi_{ij}|) = tr_R\left(\sum_{i,i',j}\sqrt{p_{ij}}\sqrt{p_{i'j}}|\psi_i\rangle|i^R\rangle\langle\psi_{i'}|\langle i'^R|\right)$$

$$= \sum_{i,i',j}\sqrt{p_{ij}}\sqrt{p_{i'j}}|\psi_i\rangle\langle\psi_{i'}|\langle i^R \mid i'^R\rangle$$

$$= \sum_{i,j}p_{ij}|\psi_i\rangle\langle\psi_i|$$

$$= \sum_i p_i|\psi_i\rangle\langle\psi_i|$$

$$= \sigma \tag{20}$$

We get $|\psi_{ij}\rangle$ is the purification of σ. In other words, σ_{ij}^A is the purification of σ^A. In terms of system A, the post-measurement state is $\rho_{ij}^A = tr_B(|\psi_{ij}\rangle\langle\psi_{ij}|) = tr_B(\sigma_{ij}) = \sigma_{ij}^A$. Similarly, ρ_{ij}^A is the purification of σ^A. Thus Bob can convert the state ρ_{ij}^A into σ_{ij}^A by performing unitary transformations. It is equal to convert the state $|\psi_1\rangle$ into $\{p_i, |v_i^A\rangle|v_i^B\rangle\}$ by performing unitary transformations for Bob. Which completes the proof.

Motivated by the fact that the perfect distinguishability of quantum states is completely characterized by the orthogonality, we have discussed the local distinguishability of two bi-partite orthogonal quantum states by majorization inequalities.

4 Conclusion

In summary, we have investigated the connection between majorization inequalities and quantum mechanics. We have proved the sufficient condition for bi-partite pure states entanglement transformation from the control to the controlled states, and a sufficient and necessary condition that an 2×2 entangled assisted state exists between 4×4 states. Also we have displayed the necessary and sufficient condition to give rise to the same given ensemble for a group of vectors. At last, it has been established local distinguishability of two bi-partite orthogonal quantum states by using ensembles of majorization in LOCC. We hope people to pay more attention to the quantum distinguishability.

References

1. Bennett, C.H., Brassard, G., Crepeau, C., Jozsa, R., Peres, A., Wootters, W.K.: Teleporting an unknown quantum state via dual classical and Einstein-Podolsky-Rosen channels. Phys. Rev. Lett. **70**, 1895 (1993)
2. Duan, R., Feng, Y., Ying, M.: Entanglement is not necessary for perfect discrimination between unitary operations. Phys. Rev. Lett. **98**(10), 100503 (2007)
3. Duan, R., Feng, Y., Ying, M.: Local distinguishability of multipartite unitary operations. Phys. Rev. Lett. **100**(2), 020503 (2008)
4. Duan, R., Feng, Y., Ying, M.: Perfect distinguishability of quantum operations. Phys. Rev. Lett. **103**(21), 210501 (2009)
5. Dou, Z., Xu, G., Chen, X.B., Yuan, K.: Rational non-hierarchical quantum state sharing protocol. Comput. Mater. Contin. **58**(2), 335–347 (2019)
6. Feng, Y., Duan, R., Ying, M.: Catalyst-assisted probabilistic entanglement transformation. IEEE Trans. Inf. Theory **51**(3), 1090–1101 (2005)
7. Henderson, L., Vedral, V.: Classical, quantum and total correlations. Phys. Rev. A **34**(35), 6899–6905 (2001)
8. Jonathan, D., Plenio, M.B.: Entanglement-assisted local manipulation of pure quantum states. Phys. Rev. Lett. **83**(17), 3566–3569 (2012)
9. Jozsa, R., Linden, N.: On the role of entanglement in quantum-computational speed-up. Proc. Roy. Soc. A-Math. Phys. **459**(2036), 2011–2032 (2003)
10. Li, Y., Qiao, Y., Wang, X., Duan, R.: Tripartite-to-bipartite entanglement transformation by stochastic local operations and classical communication and the structure of matrix spaces. Commun. Math. Phys. **358**(2), 791–814 (2018). https://doi.org/10.1007/s00220-017-3077-5
11. Marshall, A.W., Olkin, I.: Inequalities: Theory of Majorization and Its Applications. Academic Press, Cambridge (1979)

12. Nielsen, M.A.: Conditions for a class of entanglement transformations. Phys. Rev. Lett. **83**(2), 436–439 (1998)
13. Nielsen, M.A.: An introduction to majorization and its applications to quantum mechanics. Lecture Notes (2002)
14. Nielsen, M.A., Chuang, I.L.: Quantum Computation and Quantum Information, 10th Anniversary edn. Cambridge University Press, Cambridge (2011)
15. Partovi, M.H.: Majorization formulation of uncertainty in quantum mechanics. Phys. Rev. A **84**(5), 13724–13731 (2011)
16. Peres, A.: Separability criterion for density matrices. Phys. Rev. Lett. **77**(8), 1413 (1996)
17. Sun, X., Duan, R., Ying, M.: The existence of quantum entanglement catalysts. IEEE Trans. Inf. Theory **51**(1), 75–80 (2005)
18. Xiao, H., Zhang, J., Huang, W.H., Zhou, M., Hu, W.C.: An efficient quantum key distribution protocol with dense coding on single photons. Comput. Mater. Contin. **61**(2), 759–775 (2019)
19. Walgate, J., Short, A.J., Hardy, L., Vedral, V.V.: Local distinguishability of multipartite orthogonal quantum states. Phys. Rev. Lett. **85**(23), 4972–4975 (2000)
20. Zhang, S.B., Chang, Y., Yan, L., Sheng, Z.W., Yang, F.: Quantum communication networks and trust management: a survey. Comput. Mater. Contin. **61**(3), 1145–1174 (2019)

The Numerical Results of Binary Coherent-State Signal's Quantum Detection in the Presence of Noise

Wenbin Yu[1,3]([✉]), Zijia Xiong[1], Zangqiang Dong[2], Yinsong Xu[1], Wenjie Liu[1], Zhiguo Qu[1], and Alex X. Liu[1,3]

[1] Jiangsu Collaborative Innovation Center of Atmospheric Environment and Equipment Technology (CICAEET), Jiangsu Engineering Center of Network Monitoring, School of Computer and Software, Nanjing University of Information Science and Technology, Nanjing 210044, Jiangsu, People's Republic of China
ywb1518@126.com
[2] The Department of Computer Science and Application, Zhengzhou Institute of Aeronautical Industry Management, Zhengzhou 450015, Henan, People's Republic of China
[3] Department of Computer Science and Engineering, Michigan State University, East Lansing, MI 48824-1226, USA

Abstract. In the optical frequency level communication, quantum effect becomes the source of significant communication errors. One alternative to classical detection is orthogonal projection measurement. In this paper, based on the coherent state's quantum measurement and discrimination, the quantum optimal detection of binary coherent state signal in the presence of thermal noise was studied. Our numerical results showed the error probabilities of quantum detection under the OOK and BPSK modulation, and verified the performance of these two modulation methods in the thermal noise field background.

Keywords: Quantum measurement · Quantum detection · Coherent state · Optical communication

1 Introduction

For deep space optical communication, due to the large-scale diffraction loss, some quantum resources, such as entangled state and squeezed state, can not maintain the long-distance quantum property. However, the coherent state of the light field produced by the conventional laser transmitter can maintain coherence under the diffraction loss. If it is used as a signal carrier, it will be a better choice [1, 2]. At present, the classical optical receiver for deep space communication uses photon counting or coherent detection to detect the coherent state signal. The dolinar receiver is described in reference [3]. In reference [4], a different method proposed by Sasaki and Hirota is described. It does not use optical feedback, but can reach the optimal boundary through unitary transformation and photon counting.

© Springer Nature Switzerland AG 2020
X. Sun et al. (Eds.): ICAIS 2020, LNCS 12239, pp. 532–542, 2020.
https://doi.org/10.1007/978-3-030-57884-8_47

In the field of optical communication in free space, the coherent state signal of optical field has a series of different applications [5, 6]. It is pointed out in [7, 8] that the aerosol scattering effect caused by rain, snow and fog will reduce the performance of free space coherent state signal transmission. Reference [9, 10] describes the influence of atmospheric turbulence on the coherent signal transmission of optical field, and establishes various theoretical models to describe the signal fading caused by turbulence. On the other hand, in the classical communication model, the detection of the determined signal can be observed in the presence of channel additive Gaussian noise. This model is suitable for radio frequency communication applications, because in this case, quantum effects are not detected. However, in today's increasingly developed optical frequency level communication, quantum effect has become the dominant error source in communication, which must be considered in detection.

The detection of pure state signal (noiseless state) can be expressed as orthogonal projection measurement, which contains the optimization of detection operator in signal subspace [11, 12]. In reference [13], the physical realization scheme of generalized quantum measurement on Coherent States is given. In reference [14], an effective algorithm has been developed to realize the required optimization and applied to the high-dimensional signal set concerned by optical communication, including for pulse position modulation (PPM), as well as for dense signal sets such as quadrature phase-shift keying (QPSK) and binary phase-shift keying pulse position modulation (BPSK-PPM). Additionally, paper [15–17] present three novel quantum algorithms, which are applied to the quantum secure communication and quantum signature scheme respectively.

In this context, we discussed the detection method which can minimize the average probability of detection error in the thermal noise field. Based on the objective of verifying its actual detection performance, we studied the numerical results of quantum optimal detection of coherent state signal in both the OOK and BPSK cases.

2   Coherent State Signal in the Presence of Thermal Noise

In the optical frequency level communication, in order to study the communication characteristics of the optical field, it is necessary to quantize the electromagnetic field. In this case, the coherent states are used to characterize the two-dimensional and higher dimensional optical signal sets [3]

$$|\alpha> = e^{-(1/2)|\alpha|^2} \sum_{n=0}^{\infty} \frac{\alpha^n}{\sqrt{n!}} |n> \tag{1}$$

Where α is a complex number and n is the number of photons, we can know from the expression that different coherent states are always non orthogonal.

In this case, the signal carrier of the communication system is the coherent state of the optical field. The coherent signal transmitted by the transmitter has different complex envelope, so the receiver can distinguish different signal values through this parameter. According to this property, we can give the coherent state form of binary signal. Among them, the form of the signal modulated by the binary signal of coherent state OOK (on off keying) is [18]

$$\left. \begin{array}{l} H_0 : |0> \\ H_1 : |\alpha> \end{array} \right\}$$

Among them, s and X correspond to two prior hypotheses of the OOK signal set, i.e. 0 and 1, which are respectively carried by vacuum state $|0>$ and coherent state $|\alpha >$.

On the other hand, the BPSK (binary phase shift keyed) modulation of coherent state binary signals can be expressed as

$$\left.\begin{array}{l} H_0 : |-\alpha > \\ H_1 : \ |\alpha > \end{array}\right\}.$$

Here, the coherent states $|-\alpha >$ and $|\alpha >$ are used as signal carriers.

When there is no background noise, the above coherent state signal can be described by Eq. (1). However, considering the thermal noise of optical field in optical communication system, the pure state coherent state signal will be affected by the thermal noise and transformed into mixed coherent state [19, 20]

$$\rho = (\pi N)^{-1} \int \exp\left(-|\alpha - \mu|^2 / N\right) |\alpha\rangle\langle\alpha| d^2\alpha \tag{2}$$

Where μ is the complex envelope of coherent state signal except noise, and N is the average photon number of noise. Since the energy of each photon is the same, N also represents the average energy of noise.

3 Quantum Optimal Detection

A priori assumption of pure binary signal in noise free background is

$$\left.\begin{array}{l} H_0 : |\psi_0 > \\ H_1 : |\psi_1 > \end{array}\right\}.$$

In order to detect the above signals, the method of orthogonal measurement should be used.

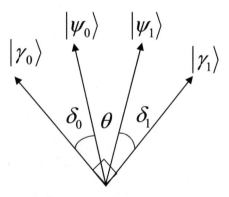

Fig. 1. The quantum detection of pure states

As described in Fig. 1, the measurement basis is a set of orthogonal states, and the orthogonal states and the measured states will form a certain angle in their expanded space. When the angles between the orthogonal measurement basis $|\gamma_0>$, $|\gamma_1>$ and the measured state $|\psi_0>$, $|\psi_1>$ are symmetric and equal, that is $\angle\delta_0 = \angle\delta_1$. At this time, the detection method is quantum optimal detection, and $|\gamma_0>$, $|\gamma_1>$ are the optimal orthogonal measurement basis [21].

Due to the common optical field thermal noise in the coherent state communication system, the aforementioned pure state coherent state signals $|\psi_0>$ and $|\psi_1>$ will be transformed into mixed state signals ρ_0 and ρ_1 with $|\psi_0>$ and $|\psi_1>$ as the energy centers under the action of thermal noise. Here, it is assumed that for noisy signals, the orthogonal measurement operators are O_0 and O_1. The problem of finding the optimal O_0 and O_1 is equivalent to the problem of finding the minimum error probability of detection.

According to the above conditions, we can deduce that the probability that the detection result is correct is

$$P(C|H_1) = tr[\rho_1 O_1]$$

$$P(C|H_0) = 1 - tr[\rho_0 O_1]$$

Because $P(H_0) = P(H_1) = 1/2$, the error probability of detection can be expressed as

$$P(E) = 1 - P(C) = \frac{1}{2}\{1 - tr[(\rho_1 - \rho_0)O_1]\}.$$

When ρ_0 and ρ_1 are known, the error probability is the lowest if and only if O_1 happens to be the nonnegative eigen decomposition of operator $(\rho_1 - \rho_0)$. If O_1 is any other projection operator, the error probability of detection will increase. We express the optimal detection operator as [3]

$$O^* = \sum_{k:\lambda_k \geq 0} |\lambda_k\rangle\langle\lambda_k|.$$

Where λ_k is the eigenvalue of $(\rho_1 - \rho_0)$. According to the above formula, the minimum error probability of detection can be further expressed as

$$
\begin{aligned}
P^*(E) &= \frac{1}{2}\{1 - tr[(\rho_1 - \rho_0)O^*]\} \\
&= \frac{1}{2}\{1 - \sum_k \langle\lambda_k|(\rho_1 - \rho_0)O^*|\lambda_k\rangle\} \\
&= \frac{1}{2}\{1 - \sum_k \lambda_k \langle\lambda_k|O^*|\lambda_k\rangle\} \\
&= \frac{1}{2}\{1 - \sum_k \lambda_k \sum_{m:\lambda_m>0} \langle\lambda_k|\lambda_m\rangle\langle\lambda_m|\lambda_k\rangle\} \\
&= \frac{1}{2}\{1 - \sum_{k:\lambda_k>0} \lambda_k\}.
\end{aligned}
$$

For the coherent signals in the background of thermal noise, their density operators are expressed by formula (2), and their forms are continuous. In order to calculate the error probability of the optimal detection and consider that the coherent state signal can be represented discretely on another set of physical quantities, it is pointed out in [3] that the integral of the continuous density operator can be transformed into the matrix form of discrete summation by using Laguerre polynomials. Each element in the matrix is

$$\langle n|\rho_1|m\rangle = \left(\frac{1}{N+1}\right)\left(\frac{n!}{m!}\right)^{\frac{1}{2}}\left(\frac{N}{N+1}\right)^m\left(\frac{u^*}{N}\right)^{m-n}$$
$$\times \, exp\left(-\left(\frac{1}{N+1}\right)|\mu|^2\right)L_n^{m-n}\left(-\frac{|\mu|^2}{N(N+1)}\right), m \geq n$$
$$\langle n|\rho_1|m\rangle = \langle n|\rho_1|m\rangle^*, m < n \qquad (3)$$

$|\mu|^2$ represents the average photon number of coherent state signal, n represents the average photon number of noise, and $L_n^{m-n}()$ is Laguerre polynomial. If the dimension of the above matrix is M, when M tends to infinity, the upper matrix will approach the integral form of the density operator of formula (2).

When the coherent state signal is limited to binary OOK modulation, ρ_0 can be given by the following formula.

$$\rho_0 = \sum_{n=0}^{\infty}\left(\frac{1}{N+1}\right)\left(\frac{N}{N+1}\right)^n|n\rangle\langle n|$$

In addition, by substituting $-\mu$ into formula (3), we can get the expression when the signal is BPSK modulated.

4 Numerical Results

Next, we will give the numerical results of the error probability of the optimal detection. The signal and channel environment parameters are set as follows: the average energy of coherent state signal is $K_S = |\mu|^2$, and the average energy of optical field thermal noise is N. We will examine the error probability curve of the optimal detection through the change of the two.

When the coherent state signal is OOK modulated, the element value of the matrix can be obtained by calculating formula (3). The calculation results of the first 4×4 dimensions of the matrix ρ_1 are given below

$$[\langle m|\rho_1|n\rangle] = \begin{pmatrix} 0.6766\ 0.2819\ 0.0831\ 0.0199 \\ 0.2819\ 0.2302\ 0.1011\ 0.0323 \\ 0.0831\ 0.1011\ 0.0681\ 0.0301 \\ 0.0199\ 0.0323\ 0.0301\ 0.0186 \\ \vdots \end{pmatrix} \cdots .$$

Here, the average photon number of signal $K_S = 0.5$ and the average photon number of noise $N = 0.2$. Further ρ_0 is

$$[\langle m|\rho_0|n\rangle] = \begin{pmatrix} 0.8333 & 0 & 0 & 0 \\ 0 & 0.1388 & 0 & 0 \\ 0 & 0 & 0.0231 & 0 \\ 0 & 0 & 0 & 0.0038 \\ & & \vdots & \end{pmatrix} \cdots .$$

By subtracting noise matrix B from additive noise signal matrix A, we can get

$$[\langle m|\rho_1 - \rho_0|n\rangle] = \begin{pmatrix} -0.1567 & 0.2819 & 0.0831 & 0.0199 \\ 0.2819 & 0.0013 & 0.1011 & 0.0323 \\ 0.0831 & 0.1011 & 0.0450 & 0.0301 \\ 0.0199 & 0.0323 & 0.0301 & 0.0148 \\ & & \vdots & \end{pmatrix} \cdots .$$

After diagonalizing the above matrix, we get

$$[\langle \lambda_k|\rho_1 - \rho_0|\lambda_k\rangle] = \begin{pmatrix} -0.3413 & 0 & 0 & 0 \\ 0 & 0.3413 & 0 & 0 \\ 0 & 0 & -0.0185 & 0 \\ 0 & 0 & 0 & 0.0185 \\ & & \vdots & \end{pmatrix} \cdots .$$

The specific value of error probability can be obtained by summing the non negative diagonal elements in the above formula.

Using the same method, we can get the value of OOK signal under different K_S and N.

For BPSK modulation, when $K_S = 0.5$ and $N = 0.3$, we can substitute them into formula (3) to get the following matrices

$$[\langle m|\rho_1|n\rangle] = \begin{pmatrix} 0.6346 & 0.2441 & 0.0663 & 0.0147 \\ 0.2441 & 0.2403 & 0.1052 & 0.0322 \\ 0.0663 & 0.1052 & 0.0841 & 0.0385 \\ 0.0147 & 0.0322 & 0.0385 & 0.0279 \\ & & \vdots & \end{pmatrix} \cdots ,$$

$$[\langle m|\rho_0|n\rangle] = \begin{pmatrix} 0.6346 & -0.2441 & 0.0663 & -0.0147 \\ -0.2441 & 0.2403 & -0.1052 & 0.0322 \\ 0.0663 & -0.1052 & 0.0841 & -0.0385 \\ -0.0147 & 0.0322 & -0.0385 & 0.0279 \\ & & \vdots & \end{pmatrix} \cdots ,$$

$$[\langle m|\rho_1 - \rho_0|n\rangle] = \begin{pmatrix} 0 & 0.4881 & 0 & 0.0294 \\ 0.4881 & 0 & 0.2104 & 0 \\ 0 & 0.2104 & 0 & 0.0770 \\ 0.0294 & 0 & 0.0770 & 0 \\ & \vdots & & & \ddots \end{pmatrix},$$

$$[\langle \lambda_k|\rho_1 - \rho_0|\lambda_k\rangle] = \begin{pmatrix} -0.5351 & 0 & 0 & 0 \\ 0 & 0.5351 & 0 & 0 \\ 0 & 0 & -0.0641 & 0 \\ 0 & 0 & 0 & 0.0641 \\ & \vdots & & & \ddots \end{pmatrix}.$$

Using similar methods to calculate the corresponding error probability one by one, we can get the following error probability curve.

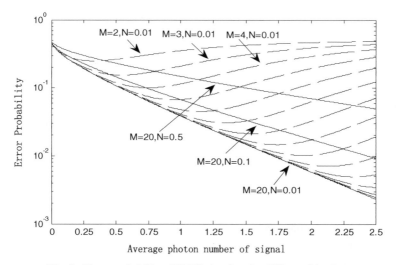

Fig. 2. Error probability of OOK signal under different M values

In Fig. 2, when the noise average energy $N = 0.01$, the matrix dimension M increases from 2 to 20 integers; when $N = 0.1$ and 0.5, M is 20, and the average photon number of the signal changes from 0 to 2.5. In Fig. 3, Ks = 1.5 and M increases from 4 to 20. It can be seen that with the increase of M, the error probability will decay faster with the increase of the average number of photons in the signal. This is because the higher dimension of matrix can make the numerical results closer to the actual situation.

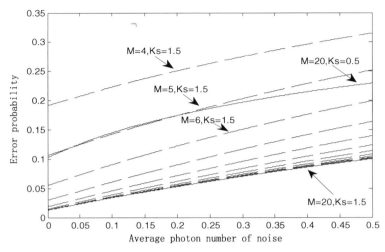

Fig. 3. Error probability of OOK signal under different M values (noise coordinate)

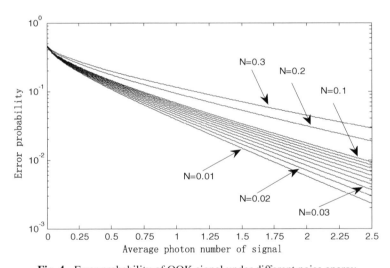

Fig. 4. Error probability of OOK signal under different noise energy

Figure 4 describes the error probability curve of OOK modulation signal when the matrix dimension M = 40, N is 0.01 to 0.1 uniform change and 0.2 and 0.3 respectively. Comparing Fig. 3 with Fig. 4, it can be seen that after the matrix dimension exceeds 20, the curve has basically converged, indicating that the performance is close to the original density operator.

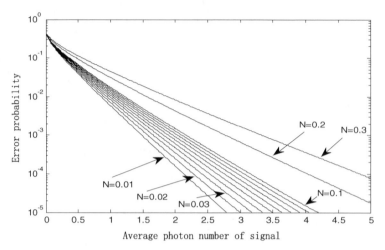

Fig. 5. Error probability of BPSK signal under different noise energy

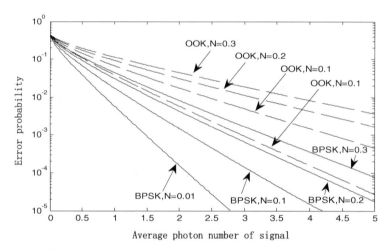

Fig. 6. Comparison of error probability between OOK and BPSK

Figure 5 shows the error probability curve of BPSK modulation signal under the same conditions as Fig. 4. It can be seen that when the average energy of noise increases gradually, the attenuation degree of error probability decreases in varying degrees. It can be seen from Fig. 6 and Fig. 7 that BPSK modulation signal has better error performance than OOK modulation signal under the same conditions.

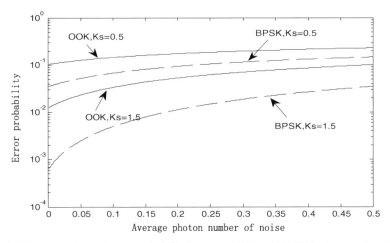

Fig. 7. Comparison of error probability between OOK and BPSK (noise coordinate)

5 Conclusion

In this paper, on the basis of the decomposition of the coherent states in the optical field, considering the influence of the quantum effect on the signal transmission at the optical frequency level, we the quantum orthogonal measurement method to detect the coherent state signal. We discussed the optimal detection of the pure state and mixed state and the quantum measurement of the coherent state signal in the background of thermal noise. In order to show the numerical results of the optimal error detection performance, we used the method of simplifying the density operator to the discrete matrix representation. Through numerical verification, we could see that as long as the dimension of matrix representation is more than 20, it was able to be close to the performance of density operator. At the same time, the numerical curve also verified the error performance of quantum optimal detection in the binary signal case, and presented the result that BPSK was much better than OOK modulation under the same detection method. In addition, the result of ternary signal and even higher dimension signal set will be our next work.

Acknowledgement. Supported by the National Natural Science Foundation of China under Grant Nos. 61501247, 61373131 and 61702277, the Six Talent Peaks Project of Jiangsu Province (Grant No. 2015-XXRJ-013), Natural Science Foundation of Jiangsu Province (Grant No. BK20171458), the Natural Science Foundation of the Higher Education Institutions of Jiangsu Province (China under Grant No. 16KJB520030), the NUIST Research Foundation for Talented Scholars under Grant Nos. 2015r014, PAPD and CICAEET funds. Partially supported by the China-USA Computer Science Research Center.

References

1. Vilnrotter, V.A.: Quantum receiver for distinguishing between binary coherent-state signals with partitioned interval detection and constant intensity local lasers. The Interplanetary

Network Progress Report, IPN Progress Report 42-189, pp. 1–22. Jet Propulsion Laboratory, Pasadena, California (2012)

2. Vilnrotter, V.A., Lau, C.-W.: Quantum detection of binary and ternary signals in the presence of thermal noise fields. The Interplanetary Network Progress Report, IPN PR 42-152, pp. 1–13. Jet Propulsion Laboratory, Pasadena, California (2002)

3. Helstrom, C.W.: Quantum Detection and Estimation Theory, Mathematics in Science and Engineering. Academic Press, New York (1976)

4. Sasaki, M., Hirota, O.: Optimum decision scheme with a unitary control process for binary quantum-state signals. Phys. Rev. A **54**(4), 2728–2736 (1996)

5. Kahn, J., Katz R., Pister, K.: Mobile networking for smart dust. In: Proceedings of The ACM/IEEE International Conference on Mobile Computing and Networking (MobiCom 1999), Seattle, WA, pp. 495–499 (1999)

6. Chu, P., Lo, N., Berg, E., Pister, K.: Optical communication using micro corner cuber reflectors. In: Proceedings of the IEEE MEMS Workshop, Nagoya, Japan, pp. 350–355 (1997)

7. Sadot, D., Melamed, A., Kopeika, N.: Effects of aerosol forwardscatter on the long- and short-exposure atmospheric coherence diameter. Waves Random Media **4**(4), 487–498 (1994)

8. Arnon, S., Kopeika, N.: Effect of particulates on performance of optical communication in space and an adaptive method to minimize such effects. Appl. Opt. **33**(21), 4930–4937 (1994)

9. Zilberman, A., Kopeida, N., Sorani, Y.: Laser bean widening as a function of elevation in the atmosphere for horizontal propagation. In: Proceedings of the SPIE Laser Weapons Technology II, vol. 4376 (2001)

10. Andrews, L., Phillips, R.: Laser Beam Propagation Through Random Media. SPIE Optical Engineering Press (1998)

11. Helstrom, C.W., Liu, J.W.S., Gordon, J.P.: Quantum mechanical communications theory. Proc. IEEE **58**(10), 1578–1598 (1970)

12. Yuen, H.P., Kennedy, R.S., Lax, M.: Optimum testing of multiple hypotheses in quantum detection theory. IEEE Trans. Inf. Theory **21**(2), 125–134 (1975)

13. Becerra, F.E., Fan, J., Migdall, A.: Implementation of generalized quantum measurements for unambiguous discrimination of multiple non-orthogonal coherent states. Nat. Commun. **4**, 2028 (2013)

14. Mingming, W., Chen, Y., Reza, M.: Controlled cyclic remote state preparation of arbitrary qubit states. CMC Comput. Mater. Contin. **55**(2), 321–329 (2018)

15. Faguo, W., Xiao, Z., Wang, Y., Zhiming, Z., Lipeng, X., Wanpeng, L.: An advanced quantum-resistant signature scheme for cloud based on Eisenstein ring. CMC Comput. Mater. Contin. **56**(1), 19–34 (2018)

16. Liu, W., Chen, Z., Liu, J., Su, Z., Chi, L.: Full-blind delegating private quantum computation. CMC Comput. Mater. Contin. **56**(2), 211–223 (2018)

17. Lau, C.-W., Vilnrotter, V.A.: Quantum detection and channel capacity using state-space optimization. The Interplanetary Network Progress Report 42-148, pp. 1–16. Jet Propulsion Laboratory, Pasadena, California (2001)

18. Liu, J.W.S.: Reliability of quantum mechanical communications systems. Technical report 468, Massachusetts Institute of Technology, Research Laboratory of Electronics, Cambridge, Massachusetts (1968)

19. Karp, S., O'Neill, E.L., Gagliardi, R.M.: Communication theory for the free-space optical channel. Proc. IEEE **58**(10), 1611–1626 (1970)

20. Barnett, S.M., Radmore, P.M.: Methods in Theoretical Quantum Optics. Oxford Science Publications, Clarendon Press, Oxford (1997)

21. Nielsen, M.A., Chuang, I.L.: Quantum Computation and Quantum Information. Cambridge University Press, Cambridge (2000)

An Adaptive Parameters Density Cluster Algorithm for Data Cleaning in Big Data

Xiaopeng Zhang[1], Ruijie Lin[2], and Haitao Xu[1($\boxtimes$)]

[1] University of Science and Technology Beijing, Beijing 100083, China
alex_xuht@hotmail.com
[2] China Telecommunication Technology Labs-Terminals, China Academy of Information and Communications Technology, Beijing 10083, China

Abstract. We have entered the era of information explosion, data has become an important driving force for the development of the industry. The huge wealth hidden in the data, enterprises can obtain a lot of useful information from business management, market analysis, scientific exploration and other aspects to support the development decisions of enterprises. However, the actual data is often intricate with different structures and types such as erroneous data, invalid data and missing duplicate data, greatly increase the difficulty of data analysis. These dirty data can greatly affect the results of data analysis, resulting in inaccurate results or even bad information. Most of the early data cleaning requires human involvement, however the exploding large-scale data is far from being able to meet with the human intervention, requiring smarter and more automated cleaning methods. The rapid development of artificial intelligence in recent years, the emergence of machine learning, provides a new opportunity for the development of data cleaning. Nowadays, there are lots of methods that have been applied to the field of data mining, including popular machine learning, deep learning, as well as classical clustering, bayesian networks, decision trees, and so on. Both provide a large number of solutions for our data cleaning. In this paper, we present a data cleaning method which use adaptive parameters density cluster algorithms based on the DBSCAN.

Keywords: Data cleaning · Clustering · Data mining

1 Introduction

Data cleaning has developed rapidly and has been applied in many ways in our data process in recent years. Because of the huge amount of data [1, 2], manual processing has become impossible. Therefore, automated data cleaning is the direction of development in the future.

In Reference [3] the author summarized the main problems of data cleaning, and generalized some solving method to different types of dirty data. Reference [4] proposed a visual data cleaning framework. The user can perform interactive operations in the graphical interface to complete the cleaning of the data. NADEEF [5] is an extensible,

© Springer Nature Switzerland AG 2020
X. Sun et al. (Eds.): ICAIS 2020, LNCS 12239, pp. 543–553, 2020.
https://doi.org/10.1007/978-3-030-57884-8_48

generalized and easy-to-deploy data cleaning platform. ESP [6, 7] is a cleaning platform that mainly deals with sensor data. In [8] the statistical characteristics of the dataset are analyzed, and the abnormal data existing in the data is analyzed by statistical algorithms to clean the abnormal data. In [9] the authors studied in depth data cleaning for RFID and WSN integration. Reference [10] present a fuzzy matching method which makes a more accurate similarity comparison of character records. In [11] the author analyzes the limitations of the existing similarity calculation between characters, and proposes a similarity matching algorithm for strings, which greatly improves the accuracy and stability of the similarity calculation between strings. CFDs [12] analyze the consistency of the data using the functional dependencies between data, then analyze the abnormal data in the data to clear the unreasonable data. In [13] the authors analyze the problem of low precision of the current data cleaning methods. The authors proposes a data cleaning method based on the KBs which greatly improves the accuracy and efficiency of data cleaning. ActiveClean [14], the author analyzes the statistical characteristics of the data, and proposes a statistical-based cleaning model, iteratively analyzes the cleaning data results, and gradually improves the accuracy of the model.

The most important part in data cleaning is to categorize the data so that similar data is grouped together. Clustering is one of the most important implementations. The purpose of clustering is that the similarity of objects of the same cluster is as large as possible, and the similarity between objects of different cluster is as small as possible. Clustering is an unsupervised classification method that does not require any prior knowledge. Clusters can be roughly divided into hierarchical clustering methods, partitioned clustering methods, and density-based and grid-based clustering algorithms. Clustering methods are widely used in the field of data cleaning. For example, when cleaning duplicate data, we can cluster the data, and finally similar data will be classified into the same cluster. Then we can design our own Rules that merge duplicate data. Similarly, erroneous data or missing data can often be corrected and populated based on the characteristics of the data in the same cluster. In fact, no clustering technique can be universally applied to data sets of various dimensions and structures. We need to choose the appropriate method according to the characteristics of data set that needs to be processed. In this paper, a density-based data cleaning method is proposed. The density-based clustering algorithm is widely used in data cleaning. Because the density-based clustering algorithm regionalizes high-density points into a cluster, it can effectively filter out low-density points that is noise data, and clustering of arbitrary shapes can be realized at the same time [15, 16].

The most classic of the density-based clustering algorithms is the DBSCAN clustering algorithm. DBSCAN mainly describes the distribution of sample points by two parameters, Eps and *MinPts*. However, the DBSCAN algorithm is very sensitive to Eps and *MinPts*. The effect is worse and even the wrong clustering result is obtained because of improper parameter value. Therefore, this paper proposes a clustering method for adaptively acquiring Eps and *MinPts*, and applies it to data cleaning. Reference [17] designed a data cleaning framework was, mainly for duplicate data and inconsistent data. In [17] the author proposes a new density clustering algorithm, and proposes the cluster density coefficient for the spatial and temporal characteristics of the data, which greatly improves the efficiency of data clustering. In [19] the authors investigate the strong dependence of DBSCAN algorithm on data parameters, so the author combines DSets with DBSCAN to determine the parameters of the clustering algorithm through DSets, which greatly improves the ease of use of the algorithm. In [20] The author builds an indexing method based on graph theory and applies it to multi-dimensional data structure cleaning, which greatly improves the cleaning efficiency of multidimensional data.

2 System Model and Problem Statement

2.1 DBSCAN: Density-Based Clustering Model

The DBSCAN [21] model measures the density of the density using the minimum sample point required in the neighborhood with a sample point size of Eps. Eps refers to the distance threshold of the Eps neighborhood, and the sample point whose distance from the sample point exceeds Eps is not in the Eps neighborhood. In the actual clustering process, we need to select the appropriate Eps value. If the Eps is too large, more sample points will fall within the Eps neighborhood, which will result in a decrease in the number of categories we eventually produce. is the threshold of the number of samples in the Eps neighborhood where the sample point is to be the core object. If the Eps is fixed, if the is too small, the resulting sample types will be too small. Below we introduce the terminology related to DBSCAN. Suppose sample set is $D = \{x_1, x_2, \cdots x_m\}$

Core point: For any sample, if its ε neighbor corresponding $x_j \in D$ contains at least $MinPts$ samples, if $|N_\varepsilon(x_j)| \geq MinPts$, then x_j is the core object.

Directly density-reachable: If x_i is in the ε neighborhood of x_j, and x_i is the core object, then x_i is said to be directly reached by x_j density. Note that the opposite is not true, that is to say, x_j cannot be said to be directly transmitted by x_i density, unless x_i is also the core object.

Density-reachable: For x_i and x_j, if there is sample sequence $p_1, p_2, \cdots, p_T$ satisfying $p_1 = x_1, p_T = x_j$, and p_{t+1} is directly reached by p_t density, then x_j is said to be reachable by x_i density. In other words, the density can reach the transferability. At this point, the transfer samples $p_1, p_2, \cdots, p_T$ in the sequence are core objects, because only the core objects can make other sample densities directly.

Density-connected: For x_i and x_j, if there is a core object sample x_k such that both x_i and x_j are reachable by x_k density, then x_i and x_j are said to be connected in density. The density connection is to satisfy the symmetry.

The DBSCAN can deal with any data set of any shape. In contrast, clustering algorithms such as K-Means are generally only applicable to convex data sets. Outliers can be found at the same time as clustering, and are not sensitive to outliers in the dataset.

2.2 Levenshtein Distance

Edit distance (also known as Levenshtein distance) refers to the number of minimum edit operations (insert, delete, replace) required between two strings. T Licensed editing operations include replacing one character with another, inserting one character, and deleting one character.

With two string S and T: $S = s_0 s_1 \cdots s_m$, $T = t_0 t_1 \cdots t_n$ Establish a $(m + 1) \times (n + 1)$ hierarchical relationship matrix LD:

$$LD_{(m+1) \times (n+1)} = \{d_{ij}\}(0 \leq i \leq m, 0 \leq j \leq n)$$

Where the matrix element is represented as d_{ij}

$$d_{ij} = \begin{cases} i & j = 0 \\ j & i = 0 \\ min(d_{(i-1)(j-1)}, d_{(i-1)j}, d_{i(j-1)}) + a_{ij} & i, j > 0 \end{cases}$$

Whether a_{ij} you need to insert, delete, replace:

$$a_{ij} = \begin{cases} 0 \; s_i = t_j \\ 1 \; s_i \neq t_j \end{cases} (i = 0, 1, \cdots, m; j = 0, 1, \cdots, n)$$

The final element is the distance between two strings, referred to as the LD distance. Represents the minimum number of edit operations required for a string to change to a string.

Based on the edit distance we calculate the similarly of the two string in:

$$sim(s1, s2) = \frac{ld}{m + n}$$

2.3 Problem of Existing Approaches

If the density of the dataset is not uniform the quality of clustering will poor. Compared with the K-Means clustering, the DBSCAN algorithm is slightly more complicated. It mainly needs to adjust the distance threshold and the neighborhood sample number. Different parameter combinations have a great influence on the final clustering effect. Different parameter combinations have a great influence on the final clustering effect (Fig. 1).

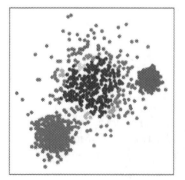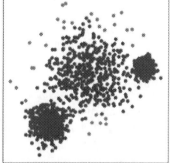

Fig. 1. A figure for different parameters results

As shown in Fig. 2, the density of cluster C_1 and cluster C_2 is different, and the distance between each sample point in C_1 from its nearest sample point is greater than the distance between noise point O_2 and cluster C_2. So we can't find the right Eps. Because if Eps is less than the distance between O_2 and C_2, then the next object in C_1 will be turned into a noise point. If the value of Eps is greater than the distance between O_2 and C_2, the noise point will be divided into cluster C_2.

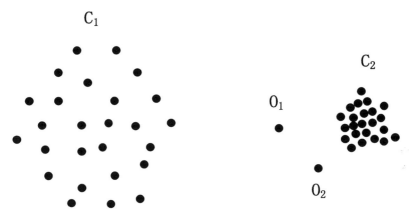

Fig. 2. A figure for different density cluster with noise

3 Data Cleaning Based on Clustering

After the text edit has been completed, the paper is ready for the template. Duplicate the template file by using the Save As command, and use the naming convention prescribed by your conference.for the name of your paper. In this newly created file, highlight all of the contents and import your prepared text file. You are now ready to style your paper; use the scroll down window on the left of the MS Word Formatting toolbar.

3.1 Dataset Analysis

Although the algorithm has the advantage of fast clustering and the ability to find clusters of arbitrary shapes in the detection of similar repeated records, the clustering algorithm also has some shortcomings. In general, the record field contains two data types, numeric and character. When the numeric data is mapped to a point in the high-dimensional space, the value in each dimension can be used with the corresponding data, but for the character field, when mapping to the midpoint of the space, the corresponding field information is lost. The general solution is to convert the string to the code value of the corresponding character, and then add the code values of the characters. In this case, there is a problem, such as string and two strings are different, but find After each character is added, the values of the two strings are the same, so that two records that are not repeated may be clustered in the same class, so after using the clustering algorithm, the obtained clustering result is recorded in the cluster. Can not be completely similar to the repeated records, and because sometimes a string contains more characters, so the value after the sum will be large, there is a certain impact on the cluster, so the first value of each character is used in the text, subtracted A relatively large value, i.e. using normalizing each character, such clustering is the inevitable errors.

A large database generally contains multiple data types. The metrics of different attributes of a record are also different. Therefore, when calculating the distance or similarity between records, we need to standardize and normalize different attributes for different attributes. The variables are scaled. Data can be divided into two data types, numeric and character. For different data types, we have to use different methods to analyze the distance and similarity between data. At the same time, in the actual processing, the influence of different attributes on the records is also different, and the influence degree of some attributes is obviously higher than other attributes, so we need to assign different weights to different attributes.

For numerical data, we use the commonly used standard Euclidean distance to measure the similarity between attributes. For character data, we use the edit distance to measure the distance between data, and set the appropriate weights for different attributes. We also normalize the numerical datasets with the Z-score

$$z(x_i) = \frac{x_i - \mu}{\sigma}$$

Let the entire data set be D, there is missing data in the D data set, and each sample has n attributes. Use the first attribute value of the sample to indicate the missing state of the attribute. If the value of the first attribute is missing, the attribute indicates that the variable is otherwise 1. The calculation formula is as follows:

$$\varepsilon_i = \begin{cases} 0 & \text{The i - th attribute value is missing} \\ 1 & \text{no value is missing} \end{cases}$$

Then, the expression of the standard distance is: the weight w set for each attribute by the user according to experience and actual demand, and the standard deviation of the attribute value of the first item. If X and Y presents the two records of the dataset. Then the distance of the two records is as follows.

3.2 Adaptive Eps and MinPts

Since Eps and *MinPts* in DBSCAN will directly affect the final clustering result, the selection of general parameters needs to be obtained by experience which does not conform to the clarity of the algorithm, so we proposed an algorithm to eliminate the parameter according to the statistical features of the dataset. The K nearest neighbor method and the mathematical expectation method are used to generate the Eps list. The K nearest neighbor method is an extension of the average nearest neighbor method. The basic idea of the algorithm is to calculate each data record in the data set D. The K-nearest neighbor distance from its Kth nearest neighbor data point, and the K-nearest. neighbor distance of all data points is averaged to obtain the K-average nearest neighbor distance of the data set. All K values are calculated to obtain a K-average nearest neighbor distance vector. The basic steps of the algorithm are as follows:

1) Calculate the distance distribution matrix of dataset D:

$$D_{n \times n} = \{Dist(i, j) | 1 \leq i \leq n, 1 \leq j \leq n\}$$

$D_{n \times n}$ is a real symmetric matrix of $n \times n$; n is the number of objects included in data set D; $Dist(i, j)$ is the distance from the i-th record to the j-th record in data set D.

2) We sort each row element of the distance matrix $D_{n \times n}$. Then the elements in column K represent the K-nearest neighbor distance.

3) By averaging the elements in the vector Dk, the K-mean nearest neighbor distance Dk of the vector Dk is obined and used as the candidate Eps parameter. Calculate all the K values and get the Eps parameter D_{Eps}, expressed as:

$$D_{Eps} = \{\overline{D_k} | 1 \leq K \leq n\}$$

4) For a given list of Eps parameters, the number of Eps neighborhood objects corresponding to each Eps parameter is sequentially determined, and the mathematical expectation value of the number of Eps neighborhood objects of all objects is calculated as the neighborhood density threshold *MinPts* parameter of the data set D, for:

$$MinPts = \frac{1}{n} \sum_{i=1}^{n} P_i$$

where P_i is the number of *Eps* neighborhood objects for the i-th object, and n is the total number of objects in data set D.

5) The elements in the set D_{Eps} are selected as the candidate Eps parameters and the *MinPts* parameters are obtained from the formula. The DBSCAN algorithm is used to cluster the data sets to obtain the number of clusters generated under different K values. When the number of clusters generated is the same for three consecutive times, this paper considers that the clustering result tends to be stable, and the number of clusters N is the optimal number of clusters. We choose the current Eps and minPts as the optimal parameters.

3.3 Data Cleaning Reprocessing

We cluster the dataset with the DBSCAN to get the classified result. For the missing data, according to the cluster in which the sample is assigned, we fill the data with the mean of the attribute which it located for the numeric type. For character data, the record with the highest frequency of occurrence can be selected as the missing data. For duplicate data, we set the repeat data similarity threshold. If the similarity of the sample data in the cluster is greater than the threshold, merge the two data record.

Table 1. Cleaning results

Record number	Actual duplicate record	Test result	Correct detection	Accuracy (%)	Recovery rate (%)
5000	471	402	346	86	85
9000	1254	1054	875	83	84
15000	2451	2015	1612	80	82
20000	3251	2766	2157	78	85

4 Experiment

This article uses an authoritative data, taken from a test set of the University of California, including a field, including name, origin, major, and individual scores. From the data table, the basic information of the student is taken out, including some artificial error records, so as to repeat similarly with some records, and the selection of parameters and thresholds is based on past experience. The figures for the cleaning results are given from Figs. 3, 4, 5, 6 and 7 (Table 1).

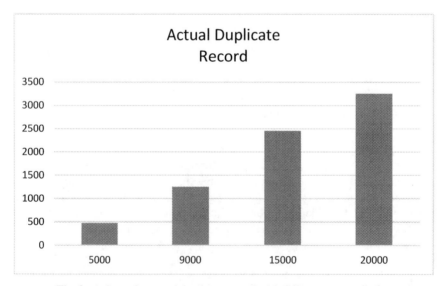

Fig. 3. A figure for actual duplicate record with different record number

Through the proposed density clustering algorithm combined with different cleaning strategies, the accuracy and recovery rate of our final cleaning results satisfy the cleaning requirements, and have a better results than the traditional k-mans clustering algorithm.

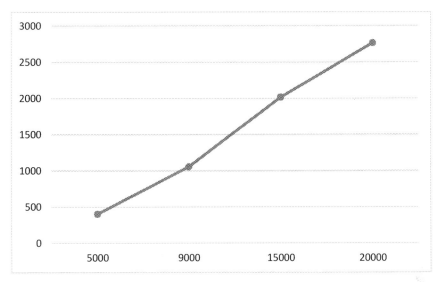

Fig. 4. A figure for test result with different record number

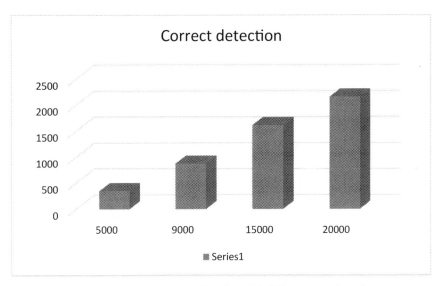

Fig. 5. A figure for correct detection with different record number

5 Conclusions

In this paper, we propose a data cleaning method based on improved DBSCAN density clustering. We introduce a method for adaptively obtaining DBSCAN parameters eps and minPts. For numerical data, we use standard Euclidean distance to calculate the similarity, and the distance between character data is calculated with edit distance. The improved clustering algorithm clusters the cleaning data. Finally we selects different

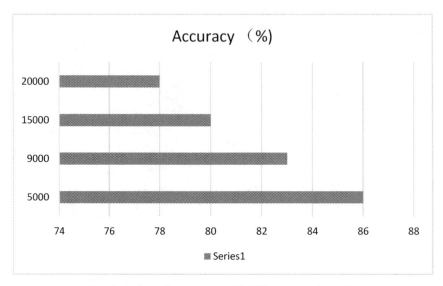

Fig. 6. A figure for accuracy with different record number

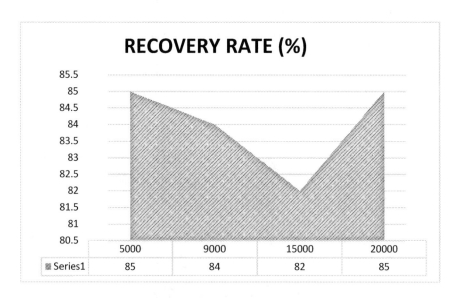

Fig. 7. A figure for recovery rate with different record number

cleaning strategies for different types of dirty data within the cluster. It has been verified that the improved cleaning algorithm can meet the requirements of data cleaning.

References

1. Centonze, P.: Security and privacy frameworks for access control big data systems. Comput. Mater. Continua **59**(2), 361–374 (2019)
2. Zhang, Z., et al.: A scalable method of maintaining order statistics for big data stream. Comput. Mater. Continua **60**(1), 117–132 (2019)
3. Rahm, E., Do, H.H.: Data cleaning: problems and current approaches. IEEE Data Eng. Bull. **23**(4), 3–13 (2000)
4. Raman, V., Hellerstein, J.M.: Potter's wheel: an interactive data cleaning system. In: VLDB, vol. 1, pp. 381–390 (2001)
5. Dallachiesa, M., Ebaid, A., Eldawy, A., et al.: NADEEF: a commodity data cleaning system. In: Proceedings of the 2013 ACM SIGMOD International Conference on Management of Data, pp. 541–552. ACM (2013)
6. Jeffery, S.R., Alonso, G., Franklin, M.J., Hong, W., Widom, J.: Declarative support for sensor data cleaning. In: Fishkin, Kenneth P., Schiele, B., Nixon, P., Quigley, A. (eds.) Pervasive 2006. LNCS, vol. 3968, pp. 83–100. Springer, Heidelberg (2006). https://doi.org/10.1007/11748625_6
7. Jeffery, S.R., Alonso, G., Franklin, M.J., et al.: A pipelined framework for online cleaning of sensor data streams. Comput. Sci. (2005)
8. Hellerstein, J.M.: Quantitative data cleaning for large databases. United Nations Econ. Comm. Eur. (UNECE) (2008)
9. Wang, L., Da Xu, L., Bi, Z., et al.: Data cleaning for RFID and WSN integration. IEEE Trans. Ind. Inf. **10**(1), 408–418 (2014)
10. Chaudhuri, S., Ganjam, K., Ganti, V., et al.: Robust and efficient fuzzy match for online data cleaning. In: Proceedings of the 2003 ACM SIGMOD International Conference on Management of Data, pp. 313–324. ACM (2003)
11. Chaudhuri, S., Ganti, V., Kaushik, R.: A primitive operator for similarity joins in data cleaning. In: 22nd International Conference on Data Engineering (ICDE'06), p 5. IEEE (2006)
12. Bohannon, P., Fan, W., Geerts, F., et al.: Conditional functional dependencies for data cleaning. In: 2007 IEEE 23rd International Conference on Data Engineering, pp. 746–755. IEEE (2007)
13. Chu, X., Morcos, J., Ilyas, I.F., et al.: Katara: a data cleaning system powered by knowledge bases and crowdsourcing. In: Proceedings of the 2015 ACM SIGMOD International Conference on Management of Data, pp. 1247–1261. ACM (2015)
14. Krishnan, S., Wang, J., Wu, E., et al.: ActiveClean: interactive data cleaning for statistical modeling. Proc. VLDB Endowment **9**(12), 948–959 (2016)
15. Hui, H., Zhou, C., Shenggang, X., Lin, F.: A novel secure data transmission scheme in industrial internet of things. China Commun. **17**(1), 73–88 (2020)
16. Jingtao, S., Lin, F., Zhou, X., Xing, L.: Steiner tree based optimal resource caching scheme in fog computing. China Commun. **12**(8), 161–168 (2015)
17. Al-janabi, S., Janicki, R.: A density-based data cleaning approach for deduplication with data consistency and accuracy. In: 2016 SAI Computing Conference (SAI), pp. 492–501. IEEE (2016)
18. Birant, D., Kut, A.: ST-DBSCAN: An algorithm for clustering spatial–temporal data. Data Knowl. Eng. **60**(1), 208–221 (2007)
19. Hou, J., Gao, H., Li, X.: DSets-DBSCAN: a parameter-free clustering algorithm. IEEE Trans. Image Process. **25**(7), 3182–3193 (2016)
20. Kumar, K.M., Reddy, A.R.M.: A fast DBSCAN clustering algorithm by accelerating neighbor searching using groups method. Pattern Recogn. **58**, 39–48 (2016)
21. Ester, M., Kriegel, H.P., Sander, J., et al.: A density-based algorithm for discovering clusters in large spatial databases with noise. In: Kdd, vol. 96, no 34, pp. 226–231 (1996)

Relevance and Time Based Collaborative Filtering for Recommendation

Aobo Zhang[1], Ruijie Lin[2], and Haitao Xu[1(✉)]

[1] University of Science and Technology Beijing, Beijing 100083, China
alex_xuht@hotmail.com
[2] China Telecommunication Technology Labs-Terminals, China Academy of Information and Communications Technology, Beijing 10083, China

Abstract. In our daily life, many recommendation systems are designed based on collaborative filtering algorithm. For example, video recommendations, shopping recommendations, music recommendations, and so on. However, the traditional collaborative filtering algorithm has problems such as low recommendation accuracy and poor real-time performance. In this paper, an enhanced collaborative filtering algorithm considering the relevance factor and time factor is proposed to improve the accuracy and timeliness of recommendations. Meantime, the average absolute error is utilized as an indicator to measure the recommended effect. Simulations are given to prove the effeteness of the proposed algorithm.

Keywords: Collaborative filtering · Recommended system · Relevance factor · Time factor

1 Introduction

Collaborative Filtering (CF) recommendation is mainly through the unrated project scoring prediction to achieve a recommended algorithm [1, 2]. Generally, in order to achieve recommendation, collaborative filtering algorithms are matched with Top-N calculations to implement its recommendation process. Currently, collaborative filtering has been used in many systems, such as Amazon, Group Lens, Movie Lens, Ringo, and so on. User-based collaborative filtering recommendation and project-based collaborative filtering recommendation are two kinds of CF algorithms that are two well-known by academics and researchers.

The theoretical basis of recommendation algorithm mainly comes from information retrieval [3, 4]. The information retrieval method is implemented by decomposing the item into feature vectors and using the feature vectors according to the users' behavior records. Then, the similarity between the feature vectors of item and the feature vectors of user is calculated, and the project with higher similarity will be recommended to the users. The recommendation algorithm has some advantages, such as strong stability of recommendation results, and easy to understand the recommendation results. However, the algorithm needs to extract features from projects and users based on the feature

© Springer Nature Switzerland AG 2020
X. Sun et al. (Eds.): ICAIS 2020, LNCS 12239, pp. 554–563, 2020.
https://doi.org/10.1007/978-3-030-57884-8_49

extraction methods such as information gain [5], mutual information [6], and CHI statistics [7]. The recommendation algorithm only recommends the content similar to the items that the user is interesting in, but cannot tap the potential interest of the users.

At present, a large number of scholars have made relevant improvements to the CBF algorithm. In [8], the author obtained a more diverse recommendation result by mixing the algorithm with collaborative filtering. Literature [9] integrates the relationships among the people in social networks into the news recommendation system based on the content semantics to better recommendation effect. The content-based recommendation algorithm is rarely used in multimedia recommendation algorithms. The main reason is that when extracting features of multimedia data, such as music, images, movies and televisions, it is also necessary to combine related technologies in the field of multimedia content analysis. However, some scholars have tried to apply this method to multimedia recommendation or retrieval systems. Zhang Erfen et al. [10] attempted to describe the music content by using the mixed Gaussian model, and pointed out that this method is also applicable to content-based music Search field. Wang Yin [11] systematically elaborated on the related concept technology based on multimedia content retrieval.

In this paper, we add two more factors into the collaborative filtering algorithm to improve the accuracy and timeliness of recommendation system. Meantime, the average absolute error is utilized as an indicator to measure the recommended effect. The whole paper is organized as follows. The traditional collaborative filtering algorithm is analyzed in Sect. 2. An optimization algorithm to the recommendation problem is given in Sect. 3. Section 4 is the numerical simulations, and it is concluded in Sect. 5.

2 Traditional Collaborative Filtering Algorithm

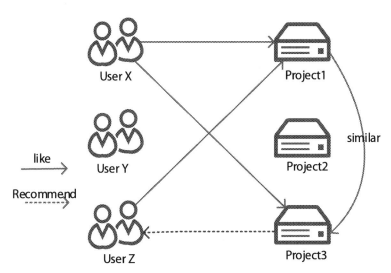

Fig. 1. Collaborative filtering algorithm based on projects

The collaborative filtering algorithm is proposed by Sarwar [12]. The basic principle of the recommendation algorithm is shown in Fig. 1. In the collaborative filtering algorithm, the authors assume that users tend to score higher on items of interest. When users are interested in one project, they are often interested in the similar projects. Then the algorithm will recommend the similar projects to these users (Table 1).

The following notations will be used in whole paper,

Table 1. Notations

Notations	Meanings
U_{ij}	Users Evaluated of project i and project j
sim_{ij}	Similarity between projects i and projects j
r_{ui}	User u's rating on project i
r_{uj}	User u's rating on project j
$\overline{r_i}$	Average rating of project i
$\overline{r_j}$	Average rating of project j

The main process of collaborative filtering algorithm is as follows. First, the scoring matrix of user for all items is listed. The similarity between the items is usually calculated based on the Pearson correlation coefficient method, as given in (2-1). After calculating the similarity between the two items, the project-item similarity matrix is listed.

$$sim_{ij} = \frac{\sum_{u \in U_{ij}} (r_{ui} - \overline{r_i})(r_{uj} - \overline{r_j})}{\sqrt{\sum_{u \in U_{ij}} (r_{ui} - \overline{r_i})^2} \sqrt{\sum_{u \in U_{ij}} (r_{uj} - \overline{r_j})^2}} \tag{2-1}$$

After the calculation of similarity matrix, we need to select the similar neighbors for the project to be predicted using the nearest neighbor method. Based on the method, K projects that are most similar to the target project can be selected from the similarity matrix as the nearest neighbor set.

After selecting the K nearest neighbors set of the target project, the predicted scores of the active users for different projects can be calculated according to (2-2), and then the rankings of the projects are ranked according to the active users' rankings from the highest to the lowest. Finally the TOP N projects are selected.

$$R_{ui} = \overline{r_i} + \frac{\sum_{j \in N} sim_{ij} \times (r_{uj} - \overline{r_j})}{\sum_{j \in N} |sim_{ij}|} \tag{2-2}$$

3   Relevance and Time Factor Based Algorithm

3.1   Relevance Based Optimization Algorithm

The traditional project-based collaborative filtering algorithm calculates the score based on user's rating information and determines the degree of similarity between the projects based on the similarity of the scores. Assuming a project is a sequel to another one. Although the scores are not the same, the content is often coherent and should be recommended to users based on the content considerations. Based on these, we ignore the intrinsic links between project in this paper. Introducing frequent items of frequent item sets, we establish a continuous matrix between projects, and try to optimize the content-based similarity between projects to improve the quality of recommendations.

Definition 3-1 (Degree of Association). The degree of association indicates the strength of continuity between items [13]. It means the probability that the user will view another item after viewing one item. The degree of association is represented by g_{ij}, indicating that association between project i and project j. The degree of association is defined as the number of users browsing project i and project j at the same time.

Proportion 3-1. The number of users browsing only project i is given as follows,

$$g_{ij} = \frac{N_i \cap N_j}{N_i} \tag{3-1}$$

where N_i is the number of users who rated item i, N_j is the Number of users who rated item j.

According to the (3-1), the degree of association between the projects can be calculated, and then the correlation degree matrix G is established as follows,

$$G_{n \times n} = \begin{bmatrix} 0 & g_{12} & g_{13} & \cdots & g_{1n} \\ g_{21} & 0 & g_{23} & \cdots & g_{2n} \\ g_{31} & g_{32} & 0 & \cdots & g_{3n} \\ \cdots & \cdots & \cdots & \cdots & g_{4n} \\ g_{n1} & g_{n2} & g_{n3} & \cdots & 0 \end{bmatrix}$$

where g_{ij} (i, j = 1, 2...n) is the degree of association between project j and project i. The degree of association between projects is generally different and asymmetrical. Scilicet $g_{ij} \neq g_{ji}$. On the basis of the traditional project-based collaborative filtering algorithm, the relevance degree factor is added, and the correlation between the projects makes the traditional similarity and the newly added relevance degree jointly determined. This kind of thinking takes into account the user's rating information, and considers the continuity of the project's content, and is more accurate and scientific in terms of recommendation.

A collaborative filtering algorithm for traditional correlation coefficient prediction scores is shown in (3-2),

$$sim_{ij} = \frac{\sum_{u \in U_{ij}} (r_{ui} - \overline{r_i})(r_{uj} - \overline{r_j})}{\sqrt{\sum_{u \in U_{ij}} (r_{ui} - \overline{r_i})^2} \sqrt{\sum_{u \in U_{ij}} (r_{uj} - \overline{r_j})^2}} \tag{3-2}$$

We combine the similarity with the above relevance degree to get the total similarity (3-3),

$$sim(i, j)_{sum} = \partial g_{ij} + (1 - \partial)sim(i, j)$$

$$= \partial \frac{N_i \cap N_j}{N_i} + (1 - \partial) \frac{\sum_{u \in U_{ij}} (r_{ui} - \overline{r_i})(r_{uj} - \overline{r_j})}{\sqrt{\sum_{u \in U_{ij}} (r_{ui} - \overline{r_i})^2} \sqrt{\sum_{u \in U_{ij}} (r_{uj} - \overline{r_j})^2}} \quad (3\text{-}3)$$

Substituting (3-3) into (2-2), a scoring method based on the degree of association weight can be obtained, as shown in the (3-4),

$$\begin{aligned} R_{ui} &= \overline{r_i} + \frac{\sum_{j \in N} sim(i,j)_{sum} \times (r_{uj} - \overline{r_j})}{\sum_{j \in N} |sim(i,j)_{sum}|} \\ &= \overline{r_i} + \frac{\sum_{j \in N} (\partial g_{ij} + (1 - \partial) sim(i,j)) \times (r_{uj} - \overline{r_j})}{\sum_{j \in N} |sim(i,j)_{sum}|} \end{aligned} \quad (3\text{-}4)$$

3.2   Time Factor Based User Interests Weight

In the real life, people's interests and hobbies are constantly changing with time. Based on this, in this paper, the project's score is assumed to be time-weighted by the fitted forgetting curve, and the closer the sampling time is, the greater the weight is given [14–16]. Therefore, the exponential decay function can be defined according to the fitted forgetting curve, to represent the change of the user's interest, and the time-based user interest weight function is given as follows,

$$w_t = e^{-\frac{t_{ui} - t_0}{T}} \quad (3\text{-}5)$$

where t_{ui} is the user u's rating time for project i, t_0 is target user sampling time, and T is the time span of the entire data set

The time-based user interest weight is added to the predicted score, and the calculation method is as shown in the formula (3-6).

$$R_{ui} = \overline{r_i} + \frac{\sum_{j \in N} sim_{ij} \times (r_{uj} \times w_t - \overline{r_j})}{\sum_{j \in N} |sim_{ij}|} \quad (3\text{-}6)$$

3.3   Relevance and Time Based Optimization

Substituting (3-3) into (3-6), we can have (3-7) which means the traditional collaborative filtering algorithm considering both the association degree weighting and the time weighting. It can be used to improve the coherence of the project content, and to solve the problem of real-time performance of the traditional collaborative filtering algorithm.

$$R_{ui} = \overline{r_i} + \frac{\sum_{j \in N} sim(i,j)_{sum} \times (r_{uj} \times w_t - \overline{r_j})}{\sum_{j \in N} |sim(i,j)_{sum}|} \quad (3\text{-}7)$$

3.4 Relevance and Time Based Optimization Algorithm

The collaborative filtering algorithm flow based on the relevance and time factor is shown in Algorithm 3-1.

Algorithm 3-1

Input: Training set and test set in the data set, nearest neighbors K

Output: User u's predicted score for item i

Step1: The user-item scoring matrix $R_{m \times n}$ and time matrix $T_{m \times n}$ are obtained from the training set

Step2: Calculate the matrix $G_{n \times n}$ according to formula (3-1)

Step3: Calculate the similarity matrix $sim(i, j)_{sum}$ between different items according to formula (3-3)

Step4: Select the K nearest neighbor sets $\{sim(i, j_1)_{sum}, sim(i, j_2)_{sum}, \ldots \ldots sim(i, j_k)_{sum}\}$ with the highest similarity to the project i.

Step5: Obtain the time weight w_t according to formula (3-5).

Step6: Substituting the formula (3-3) and the formula (3-5) into the formula (3-6) to obtain the prediction score R_{ui} of the user u for the project i.

---

4 Experimental Results

In this paper, the MovieLens dataset is selected to evaluate the proposed algorithm in Matlab. The dataset contains 943 users' scores of 1682 movies for 7 consecutive months. The score ranges from 1 to 5, where 1 means "very bad" and 5 means "very good." The MovieLens dataset provides five sets of randomly divided training sets and test sets. The experiments are performed on the five sets of data, and the final experimental results are the arithmetic mean of the five results.

In the simulation, we mainly uses MAE indicators to measure the accuracy of scoring prediction. MAE can reflect the actual situation of prediction error. We specify that the user's predicted score set on the training set is denoted by $p = \{p_{u,1}, p_{u,2}, \ldots, p_{u,n}\}$, the actual score set is denoted by $r = \{r_{u,1}, r_{u,2}, \ldots, r_{u,n}\}$, and the recommended set is denoted by N. Then the recommended performance is evaluated by calculating the difference between the two sets of scores. The MAE definition is as follows,

$$MAE = \frac{\sum_{u,i \in N} |r_{u,i} - p_{u,i}|}{N} \tag{4-1}$$

First, we must determine the value of ∂. The value of ∂ ranges from 0–1. In the experiments, we choose the nearest neighbor number K to be 20. The effect of the balance factor on MAE is shown in Fig. 2. It is found that when $\partial = 0.2$, the value of MAE is the smallest. At this time, not only the degree of correlation but also the influence of time on the similarity calculation is considered.

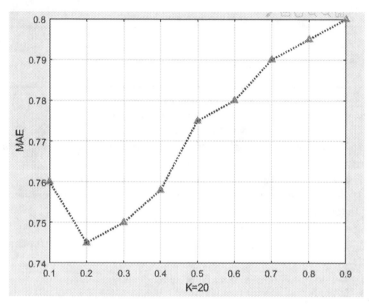

Fig. 2. Effect of balance factor ∂ on MAE

Taking the value of ∂ as 0.2, According to (3-4), we compare the results of the correlation degree algorithm and the traditional collaborative filtering algorithm. The experimental horizontal coordinate data is the nearest neighbor number K of the project, and the ordinate is the average absolute error value MAE. The experimental results are shown

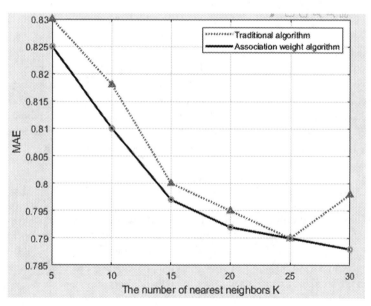

Fig. 3. The results of improve the association algorithm and the traditional algorithm

in Fig. 3. Through the experiments, we found that adding relevance weights to the traditional collaborative filtering algorithm will increase the accuracy of recommendations and optimize the performance of the recommendation system.

In order to know whether the improved time-weighting algorithm will increase the accuracy of the recommendation, we compare the two algorithms through experiments. The experiment yields the value of MAE in different situations by constantly changing the number of nearest neighbors K. The experimental results are shown in Fig. 4. From the experimental results, we can know that under the same K value, the MAE value of the improved time-weighted algorithm is smaller than the K value of the traditional collaborative filtering algorithm. Explain that this method can recommend the accuracy of the results and improve the performance of the recommended system.

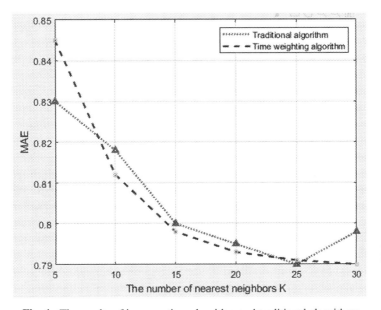

Fig. 4. The results of improve time algorithm and traditional algorithms

According to (3-7), we can comprehensively weight the degree of association and time. Then the integrated weighted algorithm is combined with the traditional collaborative filtering algorithm, the association degree weighting algorithm and time weighting algorithm are compared to observe the recommended effect of the integrated weighting algorithm. Among them, the abscissa represents the nearest neighbor number K, and the ordinate represents the mean absolute error MAE. The experimental results are shown in Fig. 5. The experimental results show that the MAE value of the association time integrated weighting algorithm is the smallest in the case of the same number of nearest neighbors K. This shows that it is also feasible to combine the correlation weight algorithm and the time weight algorithm, and the recommendation accuracy is increased, and the recommendation performance is improved.

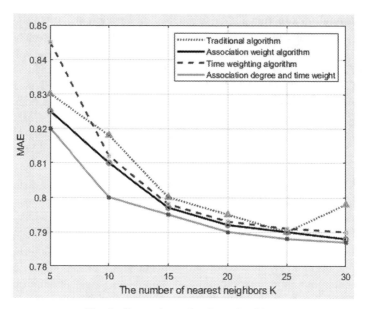

Fig. 5. Comparison of various algorithms

5 Conclusion

In this paper, we analyze the shortcomings of the traditional collaborative filtering algorithm in terms of relevance and timeliness. In the proposed algorithm, we optimize from these two aspects. Firstly, the weight of the association degree is added to the traditional collaborative filtering algorithm, and the optimal correlation coefficient is determined according to the experimental result. Time weighting is then performed on traditional collaborative filtering algorithms. By comparing the experiments, the conclusion of time-weighted feasibility is obtained. On the basis of the correlation weighting and time weighting, we combine the two to obtain the correlation time-weighted collaborative filtering algorithm. Through experiments, the algorithm is compared with the correlation degree weighting algorithm, the time weighting algorithm and the traditional collaborative filtering algorithm to verify the recommendation effect of the correlation time combined with the weighting algorithm.

Funding. This research was funded by National Key R&D Program of China (No. 2018YFB1003905).

References

1. Yan, Y., Wei, C., Xiangwei, Y.: Hybrid collaborative filtering recommendation based on project score prediction. Mod. Libr. Inf. Technol. **31**(6), 27–32 (2015)
2. Wang, G., Liu, M.: Dynamic trust model based on service recommendation in big data. Comput. Mater. Continua **58**(3), 845–857 (2019)

3. Baeza-Yates, R., Ribeiro-Neto, B.: Modern Information Retrieval: Addison Wesley[J]. Computer Science & Information Technology (CS & IT, 1999.)
4. Liu, G., Meng, K., Ding, J., Nees, J.P., Guo, H., Zhang, X.: An entity-association-based matrix factorization recommendation algorithm. Comput. Mater. Continua **58**(1), 101–120 (2019)
5. Elahi, S., Cahue, S., Felson, D.T., et al.: The association between varus–valgus alignment and patellofemoral osteoarthritis. Arthritis Rheumatol. **43**(8), 1874 (2000)
6. Penrose, J.M., Holt, G.M., Beaugonin, M., et al.: Development of an accurate three-dimensional finite element knee model. Comput. Methods Biomech. Biomed. Eng. **5**(4), 291–300 (2002)
7. Hull, D.A., Pedersen, J.O.: A comparison of classifiers and document representations for the routing problem. In: International ACM SIGIR Conference on Research and Development in Information Retrieval, pp. 229–237. ACM (1995)
8. Yang, W., Tang, R., Lu, L.: News recommendation method based on content-based recommendation and collaborative filtering. Comput. Appl. **36**(2), 414–418 (2016)
9. Fengxia, Y.: Research on news recommendation method based on content semantics in social network. Comput. Technol. Dev. **10**, 253–257 (2013)
10. Zhang, D., Xu, H.: Content-based music semantic feature description method. Electron. Des. Eng. **21**(1), 31–33 (2013)
11. Yin, W.: Content-based multimedia data query and retrieval. Research **45**, 00084–00085 (2015)
12. Sarwar, B., Karypis, G., Konstan, J., et al.: Item-based collaborative filtering recommendation algorithms. In: Proceedings of the 10th International World Wide Web Conference, pp. 285–295 (2001)
13. Kong, W.: Research on key issues of collaborative filtering recommendation system. Central China Normal University (2013)
14. Wang, J.: Collaborative filtering theory of machine learning algorithm practice recommendation system and its application. Tsinghua University (2018)
15. Hui, H., Zhou, C., Shenggang, X., Lin, F.: A novel secure data transmission scheme in industrial Internet of Things. China Commun. **17**(1), 73–88 (2020)
16. Jingtao, S., Lin, F., Zhou, X., Xing, L.: Steiner tree based optimal resource caching scheme in fog computing. China Commun. **12**(8), 161–168 (2015)

A Displacement Data Acquisition and Transmission System Based on a Wireless Body Domain Network

Juan Guo[1], Feng Sheng[2], Jie Li[2], Chao Guo[2(✉)], and Dingbang Xie[2]

[1] Guilin University of Electronic Technology-Vocational and Technical College, Beihai 536000, Guangxi, People's Republic of China
[2] Department of Electronics and Communication Engineering, Beijing Electronic Science and Technology Institute, Beijing 100070, People's Republic of China
guo99chao@163.com

Abstract. Wireless body domain network can monitor and collect the displacement data of the human body in real-time through the application of wireless sensor technology with the human body as the carrier, and upload the collected data to the displacement monitoring center for further analysis of the displacement data. In this paper, based on the wireless body domain network technology, to human body displacement data acquisition and transmission, through displacement data were collected for the human body gesture recognition and sports training, and so on to provide a series of technical support, the design of portable wearable human body displacement data acquisition unit, implements the multi-sensor wireless transmission, and application of Kalman filtering algorithm, the acquisition unit collected by the state variables and the quantity of observation data fusion and filtering, minimize noise interference, at the same time of six-axis gyro attitude algorithm by using three-axis attitude Angle of the body. To realize the real-time monitoring of human movement.

Keywords: Wireless body domain network · Human displacement data acquisition unit · Kalman filter algorithm

1 Introduction

Body area network (BAN), wireless LAN, also known as the human body, revolves around the body to build the wireless network of wearable computing equipment, is the sensor network technology and biomedical engineering, new materials, connected devices and the product of computing power, the advantage of the characteristics of the new material, combined with the feature of human body engineering, it builds a sensing, connection, calculation, and local network world with human interaction [1]. As a very small LAN on the body, body LAN BAN has the characteristics of LAN, Ethernet, cloud and other networks. The BAN system can use WPAN wireless technology as a gateway to go further and link to wearable devices connected to the Internet via the gateway device, which allows the backend device to obtain real-time health data online. The

© Springer Nature Switzerland AG 2020
X. Sun et al. (Eds.): ICAIS 2020, LNCS 12239, pp. 564–574, 2020.
https://doi.org/10.1007/978-3-030-57884-8_50

adoption of BAN as a unified standard network has had a huge impact on the Shared use of communications equipment and subsequent progress [2, 3].

2 The Overall Hardware Design of Displacement Data Acquisition and Transmission System

2.1 Demand Analysis

The system needs to meet the following requirements when working normally:

1. Acceleration data collection: six-axis gyroscope and accelerometer need to capture three axial acceleration data and axial angular velocity data of human body to obtain three-axis posture angle data of human body.
2. Wireless transmission: the function of the Bluetooth transmission module is to transfer the displacement data captured by the six-axis acceleration sensor to the Android terminal and the PC terminal for centralized processing. The wireless transmission module needs to ensure the security and stability of data transmission, as well as excellent transmission rate and low power consumption.
3. Supporting upper computer software: this design requires an Android terminal APP to realize the Bluetooth step counting function and calculate the calories burned during human activities. At the same time, the PC upper computer is also required to record the triaxial acceleration and triaxial angular velocity data of the human body in real-time, and display the human body movement posture in real-time.
4. Power supply: sensors are generally battery powered and once they exhaust their energy, they will be unserviceable. Nodes located around the sink consume more energy for information forwarding and that is so called "hot spot" phenomenon which resulting in a short lifetime to the system [4]. As the power supply device of the system, the power supply needs to guarantee the continuous operation of the system. As the power supply of the wearable device, the power supply needs to ensure that there is no safety risk [5].

This system uses data acquisition unit, Bluetooth transmission module, and data processing unit and power supply to realize the above function.

2.2 Overall System Architecture

The hardware design of this system consists of a pedometer data acquisition unit and the human body attitude acquisition unit. The human posture acquisition module and pedometer data acquisition module are respectively worn on the big arm and waist of the human body and are responsible for collecting the displacement data of the human body. The combination of a Bluetooth transmission module and the acquisition module constitutes the displacement data acquisition unit. The displacement data acquisition unit transmits the collected six-axis data to the PC terminal and Android terminal APP through a Bluetooth communication protocol. The PC and Android apps are responsible for further analysis of the data and feedback of the results to users.

The acquisition unit USES the MPU6050 six-axis acceleration sensor, which is used to collect the three-axis acceleration data and three-axis angular velocity data when the human body is moving, and then the collected data is output as a digital signal. Use ARDUINO UNO development boards. The function of the development board is to transmit the data collected by the MPU6050 six-axis acceleration sensor to the upper computer for processing through the Bluetooth communication module, which is used to analyze the posture of the human body, count steps and calculate the calories consumed in motion. The Bluetooth communication module USES the hc-06 module [6].

At present, for extending the lifetime of nodes, they are equipped with recharge-able technologies, which converts sources (e.g., body heat operation, a sensor no decan operate perpetually when the energy expenditure rate is lower than the harvested energy rate. In such Rechargeable WSNs (R-WSNs) or Energy Harvesting WSNs (EH-WSNs), although their lifetime is less of an issue, the amount of energy harvested by a sensor is limited due to the size of generation elements and limited battery capacity [7].

The overall structure of the system is shown in Fig. 1 below.

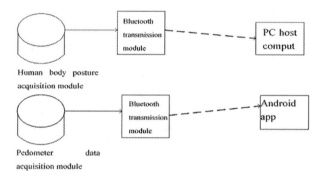

Fig. 1. Overall structure of displacement data acquisition and transmission system

2.3 Overall Hardware Design

The MPU6050 is a microelectromechanical system (MEMS) consisting of a three-axis accelerometer and a three-axis gyroscope. The module also has a (DMP) digital motion processor that is powerful enough to measure the acceleration, velocity, and direction of a system or object, displacement and many other motion-related parameters. Any change in motion is reflected in the mechanical system, which in turn changes the voltage. The IIC is the module that has a 16-bit ADC that accurately reads the voltage change and stores it in the FIFO buffer, which causes the INT (interrupt) pin to go high. So we can use MUC to read data from this FIFO buffer via IIC communication. The ARDUINO platform can use the off-the-shelf library to immediately start using the module and convert the measured analog quantities into output able, understandable digital quantities.

The coordinates of the MPU6050 chip are defined by pointing the chip surface toward itself and converting the surface to the correct Angle. The horizontal to the right is the X-axis, the vertical is the Y-axis, and the vertical is the z-axis See Fig. 2.

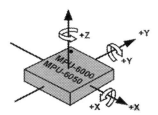

Fig. 2. MPU6050 chip coordinate system

The MPU6050 sensor digital output 6-axis or 9-axis rotation matrix, EulerAngle-forma fusion calculation data. DMP: The Digital Motion Processing engine reduces the burden of complex fusion calculus data, sensor synchronization, and gesture awareness. The motion processing database supports Android, Linux and Windows built-in operational time offset and magnetic sensor calibration algorithms without additional calibration.

The system uses ARDUINO microcontroller to realize sensor control, system control, and Bluetooth transmission and acceleration data storage in the acceleration acquisition unit. In general, ARDUINO has the following characteristics: 1. System design can be performed quickly and efficiently; 2. ARDUINO has a large number of open-source resources; 3. ARDUINO development board can directly update programs, communication, and interaction after connecting to the host computer; 4. ARDUINO development The board is suitable for a variety of power sources; 5. The development board is small in size and convenient for porting.

The HC-06 Bluetooth module used in this system is a PCBA board integrated with Bluetooth function, which is composed of PCB, chip and peripheral components. It has the characteristics of low power consumption, intelligence, and low cost. It is used to replace data cable for short-range wireless communication. Bluetooth Low Energy (BLE) is first introduced in 2010 as part of the Bluetooth specification 4, which has a remarkable capability to keep devices working for a long time (months to years [Nick (2015)]). Comparing to Classic Bluetooth, which is used usually to establish shortrange and continuous wireless connection, BLE is usually used for short bursts of some long range radio connection [8]. Hc-06 uses FHSS to avoid interference with other devices and realize full-duplex transmission. The operating frequency range of the device is from 2.402 ghz to 2.480 ghz. Hc-06 has the following advantages:

1. The module is suitable for wireless communication less than 100 m;
2. The module is easy to interface and communicate;
3. The module is one of the cheapest solutions in all types of wireless communication on the market;
4. The operation energy consumption of the module is very low, and it can be used in battery-driven mobile systems;
5. Because the module uses the UART interface, it can be used with almost all control systems Controller or processor interface.

3 Data Acquisition Unit

Before the data collection unit starts to work, we need to set up MPU6050 to ensure that the data collection unit can complete the data collection task. Before writing data to the six-axis acceleration sensor, the device should be opened first. The first step is to set the six-axis acceleration sensor into the transmission form of Wire, and then complete the selection of bus address. The fixed bus address of the six-axis acceleration sensor is 0×68. After completion, a byte is written into the registered address of the six-axis acceleration sensor, and the byte is written into the registered address to complete the data of arbitrary data length. Previously written data are sequentially put into the previous register address, and when the register runs out of space, the data is automatically entered into the next register. This completes the Wire mode. When the six-axis acceleration sensor reads the data, it is also necessary to set the sensor as Wire and transmit the data after the fixed register address is written into a byte. The data in the registered address can be read from the cache area in the Wire library. When reading the data of the six-axis acceleration sensor. The data for each axis is stored in two bytes or registers whose addresses we can see from the sensor's data table. Starting with the first register, and using the request from function, we request that all six registers on the X, Y, and Z axes be read [9]. The data is then read from each register, and since the output is a binary complement, you need to combine them accordingly to get the correct value.

3.1 Concept of Kalman Filter

Kalman filter, also known as linear quadratic estimation (LQE), is an algorithm. It uses a series of measurements observed over time, including statistical noise and other inaccuracies, to estimate the joint probability distribution of variables in each time frame, which is more accurate than those based on a single measurement.

To ensure the accuracy and completeness of the data, the human body gesture acquisition unit body using Kalman filter algorithm to attitude algorithm to get the three-axis attitude Angle data of human body [10].

Kalman gain is a relative weight given to measurement and current state estimation and can be "tuned" to achieve specific performance. In the case of high gain, the response is higher. At low gain, the filter follows the model prediction more closely. Kalman filtering algorithm is completed by five core equations in two stages.

1. In the prediction stage, prior estimation is calculated;
2. In the correction stage, calculate the posterior estimate to update the prior estimate covariance;

3.2 Human Posture Data Collection

The center of gravity of the human body as the origin to establish a three-axis coordinate system. When the human movement is offset on the X-axis, the offset Angle is set as Roll; when the human movement is offset on the Y-axis, the offset Angle is set as Pitch; when the human movement is offset on the z-axis, the offset Angle is set as Yaw.

Through six-axis acceleration sensor can be gathered by the data acquisition of the three-axis acceleration and angular velocity data of three-axis through six-axis data can be set up for the origin of three-axis coordinate system to the human body, with the earth again in order to establish the second coordinate system origin, through comparing the gap between the two coordinate system, we can complete the body posture data decoding. When the six-axis acceleration sensor is worn on the body, the three-axis coordinate system of the sensor itself will coincide with the previous human coordinate system. According to the triaxial acceleration of the six-axis acceleration sensor, we can calculate an acceleration vector an (x, y, z). When the human body is moving, we can use the formula

$$|a| = g = \backslash \mathrm{SQRT}\{x^2 + y^2 + z^2\}$$

to calculate the length of the magnitude of the vector. When the human body rotates, there will be an Angle between the Z-axis of the coordinate system based on the human body and the Z-axis of the earth coordinate system. See Fig. 3.

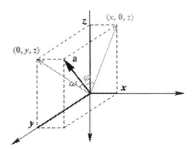

Fig. 3. The deviation between human coordinate system and the standard coordinate system

Through the formula, Roll Angle and Pitch Angle can be further calculated:

$$\mathrm{Pitch} = \mathrm{at\ an} = (x/\mathrm{SQRT}((y*y) + (z*z)));$$
$$\mathrm{Roll} = \mathrm{at\ an} = (y/\mathrm{SQRT}((x*x) + (z*z))).$$

Then Pitch was converted by formula, and Roll Angle unit:

$$\mathrm{Pitch} = \mathrm{Pitch}*(180.0/3.14);$$
$$\mathrm{Roll} = \mathrm{Roll} * (180.0/3.14).$$

At this point, Pitch and Roll Angle units are converted into angles. Due to the problem of the coordinate system, the Yaw Angle data cannot be obtained, and only the measurement of Yaw Angle transformation Angle, namely the angular velocity in the direction of the Z-axis, can be completed. The above formula can also be used to calculate the real-time Yaw Angle data of Z. However, after the six-axis acceleration sensor works for a long time, the Yaw Angle data will drift, resulting in certain errors in the data.

3.3 Pedometer Data Acquisition Unit

Pedometer specific task of data acquisition unit is in real-time acquisition to the human body three-axis direction of the acceleration value, through Kalman filter algorithm to simple filtering of data, then by connecting the data sent to the Bluetooth module on the ARDUINO Android end PC again by Android end PC by step gauge algorithm of three-axis acceleration data analysis, calculate the number of human movement [11, 12].

After the MPU6050 six-axis acceleration sensor collects the human body's acceleration in the direction of three axes, it sends the data to the Android APP through the Bluetooth module. After receiving the data, the Android terminal detects the recorded three-axis acceleration data and time stamp. After that, the acceleration data of the three axes are simultaneously drawn into the original sensor data graph as shown in Fig. 4.

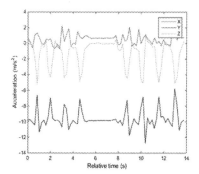

Fig. 4. Triaxial acceleration data graph   **Fig. 5.** Triaxial acceleration scalar diagram

And convert the triaxial acceleration vector at each point in time to a scalar value, where the acceleration scalar is not zero and so you have to subtract the mean to eliminate constant effects like gravity. The drawn data is centered on zero, clearly showing the magnitude of the acceleration peak. Each peak corresponds to the number of steps taken by the human body. The data graph after eliminating the mean value is shown in Fig. 5.

The find peaks function is used to find the local maximum of acceleration. Only the peak with the minimum height above a standard deviation can be considered as a step. The acceleration amplitude data is then used to visualize the peak position, as shown in Fig. 6.

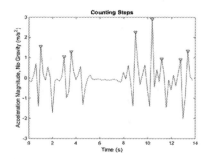

Fig. 6. Peak acceleration figure

4   System Debugging and Data Display

4.1   Test Results of Human Posture Data Acquisition Unit

The human body posture data acquisition unit is worn on the waist of the human body to start collecting human body displacement data. The human attitude data acquisition unit has carried out three experiments respectively:

1. When the human body is still, the triaxial displacement data and attitude data of the human body are collected, as shown in Fig. 7.

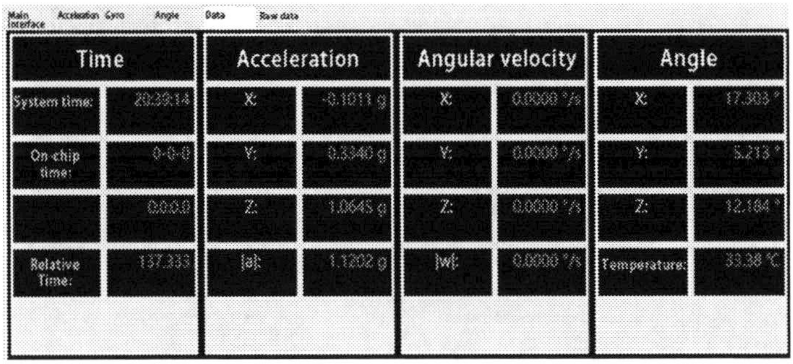

Fig. 7. Triaxial displacement and attitude data of the human body at rest

2. Triaxial displacement and attitude data of the human body when the human body moves from stationary to normal walking, as shown in Fig. 8.

time	ax	ay	az	vx	vy	vz	angx	angy	angz	T(°)
58:42.	0.1265	-0.8203	0.021	0.2441	146.4233	-11.1084	-81.1615	8.9319	-46.1755	37.1653
58:42.	0.0508	-0.7661	0.022	1.6479	143.3105	-10.8032	-81.3702	9.0417	-47.6257	37.1653
58:42.	-0.0439	-0.7178	0.0552	-1.0986	138.916	-13.3057	-81.601	9.1187	-49.032	37.1565
58:42.	-0.1006	-0.6938	0.0757	-5.6152	137.8174	-18.5547	-81.8811	9.1406	-50.4437	37.1565
58:42.	-0.0942	-0.686	0.083	-7.5684	143.0054	-24.0479	-82.1887	9.1022	-51.9104	37.1565
58:42.	-0.0391	-0.6641	0.0996	-6.6528	153.4424	-27.4048	-82.5073	9.0363	-53.4869	37.1741
58:42.	0.02	-0.6274	0.1079	-6.897	164.3677	-28.5645	-82.8424	8.9703	-55.1733	37.1859
58:42.	0.0864	-0.584	0.1055	-10.8643	172.6685	-28.6865	-83.2269	8.8989	-56.9476	37.1771
58:42.	0.1523	-0.5342	0.0977	-16.5405	179.3823	-27.771	-83.6719	8.8385	-58.7823	37.1741
58:42.	0.1797	-0.499	0.0874	-20.813	186.2793	-25.0854	-84.1772	8.7946	-60.683	37.1653

Fig. 8. Triaxial displacement and attitude data of the human body when the human body is moving from stationary to normal walking

Acceleration in the direction of three axes of the human body, as shown in Fig. 9.

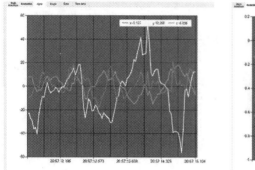

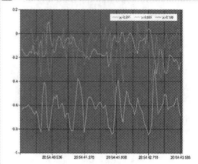

Fig. 9. Triaxial acceleration   **Fig. 10.** Static to normal angular velocity

The angular velocity diagram of the human body in the triaxial direction from stationary to normal walking, as shown in Fig. 10. The triaxial posture data diagram of the human body from stationary to normal walking, as shown in Fig. 11.

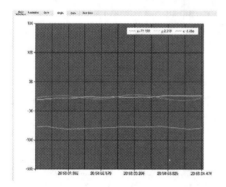

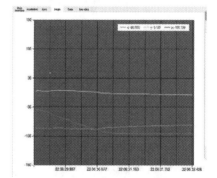

Fig. 11. Resting to normal triaxial position   **Fig. 12.** Change the direction of the three axes

3. Triaxial attitude data when the human body changes its forward direction, as shown in Fig. 12.

4.2   Pedometer Data Acquisition Unit Test Results

The pedometer data acquisition unit is worn on the human arm for testing. When starting to walk, the acceleration data graph of the human body in the direction of three axes displayed by the Android terminal APP is shown in Fig. 13.

When the human body moves forward, the Android terminal APP starts to count steps, as shown in Fig. 14.

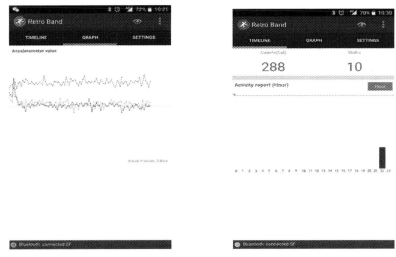

Fig. 13. Triaxial acceleration diagram of the human body displayed by Android terminal APP

Fig. 14. Android terminal APP step counting display diagram

5 Conclusion

This paper analyzes the characteristics and requirements of wireless body domain networks and designs a human displacement data acquisition, analysis and transmission system that can meet its requirements. The whole system is divided into three parts: data collection, data analysis, and data transmission. Human displacement data were determined through the collection of human posture data and pedometer data, and the data were processed through a Kalman filter. The ARDUINO platform converted the measured analog amount into an output digital amount, and finally transmitted the data to the intelligent device through the Bluetooth communication module. As a human wearable device, it should be as light and small as possible to eliminate the sense of wearing a human body. However, due to technical reasons, this system does not meet the requirements of small wearable devices in terms of size and weight, so it needs to be further improved in terms of size and weight. Although the system designed in this paper has simply realized the collection and analysis of all kinds of data when the human body is moving, it has only realized the function. Combined with the future trend of intelligent wearable devices based on a body domain networks, the system still needs to realize intelligent information interaction with the human body in the future, instead of simply collecting human body information. Only through information interaction with the human body can we better serve human beings and realize the combination of people and things to shape wearable devices into an indispensable part of the human body.

Acknowledgment. We gratefully acknowledge anonymous reviewers who read drafts and made many helpful suggestions. This work is supported by Guangxi Vocational Education Teaching Reform Research Project (GXGZJG2018A040), Fundamental Research Funds for the Central Universities of China (328201911), Higher Education Department of the Ministry of Education

Industry-university Cooperative Education Project and Education and Teaching Reform Project of Beijing Electronic and Technology Institute.

References

1. Akyildiz, I.F., Su, W., Sankarasubramaniam, Y., Cayirci, E.: Wireless sensor networks: a survey. Comput. Netw. **38**, 393–422 (2002)
2. Lin, F., Zhou, Y., You, I., Lin, J., An, X., Lü, X.: Content recommendation algorithm for intelligent navigator in fog computing based IoT environment. IEEE Access **7**, 53677–53686 (2019)
3. Lin, F., Lü, X., You, I., Zhou, X.: A novel utility based resource management scheme in vehicular social edge computing. IEEE Access **6**, 66673–66684 (2018)
4. Wang, J., Gao, Y., Liu, W., Wu, W., Lim, S.-J.: An asynchronous clustering and mobile data gathering schema based on timer mechanism in wireless sensor networks. Comput. Mater. Contin **58**, 711–725 (2019)
5. Taleb, T., Bottazzi, D., Nasser, N.: A novel middleware solution to improve ubiquitous health-care systems aided by affective information. IEEE Trans. Inf Technol. Biomed. **14**, 335–349 (2010)
6. Kang, J., Zhang, Y., Nath, B., Yu, S.: Adaptive resource control scheme to alleviate congestion in sensor networks. In: Proceedings of IEEE Workshop on Broadband Advanced Sensor Networks (2004)
7. Gao, D., Zhang, S., Zhang, F., Fan, X., Zhang, J.: Maximum data generation rate routing protocol based on data flow controlling technology for rechargeable wireless sensor networks. CMC-Comput. Mater. Contin. **59**, 649–667 (2019)
8. Zhang, Q., Liang, Z., Cai, Z.: Developing a new security framework for bluetooth low energy devices. CMC-Comput. Mater. Contin. **59**, 457–471 (2019)
9. Zhang, J.: Research on human motion behavior recognition based on acceleration sensor. Autom. Instrum., 228–229 (2016)
10. Lin, F., Zhou, Y., An, X., You, I., Choo, K.-K.R.: Fair resource allocation in an intrusion-detection system for edge computing: ensuring the security of Internet of Things devices. IEEE Consum. Electron. Mag. **7**, 45–50 (2018)
11. Wang, L.: Research on Human Behavior Recognition Technology based on Wearable Sensor Network. Nanjing University, Nanjing (2014)
12. Paek, J., Govindan, R.: RCRT: rate-controlled reliable transport for wireless sensor networks. In: Proceedings of the 5th International Conference on Embedded Networked Sensor Systems, pp. 305–319 (2007)
13. Banimelhem, O., Khasawneh, S.: GMCAR: grid-based multipath with congestion avoidance routing protocol in wireless sensor networks. Ad Hoc Netw. **10**, 1346–1361 (2012)
14. Polastre, J., Hill, J., Culler, D.: Versatile low power media access for wireless sensor networks. In: Proceedings of the 2nd International Conference on Embedded Networked Sensor Systems, pp. 95–107 (2004)
15. Su, J., Lin, F., Zhou, X., Lu, X.: Steiner tree based optimal resource caching scheme in fog computing. China Commun. **12**, 161–168 (2015)
16. Wen, Y.: Research on a Real-Time Fall Posture Detection and Heart Rate Monitoring System. Zhejiang University, Zhejiang (2008)

Data Cleaning Algorithm Based on Body Area Network

Juan Guo[1], Yuzhuo Zong[2], Fang Chen[2], Chao Guo[2(✉)], and Dingbang Xie[2]

[1] Guilin University of Electronic Technology-Vocational and Technical College, Beihai 536000, Guangxi, People's Republic of China
[2] Department of Electronics and Communication Engineering, Beijing Electronic Science and Technology Institute, Beijing 100070, People's Republic of China
guo99chao@163.com

Abstract. With the continuous development of science and technology, especially the development of medical technology, electronic information technology, biotechnology and related materials technology, and the social reality that people's demand for centralized and managed medical services in traditional medical institutions has gradually changed to the demand for new medical services of individualized treatment, disease prevention, and health management, wireless body area network has emerged as the times require, and it provides more perfect supervision and management for people's health. Although the amount of human perception data is large, its value density is very low. Human perception data is very vulnerable to the influence of perception devices, external environment and human psychology and physiology, which results in large fluctuation of perception data and cannot fully reflect the real state of human health. Therefore, data cleaning for biological information in the body area network is particularly important. For the X Y Z triaxial displacement data of biological activity location, a statistical mathematical model represented by a box diagram is proposed to clean the data. The final results show that the scheme improves the accuracy of data in the experiment of motion detection.

Keywords: Wireless body area network · Data cleaning · Box diagram method

1 Introduction

The body network, also known as the body sensation network. All kinds of sensors with communication function attached to the body of the human body and a Body Area Network (BAN) communicate through wireless technology, so it is also called wireless body area network, used for collect all kinds of information about the human body [1, 2]. Due to the increasing desire for health, the lack of medical resources and the rising medical costs in traditional medical institutions have led to the urgent need for new, cheap and convenient medical systems. Wireless body area networks came into being. The role of the wireless body area network is based on the amount of human body data perception. Although the amount of data is huge, its value density is relatively low, mainly because the human body perception data is highly susceptible to equipment, environment and

© Springer Nature Switzerland AG 2020
X. Sun et al. (Eds.): ICAIS 2020, LNCS 12239, pp. 575–585, 2020.
https://doi.org/10.1007/978-3-030-57884-8_51

personal physical and mental state [3]. The data is unstable and has a large amount of error data. It cannot correctly represent the actual state of the individual, and is prone to errors in judgment, resulting in a waste of resources. Through the process of data cleaning, the collected data can be inspected and verified, the duplicate information is deleted and the existing errors are corrected, so that more accurate data can be obtained for correct analysis.

At present, the research on data cleaning based on body area network in foreign countries is mainly reflected in the clearing of registry attributes of data sets, the cleaning of repeated records in data sets, the application of data cleaning in data warehouses, and the independent cleaning algorithm of search domains [4, 5]. The application of data cleaning in the data warehouse is relatively mature in the research, and the other algorithms are either limited or difficult to implement, or the data collection is too specific. However, there are few domestic researches on this aspect, and it is relatively late. There are few data cleaning software for Chinese data.

Aiming at the above problems, this paper designs a data cleaning algorithm based on the body area network. The algorithm is based on the IDLE data computing environment, and it is used to calculate the most common human XYZ triaxial displacement in the domain. The data under the actual situation is statistically analyzed to study the data distribution characteristics of the scorpion domain network [6]. According to the characteristics of various databases and various data cleaning algorithms of the existing body area network, the deviation is greatly discarded, and the data with the small deviation is used to perform the interpolation of the data in the database. A data cleaning algorithm for the biometric information of the domain is obtained.

Section 1 introduces the research background of the physical domain network and data cleaning algorithms, the research status at home and abroad, and systematically introduces the development details of the physical domain network and data cleaning algorithms. Section 2 focuses on the description of the most common methods used in cleaning and provides a reference for the following design. Section 3 introduces the unified statistics model applied in the data cleaning of the physical domain network, laying a theoretical foundation for the proposal of the scheme. Section 4 studies the data cleaning of the biological information of the scorpion domain, and obtains the data cleaning algorithm for the biometric information of the biological information of the scorpion domain. Section 5 runs the python program through IDLE, and the algorithm is written and the output of the distribution map is executed. Section 6 mainly summarizes and proposes ideas for the areas that need to be improved and explores future research directions.

2 Preliminary

This paper mainly introduces the use of related techniques such as statistics, data mining or pre-defined definition of clean-up rules to convert the viscous data into data that meets the data quality requirements [7, 8]. As the design of the data cleaning process becomes more and more difficult as the amount of data increases, the data model becomes more complex, and the user needs change, it is more and more difficult. Therefore, the purpose of this research is to fully study data cleaning technology. On the basis of theory, a practical data cleaning framework is designed based on the characteristics of existing systems and combined with specific cleaning algorithms, which will integrate data uploading into data cleaning and data synchronization to provide reliable data quality assurance for decision making. Ultimately achieve effective control of data quality [9–11].

This paper mainly studies from the following four aspects: research and improvement of repeated data cleaning algorithms, correlation relationship cleaning, dictionary data cleaning, and overall design of the cleaning framework [12]. The data cleaning flowchart is shown in Fig. 1.

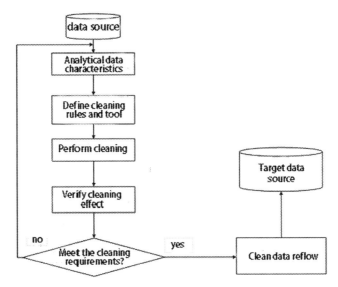

Fig. 1. Data cleaning process

2.1 Mode Layer Data Cleaning Method

The number of model layers is inherently unreasonable. There is a lack of integration constraints between the design and the attributes. It mainly includes naming conflicts and structural flaws [13]. The former means that the same name represents different pairs of objects, while the latter means that different names represent the same object, causes different objects in different data sources to be represented.

2.2 Example Layer Data Cleaning Method

The dirty data cleaning method of the instance layer is mainly divided into attribute cleaning and repeated record cleaning. Among them, the methods commonly used for the missing data values are as follows:

Delete missing values: When the number of samples is large, and the sample with missing values is relatively small in the whole sample, in this case, the samples with missing values can be directly discarded. This is a very common method [14].

Mean padding method: divide the data into several groups according to the attribute with the largest attribute correlation coefficient of missing values, and then calculate the mean value of each group separately, and put these average values into the missing values [15].

Hot card padding: How do I fill a hot card for a variable that contains missing values? For variables that contain missing values, the method of populating the hot card is to look up similar objects in the database and then use the values of the objects to be filled, most commonly using a matrix of correlation coefficients to determine variables with missing values (e.g. X variables) The most relevant variable (such as the Y variable), then, all variables are sorted by the value of Y, and then the missing value of the X variable can be replaced with the state data before the missing value [16].

There are similar methods such as the nearest distance decision filling method, the regression filling method, the multiple filling method, the K-nearest neighbor method, the ordered nearest neighbor method, and the Bayesian-based method.

For the processing of outliers, the following methods are commonly used. Simple statistical analysis: A simple descriptive statistical analysis of the data can be performed at the outset. For example, the maximum and minimum values can be used to determine whether the value of this variable exceeds a reasonable range [17].

3∂ Principle: If the data follows a normal distribution, under the principle of 3∂, the outlier is a value of a set of measured values that deviates from the mean by more than 3 standard deviations. If the data follows a normal distribution, the probability of a value other than the average value of 3∂ is P $(|x-u| > 3\partial) <= 0.003$, which is a very small probability event. If the data does not obey the normal distribution, it can also be described by how many times the standard deviation is far from the average value.

Model-based detection: First, establish a data model. The anomalies are those that cannot fit perfectly with the model. If the model is a collection of algebraic clusters, the exceptions are objects that do not significantly belong to any cluster; when using regression models, the anomalies are relative to an object far from the predicted value. (a popular algebraic cluster is a set of common zeros defined by a number of multivariate polynomial equations. If the algebraic cluster can be defined by an equation, it is called a hypersurface) [18, 19].

Based on distance: You can usually define proximity metrics between objects, which are objects that are far away from other objects.

Density-based: When a point's local density is significantly lower than most of its neighbors, it is classified as an outlier, suitable for non-uniformly distributed data [20].

Based on clustering: cluster-based outliers, one object is an outlier based on clustering. If the object does not strongly belong to any cluster, the effect of outliers on the initial clustering: if clustering is used to detect outliers Because the outliers affect clustering, there is a problem: whether the structure is valid. In order to deal with the problem, the following methods can be used: object clustering, deleting outliers, and clustering again (this does not guarantee optimal results) [21, 22].

3   The Statistical Model Applied in Body Area Network Data Cleaning

3.1   Noise Data (Outliers) Cleaning

Identify data in a single-dimensional dataset (outliers) using data distribution features and box plot methods.

Suppose a set of data is as follows:

Serial number: 1 2 3 4 … n

Data: E1E2E3E4…EnE1E2E3E4…En

Generally, for data sources that are not very discrete, the data itself will be distributed in a certain area, so the data itself is used to identify the noise data, and the outliers and outliers are identified in the data center domain according to the box plot method. First, the data set is divided into αn intervals (α can take 1, 10, 100, 1000), Interval size is $\theta = (Max\{E_1, E_2, \ldots, E_n\} - Min\{E_1, E_2, \ldots, E_n\})/\alpha n$

Intercept the interval in the data distribution set as the data concentration domain, and find the data concentration domain to form a new data group E as shown in Fig. 2.

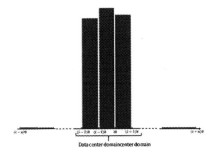

Fig. 2. Formation of the data center domain

Use the box diagram (as shown in Fig. 3) to eliminate the outliers of the new dataset, obtain the non-outlier data set [Q1−3IQR, Q3 + 3IQR], and then obtain the non-abnormal data set [Q1−1.5 IQR, Q3 + 1.5IQR], get the target data. Q1: first quartile, Q3: third quantile, IQR; interquartile range IQR = Q3−Q1;

3.2   Out of Stock Failure Value Cleaning

In the most ideal case, each piece of data in the data set should be complete. However, most of the existing data sets contain noise data. The reasons for data loss in different data sets are also various, such as human input. Carelessness or unwillingness to disclose in the case of certain special circumstances. In the data set, if the attribute value of a piece of data is marked as blank or "−", etc., indicating that it is meaningless, it is considered to be the existence of data is a missing value, this data is an incomplete data. Here we are going to use the algorithm based on the k-NN neighbor filling technique to process the missing data.

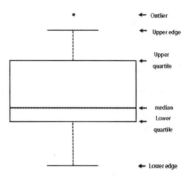

Fig. 3. Box diagram icon

The first is the k-NN classification. First, the sample is taken by the random sampling algorithm, and each data of the sample is described by n-dimensional numerical attributes. Each data represents a point of the n-dimensional, space, so that all sample samples are stored in In the n-dimensional model space, an unknown sample is given, and then the entire n-dimensional space is searched by the k-NN classification method to find the k sample samples closest to the unknown sample, so by definition, this The k sample samples are k "neighbors" of the unknown sample. After the unknown sample is found, the "proximity" of the neighbor samples is defined by the Euclidean distance, according to the definition. The Euclidean distance of any two points $X = (x\_1, x\_2,..., x\_n)$ and $Y = (y\_1, y\_2,..., y\_n)$ in the n-dimensional space is:

$$d(X, Y) = \sqrt{(X - Y)^2} \tag{1}$$

Let z be the unknown sample that needs to be tested, $z = (x'', y')$, all training samples $(x, y) \in D$, the nearest sample of the unknown sample set to D_z, at k-NN In the classification algorithm, k is the number of nearest samples, and D is the sample set of neighbor samples. After understanding the above, the data is normalized to eliminate the inaccurate data cleaning caused by different units.

$$f(x') = (x' - \min(x\_1, x\_2, x\_3, \ldots, x\_n))/(\max(x\_1, x\_2, x\_3, \ldots, x\_n) - \min(x\_1, x\_2, x\_3, \ldots, x\_n))$$

Then, calculate the distance d between the unknown sample z and each training sample (x, y), and put the k training samples closest to the sample z into the sample set D_z just preset. In the actual operation, if the test sample is determined and the k "neighbors" of the test sample are determined, the mean value corresponding to the k neighbors can be used to replace the missing value of the test sample.

3.3 Repeated Value Cleaning

Field Similarity Definition: A metric that represents the similarity of two fields based on the content of two fields. The range is 0–1, the closer to 1, the greater the similarity.

Boolean field similarity calculation method: For Boolean fields, if the two fields are equal, the similarity between the two is 0. If they are different, the similarity between the two is 1.

Numeric field similarity calculation method: For numerical fields, the relative difference of the calculated numbers can be used. Use the formula:

$$S(s\_1, s\_2) = |s\_1 - s\_2|/(\max(s\_1, s\_2)) \tag{2}$$

Character field similarity calculation method: For character fields, the simpler method is to divide the number of characters matching each other in the two strings by the number of characters in the two strings, using the formula:

$$S(s\_1, s\_2) = |k|/(((|s\_1| + |s\_2|)/2) \tag{3}$$

The k in the formula refers to the number of characters matched. For example, the string s1 = "apple", the string s2 = "apple pen" uses the above character type field similarity calculation formula to get his similarity as

$$S(s\_1, s\_2) = 5/(((|5| + |8|)/2) \tag{4}$$

Set the threshold value. When the field similarity is greater than the threshold value, identify the duplicate field and issue a reminder, and then delete or retain it according to the actual situation.

4 Data Cleaning Algorithm Preparation and Implementation

First, when importing, first import the database. In order to facilitate the operation and writing the program, first import the data in text format into Excel, as shown in Fig. 4.

A01	010-00CY	6.3379E+	27.05.	4.0629310	1.8924342	0.5074254	walking
A01	020-000-C	6.3379E+17	27.05.20C	3.203582525	1.513172626	1.083313942	falling
A01	020-000-C	6.3379E+17	27.05.20C	3.258224487	1.424609661	1.176932335	falling
A01	020-000-C	6.3379E+17	27.05.20C	3.419891119	1.463585615	1.195048332	falling
A01	020-000-C	6.3379E+17	27.05.20C	3.666965961	1.400693417	1.070500493	falling
A01	020-000-C	6.3379E+17	27.05.20C	4.500741005	1.371175289	1.271560907	falling
A01	020-000-C	6.3379E+17	27.05.20C	4.153721809	1.253938437	0.619487166	falling

Fig. 4. Importing the sensor database of Excel

The database format is as follows:

User ID	Sensor ID	Time-stamp	Time	X	Y	Z	States

Import the database and process the data in the three columns e, f, and g.

The second is to construct a distribution map of the raw data, where the horizontal axis represents displacement and the vertical axis represents frequency.

A data distribution map of the X, Y, and Z triaxial coordinate data and a distribution map in the three-dimensional coordinate system are obtained. As shown in Fig. 5, 6, 7 and 8.

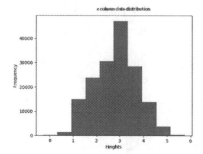

Fig. 5. e column (X-axis) raw data distribution

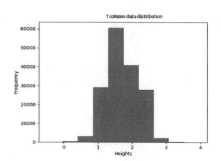

Fig. 6. f column (Y-axis) raw data distribution

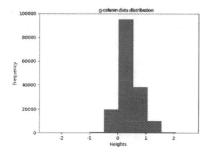

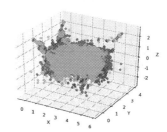

Fig. 7. g column (Z-axis) raw data distribution

Fig. 8. (x, y, z) three-axis coordinates of the original data distribution in the coordinate system

Finally, the cleaned data distribution map is drawn and the cleaned data is still output in the form of an Excel spreadsheet. See Fig. 9, 10, 11 and 12.

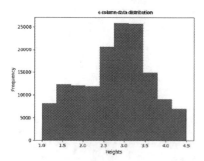

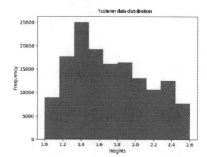

Fig. 9. Distribution of data after cleaning in column e (X-axis)

Fig. 10. Distribution of data after f column (Y-axis) cleaning

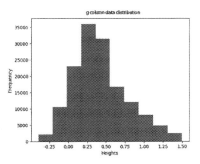
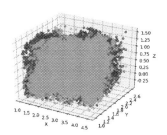

Fig. 11. Data distribution after g column (Z-axis) cleaning

Fig. 12. Data distribution of (x, y, z) three-axis coordinates after cleaning of the coordinate system

5 Achieved Purpose

1. The dirty data in the database is reduced, and the Excel file memory is reduced from 12.7 M to 11.6 M.
2. When the database is not cleaned, the detection accuracy of the input fall detection algorithm is about 50%–60%, and it can reach 80%–90% after cleaning. As shown in Fig. 13–14.

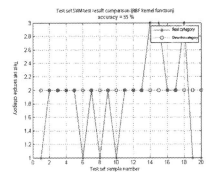
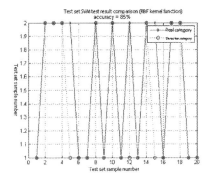

Fig. 13. Accuracy of data without data cleaning

Fig. 14. Accuracy of data after data cleaning

6 Conclusion

Based on the characteristics of various databases and various data cleaning algorithms in the existing body area network, based on the research of existing data cleaning algorithms, this paper has completed the various targets of the biological activity positioning data of the body and the area network. The established content.

The data cleaning algorithm designed in this paper is based on the existing data cleaning algorithm and combines the statistical related knowledge. The traditional data cleaning algorithm is mostly based on the deletion of data, and the deviation is extremely discarded. The combination of small deviation is small. Use the data in the database to perform the interpolation. A data cleaning algorithm for the biometric information of the domain is obtained.

Acknowledgment. We gratefully acknowledge anonymous reviewers who read drafts and made many helpful suggestions. This work is supported by Guangxi Vocational Education Teaching Reform Research Project (GXGZJG2018A040), Fundamental Research Funds for the Central Universities of China (328201911), Higher Education Department of the Ministry of Education Industry-university Cooperative Education Project and Education and Teaching Reform Project of Beijing Electronic and Technology Institute.

References

1. Chen, J.C., Cheng, C.-H., Huang, P.B.: Supply chain management with lean production and RFID application: a case study. Exp. Syst. Appl. **40**, 3389–3397 (2013)
2. Jeffery, S.R., Garofalakis, M., Franklin, M.J.: Adaptive cleaning for RFID data streams. In: Vldb, pp. 163–174. Citeseer (2006)
3. Hamad, F., Zraqou, J., Maaita, A., Taleb, A.A.: A secure authentication system for epassport detection and verification. In: 2015 European Intelligence and Security Informatics Conference, pp. 173–176. IEEE (2015)
4. Antoniolli, P.D.: Information technology framework for pharmaceutical supply chain demand management: a Brazilian case study. Braz. Bus. Rev. **13**, 27–55 (2016)
5. Chiou, J.-C., Hsu, S.-H., Kuei, C.-K., Wu, T.-W.: An addressable UHF EPCGlobal Class1 Gen2 Sensor IC for wireless IOP monitoring on contact lens. In: VLSI Design, Automation and Test (VLSI-DAT), pp. 1–4. IEEE (2015)
6. Lin, F., Su, J.: Multi-layer resources fair allocation in big data with heterogeneous demands. Wirel. Pers. Commun. **98**, 1719–1734 (2018)
7. Agarwal, M., Sharma, M., Singh, B.: Smart ration card using RFID and GSM technique. In: 2014 5th International Conference-Confluence the Next Generation Information Technology Summit (Confluence), pp. 485–489. IEEE (2014)
8. Jeffery, S.R., Alonso, G., Franklin, M.J., Hong, W., Widom, J.: A pipelined framework for online cleaning of sensor data streams. In: 22nd International Conference on Data Engineering (ICDE'06), pp. 140–140. IEEE (2006)
9. Lin, F., Zhou, Y., You, I., Lin, J., An, X., Lü, X.: Content recommendation algorithm for intelligent navigator in fog computing based IoT environment. IEEE Access **7**, 53677–53686 (2019)
10. Zhou, Y.: Research and Application of Data Cleaning Algorithm. Qiingdao University, Qingdao (2005)
11. Fan, J.: Research on Isolated Point Detection Algorithms in data mining. Central South University, Changsha (2009)
12. Park, S., Kim, Y.: User profile system based on sentiment analysis for mobile edge computing. Comput. Mater. Contin. **62**, 569–590 (2020)
13. Xia, W.: Research and Improvement of Job Scheduling Algorithm under Hadoop Platform. South China University of Technology, Guangzhou (2010)
14. Zhou, Z.: Research on data Cleaning Method Based on the Data Warehouse. Donghua University, Shanghai (2004)

15. Li, L.: A data cleaning method based on data quality dimensions. Sci. Technol. Innov. Appl. **21**, 1–5 (2017)
16. Lin, F., Zhou, Y., Pau, G., Collotta, M.: Optimization-oriented resource allocation management for vehicular fog computing. IEEE Access **6**, 69294–69303 (2018)
17. Wang, Y., Subhan, F.: Fuzzy-based sentiment analysis system for analyzing student feedback and satisfaction. Comput. Mater. Contin. **62**, 631–655 (2020)
18. Luo, Q., Wang, X.: Analysis of data cleaning technology in data warehouse. Comput. Program. Skills Maintenance **2** (2015)
19. Cao, J., Diao, X., Chen, S., Sun, Y.: Data cleaning and its general system framework. Comput. Sci. **39**, 207–211 (2012)
20. Sun, H., Chen, C., Ling, Y., Yang, M.: Cooperative perception optimization based on self-checking machine learning. Comput. Mater. Contin. **62**, 2747–2761 (2020)
21. Guo, Z., Zhou, A.: A review of data quality and data cleaning research. J. Softw. **13**, 2076–2082 (2002)
22. Su, J., Lin, F., Zhou, X., Lu, X.: Steiner tree based optimal resource caching scheme in fog computing. China Commun. **12**, 161–168 (2015)

Improving Performance
of Colour-Histogram-Based CBIR Using Bin
Matching for Similarity Measure

Martey Ezekiel Mensah[1]([⊠]), Xiaoyu Li[1], Hang Lei[1], Appiah Obed[2],
and Ninfaakanga Christopher Bombie[2]

[1] University of Electronic Science and Technology of China, Chengdu 611731, China
martey003@std.uestc.edu.cn
[2] University of Energy and Natural Resources, P.O. Box 214, Sunyani, Ghana

Abstract. Content-Based Image Retrieval (CBIR) systems work by searching
huge databases for similar images that match a query image. The CBIR systems
depend on computing similarity between two images to retrieve images of interest.
The choice of suitable similarity measuring tool is key for effective and efficient
retrieval of images. Predominantly similarity metrics such as Euclidean, Manhattan, City block distances among others are used extensively to compute the
images similarity measure when the conventional colour histogram is used for
image indexing. However, each of these similarity metrics suffers from issues of
non-similar images having the same histogram and outliers in the distribution of
colour content in images. In this paper, a proposed bin-by-bin inspections and
classification for the measurement of similarity is presented. The approach distinguishes between the queried image and the target image, to obtain a more robust
outcome. The outcomes stood superior to other state-of-the-art similarity metrics
in respect to retrieval accuracy.

Keywords: Similarity measure · Distance metrics · Colour histogram · Bin
matching · Content Based Image Retrieval

1 Introduction

The Content-Based Image Retrieval (CBIR) has been an active research area for the last
three decades [5]. The concept of "Content-based" means that the search analyzes the
contents of the images rather than the metadata such as keywords, tags, or descriptions
associated with the images. CBIR systems normally store the images' characteristics in
signature files that are actually used to perform the search. This improves computational
time associated with the search process, however, the time is still high for retrieval when
several features are used to generate signature file [15, 16]. This challenge has eventually
led to smaller devices outsourcing on the cloud for high retrieval performance [9]. Thus
perfect image retrieval and retrieval time are still prominent challenges in the CBIR
domain [22].

© Springer Nature Switzerland AG 2020
X. Sun et al. (Eds.): ICAIS 2020, LNCS 12239, pp. 586–596, 2020.
https://doi.org/10.1007/978-3-030-57884-8_52

Histogram based methods have proven to be simple and fast to index and retrieve images from the database with satisfactory results [4]. Specifically, Conventional Colour Histogram has proven to be effective for small databases or dataset. However, this indexing method fails at some point to retrieve images with similar semantic features due to failure to integrate other image descriptor such as texture and edges [14]. Various approaches to improve the performance of CCH based CBIR have been proposed in order to fix this challenge related to it. For example, the use of fuzzy histograms for indexing has proven better than ordinary CCH [1–3]. Generally, for the CCH for CBIR task or application, color histograms are used to represent images as points in a low-level visual feature space, and similarity measures are used to retrieve images with smaller distances as best matches for the search operation [8–15]. Given a distance metric, two images with shorter distance are deemed to be more similar than images that are far apart [6]. Good distance measure should not exaggerate image noise or outliers [20]. The method used to determine the distance between two images therefore become crucial for Histogram based CBIR.

The well-known problem with these similarity measures is the semantic gap [16]. Two images separated by large distance could share the same semantic content as well as two images separated by a very small distance could have different semantic content. This may be due to the fact that diverse images in a given image database or dataset may have similar colour histogram. This work seeks to tackle such influences of large difference between two bins which leads to the semantic gap.

The main contribution of our work is proposing a new bin by bin inspection and classification for the measurement of similarity. This approach discriminates between the queried image and the target image, which leads to a more robust result outperforming the state of art similarity measures.

The rest of the paper is organised as follows; we present some research trend analysis in Sect. 2, then proposed method, in Sect. 3. Next, we present some details of our experiment, of our study. We present results and discussions of this study in Sect. 4. We conclude our work and present scope for future research in the last section.

2 Related Works

Histogram Euclidean Distance (HED), Histogram Intersection Distance (HI), Histogram Manhattan Distance (HMD) and Histogram Quadratic Distance Measure (HQDM) appear to be prominent in the domain of CBIR as a result of satisfactory output. Several works have used the HED, but have performed generally poorly as compared to the city block [20]. The challenge posed by the selection of a particular similarity measure used by histogram based CBIR can easily be seen with results of various research works. For example, [22] reported that after running CBIR algorithm based on wavelet and using the three major distance calculation techniques, the final results indicate that Euclidean Distance and City block distance provided good results [21]. In an attempt to solve the issues related to the Minkowski base distance measure that are used as similarity measure, authors [2–13] proposed new distance measures with various degrees of success. For example, the Integrated Histogram Bin Matching (IHBM) proposed by [7] works slightly better than the HI (City block) and the QB methods. Some of the results in their work indicated that for retrieved images to match the top 5 and 10 images from a dataset, the HI and HQDM worked better than IHBM and HED. The IHBM method proved as metric which satisfies non-negativity, commutative and triangle inequality properties as compared with three existing methods i.e. HI, HED and HQDM. IHBM works well when considering the top 15 to 20 images retrieve with low semantic difference between queried image and retrieved images.

The challenge of semantic gap continues to be a major issue in the domain of CBIR or QBIC as a result of the limitations of various statistical models that are applied in the domain for similarity measure. The mathematical estimation of similarity based on distance measure between two images is popular and various distance measurement approaches such as Euclidean, Manhattan, Canberra, Bray-Curtis, Square Cord, and Square-Chi have been applied in order to determine the similarity among images. However, the L2 norm has been widely used as seen in research works [13–22]. Interestingly, the key assumption of similarity function remains unchanged as indicated by [1], in that the similarity between two images is the inverse to their distance. Evidently, this assumption may not always hold in reality, which always leads to the semantic gaps exhibited by various CBIR methods. The gap is widely projected by histogram based CBIR as a result of the fact the diverse images may have the same histogram signature that can lead to wrong selection of images with respect to image colour features [16].

In an attempt to fix the issue of similarity between two images, the paper Similarity Beyond Distance Measurement, [1] established two factors for the determination of similarity between two images. These are:

Separation Distance: the distance of two images in the space of low-level visual features. The closer the two images in low-level visual features, the more similar they are.

Clustering Likelihood: the likelihood for the two images to be clustered together. The more likely the two images are to be clustered together, the more similar they are.

The use of L2 norm distance for CCH has proven not to be efficient always with image retrieval. This has led to a complete redesigning of image signature and retrieval approaches that deviate from the traditional histograms (colour, texture, etc.). [23] proposed a new method of measuring similarity between two images histogram vectors by considering each individual bins. In their work, each colour bin from the queried and targeted is compared and it is accepted as similar bin or rejected based on a defined acceptable deviation ratio between the two bins. The acceptable deviation must however be determined by the user and its value generally affect the result of the queried images. We propose a method that also uses the bins comparisons since it provides a better comparison of signature vectors. This we believe can narrow the semantic gap that is found with the use of the Minkowski based similarity measure.

a. Statistical Distance Metrics
The basic principle behind the estimation of distances between two given vectors or matrices is to determine how the two can be classified as the same, close or far apart. This concept has largely been applied in the domain of CBIR with acceptable results. Various distance metric approaches have been proposed with the Minkowski being very

prominent in the determination of CBIR. In the special case of p = 1, the Minkowski metric give the City Block metric, for the special case of p = 2, the Minkowski metric gives Euclidean distance, and for the special case of p = ∞, the Minkowski metric gives the Chebychev distance metric.

Minkowski distance

$$dist = (\sum_{k=1}^{n} |p_k - q_k|^p)^{\frac{1}{p}} \tag{1}$$

a. City Block (Manhattan, Taxicab, L1 Norm) Distance

The city block is a Minkowski formula with p = 1 and eliminate the squaring of the difference of two values. One of its advantages may be the low computational time associated with it as compared to the rest of the special cases.

$$dist = (\sum_{k=1}^{n} |p_k - q_k|) \tag{2}$$

b. Eucliden Distance Metric This method is quick robust for colour histogram based CBIR with good precision and recall performance. The Euclidean distance is a special case of Minkowski with p=2.

$$dist = \sqrt{\sum_{k=1}^{n} (p_k - q_k)^2} \tag{3}$$

C. Chebychev Distance Special case of p = ∞, gives the Chebychev Distance

$$dist = \max\{|x-a_i|, \ |y-b_i|\} \tag{4}$$

When the Minkowski is used as similarity measure for CCH based CBIR, its effect of summation affect the output immensely. That is, the outlier found in the distribution of data always influences the accuracy of the result. The use of statistical mean or average for prediction of a value in a given dataset may suffer from poor performance as a result of the nature of distribution of pixel values.

b. Weighted Fuzzy Correlation for Similarity Measure of Colour-Histograms

In order to improve the retrieval accuracy of CBIR, [23] introduced the concept of analyzing each bins in both the queried and the targeted images before their cumulative grading was done. All retained colour matched peaks are in the histograms examined and their fuzzy similarity is expressed by the following membership function as illustrated in Eq. 5:

$$\mu_s(h_l, h'_{l'}) = \frac{\min(h_l, h'_{l'})}{\max(h_l, h'_{l'})} \tag{5}$$

Again a fuzzy theory operation of "α-cut defuzzification" is conducted, to distinguish the colour-matched peaks into height-matched or not height-matched. For membership function values larger than α, $\mu_s(h_l, h'_{l'}) = 1$, which refers to peaks of the compared images being height-matched, otherwise, $\mu_s(h_l, h'_{l'}) = 0$.

R_h, of the fuzzy correlation function for the two discrete curves, namely the two peak distributions in the two colour histograms, can be calculated as the membership function of histogram similarity

$$R_h = \sum_{l,l'=1}^{M} H_l \mu_{\alpha2} \left(h_l, h'_{l'} \right) \tag{6}$$

The formula used in Eq. 6 does not necessary make a distinction between the queried and target images. That is by using $\mu = min(h_i, h_i')/max(h_i, h_i')$, the actual deviation or aberration as suggested by the paper may not work well trying to estimate how one image is closer to another. The Fig. 1 illustrates the concept of distances between three points on a graph.

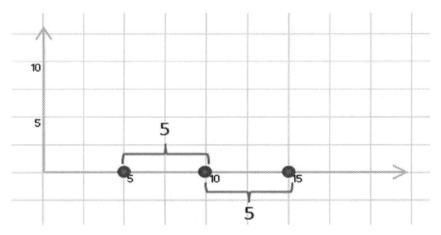

Fig. 1. Three points (5,0), (10,0), (15,0)

Given that 5, 10, and 15 are values for three (3) images (A, B, C respectively) for a particular bin, the above proposed method will generate results that suggests image C is closer to image A. When using B as a query image, $\mu(B,A) = 5/10 = 0.50$, and $\mu(B,C) = 10/15 = 0.6$ will suggest that the distance between B and A is closer than B and C. This can lead to poor performance after the estimation of μ and the use of the α-cut to determine whether bins are similar or not. However, a good selection of α can also lead to the method performing better than the HED as the results in the paper indicated.

This work addresses the semantic gap problem caused by the huge difference between the two bins when the Euclidean distance is used to estimate similarity value. Our approach is similar to what [18] proposed, but we differentiate between the queried image and the target image, to attain a more reliable result.

3   Proposed Method

The bin by bin distance function compares the matching bin of two feature vectors. Let hq = {hq₁, hq₂, hq₃, ….. hqₖ} be feature vector for queried image and ht = {ht₁, ht₂, ht$_3$,..., ht$_k$} be the feature vector for target image, then the bin by bin distance function compares hq$_i$ only with ht$_i$ for all i.

The method can be simplified as follows

- Determine the deviation percentage of two given bins. This can be done by $d_i = (|ht_i - hq_i'|/hq_i)$.

  - di : *deviation ratio between two histogram bins*
  - ht_i : *bin i of target image histogram*
  - hq_i : *bin i of query image histogram*

- Equation 7 is used to classify the percentage deviation

$$C_i = \begin{cases} 1 & \text{if } d_i < \text{T1 Two bins are similar} \\ 0 & \text{if } d_i >= \text{T1 and } d_i <= \text{T2 no judgment} \\ -1 & \text{if } d_i > \text{T2 Two bins are dissimilar} \end{cases} \tag{7}$$

C_i : Classification of bin i for histograms
T1: Threshold for classifying two bins as similar.
T2 : Threshold value for classifying two bins as dissimilar.

$$\text{Similarity Measure (SM)} = \sum_{i=1}^{n} Ci \tag{8}$$

In this method, the higher the SM value, the similar the images are. The method proposed three concepts for the comparison of two bins. The outcome of the comparison must be classified as similar bins, non-similar and no judgment. The similar bins are assigned 1, while dissimilar ones are assigned -1. This similarity measure does not solely focus on similar bins to make judgment, but dissimilar bins are also factored into the decision process. T1 and T2 thresholds values need to be determined in order to classify bins into respective blocks. Our experiment used thresholds of 0.25 for T1 and 0.75 for T2.

When image A, B, and C in the example in Fig. 1 are applied to our proposed method, $\mu(A,B) = (|5 - 10|/10) = 0.5$ and $\mu(C,B) = (|15 - 10|/10) = 0.5$. This suggests that the two images are equal in terms of distance and therefore their similarity measure been the same.

3.1 Test Data and Features

The image set used in our evaluation is the Corel Image dataset (http://www.ci.gxnu.edu.cn/cbir/Dataset.aspx-Corel10K). The dataset contains 10,000 images with 100 semantics categories. Each semantic category is made up of 100 images. The images in the dataset are save using the format: CategoryNumber_SerailNumber. You can easily determine an image's category or group by extracting the CategoryNumberfrom the filename of the image. 10 images from each semantic category as defined by the corel image dataset were selected for the experiment totaling 1,000 in all for the experiments.

RGB image format was used for creation of indexing for images in the database. RGB image format which use 8 bits for each colour channel were converted to another RGB image format which uses 2 bits for each colour channel. 6 bits RGB colour images were generated from 24 bits RGB images from the database. The number of colours in the original RGB images was therefore reduced to 64 colours, which was used to generate the colour histogram. The histograms were then normalized and used for indexing an image in the database. That is, the Conventional Colour Histogram of 64 bits was used for indexing all images in the database. The normalized histograms were then stored in the database and used for the evaluation.

3.2   Testing of Algorithm

The traditional method used by most CBIR to evaluate a retrieval measure is the Recall and Precision. The formula can be illustrated below and has been effectively used by a lot of works.

$$RR = Number\ of\ Relevant\ Images\ Retrieved$$

$$TR = Total\ Number\ of\ Images\ Retrieved$$

$$RD = Number\ of\ Relevant\ Images\ in\ the\ Database$$

$$Precision = RR/TR$$

$$Recall\ = RR/RD$$

The accuracy method for the evaluation of retrieval measure has also seen a lot of application in the CBIR domain. [6] also used a different approach to evaluate the strength of retrieval method in their work "similarity beyond metric". The accuracy method was used to evaluate the performance of the proposed method against the popular Minkowski distances used for the similarity measure. The accuracy method can be illustrated below.

$$Accuracy\ =\ (Relevant\ Images\ Retrieved\ in\ top\ T\ Returns)/(T)$$

To determine the accuracy of each similarity measurement.

1.Select a query image from the Corel Image dataset.

2.Determine the category the selected image falls in. This can be done by using the image's filename as explained earlier.

3.Generate the index for that image using the described method for generation of indexing.

4.Request from the user the number of best or top matches the algorithm is to return. This is the T variable in the accuracy formula.

5.For all the images returned at step 4, determine how many of them match the category of the queried images which was determined at step 2.

6.Calculate the accuracy of a similarity measure by using the values generated by step 5 divided by T.

4   Results

The 1,000 selected images for the evaluation of the City block, Euclidean, Weighted Fuzzy Correlated and the Proposed Methods used the above steps to calculate their accuracies.

The second experiment performed highest ranking count. Here the maximum accuracy for each experimental test case or query image is determined and a count of all the similarity measure methods that were able to match the maximum accuracy is done.

4.1   Accuracy Percentages

Figure 2: illustrates the chart of the average accuracy for the 1000 selected query images for five (5) T values. The T values were 10, 20, 30, 40 and 50. The result indicates that the proposed method performs slightly better than that of the benchmark method.

It was also observed that as T increases, the performance of the proposed declined and approached that of the City block. Figure 3 illustrates a chart for the result when ten (10) T values (10, 20, 30, 40, 50, 60, 70, 80, 90 and 100) were used.

4.2   Top Matching Similarity Measure Count

The next analysis counted the number of times each method topped the accuracy values or equaled the maximum accuracy value of the four (4) similarity measurement methods for each of the 1000 retrieval test cases. It was done by simply selecting the highest accuracy for each case and identifying the methods that generated accuracy level equal to the highest accuracy determined. It is possible for all methods to be counted for a case when all their accuracy values are equal. Figure 4 is a chart that illustrates results of the counting experiment

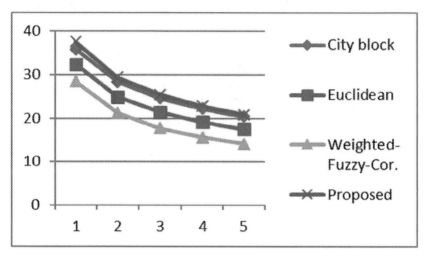

Fig. 2. A chart representing result of average accuracy for the 1000 selected query images for five (5) T values

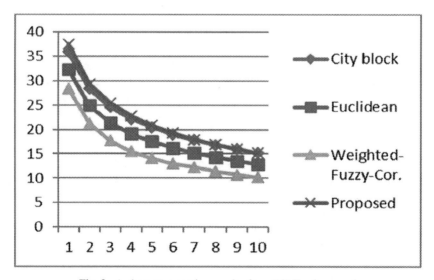

Fig. 3. A chart representing result of ten (10) T values used

4.3 Discussion

The results of the initial experiment indicated that the performance of the proposed method was about 2% better than the city block distance metric which generated second best results. The second experiment also indicates that for about 62% of the times, the proposed method selected the highest number of relevant images or among the similarity measures that selected the highest number of relevant images in all the 1000 test cases. The city block registered about 53%. It must also be indicated that the selected threshold

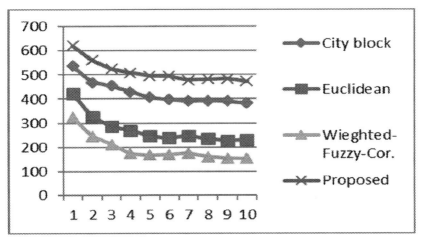

Fig. 4. Chart of maximum accuracy matching count for each query case

(T1 and T2) values for the proposed method affect its results. However, the selected thresholds used for these experiments indicate that the proposed method can be used in place of all the other three (3) methods when using the conventional colour histogram for indexing colour images in CBIR task.

5 Conclusion

This paper proposes a bin-by-bin matching for histogram based CBIR similarity measure. Experimental results indicate that the proposed method performs relatively better than most of the distances metrics used for similarity measure with the Conventional Colour Histogram. The proposed method can therefore be used in place of the city block and the Euclidean distance metrics that have been widely used as a means to measure similarities of images. Future work considerations will involve implementing adaptive autogenerate of thresholds in order to improve the retrieval accuracy of our method.

Acknowledgement. This work is supported by the Sichuan Provincial Health and Family Planning Commission of China under the name Establishment and practice of an auxiliary intelligent decision making system for tumor patient evaluation. Grant number: 19ZDYF.

References

1. Aït-Younes, A., Truck, I., Akdag, H.: Image retrieval using fuzzy representation of colours. Soft. Comput. **11**(3), 287–298 (2007)
2. Chakarvarti, R., Meng, X.: A study of colour histogram based image retrieval. In: Sixth International Conference on Information Technology: New Generations. IEEE (2009)
3. Han, J., Ma, K.K.: Fuzzy colour histogram and its use in colour image retrieval. IEEE Trans. Image Process. **11**(8), 944–952 (2002)

4. Huang, J., Ravi, S.K.: Image indexing using colour correlograms. In: Proceedings of the IEEE Conference, Computer Vision and Pattern Recognition, Puerto Rico, June 1997
5. Kakade, V.M., Keche, I.A.: Review on content based image retrieval (CBIR) technique. Int. J. Eng. Comput. Sci. **6**(3), 20414–20416 (2017). ISSN: 2319-7242
6. Kang, F., Jin, R., Hoi, S.C.: Similarity beyond distance measurement, large-scale semantic access to content (text, image, video and sound). In: Proceedings of RIAO 8th Conference 2007, 30 May–1 June 2007, Pittsburgh, PA, pp. 449–460 (2007)
7. Kumar, V.V., Rao, N.G., Rao, A.L., Krishna, V.: IHBM: integrated histogram bin matching for similarity measures of colour image retrieval. Int. J. Sig. Process. Image Process. Pattern Recogn. **2**(3), 109 (2009)
8. Lin, C.H., Chen, R.T., Chan, Y.K.: A smart content-based image retrieval system based on colour and texture feature. Image Vis. Comput. **27**, 658–665 (2009)
9. Lin, G., Liu, B., Xiao, P., Lei, M., Bi, W.: Phishing detection with image retrieval based on improved texton correlation descriptor. Comput. Mater. Continua **57**(3), 533–547 (2018)
10. Liua, Y., Zhanga, D., Lua, G., Wei-Ying, M.: A survey of content-based image retrieval with high-level semantics. Pattern Recogn. **40**, 262–282 (2007)
11. Stricker, M., Dimai, A.: Colour indexing with weak spatial constraints. In: IS&T/SPIE Conference on Storage and Retrieval for Image and Video Databases IV, vol. 2670, pp. 29–40 (1996)
12. Marín-Reyes, P.A., Lorenzo-Navarro, J., Castrillón-Santana, M.: Comparative study of histogram distance measures for re-identification. arXiv preprint arXiv:1611.08134 (2016)
13. Mustikasari, M., Madenda, S., Prasetyo, E., Kerami, D., Harmanto, S.: Content based image retrieval using local colour histogram. Int. J. Eng. Res. **3**(8), 507–511 (2014)
14. Pass, G., Zabih, R., Miller, J.: Comparing images using colour coherence vectors. http://www.cs.cornell.edu/home/rdz/ccv.html. Accessed Mar 2018
15. Pass, G., Zabih, R.: Comparing images using joint histograms. Multimedia Syst. **7**(3), 234–240 (1999)
16. Pass, G., Zabih, R.: Refinement histogram for content-based image retrieval. In: IEEE Workshop on Application of Computer Vision, pp. 96–102 (1996)
17. Picard, R.W., Minka, T.P.: Vision texture for annotation. Multimedia Syst. **3**, 3–14 (1995)
18. Singha, N., Singhb, K., Sinha, A.K.: A novel approach for content based image retrieval. Proced. Technol. **4**, 245–250 (2012)
19. Suhasini, P.S., Krishna, K.R., Krishna, I.V.M.: CBIR using colour histogram processing. J. Theoret. Appl. Inf. Technol. **6**(1), 116–122 (2009)
20. Tyagi, V.: Content-Based Image Retrieval. Springer, Singapore (2017)
21. Xia, Z., Lu, L., Qiu, T., Shim, H.J., Chen, X., Jeon, B.: A privacy-preserving image retrieval based on AC-coefficients and color histograms in cloud environment. Comput. Mater. Continua **58**(1), 27–44 (2019)
22. Yadav, O., Suryawanshi, V.: CBIR evaluation using different distances and DWT. Int. J. Comput. Appl. **93**(16), 36–40 (2014)
23. Yu, S., Niu, D., Zhang, L., Liu, M., Zhao, X.: Colour image retrieval based on the hypergraph combined with a weighted adjacent structure. IET Comput. Vis. **12**(5), 563–569 (2018)
24. Yuvaraj, D., Sivaram, M., Karthikeyan, B., Abdulazeez, J.: Shape, color and texture based CBIR system using fuzzy logic classifier. CMC-Comput. Mater. Continua **59**(3), 729–739 (2019)
25. Zabih, R., Miller, J., Mai, K.: A feature-based algorithm for detecting and classifying scene breaks. In: ACM Multimedia Conference, pp. 189–200, November 1995
26. Zhai, H., Chavel, P., Wang, Y., Zhang, S., Liang, Y.: Weighted fuzzy correlation for similarity measure of colour-histograms. Opt. Commun. **247**, 49–55 (2005)

A Survey on Risk Zonation of Lightning

Yunfei Wang[1(✉)] and Meixuan Qu[2]

[1] Nanjing University of Information Science and Technology, Nanjing 210044, China
wangyunfei666@qq.com
[2] Hunan Meteorological Disaster Prevention Technology Center, Hunan 410007, China

Abstract. This paper not only reviews a series of studies on disaster risk zoning in the field of lightning disaster prevention in the past decade, but also combs and evaluates the results and methods of this research direction. It aims to summarize the work of lightning disaster risk zoning in recent years. Research on lightning risk regionalization can be divided into three stages: constructing lightning risk assessment model, classification of lightning disaster level, and classification of lightning disaster area. On the construction of lightning disaster risk assessment model is divided into two main categories: weighted comprehensive evaluation method and the combination evaluation method, the weighted comprehensive evaluation method for weight assignment include: the sampling method, that is simple, cumbersome, subjective, and the average weight method, that is simple to calculate but difficult to distinguish the difference of indicators, and the analytic method with certain objectivity but lack of flexibility in index selection. The combination evaluation method also includes selection of principal component analysis method, entropy method, and cluster analysis method. In the method of lightning disaster classification, most studies mainly adopt two classification methods: one is to classify according to the meteorological disaster classification statistical method. Such methods have certain statistical significance, but the results are rough and cannot be refined to reflect the spatial difference of disaster vulnerability risk, the second is to use the natural breakpoint classification method in GIS to classify, such methods can make the classification level have certain accuracy. In the lightning risk zoning, most studies basically use administrative districts as basic zoning units.

Keywords: Analytic hierarchy process · Entropy method · Principal component analysis · Cluster analysis · Natural breakpoint grading

1 Introduction and Background

Lightning is a phenomenon of atmospheric discharge. With the development of social economy, it has caused more and more disasters, and the scope of its spread has become wider and wider. Lightning disasters have been listed as one of the 10 most serious natural disasters by relevant UN organizations, IEO. It has been identified as a major public hazard in the electronic age. Therefore, research on lightning disaster prevention is of great significance. In recent years, scientific protection of lightning disaster has been paid more and more attention. Due to China's vast territory and great differences

© Springer Nature Switzerland AG 2020
X. Sun et al. (Eds.): ICAIS 2020, LNCS 12239, pp. 597–606, 2020.
https://doi.org/10.1007/978-3-030-57884-8_53

in climate, lightning disaster process is not the same, the regional lightning disaster risk regionalization research is very plentiful in recent years. At the same time, in the research of lightning zoning, the research direction mainly focuses on the improvement of the method of assigning the index weight value and dividing the disaster grade, and the selection of the basic dividing area.

In most of the research literature, the research on lightning disaster risk zoning is mainly divided into three stages: lightning risk assessment, classification of lightning disaster level, and lightning disaster regional division. The flow chart is shown in Fig. 1.

In the lightning risk assessment phase, the main task is to establish a risk assessment model and empower the assessment indicators. Most of the early research methods used weighted comprehensive evaluation methods for the weighting methods of evaluation models. In the assignment of weights, such as Min Wang et al. [4], Yuan Guo et al. [9], Xiangling Yuan et al. [10] and others take the analytic hierarchy process, such as Hu Guo et al. [6], Yanting Tian et al. [7], Jiaqi Li et al. [25], Cailian Li et al. [32] and others used the method of average weight, or such as Yichang Li et al. [2] to take extensive research and sample questionnaires. Based on the early evaluation model, the weighting method of the lightning risk assessment model is continuously improved. In the middle and late risk assessment, Xun Cui et al. [13] combined a variety of highly relevant assessment methods to empower the assessment. The model is more comprehensive. Gao et al. [11] adopted a cluster analysis method to select the threshold between the index variable and the rank, which enhanced the flexibility of classification.

In the stage of lightning disaster classification, the main task is to determine the threshold of the disaster classification level and the disaster level limit. The early litera-tures classify the lightning disaster level according to the statistical method of meteoro-logical disasters, and are mainly divided into 4 and 5 levels. In most of the research in the middle and late period, the natural breakpoint classification method in GIS is used to classify the classification, which makes the classification level have certain accuracy.

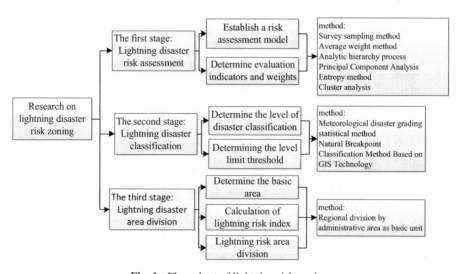

Fig. 1. Flow chart of lightning risk zoning

There are also studies that use the steepness of the aggregation coefficient to determine the classification number and then conduct classification according to the rank of risk assessment, which enhances the flexibility of classification.

In the stage of lightning disaster area division, the main work is to determine the basic division area, calculate the corresponding lightning risk index, and combine the stage division threshold of stage 2 to divide the lightning risk area. In the determination of the basic division area, most studies use the administrative area such as county (city, district) as the basic area division unit.

2 Model

In the establishment of the lightning risk assessment model, most studies use the weighting of indicators: average method, analytic hierarchy process, principal component analysis and entropy method. This paper will briefly introduce and evaluate these several weighting models in combination with specific research examples.

2.1 Average Weight Method

The average method assigns an average value to all selected indicators, and evaluates the importance of each indicator equally for the lightning risk assessment model. This method is more practical and less cumbersome than the survey sampling in terms of computational simplicity. For example, Hu Guo et al. [6], Yanting Tian et al. [7], selected 4 indicators, the weight of each indicator is given 1/4, and a lightning risk assessment model is constructed.

2.2 Analytic Hierarchy Process

The analytic hierarchy process has the characteristics of simple, flexible and practical multi-criteria for the evaluation of the weight of the lightning risk assessment model. In the lightning disaster risk assessment model based on analytic hierarchy process, the technical key is to reasonably select certain disaster risk assessment indicators according to the impact and damage degree caused by lightning disasters and the vulnerability and vulnerability characteristics of regional disaster-tolerant bodies. For example, Xiangling Yuan et al. [10] selected four indicators: thunderstorm days (M), lightning disaster frequency (P), economic (GDP) vulnerability modulus (D), and life vulnerability modulus (L) to conduct a lightning disaster risk assessment.

Step1: Establish a hierarchical model. The model is divided into three layers according to the requirements of the risk assessment. That is: the first layer is the target layer of lightning disaster risk; the second layer is the criterion layer of disaster formation conditions; the third layer is the indicator layer of influencing factors.

Step2: Establish a judgment matrix for the weight of lightning disaster risk assessment indicators. The judgment matrix represents a comparison of the relative importance of each indicator related to the previous level to a certain indicator. The judgment matrix

T is established according to the selected evaluation index and the hierarchical model.

$$T = \begin{pmatrix} 1 & 2 & 3 & 5 \\ \frac{1}{2} & 1 & 2 & 4 \\ \frac{1}{3} & \frac{1}{2} & 1 & 3 \\ \frac{1}{5} & \frac{1}{4} & \frac{1}{3} & 1 \end{pmatrix} \tag{1}$$

Step3: The relative weights of the indicators are calculated by the judgment matrix. For the constructed judgment matrix, the maximum eigenvalue λmax $=$ 4.0511 and the eigenvector are calculated by Matlab software. Then normalize the feature vector, and the normal weight of the feature vector is the relative W $=$ $[0.472850.284380.169920.07285]^T$ are weight of each indicator.

Step4: Judging the consistency check of the matrix. Because CR $= 0.017 < 0.10$, the article considers that the consistency of the judgment matrix T is acceptable.

2.3 Principal Component Analysis

Principal component analysis method can be used to evaluate the correlation between multiple lightning risk assessment indicators for the lightning risk assessment model. The internal structure of multiple lightning risk assessment indicators is revealed by a few principal components. The linear combination of indicators, as a new comprehensive indicator, makes the multi-index lightning risk assessment model not only guarantees objectivity, but also achieves dimensionality reduction and simplifies the calculation effect. Take the weight assignment steps of the lightning risk assessment model based on Principal Component Analysis in [13] by Xun Cui et al. as an example.

Step1: The lightning indicators of eight indicators including thunderstorm days, lightning disasters, ground flash density, lightning current intensity, gross domestic product (GDP), population density, personal injury and death, and economic loss were standardized.

Step2: Calculate the correlation coefficient:

$$R = \begin{bmatrix} r_{11} & \cdots & r_{18} \\ \vdots & \ddots & \vdots \\ r_{81} & \cdots & r_{88} \end{bmatrix} \tag{2}$$

The above r_{ij} for variables x_i and y_j correlation coefficient.

Step3: Calculate eigenvalues and eigenvectors

The eigenvalue λ_i can be obtained by Jacobian method and arranged from small to large: $\lambda_1 > \lambda_2 > \lambda_3 > \ldots \lambda_8 > 0$, and then the feature vector e_i corresponding to the eigenvalue λ_i is respectively obtained, where e_i represents the jth of e_i Component.

Step4: Calculate the principal component contribution rate and the cumulative contribution rate

The contribution rate of the principal component z_i is:

$$\frac{\lambda_i}{\sum_{k=1}^{p} \lambda_k} i = 1, 2 \ldots 8 \tag{3}$$

Cumulative contribution rate:

$$\sum_{i=1}^{p} \frac{\lambda_i}{\sum_{k=1}^{p} \lambda_k} \quad i = 1, 2, \ldots 8 \tag{4}$$

Take the principal components of which the cumulative contribution rate reached 85%–95%, and determine the indicators and their weights: thunderstorm day (0.052), lightning disaster number (0.211), ground flash density (0.089), lightning current intensity (0.068), GDP (0.158), population density 0.094), personal injury (0.105), economic loss (0.224)

2.4 Entropy Method

The entropy method can be used to determine the impact of a certain lightning risk assessment index on the lightning risk assessment by determining the discrete degree of a lightning risk assessment index, that is, using the entropy value to judge the degree of influence of an indicator. Take the weight assignment steps of the lightning risk assessment model based on the entropy method established by Cui Xun et al. [13] as an example.

Step1: Three methods for determining lightning risk weights and eight indicators for lightning risk assessment were selected. x_{ij} was set as the i-th method and the value of the j-th index.

Step2: To standardize x_{ij}, the absolute value of the indicator is converted into a relative value. For the sake of convenience, the normalized data is still recorded as x_{ij}.

Step3: Calculate the proportion of the i-th method under the j-th indicator

$$P_{ij} = \frac{x_{ij}}{\sum_{i=1}^{n} x_{ij}} \tag{5}$$

Step4: Calculate the entropy of the j-th indicator

$$e_j = -k \sum_{i=1}^{n} p_{ij} \ln(p_{ij}) \tag{6}$$

Where $k = \frac{1}{ln(n)}$, satisfying $e_j \geq 0$.

Step5: Calculate information entropy redundancy

$$d_j = 1 - e_j \tag{7}$$

Step5: Calculate the weight of each indicator

$$p_{ij} = \frac{d_j}{\sum_{j=1}^{n} d_j} \tag{8}$$

Obtained indicators and weights: thunderstorm day (0.717), lightning disaster number (0.045), ground flash density (0.377), lightning current intensity (0.192), GDP (0.66), population density (0.022), personal injury (0.156), economic loss (0.039)

3   Analysis and Comparison

In the introduction, this per simply divides the research on lightning disaster risk zoning in recent years into three stages: lightning risk assessment, lightning disaster classification and lightning disaster regional division. The following three aspects will be used to compare and analyze various methods of lightning disaster risk zoning in recent years.

In the lightning risk assessment, the core of the technology lies in the construction of the evaluation index system and the weighting, as shown in Table 1. For the determination of the weight of the risk assessment model, from the perspective of the weighting attribute, the research in recent years has chosen two types of methods, one of which is directly assigned by work experience and questionnaire. For example, Yichang Li et al. [2] used extensive surveys and sample survey questionnaires to determine the weight of impact factors. Xiangyang Cheng et al. [15] used disaster analysis and expert scoring to determine the weights of the evaluation models. Evaluation models obtained in this way are highly subjective. In contrast, the other is the use of analytic hierarchy process, grey relational analysis, cluster analysis, entropy method, etc., to avoid artificial subjective arbitrariness. For example, Meng Cheng et al. [12] adopted the analytic hierarchy process to determine the weight, and conducted the zoning of lightning disaster vulnerability in southwestern Shandong. Guangchang Chen et al. [14] used the gray correlation analysis method to determine the weight and finished the work of quantitative assessment and zoning of lightning disaster risk in Jiangsu Province.

Yin et al. [26] used principal component analysis to assign weights, and Gao et al. [11] used cluster analysis to determine the weight for conducting the risk zoning of lightning disaster vulnerability in Hainan Island. Cui Xun et al. [13] introduced the entropy method into the combined evaluation method to carry out the lightning disaster risk zoning in Jiangsu Province. The weighting methods selected in these studies all have certain objectivity, and the evaluation model constructed is also more scientific. From the perspective of weighting calculation, the analytic hierarchy process is more likely to show the difference in the importance of impact factors than the average weight method adopted by Yuan Xiangling [16] and Tian Yanting [7]. The cluster analysis method selected by Gao et al. [11] has more flexibility and pertinence in selecting the threshold between indicator variables and ranks than Zhao Wei et al. [20], adopted the analytic hierarchy method.

In the classification of lightning disaster levels, as shown in Table 2. The early research is based on the classification of statistical methods of meteorological disasters. It is divided into 4 levels and 5 levels mostly, which is statistically significant. However, once the classification of the vulnerability zoning is determined, it cannot be changed, and there are certain limitations and subjectivity for the classification of the grading. In contrast, most of the research in the middle and late stages used the natural breakpoint grading method in GIS technology [35] for classification. For example, Guangchang Chen et al. [14] used GIS technology to carry out lightning disaster risk zoning in Jiangsu Province. Li Hao et al. Ruili Xu et al. [29] and [39] applied GIS technology to the mountainous areas of Zhejiang and Shandong. Conducting lightning risk zoning, Li Changyu et al., Zhun Li et al. used the four-level division in [30] and [40] to conduct lightning risk zoning in Xining and Nanchang areas of Qinghai, and Xiangyang Cheng

Table 1. Comparison table of weighting methods for lightning risk assessment and evaluation index system

Weight determination method	Parameter selection	consider the correlation between indicators	Subjective/Objective	Characteristics of the weighting method of lightning risk assessment and evaluation system		
				Calculation method is simple	Show indicator differences	The level of threshold is flexible
Work experience, questionnaire			Subjective	✓		
Average weight method	Number of indicators n		Subjective	✓		
Analytic hierarchy process	Judgment matrix T, Maximum characteristic root λ_{max}, Feature vector W	✓	Subjective	✓	✓	
Entropy method	Entropy value of the indicator e_j, Information entropy redundancy d_j	✓	Objective	✓	✓	
Principal Component Analysis	Correlation coefficient between indicators r_{ij}, Eigenvalues λ_j	✓	Objective	✓	✓	
Cluster analysis	Group center of gravity, Canonical correlation coefficient r, Classification matrix T	✓	Objective	✓	✓	✓

et al. [15] zoning the risk of lightning disasters in Jiangsu Province according to five levels. This method makes the classification level have certain accuracy.

In the area of lightning disaster area division, most studies use administrative areas such as counties (cities, districts) as basic area division units. The accuracy of the risk area is not high, and there is a large error. The result of the zoning is also relatively

Table 2. Comparison table of lightning risk classification methods

Grading method	Dividing flexibility	Objectivity/subjectivity	Characteristics of lightning risk classification method		
			Whether it is statistically significant	Whether it is highly accurate	Whether the level is flexible
Meteorological disaster grading statistical method	Weak	Objectivity	$\checkmark$		
Natural breakpoint classification method based on GIS technology	Strong	Objectivity		$\checkmark$	$\checkmark$

extensive, and the spatial difference that reflects the vulnerability of lightning disasters in various places cannot be refined.

4 Conclusion

Based on the above analysis of recent research on lightning disaster risk zoning, it is not difficult to find out how to construct a more objective and scientific evaluation index system and the method of assigning weight for evaluation index has been accompanied by relevant research. In the index system, the weighted comprehensive evaluation method and the combined evaluation method provide a variety of objective selection types for the evaluation model. In the method of assigning weight, the analytic hierarchy process and the entropy method are used to distinguish the importance of the impact factors in different ways. In recent years, the study of lightning zoning has gradually completed the selection of assigning weight methods, but at the same time, the evaluation and testing of the partition model are not completed enough, which makes the reliability and scientificity of the model difficult to be evaluated. Moreover, most the zoning study uses the administrative district as the basic zoning unit, and the zoning results are difficult to refine, and the geographical influence factors are ignored. Therefore, in the future research trend, based on the current technical background of artificial intelligence and big data, the grid technology is introduced to finely classify the lightning data, economic data and geographic data of the geographical area. Combined the classification model construction of machine learning and deep learning, how to construct a risk assessment model for high-accuracy lightning disasters is likely to be the next research hotspot.

References

1. Tang, Q., Sun, J., Feng, Q.: GIS-based lightning disaster risk zoning in Shandong Province. In: The 32nd Annual Meeting of Chinese Meteorological Society S20 13th Lightning Protection and Disaster Reduction BBS – Lightning Physics and Lightning Protection Technology (2015)
2. Li, Y., Yu, C., Wu, L., et al.: Study on the climatic characteristics of thunder and lightning and the vulnerability of lightning disasters in Sanming City. J. Nanjing Univ. Inf. Technol. Nat. Sci. 9(2), 220–226 (2017)
3. Wu, A.: Research on risk assessment and zoning of lightning disasters in Guizhou Province. China Agric. Res. Regionalization 39(2), 88–93 (2018)
4. Wang, W., Kong, S., Wang, X.: Analysis and division of lightning disaster vulnerability in Eastern Qinghai. Meteorol. Sci. Technol. 46(2), 412–417 (2018)
5. Liu, Y., Li, Z., Cheng, X., et al.: Anhui Province lightning disaster risk zoning. J. Nanjing Univ. Inf. Technol. Nat. Sci. 6(2), 163–168 (2014)
6. Guo, H., Xiong, Y.: Analysis,: assessment and vulnerability zoning of lightning disasters in Beijing. Chin. J. Appl. Meteorol. 19(1), 35–40 (2008)
7. Tian, Y., Wu, M., Shi, F., et al.: Comprehensive assessment and zoning of lightning disaster vulnerability in Hebei Province. Meteorol. Sci. Technol. 40(3), 507–512 (2012)
8. Li, Y., Shen, Y., Chen, A., et al.: Refined lightning disaster vulnerability zoning model based on GIS graph stacking method. Meteorol. Sci. Technol. 46(1), 182–188 (2018)
9. Guo, Y., Wu, L., He, K., et al.: The risk division of guangxi thunderstorms based on analytic hierarchy process. Meteorol. Res. Appl. 39(02), 108–113 + 145 (2018)
10. Yuan, X., Ji, H., Cheng, L.: Hierarchical analysis model for lightning disaster risk zoning in Heilongjiang Province. Rainstorm Disaster 29(3), 279–283 (2010)
11. Gao, Y., Meng, X., Lao, X.: Risk zoning of lightning disaster vulnerability in Hainan Island based on cluster analysis. J. Nat. Disasters (1), 175–182 (2013)
12. Cheng, M., Wang, X.: Disaster zoning of lightning disasters in southwestern Shandong based on lightning location data. Sci. Meteorol. Environ. 40(4), 126–131 (2017)
13. Cui, X., Zhuang, Y., Wang, H.: The zoning of lightning disaster risk in Jiangsu Province based on the combination evaluation method. J. Nat. Disasters 24(6), 187–194 (2015)
14. Chen, G., Cui, X., Tian, X.: Quantitative assessment and zoning of lightning disaster risk in Jiangsu Province. Disaster Sci. 2017(01), 32–35 (2017)
15. Cheng, X., Xie, W., Wang, K., et al.: Research on lightning disaster risk zoning method and its application in Anhui Province. Meteorol. Sci. 32(1), 80–85 (2012)
16. Yuan, X., Wang, Z., Xiao, W., et al.: Analysis of potential and realistic vulnerability of lightning disasters and zoning study. Disaster Sci. 26(1), 20–25 (2011)
17. Wang, Z., Tang, Y., Zeng, X., et al.: Development of visualization analysis system for lightning disaster data. Meteorology 35(5), 97–104 (2009)
18. Zhang, Y., Jiao, X., Jiang, H., et al.: Research on risk assessment and vulnerability zoning of lightning disasters in Nanjing. In: The 28th Annual Meeting of Chinese Meteorological Society, vol. 2012, pp. 1–7 (2011)
19. Wang, H., Deng, Y., Yin, L., et al.: Analysis and division of lightning disasters in Yunnan Province. Meteorology 33(12), 83–87 (2007)
20. Zhao, W., Yang, X., Zhang, B.: Analysis and division of lightning disaster risk in Zhejiang Province. J. Trop. Meteorol. 30(5), 996–1000 (2014)
21. Jiang, Y., Kuang, M., Qi, H.: Regional vulnerability analysis, assessment and vulnerability zoning - taking Chongqing as an example. Disaster Sci. 16(3), 59–64 (2001)
22. Na, Y., Wenan, X.: Analysis, assessment and vulnerability zoning of lightning vulnerability
23. Yan, C.: Analysis of the vulnerability of lightning disasters in Jiangxi Province and its division. Jiangxi Province Sci. 24(2), 131–135 (2006)

24. Fan, Y., Luo, Y., Chen, Q.: Determination of the weights of comprehensive evaluation indexes for regional vulnerability. Disaster Sci. **16**(1), 85–87 (2001)
25. Li, J., Shen, S., Qin, J.: Comprehensive assessment and zoning of vulnerability risks of lightning disasters in Chongqing. J. Southwest Univ. Nat. Sci. Ed. **33**(1), 96–102 (2011)
26. Yin, X., Xiao, W., Feng, M., et al.: Distribution characteristics and vulnerability zoning of regional lightning disasters. Meteorol. Technol. **37**(2), 216–220 (2009)
27. Blaikei, P., Cannon, X., Davis, I., et al.: At Risk: Natural Hazard, People Vulnerability and Disasters. Routledge, Abingdon (2005)
28. Jin, C., Xiao, W., Wang, X.: Analysis and division of lightning disaster hazard in Hubei Province. Rainstorm Hazard **30**(3), 272–276 (2011)
29. Li, H., Bian, X.: Analysis of vulnerability of lightning disasters in Zhejiang Province and risk zoning. J. Nanjing Univ. Inf. Technol. Nat. Sci. **6**(4), 336–341 (2014)
30. Li, C., Zhao, W.: Vulnerability analysis and zoning of lightning disasters in Xining City, Qinghai Province. Gansu Sci. Technol. **29**(13), 47–49 (2013)
31. Hu, H., Wang, Y., Xiong, Y.: Beijing lightning disaster risk assessment based on analytic hierarchy process. J. Nat. Disasters **19**(1), 104–109 (2010)
32. Li, C., Zhao, X., Zhao, D., et al.: Analysis and assessment of vulnerability of lightning disasters in Shaanxi Province and vulnerability area. Disaster Sci. **23**(4), 49–53 (2008)
33. Yan, C., Wu, G., Zhu, J.: Empirical analysis of regional lightning vulnerability and its zoning. Chin. J. Meteorol. Environ. **23**(1), 17–21 (2007)
34. Wang, F., Ren, Z., Wei, L., et al.: Discussion on the classification of lightning disasters and the potential indicators. Meteorol. Sci. Technol. **37**(6), 744–747 (2009)
35. Zhang, Y., Feng, Z., Wang, Y., et al.: GIS-based grid lightning disaster risk assessment model and its application. Meteorol. Sci. Technol. **01**(2016), 142–147 (2016)
36. Wang, S., Zou, S., Wang, N.: Dalian city lightning disaster risk assessment system index quantification method. Meteorol. Sci. Technol. **44**(3), 510–516 (2016)
37. Hu, H., Wang, Y.: Risk assessment of urban meteorological disasters using analytic hierarchy model. Chin. Meteorol. Soc. (2007)
38. Yang, S., Hao, X., Ren, X.: Analysis of causes of lightning disasters in Shanxi Province in 2006 and Countermeasures. J. Nat. Disasters **17**(2), 116–121 (2008)
39. Xu, R., Zhu, X.: GIS-based lightning disaster risk assessment and zoning technology in southeast Shandong mountainous area. Chin. Agric. Sci. Bull. **30**(5), 292–296 (2014)
40. Li, Z., Yang, H., Lin, C.: Vulnerability analysis and zoning of lightning disasters in Nanchang Area. Meteorol. Res. Appl. **34**(3), 78–82 (2013)
41. Lopez, R.E., Holle, R.L.: Changes in the number of lightning deaths in the United States during the twentieth century. J. Clim. **11**(8), 2070–2077 (1998)
42. Doll, C.N.H.: CIESIN Thematic Guide to Night-Time Light Remote Sensing and its Applications. Center for International Earth Science Information Network of Columbia University, Palisades (2008)
43. He, L., et al.: A method of identifying thunderstorm clouds in satellite cloud image based on clustering. Comput. Mater. Continua **57**(3), 549–570 (2018)
44. Zhao, Y., Zhang, S., Yang, M., He, P., Wang, Q.: Research on architecture of risk assessment system based on block chain. Comput. Mater. Continua **61**(2), 677–686 (2019)
45. Maamar, A., Benahmed, K.: A hybrid model for anomalies detection in Ami system combining k-means clustering and deep neural network. Comput. Mater. Continua **60**(1), 15–39 (2019)

Joint Extraction of Entity and Semantic Relation Using Encoder - Decoder Model Based on Attention Mechanism

Yubo Mai[1], Yatian Shen[1(✉)], Guilin Qi[2], and Xiajiong Shen[1]

[1] School of Computer and Information Engineering, Henan University, Kaifeng 475000, Henan, China
sy602@126.com
[2] School of Computer and Engineering, Southeast University, Najing 210000, Jiangsu, China

Abstract. Attention-based encoder-decoder neural network models have recently shown promising results in machine translation and speech recognition. In this work, we propose an attention based neural network model for joint named entity recognition and relation extraction. We explore different strategies in incorporating the alignment information to the encoder-decoder framework, and propose introducing attention mechanism to the alignment-based recurrent neural networks (RNN) models. Such attentions provide additional information to relation extraction and named entity recognition. Our independent models achieve state-of-the-art named entity recognition performance on the benchmark CoNLL04 dataset. Our joint training model further obtains 0.5% F1 absolute gain on named entity recognition and 0.9% F1 absolute improvement on relation extraction over the best models.

Keywords: Encoder - decoder model · Semantic relation extraction · Attention mechanism

1 Introduction

Entity recognition and relation extraction are two serialized subtasks in traditional information extraction: named entities are first discovered, and semantic relation classification is performed between these entity pairs. This approach to the problem ignores the fact that there is strong dependency between them and they share semantic representations [1, 2]. However, traditional serialization methods cannot share information between two tasks. In addition to the great correlation between them, another important problem is that these traditional methods have to face the problem of feature engineering, and how to avoid a lot of tedious feature engineering is also very tricky.

To solve the above two problems, we use a multi-task learning framework based on sequence labeling. Similar frameworks have been applied to other tasks, such as intention detection and slot filling, part of speech annotation and multilingual machine translation. We extend the recurrent neural network structure based on encoder-decoder framework as multi-task learning model, which can share encoder among multiple tasks to complete

© Springer Nature Switzerland AG 2020
X. Sun et al. (Eds.): ICAIS 2020, LNCS 12239, pp. 607–618, 2020.
https://doi.org/10.1007/978-3-030-57884-8_54

two different tasks: named entity recognition and relation extraction. This framework of joint extraction simplifies the way of information extraction and the problem of named entity recognition and relation extraction can be solved by just training model and fine-tuning a batch of parameters.

The main idea of encoder-decoder model is to encode the input sentence sequence into a dense vector using the encoder, and then generate the corresponding output sequence from the dense vector using the decoder. In recent years, this model has been successfully applied to many tasks of natural language processing. Therefore, we use encoder-decoder model as the framework of joint extraction and use recurrent neural network for semantic encoding and decoding. Recurrent neural network is a time-series related network model, which can effectively remember and process contextual information, and its extended model, long and short memory neural network [3] can solve the problem of word sparsity and semantic long-distance dependence. And it can learn the dense representation of a word through the recurrent structure, which makes the word have a strong generalization ability and the memory structure can remember the context longer, to some extent, the problem of artificial feature extraction is solved.

In this paper, the model of sequence annotation is utilized to recognize named entity, that is, each word corresponds to one annotation, and the input and output are one-to-one correspondence of equal length. For semantic relation classification, only a dense semantic representation of the encoder is required. Considering the characteristics of the above tasks, in the encoder-decoder framework, we use two variants of the framework. In the first case, the length of encoding and decoding sequence are aligned one by one in the task of named entity recognition. In the second case, the attention mechanism is used to decode the alignment model to obtain more accurate annotation of named entities. In both models, the encoding of semantic relation is fixed. Through encoder - decoder neural network framework, we integrate named entity recognition and semantic relation classification into an overall framework, so as to achieve the goal of joint extraction. The data we used is the CoNLL04 dataset, which is a dataset for named entity and relation identification [4].

The main contributions of this paper are summarized as follows:

1) We propose a deep encoder-decoder structure based on recurrent neural network, which can integrate named entity extraction and semantic relation extraction into a framework to achieve the purpose of joint extraction.
2) We explore the alignment model of named entity recognition in encoder-decoder model. Based on the alignment model, the attention mechanism is used for better decoding, and the named entity is more accurately labeled.
3) The experiment on CoNLL04 dataset shows that the model can reach 94.2% F1 value on entity recognition task and 76.1% F1 value on semantic relation classification task. Compared with the previous best system, we have achieved improvement on both tasks.

2   Related Work

In natural language processing tasks, the recurrent neural network model takes sentences as word sequences and has been successfully applied to language models, word segmentation and phonetic understanding. While traditional recurrent neural networks only contain some information about the past, bidirectional variants combine information about the past and the future. Stacked recurrent neural network was proposed by [5], and its depth structure shows great advantages [6]. [7] studied other ways of constructing deep recurrent neural network.

Many different neural network models have been used for named entity recognition. [8] used the neural network model in unsupervised learning on unlabeled data and has achieved good results in naming and recognition tasks. [9] applied convolutional neural network on the sequence of word embeddings, and conditional random field was used on the top layer. [10] utilized manual spelling feature as input and adopts LSTM-CRF neural network model. The spelling feature and place-name index feature were applied to linear chain conditional random field, and the learning method of word embedding [11] was used, and good results were obtained.

Recurrent neural network and convolutional neural network have achieved high performance in the task of sentence level relation classification [12]. [13] proposed recurrent and recursive neural network based on dependency syntax tree to construct the representation of sentences. However, these methods can not model the internal dependencies of entities and relations.

Most of the existing joint extraction systems of entity and semantic relation are based on features [4, 14, 15], and the dependence between tags is not based on the neural network model. The method [16] is based on neural network, and establishes the dependency between tags through the data structure of the table. Multi-task learning based on neural network has been used to establish associations among various tasks and achieved good results [17]. Therefore, we propose a multi-task learning framework based on neural network for joint extraction of named entity recognition and semantic relation.

3   Model

3.1   Long Short Term Memory (LSTM)

The hidden state of the recurrent neural network depends on the previous time step at each time step, in form, given a sequence $x^{(1:n)} = \left(x^{(1)}, x^{(2)}, \ldots, x^{(t)}, \ldots, x^{(n)}\right)$, and update its hidden state $\mathbf{h}^{(t)}$ in a loop according to the following formula,

$$\mathbf{h}^{(t)} = g(\mathbf{U}\mathbf{h}^{(t-1)} + \mathbf{W}x^{(t)} + \mathbf{b}) \tag{1}$$

For the activation function of the recurrent neural network, we use the long short term memory model unit (LSTM) proposed in [18], which can better model and learn the long distance dependence of words in sentences. This results in a computational path in the expanded recurrent neural network, whose derivative product is close to 1. These paths prevent gradient back propagation from suffering from the disappearance of the gradient [19]. The tanh function is chosen because its derivative has a larger range.

3.2 Encoder-Decoder Framework for Entity and Semantic Relation Joint Extraction

In the encoder-decoder framework, we use two variants of the framework: the first case is to consider the alignment of encoding and decoding sequence length in the named entity recognition task; in the second case, the attention mechanism is used to decode the alignment model to obtain more accurate annotation of named entities.

Encoder - Decoder Framework Based on Alignment for Joint Extraction of Entity and Semantic Relation

There are two variants of this model: 1) alignment model without attention mechanism; 2) alignment model with attention mechanism. The framework of joint extraction of entity and semantic relation based on encoder - decoder model is shown in Fig. 1. We use the model of sequence annotations in the task of named entity recognition, which maps a word order $x = (x_1, x_2, \ldots, x_T)$ to the corresponding annotation sequence $y = (y_1, y_2, \ldots, y_T)$.

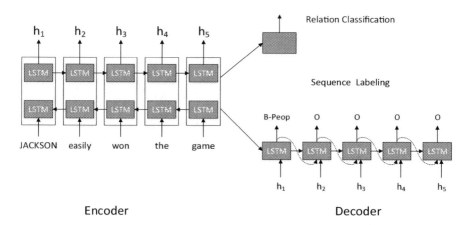

Fig. 1. Aligned framework of encoder - decoder neural network model

In the encoder stage, we use LSTM as the basic recurrent neural network unit. The forward recurrent neural network reads the word order according to the source sequence, and then generates the hidden state fh_i at each time step. Similarly, the backward recurrent neural network reads the reverse sequence and produces the hidden state of the sequence $(bh_T, \ldots, bh_1)$. The hidden state of the encoder at time step i is concatenation of forward hidden state fh_i and backward hidden state bh_i, that is $h_i = [fh_i, bh_i]$.

The final state of the forward and backward recurrent neural network encoder can encode the information of the whole sentence sequence. Similar to [20], we use the last state of the encoder as the initial hidden state of the decoder. The decoder is a one-way recurrent neural network. We use the long short term memory model as the basic recurrent neural network. At each time step i in the decoding stage, the state s_i of the decoder is calculated from the following parts: the previous decoding state s_{i-1}, the previous prediction label y_{i-1} and the aligned encoding hidden state h_i, its formula is as follows:

$$s_i = f(s_{i-1}, y_{i-1}, h_i) \tag{2}$$

For the above formula, due to the addition information of aligned encoder hidden state h_i in the decoding stage, each source input sequence and output sequence in the decoding stage are strictly one-to-one.

For the joint extraction of entity and semantic relation, we add an extra relation classifier to the model to generate the relation type distribution of a relation. The final state of the whole encoder is regarded as the encoding information of the whole sentence, the two subtasks, named entity recognition and relation classification, share the same encoder. In the training phase of the model, the model losses from the two subtasks are propagated to the encoder through the back propagation algorithm.

Aligned Encoder - Decoder Framework Based on Attention Mechanism for Joint Extraction of Entity and Semantic Relation

The structure used by the attention-mechanism-based recurrent neural network model in the joint extraction task of named entity recognition and relation classification is shown in Fig. 2. In the encoder-decoder neural network framework based on alignment, we introduce attention mechanism to consider more contextual information.

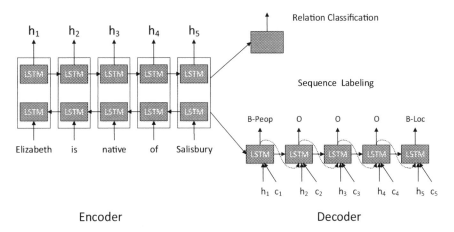

Fig. 2. Aligned framework of encoder - decoder neural network model based on attention mechanism

Attention mechanism is to pay special attention to some information and break the disadvantages of traditional information encoding or decoding, that is, to treat all information equally and indiscriminately. [21] first introduced the attention mechanism into the task of machine translation by using the attention mechanism to select the most relevant words to decoding from the source language. [22] used the attention mechanism to select the most relevant image area when generating an image description. Similarly, attention mechanisms are applied to various tasks, such as semantic implication, text classification, relation extraction, machine translation [23], image description generator [24], and semantic analysis [25].

We use the bidirectional recurrent neural network model to read the source sentences forward and backward, and use the LSTM unit as the basic recurrent neural network unit. The annotation of named entities is modeled using recurrent neural networks. Similar to the encoder-decoder structure described above, the hidden state of encoder for each time step i is a concatenation of the forward hidden state and backward hidden state, that is, $h_i = [fh_i, bh_i]$. Each hidden h_i contains information about the entire input word order, and pays more attention to the word information around the time step i. When bidirectional recurrent neural network is used for sequence annotation tasks, the hidden state of each time step encodes the whole sequence of information, but this information may gradually disappear in the process of forward and backward propagation. Therefore, in the process of named entity identification, we will not only align the information of hidden h_i with each time step, but also focus on the context information c_i whether provides additional support information, especially the information requires a longer time sequence dependence, but the ordinary encoder hidden state is not easily available.

The final state of the bidirectional recurrent neural network encoder can encode the information of the whole sentence sequence. We use the last state of the encoder as the initial hidden state of the decoder, similar to [21]. The decoder is a one-way recurrent neural network model, which uses the long short term memory model as the basic recurrent neural network. At the decoding stage of each time step i, the state s_i of the decoder is calculated from the following parts: the previous decoded state s_{i-1}, the previous prediction label y_{i-1}, the aligned encoder hidden state h_i and context vector c_i. Its formula is as follows:

$$s_i = f(s_{i-1}, y_{i-1}, h_i, c_i). \tag{3}$$

In the above formula, the value of context vector c_i is obtained from the hidden state weighted sum, and its calculation method is as follows:

$$c_i = \sum_{j=1}^{T} \alpha_{i,j} h_j. \tag{4}$$

$$\alpha_{i,j} = \frac{\exp(e_{i,j})}{\sum_{t=1}^{T} e_{i,k}}. \tag{5}$$

$$e_{i,k} = g(s_{i-1}, h_k) \tag{6}$$

g is a feed forward neural network, which is a kind of artificial neural network. During the decoding phase, the output of decoder is aligned with the corresponding input hidden state h_i. The context vector c_i provides additional information for the decoder and can be viewed as a normalized feature of hidden state $h = (h_1, \ldots, h_T)$ after weighting, and the information of c_i is the attention mechanism in the model.

For the joint extraction of named entity and semantic relation, we also add an extra relation classifier on the model to generate the distribution of relation types of a sentence, the final state of encoder as the encoding information of the sentence, the two subtasks, named entity recognition and relation classification, share the same encoder. In the training phase, the model losses from the two subtasks are propagated to the encoder through the back propagation algorithm.

4 Training Model

This model is a joint training process, in which the tasks of entity recognition and relation classification are jointly trained to learn the parameters of the model.

For named entity recognition, we use cross entropy loss as the objective function. For each sample instance (x, l), its input sequence is x and its label sequence is l. The model parameters of the neural network are learned from all the training datasets. Since the source input and output label of the model are aligned, the length of input and output is equal ($|x| = |l|$), and its target formula is as follows:

$$loss1_{ce} = - \sum_{(x,l)} \sum_{t=1}^{|x|} \sum_{l} \delta(l, \mathbf{l}_t) \log y_l^t \qquad (7)$$

In the above formula, $\delta(l, \mathbf{l}_t)$ is the Kronecker delta function, y_l^t is the output of the l tag network at time t, y_l^t is calculated by adding softmax layer to the output of every t of the network.

For relation extraction, we also use the cross entropy loss function as the target, and its target formula is as follows:

$$loss2_{ce} = - \sum_{i=1}^{N} \sum_{j=1}^{C} y_i^j \log(\hat{y}_i^j) \qquad (8)$$

y_i^j is the relation tag, $\hat{y}_i^j$ is the predicted distribution, N is the number of training samples, C is the number of classes.

The training loss on the entire dataset can be written as the following formula:

$$loss = loss1_{ce} + loss2_{ce} \qquad (9)$$

In the training stage of the model, the model loss from the two subtasks is propagated to the encoder through the back propagation algorithm, and conducting gradient propagation. We use the Adam optimization method to optimize the parameters, and its parameter settings refer to the method [26]. The gradient of the network model is calculated using BPTT method [27]. Assume that θ_j is the parameter after updating step j, η_0 is the learning rate, N is the minimum batch size, $\nabla_\theta(j)$ is the gradient of the cost function. Parameter updating is shown in Eq. (10):

$$\theta(j + 1) = \theta(j) - \eta_0(1 - \frac{j}{N}) \frac{\nabla_\theta(j)}{\|\nabla_\theta(j)\|} \qquad (10)$$

5 Experiment

5.1 Datasets and Assessments

We trained and tested our model using the CoNLL04 dataset, which is an entity extraction and relation identification dataset [4]. This dataset defines four entity types, including location, Organization, Person, and Other entity types. Relation types include Kill, Live In, Located In, OrgBased In, and Work For.

We use F value as evaluation indicator of entity and relation joint extraction performance, and use this evaluation indicator to adjust the parameters. If the entity is correctly identified and labeled with the correct type, it is considered the correct annotation. The calculation formula of F value is as follows:

$$F = \frac{(\beta^2 + 1) \times P \times R}{\beta^2 \times P + R} \tag{12}$$

In the above formula, P is the accuracy rate and R is the recall rate. The value of β is equal to 1, indicating that accuracy is as important as recall.

5.2 Hyper-parameter

For network learning, the setting of hyper-parameters is very important. According to the experimental results, the hyper-parameters we selected are shown in Table 1. The dimension of the word embedding is set to 128 dimensions and randomly initialized. The number of hidden cells is 128. The minimum batch size during training is 16. The learning rate of the model is set to 0.01. The dropout rate of the model is set to 0.5. The maximum standard value used for gradient reduction is set to 5. We use the Adam method to update the parameters of the model during training.

Table 1. Setting of model hyperparameters.

Parameter	Values
Word embedding size	128
hidden size	128
Minibatch size	16
Dropout rate	0.5
Learning rate	0.01

5.3 Joint Extraction Method of Entity and Semantic Relation Versus Single Method

In the encoder - decoder based entity and semantic relation joint extraction model, each word in the sample is pretrained as a 128-dimensional word embedding. In addition to

Table 2. Comparison of experimental results on CoNLL04 dataset by single model method and joint method.

Methods	NER			RE		
	P	R	F_1	P	R	F_1
Sperate+basic	.851	.884	.867	.613	.503	.553
+POS	.854	.892	.872	.621	.515	.563
+CF	.861	.905	.882	.644	.534	.584
+CTX	.862	.908	.884	.662	.423	.619
Pipeline+basic				.635	.512	.567
+POS				.652	.542	.592
+CF				.654	.575	.612
+CTX				.751	.628	.684
Joint entity&Relation+basic	.873	.881	.877	.658	.549	.599
+POS	.914	.921	.917	.661	.552	.602
+CF	.925	.918	.921	.703	.554	.620
+CTX	.938	.944	.941	.731	.602	.660
+(semantic dictionary of names and places)	.940	.946	.943	.862	.681	.761

the input of the basic word embedding, there are also part of speech tagging features (POS), initial character features (CF), context feature vectors (CTX), various names of people and places and other semantic dictionaries. Word embeddings are shared by two tasks of entity and semantic relation extraction. These word embeddings are randomly initialized and updated during training. We use an aligned encoder - decoder model based on attention mechanism (NN(bi) aligned inputs & attention) for comparative experiments.

Table 2 shows the results of named entity recognition and relation classification. "+" (e.g., +POS) represents the accumulation of each feature. We used ablation tests to verify the extraction results of each feature on different models. We have verified three different extraction methods. The basic model is unidirectional or bidirectional LSTM. Spearte (single method) respectively used recurrent neural network in the task of named entity recognition and relation classification. Pipeline (serial method) first identifies named entities, and then carries out semantic relation classification of sentences. Joint (Joint method) uses encoder-decoder network framework to conduct named entity recognition and relation classification joint extraction multi-task learning process.

We show the results of three different models on the named entity recognition and relation classification tasks in the table. In Table 2, we find that: (1) In all models, various features (POS, CF, CTX and various semantic dictionaries of names and places) improve the performance of the model, especially the semantic dictionaries of names and places. This shows that semantic dictionary is important for the recognition of named entities, which can fundamentally determine the type of entities, and the improvement of the accuracy of entity recognition can affect the improvement of semantic relation results, which is a process of mutual promotion. (2) Undoubtedly, in the serialization method of pipeline, since the relation classification is closely related to the task of named entity recognition, and entity recognition is conducive to relation extraction, its performance is greatly improved compared with the single method. (3) The results of our proposed joint model are better than those of single and serial models, indicating that joint training and decoding are beneficial to the task of name entity recognition and relation classification.

5.4   Comparison with Other Relevant Methods

In the experiment, we compared the aligned encoder-decoder model based on the attention mechanism (NN(bi) aligned inputs & attention) with other related methods. Kate and Mooney [14] used a structure called "card-pyramid" to conduct joint extraction of entities and relations, which could encode all possible entities and relations in sentences. Therefore, the joint extraction task is transformed into the joint annotation of the graph's node, and the experimental results show that: on the CoNLL04 dataset, the model obtained 91.7% F1 value of the entity recognition and 62.2% F1 value of the relation extraction. Using a structured machine learning method, Miwa and Sasaki [15] introduced a table structure to flexibly represent entities and relations, so as to decode the model and achieve the purpose of joint extraction of entities and relations. Gupta et al. [16] proposed the task of entity and relation joint extraction based on neural network through semantic combination of words, and simplified the task of entity and relation joint extraction into table filling and dependency between entities and relations with the recurrent neural network model of table filling. Our model uses an attention-based encoder-decoder alignment model.

Table 3. Comparative experiments with different models on various entity and relation types in the CoNLL04 dataset

Methods	Kate & Mooney			Miwa & Sasaki			Gupta et al. (2016)			Ours		
	R	P	F_1	R	P	F_1	R	P	F_1	R	P	F_1
Person	.942	.921	.932	.948	.931	.939	.988	.932	.959	.974	.935	.954
Location	.942	.908	.924	.939	.922	.930	.956	.974	.965	.964	.951	.957
Organization	.887	.905	.895	.896	.903	.899	.873	.939	.905	.925	.941	.933
Average	.924	.911	.917	.924	.924	.924	.961	.926	.943	.954	.942	.948
Live_In	.601	.664	.629	.532	.819	.644	.640	.727	.681	.645	.734	.687
OrgBased_In	.641	.662	.647	.572	.768	.654	.562	.831	.671	.573	.841	.682
Located_In	.567	.675	.583	.549	.821	.654	.553	.867	.675	.561	.874	.683
Work_For	.683	.735	.707	.642	.886	.743	.671	.945	.785	.686	.937	.792
Kill_For	.641	.916	.752	.797	.933	.858	.894	.857	.875	.901	.869	.885
Average	.607	.672	.622	.599	.837	.698	.664	.825	.737	.673	.851	.746

From the comparison in Table 3 (statistical results after removing the type "NONE"), it can be seen that our method of entity and relation joint extraction is superior to the results of Kate and Mooney [14], Miwa and Sasaki [15] and Gupta et al. (2016) [16]. The proposed joint extraction model obtained an average F value of 94.8% in the task of named entity recognition and 74.6% in the result of relation classification. Compared with the current best system [16], the proposed model has obtained an improvement of 0.5% and 0.9% F1 values respectively in the task of entity extraction and relation classification. Our results on the Location entity improve even more because many entities are identified depending on the semantic constraints of the relation.

In addition, we use the semantic dictionary of names and place names and the output of other information extraction models as additional features, which are helpful for the extraction of named entities and semantic relations. In this experiment, we improved the overall performance of the system by adding semantic dictionary and additional feature sets to our model. Therefore, for end-to-end named entity recognition and relation extraction, it is effective that the multi-task joint extraction method with additional features.

6   Conclusion

In this paper, we study encoder-decoder model based on attention mechanism on the joint task of entity recognition and relation extraction. Named entity extraction of sentences is regarded as a sequence labeling task, while relation extraction is a semantic relation classification task. We unify the two in a framework with a recurrent neural network based on attentional mechanism. Experiments on the CoNLL04 dataset show that our proposed model achieves the best results in all systems.

References

1. Nanda, K.: Combining lexical, syntactic, and semantic features with maximum entropy models for extracting relations. In: Proceedings of the ACL 2004 on Interactive Poster and Demonstration Sessions, p. 22. Association for Computational Linguistics (2004)
2. Luo, G., et al.: Joint entity recognition and disambiguation. In: Proceedings of the 2015 Conference on Empirical Methods in Natural Language Processing (2015)
3. Gers, F.: Long short-term memory in recurrent neural networks. Diss. Verlag nicht ermittelbar (2001)
4. Roth, D., Yih, W.: A linear programming formulation for global inference in natural language tasks. Illinois Univ at Urbana-Champaign Dept of Computer Science (2004)
5. El Hihi, S., Bengio, Y.: Hierarchical recurrent neural networks for long-term dependencies. In: Advances in Neural Information Processing Systems (1996)
6. Hermans, M., Schrauwen, B.: Training and analysing deep recurrent neural networks. In: Advances in Neural Information Processing Systems, pp. 190–198, (2013)
7. Pascanu, R., et al.: How to construct deep recurrent neural networks. arXiv preprint arXiv: 1312.6026 (2013)
8. Qi, Y., et al.: Combining labeled and unlabeled data with word-class distribution learning. In: Proceedings of the 18th ACM Conference on Information and Knowledge Management (2009)

9. Collobert, R., et al.: Natural language processing (almost) from scratch. J. Mach. Learn. Res. **12**, 2493–2537 (2011)
10. Huang, Z., Xu, W., Yu, K.: Bidirectional LSTM-CRF models for sequence tagging. arXiv preprint arXiv:1508.01991 (2015)
11. Passos, A., Kumar, V., McCallum, A.: Lexicon infused phrase embeddings for named entity resolution. arXiv preprint arXiv:1404.5367 (2014)
12. Zeng, D., Liu, K., Lai, S., Zhou, G., Zhao, J.: Relation classification via convolutional deep neural network. In Proceedings of COLING, pp. 2335–2344 (2014)
13. Socher, R., et al.: Semantic compositionality through recursive matrix-vector spaces. In: Proceedings of the 2012 Joint Conference on Empirical Methods in Natural Language Processing and Computational Natural Language Learning. Association for Computational Linguistics (2012)
14. Kate, R., Mooney, R.: Joint entity and relation extraction using card-pyramid parsing. In: Proceedings of the Fourteenth Conference on Computational Natural Language Learning, pp. 203–212. Association for Computational Linguistics (2010)
15. Miwa, M., Sasaki, Y.: Modeling joint entity and relation extraction with table representation. In: EMNLP, pp. 1858–1869 (2014)
16. Gupta, P., Schütze, H., Andrassy, B.: Table filling multi-task recurrent neural network for joint entity and relation extraction. In: Proceedings of COLING 2016, the 26th International Conference on Computational Linguistics: Technical Papers (2016)
17. Mesnil, G., et al.: Using recurrent neural networks for slot filling in spoken language understanding. IEEE/ACM Trans. Audio Speech Lang. Process. **23**(3), 530–539 (2014)
18. Cho, K., et al.: Learning phrase representations using RNN encoder-decoder for statistical machine translation. arXiv preprint arXiv:1406.1078 (2014)
19. Pascanu, R., Mikolov, T., Bengio, Y.: On the difficulty of training recurrent neural networks. ICML **3**(28), 1310–1318 (2013)
20. Sutskever, I., Oriol, V., Le, Q.V.: Sequence to sequence learning with neural networks. In: Advances in neural information processing systems (2014)
21. Bahdanau, D., Cho, K., Bengio, Y.: Neural machine translation by jointly learning to align and translate. arXiv preprint arXiv:1409.0473 (2014)
22. Xu, K., et al.: Show, attend and tell: neural image caption generation with visual attention. In: International conference on machine learning (2015)
23. Qiu, J., et al.: Dependency-based local attention approach to neural machine translation. Comput. Mater. Continua **58**, 547–562 (2019)
24. Qu, Z., et al.: Feedback LSTM network based on attention for image description generator. CMC-Comput. Mater. Continua **59**(2), 575–589 (2019)
25. Ling, H., et al.: Attention-aware network with latent semantic analysis for clothing invariant gait recognition. Comput. Mater. Continua **60**, 1041–1054 (2019)
26. Kingma, D.P., Ba, J.: Adam: A method for stochastic optimization. arXiv preprint arXiv: 1412.6980 (2014)
27. Graves, A., Schmidhuber, J.: Framewise phoneme classification with bidirectional LSTM and other neural network architectures. Neural Netw. **18**(5-6), 602–610 (2005)

Internet of Things

Image Segmentation of Manganese Nodules Based on Background Gray Value Computation

Ha-de Mao[1(✉)], Yu-liang Liu[1], Hong-zhe Yan[1], and Cheng Qian[2]

[1] School of Naval Architecture and Mechanical Engineering, Zhejiang Ocean University,
Zhoushan 316004, China
540493801@qq.com, 2574951395@qq.com
[2] School of Physics and Electrical Engineering, Huaiyin Normal University,
Huaian 223300, China

Abstract. Aiming at two problems from the process of target and background segmentation in the field of manganese nodule image processing, the uneven illumination and the morphological defects of manganese nodules caused by white sand cover, this paper proposes an improved method. After the image is processed by this method, the above two problems are solved and segmentable manganese nodule images are obtained. Finally, the processed image is segmented to verify the feasibility and stability of the method. The original method is only used in document images processing. It is not suitable for the processing of manganese nodule images because it can't repair the morphology of manganese nodules and reduce the contrast between the target and the background. The image of manganese nodules has the following characteristics: the pixels with higher gray value of white sand are scattered around the pixels with lower gray value of manganese nodules, and the area covered by white sand is smaller. In view of this feature, controllable artifacts are used to repair the morphology of manganese nodules. First, the image is preprocessed, and the background gray value is calculated and subtracted from the original image. Then the gray value of the image is adjusted by the block diagram. Finally, by performing the above operation again, the problem of uneven illumination in the image can be solved, and the morphology of manganese nodules in the image can be repaired without affecting the gap between them The simulation results show that, compared with the original method and other methods, several images processed by this method have significant effect in the process of segmentation, which proves that the method is effective and stable.

Keywords: Underwater image · Image segmentation · Morphological dilation and erosion · Manganese nodules

1 Introduction

Manganese nodules contain dozens of precious metal elements such as cobalt and nickel, and are widely used in aviation, aerospace, electronic information and other fields [1, 2]. It mainly distributed on the seabed of two thousand to six kilometers, we need to analyze the images obtained by autonomous underwater vehicle (AUV) to determine the

© Springer Nature Switzerland AG 2020
X. Sun et al. (Eds.): ICAIS 2020, LNCS 12239, pp. 621–634, 2020.
https://doi.org/10.1007/978-3-030-57884-8_55

reserves and distribution of manganese nodules in an area. In the process of statistics of manganese nodules, image segmentation is generally needed. In this process, the problems of uneven illumination and morphological incompleteness caused by white sand cover of manganese nodules are the main problems to be solved. Because of the bad underwater environment, Images become very blurred [3], so image enhancement [4–6] is needed first. At the same time, due to the inevitably uneven illumination of the photographed image, it is necessary to adopt effective methods to make the illumination of the whole image more uniform. As for the problem of uneven illumination, Badeka *et al.* proposed some solutions [7–10], but these methods are only used to deal with uneven illumination of document images, and can't repair morphology of manganese nodules. Retinex algorithm [11] has a good effect in dealing with uneven illumination, but it also can't repair the morphology of manganese nodules by executing the algorithm. Regarding image segmentation, the methods proposed by Diels *et al.* [12–16] are mostly done without the target being covered, but manganese nodules usually cover a layer of white sand. With regard to the recovery of object morphology, existing literature is usually processed using a corrosion expansion algorithm. However, the spacing between the manganese nodules is small, and the corrosion expansion algorithm fills the gap between the manganese nodules or causes a whole piece of manganese nodules to be divided into multiple pieces. There are few methods for processing images of manganese nodules that can simultaneously solve the problem of uneven illumination and repair the morphology of manganese nodules. Therefore, for the image segmentation processing of manganese nodules, this paper proposes a method to simultaneously solve the problem of uneven illumination and repair the morphology of manganese nodules. Firstly, color image is transformed into gray image. Secondly, the background gray value of the image is calculated and subtracted from the original image. Then the gray value of the image is adjusted according to the block diagram of this paper. Finally, repeat the above steps once. For the first time, the process solves the problem of uneven illumination. Then the image which solves the problem of uneven illumination will be executed again to repair the morphology of manganese nodules. Finally, in MATLAB simulation [17–19], several images processed by this method are segmented to verify the feasibility and stability of the proposed method.

2   Selection of Image Samples and Design of Method Flow

As shown in Fig. 1, this paper selects the manganese nodule images taken by AUV as a sample for research. It can be seen that the image processing of manganese nodules usually faces the following difficulties: a) The problem of uneven illumination inevitably exists in images due to different illumination angles of light sources. b) Manganese nodules are generally covered by white sand, and the gap between them is small. In the process of restoring morphology, the gap between the particles of manganese nodules will be filled. Therefore, it is difficult to restore the morphology of manganese nodules without affecting the gap between manganese nodules by using conventional image processing methods. In a word, the problem of uneven illumination and morphological recovery of manganese nodules in image segmentation is the core task of image processing of manganese nodules. The step of image enhancement lays a foundation for eliminating the uneven illumination and repairing the morphology of manganese nodules.

Fig. 1. Sample images taken by AUV

In order to deal with the above problems, the method proposed in this paper is shown in Fig. 2.

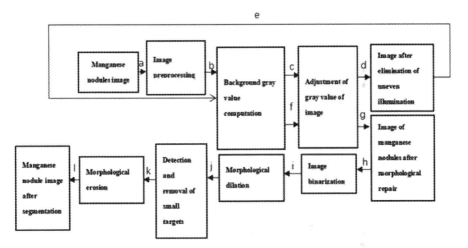

Fig. 2. Flow chart of the method of this paper

As shown in Fig. 2, after obtaining the uniform illumination image of manganese nodules through C and D steps, the morphology of manganese nodules in the image can be restored by performing the above-mentioned operations again. After binarizing the manganese nodule image, the obtained image is segmented. Then, the feasibility and stability of this method are verified by comparing the effect of image segmentation.

3 Image Preprocessing

The preprocessing of images in this paper includes image enhancement and Gaussian filtering. Among them, the enhancement of the image aims to improve the contrast between manganese nodules and the background. The image taken by the AUV is generally a color image. For the convenience of subsequent processing, we grayscale the color image and perform grayscale stretching on the obtained grayscale image. The gray histogram and image before and after grayscale stretching are shown in Fig. 3

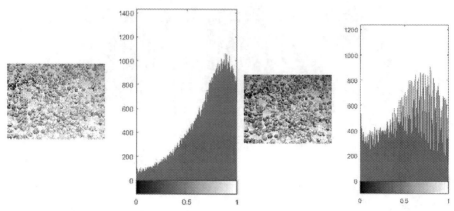

(a) gray histogram and image before grayscale stretching (b) gray histogram and image after grayscale stretching

Fig. 3. The gray histogram and image before and after grayscale stretching

It can be seen that the contrast between the enhanced manganese nodule image and the background is obviously enhanced, and the histogram is more balanced. This lays a foundation for the subsequent computation of image background gray value.

In order to reduce the influence of debris and reduce the gray value of some white sand on manganese nodules, and then lay the foundation for the calculation of background gray value and the generation of artifacts, it is necessary to filter the gray image first. The formula is as follows:

$$G(x, y) = \frac{1}{2\pi\sigma^2} e^{-\frac{x^2+y^2}{2\sigma^2}},$$ (1)

where, G (x, y) is the gray value of the filtered image at (x, y). The size of the filter's range of action depends on σ. The larger the value of σ, the more blurred the image is.

4 Elimination of Uneven Illumination and Repair of Manganese Nodule Morphology

4.1 Background Gray Value Computation

In order to eliminate the influence of uneven illumination [8, 9]. Firstly, we need to calculate the background gray value of gray image. Secondly, the background gray value is subtracted from the original image to eliminate its influence. Finally, the gray value of the image is adjusted by the block diagram, so that the manganese nodules can be clearly displayed in the image.

The first implementation of this process is to solve the problem of uneven illumination. Firstly, all the pixels in the image are traversed by a mask of $N*N$ size, and the n points with the highest gray value in the mask are recorded in turn. The recorded gray value is assigned to L_P where $1 \leq P \leq$ n, and the set L is formed in the order of large to small. Then the set L is $L = \{L_1, L_2, ..., L_n\}$. Considering that there will be interference points in the image, after removing the maximum gray value L_n and the minimum gray value L_1 before calculating the background gray value. The remaining points in the set are summed and taken as the mean of the background gray value of the central point $G\ (x,\ y)$ of the mask. The number of points is n, $n = 8$. However, as shown in Fig. 4, different sizes of masks will have a great impact on the calculation of the background, so we need to determine the value of N next.

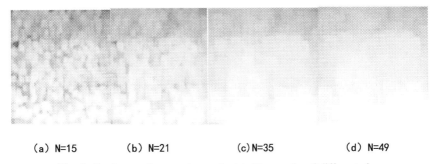

(a) N=15   (b) N=21   (c) N=35   (d) N=49

Fig. 4. Background gray values calculated by masks of different sizes

Figure 4 shows that the smaller the mask is, the more accurate the calculation of background gray value is. However, subtracting from the original image is not conducive to the elimination of illumination unevenness, and is not conducive to the subsequent steps to generate artifacts. But the size of the mask can not be too large, too large mask is also not conducive to the elimination of uneven light. In order to find a suitable mask, as shown in Table 1, we recorded the size of the mask and its running time.

Table 1. Operating time under different masks

Mask size	The duration of the first operation (S)	The duration of the second operation (S)	The duration of the third operation (S)	Average duration (S)
15*15	0.170510	0.117943	0.123959	0.137471
21*21	0.248632	0.247915	0.248542	0.248363
37*37	0.737713	0.765184	0.753215	0.752037
49*49	1.245406	1.228611	1.270246	1.248087

According to Table 1, the larger the mask, the longer the calculation time is required. So, we let $N = 21$ to calculate the gray value. Next, we subtract the background gray value from the original image to solve the problem of uneven illumination.

4.2   Solution to the Problem of Uneven Illumination

In order to solve the problem of uneven illumination, we subtract the original image from the background gray value. After subtraction, in the region where the background gray value of the image is small, the contrast between the manganese nodules and the surrounding environment in the subtracted image will decrease. Therefore, it is necessary to adjust the gray value between the manganese nodules and the background. Let the gray value of the original image and the background gray at (x, y) be X, Y, respectively, and the gray value of the image after the gray value adjustment is E. The contrast adjustment block diagram is shown in Fig. 5.

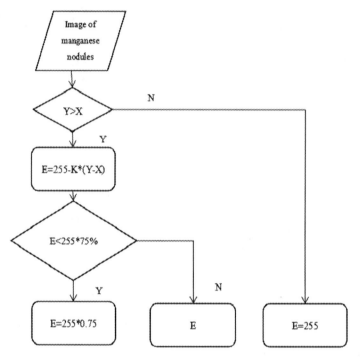

Fig. 5. Contrast adjustment block diagram

Where, K is a multiple of magnification. After pre-processing, the gray value of white sand covered with manganese nodules is lower than that of white sand in background. The original method [10] requires the suppression of artifacts, but we need to make the image generate artifacts appropriately to prepare for the morphological restoration of manganese nodules while dealing with uneven illumination. But at the same time, noise generation and image distortions should be suppressed, so the value of K is as follows:

$$\begin{cases} B_1 & g(x, y) \in [0, 20) \\ 1 + (B_1 - 1)\frac{100 - g(x,y)}{80} & g(x, y) \in [20, 100] \\ 1 & g(x, y) \in (100.200) \\ 1 + B_2\frac{g(x,y) - 220}{35} & g(x, y) \in [220, 225] \end{cases} \qquad (2)$$

where, $B1$ and $B2$ are constant, and their values are 2.6 to 3 and 1, respectively. Equally, The values of these two constants depend on the size of the region to be repaired and the contrast between the target and the background. After the contrast of the image is adjusted, the gray scale of the image is stretched. As shown in Fig. 6, the results of this method are compared with those of the Retinex algorithm and the original method.

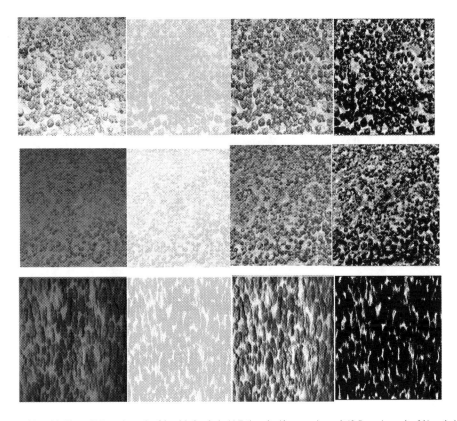

(a)   original image (b) Processing results of the original method   (c) Retinex algorithm processing results(d) Processing results of this method

Fig. 6. Comparison of the results of different methods

As can be seen from Fig. 6, this method and the other two methods can solve the problem of uneven illumination. However, the method in this paper improves the contrast between manganese nodule particles and the surrounding environment, which is conducive to image segmentation. At the same time, the advantage of this method is that the morphology of manganese nodules can be repaired by using this method again, which can not be completed by the Retinex algorithm and the original method. Although the problem of uneven illumination has been well solved, the morphology of manganese nodules has not been completely restored. Next, we need to use this method again to repair the morphology of manganese nodules, and then test the results of this paper in image segmentation.

4.3   Morphological Repair of Manganese Nodules

After solving the problem of uneven illumination, we need to repair the morphology of manganese nodules. As shown in Fig. 7, the gray value of the area covered by white sand is higher than that of the area not covered by white sand. On a single manganese nodule, the pixels with higher gray value of white sand are mostly surrounded by the pixels with lower gray value, and the distribution of covered white sand on a single manganese nodule is scattered. In view of this characteristic, we execute the method again to calculate the background gray value of the image.

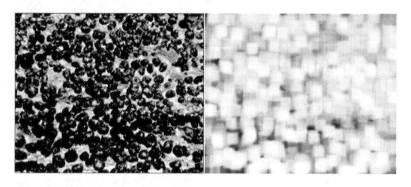

(a) Image after illumination equalization   (b) Background gray values calculated from images

Fig. 7. Background gray value calculated in the second time

Next, the image is subtracted from its background gray value, and the gray value of the image is readjusted using the flow of Fig. 5. In this way, the morphology of manganese nodules can be repaired. As shown in Fig. 8, if we use algorithm of corrosion and expansion to repair morphology directly at this time, the following problems will arise: (1) In order to restore the morphology of manganese nodules, corrosion algorithm is applied to the image. Although the morphology of manganese nodules is restored, the gap between individual manganese nodules disappears and the interference increases. After that, it is difficult to separate the manganese nodules from the interfering pixels by the expansion algorithm. (2) In order to remove the interference, the expansion algorithm is applied to the image. Although the interference is excluded, due to the white sand cover, the whole manganese tuberculosis will split into many pieces, which are removed together with the interference pixels by the small target removal algorithm.

Compared with other methods, the method in this paper not only ensures the gap between individuals, but also prevents the partial loss of manganese nodules after morphological restoration. It is shown that the method presented in this paper is applicable to the morphological restoration of manganese nodules and is a feasible and stable method. It should be noted that compared with solving the problem of uneven illumination, the size of the mask has a great influence on the results in the process of morphological restoration. As shown in Fig. 9, too large or too small masks can affect the morphological repair of manganese nodules.

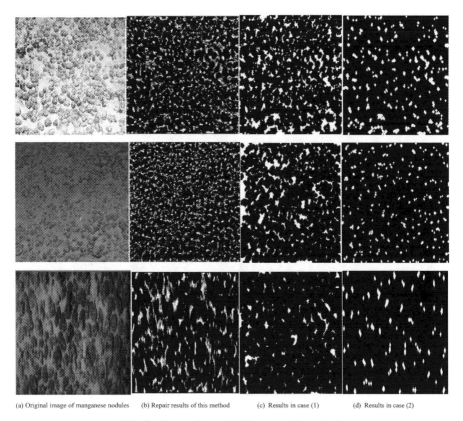

(a) Original image of manganese nodules   (b) Repair results of this method   (c) Results in case (1)   (d) Results in case (2)

Fig. 8. Comparison of different repair methods

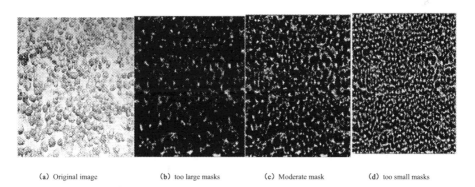

(a) Original image   (b) too large masks   (c) Moderate mask   (d) too small masks

Fig. 9. Effect of different masks on repair results

As can be seen from Fig. 9, excessive masking will lead to the disappearance of gaps between manganese nodules, and too small masking will lead to incomplete morphological repair of manganese nodules. After solving the two core problems, we need to segment the processed image to verify the stability and feasibility of this method.

5 Simulation Results

5.1 Binarization of Image

Since we need to obtain binary images containing only manganese nodules, we need to check whether the processed images can achieve the desired results in image segmentation after the above operations are completed. If the image needs to keep more details, we use Sauvola algorithm [8] to binary the image.

The principle of the algorithm is that the mean and standard variance of the gray value in the mask of r*r size are calculated to determine the threshold of the mask center (x, y). In order to eliminate a small part of the interference, we increase the r value as much as possible. The average gray value of a mask centered on (x, y) is:

$$m(x, y) = \frac{1}{r^2} \sum_{i=x-\frac{r}{2}}^{x+\frac{r}{2}} \sum_{j=y-\frac{r}{2}}^{y+\frac{r}{2}} f(i, j). \tag{3}$$

The variance in the mask is:

$$s(x, y) = \sqrt{\frac{1}{r^2} \sum_{i=x-\frac{r}{2}}^{x+\frac{r}{2}} \sum_{j=y-\frac{r}{2}}^{y+\frac{r}{2}} (f(i, j) - m(x, y))^2}. \tag{4}$$

According to Eqs. 3 and 4, the threshold of image at point (x, y) can be obtained as follows:

$$T(x, y) = m(x, y) * [1 + K * (\frac{s(x, y)}{R} - 1)], \tag{5}$$

where, the dynamic range of standard deviation is R, $R = 128$. And the correction parameter is K, $K = 0.5$.

If the image needs to be quickly segmented, the threshold can be set to 0 directly. The binary image is shown in Fig. 10.

From the comparison of the images, we can see that the morphology of manganese nodules has been greatly restored, but the interference still exists. Next, we will eliminate the interference and test the effectiveness and stability of the method by image segmentation.

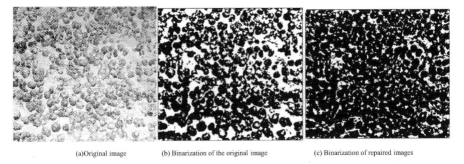

(a)Original image   (b) Binarization of the original image   (c) Binarization of repaired images

Fig. 10. Comparison of binary images

5.2 Image Segmentation

Because most of the manganese nodule particles have returned to normal shape, corrosion expansion algorithm and small target removal algorithm can be used to remove interference, thus completing image segmentation. Because the morphology of most manganese nodules has returned to normal, algorithm of corrosion and expansion and small target removal algorithm can be used to remove the interference, thus completing image segmentation and obtaining binary images containing only manganese nodules. Because the interference in the image is mostly slender and the number of pixels in the joint with manganese nodules in the image is relatively scarce, we use the expansion algorithm to disconnect the joint of manganese nodules. After experimenting with different masks, the structural element $B = [0, 1, 0; 1, 1, 1, 0, 1, 0]$ is selected, and then the binary image is expanded three times, and the small target removal algorithm is performed to remove the interference in the image. Finally, the corrosion algorithm is implemented by using the flat disc structure element with radius of 4. The image processing results are shown in Fig. 11.

As can be seen from Fig. 11, compared with the results of image segmentation in the case of direct binarization and image without repaired by this method, the image processed by this method not only solves the problem of uneven illumination, but also retains the gap between manganese nodules. The manganese nodule image processed by this method has a good effect in image segmentation, which proves that this method can be applied to the segmentation of manganese nodule image, and it is feasible and stable. Although the morphology of manganese nodules is slightly distorted, it will not have much influence on the identification of the distribution of manganese nodules in this area. However, manganese nodules buried almost entirely in the sand can not be repaired by this method.

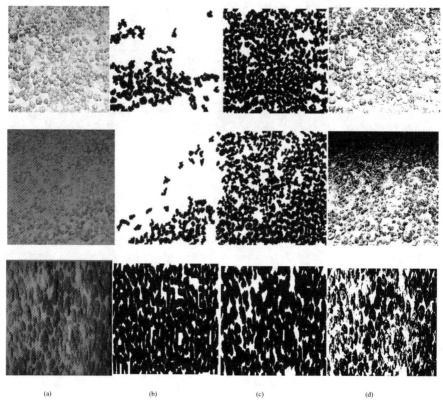

(a)   (b)   (c)   (d)

(a) Original image
(b) Image segmentation results of unrepaired manganese nodules
(c) The result of image segmentation after processing by this method
(d) The result of binarization of original image

Fig. 11. Result of image segmentation

6 Conclusion

In order to realize the segmentation of manganese nodule images, it is necessary to solve the two core problems of uneven illumination and repair the morphology of manganese nodule morphology. Therefore, according to the characteristics of manganese nodule image, this paper proposes a method and verifies its feasibility in image segmentation. Firstly, he gray image is preprocessed and the method presented in this paper is executed to get the image which eliminates the uneven illumination. After that, the morphology of manganese nodules can be repaired by performing the method again on the images which eliminate the uneven illumination. Here, the applicability and stability of this method in eliminating the uneven illumination and repairing the morphology of manganese nodules are verified by comparison with other methods. Finally, the image with only manganese nodules is obtained by image segmentation, which verifies the feasibility and stability of this method. However, the morphology of manganese nodules treated by this method is slightly distorted. In a word, the method is effective and stable by processing several images, and the manganese nodule image processed by this method has achieved good results in image segmentation. The method presented in this paper can synchronously process a large number of manganese nodules with different conditions in an image, which lays a good foundation for automatic underwater manganese nodule survey. How to make statistics on the manganese nodules is what we need to study next.

Acknowledgements. This work was supported by Open Fund Project of China Key Laboratory of Submarine Geoscience (KLSG1802), Science & Technology Project of China Ocean Mineral Resources Research and Development Association (DY135-N1-1-05).

References

1. Zhuang, W., Chen, Y., Su, J., Wang, B., et al.: Design of human activity recognition algorithms based on a single wearable IMU sensor. Int. J. Sens. Netw. **30**(3), 193–206 (2019)
2. Zhuang, W., Sun, X., Zhi, Y., et al.: A novel wearable smart button system for fall detectio. In: 2017 AIP Conference Proceedings, vol. 1838(1), pp. 020–035 (2017)
3. Awad, NH, Ali, M.Z., Suganthan, P.N.: Ensemble sinusoidal differential covariance matrix adaptation with Euclidean neighborhood for solving CEC2017 benchmark problems. In: 2017 IEEE Congress on Evolutionary Computation, San Sebastian, Spain, pp. 372–379 (2017)
4. Petit, F., Capelle-Laize, A.S., Carre, P.: Underwater image enhancement by attenuation inversion with quaternions. In: Proceedings of IEEE International Conference on Acoustics, Speech and Signal Processing, Taiwan, vol. 200, pp. 1177–1180 (2009)
5. Sahu, P., Gupta, N., Sharma, N.: A survey on underwater image enhancement techniques. Int. J. Comput. Appl. **87**(13), 19–23 (2014)
6. Huang, D., Wang, Y., Song, W., et al.: Underwater image enhancement method using adaptive histogram stretching in different color models. J. Image Graph. **23**(5), 640–651 (2018)
7. Badekas, E., Papamarkos, N.: Estimation of appropriate parameter values for document binarization techniques. Int. J. Robot. Autom. **24**(1), 66–78 (2009)
8. Sauvola, J., Pietikainen, M.: Adaptive document image binarization. Pattern Recogn. **33**(2), 225–236 (2000)
9. Hassan, N.M., Mostfa, E., Salehi, M.E.: A fast fault-tolerant architecture for sauvola local image thresholding algorithm using stochastic computing. IEEE Trans. Very Large Scale Integr. Syst. **24**(2), 808–812 (2016)
10. He, Z.: Binarization of document image captured under uneven illumination. J. Shanghai Univ. Eng. Sci. **25**(2), 164–166 (2011)
11. Wang, G., Dong, Q., Pan, Z.: Retinex theory based active contour model for segmentation of inhimogeneous images. Digit. Process. **50**, 43–50 (2016)
12. Diels, E., van Dael, M., Keresztes, J., et al.: Assessmennt of bruise volumes in apples using X-ray computed tomography. Post-Harvest Biol. Technol. **128**, 24–32 (2017)
13. Liu, Y., Zhou, T., Wang, X., et al.: High precision pose calibration method of multi-line camera system. J. Huazhong Univ. Sci. Technol. (Nat. Sci. Edn.) **46**(11), 81–86 (2018)
14. Wang, Q., Wu, K., Wang, X., et al.: Automatic detection and classification of foreign bodies of dumplings based on x-ray. J. Comput.-Aided Des. Comput. Graph. **30**(12), 2242–2252 (2018)
15. Jing, J., Guo, G.: Yarn packages hairiness detection based on machine vision. J. Text. Res. **40**(1), 148–152 (2019)
16. Kaur, R.: Modified watershed transform using convolution filtering for image segmentation. Int. J. Eng. Sci. Res. Technol. **3**(8), 449–454 (2014)

17. Qun, M., Heng, Y., Fang, C., et al.: Reversible data hiding in encrypted image based on block classification permutation. Comput. Mater. Continua **59**(1), 119–133 (2019)
18. Niu, B.H., Huang, Y.F.: An improved method for web text affective cognition computing based on knowledge graph. Comput. Mater. Continua **59**(1), 1–14 (2019)
19. Wang, Q., Yang, C.G., Wu, S.E., et al.: Lever arm compensation of autonomous underwater vehicle for fast transfer alignment. Comput. Mater. Continua **59**(1), 105–118 (2019)

A Portable Intelligent License Plate Recognition System Based on off-the-Shelf Mini Camera

Haixia Yang[1], Wei Zhuang[2], Qingfeng Zhou[2], Dong Dai[3], and Weigong Zhang[1(✉)]

[1] School of Instrument Science and Engineering, Southeast University, Nanjing 210096, China
zhzhy411@163.com, zhangwg@seu.edu.cn
[2] School of Computer and Software, Nanjing University of Information Science and Technology, Nanjing 210044, China
zw@nuist.edu.cn, 573137902@qq.com
[3] School of Cyber Science and Engineering, Southeast University, Nanjing 210096, China
daidong@seu.edu.cn

Abstract. This paper presents a portable license plate recognition system based on single chip microcomputer with the off-the-shelf mini camera. The system hardware mainly consists of STM32 single-chip microcomputer and the image acquisition sensor OV7670. A TFT color screen is adopted for real time image display and AMS1117 is used for the power conversion and supply. The system uses STM32 MCU to control the overall operation of the license plate recognition system, and controls the mini camera to perform image acquisition through the jump-point analysis and image binary processing method, and then locates the license plate area. With the plate characters successfully localized, the recognition of the license plate is finally obtained through the character segmentation and matching. The experimental results have shown that our system has the best performance under normal light, when the inclination is less than $15°$ and the character integrity is more than 85%.

Keywords: License plate recognition · Single chip system · License plate location · Character cutting

1 Introduction

With the increasingly advancing information and Internet of things technology, the continuous improvement and development of intelligent traffic management and visualization to achieve all-round vehicle monitoring and management is a hot research direction of relevant researchers in all of the world [1–4]. Automatic license plate recognition technology plays a very important role in this research, which can be applied in the field of road toll collection, vehicle inspection and tracking, and parking station vehicle management, etc. [5–8].

In many practical applications nowadays, the application of license plate recognition technology mostly relies on the fixed camera to take photos of the vehicle, and then the PC terminal is used to capture the image, screen the license plate image, and carry out complex image processing to extract the license plate characters [9]. With the continuous progress of road construction, urban road traffic management is gradually becoming tired [10]. Under the changeable environment, the existing license plate recognition methods are no longer applicable. Therefore, the design of advanced license plate acquisition and recognition technology has become an important research topic of image and information processing technology [11, 12].

In recent years, many studies on the fusion of vision and image processing have been proposed [13–15]. The license plate images collected by the visual sensor are processed in a series of ways, including license plate recognition, character cutting, character feature extraction, character template construction and character recognition. According to this series of steps, researchers finally successfully designed an automatic license plate recognition system with high recognition rate, and later proposed template matching method, adaptive threshold method, vertical projection method, boundary tracking method and other license plate image processing technologies [16]. After years of research and design, on the basis of mature theory, the license plate recognition technology has been successfully applied in practical applications [17, 18]. In recent years, Singapore, Israel, Germany, Britain, the United States and other countries successively do research on license plate recognition, and put those novel products into applications, such as VLPRS developed by Singapore Optasia company, ARGUS developed by British Alphatech, the AUTOSCOF System developed by the United States, and the ARTEM7SXI System developed by Siemens, and so on.

2   System Hardware Design

2.1   System Master Control Design

To realize the complex operation of the whole system, the selection of its master control chip is crucial. The master control chip of the system needs to control the camera sensor to scan and intercept the image, read the image data, and then conduct a series of image processing, and display the processed image on the display screen. The functions to be realized by the master control chip are relatively complicated and tedious. It can be seen that the master control chip needs strong computing power and image processing ability. Therefore, the functions of the 51 MCU cannot be realized, so it is abandoned. After multiple comparisons, the design chooses embedded microcontroller chip STM32F103RBT6, which is a 32-bit, with ARM cortex-m3 core microcontroller minimum system [7].

In the design of the whole hardware system, to realize the complete operation of the whole circuit, the microcontroller minimum system circuit is the core of the whole system, whether the circuit is normal determines whether the whole system can achieve image acquisition, processing, display and other functions. In order to make the minimum system run normally, it is necessary to design a complete and feasible minimum system circuit. The circuit diagram of the minimum system of the single chip microcomputer is shown in Fig. 1. The circuit diagram shows that the complete minimum system circuit can be composed of power supply, clock and reset circuit.

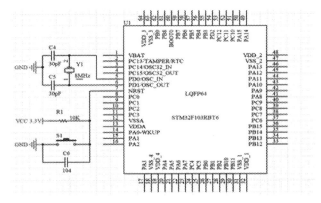

Fig. 1. Microcontroller minimum system circuit diagram

2.2 Power Circuit Design

Ams1117-3.3 can provide enough power supply for system operation and has a high cost performance. The specific power supply flow provided in this design is as follows: when the power is switched on, the system current passes through C3 and C4 capacitors and is filtered to remove signals lower than the specified voltage, so as to output a stable 3.3v dc voltage to supply power to the system. In addition to the filter, inductance is also designed, and the two are connected in parallel to form the filter circuit of LC and RC, so as to obtain a relatively ideal and stable dc voltage and filter out the voltage ripple components of the power supply [8]. The power circuit of the system is shown in Fig. 2:

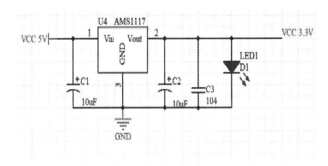

Fig. 2. System power circuit diagram

2.3 Image Acquisition Module

The image acquisition module is mainly used for the acquisition of license plate image. The main principle is to continuously scan the image of the designated area by using the image binary method. Every 15 s, when the image background conforms to the background of the license plate image, the image is intercepted and the character is segmented. The captured image is sent to the main control chip for processing. OV7670 camera is proposed to be adopted, after being compared with various cameras and being considered the matching of STM32 applicable cameras.

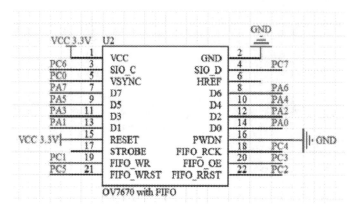

Fig. 3. Connection diagram of relevant interface between OV7670 and STM32

Image Acquisition Circuit Design. Image acquisition circuit to achieve license plate image capture, and the selection of OV7670 camera can achieve the design of the function. In OV7670, the SCCB bus pin is connected to the main control chip and controlled by it to achieve multi-mode camera shooting and multi-resolution image data output, which can output up to 30 frames/second of VGA images. STM32 also has DCMI interface of high-speed digital camera. The camera can realize automatic recognition and interception of license plate through connection with MCU, and display license plate image through MCU in the display module, and then conduct image processing and other operations. The interface connection between OV7670 camera and STM32 is shown in Fig. 3.

3   System License Plate Recognition Algorithm

3.1   Graying Algorithm

Since the camera collects color images, it is easy to overlap with the background of the license plate, and the most important way to realize automatic recognition and interception of license plate images is based on the background color, so grayscale processing should be carried out. Color image occupies a high capacity, especially in some microprocessors due to the limited capacity, color image storage load is too heavy, but the use of grayscale can reduce the load can also improve the identification efficiency and complex light conditions of the image interception. At present, the red, green and blue component model is commonly used, namely the RGB model (R is the red component, G is the green component and B is the blue component). Gray value is an important parameter in the gray processing of the model. When gray = R = G = B, gray image can be transformed. This is the simplest method, but the error is relatively large. Currently, the average method is mainly adopted: gray takes the average of R, G and B.

Define:

$$gray = \frac{R + G + B}{3} \tag{1}$$

Based on the above formula, grayscale can be further realized. Most researchers use the average weighted value to calculate, and the formula is:

H = 0.229R + 0.588G + 0.144B (H is the brightness value, and the coefficient of RGB represents the empirical weighted value). After the grayscale value is calculated, in order to make the grayscale image consistent with the color image, grayscale stretching must be carried out. The formula of grayscale transformation function is as follows:

$$f(x) = \begin{cases} \frac{y1}{x1}x \ \ldots\ldots\ x < x1 \\ \frac{y2-y1}{x2-x1}(x - x1) \ \ldots\ldots\ x1 \le x \le x2 \\ \frac{255-y2}{255-x2}(x - x2) + y2 \ \ldots\ldots\ x > x2 \end{cases} \tag{2}$$

3.2 Binary Algorithm

After the algorithm for calculating grayscale change, further step is binarization processing, which is the further extension of the gray mapping into binarization. Binarization processing transform the grey value by two values (1, 0, where 1 means white, 0 means black), so the binary image is also known as black and white image. The main method is to select the appropriate threshold, normalize the grayscale image with the grayscale value, and transform the black-and-white binary image, which can present the local features of the image. This paper adopts the limited threshold method, which is an improved algorithm. Different from the traditional threshold method, it needs to stretch the image gray level first, and then set the threshold. It can only show the background color part of the license plate image, which can realize the license plate interception better in the location area. After the color grayscale, the image can be divided into six levels of gray value, and then binarization, so the license plate background color corresponding to the set gray value reflect its background color. Due to the limited level and the limited capacity of SCM, this paper only studies the common blue license plate binarization processing.

For the binarization of blue license plate, because the color of the number on the blue license plate is white, the image can be divided into regions. We identify the general blue region, divide the blue region, and then search for the gray value corresponding to white in each blue region. The white gray value is 150 and the blue gray value is 255. We figure out the percentage of white and the percentage of the background color, then the relevant comparison can realize the binarization judgment of the license plate background color. The specific formula is: alpha = w/Qb, beta = b/Qb (alpha represents the percentage of number color, beta represents the percentage of base color, w represents the white area value, b represents the blue area value, Qb represents the blue gray value).

According to the equation, after calculating the percentage of each in the candidate area, if it is determined, the license plate can be determined as a blue plate.

Define:

$$G(X, Y) = \begin{cases} 1 \ldots\ldots \alpha + \beta \approx 1 \&\& G(x, y) = 255 \\ 0 \ldots\ldots others \end{cases} \tag{3}$$

3.3   Regional Localization Algorithm

This paper adopts the method of horizontal projection and vertical projection alternately, which can display the license plate area clearly by using reliable electronic equipment, and adopts the method of horizontal and vertical two-way projection, which can locate quickly. In the projection method, we adopts the method of extracting the corner points of the license plate to realize the positioning and make some improvements. The basic idea of this algorithm is traversal that is traversing row by row and column by column. To determine the fixed point, its reference value is P1 (x1, y1) in the upper left corner and P2 (x2, y2) in the upper right corner. The specific method is as follows: first draw up a maximum value for reference, then comprehensively slide search the license plate area, find the two points that conform to the maximum value, determine the horizontal and vertical coordinates of P1 and P2, and then extract the area between the two points.

3.4   Character Segmentation and Matching Algorithm

Character segmentation and matching is the last step and the key one in the overall algorithm research. Whether the function can be realized depends on whether the segmentation and matching algorithm is suitable. The character segmentation algorithm adopted in this paper is the vertical area projection method. According to the characteristics of Chinese license plate, its characters have a total of 8 bits, which can be divided into regions in the binarization processing time and divided successively. These 8 characters have gaps, and the character projection must be the local minimum through vertical projection.

Once the character segmentation is complete, the next step is character recognition. Character matching algorithms are commonly used in edge feature extraction, wavelet transform and template matching. Through experimental comparison, this paper chooses template matching method, which is easier for us to operate, relatively simple to implement and acquire high identification rate. There are a lot of commonly used matching methods, because of the characteristics of domestic license plate, it is not appropriate to use only one template matching to separate the letters and Numbers of Chinese characters, so this paper chooses simple binary template matching, stroke direction density characteristics method.

Chinese recognition steps are generally composed of two parts: Chinese character recognition, letter and number recognition. Its first task is to establish the corresponding template library, according to the Chinese license plate character characteristics of the statistical input, and next step is characteristics extraction and classification.

A template library is established after the split of the license plate character extraction with characteristics being recognized in turn from right to left. Through comparing operation characteristics, with the characters in the template library maximum likelihood matching, the matching degree is determined by the related equation after the operation, in which the available identified R represents the characters with template matching similarity degrees. Then we choose the one corresponding to the maximum similarity template output character, and its R calculation equation is:

$$R = \frac{\sum_{y=1}^{y}\sum_{x=1}^{x}\left[S(x,y) \times T(x,y)\right]}{\sum_{y=1}^{y}\sum_{x=1}^{x}\left[S(x,y)\right]^{2}} \tag{4}$$

Where, S represents the input character; T represents the characters in the template library; they both have size X times Y.

4 System Experiments

4.1 System Function Tests

License Plate Interception Test System. This test is to validate whether can realize license plate localization and interception function after setup of the system. The screen will display the real-time video camera images of the transmission, when the license plate area has been specified during the camera scanning. The system automatically recognize the license plate background, after capturing the timing, image binarization, and the license plate number in the picture with the blue line in turn divided into each character, which will be transformed into eight regions. Finally, the system will transmit the results to the screen in real time.

System License Plate Recognition Display Test. After image binarization being processed by the system, 8 characters are matched by character template in the main control chip, and then the matched characters are displayed. In order to verify the validity of license plate recognition for various provinces, the license plate recognition of several provinces is found in the computer, as shown in Fig. 4.

Fig. 4. License plate image binarization processing

4.2   System Performance Test and Comparison

From the above tests, it can be known that the basic functions of the design have been realized, and these are carried out in a relatively stable environment, so a comparative test should be conducted to obtain the specific performance of the system. This test is conducted in three aspects, namely illumination intensity, license plate Angle and license plate integrity.

Identification Performance Test under Light Intensity. According to the above test results, license plate can be recognized under normal, strong or weak luminosity. In order to facilitate the test, the license plate images are all from the Internet. Next, we test the results in bright light and dark.

1) For a strong light, we use a commodity lamp to simulate strong light exposure. First, testing the results under strong light intensity, as shown in Fig. 5. Under strong light, its license plate is too blurred on the display screen, which makes it unrecognizable. Its identification box cannot capture the license plate area, which makes the identification fail.

Fig. 5. Recognition results in bright light

2) This paper also simulate the environment under dark night. The test results are shown in Fig. 6. As can be seen from the identification figure, the camera can scan the license plate area, and the license plate can be seen on the screen. However, the background color of the license plate becomes dark under low light, and the background color cannot be recognized by binarization, so the recognition fails.

License Plate Angle Different Recognition Performance Test. In order to verify its efficiency, we test the influence of license plate tilt on the system and find out the tilt limit. After several tests, it was found that the license plate can be recognized when it is tilted at 15° to 20°, but the error is large. Accurate identification tilt is under 10°. The identification diagram at 15° tilt is shown in Fig. 7. As can be seen from the figure, the recognition rate is greatly reduced. The license plate number is zhejiang d.jf555, while it is displayed as zhejiang q.jfrfs. It can be seen from the blue line recognition box and license plate tilt angle is about 15°.

As shown in Fig. 8, this is the identification diagram at a tilt of 20°. It can be seen that it cannot be recognized after 20°.

Fig. 6. Low light or no light recognition results

Fig. 7. Identify results at a 15° tilt (Color figure online)

Fig. 8. Identify results at a 20° tilt

4.3   System Performance Evaluation

After several tests, the overall performance evaluation of the system can be obtained. Its performance is as shown in Table 1:

Table 1. Performance comparison diagram

Illumination intensity/recognition rate	Tilt/recognition rate	Character integrity/recognition rate
High light/50%	<15°/90%	>80%/85%
Normal light/90%	15°–25°/50%	80%–70%/50%
Weak light/0%	>25°/0%	<70%/0%

5   Conclusion

The proposed license plate recognition system has features of low cost, portable and high efficiency, and can realize plate recognition in complex environment, which plays an important role in modern traffic management intelligence. The system is of great significance in the fields of car park management and street vehicle inspection. In this paper, a complete and feasible minimum system circuit is designed, and image binary method is used to construct the image acquisition module. In addition, binary processing is carried out after the grayscale algorithm, and the key is the subsequent research on location positioning algorithm. Finally, random character segmentation is carried out to normalize the characters, and then the character template is established to send the characters to be recognized into the matching program for matching. Experimental results show that the system has the best performance under normal light, when the inclination is less than 15° and the character integrity is more than 80%.

Acknowledgements. This work was supported by the National Natural Science Foundation of China (Grant No. 61972207), Jiangsu Provincial Government Scholarship for Studying Abroad and the Priority Academic Program Development of Jiangsu Higher Education Institutions (PAPD).

References

1. Li, S., Liu, F., Liang, J., Cai, Z., Liang, Z.: Optimization of face recognition system based on azure IoT edge. Comput. Mater. Continua **61**(3), 1377–1389 (2019)
2. Fang, W., Zhang, F., Sheng, V.S., Ding, Y.: A method for improving CNN-based image recognition using DCGAN. Comput. Mater. Continua **57**(1), 167–178 (2018)
3. Xia, Z., Lu, L., Qiu, T., Shim, H.J., Chen, X., Jeon, B.: A privacy-preserving image retrieval based on AC-coefficients and color histograms in cloud environment. Comput. Mater. Continua **58**(1), 27–43 (2019)

4. Xia, X., Deng, Y.: Analysis of parking lot intelligent guidance system. Technol. Entrepreneur (11), 8 (2013)
5. Zhuang, Z., Cai, L.: Vehicle licence plate recognition using super-resolution technique. In: IEEE International Conference on Advanced Video and Signal Based Surveillance, pp. 411–412 (2014)
6. Liu, Y., Nannan, L.: Design of license plate recognition system based on the adaptive algorithm. In: 2008 IEEE International Conference on Automation and Logistics, pp. 2818–2821 (2008)
7. Xin, J., Mingyong, L., Kaixuan, Z., Jiangtao, J., Hao, Ma., Zhaomei, Q.: Development of vegetable intelligent farming device based on mobile APP. Cluster Comput. 22(4), 8847–8857 (2018). https://doi.org/10.1007/s10586-018-1979-4
8. Ma, Y., Zeng, X., Li, J.: License plate location research based on texture analysis & mathematics morphology. Adv. Mater. Res. 317, 74–77 (2011)
9. Kaburuan, E.R., Jayadi, R.: A design of IoT-based monitoring system for intelligence indoor micro-climate horticulture farming in Indonesia. Procedia Comput. Sci. 157, 459–464 (2019)
10. Amitrano, D., Arattano, M., Chiarle, M., et al.: Microseismic activity analysis for the study of the rupture mechanisms in unstable rock masses. Nat. Hazards Earth Syst. Sci. 10(4), 831–841 (2010)
11. Liu, Q., Wang, Y., Linheng, W.: Design of image acquisition and display system based on STM32. Math. Tech. Appl. 02, 94 (2012)
12. Aboura, K., Al-Hmouz, R.: An overview of image analysis algorithms for license plate recognition. Organizacija 50(3), 285–295 (2017)
13. Zhuo J, Hu Y.: Research of license plate locating method based on edge detective and projection. Bulletin of Science and Technology (2010)
14. Xia, H, Liao, D.: The study of license plate character segmentation algorithm based on vetical projection. In: 2011 International Conference on Consumer Electronics, Communications and Networks (CECNet), pp: 4583–4586 . IEEE (2011)
15. Ma, J., Mo, Y., Wang, M.: A method of license plate character recognition based on improved template matching. Minicomput. Syst. (13), 32 (2003)
16. Sarfraz, M.S., Shahzad, A., Elahi, M.A., Fraz, M., Zafar, I., Edirisinghe, E.A.: Real-time automatic license plate recognition for CCTV forensic applications. J. Real-Time Image Process. 8, 285–295 (2013)
17. Nan, L., Rui, Y., Jiakang, D., et al.: The design of wireless intelligent door lock system based on the microcontroller. Electron. Test (4), 8 (2018)
18. Zhuang, W., Chen, Y., Su, J., Wang, B., Gao, C.: Design of human activity recognition algorithms based on a single wearable IMU sensor. Int. J. Sens. Netw. 30(3), 193–206 (2019)
19. Su, J., Hong, D., Tang, J., Chen, H.: An efficient anti-collision algorithm based on improved collision detection scheme. IEICE Trans. Commun. E99-B(2), 465–469 (2016)

Mechanical Analysis and Dynamic Simulation of Ship Micro In-pipe Robot

Zhipeng Xu[1], Yuliang Liu[1(✉)], Han-de Mao[1], Sheng Chang[1], and Shuxun Li[2]

[1] School of Ship and Mechanical and Electrical Engineering, Zhejiang Ocean University,
Zhoushan 316000, China
15195802915@163.com, 2574951395@qq.com
[2] Zhejiang Jiachang Engineering Technology Co, Ltd., Zhoushan 316000, China

Abstract. A new design scheme of ship micro in-pipe robot is proposed where both complex working environment inside ship boiler flue and limitations from traditional cleaning methods. By two groups of three-claw cylinders and transmission device to coordinate the action, the peristaltic movement of the robot is realized. Multi-module design method is used to make the overall structure of the robot more compact, and to improve the working and driving ability in a small space. In this paper driving condition of the robot in vertical pipeline is mainly studied, and the regulating characteristics of the driving wheel is used to adapt to different pipe diameters. Dynamic analysis and simulation are completed based on Newton-Euler equation, and the objects include joints, guide wheels end and driving characteristics of the regulating mechanism. The results show that the regulating mechanism can help the robot to move under different pipe diameter, the guide wheels can maintain the expected contact force in the restricted direction when obstacles occur, and the driving force of the robot can meet the actual work requirements.

Keywords: Ship boiler · Pipeline robot · Peristaltic · Newton-Euler equation · Dynamics

1 Preface

In the field of auxiliary marine boiler, the design requirement of a smoke pipe is that the pipe should hold a high gas flow rate and self-cleaning ability [1]. However, long-term use of inferior fuel or poor combustion during work will result in widespread accumulation of ash in the pipe. During the voyage, usually it's necessary to blow the ash every two or three weeks [2]. After the beginning of the main engine, it is especially necessary to blew the ash. At present, it is mainly washed with water or manual [3]. The washing not only consumes a lot of fresh water, but also the water having dissolved the ash is acidic. Also, frequent washing may aggravate the corrosion of the pipe and the drum, and water infiltration into firebricks may easily damage the walls [4]. Manual ash removal is to use chisel, brush and other manual tools, the crew enters the furnace for mechanical cleaning. Due to high temperature in the furnace and the large soot, the high intensity

© Springer Nature Switzerland AG 2020
X. Sun et al. (Eds.): ICAIS 2020, LNCS 12239, pp. 646–657, 2020.
https://doi.org/10.1007/978-3-030-57884-8_57

work for a long time has caused damage to the body of the crew [5, 6]. Especially in the industry, there are many small pipes below 10 cm, which is very difficult to operate manually, and it is even more difficult to clean under the harsh environment such as shipping and chemical industry [7]. In summary, pipeline robots have good application prospects in small pipeline cleaning, we urgently need to replace the manual with tiny robots that can enter the pipeline for inspection and cleaning.

At present, there are many related researches on pipeline robots at home and abroad. The research on pipeline robots in foreign countries started earlier. As early as the 1980s, pipeline robots began to develop rapidly [8, 9]. Jointed peristaltic pipe robot developed by Alive Adria, Herman Streich, etc., adopts wheeled walking mechanism, has good flexibility and strong cornering ability, but the weight is large, the joint control ability is poor, and it is not suitable for complex environment in pipeline [10]. Young-Sik Kwon of South Korea developed a wheeled pipe robot with a small overall size and light weight which is suitable for micro-pipe environment, but because of its large driving friction, it faces when facing uneven pipes [11]. The ability is weak and the turning radius is large. The research on pipeline robots in China started relatively late in developed countries. The spiral-driven pipeline robot developed by Deng Zongquan of Harbin Institute of Technology and the large-diameter pipeline detection robot developed in succession has marked the entry of China's pipeline robot into the practical stage from theoretical research [12]. Liu Shuaimin and others have developed a miniature spiral wheeled pipeline robot using a spiral propulsion has the characteristics of simple control [13], tight structure, and large traction, but it can only adapt to pipes with small pipe diameter changes.

In this paper, the pipeline robot is designed to clean a ship boiler flue with advantages of multi-module structure and excellent obstacle surmounting ability. In addition it has good adaptability to different pipe diameters, and big road capacity when cleaning mechanism and can carry more surveying devices through a variety of complex pipeline environments such as vertical pipes.

2 Movement Principle and Structural Design of Pipeline Micro-robot

2.1 Mechanical Structure Design

The main structure of the ship boiler pipe micro-robot is shown in Fig. 1. It mainly consists of three parts: cleaning mechanism, drive mechanism and support structure. The copper brush is connected to the motor to form a cleaning structure, the support bar is fixed to the ordinary cylinder, the intermediate ring has a spring, the cleaning structure, the three-jaw cylinder 2 and the driving structure together form the front half of the robot and are connected to the three-jaw cylinder 1 of the rear half.

2.2 Principle of Motion

Assuming that the initial state of the robot is completely into the pipeline, the robot is in a completely free contraction state, at which point the three claws are not grasping

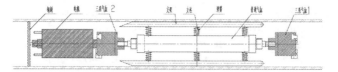

Fig. 1. Pipe robot body structure diagram.

the pipe wall. At this time, the cylinder operation is started. The following is a specific movement step, which is divided into six steps, as shown in Fig. 2: ① the three-jaw cylinder 1 works, the support unit moves, the rear claw on the cylinder grips the pipe wall, ② the rear claw keeps gripping the pipe wall state, the ordinary cylinder work, the piston motion, so that the piston rod extends, thereby driving the front half of the robot to move forward, ③ the three-jaw cylinder 2 work, the front claw grips the pipe wall, ④ the piston rod of the ordinary cylinder remains stretched, the hind claw releases the support to the pipe wall, ⑤ the ordinary cylinder enters the contracted state, thereby driving the rear portion of the robot to move upward, ⑥ when the piston rod of the ordinary cylinder is fully retracted, the rear portion of the robot moves to a specific position. The jaw cylinder 1 is operated again, and the rear claw grips the inner wall again.

The above is a complete periodic motion of the robot movement. By repeating the above actions, the robot can continue to move forward. In the case of reverse motion, the reverse sequence of actions can be achieved.

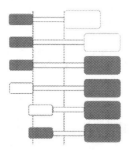

Fig. 2. Mobile schematic.

3   Overall Mechanical Analysis of the Robot

Take the robot as an example of downward movement of the vertical pipe for analysis. Assumption: 1, support structure and pipe wall do not slip; 2, three sets of support claws are equally stressed; 3, the weight of the robot is constant during the movement.

3.1   In-Tube Static Analysis

In the first step, the hind claw grip the pipe wall, and most of the robot's support is concentrated on the three claws. Since the guide wheel only plays a guiding role. When the vertical tube crawls, the supporting force of the tube wall on the wheel is so small as to be negligible compared to the supporting force of the tube wall on the support bar and the three claws, and the front claw does not touch the tube wall in the first step:

$$\begin{cases} 3\mu(N_1 + N_2 + N_3 + N_4 + N_5) = G \\ N_1 = N_2 = N_3 = 0 \end{cases}, \tag{1}$$

in the formula: μ is maximum static friction coefficient, N_1 is wall support for the front claw, N_2 is supporting force of the pipe wall to the front wheel, N_3 is supporting force of the pipe wall to the rear wheel, N_4 is supporting force of the wall to the hind claw, N_5 is tube wall pressure on the support strip, G is overall gravity of the robot.

In the third step, the front and hind claws will grasp the wall of the tube. If the wall has the same support force for the front and hind claws, at this time there are:

$$\begin{cases} 3(N_1' + N_2' + N_3' + N_4' + N_5') = G \\ N_1' = N_4' \end{cases}. \tag{2}$$

Similarly available:

$$N_2' = N_3' = 0. \tag{3}$$

3.2 Force Analysis of the Robot Moving Along the Tube

In the middle of the ordinary cylinder and the three-claw cylinder 1, the cylinder motor and the screw nut are used for transmission. In the case of the external guide rod, the output force of the cylinder is directly converted into the linear output of the guide sleeve. When ordinary cylinder works, the following relationship exists between the output thrust F_1 and pull force F_2 with the cylinder inner diameter and the piston rod diameter:

Thrust:

$$\frac{\pi}{4}D^2P\eta, \tag{4}$$

Pull:

$$\frac{\pi}{4}(D^2 - d^2)P\eta, \tag{5}$$

where D is cylinder inner diameter, d is piston rod diameter, P is used air pressure; η is cylinder load rate.

In the second step, thrust F_1 generated by ordinary cylinder must overcome the rolling friction of the pipe wall to the front wheel and the overall gravity of the robot, so that the first half of the robot moves forward. In the fourth step, the ordinary cylinder piston rod retracted, driving the latter half of the robot moves upwards, the pulling force generated by ordinary cylinder must overcome the gravity of the three-jaw cylinder 1 and the rolling friction of the pipe wall to the rear wheel. Assuming that the robot is moving at a constant speed during the ascent, there are:

$$\begin{cases} F_1 - 3f_1 - G_1 = 0 \\ F_2 - 3f_2 + G_2 = 0 \end{cases}, \tag{6}$$

when the rear half of the robot moves upwards, the front support claw should provide enough friction to prevent the robot from slipping:

$$3\mu(N_1 + N_2) > G_1 + F_2, \tag{7}$$

in the middle: F_1 is ordinary thrust, F_2 is ordinary pull, G_2 is the weight of the rear part of the robot, G_1 is the weight of the first half of the robot, f_1 is rolling friction of the pipe wall to the front wheel, f_2 is rolling friction of the pipe wall to the rear wheel.

When the cylinder thrust pushes the front and rear parts of the robot upward, the rolling friction f_1 and f_2 are too small compared to the gravity of the front half of the robot, so it can be ignored in the calculation.

3.3   Support Structure Stress Analysis

Support bar: when the robot is at rest, it is close to the pipe wall and receives pressure from the pipe wall, the spring will be compressed. When the robot moves, the adjustment of the ordinary cylinder ensures that the support bar does not come into contact with the pipe wall, reducing the sliding friction. When facing the inner wall of different pipe diameters, the amount of spring compression under the support bars is different. At this time:

$$\begin{cases} N_5 = f' \\ f' = k\Delta x \end{cases}, \tag{8}$$

in the formula: f' is spring force, K is spring elastic coefficient, Δx is spring shape variable.

When μ, G_1, G_2 are known, the simultaneous type Eqs. 1 to 3, according to the range of the spring deformation, can determine the pressure of the pipe wall against the three claws in different movement steps, and then the vertical type Eqs. 6 and 7, can obtain the thrust F_1 and pull F_2 of the cylinder. Finally, substitute F_1 and F_2 into the Eq. 4 and Eq. 5, and the main parameters of the ordinary cylinder can be obtained. Similarly, the main parameters of the two three-jaw cylinders can be obtained according to N_1, N_1', which can help us choose the right cylinder to adapt to different pipeline environments.

3.4   Force Analysis of Regulating Mechanism

The guiding mechanism of the robot has a compact structure and a large load capacity, which satisfies the basic conditions of walking in the pipeline. In the front and rear support of the guide wheel structure, all three wheels are evenly distributed in the radial direction, the front and rear parts are symmetrically distributed along the axial direction, the support points are six in total, so the shape closure condition is satisfied. When the moving mechanism walks, the three wheels are evenly distributed in the radial direction, the three points of contact are always on a cylindrical surface, so that the self-centering can be realized. Under the action of the supporting device, the guiding wheel is pressed tightly on the inner wall of the pipe, which has a strong adaptability. Simultaneously, the proper distance between the center of gravity of the robot and the axis of the cylinder's operation is guaranteed on the vertical plane of the cylinder axis, thereby ensuring the smoothness of the entire robot during operation.

The force of the regulating mechanism is shown in Fig. 3, where L_1 is the length of the support rod AO, L_2 is the distance from the fulcrum B to the center of the fixed frame O, L_3 is the length of the spring rod BC, the spring coefficient of the spring is k, the deformation of the spring is Δx, the radius of the driving wheel is R, N is the pressure of the inner wall of the pipe to the driving wheel, F is the lateral thrust of the screw to the screw nut, α, β is the angle between the support rod AO and the spring rod BC with the horizontal direction respectively. The coordinate system XOY is established centering on the O point.

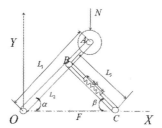

Fig. 3. Regulating mechanism force diagram.

Derived from geometric relations:

$$
\begin{cases}
y_A = L_1 \sin \alpha \\
L_3 \sin \beta = L_2 \sin \alpha \\
x_C = L_3 \cos \beta + L_2 \cos \alpha
\end{cases}
\tag{9}
$$

in the formula, y_A is the ordinate of point A, x_C is the abscissa of point C.

Differentiate the two formulas, get:

$$
\begin{cases}
\delta y_A = L_1 \sin \alpha \delta_\alpha \\
L_3 \sin \beta \delta_\beta = L_2 \sin \alpha \delta_\alpha \\
\delta x_C = L_3 \cos \beta \delta_\beta + L_2 \cos \alpha \delta_\alpha
\end{cases}
\tag{10}
$$

simplified, get:

$$
\delta x_C = -\frac{L_2}{L_1}(\tan \alpha + \tan \beta) \delta y_A.
\tag{11}
$$

From the principle of virtual work:

$$
F\delta x_C + 3(N_1 + N_2 + N_3)\delta y_A = 0.
\tag{12}
$$

There are six wheels in front of and behind the pipeline robot, and the positive pressure received by each group is approximately equal under normal operation. N_1 is the pressure on the front wheels, N_2 is the pressure on the rear wheels.

Bringing Eq. 11 into Eq. 12, get:

$$N_1 + N_2 = \frac{FL_2}{6L}(\tan\alpha + \tan\beta). \tag{13}$$

On the CD bar, from the balance equation:

$$k\Delta x \cos\beta = F. \tag{14}$$

Therefore:

$$N_1 + N_2 = \frac{k\Delta x \cos\beta L_2}{3L}(\tan\alpha + \tan\beta). \tag{15}$$

As can be seen from Eq. 15, L_1, L_2 are constant and α, β will vary according to different pipe diameters. Therefore, we can adjust the rotation of the motor to drive the screw nut and then apply different pretightening force to the spring, finally ensure the pressure N of the pipe to the guide wheel is kept within the working allowable range.

3.5   Cable Resistance Analysis

When the robot drags the cable forward, the cable will rub against the inner wall of the pipe under its own gravity, resulting in cable resistance, which requires the robot to provide driving force to overcome. The longer the travel distance, the more obvious the cable drag resistance, so this resistance cannot be ignored. Assume that there is no protrusion between the cable and the inner wall of the pipe, it is evenly contacted, and the cable expansion and contraction is not considered.

As shown in Fig. 4, when the robot is moving down on the vertical pipe, S is the total length of the cable, ds is a small unit, and dN is the supporting force of the inner wall of the pipe to the ds unit cable. F, F+dF are the tensile forces at the two ends of the ds unit cable respectively. F_1, F_2 are the pressure at both ends of the S length cable, and f is the dynamic friction factor between the inner wall of the pipe and the cable. That is, the ds unit cable force balanced is:

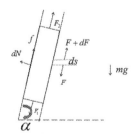

Fig. 4. Cable force diagram.

$$\begin{cases} F + mgds = F + dF + 2f \cdot dN + \rho g \cdot ds \cdot \sin\alpha \\ dN = \rho g \cdot ds \cdot \cos\alpha \end{cases}, \tag{16}$$

in the above formula, ρ is the quality of the cable per unit length.

Simplify the integration of Eq. 16:

$$F_1 = F_2 + 2\rho g f + \rho g S - mg, \tag{17}$$

Equation 17 is the resistance formula of the cable with the length S in the pipeline, F_1 is the pulling force of the robot on the cable, F_2 is the resistance of the cable in the pipeline. Which is the cable resistance is related to the inclination of the pipeline and the length of the cable.

4 Dynamics Simulation

4.1 Regulating Mechanism

The Euler-Lagrange equation is used to describe the variation of the robot system over time under constraints. Lagrange equation is as follows:

$$\begin{cases} \frac{d}{dt}\frac{\partial L}{\partial \dot{q}_i} - \frac{\partial L}{\partial q} = \tau_i \\ L = K - P \end{cases}, \tag{18}$$

where L is Lagrangian operator, K is total kinetic energy of the system, P is total potential energy of the system, $\dot{q}_i$ is Generalized coordinates of the robot, τ_i is the generalized or generalized moment acting on the system at joint i.

Taking the double support rod of the regulating mechanism as the controlled object, we get the dynamic equation:

$$M(q)\ddot{q} + C(q, \dot{q})\dot{q} + G(q) = \tau, \tag{19}$$

where q = $[q_1, q_2]$, $\tau = [\tau_1, \tau_2]$, and other arguments is defined as below.

$$\begin{cases} M(q) = \begin{bmatrix} \alpha + 2\varepsilon\cos(q_2) + 2\eta\sin(q_2) & \beta + \varepsilon\cos(q_2) + \eta\sin(q_2) \\ \beta + \varepsilon\cos(q_2) + \eta\sin(q_2) & \beta \end{bmatrix} \\ C(q, \dot{q}) = \begin{bmatrix} (-2\varepsilon\sin(q_2) + 2\eta\cos(q_2))\dot{q}_2 & (-\varepsilon\sin(q_2) + \eta\cos(q_2))\dot{q}_2 \\ (\varepsilon\sin(q_2) - \eta\cos(q_2))\dot{q}_1 & 0 \end{bmatrix} \\ G(q) = \begin{bmatrix} \varepsilon e_2\cos(q_1 + q_2) + \eta e_2\sin(q_1 + q_2) + (\alpha - \beta + e_1)e_2\cos(q_1) \\ \varepsilon e_2\cos(q_1 + q_2) + \eta e_2\sin(q_1 + q_2) \end{bmatrix} \end{cases}. \tag{20}$$

The simulation results are shown in Fig. 5:

It can be seen from Fig. 5 that when the robot passes through the nodes of different pipe diameters, the spring of the regulating mechanism will be deformed, and the angular velocities of the joints 1 and 2 will change relatively before. Then, when the robot completely passes through the node, the angular velocity will tend to be stable, and the angular velocity trace of the joint is successfully tracked, which verifies the feasibility of the regulating mechanism. Simulation results show that the robot can move under different pipe diameters.

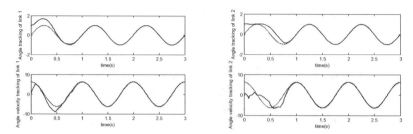

Fig. 5. Joint angular velocity and angular velocity tracking.

4.2 Guide Wheel End

For the contact work robot like pipeline cleaning robot, there is a requirement for the contact force during work. A slight positional deviation at the end of the robot may cause a huge contact force, which may cause certain damage to the robot. So, in this paper, a dynamic relationship is designed, which is established between the end force and the position of the robot, and achieves the purpose of controlling the end force by controlling the displacement of the robot, and ensure the robot maintains the desired contact force in the constrained direction. When the end of the robot guide wheel contacts the obstacle, it continues to move in the vertical direction. At this time, the dynamic equation is constructed for the end of the guide wheel:

$$D(q)\ddot{q} + C(q, \dot{q})\dot{q} + G(q) = \tau. \tag{21}$$

Among them:

$$
\begin{cases}
D(q) = \begin{bmatrix} m_1 + m_2 + 2m_3 \cos q_2 & m_2 + m_3 \cos q_2 \\ m_2 + m_3 \cos q_2 & m_2 \end{bmatrix} \\
C(q, \dot{q}) = \begin{bmatrix} -m_3 \dot{q}_2 \sin q_2 & -m_3(\dot{q}_1 + \dot{q}_2) \sin q_2 \\ m_3 \dot{q}_1 \sin q_2 & 0 \end{bmatrix}, \\
G(q) = \begin{bmatrix} m_4 g \cos q_1 + m_5 g \cos(q_1 + q_2) \\ m_5 g \cos(q_1 + q_2) \end{bmatrix}
\end{cases} \tag{22}
$$

where the m_i value is given by the formula $M = P + p_l l$, there are:

$$
\begin{cases}
M = \begin{bmatrix} m_1 & m_2 & m_3 & m_4 & m_5 \end{bmatrix}^T \\
P = \begin{bmatrix} p_1 & p_2 & p_3 & p_4 & p_5 \end{bmatrix}^T, \\
L = \begin{bmatrix} l_1^2 & l_2^2 & l_1 l_2 & l_1 & l_2 \end{bmatrix}
\end{cases} \tag{23}
$$

$$M_m(\ddot{x}_c - \ddot{x}) + B_m(\dot{x}_c - \dot{x}) + K_m(x_c - x) = F_e, \tag{24}$$

$$M_m \ddot{x}_d + B_m \dot{x}_d + K_m x_d = -F_e + M_m \ddot{x}_c + B_m \dot{x}_c + K_m x_c, \tag{25}$$

where, F_e is the contact resistance at the end, X_C is the instruction track of contact position, M_m, B_m and K_m are the matrix of mass, damping and stiffness coefficients respectively, and the simulation results are shown in Fig. 6.

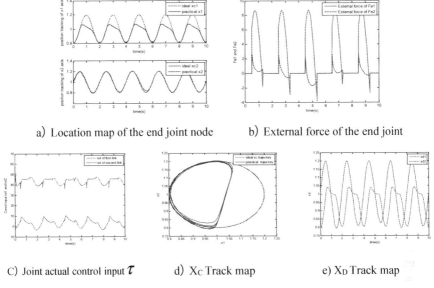

a) Location map of the end joint node   b) External force of the end joint

C) Joint actual control input τ d) Xc Track map   e) XD Track map

Fig. 6. Guide wheel end control diagram.

Analysis of the curve in the graph, can get the following conclusions:

1) When encountering an obstacle, due to the action of the regulating mechanism, the position of the end of the joint 1 has changed compared to the previous, and the position of the end of the joint 2 remains unchanged, in accordance with the theoretical analysis of the regulating mechanism. Meanwhile, due to the compression of the spring the thrust of the screw, the entire external force of the joint 2 changes compared to the normal situation, and the joint 1 remains unchanged, in accordance with the desired setting of the regulating mechanism and the guide wheel.

2) When the contact position of the end of the guide wheel encounters an obstacle, the end force can be controlled by controlling the robot's displacement to ensure that the robot maintains the desired contact force when encountering an obstacle.

4.3 Overall Motion Simulation

According to the design requirements, the speed and acceleration of the robot are simulated in the MATLAB environment. The motion function is shown in Eq. 26.

$$\begin{cases} 0 \leq t \leq \frac{v}{a}; s(t) = \frac{1}{2}at^2 \\ \frac{v}{a} < t \leq T - \frac{v}{a}; s(t) = vt - \frac{v^2}{2a} \\ T - \frac{v}{a} < t \leq T; s(t) = \frac{2avT - 2v^2 - a^2(t-T)^2}{2a} \end{cases}, \qquad (26)$$

Here v, a, T have constraints: $s(T) = \frac{avT - v^2}{a} = 1, t_a = \frac{v}{a} \leq \frac{T}{2}, a > 0$. When v, a is specified, then: $T = \frac{a+v^2}{va}, \frac{v^2}{a} \leq 1$. When v, T is specified, then: $a = \frac{v^2}{vT-1}, 1 < vT \leq 2$. When a, T is specified, then: $v = \frac{1}{2}(aT - \sqrt{a}\sqrt{aT^2 - 4}), aT^2 \geq 4$.

The simulation results are shown in Fig. 7:

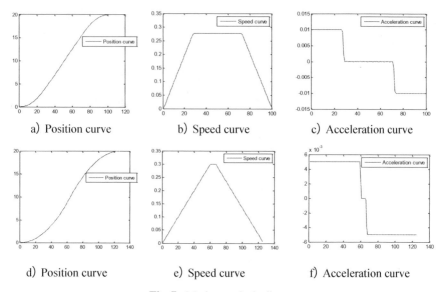

a) Position curve   b) Speed curve   c) Acceleration curve

d) Position curve   e) Speed curve   f) Acceleration curve

Fig. 7. Motion analysis diagram.

If the initial speed and the final speed are assumed to be zero, given the maximum acceleration, running time, starting angle, and ending angle, the position curve, velocity curve, and acceleration curve of the simulated robot are as shown in Fig. 7 a to c. If the initial speed and the final speed are assumed to be zero, given the starting angle, the ending angle, the speed in the uniform speed phase, and the given maximum acceleration, then the corresponding position curve, velocity curve and acceleration curve are shown in Fig. 7 d to f shown. As can be seen from the figure, the velocity curve, position curve and acceleration curve of the robot correspond to each other, and the simulation results are consistent with the analysis and solution results of the motion force inside the robot tube in Sect. 2, which verifies the correctness of the theoretical analysis and simulation analysis of the robot's walking mechanism.

5 Summary

In this paper, a new type of marine boiler tube micro-robot and a novel driving structure have been designed, where a multi-module structure is adopted. Through the expansion and contraction of the front and rear three-jaw cylinders and the screw drive, the creeping movement of the robot is realized. The robots can move in complex pipes and adapt to pipes with different diameters. After a detailed analysis of the force of each unit of the robot in the pipeline, we complete dynamic simulation of the robot. Then we use MATLAB software to simulate and analyze driving ability, regulating mechanism and the end of the guide wheel, and the motion analysis of each component of the robot is verified. The conclusion drawn in this paper can provide reference for the robot's force analysis when the robot is walking in other types of pipelines, can also provide a theoretical foundation for the subsequent robot optimization.

Acknowledgement. This work was supported by the Zhoushan Science and Technology Bureau of Zhejiang Province of China (2018C31075, 2018C31088).

References

1. Kim, H.M., Yun, S.C.: Novel mechanism for in-pipe robot based on multi-axial differential gear mechanism. IEEE/ASME Trans. Mechatron. **99**, 1 (2016)
2. Liu, Z., Wang, X.A., Sun, C.J., Lu, K.: Implementation system of human eye tracking algorithm based on FPGA. Comput. Mater. Continua **58**(3), 653–664 (2019)
3. Kolesnikov, Y., Ptitryna, A.S.: Mathematical model for pipeline control applying in-line robotic device. Indian J. Sci. Technol. **9**(11), 1–11 (2016)
4. Ren, X.X., Hu, W.T., Lv, H.X.: Research and design of a robot for pipeline inspection and cleaning. J. Electr. Power **31**(2), 142–147 (2016)
5. Xu, F.P., Deng, Z.Q.: Research on traveling-capability of pipeline robot in elbow. Robot **26**(2), 155–160 (2014)
6. Choi, Y.S., Kim, H.M., Mun, H.M.: Recognition of pipeline geometry by using monocular camera and PSD sensors. Intel. Serv. Robot. **10**(3), 213–227 (2017)
7. Kim, H.M., Suh, J.S., Choi, Y.S.: An in-pipe robot with multi-axial differential gear mechanism. In: IEEE/RSJ International Conference on Intelligent Robots & Systems, pp. 252–257. IEEE, Piscataway (2013)
8. Centonze, P.: Security and privacy frameworks for access control big data systems. Comput. Mater. Continua **59**(2), 361–374 (2019)
9. Brown, L., Watson, S., Lennox, B.: Elbow detection in pipes for autonomous navigation of inspection robots. J. Intell. Robot. Syst. **95**(2), 527–541 (2019). https://doi.org/10.1007/s10 846-018-0904-7
10. Kwon, Y.S., Yi, B.J.: Design and motion planning of a two-module collaborative indoor pipeline inspection robot. IEEE Trans. Rob. **28**(3), 681–696 (2012)
11. Nagaya, K., Yoshino, T., Ando, Y.: Wireless piping inspection vehicle using magnetic adsorption force. IEEE-ASME Trans. Mechatron. **17**(3), 472–479 (2012)
12. Lin, N., Tang, J., Li, X., Zhao, L.: A novel improved bat algorithm in UAV path planning. Comput. Mater. Continua **61**(1), 323–344 (2019)
13. Liang, L., Guo, Z.H., Zhu, Z.M.: Steering analysis of a four-spiral in-pipe robot. J. Mach. Des. **32**(7), 16–19 (2015)

LoRa Devices Identification Based on Differential Constellation Trace Figure

Xiuting Wu[1(✉)], Yu Jiang[1,2(✉)], and Aiqun Hu[1,2(✉)]

[1] Southeast University, Nanjing, China
{220184473,jiangyu,aqhu}@seu.edu.cn
[2] Purple Mountain Laboratories, Nanjing, China

Abstract. With the development of the Internet of Things, the application of LoRa devices is more and more extensive. However, the security access issues of LoRa terminals has not been paid enough attention yet. In this paper, a novel device identification method is proposed according to the modulation principle and frame format of LoRa signals, which is based on the implementation of differential constellation figure on the physical layer. Unique RF fingerprint features can be extracted from each LoRa signal frame or even the preamble part through this method, thus realizing device identification. The method principle is theoretically analyzed and then experiments are conducted with four LoRa devices using the method of pattern classification, which turns out to be a high accuracy rate device identification for every signal frame of these devices, verifying the effectiveness of the identification method.

Keywords: LoRa · RF fingerprint · Differential constellation trace figure · Device identification

1 Introduction

Internet of things (IoT) is a network technique which connects information collection equipments such as radio frequency identification (RFID) cards, sensors, infrared sensors, global positioning system and laser scanners with the Internet for information exchange and communication according to the agreed protocol, so as to realize intelligent identification, positioning, tracking, monitoring and management, etc. [1, 2]. With the continuous development of science and technology, IoT technology is widely used in daily life, urban infrastructure, agriculture and other places, and therefore the variety and the number of IoT devices increases, bringing up many addressing issues, attacks and information leaks [3, 4].

Traditional Internet-based security problems have a lot of mature and effective solutions, such as terminal weak password has a high-strength password design scheme to solve, etc. [5]. However, IoT has its own unique security problems, including weak authentication and authorization mechanism; lack of transmission layer encryption; illegal remote control after terminals online; risks brought by new nodes and local area networks introduced by the perception layer; security issues of uncontrolled environmental terminals, etc. [6], however there isn't any perfect security scheme yet on the issue of access control.

There is a contradiction between communication distance and power consumption for technologies like Wi-Fi, Bluetooth, ZigBee, 3G, LTE, etc. The emergence of Low Power Wide Area Network (LPWAN) communication technology solves this difficult problem [7]. LoRa (long range) communication technology, as a kind of wireless technology in LPWAN, has the advantages of long transmission distance and low power consumption simultaneously. In addition, LoRa equipment has high autonomy in networking, and its industrial chain is relatively mature and the commercialization application is earlier [8]. LoRaWAN is a set of communication protocol and system architecture designed by LoRa alliance based on LoRa long-distance communication network, which follows the protocol of Low-Rate Wireless Personal Area Networks (IEEE802.115.4-2011) [9].

In recent years, more and more researches show that radio frequency characteristics of the equipment can be extracted from electromagnetic waves emitted by the wireless communication system [10]. Due to the differences of electronic components in the equipments, the electromagnetic wave emitted by these equipments contains their unique RF characteristics, which can be a parameter for device identity authentication, also known as "RF fingerprint" [11]. Feature extraction based on RF fingerprints is on the physical layer of the communication system, which is not easy to be modified, so it can protect the system security from the bottom of the communication system.

At present, the most widely used method of RF fingerprint extraction is based on the transient response and steady-state response of the system [12–14]. In addition to the above two methods, in 2016, Peng et al. effectively completed the identification of ZigBee equipment based on the novel method of differential constellation trace figure (DCTF) [15]. In 2017, robyns et al. proposed and analyzed a new fingerprint recognition method based on supervised machine learning, taking the data preprocessed as the whole identification object, turning out a high recognition accuracy [16].

This paper explores how to realize the device identification based on the fingerprint information of the physical layer in the LoRa network. Theoretically we analyze the modulation mode, frame format and demodulation principle of LoRa signal. Experimentally LoRa signal is collected with a USRP device, and then after preprocessing, signal characteristics of valid data section and preamble section of every signal frame are analyzed to extract RF Fingerprint using the method of DCTF to realize LoRa device identification on the physical layer. Finally through the method of pattern classification, we realize the classification of four LoRa modules, which verifies the effectiveness of the LoRa device identification method.

2   LoRa Signal Analysis

2.1   LoRa Modulation Principle

LoRa modulation scheme is improved from chirp spread spectrum (CSS) scheme [17]. Linear frequency modulation (LFM) signal, also known as chirp signal, has a constant amplitude with the frequency changing linearly across the whole bandwidth. LoRa signal modulation mainly depends on chirp pulse to encode information.

LoRa modulation technology mainly has four key parameters: carrier frequency (f_c), bandwidth (BW), spreading factor (SF) and code rate (CR) to realize signal modulation and control of the wireless communication. The signal representation is as follows:

$$s(t) = e^{j(2\pi f_c t + 2\pi \frac{\beta}{2} t^2)} \tag{1}$$

$$\text{Where,} \quad \beta = \frac{BW}{T_{symbol}} \tag{2}$$

$$T_{symbol} = \frac{2^{SF}}{BW} \cdot CR \tag{3}$$

For LFM spread spectrum, the instantaneous frequency of chirp signal is a time-dependent linear function:

$$f(t) = f_c + \mu \cdot \frac{B}{T} \cdot t = f_c + \mu k t \tag{4}$$

Where, f_c is the carrier center frequency, μ represents the instantaneous frequency change slope of chirp signal, $\mu = 1$ represents a up-chirp, $\mu = -1$ represents a down-chirp, B represents bandwidth, k is the frequency modulation slope. A typical chirp signal is shown in Fig. 1.

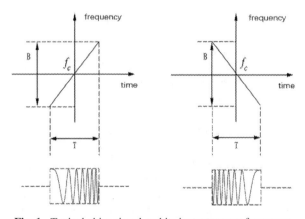

Fig. 1. Typical chirp signal and its instantaneous frequency.

LoRa frame structure is shown in Fig. 2, which starts from the preamble that is used to keep the receiver and transmitter synchronized, including synchronization word (SYN) and start of frame delimiter (SFD). And then following the LoRa physical header (PHDR) plus a header CRC, and the PHDR can explicit or implicit. PHDR includes the length of information data (only 255 bytes at most), error correction coding rate and whether load CRC is carried at the end of the frame. While these messages are known or fixed, implicit mode is recommended to improve efficiency, shorten transmission time and reduce power consumption.

The most significant difference between the protocol of LoRa and LoRaWAN is that part of the data load in LoRaWAN is encrypted by AES128 while the frame format is almost the same.

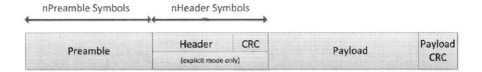

Fig. 2. LoRa frame format in explicit mode.

2.2 Modulation and Demodulation

In LoRa modulation, each chirp symbol is composed of a linear sweep signal. Time from the signal starting position to the frequency mutation is called the symbol duration (SD), up to the value of spreading factor (SF) within the frequency range of band width, determining the symbol value (SV) ranging from 0 to $2^{SF} - 1$, as shown in Fig. 3.

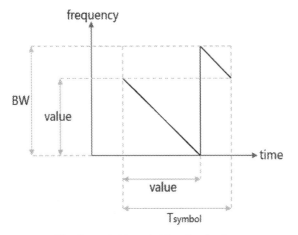

Fig. 3. Definition of chirp signal value.

$$\text{Here,} \quad SD = \frac{sv}{2^{SF}} T_{symbol} \tag{5}$$

The time-frequency diagram of the effective data section of a typical LoRa signal is shown in the Fig. 4, with 10 up-chirp symbols as the preamble part, and 2 down-chirp as SFD, and then the data load part. In the process of demodulation, we find the starting position of the effective data segment through the signal energy threshold method, and then synchronize the signal preamble using the sync word (SYN). Additionally the data synchronization is precisely calibrated based on SFD symbols, and then SD and SV can be calculated.

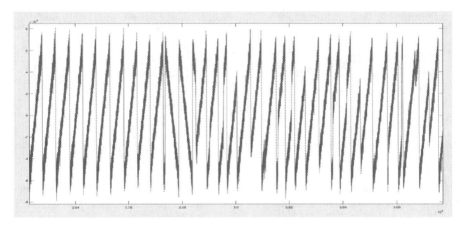

Fig. 4. Time frequency diagram of LoRa effective data segment.

3 LoRa Signal Acquisition and Preprocessing

3.1 Signal Acquisition

In the laboratory environment, USRP Mini B200 is used for data sampling, with LoRa modules placed in a fixed position. USRP is connected to the host through a USB port, and is equipped with a receiving antenna, as shown in Fig. 5.

Fig. 5. USRP Mini B200 receiving device.

Two types of LoRa modules used to generate LoRa signals, among which there are three modules based on LoRa SX1278 chip and one RAKWireless rak811 module following the LoRaWAN protocol within a SX1276 chip. The differences between these LoRa chirps are shown in Table 1.

According to different chip parameters, we set $f_c = 433$ MHz for the SX1278 modules while set $f_c = 868$ MHz for the rak811 module. As for other parameters, SF = 7, BW = 125 kHz, CR = 4/5, PHDR in implicit mode, they are set the same. Besides, all these modules are set to transmit certain content. The modules are shown in Fig. 6.

Table 1. Differences between LoRa chips.

Chip	Frequency	SF	BW	Bit rate
SX1276	137–1020 MHz	6–12	7.8–500 kHz	0.018–37.5 kbps
SX1277	137–1020 MHz	6–9	7.8–500 kHz	0.11–37.5 kbps
SX1278	137–525 MHz	6–12	7.8–500 kHz	0.018–37.5 kbps

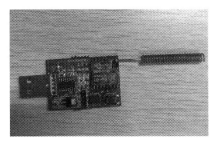

Fig. 6. LoRa SX1278 module and RAKWireless rak811 module.

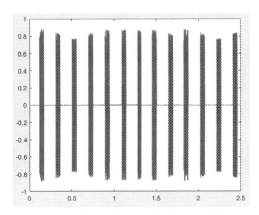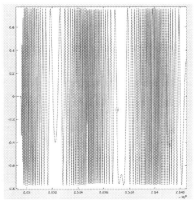

Fig. 7. The original signal and its effective data fragment.

Matlab is used to process the received signal. The original signal and its effective data fragment are shown in the Fig. 7.

After preprocessing, draw the time-frequency graph of the effective data segment of each type of module, and they are shown in the Fig. 8(a) and Fig. 8(b) respectively. It can be seen that different modules have different frame structures in the preamble. The signal of sx1278 module starts with a preamble composed of 14 up chirp signals while rakwireless rak811 module consists of 10 up-chirp preamble. This is because different LoRa modules can be designed according to different standards as long as they follow the protocol. The longer the preamble is sent, the longer the power consumption will be, and the channel resource occupation will also be affected.

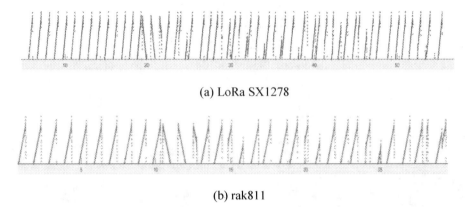

(a) LoRa SX1278

(b) rak811

Fig. 8. Time-frequency graph of the effective data segment for the two types of modules.

4 Feature Extraction and Module Recognition

4.1 Principle of Differential Constellation Trace Figure

The constellation diagram can be obtained if we draw the I/Q channel data on the coordinate plane, which reflects the characteristics and relationships between the signals and can be used for studying the digital communication system through the image with an image recognition based method. However, when the constellation diagram is used to analyze the RF characteristics directly, the receiving symbol deviates from its position due to the influence of frequency offset, and as time accumulates, a concentric ring is finally drawn, covering up valid information. By performing differential processing on the received baseband signal, the rotation of the received symbol due to the frequency offset can be eliminated.

Generally, the transmitter and the receiver have frequency offset. If the transmitter carrier frequency is f_{ct1} and the baseband signal is $X(t)$, the transmitted is:

$$S(t) = X(t)e^{-j2\pi f_{ct1}t} \tag{6}$$

For an ideal channel, there is R(t) = S(t), However the signal received is:

$$Y(t) = R(t)e^{-j(2\pi f_{ct2}t+\phi)} = S(t)e^{-j(2\pi f_{ct2}t+\phi)} \tag{7}$$

Where f_{ct2} is the receiver carrier frequency and ϕ is the received signal phase error. Since the transmitter and receiver have a deviation $\Delta f = f_{ct2} - f_{ct1}$, so

$$Y(t) = X(t)e^{j2\pi \Delta ft+\phi}. \tag{8}$$

While received signal contains a rotation factor $e^{j2\pi \Delta ft}$, in order to solve this problem and reflect the frequency offset on the picture, the data need to be differentially operated:

$$D(t) = Y(t) * Y^*(t+n) = X(t) \cdot X(t+n)e^{-j2\pi \theta n} \tag{9}$$

The result of the difference still has a rotation factor $e^{-j2\pi \theta n}$, but it is a stable value, which can be used to directly reflect the frequency offset characteristic of the signal in the constellation trace figure, making it feasible to extract RF fingerprint based on the method of subsequent differential constellation trace figure.

4.2   Feature Extraction and Module Recognition

According to the parameters set during signal acquisition in Sect. 3.1, set the rakwireless rak811 module FC = 868.1 MHz, SF = 7, BW = 125 kHz, Cr = 4/5, set the PHDR to the hidden header mode, and send a fixed section of data irregularly The difference between the parameters set by sx1278 and rak811 is that the carrier frequency fc is 433 MHz, and a fixed period of data is sent at a certain interval. The center frequency point set by USRP equipment at the receiving end is consistent with the carrier frequency of LoRa transmitting module, and the sampling frequency FS is set to 5 MHz. During sampling, USRP and LoRa modules are fixed at a distance of about 1 m, and there is no shelter between them.

According to the parameters set by LoRa module, the signal transmission rate Rs can be obtained by:

$$R_s = \frac{BW}{2^{SF}} = 976.5625 \, symbol/s \tag{10}$$

Combined with the sampling frequency, the number of sampling points of each LoRa symbol can be calculated:

$$N = \frac{f_s}{R_s} = 5120 \tag{11}$$

Taking the differential interval to 5120, the sampling data is differential processed, and the differential constellation trace figure is drawn. After visual processing, the density of symbols in different positions is represented by colors on the constellation figure. The larger the symbol density is, the closer the color is to red. The final DCTF for each device is as shown in the Fig. 9.

It can be seen that for LoRa modulated signals, the DCTF clustering center is obvious, and the symbol distribution is centralized with small noise interference. Based on the differences of these images, the methods of image processing and pattern classification can be applied to feature extraction. In this experiment, the method of pattern classification is used to realize LoRa device recognition and the specific operation is as follows:

First, in the training data set of a module I, find the maximum value of the point density in the DCTF for each effective data segment;

Second, find the location of all the qualified points reaching 0.95 of the maximum, as shown in the figure.

Then, calculate the clustering center Ci of the module I taking all the eligible points in all valid data segments of this device into consideration, and the clustering centers of other three devices are calculated by this method.

Finally, analyze the valid data segments of four modules in the test data set by calculating the average position X of all the eligible points in each valid data segment. And then calculate the Euclidean distance Di between X and each cluster center Ci, and classify this processing data segment as the category where Di is the minimum value.

In the experiment, we trained 304 pieces of valid data segments of four modules to get the clustering center of each module, and then classify 149 pieces of valid data segments of these four modules, which are all successful, as shown in the Fig. 10.

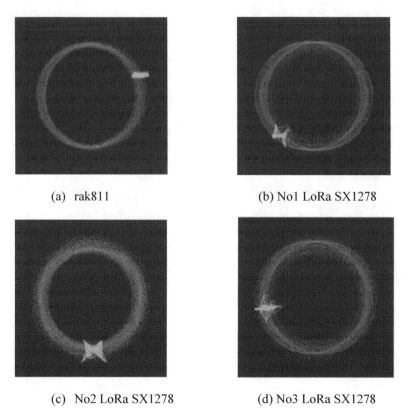

(a) rak811   (b) No1 LoRa SX1278

(c) No2 LoRa SX1278   (d) No3 LoRa SX1278

Fig. 9. Typical DCTF of a valid data segment for each device. (Color figure online)

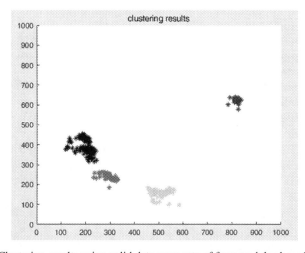

Fig. 10. Clustering results using valid data segments of four modules based on DCTF.

The above experimental results are obtained when all modules send specified data, while the data of the rak811 module is partially encrypted and can be regarded as random data. From section, it is known that preamble of each module is unchanging, therefore we try to draw the DCTF just using the preamble part, and the result is as shown in the Fig. 11, corresponding to the same module in Fig. 9.

Fig. 11. DCTF of the preamble part for each device.

It can be seen that, based on the DCTF of the preamble part, the data noise interference is basically absent, and the data is more centralized. In addition, the contour similarity between the DCTF drawn by a whole segment of data and the DCTF drawn by the preamble is very high.

For the same test data set, pattern classification is executed based on the DCTF generated from the preamble parts of all valid data segments, and the classification accuracy is 98.66%, which is shown in Fig. 12.

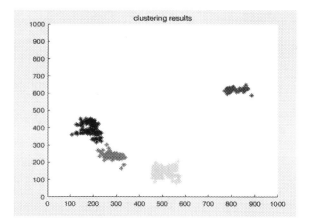

Fig. 12. Clustering results using data preamble parts of four modules based on DCTF.

From the clustering results in Fig. 10 and Fig. 12, we can conclude that the feature extracted from DCTF can achieve the purpose of module identification, and has no relation with the content transferred. Considering the high contour similarity between the DCTF drawn by a whole valid data segment of data and its preamble, so it can be considered that when there are many devices to be identified, the characteristic of DCTF contour similarity can be applied for further precise classification.

5 Conclusion

After analyzing the security situation of IoT, this paper proposes a physical layer based device identification method—DCTF for LoRa devices. According to the modulation principle and frame format of LoRa signal, DCTF can be used to extract unique RF fingerprint features of each device and experimentally realize a high accuracy rate in device identification for per valid data frame of 4 LoRa devices with the whole valid data segment or just the preamble part. The follow-up work will focus on how to uniquely identify LoRa devices according to the preamble of LoRa signals taking SNR or sample rate into account, so as to improve the stability and robustness of the device identification system.

References

1. Huang, Y., Li, G.: Descriptive models for Internet of Things. In: 2010 International Conference on Intelligent Control and Information Processing, pp. 483–486. IEEE, August 2010
2. Wang, J., Gao, Y., Liu, W., Wu, W., Lim, S.J.: An asynchronous clustering and mobile data gathering schema based on timer mechanism in wireless sensor networks. Comput. Mater. Contin. **58**, 711–725 (2019)
3. Conti, M., Dehghantanha, A., Franke, K., Watson, S.: Internet of Things security and forensics: challenges and opportunities. Future Gener. Comput. Syst. **78**, 544–546 (2018)
4. Liu, W., Luo, X., Liu, Y., Liu, J., Liu, M., Shi, Y.Q.: Localization algorithm of indoor Wi-Fi access points based on signal strength relative relationship and region division. Comput. Mater. Contin. **55**(1), 71–93 (2018)
5. Cheng, J., Xu, R., Tang, X., Sheng, V.S., Cai, C.: An abnormal network flow feature sequence prediction approach for DDoS attacks detection in big data environment. Comput. Mater. Contin. **55**(1), 95–119 (2018)
6. Mahmoud, R., Yousuf, T., Aloul, F., Zualkernan, I.: Internet of Things (IoT) security: current status, challenges and prospective measures. In: 2015 10th International Conference for Internet Technology and Secured Transactions (ICITST), pp. 336–341. IEEE, December 2015
7. Li, L., Ren, J., Zhu, Q.: On the application of LoRa LPWAN technology in Sailing Monitoring System. In: 2017 13th Annual Conference on Wireless On-demand Network Systems and Services (WONS), pp. 77–80. IEEE, February 2017
8. Neumann, P., Montavont, J., Noël, T.: Indoor deployment of low-power wide area networks (LPWAN): a LoRaWAN case study. In: 2016 IEEE 12th International Conference on Wireless and Mobile Computing, Networking and Communications (WiMob), pp. 1–8. IEEE, October 2016
9. de Carvalho Silva, J., Rodrigues, J.J., Alberti, A.M., Solic, P., Aquino, A.L.: LoRaWAN—a low power WAN protocol for Internet of Things: a review and opportunities. In: 2017 2nd International Multidisciplinary Conference on Computer and Energy Science (SpliTech), pp. 1–6. IEEE, July 2017
10. Danev, B., Zanetti, D., Capkun, S.: On physical-layer identification of wireless devices. ACM Comput. Surv. (CSUR) **45**(1), 6 (2012)
11. Honglin, Y., Aiqun, H.: Fountainhead and uniqueness of RF fingerprint. J. Southeast Univ. (Nat. Sci. Ed.) **39**(2), 230–233 (2009)
12. Demers, F., St-Hilaire, M.: Radiometric identification of LTE transmitters. In: 2013 IEEE Global Communications Conference (GLOBECOM), pp. 4116–4121. IEEE, December 2013

13. Remley, K.A., et al.: Electromagnetic signatures of WLAN cards and network security. In: Proceedings of the Fifth IEEE International Symposium on Signal Processing and Information Technology 2005, pp. 484–488. IEEE, December 2005

14. Romero, H.P., Remley, K.A., Williams, D.F., Wang, C.M.: Electromagnetic measurements for counterfeit detection of radio frequency identification cards. IEEE Trans. Microw. Theory Tech. **57**(5), 1383–1387 (2009)

15. Peng, L., Hu, A., Jiang, Y., Yan, Y., Zhu, C.: A differential constellation trace figure based device identification method for ZigBee nodes. In: 2016 8th International Conference on Wireless Communications & Signal Processing (WCSP), pp. 1–6. IEEE, October 2016

16. Robyns, P., Marin, E., Lamotte, W., Quax, P., Singelée, D., Preneel, B.: Physical-layer fingerprinting of LoRa devices using supervised and zero-shot learning. In: Proceedings of the 10th ACM Conference on Security and Privacy in Wireless and Mobile Networks, pp. 58–63. ACM, July 2017

17. Vangelista, L.: Frequency shift chirp modulation: the LoRa modulation. IEEE Signal Process. Lett. **24**(12), 1818–1821 (2017)

Improved Helmet Wearing Detection Method Based on YOLOv3

Peizhi Wen[1], Mingyang Tong[1], Zhenrong Deng[1(✉)], Qinzhi Qin[2],
and Rushi Lan[1]

[1] School of Computer Science and Information Security,
Guilin University of Electronic Technology, Guilin 541004, China
zhrdeng@guet.edu.cn
[2] Guangxi Construction Industry Mostly Lease Co. Ltd., Guilin 530000, China
qqzhome@163.com

Abstract. In order to monitor the wearing of safety helmet in dangerous workplace in real-time and ensure the safety of production, this paper proposes a method based on the yolov3 algorithm to detect the wearing of safety helmet. Firstly, K-means algorithm is used to cluster the target boxes on the self-made data set, so that the prediction boxes can fit the target better that in the data set. At the same time, the network is pre-trained on the voc2007 data set to make the model parameters more accurate and reduce the training time. Secondly, multi-scale feature extraction and multi anchor box mechanism are used to improve the accuracy of small object detection. Finally, optimizing the non-maximum suppression (NMS) algorithm with Gaussian function that can improve the detection accuracy of the occluded target. Experimental results show that the algorithm has better detection effect while meeting the real-time helmet wearing detection.

Keywords: Helmet detection · Non-maximum suppression · Object detection · YOLOv3

1 Introduction

In many complex production environments, such as construction sites, mining and other places, there are various risk factors threatening personal safety. As the head is the most critical part of the human body and the most vulnerable to fatal injury, wearing a helmet in such workplace is an effective guarantee for the life safety of personnel in the production area. Therefore, monitoring whether the personnel in the production area wear safety helmet in real-time is an important means to ensure the safety of production.

This work is supported in part by the National Natural Science Foundation of China (Nos. 61702129, 61772149, and U1701267), Innovation Project of Guet Graduate Education (No. 2019YCXS050), and Guangxi Natural Science Foundation (Nos. 2018GNS-FAA138132, AD18216004, and AD18281079).

X. Sun et al. (Eds.): ICAIS 2020, LNCS 12239, pp. 670–681, 2020.
https://doi.org/10.1007/978-3-030-57884-8_59

Traditional helmet detection methods are primarily implemented by hand-designing features. Jia, et al. proposed a feature vector composed of a block-based local binary histogram based on the variability component model and the gradient direction histogram and color features, using the support vector machine (SVM) to achieve the helmet detection [7]. Liu, et al. used a skin color detection method to locate the face region, using the Hu matrix of the face and above region as the feature vector of the image, training the model with BP neural network and SVM to implement helmet detection [9]. Feng, et al. used the mixed Gaussian model to detect the foreground, judged whether it belongs to the human body through the connected region, and finally positioned the human head region to realize the automatic identification of the helmet [3]. These detection methods have vulnerable robustness and low accuracy.

With the development of deep learning, convolutional neural networks have become the focus of research. At the same time, the application of convolutional neural network in many aspects has achieved good results, Zhong Liu et al. used neural network to achieve the automatic detection of arrhythmia [10]. Zhenguo Gao et al. proposed a real-time target tracking algorithm combining shape and color features [4]. Cirshick proposed a regional convolutional neural network model (R-CNN) [6] based on region proposal, and applied convolutional neural networks to object detection for the first time. Afterwards, Crishick and Ren proposed fast regional convolutional neural networks (Fast R-CNN) [5] and faster regional convolutional neural networks (Faster R-CNN) [15], respectively. Ruohan Meng present an approach base on Faster R-CNN which corresponds multiple steganographic algorithms to complex texture objects was presented for hiding secret message [11]. Compared to R-CNN in accuracy and speed has improved. Redmond [12] and others first proposed a grid based YOLO algorithm, which greatly improved the detection speed. Based on YOLO, Redmond successively proposed the YOLOv2 [13] and YOLOv3 [14] detection algorithms. Compared with YOLO, the detection speed and accuracy of YOLOv2 have been significantly improved. Although YOLOv3 is inferior to YOLOv2 in terms of speed, the detection accuracy is greatly improved without affecting real-time detection. Since feature extraction does not require manual design, but relies on convolutional neural networks, greatly saved time and improved usability. Fang, et al. borrowed the idea of densely connected networks and added dense blocks to the original YOLOv2, achieving multi-layer feature fusion and balance of spatial information and semantic information, improving sensitivity to small target detection, and effectively improving the accuracy of the test was detected, but the practical requirements were still not met [2]. Lin, et al. used YOLOv3 to test the helmet, and the detection accuracy on the self-made data set reached 98.7%, but only the single-category detection method could not meet the matching problem between the person and the helmet [8]. Xu, et al. used multi-scale and increased number of anchor box on the original Faster RCNN to enhance the robustness of the network to detect different scale targets, at the same time use multi-component combination to improve the accuracy of category determination [16]. This method can detect many categories well, but the detection speed cannot meet the requirements of real-time detection.

Aiming at the problems of low detection accuracy, slow detection speed and single category detection that do not meet the requirements of actual working scenes, this paper proposes a helmet wearing detection method. The multi-anchor box mechanism is used to enhance the detection effect of the YOLOv3 on small targets. On the basis of YOLOv3 algorithm, NMS method is optimized by Gaussian function (soft-NMS) [1]. Which can reduce the phenomenon of missing detection caused by mutual occlusion of target with same label, and improve the accuracy of objects detection.

The remainder of the paper is organized as follows. In Sect. 2, we introduce the basic work used in this paper; Sect. 3 introduce the methods we proposed in detail; Sect. 4 discusses the experiment setup and results. Section 5 concludes the work.

2   Basic Work

2.1   Feature Extraction Network of YOLOv3

YOLOv3 uses DarkNet-53 as a feature extraction network. DarkNet-53 consists mainly of 53 full convolution layers and 24 residual units. The full convolution layer uses a 3×3, 1×1 convolution kernel for convolution operations, the 3×3 convolution kernel is responsible for increasing the number of feature map channels, and the 1×1 convolution is responsible for compressing the characteristic representation obtained after 3×3 convolution. Five residual layers with different scales and depths are used to calculate residual operations between different convolution layers. The residual layers can prevent gradient disappearance and gradient explosion in deep convolutional networks, enabling deeper networks to obtain better feature maps and improve target detection accuracy.

2.2   Feature Pyramid Network

Yolov3 uses the feature pyramid network to extract multi-scale features [14]. In the object detection process, spatial information is needed to locate the target, and semantic information is needed to classify the target. However, in the convolution process, spatial information will be continuously lost. The lower layer features have more spatial information, and the higher layer features have richer semantic information. The feature pyramid network can retain the spatial information of lower layer features and rich semantic information of higher layer features at the same time.

2.3   NMS Algorithm

NMS is a method often used in target detection. The purpose is to remove redundancy. The core formula of the algorithm is as follows:

$$S_{confi} = \begin{cases} S_{confi}, I\left(M, b_i\right) < N_{It} \\ 0, I\left(M, b_i\right) < N_{It} \end{cases} \tag{1}$$

Where S_{confi} is the confidence score of prediction box i, M is the prediction box corresponding to the maximum confidence score, N_{It} is the Intersection-over-Union (IOU) threshold of non-maximum suppression, and $I(M, b_i)$ is the IOU of prediction box M and b_i. The formula for calculating the IOU is as follows:

$$I(M, b_i) = \frac{M \cap b_i}{M \cup b_i} \tag{2}$$

The specific steps of the algorithm are as follows:

(1) The score set S and the predicted box set B are sorted according to the confidence score.
(2) Select the prediction box with the largest confidence score as M to add it to the set D and remove M from the set B.
(3) Calculate the IOU of M and the prediction box in the set B. If the IOU is greater than the set IOU threshold, delete the prediction box b_i and the corresponding confidence score s_i.
(4) Determine whether s_i is less than the set score threshold N_{St}, and if s_i is smaller than N_{St}, delete s_i and b_i.
(5) Repeat steps 2 through 4 until set B is empty, returning sets D and S. Getting the prediction box D with the highest score.

3   An Improved Detection Algorithm of Helmet Wearing Based on YOLO

Aiming at the problems of low detection accuracy, slow detection speed and single category detection that do not meet the requirements of actual working scenes, this paper proposes a helmet wearing detection method, which uses the multi-anchor box mechanism to enhance the detection effect of the YOLOv3 on small targets. Using Gaussian function to optimize NMS based on YOLOv3 algorithm, to reduce the phenomenon of missed detection caused by occlusion of object with same label. The training and detection process of this method is shown in Fig. 1.

The helmets in the data set come in a variety of forms to meet the requirements of helmet detection in a variety of work environments. Using voc2007 to pre-train the network, pre-training can make the model parameters more accurate. Then use the self-made data set to train the network, the loss can be faster convergence. That can reduce the training time and improve the accuracy of the model. Using multi-scale feature extraction and anchor box mechanism can effectively detect small objects, and soft-NMS algorithm can reduce the phenomenon of missing detection caused by occlusion of objects with the same label.

3.1   Multi Scale Feature Extraction for Small Object Detection

Multi-scale feature extraction and feature pyramid network can make large-scale feature maps have more semantic features and improve the accuracy of multi-label classification. The K-means algorithm cluster the data set detection boxes

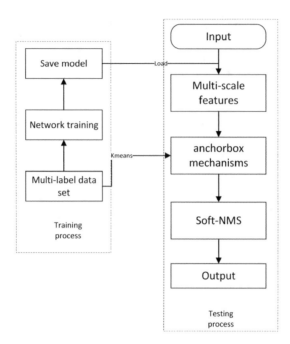

Fig. 1. Flow chart of model training and detection. The left dotted box is the data preprocessing and network training process. The algorithm flow in the detection process is shown in the dotted box on the right

to obtain 9 anchor boxes to match the target size. In the detection process, multiple anchor box mechanisms are used to detect feature maps of different scales. The larger scale feature map uses a small anchor box to improve the detection accuracy of small targets. Using the soft-NMS to filter the detection boxes can reduce the missed detection caused by the same target occlusion and improve the detection accuracy. The multi-scale feature detection process is shown in the Fig. 2.

First, the feature map obtained by the 24th residual layer is convoluted to obtain a feature map Y3 with a size of 13 × 13, and the Y3 is upsampled and combined with the feature map obtained by the 23rd Residual layer to perform a convolution operation to obtain a characteristic map Y2 with a size of 26 × 26. The upsampling of Y2 is combined with the feature map obtained by the 19th Residual layer, and a convolution operation is performed to obtain a feature map Y1 having a size of 52 × 52. The feature maps Y1, Y2 have both detailed positioning information at the lower level and semantic information at the higher level.

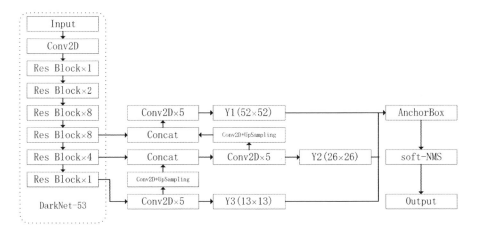

Fig. 2. Multi-scale feature detection flowchart. The feature extraction network is inside the dotted box on the left. The right is a pyramid feature extraction model

3.2 Using Soft-NMS to Solve the Occlusion Problem of the Same Label Target

In reference [1], soft-NMS is proposed to solve the problem of missing detection caused by the same label occlusion. This paper uses the method to optimize the YOLOv3 model to solve the problem of missing detection caused by personnel occlusion in the process of helmet wearing detection. Soft-NMS attenuation the confidence score using Gaussian function to optimize the NMS This algorithm can effectively solve the problem of NMS of missed detection caused by violent deletion of prediction boxes. When the value of I(M, b_i) is greater than the threshold N_{It}, the confidence score S is processed with Gaussian attenuation, which is shown in Formula 3:

$$S_{confi} = \begin{cases} S_{confi}, I\left(M, b_i\right) < N_{It} \\ S_{confi} e^{-\frac{I\left(M, b_i\right)^2}{\sigma}}, I\left(M, b_i\right) < N_{It} \end{cases} \tag{3}$$

The pseudo code of the improved algorithm is as Algorithm 1.
The specific steps of the algorithm are as follows:

(1) The score set S and the predicted box set B are sorted according to the confidence score.
(2) Select the prediction box with the largest confidence score as M to add it to the set D and remove M from the set B.
(3) Calculate the IOU of M and the prediction box in the set B. If the IOU is greater than the set IOU threshold, The confidence score s_i is Gaussian decayed and returned to s_i, and if the IOU is less than the set threshold, s_i remains unchanged.
(4) Determine whether s_i is less than the set score threshold N_{St}, and if s_i is smaller than N_{St}, delete s_i and b_i corresponding thereto.
(5) Repeat steps 2 through 4 until set B is empty, returning sets D and S.

4   Experimental Analysis

Experimental environment configuration: GPU: NVIDIA Quadro k620, cuda10.1, Ubuntu 16.04, memory 16 GB. The experimental code uses the keras framework.

4.1   Data Set Production

Using self-made data set to complete the experiment. The pictures in the data set are mainly obtained from online picture crawling and construction site collection, totaling 5000. Dataset image scenes mainly include construction sites, factory floors, and aerial work. There are many angles for shooting, such as looking down, looking up and looking up. There are also close-range acquisitions, as well as distant scenes, daytime shooting and nighttime shooting. It basically covers all situations that may be encountered in the helmet wearing detection. Images are annotated using the LabelImg tag dataset authoring tool and finally converted to the VOC2007 dataset format. Some scenes in the dataset are shown as Fig. 3.

4.2   Experiment Results Analysis

The experimental process is mainly divided into three parts: basic network training, threshold setting and adjustment and comparison experiment.

Fig. 3. Part of the scene in the data set. The picture in the upper left corner is night scene, the indoor scene in the upper right corner, the dense crowd scene in the lower left corner, and the small target scene in the lower right corner.

Training Parameter Setting. Pre-training uses the voc2007 data set, that can get better initialization network parameters, speed up the convergence of the network during training, and effectively prevent over-fitting. The voc2007 data set is divided into training set and verification set according to the ratio of 8:2. Bench size is set to 32, epoch is set to 50, and learning rate is set to 0.001. The loss of the verification set is taken as the tracking target. If the loss of 3 consecutive epochs does not decrease, the learning rate will be reduced to 10% of the original learning rate. If the loss of 10 consecutive epochs does not decrease, the training will be stopped and the final network model parameters will be saved.

Load the pre-training model, use the self-made helmet data set to train the network again. The epoch is changed to 100, and other parameter settings are the same as the pre-training, the training model is saved. The accuracy of the model is 87%. Compared with yolov3, the accuracy of VOC data set performance is improved slightly.

Comparison of Experimental Results. In the experiment, precision and recall are used to measure the effectiveness of the method. The formula is as follows:

$$precision = \frac{TP}{TP + FP} \tag{4}$$

$$recall = \frac{TP}{TP + FN} \tag{5}$$

TP (true positive) indicates that the model is correct to predict the correct positive sample; FP (false positive) indicates the positive sample predicted by the model error; FN (false negative) indicates the negative sample predicted by the model error.

Compare Experiments with Different Parameters. The test set was verified using the trained network and found to be false detection and missed detection. Mainly in (1) target gives two different labels. (2) The occluded target cannot be detected. The main reason for false detection and missed detection is due to the setting of score threshold and IOU threshold when removing redundant prediction boxes with NMS in the detection process. When the score is set too low, the false detection box with low confidence score also remained, resulting in false detection. When the score value is set too high, the correct detection box with lower confidence score will be deleted, resulting in missed detection. When the IOU threshold is set too low, the detection effect of the occluded object is not good, resulting in a missed detection phenomenon. Set different parameter sets for comparison experiments and select the best parameter set.

Table 1. Comparison of threshold setting

Score	IOU	Precision(%)
0.1	0.1	85.5
0.2	0.1	86.3
0.2	0.2	87
0.3	0.2	87.4
0.3	0.3	88.3

As shown in Table 1, when the confidence score is 0.2 and the IOU value is 0.1, the detection effect is the same as that of yolov3 on common data sets. Due to the difference of data sets, there are occlusion, big difference between object sizes, and fuzzy targets. Adjusting the parameters can make the model fit the data set better. When the score is 0.3 and IOU is 0.3, the error detection rate of the model is 10.4%, and the accuracy rate is 88.3%. It has achieved good results in data set.

As shown in the Fig. 4, when the confidence is 0.1, there are three objects with two prediction boxes in the figure, in which the safety helmet wearing personnel with confidence of 0.65, 0.72 and 0.62 are detected. At the same time, three error prediction boxes with confidence of 0.2, 0.12 and 0.13 are marked out, resulting in false detection. When the confidence threshold is set to 0.3, three prediction boxes with confidence scores less than the threshold are deleted, and the false detection phenomenon disappears. The selection of appropriate threshold parameters can effectively improve the detection efficiency and accuracy.

Comparative Experiments with Different Detection Methods. Model training on YOLOv2, YOLOv3, Reference [8] and this method, respectively. Compare the accuracy, recall, and detection speed of each model on the same test set. The contrast reflects the advantages between each model (Table 2).

Fig. 4. Test results of different parameter combination. The left figure shows the test results of parameter group [0.1, 0.1], and the right figure shows the test results of parameter group [0.3, 0.3].

Table 2. Comparison of test results.

Method	Precision(%)	Recall(%)	fps
YOLOv2	72.4	78.2	46
YOLOv3	88.3	89.7	36
Reference [8]	90.3	90.8	36
Our	90.7	91.2	31

Compared with yolov2, yolov3 has higher detection accuracy and recall rate. At the same time, yolv3 uses more size of anchor box, which makes the detection accuracy of yolv3 for multi-object and multi-size target improved significantly. At the same time, the matching degree between the detection frame and the target is also significantly improved.

Figure 5 shows the result of YOLOv2 test, there is obvious missed detection behavior. The construction personnel wearing the helmet are not detected, and the detection box of the hand-held camera object was too large, and large distance between the center of the detection box and the object. There is still a missing inspection phenomenon in the right figure, the staff wearing the helmet is not detected, but the helmet is correctly tested. At the same time, the center of the detection box is closer to the object center, and the frame selection range is more accurate.

Compared with YOLOv2, YOLOv3 has a significant improvement in detection accuracy, but the detection accuracy of yorov3 for occluded objects is still insufficient, and there are still false detection and missing detection. This is because the detection model recognizes two objects as one when two detection objects with similar features occlude each other, or the detection box of occluded objects is deleted violently due to NMS, resulting in the phenomenon of missing detection.

Fig. 5. On the left is the test result of YOLOv2, on the right is the test result of YOLOv3. YOLOv3 has higher detection accuracy than YOLOv2

Fig. 6. On the left is the test result of Reference [8], on the right is the test result of Our method.The comparison shows Our method effectively improves the detection accuracy of the occluded object

Figure 6 shows the detection results of Reference [8], in which the human target B is blocked by the human target A, and the safety helmet B is blocked by the safety helmet A, resulting in the human target B and the safety helmet B are not detected, resulting in the phenomenon of missing detection. The right figure shows the detection results of Our method, in which both helmet B and person B are detected. It effectively improves the detection accuracy of the occluded object, and is more practical.

5   Conclusion

This paper optimizes the YOLOv3 model using the soft-NMS algorithm. From the experimental results, the improved YOLOv3 algorithm can effectively detect the occluded target and improve the accuracy and practicability of the detection. However, the method described in this paper still has certain limitations. The target detection effect of the occlusion rate exceeding 60% is not ideal, and the target miss detection rate is high for small size in occlusion. This is an issue that needs to be addressed in future research.

References

1. Bodla, N., Singh, B., Chellappa, R., Davis, L.S.: Soft-NMS-improving object detection with one line of code. In: Proceedings of the IEEE International Conference on Computer Vision, pp. 5561–5569 (2017)
2. Fang, M., Sun, T., Shao, Z., et al.: Fast helmet wearing detection based on improved YOLOv2. Opt. Precis. Eng. **27**(5), 1196 (2019)
3. Feng, G., Chen, Y., Chen, N., Li, X., Song, C., et al.: Research on automatic recognition technology of safety helmet based on machine vision. Mach. Des. Manuf. Eng. **10**, 39–42 (2015)
4. Gao, Z., et al.: Real-time visual tracking with compact shape and color feature. Comput. Mater. Continua **55**(3), 509–521 (2018)
5. Girshick, R.: Fast R-CNN. In: Proceedings of the IEEE International Conference on Computer Vision, pp. 1440–1448 (2015)

6. Girshick, R., Donahue, J., Darrell, T., Malik, J.: Rich feature hierarchies for accurate object detection and semantic segmentation. In: Proceedings of the IEEE Conference on Computer Vision and Pattern Recognition, pp. 580–587 (2014)
7. Jia, J., Qingjie, B., Huiming, T.: Safety helmet wear detection based on deformable component model. Appl. Res. Comput. **33**(3), 953–956 (2016)
8. Lin, J., Dang, W., Pan, L., Bai, S., Zhang, R.: Fast helmet wearing detection based on improved YOLOV3. Comput. Syst. Appl. **28**(9), 174–179 (2019)
9. Liu, X., Xining, Y.: Application of skin color detection and hu moment in helmet recognition. J. East Chin Univ. Sci. Technol. **40**(3), 365–370 (2014)
10. Liu, Z., Wang, X., Kuntao, L., David, S.: Automatic arrhythmia detection based on convolutional neural networks. Comput. Mater. Continua **63**(2), 1079–1079 (2020)
11. Meng, R., Rice, S.G., Wang, J., Sun, X.: A fusion steganographic algorithm based on faster R-CNN. Comput. Mater. Continua **55**(1), 1–16 (2018)
12. Redmon, J., Divvala, S., Girshick, R., Farhadi, A.: You only look once: unified, real-time object detection. In: Proceedings of the IEEE Conference on Computer Vision and Pattern Recognition, pp. 779–788 (2016)
13. Redmon, J., Farhadi, A.: YOLO9000: better, faster, stronger. In: Proceedings of the IEEE Conference on Computer Vision and Pattern Recognition, pp. 7263–7271 (2017)
14. Redmon, J., Farhadi, A.: YOLOv3: an incremental improvement. arXiv preprint arXiv:1804.02767 (2018)
15. Ren, S., He, K., Girshick, R., Sun, J.: Faster R-CNN: towards real-time object detection with region proposal networks. In: Advances in Neural Information Processing Systems, pp. 91–99 (2015)
16. Shoukun, X., Wang, Y., Yuwan, G., Li, N., Zhuang, L., Shi, L.: Research on safety helmet wear detection based on improved fast RCNN. Appl. Res. Comput, **28**(9), 1–6 (2019)

MobiMVL: A Model-Driven Mobile Application Development Approach for End-Users

Zhongyi Zhai[1,2], Ke Xiang[1], Lingzhong Zhao[1(✉)], and Junyan Qian[1]

[1] Guangxi Key Laboratory of Trusted Software, Guilin University of Electronic Technology, Guilin 541004, Guangxi, China
zhaizhongyi@guet.edu.cn, hisangke@163.com, zhaolingzhong163@163.com, qjy2000@gmail.com
[2] State Key Laboratory of Networking and Switching Technology, Beijing University of Posts and Telecommunications, Beijing 100876, China

Abstract. With the popularity of mobile Internet, many mobile users begin to create their own applications by using end-user development tools in the Web 2.0 era. These tools not only require users that develop applications equipped with more or less programming skills, but also focus on the type-specific mobile applications. To address these issues, we propose a model-driven development approach, called MobiMVL, for conducting end-users to develop mobile applications. The MobiMVL provides a full paradigm for mobile applications, including unified service model, business logic model and GUI model. These models formalize respectively the application's component domain, business interactions, and graphical interface together with application behavior. We also implemented an integrated development platform that can facilitate end-users to develop mobile applications following the MobiMVL. Finally, performance evaluation are conducted to evaluate our platform.

Keywords: Mobile applications · Model-driven development · End-user development · Mashup · Web service

1 Introduction

Owing to the popularity of wireless network and mobile devices, mobile applications [1–3] have been widely used in people's daily life. Nowadays, more than 1.4 million apps in Google's open market and 1.2 million apps in Apple's market are offered to users [4]. However, many end-users often cannot find adequate applications from the app store to meet their requirements. This is because that many

This work was supported by the National Nature Science Foundation of China (Nos. 61562015, 61572146, U1711263, 61862014, 61902086), Guangxi Natural Science Foundation of China (No. 2018GXNSFBA281142), Innovation Project of young talent of Guangxi (AD18281054), Guangxi Key Laboratory of Trusted Software (kx201718), the Innovation Project of Guet Graduate Education (2019ycxs049).

requirements of end-users are personalized, but most of mobile applications are not able to meet users' individual requirements [5]. End-user development (EUD) provides an effective solution to deal with the personalized application requirements [6,7]. EUD commonly provides a lightweight development environment and method, which enable ordinary users who are non-professional programmers to develop applications with their domain experience [8,9]. Mashup is the most successful technology for EUD. It enables mobile users to combine web services together to rapidly create mobile applications according to their requirements. With the evolution of Web 2.0, many web services have been published on the Internet that can be used easily for such development [10,11].

Presently, some end-user-oriented mobile development tools have used mashup technology to facilitate end-users to develop mobile applications. The MIT App Inventor [12,13] is a visual development environment for designing and developing mobile applications. This tool follows a reasonable development pattern, in which the user interface can be designed and created through a Designer Environment, and the application behavior can be developed through a Block Editor. However, the Block Editor adopts a traditional programming style which is not suitable for end-users. Some other tools are also provided for specific type of mobile applications. For example, MobiMash [14] only focuses on the simple mobile apps which do not involve business interactions. Overall, the existing end-user development tools have some challenges for developing mobile applications.

1) Most of these tools only support for type-specific development that limit the scope of mobile applications, for example, the App Inventor only be used to develop local applications.
2) These tools lack a unified abstract model to coordinate web resources.
3) They commonly demonstrate the development tasks by examples, and lack a well-constructed paradigm of mobile applications to conduct end-user development systematically.

To solve those issues, we present a model-driven development (MDD) approach for mobile applications based on a common software architecture, namely MobiMVL. It provides a triple paradigm for modeling mobile applications. Within our MobiMVL, a mobile application is modeled using three interrelated models, i.e. unified service data model, GUI model and business logic model.

This rest of this paper is organized as follows. Section 2 introduces some related work. Section 3 defines the MobiMVL framework, and specifies the three models and its implementation method. In Sect. 4, the development platform is implemented. Then, our platform is evaluated in Sect. 5. Finally, we draw some conclusions in Sect. 6.

2   Related Work

The related work can be generally classified into two categories: (1) Model-driven development approach, (2) Mashup development for mobile applications.

2.1 Model-Driven Development Approach

Model-driven Development (MDD) is a software development approach that focuses on the abstract models as the primary artifacts in the development process, and adopts the model transformation to automatically create the concrete implementations [15]. Instead of requiring developers to code every detail of a system's implementation using a programming language, MDD lets them model what functionality is needed and what overall structure the system should have. Presently, some research institutions have provided several specific implementation of MDD, such as OMG's MDA [16], Software Factories of Microsoft [17], and Domain Specific Modeling (DSM) [18,19].

The reference [20] provides a model-driven development approach that can transform the abstract description of embedded systems (i.e. UML) into its low-level implementation details (i.e. VHDL). Similarly, the reference [21] presents a model-driven methodology for developing secure data-management applications. It adopts the improved UML to describe the components, authorization, and GUI, and automatically generates the secure applications by a code generator.

The reference [22] proposes a model-driven development patterns for IoT applications. It defines some semi-formal model (i.e. PIMs) and its design patterns to facilitate domain people to develop IoT applications.

2.2 Mashup Development for Mobile Applications

Presently, many mashup tools have been used to facilitate end-users to develop mobile applications based on Web resources. These tools mainly involve Web Mashup and Service Mashup.

The Web Mashup is commonly used to create the GUI of mobile applications based on graphical development environment. For example, MobiDev [23] allows end-users to build mobile user interface by using the Web Mashup mode. Cabana [24] and LightApp [25] provides a lightweight mashup environment by composing UI elements and web services.

The Service Mashup is mainly used to create the business interactions of mobile applications. For example, MicroApp [26] enables end-users to develop their personal applications through the composition of existing web services. In MicroApp, the user interface is created through combining some sub-interfaces of services that is not flexible. The Appgyver [27] provides a combination mechanism of dataflow and controlflow that can be used to design the business processes based on web services.

3 Model-Driven Development Pattern

3.1 MobiMVL Framework

Figure 1 shows the MobiMVL development framework that follows the MDA. In the framework, the application involves three layers: presentation layer, component layer and logic layer. The presentation layer takes charge of the creation of interaction interface. The component layer provides necessary feeds of applications. The logic layer is in charge of the function innovation based on existing services. According to these three layers, the application can be abstracted as three elements service model, GUI, business logic. Differ from the traditional MVC, which provides a more intuitive paradigm that can facilitate technical programmers to organize programs; MobiMVL can facilitate end-users to design applications according to their natural structure. MobiMVL minimizes the coupling among modules by separating user interfaces, resources and business logic. Every element of MobiMVL corresponds to the object in the real world, and it can be easily designed by end-users. To support the end-user development based on MobiMVL, some model-driven theories should be constructed for every element as follow.

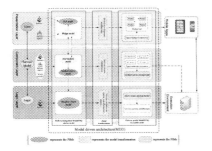
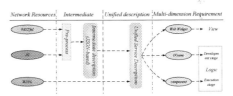

Fig. 1. The MobiMVL framework for mobile applications.

Fig. 2. The multi-dimension service adaptation.

1) In the Model element, we provide unified service model that can access heterogeneous web services as basic feeds of applications; we also need to built a multi-dimension service adaptation method that can not only provide available service representation for View and Logic, but also implement an automatic execution mechanism of unified service component.

2) In the View element, we provide an adaptive widget model that can be easily used by end-users to compose interaction interface, i.e. GUI model. This model formalizes both UI structure and configuration of behavior. For the widget model, an adaptive development framework should be constructed to link the underlying functions automatically.

3) In the Logic element, we design a dataflow-based mashup model to coordinate services to provide value-added functions. To conduct end-user to complete the business logic, we also introduce a data-driven development pattern. To realize the seamless connection between abstract model and executable script, a hierarchical adaptation should be built for the dataflow-based model.

3.2   Unified Service Data Model and Its Implementation

Enabled by web service standards, such as WSDL [28] and RESTful [29], a large number of web services have been published on the Internet. However, these services cannot be invoked by end-users without barriers because that they are heterogeneous, and they all have a type-specific interface. To complete the compatibility of open services, the Model element provides a multi-dimension service adaptation method for web services, as shown in Fig. 2. In this method, services should be redescribed in a unified way. Here, we adopt a JSON-based template to extract the service type, usage, operation address, etc. The RESTful service and JS service have not provided a standard description so that they need to be preprocessed into JSON-based intermediate data and then be transform into the standard format. These uniform services are finally processed into available services for View and Logic. The widget transformation of services is based on the widget model, so, it will be introduced in next section. Based on the adaptation mechanism, we implemented a service middleware that can directly parse the unified service components. This middleware is the same as the model transformation device of MDD. To implement the model-driven architecture, we define the unified service data model (SDM) to adapt heterogeneous service resource to participate in business development.

Definition 1. (SDM) An SDM is defined by a quintuple (S_D, A_S, T, V, E), where:

1) S_D is a text description about the service functions and usages.
2) A_S is the attribute set of a service, each attributes in the service is a data object in the interaction. Attributes of a service are roughly classified into three parts: input attributes, output attributes, and extension attributes.
3) T is the operation set that can provides relevant service's functions. Formally, the transformation T in SDM is a function set for processing data, denoted as $\{f_1(A_{I1}), f_2(A_{I2}), \ldots, f_n(A_{Ik})\}$.
4) V is describes the visual operations that can display the visual elements of service. Here, services visualization involves basic service framework S_F and specific service function S_P. For S_F, some common operations V_{PC} can be used to display uniform graphical framework of services, like a widget-based representation. For S_P, the service providers should provide specific operations V_{PS} to display the data or virtual effects of service functions.
5) E describes the extension of service function.

Via SDM, the repackaged service provides functionalities that can be used in a data-driven way. The relevant functionalities in the T can be provided by different types of web service, such as SOAP service and RESTful service.

In the business development, the SDM should be implemented as two forms for the development and execution respectively. These two implementations all follow the SDM specification. Here, we just illustrate the IFrame implementation for SDM, as shown in Fig. 3. As the IFrame instance is quite verbose, we use a comprehensive template to show how it can be developed. Some annotations have been added to this template to show which piece of code corresponds to the element in the SDM. In what follows, we will discuss the template in detail.

At the beginning of the IFrame template, the dependency library should be invoked according to the service category. Each dependency library corresponds to a combination of S_F and S_P. We have preliminarily developed several dependency libraries based on the pre-categorized services. When the dependency library is selected, the IFrame will take a visual shape and a graphical interface for the service.

The next part is a group of configuration information, which needs to be reprocessed by relying on specific service requirement. This part includes three elements: setinfo, addInputs & addOutputs, and transformation. Setinfo contains some basic service information, such as author, service title and description. Utilizing this information, end-users could understand the service function, characteristics and usage quickly. AddInputs & addOutputs is the core of the IFrame template, which corresponds to the T in the definition of SDM. For the AddInputs, the service provider should set the inputs by configuring the attributes names, and then construct the service operations behind every input attribute. For the addOutputs, the service provider only needs to set the outputs. The inputs and outputs are considered as the interface of data interaction between services, while the T takes charge of data processing from inputs to outputs. In our platform, we have implemented an adaptation mechanism of heterogeneous services for the T, so that it can be realized by invoking service interface directly. Generally, the transformations can be based on the SOAP services, RESTful services or JS services. Service providers just need to configure some basic information, such as operation name, interface type and service address.

The third part is optional. If there is not a proper web service for producing T, one can develop some local operations to process the data. Besides that, if service providers want to extend services, they also can develop some additional operation for the T.

The rest two parts are respectively page elements and renderscript, which provides basic elements for realizing the interactive interface of Iframe. If the service provider wants to present the data or virtual function of services specially, he/she can configure necessary elements in these two parts.

3.3   Cross-Platform GUI Model and Its Implementation

Cross-platform GUI development is an important issue for end-user development tools. Here, we propose a widget-based generation approach for GUI. The generated GUI can automatically adapt different mobile platform, and is finally created as mobile installation package, which is an easy-to-use form for end-users. In the approach, we construct an adaptive widget model (AWM) for the basic visual element of GUI.

Definition 2. (AWM) An AWM is defined by a quintuple (W_D, W_F, O_P, E_S, A_{op}), where:

688   Z. Zhai et al.

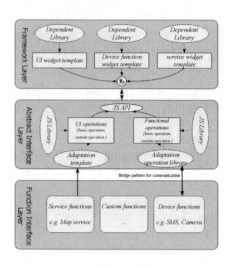

Fig. 3. The Iframe implementation of service.

Fig. 4. The three-layer development framework for widget.

1) W_D is a description about the widget functions and usages.
2) W_F represents the graphical frame of the widget. In the widget library, it is classified into three types: UI elements, device functions and interactive services. We have respectively provided multiple widget frames for these widgets.
3) O_P represents abstract operations for the widget that manages the widget, manipulate the widget, and connect to the underlying functional operations.
4) E_S represents the event set related to the widget, which is used to trigger actions by invoking corresponding operations.
5) A_{OP} represents the adaptation operations for different mobile platform.

Based on the AWM, widgets not only can be used as feeds to create GUI, but also can be used to connect the business functions. Moreover, the AWM provides a cross-platform adaptation mechanism for the device functions. To facilitate the widget development following the AWM, we propose a three-layer development framework, as shown in the Fig. 4. In the framework layer, the widget framework can be quickly created by invoking the widget template. The developer can choose proper widget template according to the type of widget. In the template, it have provides some basic operation of widget, e.g. the management of widget lifecycle. If user wants to extend widget's capability, he/she can customize some abstract operations for the widget at abstract interface layer. If a widget represents a functional component, it should be provided functional operations at function interface layer, and also be abstracted corresponding operations at abstract interface layer. The development framework of AWM also provides a bottom-up adaptation mechanism between different levels. This mechanism has been implemented as model transformation of GUI model through a mobile development middleware.

Many UI elements, device functions and web services have been integrated into our library by using some open project. We also allow users to add some new widgets following the above development method. GUI consists of a set of widgets and a set of trigger events.

Definition 3. The GUI model is defined as a by a quintuple $(GUI_N, W_S, O_{GUIS}, E_{GUIS}, R_{(E \to O)})$, where:

1) GUI_N is a GUI name about the application.
2) W_S represents the widget set related to the GUI of application, $W_S = \{W_1, W_2, \ldots, W_n\}$.
3) O_{GUIS} represents the operation set related to the GUI, $O_{GUIS} = O_P(W_1) \cup O_P(W_2) \cup \ldots \cup O_P(W_n)$.
4) E_{GUIS} represents the event set related to the GUI, $E_{GUIS} = E_S(W_1) \cup E_S(W_2) \cup \ldots \cup E_S(W_n)$.
5) $R_{(E \to O)} = e_i \to o_j | e_i \in E_{GUIS}, o_j \in O_{GUIS}$ represents the relationships of events and operations.

According to the GUI model, end-users can utilize these widgets to design the GUI of applications by a WYSIWYG (what you see is what you get) mode, and configure some functions for the GUI in a visual manner. If end-users cannot well design a GUI, they can publish their task on the crowdsourcing platform, collect the returned results and choose an optimal one as the solution.

3.4 Dataflow-Driven Business Logic Design

In the MobiMVL, we adopt a dataflow-based approach to develop the business logic. Here, we defined a structured dataflow model, called service relation model (SRM), which can facilitate the business logic development by refining the data interaction between services.

Definition 4. (SRM) An SRM is defined by a triple (f, M_E, M_S), where:

1) $f = < S_j, a_{Oi}, a_{Ii}, S_i >$, S_i and S_j represent two services, a_{Oi} is a output attribute of S_j, a_{Ii} is a output attribute of S_i.
2) $\{ME = < a_{Oi}, a_{Ii} > \in f$, a_{Oi} is type strong matching with aIi, or a_{Oi} is type weak matching with $a_{Ii}\}$.
3) $MS = \{< a_{Oi}, a_{Ii} > \in f$, a_{Oi} is semantics strong matching with a_{Ii}, or a_{Oi} is semantics weak matching with $a_{Ii}\}$.

In our framework, we have introduced a basic set of attribute types-TA: Integer, Float, String, Boolean, Image, Audio, JSON, and permit service providers to extend attribute type based on these basic types. For the attribute types, we construct the dependency relationship between these types, for example, $Integer \angle Float$ represents the value of Integer attributes that can be accepted by the Float attributes. If an attribute a_i has a same data type with attribute a_j, then a_i is type strong matching with a_j; if an attribute a_i has a same data type with attribute a_j, then a_i is type weak matching with a_j. Similarly, we construct a semantic tree for the attributes of services. Then, the four hierarchical relationships of attributes can be extracted based on the semantic tree.

1) IstheSame($\sigma(a_1)$, $\sigma(a_2)$): The concept of attribute a_1 is the same with the attribute a_2.
2) IsSubject($\sigma(a_1)$, $\sigma(a_2)$): The concept of attribute a_1 is subject to the attribute a_2.
3) IsSameSource($\sigma(a_1)$, $\sigma(a_2)$): The concept of attribute a_1 is have a same source concept with the attribute a_2.
4) IsIrrelevant($\sigma(a_1)$, $\sigma(a_2)$): The concept of attribute a_1 is irrelevant to the attribute a_2.

If there exists IstheSame($\sigma(a_1)$, $\sigma(a_2)$), then a_i is semantics strong matching with a_j. If there exists IsSubject($\sigma(a_1)$, $\sigma(a_2)$), then a_i is semantics weak matching with a_j.

Compared to traditional dataflow approach [30], SRM is a structured dataflow that not only can be as the basic unit of business logic, but also to improve the accuracy of data connections. Based on SRM, the business logics can be described as dataflow graphs. We call such dataflow graph as "Service Process Graph (SPG)".

Definition 5. (SPG) The SPG can be represented as a triple: $< V, E, I >$, where:

1) $V = \{S_1, \ldots, S_n\}$ describes the involved services in the SPG.
2) $E = \{SRM_1, \ldots, SRM_n\}$ is the set of dataflows.
3) $I = a_{I1}, a_{I2}, \ldots, a_{Ii}$ is the minimal attribute set that can drive the SPG.

The design of SPG is actually a graph creation process that relies on service attributes (i.e. data) and service dependencies. We propose a data-driven development pattern that can facilitate end-users to create business processes. We firstly introduce a service creation model (SCM) as the basic framework of the design of SPG.

Definition 6. (SCM) The SCM is defined by a quad: $< D_R, S_R, SD_G, SPG_i >$, where:

1) D_R describes the related data entities of the business logic.
2) S_R shows the alternative services that might be used by the business logic.
3) SD_G describes a service dependency graph that implies the interaction relationships of services in the business logic.
4) SPG_i describes a dataflow graph of the business logic, which is the outcome of SCM.

Based on the SCM, the generation process of SPG is as follow:
Step 1: According to requirement of business logics, the end-user needs to analyze the related data entities (i.e service attributes) for D_R. If there have some new services selected for S_R in the step 2, the developer can also extract related new data entities from their service attributes iteratively.
Step 2: The end-user should search and add related services for S_R according to the new data entities in D_R, and construct the service dependency graph for

S_R. If an added service is selected by new data entity of another service, the dependency relationship between these two services need to be established.

Step 3: The service dependency graph need to be checked whether it forms a complete dependency pathway from the initial services to the final services, by means of end-users' domain knowledge or real-world experience. If there are some missing service attributes in the dependency graph, corresponding data entities will be added in the D_R, and the end-user needs to repair the dependency graph from step 1 again.

Step 4: According to the dependency graph, the user should complete all the data connections among services, and simulate the execution effect of SPG.

Through the above steps, ordinary users can develop a new service in a natural way. In the whole process, the focus is on the "data", i.e. service attributes, which are easily understood and acquired by ordinary users.

4 Platform Implementation

Figure 5 shows the overview of mobile development environment. The design goal of this platform is to provide a comprehensive development ecosystem to facilitate end-users to create full mobile applications. Three sub-systems of this platform can cooperate with each other in eliminating obstacles to EUD. ServiceAccess provides a registration interface of web services, and enables to export available services to EasyApp and LSCE automatically. EasyApp allows end-users to design the GUI through a WYSIWYG mode. Once a GUI is completed, the installation package of mobile app (e.g. .apk or .ipa) is created through a one-click way. If a mobile application involves some application logics, the end-users should develop and deploy these logics by using LSCE, and then, the open interfaces (e.g. RESTful API) of them would be published for the corresponding app. This tool suite provides a systematic development environment for mobile applications. The end-users are the most important participants in such development process, who have the capacity to engage in every task in the development.

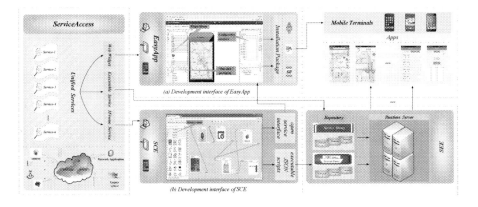

Fig. 5. The overview of mobile development environment

5   Performance Evaluation

In our platform, the mobile development takes MobiMVL as a new paradigm of applications. The mobile application of MobiMVL involves two main parts: the GUI is developed as a mobile app on the mobile device, and the service processes are designed as business logics on the execution environment. To evaluate the efficiency of MobiMVL and our platform, experiments are designed.

In this experiment, we measure the respond time of two business processes at three different concurrent situations—100 process instances, 500 process instances and 1000 process instances, as shown in Fig. 6.

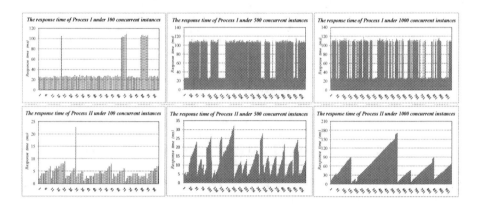

Fig. 6. The respond time of service processes under different concurrent instances

The Process I was designed for computing the value of Pi and outputting the result to another service. For this process, we implemented some native components as required services. The Process II was designed for inquiring weather service and transmitting this information to subscribers. For the limitation of concurrency of open web services, we had constructed some virtual services for the Process II. As can be seen from Fig. 6 the respond time of Process I increases rapidly from the 100 instances to the 500 instances. That is due to the higher calculation load of components in Process I, which will consume more time to wait for the release of system resources with the increase of process instances. When concurrent instances of Process I arrive at 1000, the peak value of respond time stays around 110ms. For Process II, the respond time is very short under 500 concurrent instances. When concurrent instances of Process II arrive at 1000, the average respond time increases more. That is due to the respond time is limited by some external resource, such as network transmission, network bandwidth and processing power of public web services. Similarly, for a process invoke third-party web services on the Internet, the respond time depend heavily on the handling capability of third-party services and network situation. Overall, LSCE supports high concurrency of execution of service processes.

6 Conclusion

In this paper, we present a model-driven development approach for mobile applications, called MobiMVL. The MobiMVL not only provides a full application paradigm at abstract layer, but also can facilitate end-users to create mobile applications systematically. To facilitate end-user development, we implemented an integrated development platform. This platform includes three sub-systems: ServiceAccess, EasyApp and LSCE, which can be used to design the three element of MobiMVL respectively.

In the future work, we intend to construct an approach that facilitates end-users to create services if they cannot find available services from Internet.

References

1. Wei, S., Hongji, D., et al.: Traffic sign recognition method integrating multi-layer features and kernel extreme learning machine classifier. Comput. Mater. Continua **60**(1), 147–161 (2019)
2. Jiancheng, Z., Zhengzheng, L., et al.: Super-resolution reconstruction of images based on microarray camera. Comput. Mater. Continua 60(1), 163–177 (2019)
3. Jiwei, Z., Yueying, L., et al.: Improved fully convolutional network for digital image region forgery detection. Comput. Mater. Continua **60**(1), 287–303 (2019)
4. Choi, H., Kim, J., Hong, H., Kim, Y., Lee, J., Han, D.: Extractocol: automatic extraction of application-level protocol behaviors for android applications. In: SIG-COMM 2015, London, United Kingdom (2015)
5. The App Stores are not "long tail" (2014). http://www.rudebaguette.com/2014/02/25/app-stores-long-tail/
6. Ko, A.J., Abraham, R., Beckwith, L., et al.: The state of the art in end-user software engineering. ACM Comput. Surv. **43**(3), 1–44 (2011)
7. Namoun, A., Daskalopoulou, A., Mehandjiev, N., Xun, Z.: Exploring mobile end user development: existing use and design factors. IEEE Trans. Softw. Eng. Early Access (2016)
8. Lizcano, D., Alonso, F., Soriano, J., et al.: A web-centred approach to end-user software engineering. ACM Trans. Softw. Eng. Methodol. **22**(4) (2013)
9. Chudnovskyy, O., Nestler, T., et al.: End-user-oriented telco mashups: the OMELETTE approach. In: Proceedings of the 21st Annual Conference on World Wide Web (WWW), Lyon, France, pp. 235–238 (2012)
10. Cappiello, C., Matera, M., et al.: A UI-centric approach for the end-user development of multidevice mashups. ACM Trans. Web (TWEB) **9**(3) (2015)
11. Cheng, B., Zhai, Z., et al.: LSMP: a lightweight service mashup platform for ordinary users. IEEE Commun. Mag. **55**(4), 116–123 (2017)
12. App Inventor, MIT Center for Mobile Learning (2016). http://appinventor.mit.edu/explore
13. Crawford Pokress, S., Veiga, J.J.D.: MIT app inventor: enabling personal mobile computing. In: PROMOTO 13, Indianapolis, IN, USA (2013)
14. Cappiello, C., Matera, M., Picozzi, M., Caio, A., Guevara, M.T.: MobiMash: end user development for mobile mashups. In: WWW 2012, Lyon, France (2012)
15. Atkinson, C., Kühne, T.: Model-driven development: a metamodeling foundation. IEEE Softw. **20**(5), 36–41 (2003)

16. OMG's MDA Guide Version (2003). http://www.omg.org/mda/mda-files/MDA-Guide-Version1-0.pdf
17. Greenfield, J., Short, K.: Software factories: assembling applications with patterns, models, frameworks, and tools. In: International Conference on Software Product Lines (2004)
18. Liu, X., Xu, M., Teng, T., Huang, G., Mei, H.: MUIT: a domain-specific language and its middleware for adaptive mobile web-based user interfaces in WS-BPEL. IEEE Trans. Serv. Comput. Early Access Article (2016)
19. Karsai, G., Sztipanovits, J., Ledeczi, A., Bapty, T.: Model- integrated development of embedded software. Proc. IEEE **91**, 145–164 (2003)
20. Wood, S.K., Akehurst, D.H., Uzenkov, O., et al.: A model-driven development approach to mapping UML state diagrams to synthesizable VHDL. IEEE Trans. Comput. **57**(10), 1357–1371 (2008)
21. Basin, D., Clavel, M., Egea, M., et al.: A model-driven methodology for developing secure data-management applications. IEEE Trans. Softw. Eng. **40**(4), 324–337 (2014)
22. Cai, H., Yizhi, G., Vasilakos, A.V., Boyi, X., Zhou, J.: Model-driven development patterns for mobile services in cloud of things. IEEE Trans. Cloud Comput. Early Access Article (2016)
23. Seifert, J., Pfleging, B., Bahamóndez, E.C.V., Hermes, M., et al.: MobiDev: a tool for creating apps on mobile phones. In: MobileHCI 2011, Stockholm, Sweden (2011)
24. Dickson, P.E.: Cabana: a cross-platform mobile development system. In: SIGCSE 2012, Raleigh, North Carolina, USA (2012)
25. Baidu LightApp (2016). http://qing.baidu.com/
26. Francese, R., Risi, M., Tortora, G., et al.: Visual mobile computing for mobile end-users. IEEE Trans. Mob. Comput. **15**(4), 1033–1046 (2016)
27. Appgyver (2017). https://www.appgyver.eu/
28. Roriguez, J.M., Mateos, C., et al.: Assisting developers to build high-quality code-first web service APIs. J. Web Eng. **14**(3–4), 251–285 (2015)
29. Bellido, E., Alarcon, R., Pautasso, C.: Control-flow patterns for decentralized RESTful service composition. ACM Trans. Web **8**(1) (2013)
30. Wang, G., Han, Y., Zhang, Z., Zhang, S.: A dataflow-pattern-based recommendation framework for data service mashup. IEEE Trans. Serv. Comput. **8**(6), 889–902 (2015)

Analysis on Buffer-Aided Energy Harvesting Device-to-Device Communications

Guangjun Liang[1,2], Qun Wang[1,2(✉)], Jianfang Xin[3], Lingling Xia[1,2],
Xueli Ni[1,2], and Jie Xu[1,2]

[1] Department of Computer Information and Network Security,
Jiangsu Police Institute, Nanjing, China
{liangguangjun,wqun}@jspi.cn
[2] Jiangsu Provincial Public Security Department Key Laboratory of Digital
Forensics, Nanjing, China
[3] Electrical Engineering School, Anhui Polytechnic University, Wuhu, China
xinjfang@163.com

Abstract. We address the performance analysis issue of the buffer-aided energy harvesting device-to-device communication system. A Coupled Processor Queuing Model with data packet and energy packet poisson arrival but interactive departure is proposed. The Quasi-Birth and Death method is adopted to obtain the steady state transition probability of proposed queuing model. Then the expressions of throughput, delay and packet drop rate for both data queue and energy queue are derived. The simulations verify the accuracy of the theoretical derivation results.

Keywords: Buffer-aided · Energy harvesting · Device-to-Device

1 Introduction

With the increasing number of massive intelligent terminal devices, up-and-coming communication services that meet the specific needs of users are also exploding. The contradiction between the explosive growth of data traffic carried by mobile communication and the shortage of wireless spectrum resources is increasingly emerging. Therefore, how to effectively heighten network capacity, enhance wireless spectrum utilization, and improve the end user experience in different communication scenarios becomes an urgent task. As one of the key candidate technologies for 5G, Device-to-Device (D2D) communication technology will be used in 5G mobile communication based on the characteristics of ultra-high speed, large bandwidth, ultra-large-scale access capability and large data processing capability [1,2].

This research has been supported by the National Natural Science Foundation of China (Grant No. 61802155), the High-level Introduction of Talent Scientific Research Start-up Fund of Jiangsu Police Institute (2019) and the General Research Project of Anhui Higher Education Promotion Plan(Grant TSKJ2015B18, KZ00215021, KZ00215022).

© Springer Nature Switzerland AG 2020
X. Sun et al. (Eds.): ICAIS 2020, LNCS 12239, pp. 695–706, 2020.
https://doi.org/10.1007/978-3-030-57884-8_61

Recently, Buffer-aided D2D communication which allows D2D terminals to be equipped with buffers has caused great concern. Objectively speaking, adding buffers can improve system performance, such as system throughput, power saving and communication quality, but it also inevitably increases the related challenges of transmission delay and optimal bandwidth allocation. The authors in [3] analyze the basic effects of data buffer constraints on D2D communication through an optimization framework. They also point out the opportunities created by buffer-aided D2D communication for system bandwidth protection which caused by the positive correlation between buffer size and overall system performance. [4,5] propose two relay resource management methods to improve the performance of D2D communication, respectively. The authors in [4] propose a relay selection method based on social aware energy efficient which takes into account the transmission rate, relays available residual energy and buffer status. Each time slot, the scheduling rule is dynamically calculated by the relay selection method to decide which link is activated to transmit the data packet. The authors in [5] propose a relay selection strategy in C-RAN-based D2D communication systems. First, the transmission success probability in one-hop D2D communication and the average transmission delay are derived. Then, considering the similarity of interest between users, the remaining energy of the candidate relay users, the buffer status, and different channel conditions, a relay selection mathematical utility model is given. Without considering energy factors, [3–5] analyze the impact of data buffer settings on the performance of D2D communication networks.

Energy harvesting (EH) technology [6] capable of collecting energy from the surrounding environment, especially suitable for future green wireless communication scenarios. [7–11] discuss the joint resource allocation problems in the energy harvesting D2D communications. [7] addresses the dynamic resource allocation strategy based on wireless power supply D2D communication mode. A joint optimization algorithm for resource blocks and power allocation is proposed in an energy EH D2D two-layer heterogeneous network scenario in [8]. Without excessively reducing the throughput of MU, a low complexity method to optimize D2D transmission power is also raised. Considering the D2D downlink resource reuse and EH problem, two joint resource allocation algorithms for offline (non-causal) and online (causal) energy harvesting profile at the D2D transmitter are proposed in [9]. The authors in [10] discuss the power allocation problem of EH D2D communication for the goal of maximizing service quality. A joint power allocation and subcarrier pairing algorithm with the constraints of subcarrier multiplexing and energy collection is given. Considering service quality (QoS) and available energy constraints, a joint resource allocation scheme which aims at rate maximization is proposed in [11]. The above five references do not consider the impact of D2D user mobility in the D2D communication system.

In this paper, we concentrate on a buffer-aided energy harvesting device-to-device communication networks. Firstly, we profile the data packet and energy packet poisson arrival but interactive departure scenario as a Coupled Processor

Queuing Model (CPQM). Secondly, we apply the Quasi-Birth and Death (QBD) method the obtain the steady state distribution(SSD) of proposed CPQM. Then, we derive the expressions of data throughput, data delay and data packet drop rate. Thirdly, the simulations verify the accuracy of the theoretical derivation results.

The main contributions of this paper are summarized as follows:

- Compared with traditional D2D communications scenario, we consider not only the data poisson arrival but also the energy poisson arrival in the buffer-aided EH D2D communications networks. Each D2D transmitter is equipped with a data buffer and an energy buffer.
- We consider the tradeoff between data throughput and energy consumption. The D2D scenario of data packet and energy packet independent arrival but interactive departure are profiled as a CPQM. The expressions of CINR interval division which contains the interferences from CU and the other D2D users is raised.
- By Quasi-Birth and Death method to acquire steady state transition probability, we derive the expressions of data throughput, data delay and data packet drop rate of Coupled Processor Queuing Model.

The rest of this article is organized as follows. Section 2 describes the system model, the queue model and main assumption. Section 3 applies QBD method to solve the CPQM. Section 4 analyzes the performance parameters of data queue, such as data throughput, data delay and data packet drop rate. The simulations verify the accuracy of the theoretical derivation results in Sect. 5. In Sect. 6, there is a conclusion for this article.

2   System Model and Queue Model

2.1   System Model and Assumption

As shown in Fig. 1, the D2D communication heterogeneous cellular network that allows a group of D2D users to multiplex the cellular user's uplink channel is considered. It is assumed that the BS is arranged according to a hexagonal grid, and the BS density per unit area is λ_B. To simplify the expression, we approximate the coverage radius of the base station to $R = \frac{1}{\sqrt{\pi \lambda_B}}$ [12]. A cluster of D2D communication pairs multiplex a cellular user's uplink channel and the bandwidth is B_0. The cellular user obeys the PPP process Φ_C with the density parameter λ_C, and the D2D users obeys the PPP process Φ_D with the density parameter λ_D. The distance between the cellular user and the serving base station is defined as $r, r \in (0, R)$. For the convenience of analysis, we assume that the distance of each D2D user pair is fixed, and its probability density function of r is $\frac{2r}{R^2}$.

The D2D transmitter adaptively adjusts the transmission power and transmits data according to the channel state and collecting energy. Since the variety of channel state is a continuous process, it is not easy to analyze and adaptively

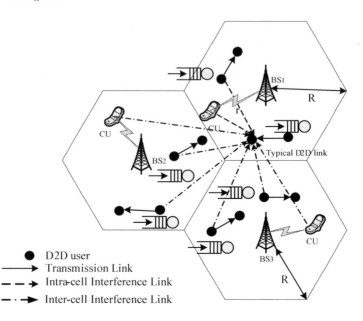

Fig. 1. Cellular wireless networks with EH relaying system.

adjust the transmission power. In this paper, we divide the channel into $M + 1$ intervals. Except for the cutoff transmission interval, the remaining M channel state intervals correspond to M transmit powers $P_{k,i}^D$, $i = 1, 2, ..., M$ respectively, where k represents the typical D2D transmission pair to be studied. The Shannon capacity of a typical D2D k receiver is given as

$$R_i = B_0 \log \left(1 + \frac{P_{k,i}^D h_{k,k} d_{k,k}^{-\alpha_D}}{I_C + I_D + \sigma^2} \right), \quad i = 1, 2, ..., M. \tag{1}$$

where $h_{k,k}$ and $d_{k,k}$ indicate small-scale fading and transmission distance between pairs of typical D2D users, respectively. σ^2 is the noise power. α_D is the path loss factor for D2D user transmission link path loss index. I_C and I_D represents interference from cellular users and other D2D users on the same channel, respectively. Considering the channel interference I_C and I_D, we define the channel interference to noise ratio(CINR) of typical D2D k as $CINR_k$.

$$CINR_k = \frac{h_{k,k} d_{k,k}^{-\alpha_D}}{I_C + I_D + \sigma^2} = \frac{h_{k,k} d_{k,k}^{-\alpha_D}}{\sum\limits_{n \in \Phi_C} P_n^C h_{n,k} d_{n,k}^{-\alpha_C} + \sum\limits_{m \in \Phi_D \setminus \{k\}} P_m^D h_{m,k} d_{m,k}^{-\alpha_D} + \sigma^2} \tag{2}$$

2.2 Queue Model and Assumption

As described above, there is a coupling relationship between the data arrival and energy harvesting in the EH D2D cellular networks. Different from traditional queue model, we consider the joint resource scheduling problem with data and energy coupling. Then a data and energy Coupled Processor Queuing Model is modeled. In order to describe the CPQM, a two-state Markov process (i, j), $i \in \{0, 1, 2, ..., N_d\}$, $j \in \{0, 1, 2, ..., N_e\}$ is defined, where i and j denote the data queue length and energy queue length, respectively. The buffer size of the data queue is N_d, and the buffer size of the energy queue is N_e.

Since energy harvesting is a continuous process that is not easy to analyze, we discretize this continuous energy harvesting process. We define the size of an energy pack to E_0 and fix the D2D communication transmission slot length to T_0. The M transmit powers satisfy the following relationship

$$P_{k,1}^D = \frac{1}{2} P_{k,2}^D = ... = \frac{1}{i} P_{k,i}^D = ... = \frac{1}{M} P_{k,M}^D \tag{3}$$

According to the research results in [13], if the energy consumed E_0 is fixed, the stable transmit power in the whole transmission slot is the highest energy efficiency type.

$$E_0 = P_{k,1}^D T_0 = \frac{1}{2} P_{k,2}^D T_0 = ... = \frac{1}{i} P_{k,i}^D T_0 = ... = \frac{1}{M} P_{k,M}^D T_0 \tag{4}$$

The unit data packet size is fixed to L_0, we have

$$L_0 = B_0 T_0 \log \left(1 + P_{k,i}^D CINR_{k,i}\right), \ i = 1, 2, ..., M. \tag{5}$$

Thus we can divide the intervals of different transmit powers as follow

$$CINR_{k,1} = iCINR_{k,i}, i = 2, 3, ..., M \tag{6}$$

Considering more general data and energy arrivals, data arrival and energy arrival are subject to Poisson distributions with parameters of λ_d and λ_e. p_a^d indicates the probability that a packet arrives in the current transmission slot, and p_b^e indicating the probability that b energy packet arrives in the current transmission slot.

$$p_a^d = ..., a = 0, 1, ..., N_d \tag{7}$$

$$p_b^e = ..., b = 0, 1, ..., N_e \tag{8}$$

Thus $p_{N_d+1}^d = p_{N_d+2}^d = ... = 0$ and $p_{N_e+1}^e = p_{N_e+2}^e = ... = 0$.

$$p_{N_d}^d = 1 - \sum_{x=0}^{N_d-1} p_x^d \tag{9}$$

$$p_{N_e}^e = 1 - \sum_{x=0}^{N_e-1} p_x^e \tag{10}$$

The D2D transmitter adaptively adjusts the transmitting power according to the channel state through the collected energy, and transmits a fixed-size data packet L_0 in a fixed time slot T_0. Consider the PPP distribution of D2D users and CU users, we divide the $M + 1$ channel state intervals according to channel gain as $p_x^{ch}, x = 0, 1, 2, ..., M$. p_x^{ch} indicates the probability that the current transmission slot channel is in the first state x.

3   QBD Method for Solving CPQM

First, we describe the two-dimensional Markov process with two one-dimensional processes. The state space is redefined as (i, j), $i \in \{0, 1, 2, ..., N_d\}$, $j \in \{0, 1, 2, ..., N_e\}$. For the two-dimensional Markov process (i, j), there is a possible state of the $(N_d + 1)^2 (N_e + 1)^2$, and as the maximum buffer size N_d and N_e increase, the state will be more complicated. In order to reduce the computational complexity, we use the QBD method to solve the CPQM. We define the subset of states of CPQM as $L_i = \{(i, j) | 0 \leq j \leq N_e \}$, $0 \leq i \leq N_d$, and the transition probability matrix is expressed as

$$\mathbf{P} = \begin{bmatrix} B_0 & C_0^1 & \cdots & \cdots & \cdots & & \cdots \\ A_1 & B_1 & C_1^2 & \cdots & C_I^J & & \cdots \\ 0 & A_2 & B_2 & C_2^3 & \cdots & & \cdots \\ \cdots & 0 & \cdots & \cdots & \cdots & & \cdots \\ \cdots & \cdots & \cdots & \cdots & \cdots & C_{N_d-1}^{N_e} & \\ 0 & \cdots & \cdots & 0 & A_{N_d} & B_{N_d} \end{bmatrix}, \quad \begin{array}{l} I = 0, 1, ..., N_d - 1 \\ J = 1, 2, ..., N_e \end{array} \quad (11)$$

where the elements on the secondary diagonal A_i, $i = 1, 2, ..., N_d$ and the elements on the main diagonal B_i, $i = 0, 1, ..., N_d$ are sub-matrices of the transition probability.

Different the Bernoulli arrival of data and energy in [14], the Poisson arrival process is more complicated and close to the actual situation, so the boundary conditions of the transition probability matrix of CPQM are more complicated. After careful deduction, the elements in the transition probability matrix P are zero, except for the main diagonal, the N_e diagonals in the upper triangle, and the one diagonal in the lower triangle. The nonzero elements in the lower triangle A_i, $i = 1, 2, ..., N_d$, indicates that the state subset L_{i-1} transfer to the state subset L_i. The nonzero elements in the main diagonal B_i, $i = 0, 1, ..., N_d$, indicates that the state subset L_i transfer to the state subset L_i. The nonzero elements in the upper triangle C_I^J, $\begin{array}{l} I = 0, 1, ..., N_d - 1 \\ J = 1, 2, ..., N_e \end{array}$, indicates that the state subset L_i transfer to the state subset L_{i+1}.

4   Queuing Stationary Analysis and Performance Parameter

4.1   Queuing Stationary Analysis

The queuing system of the steady state distribution of the CPQM can be obtained after long-running process. Therefore, the SSD of the CPQM is defined as $\Pi = [\pi(i,j)]$, $i = 0,1,...,N_d$, $j = 0,1,...,N_e$ when the time shaft is infinity. The SSD of data and energy queue length are represented as follow

$$
\begin{cases}
\Pi_d = [\pi_d(0), \pi_d(1), ..., \pi_d(N_d)] \\
\pi_d(i) = \sum_{j=0}^{N_e} \pi(i,j)
\end{cases}
\tag{12}
$$

$$
\begin{cases}
\Pi_e = [\pi_e(0), \pi_e(0), ..., \pi_e(N_e)] \\
\pi_e(j) = \sum_{i=0}^{N_d} \pi(i,j)
\end{cases}
\tag{13}
$$

where Π_d and Π_e are the SSD of data and energy queue length, respectively.

According to normalized to the sum of probability, the SSD of the CPQM is solved by the classic queuing theory as follow

$$
\begin{cases}
\Pi = \Pi P \\
\sum_{i=0}^{N_d} \sum_{j=0}^{N_e} \pi(i,j) = 1
\end{cases}
\tag{14}
$$

In order to gain the SSD Π and transition probability matrix P, we provide a fixed point iteration algorithm to solve (14) as [12].

4.2 Performance Metrics

After obtaining the SSD of the CPQM by the fixed point iteration algorithm, we can get some system performance parameters, for instance average data queue length, average energy queue length, average data throughput, average energy consumption, average data delay, average energy consumption delay, data drop probability and energy drop probability.

The expressions of average data queue length $\mathrm{E}\,[Q_{data}]$ and average energy queue length $\mathrm{E}\,[Q_{energy}]$ can be acquired as follow

$$
\mathrm{E}\,[Q_{data}] = \sum_{i=0}^{N_d} i \left(\sum_{j=0}^{N_e} \pi(i,j) \right)
\tag{15}
$$

$$
\mathrm{E}\,[Q_{energy}] = \sum_{j=0}^{N_e} j \left(\sum_{i=0}^{N_d} \pi(i,j) \right)
\tag{16}
$$

Given the channel state x probability of current transmission slot p_x^{ch}, $x = 0,1,...,M$, the average data throughput $\mathrm{E}\,[Thr_{data}]$ and average energy consumption $\mathrm{E}\,[Thr_{energy}]$ can be acquired as

$$
\mathrm{E}\,[Thr_{data}] = (1 - \pi_d(0)) \left(\sum_{i=1}^{M} p_i^{ch} \sum_{j=i}^{N_e} \Pi_e(j) \right)
\tag{17}
$$

$$\mathrm{E}\left[Thr_{energy}\right] = (1 - \pi_d(0)) \left(\sum_{i=1}^{M} i p_i^{ch} \sum_{j=i}^{N_e} \Pi_e(j)\right) \tag{18}$$

Then, the average data delay $\mathrm{E}\left[Del_{data}\right]$ and average energy consumption delay $\mathrm{E}\left[Del_{energy}\right]$ are obtained by Little's formula as follow

$$\mathrm{E}\left[Del_{data}\right] = \frac{\sum_{i=0}^{N_d} i \left(\sum_{j=0}^{N_e} \pi(i,j)\right)}{(1 - \pi_d(0)) \left(\sum_{i=1}^{M} p_i^{ch} \sum_{j=i}^{N_e} \Pi_e(j)\right)} \tag{19}$$

$$\mathrm{E}\left[Del_{energy}\right] = \frac{\sum_{j=0}^{N_e} j \left(\sum_{i=0}^{N_d} \pi(i,j)\right)}{(1 - \pi_d(0)) \left(\sum_{i=1}^{M} i p_i^{ch} \sum_{j=i}^{N_e} \Pi_e(j)\right)} \tag{20}$$

Further, we can also write the closed-form expressions of data drop probability p_{data}^{Drop} and energy drop probability p_{energy}^{Drop} as follow

$$p_{data}^{Drop} = \sum_{j=0}^{N_e} \pi(N_d, j) \tag{21}$$

$$p_{energy}^{Drop} = \sum_{i=0}^{N_d} \pi(i, N_e) \tag{22}$$

5   Simulation and Analysis Results

In this section, we verify the accuracy of the theoretical derivation results by Monte Carlo simulation, which gives a few practical guidance for the real system performance with physical significance. The curve CPT $M = 1$ indicates that the D2D terminal uses the CPT strategy as a reference to show the performance gain considering the APT strategy. A curve with APT $M = 2, 4, 6$ indicates that the D2D terminal employs an APT strategy that allows more energy packets to be sacrificed in the event of poor channel conditions to achieve better system performance due to insufficient energy. The Bernoulli process parameter λ_d for data arrival satisfies $0 \le \lambda_d \le 1$. The Bernoulli process parameter λ_e for energy arrival satisfies $0 \le \lambda_e \le 1$.

Figure 2 depicts the average queue length performance of CPQM with average data arrival and energy arrival when $M = 2$. As shown in the figure, the average data queue length begins to grow, and then gradually becomes saturated as the average data arrival rate increases and the energy arrival rate decreases. In

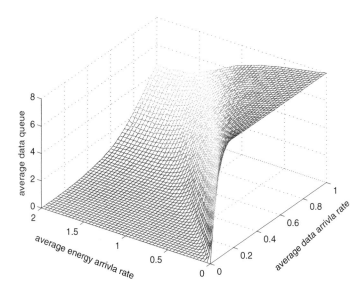

Fig. 2. The average data queue length performance of CPQM with average data arrival and energy arrival.

particular, when the energy arrival rate is low, the data queue quickly saturates as the average data arrival rate increases. If the average energy arrival rate increases, meaning that the rich energy can better satisfy the data transmission, the data queue will slowly increase and then reach saturation.

Figure 3 illustrate the average data throughput as functions of average data arrival when $\lambda_e = 1.6$. We can see that the deduced derivation values are consistent with the computer simulation values which further verifies the correctness of the numerical derivation. As mentioned earlier, when the channel status is not good, using a larger M value and consuming more energy packets to send a packet may reduce the data delay in the buffer, but it does not always achieve the best data throughput performance. Shown in Fig. 3, the average data throughput begins to increase and then saturates as the average data arrival increases, $M = 2$ is the best choice when $M = 1, 2, 4, 6$. Specially, the optimal system throughput performance is $M = 2$, which provides the biggest data throughput but moderate energy consumption.

Shown in the Fig. 4, the theoretical value is consistent with the computer simulation value, indicating the correctness of the theoretical derivation. Figure 4 illustrates that the average data delay of CPQM begins to increase and then saturates as the average data arrives for $M = 1, 2, 4, 6$. Seen from the figure, when the channel status is not good, using a large M value and consuming more energy packets to send a data packet does not always achieve the lowest data delay performance, because the rich energy consumption strategy may cause the next few data to be absent, which in turn increases the data delay. Figure 4 profile the average data delay as functions of average data arrival when $\lambda_e = 1.6$.

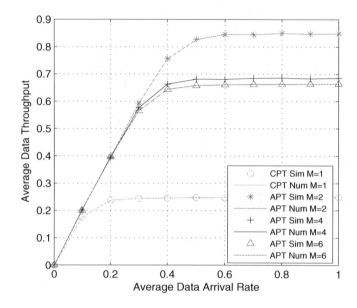

Fig. 3. The average data throughput performance of CPQM.

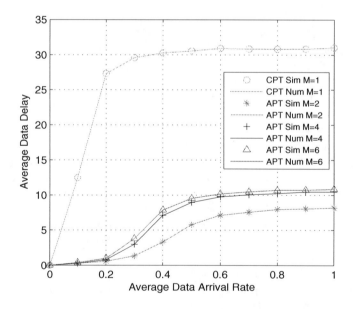

Fig. 4. The average data delay performance of CPQM.

When the energy arrival is fixed, all the average data delay curves with different M value when the average data arrival increases. The $M=2$ has the lowest delay, $M=4, 6$ are second, $M=1$ has the worst delay.

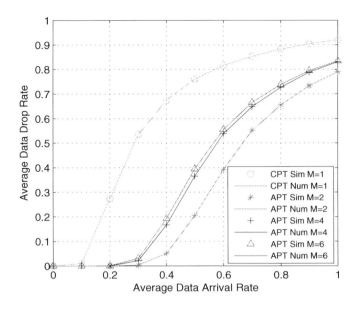

Fig. 5. The average data drop rate performance of CPQM.

Figure 5 depicts the average data drop as functions of average data arrival when $M = 1, 2, 4, 6$. The deduced derivation values are consistent with the computer simulation values which further verifies the correctness of the numerical derivation. Figure 5 illustrate the average data drop as functions of average data arrival with different M value when $\lambda_e = 1.6$. When $\lambda_d < 0.1$ in which energy resources are richer than data resources, the performance gap of data drop with $M = 1, 2, 4, 6$ are negligible. When $\lambda_d > 0.1$ in which data resources are more and more, the performance gap of data drop with different M value are get bigger. The strategy $M = 2$ is the best, the strategy $M = 4, 6$ are the second, and the strategy $M = 1$ is the worst.

6   Conclusions

In this paper, we focus on a buffer-aided energy harvesting device-to-device communication networks. We describe the packet and energy packet Poisson arrival but the interactive leaving scenario. Then the system model is described as the Coupling Processor Queueing Model. Applying Quasi-Birth and Death method, we solve the steady-state distribution of the proposed Coupling Processor Queueing Model, and derive a few system performance parameter. The simulation verifies the accuracy of the theoretical derivation results.

References

1. Ansari, R.I., Chrysostomou, C., Hassan, S.A., et al.: 5G D2D networks: techniques, challenges, and future prospects. IEEE Syst. J. **PP**(99), 1–15 (2018)

2. Zhao, C., Wang, T., Yang, A., et al.: A heterogeneous virtual machines resource allocation scheme in slices architecture of 5G edge datacenter. Comput. Mater. Continua **61**(1), 423–437 (2019)
3. Zhang, H., Li, Y., Jin, D., et al.: Buffer-aided device-to-device communication: opportunities and challenges. IEEE Commun. Mag. **53**(12), 67–74 (2015)
4. Li, Y., Zhang, Z., Wang, H., et al.: SERS: social-aware energy-efficient relay selection in D2D communications. IEEE Trans. Veh. Technol. **PP**(99), 1 (2018)
5. Gui, J., Li, Z., Zeng, Z.: Improving energy-efficiency for resource allocation by relay-aided in-band D2D communications in C-RAN-based systems. IEEE Access **7**, 8358–8375 (2019)
6. Beeby, S.P., ODonnell, T.: Electromagnetic energy harvesting. In: Priya, S., Inman, D.J. (eds.) Energy Harvesting Technologies, pp. 129–161. Springer, Boston (2009). https://doi.org/10.1007/978-0-387-76464-1_5
7. Feng, L., Yang, Q., Kim, K., et al.: Two-timescale resource allocation for wireless powered D2D communications with self-interested nodes. IEEE Access **7**, 10857–10869 (2019)
8. Gupta, S., Zhang, R., Hanzo, L.: Energy harvesting aided device-to-device communication in the over-sailing heterogeneous two-tier downlink. IEEE Access **6**, 245–261 (2017)
9. Shanshan, Y., Liu, J., Zhang, X., Shangbin, W., et al.: Social-aware based secure relay selection in relay-assisted D2D communications. Comput. Mater. Continua **58**(2), 505–516 (2019)
10. Saleem, U., Jangsher, S., Qureshi, H.K., et al.: Joint subcarrier and power allocation in the energy-harvesting-aided D2D communication. IEEE Trans. Ind. Inf. **14**(6), 2608–2617 (2018)
11. Luo, Y., Hong, P., Su, R., et al.: Resource allocation for energy harvesting-powered D2D communication underlaying cellular networks. IEEE Trans. Veh. Technol. **66**(11), 10486–10498 (2017)
12. Xin, J., Zhu, Q., Liang, G.: Performance analysis of D2D underlying cellular networks based on dynamic priority queuing model. IEEE Access **7**, 27479–27489 (2019)
13. Yang, J., Ulukus, S.: Optimal packet scheduling in an energy harvesting communication system. IEEE Trans. Commun. **60**(1), 220–230 (2012)
14. Han, Y., Zheng, W., Wen, G., Chu, C., Jian, S., Zhang, Y.: Multi-rate polling: improve the performance of energy harvesting backscatter wireless networks. Comput. Mater. Continua **60**(2), 795–812 (2019)

Research on Traceability of Agricultural Products Supply Chain System Based on Blockchain and Internet of Things Technology

Hongyi Jiang[1], Xing Sun[2(✉)], and Xiaojun Li[1]

[1] Hebei Earthquake Agency, Shijiazhuang 050021, China
[2] School of Information Science and Engineering, Hebei University of Science and Technology, Shijiazhuang 050000, China
291948835@qq.com

Abstract. In this paper, the chain and the block chain thinking are introduced. Making use of the advantages of blockchain technology, this paper applies blockchain technology to the traditional agricultural product supply chain based on Internet of Things Technology, and designs the concept and reference framework of the agricultural product supply chain system based on blockchain and Internet of things technology. It effectively solves the problem of poor traceability in the agricultural product supply chain, and realizes the information management of the whole agricultural product supply chain. Through the combination of agricultural products and block chain, the data of the main bodies of the whole agricultural product supply chain are completely preserved, which ensures that the data is open, transparent, traceable and insurable, which makes the agricultural product supply chain system more automatic, more secure and reliable, and convenient for supervision and traceability. It is helpful to the quality and safety control of agricultural products.

Keywords: Blockchain · Internet of things · Supply chain · Traceability

1 Introduction

In recent years, food safety issues have been highly valued by people. Diet and health issues are closely related. Real-time supervision of the production and processing of agricultural products [1] can effectively control the abuse of pesticides, feed, additives and other problems by producers and operators. At present, the rapid development of Internet of Things, cloud computing, big data, artificial intelligence, radio frequency identification, and other technologies provide a comprehensive supervised and traceable technical basis for the whole process of agricultural product supply. More and more enterprises have established supply chain systems to ensure the sharing of information on agricultural products from production to processing to sales, and to achieve lean management levels. The agricultural product traceability system is a system that tracks

© Springer Nature Switzerland AG 2020
X. Sun et al. (Eds.): ICAIS 2020, LNCS 12239, pp. 707–718, 2020.
https://doi.org/10.1007/978-3-030-57884-8_62

agricultural products (including food, feed, etc.) into each stage of the market (the entire process from production to distribution), and is helpful to the quality and safety control of agricultural products.

The agricultural product supply chain based on the Internet of Things has developed relatively mature, to realize the supply chain informatization, in order to meet the business needs of each entity in the supply chain, to make the operation process and information system closely cooperate, so as to realize the seamless link between the information of the participants and each link process. It usually covers the whole business chain of agricultural products from production, processing, circulation, testing, sales to consumption, generally including production bases (farmers, farms or planting bases), processing and packaging links (processing workshops, wholesale markets or logistics centers) and sales links (farmers' markets, supermarkets, online stores or specialty stores). Promoting entities to form a community of interests through the supply chain is conducive to strengthening the power of accountability and supervision, improving the overall automated production efficiency, improving profits and reducing costs. However, in the current supply chain of agricultural products based on the Internet of Things, the lack of transparency in the process of production, sales, and consumption of agricultural products is easily falsified. The traceability ability is poor, the time-consuming and costly, and the security and reliability of traceability information can not be verified [3].

This paper takes blockchain technology as the starting point and uses blockchain technology and Internet of Things technology to solve the above-mentioned traceability problem [9]. The untamperable and traceable function of blockchain can effectively solve the problem of food traceability. On the basis of the Internet of things technology, combined with the block chain technology, the system designs and develops the agricultural product supply chain traceability system based on the blockchain and the Internet of things. The whole supply chain information of agricultural products is recorded completely and is transparent and open within the whole supply chain. In the case of problems with food quality and safety, it can quickly and effectively query production problems or processing links, recall products when necessary, and implement targeted punishment measures to improve the quality of agricultural products.

2   Research Status

The issue of agricultural product safety has become the focus of attention of all countries in the world. In order to grasp the information on the production and circulation of agricultural products, the traceability system of agricultural products has emerged as the times require. The research progress is as follows: T. Moe [6] believes that traceability is divided into two parts: internal traceability and external traceability, which mainly focuses on the recording and tracking of the information experienced by the commodity in the production, while the external traceability of the commodity is mainly on the ability of the commodity to trace the information or record except to the production link; Jansen Vullers [7] believes that the traceability is the ability to identify the materials consumed by the product and the relationship between them, identify the relationship between the upstream and downstream products, and determine the source of the product; Liddell & Bailey [8] distinguished the traceability, transparency and quality assurance of livestock traceability systems, and compared the livestock traceability systems of six major pork exporting and importing countries; Hobbs [4] and the like consider that the traceability system has three functions, One is the recall of foods with hidden safety hazards, which reduces public costs. The other is to clarify the responsible body of agricultural product safety accidents. The third is to reduce the information cost of consumers purchasing agricultural products; Yan Bo [10], etc. designed and developed five subsystems of breeding, processing, distribution, sales, and query management based on the RFID and the electronic code of the Internet of Things from the perspective of node companies, consumers and government regulators in the supply chain The tilapia supply chain traceability platform, with a detailed design of the platform's object name service and EPC information service, is used to realize the full tracking and tracing of aquatic products from breeding, processing, distribution to sales. Mario [2] proposed a wine product information traceability system, which stores all the storage and processing information of the whole supply chain of wine from the vineyard to the hands of consumers. With the help of this system, the whole process of the supply chain can be analyzed effectively and product quality management can be realized.

By analyzing and combing the existing domestic and foreign literature, this article believes that the Internet of Things and related technologies still have some shortcomings in the research and application of traceability of agricultural product supply chains. (1) The security of agricultural product information traceability system is low. The traditional agricultural product information traceability system adopts the central data storage mode. Once the central system is attacked, the product information will be leaked or even tampered with, so it is difficult to guarantee the security and reliability of the information. (2) Information sharing cannot be achieved in all links of the agricultural product supply chain. Although the traditional information traceability model can realize the collection and recording of agricultural product information, participants in each link of the supply chain cannot obtain product information equally and without discrimination, and there is a product information hierarchy. (3) The cost of agricultural product information traceability system and anti-counterfeiting technology is high. Advanced agricultural product information traceability system and anti-counterfeiting technology means that production enterprises not only need to build a large number of hardware infrastructure, but also need to invest a lot of manpower and material resources to maintain the system, which will increase the production cost of products.

Combined with the existing research results, this paper proposes an agricultural product supply chain system based on block chain and Internet of Things Technology, which uses block chain technology to realize the traceability function of agricultural product information, and then realizes the information management of the whole agricultural product supply chain.

3 Brief Introduction of Blockchain

Blockchain technology refers to the collective maintenance of a reliable and growing database by decentralized and distrustful means. Its characteristics are not easy to tamper, difficult to forge, traceable, block chain records all the transaction information, the process is efficient and transparent, and the data is highly secure [5]. It is more suitable for the agricultural product supply chain, which has multiple participants, and needs to establish an open, transparent and mutual trust mechanism.

3.1 Data Structure of Block Chain

Blockchain is a distributed database that divides data into different blocks. Each block is linked to the back of the previous block through specific information. Each newly generated block advances strictly in chronological order to form an irreversible chain. The algorithm used in blockchain is not a simple calculation problem, but a Hash algorithm. Hash is a classical technique in cryptography, which can be used to verify that no one tampered with the data content. Block chain includes block header and block. The block chain uses the hash value of the block as the unique identification of the block, which is stored in the "hash value of the previous block" field of the next block. By recording the hash value of the block, a data chain from the latest block to the first block is formed, which ensures that the supply chain system can trace the data of each subject at each time stage.

In addition, the "Hash value of data" field of the block records the hash value of each data of each subject in the time stage of the block, which ensures that each data of each subject in the supply chain system can be traced back, and through these two parts of the block chain, the data of the supply chain system can be traced back, which is convenient for accountability and supervision. The "root node hash value of the Merkle tree" field in the block is to iteratively calculate the hash value of the data in the block and record the final hash value to the block head. The fields of "hash value of previous block" and "root node hash value of Merkle tree" make use of the unidirectional and anti-conflict property of hash algorithm to ensure that every data can not be tampered with, and once tampering can be located quickly, the reliability and reliability of supply chain system data are improved. The data structure of a single block in the blockchain is shown in Fig. 1.

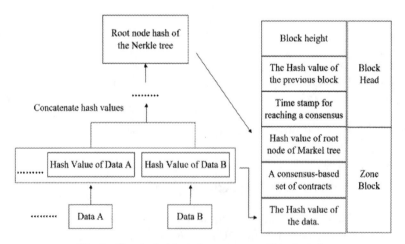

Fig. 1. The single block data structure of block chain

3.2 Application of Block Chain

Block chain 1.0 is represented by Bitcoin, which solves the problem of decentralized currency and payment means. Block chain 2.0 era stays at the conceptual level, block chain 3.0 era block chain application brings us more possibilities. At the application level, the blockchain has the following important characteristics:

(1) The blockchain does not have any central agency or central server, and all transactions occur in client applications installed on the computers or mobile phones of each subject in the blockchain;

(2) The contract or legal document loaded on the blockchain is an executable program. When the conditions are met, legal affairs will be automatically generated. This is the so-called "on-chain code", which is also called "smart contract" in Ethereum.

(3) Each node stores the same information while working independently and has the same rights to achieve point-to-point direct interaction;

(4) Block chain can be understood as a technical scheme of public accounting, the system is completely open and transparent and low cost, the books of account are open to the main platform of the supply chain, and the data and information are shared, and all the subjects in the block chain can trace to the source of the "account";

(5) Each newly generated block in the block chain is advanced strictly according to the time linear order, and the irreversibility of the time leads to any behavior that attempts to invade and tamper with the data information in the block chain is easy to be traced, resulting in the exclusion of other nodes, and the cost of counterfeiting is extremely high, which can limit the occurrence of such behaviors as tampering with the data.

3.3 Consensus Mechanism and Smart Contracts

In the blockchain system, consensus mechanism [13] and intelligent contract ensure the authenticity of data and contract execution, and realize "decentralized".

Consensus Mechanism. The consensus mechanism is the core of blockchain technology. In distributed computing, different computers reach a consensus through communication and exchange of information and act according to the same set of cooperative strategies. The consensus support ability of block chain includes the selection and application of the formula algorithm. A good consensus mechanism is the basis of the trust of each subject in the system. In the process of consensus formation, the relevant nodes will distribute the legitimacy of the transaction and agree with the consensus of other nodes.

The consensus mechanism is essentially a programmable protocol. Such a protocol is also called a consensus algorithm. The most well-known consensus algorithm is the Proof-of-Work Algorithm (PoW) adopted by Bitcoin, which can mathematically prove that the blockchain is safe and reliable when more than half of the participants adhere to the consensus mechanism.

Intelligent Contracts. The smart contract was put forward by Nick Szabo, a legal scholar. His definition is as follows: "an intelligent contract is a set of commitments defined in digital form, including agreements on which contract participants can implement these commitments." Generally speaking, an intelligent contract is a piece of computer program code that specifies in advance the rights and obligations between two or more participants, which are often executed not immediately, but when a series of conditions are met in the future.

Based on the requirements of scalability and applicability, supporting intelligent contracts should be the basic capability requirement of blockchain. Distributed computing in blockchain is embodied in the consensus process [15] and intelligent contract execution process. The process of intelligent contract execution is the distributed independent execution of the contract by the relevant nodes, and the implementation results are agreed upon. The intelligent contract is divided into two parts: contract generation and contract execution. Contract generation is responsible for submitting the intelligent contract code to the blockchain module for storage [11], and contract execution is responsible for running the intelligent contract code to realize the function of storing or querying data from the block chain module.

4   Architecture Design of Agricultural Product Supply Chain System Based on Block Chain and Internet of Things Technology

The system is based on the Internet of Things Technology, and the producers [14], processors, transporters, and sellers in the agricultural product supply chain are set as the nodes in the blockchain. The traceability system of agricultural product supply chain based on Internet of Things and block chain is constructed [12]. The system architecture is divided into four levels: perception layer, network layer, application support layer, and application layer. 1) The sensing layer adopts a state sensor, QR code, Radio Frequency Identification (RFID) and other Internet of Things Technology to realize the fast reading and writing function of information, which is the data source of the whole supply chain traceability system. 2) The network layer adopts point-to-point communication P2P, Distributed Internet technologies such as CDN, a connected storage NAS, content distribution network, provide the system with greater network carrying capacity and faster response speed and efficiency. 3) The application support layer is the core technical part of the whole system, including the Internet of things block chain support platform, which mainly provides the support capacity of storage, consensus, contract, and identity. 4) The application layer will cover the agricultural product supply chain from the production link, the information intelligence of processing link, distribution link to sales link, as well as the quality and safety management and traceability function of the system. The system architecture figure is shown in Fig. 2 below.

Perceptual layer: It provides the reading and writing function of agricultural product supply chain information. The main nodes in the chain of agricultural products supply chain are uploaded to the blockchain by Internet of Things Technology through state sensors, QR codes, Radio Frequency Identification (RFID) and other equipment, including the planting and feeding mode of agricultural products, the mode of processing and

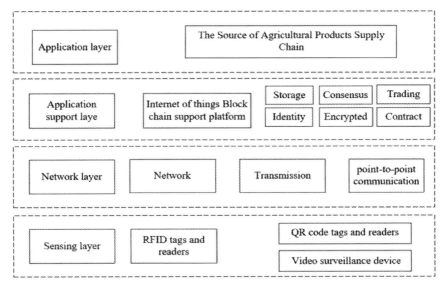

Fig. 2. System architecture diagram

additives, the place and time of transportation, the location, and storage mode of sales links, and so on. Only by reading the information accurately and quickly can we ensure the block building speed of the block chain and publish the blocks carrying the latest data information to all the main nodes in the chain. While uploading information, the main body in the chain can read all the information before agricultural products through RFID and other devices.

Network layer: P2P, NASN, CDN and other distributed Internet technologies used in the Internet of things are compatible with blockchain technology. Distributed Internet of things network can solve many of the problems mentioned above. The standardized point-to-point communication model is used to process the information between thousands of devices in the agricultural supply chain, which effectively reduces the cost, including the cost of deploying and maintaining large data centers, and the computing requirements and storage requirements can be decentralized through thousands of Internet of things devices. This will avoid the collapse of the whole network caused by the failure of one node, and provide the network foundation for the information transmission of the blockchain.

Application support layer: It mainly includes the Internet of things block chain support platform, which is the core technology of the system. The platform gives full play to the advantages of Internet of things and blockchain technology and realizes the complementary technical advantages. The Internet of Things can collect the raw material information of products, as well as information from various links such as processing, warehousing, logistics, and transactions, and feed it back to the blockchain to ensure the authenticity of the original data. The chain connection and its smart contract and consensus mechanism also ensure the traceability of the data and tamper-proof characteristics.

Application layer: The system can be applied to every link of the agricultural product supply chain. Here we focus on the traceability of the system. Traceability ability is embodied in quality and safety management and traceability in the agricultural product supply chain. The main purpose is to achieve two objectives 1) The traceability of agricultural supply chain nodes: In order to ensure the traceability management of the whole agricultural product supply chain, it is necessary to realize the production of agricultural products from the original production base. The traceability of each participating unit involved in the whole process of processing and circulation to the final consumer. 2) Traceability of quality and safety information of agricultural products: every key step or key procedure of each link in the process of the agricultural product supply chain may lead to food safety problems. Therefore, the traceability of each specific batch of agricultural products is another mainline to realize the traceability management of quality and safety. Product traceability mainly involves the traceability of the key indicators affecting product safety and the tracking and monitoring of the position of the product in the supply chain.

4.1   Technical Analysis of Application Support Layer

The application of the supporting layer is the key to the whole system, and the technology of the Internet of things and the technology of block chain are mainly completed. The traditional Internet of Things Technology provides timely and accurate data of various main bodies in the supply chain, such as processing time, processing materials, processing methods and the like in the processing link. The data information is transmitted to the block chain, a new block is established in the agricultural product information, and all the nodes in the chain are distributed so that each participant in the agricultural product supply chain shares the agricultural product information in real-time. The Internet of Things, block chain technology support platform model is shown in Fig. 3.

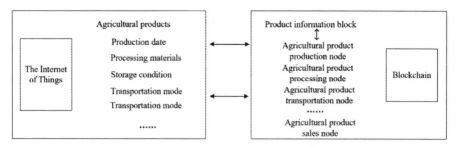

Fig. 3. Internet of things, Block chain Technology supporting platform Model

4.2   Analysis of Traceability Ability of Application Layer

To analyze the traceability ability of the agricultural product supply chain is to analyze the integrity and accuracy of the whole supply chain information record. The biggest difference between the system and the traditional agricultural product supply chain based on Internet of things technology is that the recording process of the whole system makes use of the consensus mechanism in the block chain, point-to-point transmission, hash algorithm and other technologies to ensure the consistency of the data, prevent usurpation, security, ensure that the data of each node is open and transparent to other nodes, and enhance the trust between the main bodies of the supply chain system. Therefore, from the point of view of blockchain technology, the traceability of the system is analyzed.

The recording process of the system is to store the newly generated data of each subject into the newly generated block and to ensure that the data of each main body of agricultural products as nodes are consistent. The specific process is as follows: 1) The system uses the consensus mechanism of blockchain to select the authorized node among the nodes of the block chain; 2) The new data generated by the nodes of each subject is transmitted to the authorized node through point-to-point transmission; 3) The authorized node generates a new block through the consensus mechanism; 4) The authorized node uses the hash algorithm to process the new data and adds the fields such as "previous block hash value" and "agreed timestamp" to fill the new block together. 5) The authorized node uses the point-to-point transmission mechanism to transmit the new block to the whole chain. 6) After each main node receives the new block, the hash algorithm is used to verify it, and after the verification is passed, the new block is added to the end of the chain of the existing block chain. The data recording process of the system from the perspective of blockchain is shown in Fig. 4.

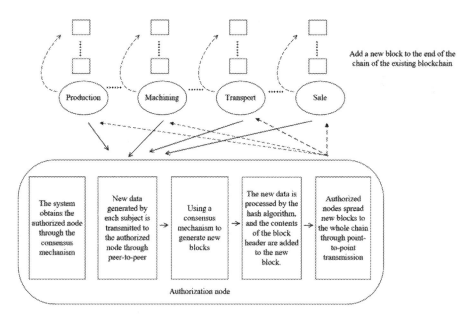

Fig. 4. Data recording flow of the system from the perspective of block chain

On the basis of the recording process, the process of tracing back to the source is relatively simple. The main body of the supply chain sends a query request, that is, the hash value query of the data recorded in the blockchain can be used to obtain the results. When there are quality and safety problems in the agricultural products purchased by consumers, consumers can quickly and accurately trace back the flow direction of all products related to the problem products, and the relevant enterprises can recall all the problem products quickly and accurately to avoid the further expansion of the harm.

5 Case Analysis

Whether it is the traceability of the participating units of the agricultural product supply chain or the traceability of the quality and safety information of the batch agricultural products, it must be based on the analysis of the specific agricultural product supply chain process. Taking chicken product supply chain system as an example, this paper introduces the model of chicken product information traceability. The specific process is as follows:

Step 1: The system is equipped with a chip for each live chicken on the farm, which will record the physical condition of each chicken from hatching to entering the slaughterhouse.
Step 2: The intelligent production line is used to collect and organize the product information and generate a QR code, which is marked on the product label.
Step 3: Upload the processing information of each batch of products and the corresponding QR code to the product information blockchain.
Step 4: The storage conditions and transportation process of chicken products are monitored in real-time by using devices such as the Internet of things sensors, and the information is passed into the block chain.
Step 5: Consumers buy chicken products through sales channels, and the relevant transaction records are kept in their own accounts and stored in the block chain.
Step 6: The consumer of the system provides the query client, while the main body of the supply chain maintains the core client to prevent the leakage of all the business information of the supply chain. Consumers who have query clients but have not purchased chicken products can query the production and logistics information of chicken products; consumers who have query clients and have purchased chicken products can not only query the production and logistics information of chicken products but also query the processing raw materials, processing methods and other information of chicken products.

Figure 5 shows the information traceability process based on the chicken product supply chain.

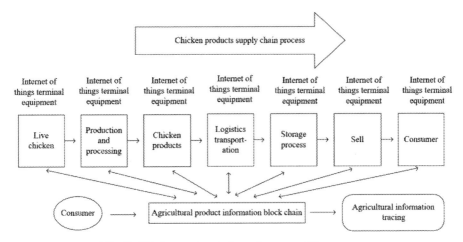

Fig. 5. Chicken product information traceability process

6 Conclusion

Through the above analysis, we conclude that the application of blockchain technology to the traditional agricultural product supply chain based on the Internet of things technology can effectively solve the problem of poor traceability in the agricultural product supply chain. The application of block chain and Internet of things to each link of the agricultural product supply chain can not only facilitate the comprehensive and detailed reading of all kinds of information into the management information system but also ensure the traceability of supply chain and the security and reliability of traceability information. The process of agricultural product supply chain mainly covers the basic links of production, processing, finished product distribution and finished product retail, as well as warehousing, transportation, loading and unloading, handling and other logistics activities running through each link and in the upstream and downstream circulation of the supply chain. The data information obtained from each activity link will be completely stored in the block chain.

Funding. This research was funded by the Science and Technology Support Plan Project of Hebei Province, grant number 17210803D.

This research was funded by the Science and Technology Support Plan Project of Hebei Province, grant number 19273703D.

This research was funded by Science and Technology Spark Project of the Hebei Seismological Bureau, grant number DZ20180402056.

This research was funded by Education Department of Hebei Province, grant number QN2018095.

References

1. Sun, X.: Study on the Design and Application of Fresh Agricultural Products Supply Chain Traceability System Based on Near Field Communication (NFC) Technology. College of Biological and Agricultural Engineering, Jilin University, Jilin (2016)

2. Cimino, Mario G.C.A., Marcelloni, F.: Enabling traceability in the wine supply Chain. In: Anastasi, G., Bellini, E., Di Nitto, E., Ghezzi, C., Tanca, L., Zimeo, E. (eds.) Methodologies and Technologies for Networked Enterprises. LNCS, vol. 7200, pp. 397–412. Springer, Heidelberg (2012). https://doi.org/10.1007/978-3-642-31739-2_20

3. Jiang, X., Liu, M.Z., Yang, C., Liu, Y.H., Wang, R.L.: A blockchain-based authentication protocol for WLAN mesh security access. Comput. Mater. Continua. **58**(1), 45–59 (2019)

4. Hobbs, J.E., Bailey, D., Dickinson, D.L., Haghiri, M.: Traceability in the Canadian red meat sector: do consumers care. Can. J. Agr. Econ. **53**(1), 47–65 (2005)

5. Aitzhan, N.Z., Svetinovic, D.: Security and privacy in decentralized energy trading through multi-signatures, block-chain and anonymous messaging streams. IEEE Trans. Dependable Secure Comput. **15**(5), 840–852 (2018)

6. Moe, T.: Perspectives on traceability in food manufacture. Trends Food Sci. Technol. **9**(5), 211–214 (1998)

7. Jansen, V.M.H., van Dorp, C.A., Beulens, A.J.M.: Managing traceability information in manufacture. Int. J. Inf. Manage. **23**(5), 395–413 (2003)

8. Liddell, S., Bailey, D.V.: Market opportunities and threats to the U.S. Pork industry posed by traceability system. Int. Food Agribusiness Manage. Rev. **4**(3), 287–302 (2001)

9. Ye, X.R., Shao, Q., Xiao, R.: A supply chain prototype system based on blockchain, smart contract and Internet of Things. Sci. Technol. Rev. **35**(23), 62–69 (2017)

10. Yan, B., Shi, P., Huang, G.W.: Development of traceability system of aquatic foods supply chain based on RFID and EPC Internet of Things. Trans. Chin. Soc. Agric. Eng. **29**(15), 172–183 (2013)

11. Deng, Z.L., Ren, Y.J., Liu, Y.P., Yin, X., Shen, Z.X., Kim, H.J.: Blockchain-based trusted electronic records preservation in cloud storage. Comput. Mater. Continua **58**(1), 135–151 (2019)

12. Song, R., Song, Y.B., Liu, Z.M., Tang, M., Zhou, K.: GaiaWorld: a novel blockchain system based on competitive PoS consensus mechanism. Comput. Mater. Continua **60**(3), 973–987 (2019)

13. Watanabe, H., Fujimura, S., Nakadaira, A., et al.: Blockchain contract: a complete consensus using blockchain. In: 2015 IEEE 4th Global Conference on Consumer Electronics (GCCE), Osaka, Japan, pp. 577–578. IEEE (2016)

14. Huh, S., Cho, S., Kim, S.: Managing IoT devices using blockchain platform. In: 2017 19th International Conference on Advanced Communication Technology (ICACT), Bongpyeong, South Korea, pp. 464–467. IEEE (2017)

15. Christidis, K., Devetsikiotis, M.: Blockchains and smart contracts for the Internet of Things. IEEE Access **4**, 2292–2303 (2016)

Design of Remote-Control Multi-view Monitor for Vehicle-Mounted Mobile Phone Based on IoT

Qian Ni[1], Meixia Ji[1], Ning Cao[2(✉)], Shouxu Du[1], and Wei Huo[2]

[1] Qingdao Binhai University, Qingdao 266555, China
[2] School of Internet of Things and Software Technology, Wuxi Vocational College of Science and Technology, Wuxi, China
ning.cao2008@hotmail.com

Abstract. In order to improve the deficiency of traditional vehicle mounted monitor and reduce the incidence of traffic accidents, a remote control multiple view monitor for the vehicle mounted mobile phone which is based on the technology of Internet of things is designed. The monitor includes the circuit board module, the camera module and the lift rod module. The monitor controls the operation of the multiple steering gears through the embedded circuit, and then controls the rotation of the lifting rod as well as the cloud deck which carries the camera; the router is used to generate the wireless signals which connect the mobile phone, and the mobile phone is used to send instructions and receive the collected signals. The monitor can realize the lifting and rotation of the camera through the remote control with the mobile phone, implementing acquisition of the scenes in multiple directions. With this monitor, it is possible to carry out the auxiliary observations while the vehicle is moving forward, backing up or perform the side parking or in the case that there is any high obstacle in front of the vehicle, and this will be helpful in avoid traffic accidents in an effective manner. This monitor can also be used for the household monitoring, the commercial monitoring as well as the monitoring in the public security system and the traffic system.

Keywords: Internet of Things · Remote control of mobile phone · Vehicle monitor system

1 Introduction

The Internet of things, being an information carrier of the Internet, the traditional telecommunications network and so on, is a network which enables all ordinary objects that can exercise independent functions to implement the interconnection. With the rapid development of the technology of Internet of things, it has found extensive applications in the transportation industry, such as GPS global positioning technology, the traffic road monitoring technology and the traffic recording technology (Faezipour et al. (2012)). At present, the traditional vehicle monitor (automobile data recorder) on the market has some shortcomings, such as incomplete monitoring and so on, which are mainly manifested in the following aspects:

© Springer Nature Switzerland AG 2020
X. Sun et al. (Eds.): ICAIS 2020, LNCS 12239, pp. 719–728, 2020.
https://doi.org/10.1007/978-3-030-57884-8_63

(1) The existing vehicle camera equipment can only shoot the scene in one direction of the car. Even if it can rotate, the rotation angle is very limited. For a vehicle which is performing backing up or parking, it has no auxiliary effect. So it is necessary to install additional cameras to provide assistance while the vehicle is performing backing up or parking (Yap and Chong (2018)).

(2) Generally, the existing vehicle monitor is fixed on the front windshield of the vehicle. When there is a barrier (such as a higher vehicle) which has blocked the line of sight, the camera can not capture the road condition in front of the vehicle and, in particular, the driver can not accurately judge the traffic lights and traffic conditions at the crossroads. This will cause some unnecessary traffic accidents (Li et al. (2013)).

(3) For the existing vehicle monitor, when the vehicle owner is not in the car, it can not monitor the surrounding situation of the vehicle (Yuan (2013)). The common practice is to read the recorded images through the memory card. As a result, it will be impossible to find any special situation around the vehicle in time.

In view of the shortcomings of the existing vehicle recorders in the market, a multiple view monitor remotely controlled by the vehicle mobile phone is designed by using the Internet technology. The main innovative functions of the monitor are as follows: The camera can perform 360° rotation in the horizontal position and 180° rotation in the vertical position as well as multiple angle shooting and feedback the images to the mobile phone under the remote control of the mobile phone APP software, so the auxiliary observation can be carried out when the vehicle is performing the forward moving, backing back and parking; in case that there is any barrier in front of the camera, the mobile phone can control the lifting rod in the monitor to drive the camera to rise, and the camera can rise above the barrier to photograph the road condition in front of the camera, so that the driver can observe the road condition in a better manner and effectively avoid the traffic; the router in the monitor can send the surrounding scenes to the user's mobile phone remotely and in real time through the network, and if any special circumstance has been found, it will be possible to use the mobile phone to remotely control the monitor to send out an alarm (Wang et al. (2014)).

2 The Overall Design

2.1 The Overall Structural Design

The multiple view monitor remotely controlled by the mobile phone is installed outside the vehicle roof, as shown in Fig. 1. It includes the circuit module, the camera module and the lift rod module. The main components of each module are as follows.

(1) The circuit board module includes the embedded control module, the router module and the power supply module; the core control module includes the embedded control module which controls all the modules to perform the unified and coordinated operation; the router module is used to receive the instructions which are sent by the mobile phone and transmit the video signals which are collected by the camera to

the terminal equipment; the power module supplies the electric power to the whole monitor system.

(2) In the camera module, the camera cloud deck is used to carry the camera. It enables the camera to reversion its angle. The cloud deck is fixed at one end of the lifting rod, and the camera is driven by the lifting rod to move up and down.

(3) The lifting rod module comprises two steering gears and one connecting rod. Moving up and down as well as rotating of the connecting rod are realized through the rotation of the steering gears.

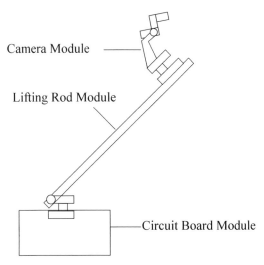

Fig. 1. A diagram of the overall structure of the multiple view monitor which is remotely controlled by the mobile phone

2.2 Network Structure Design

The design of the vehicle mounted multiple view monitor which is remotely controlled by the mobile phone combines the technology of Internet of Things (Li et al. (2013); Shi (2018)). The network structure of the vehicle mounted multiple view monitor which is remotely controlled by the mobile phone is as shown in Fig. 2, from bottom to top are the sensing layer, the network layer and the application layer, respectively. In the sensing layer, the videos of the scenes shot by the camera are transmitted to the embedded circuit for processing. In the network layer, the router in the monitor, through transmitting the wireless signals, transmits the video signals to the application layer or transmits the instructions from the application layer to the control circuit in the monitor; the monitor can be applied to the close range control as well as the remote monitoring, The mobile phone is connected with the free WiFi signal generated by the router at the time of the close range control, and the mobile phone can access the specified network address at the time of the remote monitoring so as to perform the remote monitoring on the surroundings of the equipment in real time (Gong et al. (2019)). In the application layer, the mobile phone APP software is used to receive the signals which are sent by the network layer or to transmit these signal to the monitor design through the network layer (Ding and Zhang (2014)).

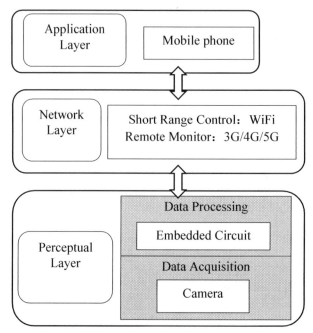

Fig. 2. A structure chart of Internet of Things applications of the vehicle mounted multiple view monitor which is remotely controlled by the mobile phone

3   The Detailed Design of the Camera Module and the Lifting Rod Module

3.1   Design of the Camera Module

The camera is carried by a cloud deck with two-degree of freedom, which is made up of the two miniature steering gears and their supports, and there are two types of steering gear, as shown in Fig. 3. The steering gears which can perform 360° rotation and the steering gears which can perform 180° rotation. Because the 360° steering gear can only control the direction and the speed, and the 180° steering gear can control its angle. So the 180° steering gear is selected in this design. The angle of the output shaft in the 180° steering gear changes from −90° to 90° in a linear manner. In the top part of Fig. 3, the steering wheel of the A steering gear is fixed on the side support, which drives the camera to perform 180° rotation up and down in the vertical direction; in the lower part, the steering wheel of the B steering gear is fixed on the base at one end of the lifting rod module, which drives the camera to perform 180° rotation rightward and leftward in the horizontal direction (Cao and Li (2019)).

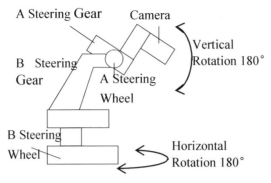

Fig. 3. A diagram of the structure of the camera module in the monitor

3.2 Design of the Lift Bar Module

With only the A steering gear and the B steering gear, it will be not possible for the camera in this invention to realize the shooting in the range of 180° in the rear, and as shown in Fig. 4. The multiple view monitor in this invention is provided with two steering gears at the lower end of the lifting rod module. The lifting and the rotation of the connecting rod are realized by the rotation of the steering gear. Wherein the steering wheel of the C steering gear is connected with the bottom end of the lifting rod to realize the lifting function of the camera; the D steering gear is embedded in the casing of the circuit board module, and the steering wheel of the D steering gear is fixed on the C steering gear and drive the C steering gear to perform rotation in the horizontal direction. The lifting rod in this invention is a hollow tube, and the connecting wire of the camera module passes through this tube and is connected to the circuit board module.

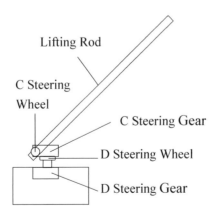

Fig. 4. A diagram of the structure of the lifting rod module in the monitor

3.3 The Operating Relationship

The invention mainly relies on the two steering gears in the lifting rod module and the two steering gears in the camera head unit, which enable the camera to have the multiple angle viewing. The operating relationship in each direction of the camera is as follows:

(1) The lifting rod is in the forward state in the horizontal direction

This state is applicable to observing the road condition and recording the driving condition while the vehicle is performing its normal driving operation. When the lifting rod is in the horizontal direction, it is possible to cause the shooting angle of the camera to be at 180° (rightward and leftward) in the horizontal direction through rotating the B steering gear on the cloud deck and to cause the shooting angle of the camera to be at 180° (upward and downward) in the vertical direction through rotating the A steering gear.

(2) The lifting rod is in the going up state

This state is applicable to the occasion where there is any obstacle in the front: the angle between the lifting rod and the horizontal line can be adjusted by means of the C steering gear; the camera will go up to the highest point when this angle is 90°, enabling the camera to perform 180° shooting in the front -> up -> rear direction through the A steering gear on the cloud deck.

(3) The lifting rod is in the backward state

The C steering gear is driven to rotate through the rotating of the steering wheel of the D steering gear, so that the camera on the lifting rod is driven to rotate by 180° from the front to the rear. In this state, it is possible to cause the camera to face towards the rear of the vehicle. When the camera is arranged at the top of the vehicle, it can assist the automobile to perform backing up. In addition, when the camera is in the backward state, it will be possible to the lift the lifting, so as to observe the scenes in multiple directions.

4 Detailed Design of the Core Control Module

4.1 Design of the Embedded Circuit

The core control module is controlled mainly by the embedded circuit, as shown in Fig. 5. The steering gear, the camera, the router, the alarm and the like are under the control of the core processor.

For this design, it is appropriate to employ the single chip microcomputer MEGA328P as the core processor. This is a high performance and low power 8-bit AVR single chip computer, which supports SPI, IIC and UART serial communication. It has 6 output ports for PWM signals, 8 channels of 10-bit ADC, and so on. The four PWM

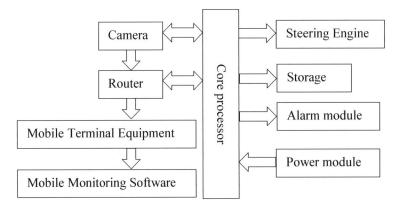

Fig. 5. A diagram of the structure of the core control module

(pulse width modulation) signal output ports of the MEGA328P single-chip microcomputer respectively control the A steering gear and the B steering gear in the camera module as well as the C steering gear and the D steering gear in the lifting rod module; the I/O port is used for data acquisition and control of the alarm module of the camera; the RXD/TXD port is used to connect to the router.

4.2 Design of the Steering Gear Drive Module

The rotation angle of the steering gear output shaft is controlled by the pulse signal with a period of 20 ms (that is, PWM signal). The pulse width of the pulse signal varies from 0.5 ms to 2.5 ms, and the angle of the steering gear output shaft changes accordingly from $-90°$ to $90°$ in a linear manner. Table 1 shows the relationship between the pulse width of the input signal and the output rotation angle of the steering gear.

Table 1. The congruent relationship between the pulse width of the input signal and the output rotation angle of the steering gear

The width of the input positive pulse	The number of the output rotation angle of the steering gear
0.5 ms	$-90°$
1.0 ms	$-45°$
1.5 ms	$0°$
2.0 ms	$45°$
2.5 ms	$90°$

The <servo.h> library of MEGA328P provides several functions for controlling of the steering gears. With these functions, it will be possible to control the angle of the steering gear with a simple program language. For example, the role of the function

attach() is to set the interfaces of the steering gear, and the number parameter of the interface should be added within the brackets of function at the time calling the function; the role of the function **write()** in is to set the rotation angle of the steering gear, and the degree parameter of the function should be added within the brackets at the time calling the function; the role of the function **read()** is to read the angle of the steering gear; the role of the function **attached()** is to determine whether the steering gear parameters have been sent to the steering gear interface.

4.3 Design of the Camera Control Module

For the camera, it is appropriate to adopt OV7670 sensor which is equipped with FIFO (First Input First Output) memory chip. OV7670 sensor has the built-in photosensory array, the timing generator, the AD converter, the analog signal processor, the digital signal processor and so on. FIFO is used as the data buffer, which can simplify the data acquisition process to a great extent. The process of image access is as follows:

(1) Storing of the image data (OV7670 writes data to FIFO)
    First, wait for synchronization signal from OV767, and then reset the write pointer of FIFO, at this point, FIFO write is enabled, and then wait for the second synchronization signal from OV767, and then disable the FIFO write. Through the above steps, the storage of an image is completed.
(2) Reading of the image data (MCU reads data from FIFO)
    First reset the read pointer of FIFO, then send a read clock (FIFO RCLK) to FIFO and read the first pixel high byte; then send a read clock (FIFO RCLK) to FIFO and read the first pixel low byte; once again send a read clock (FIFO RCLK) to the FIFO and read the second pixel high byte,... Do the loop for reading the remaining pixels (Zou et al. (2019)).

4.4 Design of the Camera Control Module

The monitor uses the TL-WR703N miniature 3G wireless router as the module for processing the WiFi signals and the remote monitoring. Based on IEEE 802.11n standard, it can expand the range of wireless network and provide stable transmission up to 150Mbps. At the same time, it is compatible with IEEE 802.11b and IEEE 802.11 g standard to convert 3G network into WiFi signal, so that an user who is using a smart phone with a GSM card which does not have the 3G function can share the 3G network through WiFi..

The monitor is mainly intended for usages in two occasions: the close range control and the remote monitoring.

(1) The control method for the close-range WiFi signal

The mobile phone is connected with the free WiFi signals which are generated by the router and sends the commands to the single-chip microcomputer by using the mobile phone APP software. The single-chip microcomputer controls the operation of the four steering gears, thereby enabling the camera to perform the multiple position acquisition of the scenes; at the same time, the shot scenes are fed back to the mobile phone through the WiFi signals.

The steps for setting the WiFi signals are as follows: 1) install the OpenWrt operating system to the router; 2) wait for the operating system to be successfully installed in the router and to start normally; 3) complete the configuration of the WiFi network and causethe WiFi signal to be sent out; 4) open the wireless signal switch of the mobile terminal of the mobile phone to search the name of the WiFi hot spot which has been set by the current router, and input the name of the current WiFi hot spot.

The router is equivalent to a microcomputer. After being connecting with the USB camera, the router can monitor the images of the vehicle. Connect the router to the single chip microcomputer, then the wireless WiFi connection can be realized through the mobile phone terminal equipment. The upper computer software of the mobile phone performs the control operations. After the control data is transmitted by WiFi and processed by the single chip microcomputer, the control of the camera is implemented. This control method does not need consuming the mobile phone traffic.

(2) The remote monitoring method

Set up the router to connect to the network at the specified address, and the mobile phone uses the network card to perform remote monitoring on the scenes around the equipment by accessing the Web pages (Hu et al. (2019)).

In the remote monitoring, the network card is inserted into the network card socket on TL-WR703N. After the corresponding settings are completed, the camera system can be directly connected to the network using the general network interface and the unique MAC address. The user can access the MAC of the camera at any time through the network, so the rotation of the camera and alarming of the alarm module will be under control. At the same time it will be possible to receive the real-time image information which is sent by this equipment. For example, set the MAC address to 192.168.1.108 and enter the URL http://192.168.1.108:8080/?action=stream in the viewer of the mobile phone or the computer, the user can see the remote shot videos.

5   Conclusion

The multiple view monitor is mainly used for the automobile monitoring. while the vehicle is performing the normal driving operation, it can have the function of an ordinary automobile data record and record the driving condition of the vehicle to the memory card. In case that there is any obstacle in the front or on the side, which blocks the line of sight, the lifting rod can be raised so as to observe the road conditions. At the time when the vehicle is perform backing up, the camera can be turned back for the purpose of observing the rear road conditions and assisting the driver to back up the vehicle. In the event that the driver is not in the vehicle, it will be possible to moved the device into the vehicle. The user can observe the scenes around the vehicle remotely through the network and cause the equipment to send out the alarm sound through the remote control. The multiple view monitor can be controlled by the mobile phone, so it is more convenient and simple to use.

This monitor can also be used for the household monitoring, the commercial monitoring as well as the monitoring in the public security system and the traffic system.

References

Faezipour, M., Nourani, M., et al.: Progress and challenges in intelligent vehicle area networks. Commun. ACM **36**, 90–100 (2012)

Yap, K.-L., Chong, Y.-W.: Software-defined networking techniques to improve mobile network connectivity: technical review. IETE Tech. Rev. **35**(3), 292–304 (2018)

Li, J., Zhang, Y., et al.: Mobile phone based WSN infrastructure for IoT over future internet architecture. In: IEEE International Conference on Green Computing and Communications and IEEE Internet of Things and IEEE Cyber, Physical and Social Computing (2013)

Yuan, C.: Design and implementation of an intelligent video surveillance system based on Android phone. In: Advanced Materials Research, vol. 816 (2013)

Wang, C., Rong, S.C., et al.: An intelligent video surveillance system for Android smart phone. In: Advanced Materials Research, vol. 850 (2014)

Li, N., Hao, X., et al.: Design and implementation of the smart home security control system based on the Android platform. In: Advanced Materials Research, vol. 753 (2013)

Shi, C.: A novel ensemble learning algorithm based on D-S evidence theory for IoT security. Comput. Mater. Continua **57**(3), 635–652 (2018)

Gong, S., Li, K., Tong, E., et al.: Smart factory solutions based on cellular industrial Internet of things. Chin. J. Internet Things **3**(02), 108–114 (2019)

Ding, M., Zhang, J.: Application of IoT technology in virtual instrument course. In: Advanced Materials Research on Electronics Education, vol. 43(11) (2014)

Cao, H., Li, Q.: Intelligent monitoring system for safe electricity use based on Internet of Things. Comput. Sci. Appl. **9**(08), 1591–1603 (2019)

Zou, J., Li, Z., Guo, Z., Hong, D.: Super-resolution reconstruction of images based on microarray camera. Comput. Mater. Continua **60**(1), 163–177 (2019)

Bing, H., Xiang, F., Fan, W., Liu, J., Sun, Z., Sun, Z.: Research on time synchronization method under arbitrary network delay in wireless sensor networks. Comput. Mater. Continua **61**(3), 1323–1344 (2019)

Super-Resolution Reconstruction of Electric Power Inspection Images Based on Very Deep Network Super Resolution

Bowen Shi[1] , Jilong Gao[1] , Zhentao He[2] , Tian Zhang[3] , Tie Zhong[1] ,
and Haipeng Chen[1](✉)

[1] Key Laboratory of Modern Power System Simulation and Control &
Renewable Energy Technology, Ministry of Education, Northeast Electric Power University,
Jilin City 132012, Jilin, China
haipeng0704@126.com
[2] College of Electrical Engineering, Northeast Electric Power University, Jilin City 132012,
Jilin, China
[3] School of Physics, Northeast Normal University, Changchun 130024, China

Abstract. Aiming at the problem of getting low resolution image and blurred image during unmanned aerial vehicle (UAV) inspection, a super-resolution reconstruction algorithm based on very deep network super resolution (VDSR) is proposed. The algorithm model is composed of deep convolution neural network and residual structure, which is improved on the basis of super resolution convolutional neural network (SRCNN). By deepening the network to 20 layers, the receptive field can be expanded, and the residual structure can be combined to obtain better reconstruction effect. The experimental results show that the proposed super-resolution method based on VDSR has richer texture and more realistic visual effect, with 2.95 dB and 0.037 improvement in PSNR and SSIM compared with the super-resolution methods based on Bicubic Interpolation, Sparse Coding and SRCNN. The proposed algorithm further promotes the theoretical research of inspection image super-resolution, and can be effectively applied to the practical application of power inspection.

Keywords: Deep learning · Super-resolution · VDSR · Inspection image

1 Introduction

In recent years, with the application upsurge of UAV [1] at home and abroad, UAV inspection technology has been widely concerned by major power grid companies. However, in the process of UAV inspection, due to some external factors, such as the disturbance of the UAV unable to maintain stable flight caused by the wind speed during the shooting, some blurred images will be taken, which will seriously affect the identification of electrical components in the inspection image, so it is necessary to preprocess the blurred images.

© Springer Nature Switzerland AG 2020
X. Sun et al. (Eds.): ICAIS 2020, LNCS 12239, pp. 729–738, 2020.
https://doi.org/10.1007/978-3-030-57884-8_64

Super-resolution reconstruction refers to the reconstruction of one or more low-resolution images (LR) by image processing algorithm to obtain the corresponding high-resolution images (HR). It has been widely used in the field of satellite remote sensing, medical image, video monitoring and image compression. At present, interpolation based, reconstruction based and learning based super-resolution reconstruction algorithms are the three main directions of super-resolution reconstruction. In recent years, experts and scholars have carried out in-depth research on super-resolution reconstruction technology. In reference [2], a super-resolution method based on bicubic interpolation is proposed. By finding out the influence factors of 16 pixels in the neighborhood on the pixels to be interpolated, the corresponding pixel values are finally obtained, and a better reconstruction effect is achieved. In reference [3], a super-resolution method based on sparse representation is proposed. By using joint dictionary learning, the dictionaries and sparse coefficients of high-resolution and low-resolution images are learned, and then the reconstructed images are obtained by gradient descent. In reference [4], a super resolution convolutional neural network (srcnn) method is proposed. This method uses three-layer network, and its functions are feature extraction, nonlinear mapping and reconstruction. Compared with the previous two traditional methods, this method has greatly improved the reconstruction effect. Although srcnn has achieved high performance in image super-resolution reconstruction, it still has limitations in the following aspects: (1) it depends on the context information of small image area; (2) the training convergence is not fast enough; (3) it can only be applied to single scale image.

In this paper, the very deep network super resolution (VDSR) algorithm is used to train the low-resolution image captured by UAV power inspection. The VDSR algorithm makes up for the above shortcomings of srcnn by building multi-layer neural network, applying residual learning, improving the learning rate, and improving the gradient descent method [5]. The experimental results show that the algorithm can quickly and effectively achieve the super-resolution reconstruction of low-resolution image. The reconstructed image has obvious details, clear edges, good visual effect, great improvement in image quality, and significant improvement in PSNR and SSIM.

2   Inspection Image Preprocessing

In order to generate the corresponding low-resolution image from the high-resolution image, 3×3 convolution is used to check the high-resolution clear image before the training to generate the low-resolution blurred image. There are many ways to judge whether the image is clear or blurred. In this paper, tenengrad gradient function is used to judge and classify the clear image and blurred image. Tenengrad gradient function uses Sobel operator to extract gradient values in horizontal and vertical directions respectively, which can highlight areas with fast density changes in the image, so it is commonly used in edge detection [6]. In image processing, it is generally believed that the image with good focus has a sharper edge, so with a larger gradient value, the image will be clearer; on the contrary, the more blurred the image, the smaller the gradient value. By setting a threshold value in the training process, if the gradient value of an image is greater than this threshold value, we can think it is clear, if the gradient value is less than this threshold value, we can think it is blurred, so as to realize the classification of clear image and blurred image.

3   Network Structure of VDSR

VDSR consists of 20 network layers, each of which is composed of convolution layer and nonlinear layer. The first layer is used to process the input image, and the last layer is used to reconstruct the image. The first layer and the last layer contain a 3 × 3 × 64 filter, that is, the 64 dimension feature image is converted into a dimension residual image; except for the first layer and the last layer, each layer is composed of 64 3 × 3 × 64 filters, that is, each filter has a 3 × 3 convolution operation on 64 channels of the input characteristic image [7].

VDSR is an improved network based on srcnn. First of all, because srcnn uses the traditional convolution method, the image size will be smaller and smaller through gradual convolution; second, if the traditional convolution operation is adopted, the boundary pixels will have poor results due to the small receptive field. The traditional method is usually to cut the convolution results to cut off the boundary area, so that the convolution output image size will become smaller after being cut. Therefore, in order to keep the size of the image, VDSR will fill the image with 0 before convolution operation, which solves the problem of network depth. A deeper network means a larger receptive field. VDSR uses a convolution kernel of 3 × 3, so that the network with a depth of D has a receptive field of (2D-1) × (2D-1), so that the output pixels can be inferred from more input pixels.

However, with the increase of network layer, layer weights also will increase, in the process of training a large number of parameters are calculated explosion could lead to a gradient, so the VDSR network structure, introduces the study of residual image is about to enter the low resolution of images compared with the reconstruction of the network output image to add again after the operation, finally get the high resolution image. Compared with SRCNN, VDSR greatly accelerates the convergence speed of the network and reduces the learning difficulty.

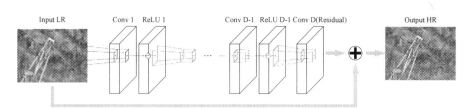

Fig. 1. Network structure of VDSR.

4   VDSR Network Training and Evaluation Index

4.1   Datasets

Since there is currently no large public data set of electrical components, this paper adopts the inspection images taken by uav as the training and test data set. The data set contains 630 images, including 480 images in the training set and 150 images in the test set. In order to avoid the problem of over-fitting that may occur in the network due to the small amount of data, some data in the training set are processed by different degrees of resize and expanded by horizontal or vertical flips. In this way, the constructed data set contains multiple sets of data with different magnification multiples and different directions. With a small amount of data, the diversity of training data can be guaranteed to the greatest extent, and the model can be fully trained to prevent overfitting, so as to obtain better super-resolution reconstruction results.

4.2 Some Key Network Parameters

Learning Rate. Due to the introduction of residual structure in the network, VDSR can use a higher learning rate for training to avoid a long training time. Although using a high learning rate can improve the convergence rate in training, simply setting a high learning rate can also lead to the disappearance of the gradient or explosion. Therefore, VDSR USES an adjustable gradient clipping method to maximize the training speed and suppress gradient expansion at the same time.

Weight Decay. L2 regularization aims to attenuate the weight to a smaller value and reduce the problem of overfitting the model to some extent. The formula is to add a regularization term after the cost function:

$$C = C_0 + \frac{\lambda}{2n} \sum_w w^2 \tag{1}$$

Where, C_0 represents the original cost function, and the latter term is the L2 regularization term, which is the sum of the ownership weight parameter w squared, divided by the sample size of the training set n. λ is the regular term coefficient, which weighs the regular term against the C_0 term.

Loss Function. Since the structure of the input and output images are basically similar, the residual image is defined as $r = y - x$, where x is a low-resolution image and y is a high-resolution image. Loss function is defined as:

$$Loss = \frac{1}{2} \|r - f(x)\|^2 \tag{2}$$

Where f(x) is network prediction [9]. The loss layer has three inputs: residual estimation, network input (ILR image) and baseline HR image.

The network needs to be initialized and corresponding parameters set before training. The network with a depth of 20 layers was used, the convolution kernel size was 3 × 3, the batch size was set as 64, the initial learning rate was set as 0.1, the momentum attenuation coefficient and the weight attenuation coefficient were set as 0.9 and 0.0001, respectively. The experimental training was 80 cycles, and the learning rate decreased by 10 times after 20 cycles.

4.3   Objective Evaluation Index of Image Quality

Peak Signal-to-Noise Ratio (PSNR). PSNR is an important indicator of image quality. The physical meaning is the ratio of the energy of peak signal to the average energy of noise. The unit is usually expressed as dB. PSNR is often used to represent the distortion degree of image. The calculation formula of PSNR is as follows:

$$PSNR = 10\log_{10}(\frac{MAX^2}{MSE}) = 10\log_{10}(\frac{MAX^2}{\frac{1}{M \times N}\sum\limits_{i=0}^{M}\sum\limits_{j=0}^{N}(xij - yij)}) \qquad (3)$$

Where, *MSE* is mean-square error, representing the mean-square error between the original image and the reconstructed image. *M* and *N* represent the length and width of the image respectively. x_{ij} and y_{ij} represent the pixels in the *i*th row and the *j*th column of the original image and the reconstructed image respectively. *MAX* is the maximum pixel value of the image. If each pixel is represented by 8-bit binary, then it is 255 [10]. In general, if the pixel value is represented by the B-bit binary, then $MAX = 2^B - 1$. According to formula (2), the smaller MSE is, the larger PSNR will be. The larger the PSNR, the better the quality of the reconstructed image.

Structural Similarity Index (SSIM). SSIM is an index to measure the similarity of two images. The value range is 0 to 1. The closer to 1, the better the similarity of two images is. SSIM measured luminance, contrast and structure, and used statistical parameters such as mean and variance to reflect the similarity between the two images [11]. If the original image is x and the reconstructed image is y, the representation of brightness, contrast and structure is as follows:

$$l(x, y) = \frac{2\mu_x\mu_y + C_1}{\mu_x^2 + \mu_y^2 + C_1} \qquad (4)$$

$$c(x, y) = \frac{2\sigma_x\sigma_y + C_2}{\sigma_x^2 + \sigma_y^2 + C_2} \qquad (5)$$

$$s(x, y) = \frac{\sigma_{xy} + C_3}{\sigma_x\sigma_y + C_3} \qquad (6)$$

$$SSIM(x, y) = [l(x, y)]^\alpha \cdot [c(x, y)]^\beta \cdot [s(x, y)]^\gamma \qquad (7)$$

Where, μ_x and μ_y represent the mean of *x* and *y* respectively; σ_x^2 and σ_y^2 represent the variance of *x* and *y* respectively; σ_{xy} represents the covariance of *x* and *y*; $C_1 = (k_1 MAX)^2$ and $C_2 = (k_2 MAX)^2$ are two constants to avoid dividing by zero. *MAX* is the maximum pixel value, k1 = 0.01 and k2 = 0.03 are the default values.

In order to make the expression more convenient, formula (8) can be obtained by generally taking $\alpha = \beta = \gamma = 1$.

$$SSIM\,(x, y) = \frac{(2\mu_x\mu_y + C_1)(2\sigma_{xy} + C_2)}{(\mu_x^2 + \mu_x^2 + C_1)(\sigma_x^2 + \sigma_x^2 + C_2)} \tag{8}$$

In addition, since the input image is grayscale when calculating SSIM, it is necessary to convert the color image to grayscale before the calculation. During each calculation, an N × N window was taken from the picture, and then the window was continuously moved for calculation. Finally, the average value was taken as the global SSIM.

5 Experimental Simulation and Result Analysis

5.1 Experimental Environment

The experimental environment is basically configured as follows: Intel (R) Xeon (R) CPU e5-8276, main frequency 4.0 GHz, NVIDIA Tesla K80 as the graphics card, dual GPU accelerator, 24 GB GDDR4 memory, 128 GB memory. The experimental platform is Ubuntu 16.04 operating system, and the algorithm programming is implemented by Tensorflow deep learning framework.

5.2 Experimental Design and Result Analysis

In order to verify the super-resolution reconstruction effect of VDSR algorithm, Bicubic, sparse coding (SC) algorithm and SRCNN algorithm were selected for simulation experiments. Four representative and richly textured inspection images were selected for comparison. The experimental results are shown in Fig. 2.

In order to objectively display the super-resolution reconstruction effect of VDSR algorithm and the algorithms compared with it [12], PSNR, SSIM and reconstruction time were compared for the reconstructed images of the four algorithms. The test results are shown in Table 1.

It can be seen from Table 1 that the PSNR and SSIM of the four images processed by VDSR algorithm are all higher than Bicubic, SC and SRCNN algorithms. However, because the network depth of the VDSR algorithm model is relatively deep, the reconstruction time is slightly longer than [13] other algorithms.

In order to display the reconstruction effect of VDSR algorithm and other three algorithms more intuitively, this paper obtained the reconstruction effect comparison diagram of each algorithm model in the training process of algorithm model. Comparison of reconstruction effects is shown in Fig. 2. The horizontal axis is Backprops and the vertical axis is PSNR.

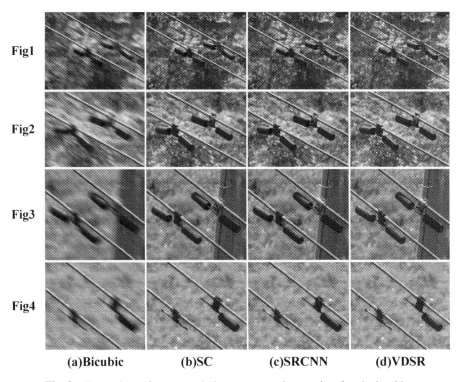

Fig1

Fig2

Fig3

Fig4

(a)Bicubic   **(b)SC**   **(c)SRCNN**   **(d)VDSR**

Fig. 2. Comparison of super resolution reconstruction results of each algorithm.

Table 1. Table captions should be placed above the tables.

Figures	Algorithms	PSNR/dB	SSIM	Time/s
Figure 1	Bicubic	29.25	0.7022	0.10
	SC	30.32	0.7163	0.13
	SRCNN	31.95	0.8653	0.30
	VDSR	34.87	0.8933	0.32
Figure 2	Bicubic	29.56	0.7122	0.13
	SC	30.69	0.7455	0.15
	SRCNN	32.27	0.8548	0.33
	VDSR	35.12	0.8743	0.40
Figure 3	Bicubic	30.12	0.7442	0.11
	SC	30.95	0.7436	0.14
	SRCNN	32.45	0.8333	0.32
	VDSR	35.38	0.8799	0.38
Figure 4	Bicubic	29.67	0.7233	0.13
	SC	30.51	0.7566	0.15
	SRCNN	32.16	0.8323	0.34
	VDSR	35.24	0.8875	0.36

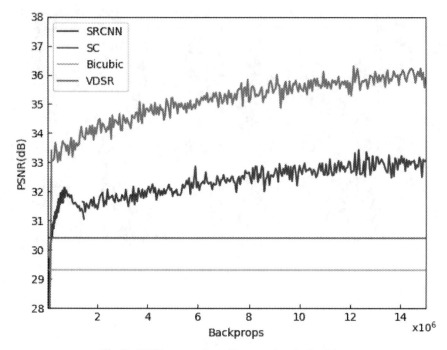

Fig. 3. PSNR comparison diagram of each algorithm.

As can be seen from Fig. 3, with the increase of the number of training iterations, the reconstruction effect of VDSR algorithm and SRCNN algorithm is gradually improved. In addition, compared with the traditional algorithm and SRCNN, the PSNR of the VDSR algorithm has a more significant improvement. Compared with the SRCNN algorithm, the PSNR is improved by 1–2 dB, which is 3–5 dB higher than the traditional algorithm [14].

In order to verify the relationship between the number of iterations and the reconstruction effect, the reconstruction images when the number of iterations was 4000, 8000 and 12000 were respectively shown, as shown in Fig. 4. It can be seen that with the increase of the number of iterations, the reconstruction effect is gradually improved.

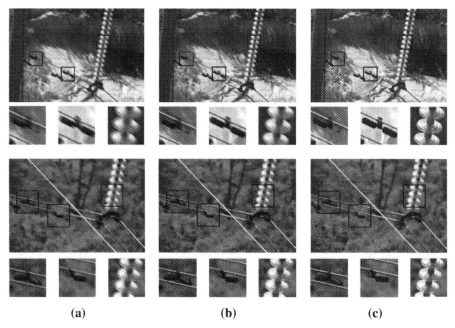

(a)　　　　　　　　**(b)**　　　　　　　　**(c)**

Fig. 4. Super resolution reconstruction process using VDSR. (From left to right: (a) 4000 iterations; (b) 8000 iterations; (c) 12000 iterations)

6 Conclusion

Super resolution reconstruction of power inspection image is of great significance to power inspection. In order to solve the problem of blurred image, this paper proposes a super resolution reconstruction algorithm of inspection image based on VDSR. By applying VDSR model to power inspection, the blurred image can be reconstructed into a clear one. The experimental results show that the image reconstruction based on VDSR algorithm has a very small distortion degree and is very consistent with the original image, and it is proved that the image reconstruction effect of VDSR algorithm is significantly better than that of traditional algorithm and SRCNN algorithm, which provides a certain theoretical basis for the intelligent inspection and identification of electrical components.

Acknowledgements. National Natural Science Foundation of China (Grant No. 11905028).

References

1. Hu, J., Li, L., Xie, Q., Zhang, D.: A novel segmentation approach for glass insulators in aerial images. J. Northeast Electr. Power Univ. **38**(2), 87–92 (2018)
2. Xing, X., Sun, Q., Zhang, P., Li, M.: Research on distribution network fault recovery and reconstruction based on deepfirst search and colony algorithms. J. Northeast Electr. Power Univ. **39**(3), 38–43 (2019)
3. Lin, C., Tong, H., Yu, W., Xin, Y., Zhang, R.: Distribution grid intelligent state monitoring and fault handling platform based on ubiquitous power internet of things. J. Northeast Electr. Power Univ. **39**(4), 1–4 (2019)
4. Yang, M., Huang, B., Huang, B., Lin, S.: Real-time prediction for wind power based on Kalman filter and support vector mahines. J. Northeast Electr. Power Univ. **37**(2), 45–51 (2017)
5. Li, P., Shi, Z., Liu, X.: Application of fuzzy C-means clustering algorithm based on immune genetic algorithm. J. Northeast Electr. Power Univ. **38**(3), 79–83 (2018)
6. Pech-Pacheco, J., Cristóbal, G., Chamorro-Martinez, J., Fernandez-Valdivia, J.: Diatom autofocusing in brightfield microscopy: a comparative study. In: Proceedings 15th International Conference on Pattern Recognition, vol. 3, pp. 314–317. IEEE, Barcelona (2000)
7. Tai, Y., Yang, J., Liu, X.: Image super-resolution via deep recursive residual network. In: Proceedings of the IEEE Conference on Computer Vision and Pattern Recognition, pp. 3147–3155 (2017)
8. Krogh, A., Hertz, J.: A simple weight decay can improve generalization. In: Advances in Neural Information Processing Systems, pp. 950–957 (1992)
9. Moist, L., Port, F., Orzol, S.: Predictors of loss of residual renal function among new dialysis patients. J. Am. Soc. Nephrol. **11**(3), 556–564 (2000)
10. Huynh-Thu, Q., Ghanbari, M.: Scope of validity of PSNR in image/video quality assessment. Electron. Lett. **44**(13), 800–801 (2008)
11. Hore, A., Ziou, D.: Image quality metrics: PSNR vs. SSIM. In: 2010 20th International Conference on Pattern Recognition, pp. 2366–2369. IEEE, Istanbul (2010)
12. Xiao, D., Liang, J., Ma, Q., Xiang, Y., Zhang, Y.: Capacity data hiding in encrypted image based on compressive sensing for nonequivalent resources. Comput. Mater. Continua **58**(1), 1–13 (2019)
13. Luo, M., Wang, K., Cai, Z., Liu, A., Li, Y., Cheang, C.: Using imbalanced triangle synthetic data for machine learning anomaly detection. Comput. Mater. Continua **58**(1), 15–26 (2019)
14. Xia, Z., Lu, L., Qiu, T., Shim, H., Chen, X., Jeon, B.: A privacy-preserving image retrieval based on ac-coefficients and color histograms in cloud environment. Comput. Mater. Continua **58**(1), 27–43 (2019)

Single Color Image Dehazing Based on Vese-Osher Model and Dark Channel Prior Algorithm

Wei Weibo[⊠], Zhao Hui, and Gao Zhuzhu

Qingdao University, Qingdao 266071, China
njustwwb@163.com

Abstract. A new image dehazing method based on Vese-Osher model and dark channel prior algorithm is proposed in this paper. Vese-Osher model is used to decompose a hazy image into texture part and structure part. According to many experiment results, it can be seen that the haze is mainly distributed in structure part. Dark channel prior algorithm is used to dehaze structure part, and then fuse the dehazed structure part and orginal texture part to obtain final haze-free image. The simulation results show that the proposed method could get clear natural images and maintain edges and texture features. The evaluation indicators are also better.

Keywords: Vese-Osher model · Structure part · Texture part · Dark channel prior

1 Introduction

Due to the scattering effect of tiny particles in the atmosphere in hazy or dusty weather conditions, contrast of outdoor images is usually reduced. Color and saturation are also reduced. It is also a major influence to the reliability of many vision understanding applications, such as video surveillance, intelligent vehicles, satellite imaging, and aerial imagery [1–3]. Therefore, it is of great significance for the application of computer vision and image processing to study how to obtain clear haze-free images in hazy conditions.

At present, image dehazing methods could be divided into two categories, which are enhancement methods based on image processing and restoration methods based on physical model. Enhancement methods can improve perceptual quality of scene appearance [4, 5], enhance degraded images, improve image quality, and achieve a better effect of thin haze removal [6, 7]. However, these methods do not consider about the differences of haze thickness. Since there is no objective reason for the formation of haze, fundamentally, it is impossible to achieve dehazing. To tackle this problem, restoration methods based on physical model establish atmospheric scattering model [8, 9]. By analyzing scattering effect of atmospheric suspended particles on light, these methods can recover original image according to physical mechanism of image degradation. They could get an obvious and natural dehazing effect on hazy images under different conditions and maintain detail features. Tan [10], Fattal [11], Tarel [12], He [13] are representative

© Springer Nature Switzerland AG 2020
X. Sun et al. (Eds.): ICAIS 2020, LNCS 12239, pp. 739–750, 2020.
https://doi.org/10.1007/978-3-030-57884-8_65

algorithms. In particular, dark channel prior(DCP) algorithm [13] has become one of the most popular and successful physics-based dehazing methods [14–16]. It assumed that most local non-sky patches in haze-free images have pixel values close to zero. DCP-based dehazing performance has been further improved [17, 18]. However, sometimes, texture details of these methods are not effective particularly.

Du et al. [19] first decomposed high-resolution satellite images into different spatial layers, then utilized the frequency characteristics to achieve dehazing. [20] proposed a method to divide images into low frequency part and high frequency part [21, 22] in the frequency domain, then dehaze in low frequency part and denoise in high frequency part. The proposed approach aims to significantly increase perceptual visibility of haze scene. However, this method used wavelet transform to transform images from spatial domain to frequency domain. The process is relatively complex.

Fattal [11] assumed that transmittance of local area of the image was a constant vector, and chromaticity of object surface had a statistical irrelevance with medium propagation. This method required a large amount of physical color information. However, under the condition of dense haze, the image lost a large amount of color information. Gibson [1] used median filter to estimate atmospheric scattering function, achieving image restoration. However, this method is prone to edge information loss and dehazing image distortion. Nishino [23] utilized bayesian posterior probability model to fully excavate potential statistical features in the image for dehazing. This method can better deal with dense fog, but when dealing with thin haze, its color is too bright and lacks a sense of reality. Kratz [24] adopted a factorial MRF to model a hazy image and the aesthetically pleasing results can be obtained. However, this method often tended to underestimate transmission and thus induced some halo artifacts. Tan [10] assumed that local ambient light is constant, and contrast intensity is significantly enhanced, in the framework of Markov random field model, structure of edge strength cost function, using graph theory to estimate optimal segmentation illumination, but resulting image color distortion of supersaturation is inevitable.

Tarel [12] assumed that the atmospheric dissipation function changed slowly locally, so median filter was used to estimate medium transmission coefficient. However, hazing effect was not noticeable. Ancuti [25] put forward a dehazing algorithm based on image fusion. Image fusion of the two input images are derived from original image, the method also eliminated halo effect through multiscale processing, however, the method have appeared on the tower image texture details loss phenomenon. Kopf [26] obtained the depth information based on the known 3D texture of the scene. This kind of method could effectively restore the image, but texture details of the obtained dehazing image are also lost. DEFADE [27] described a prediction model of perceptual haze density called FADE and a perceptual image dehazing algorithm, both based on image NSS and haze aware statistical features. However, this method may also fail to maintain texture details effectively.

In this paper, we uses Vese-Osher (VO) decomposition model [28, 29] to decompose hazy images into low-frequency structure part and high- frequency texture part. And structure part is used to be dehazed. Then, the dehazed structure part and texture part are fused to obtain a final haze-free picture. Finally, the vailidity of proposed method is proved in experiment results.

2 Background and Observation

2.1 Dark Channel Prior

Based on McCarteny's scattering model and ambient light model, Schechner et al. [30] proposed a monochromatic atmospheric scattering model to simulate image degradation process under hazy conditions, as follows:

$$f(x) = u(x) \cdot t(x) + A(1 - t(x)) \tag{1}$$

Where, $x = (x, y)$ stands for index pixel position, f is the image to be dehazed; u is the image after dehazing; A is atmospheric light intensity, represents light intensity of different environments in the image, and its value can be obtained by virtue of dark channel prior knowledge. t is atmospheric transmission function, namely transmission rate, which reflects the depth of field information of the scene object $t(x) = e^{-\beta d(x)}$; β is atmospheric scattering coefficient; $d(x)$ is the scene depth.

Based on Eq. (1), the goal of image dehazing is to recover $u(x)$. This requires us to estimate transmission function $t(x)$ and global atomospheric light A. Once A and $t(x)$ are known, u could be well recovered by inverting atmospheric scattering model:

$$u(x) = \frac{f(x) - A}{\max(t(x), t_0)} + A \tag{2}$$

Where t_0 is a constant set to avoid divisor being zero, generally 0.1

2.2 Vese-Osher Model

Meyer [28] established oscillation function modeling theory of texture image on the basis of TV model, and took oscillation function space (G space) dual to the function space of bounded variation BV function space as function space of texture image, and put forward E space and F space. However, Meyer did not give implementation method. Vese and Osher [28, 29] combined Meyer's theory of oscillation function modeling and TV model, then used L^P approximate $\|\cdot\|_G$ norm to propose a new structure-texture decomposition model (Vese-Osher model). Energy functional of VO model of vector image is as follows:

$$G_p(u, g_1, g_2) = \int_\Omega |\nabla u| dx dy + \lambda \int_\Omega |f - u - \nabla \cdot \vec{g}|^2 dx dy + \mu \left[\int_\Omega \left(\sqrt{g_1^2 + g_2^2} \right)^p dx dy \right]^{\frac{1}{p}} \tag{3}$$

Where, $\lambda, \mu > 0$ is penalty parameter, and $p \to \infty$. In formula (3), the first term on the right side of the equation ensures that $\mu \in BV(R^2)$; The second term guarantees that $f \approx u + div\xi$ and $g = (g_1, g_2)$; The third term is penalty term for the norm of function v in G space

2.3 Observations of Haze Analysis in Frequency Domain

In general, surface reflections and light changing will cause rapid spatial change, while airlight is modeled by slow spatial variation corresponds to atmospheric scattering. Spatial variation of haze distribution will be very slow and smooth because the haze is mainly produced by atmospheric scattering and spreads widely. Concerning image processing in the frequency domain, it is reasonable to assume that the haze almost exists in the low-frequency component.

Figure 1 is original hazy images, texture part and structure part decomposed by VO decomposition model (Because grayscale value of texture image is too small to display clearly, brightness enhancement processing is carried out on it, that is, brightness of whole image increases by 150 uniformly, making texture part relatively obvious). It can be clearly seen that the haze is mainly distributed in the structure part. Therefore, it is quite natural to remove the haze in the structure part.

Fig. 1. VO decomposition model applied to hazy images. The first line is hazy images, the second line is corresponding structure part, and the third line is texture part.

3 Proposed Dehazing Method

3.1 VO Model of Color Hazy Image Decomposition and Its Split Bregman Method

Meyer [28] proposed to represent spatial function of v by G space dual to BV space in a certain sense, to better decompose the u part of BV space and the v part of oscillation function space from image f, Where u represents structure part and v represents texture part. In general, G space proposed by Meyer [28] can be defined as:

$$G = \{v_i | v_i = \partial_x g_{i,1}(x, y) + \partial_y g_{i,2}(x, y), g_{i,1}, g_{i,2} \in L^\infty(\Omega)\} \tag{4}$$

$$\|v_i\|_* = \inf_{g=(g_{i,1}, g_{i,2})} \{\sqrt{g_{i,1}^2 + g_{i,2}^2}_{L^\infty} | v_i = \partial_x g_{i,1} + \partial_y g_{i,2}\} \tag{5}$$

VO model of color hazing image is shown as follows:

$$G_p(u, g) = \int_\Omega \sqrt{\sum_{i=1}^n |\nabla u_i|^2} dxdy + \lambda \int_\Omega \sum_{i=1}^n |f_i - u_i - \nabla \cdot \vec{g}_i|^2 dxdy$$

$$+ \mu \left[\int_\Omega \sum_{i=1}^n \left(\sqrt{|g_{i,1}|^2 + |g_{i,2}|^2} \right)^p dxdy \right]^{\frac{1}{p}} \tag{6}$$

Where, $\lambda, \mu > 0$ is penalty parameter, and $p \to \infty$, $u = (u_1, u_2, u_3)$, $g = (g_{i,1}, g_{i,2})$ where $i = 1, 2, 3$. For Eq. (6), variational method is used to find the minimum value of its energy functional, and Euler-Lagrange equation for solving u is obtained as follows:

$$u_i = f_i - \partial_x g_{i,1} - \partial_y g_{i,2} + \frac{1}{2\lambda} div \left(\frac{\nabla u_i}{|\nabla u|} \right) \tag{7}$$

The solution of $div \left(\frac{\nabla u_i}{|\nabla u|} \right)$ in Eq. (7) is complex and inefficient. To solve this problem, Split Bregman method for solving minimum problem is designed. By introducing auxiliary variable $w = (w_1, w_2, w_3)^T$, and $w \approx \nabla u$, and Bregman's iteration parameter $b = (b_1, b_2, b_3)^T$, the model is as follow:

$$G_p(u, g) = \int_\Omega \sqrt{\sum_{i=1}^3 |w_i|^2} dxdy + \lambda \int_\Omega \sum_{i=1}^3 |f_i - u_i - \nabla \cdot \vec{g}_i|^2 dxdy$$

$$+ \mu \left[\int_\Omega \left(\sqrt{|g_1|^2 + |g_2|^2 + |g_3|^2} \right)^p dxdy \right]^{\frac{1}{p}}$$

$$+ \frac{1}{2\theta} \int_\Omega \sum_{i=1}^3 \left(w_i - \nabla u_i - b_i^{k+1} \right)^2 dxdy \tag{8}$$

$$b_i^{n+1} = b_i^n + \nabla u_i^n - w_i^n \tag{9}$$

Experimental data of Vese and Osher proved that the value of parameter P within the range of [1, 10] is not particularly significant for image decomposition. Compared with other values, iterative operation is faster when P is 1. The following is the idea of alternating iteration when $P = 1$. First fix w to solve u, then fix u to solve w, and corresponding solution is:

$$u_i = f_i - \partial_x g_{i,1} - \partial_y g_{i,2} + \frac{1}{2\lambda\theta} \left(\Delta u_i + \nabla \cdot b_i^{n+1} - \nabla \cdot w_i \right) \tag{10}$$

$$w_i^{n+1} = \left(\nabla u_i + b_i^{n+1} \right) - \theta \frac{w_i^{k+1}}{\sqrt{\sum_{i=1}^n |w_i^{k+1}|^2}} \tag{11}$$

The analytical solution can be obtained through generalized soft threshold formula:

$$w^{n+1} = Max \left(\left| \nabla u_i + b_i^{n+1} \right| - \theta, 0 \right) \frac{\nabla u_i + b_i^{n+1}}{|\nabla u + b^{n+1}|}, 0\frac{0}{0} = 0 \tag{12}$$

$$\mu \frac{g_{i,1}}{\sqrt{g_{i,1}^2 + g_{i,2}^2}} = 2\lambda \left[\frac{\partial}{\partial x}(u_i - f_i) + \partial_{xx}^2 g_{i,1} + \partial_{xy}^2 g_{i,2} \right] \tag{13}$$

$$\mu \frac{g_{i,2}}{\sqrt{g_{i,1}^2 + g_{i,2}^2}} = 2\lambda \left[\frac{\partial}{\partial x}(u_i - f_i) + \partial_{xy}^2 g_{i,1} + \partial_{yy}^2 g_{i,2} \right] \tag{14}$$

Split Bregman method can be applied to VO model to achieve simple solution and fast calculation speed. It can better extract texture part and structure part from f at the same time.

3.2 Structure Part Dehazing and Image Fusion

In this paper, guidance filtering algorithm [31] is adopted to refine transmittance t in Eq. (2). This method mainly relies on simple box fuzzy effect, which is fast and practical.

Equation (1) is deformed, and Eq. (15) is dehazing process. After dehazing structure part through Eq. (2), texture part and structure part need to be fused to obtain final dehazing image. Image dehazing is not a linear process, so processed structure part and texture part cannot be simply superimposed [32]. Denotes dehazing process of Eq. (15) as operator $F(\cdot)$,

$$u(x) = \frac{f(x)}{t(x)} + A(1 - \frac{1}{t(x)}) \tag{15}$$

Then $F(f(x)) = F(f(x)_s + f(x)_T)$. f is a hazy image, f_s is structure part, and f_T is texture part. Since the proportion of texture part in original image with haze is much smaller than that of structure part, according to Taylor equation:

$$
\begin{aligned}
u(x) = F(f(x)) &= F(f(x)_S + f(x)_T) \\
&\approx F(f(x)_S) + F(f(x)_S)' f(x)_T \\
&= S(x) + \frac{1}{t(x)} f(x)_T
\end{aligned}
\tag{16}
$$

Where, $S(x)$ represents structure part after dehazing, and $F(\cdot)$ represents the first derivative of operator $F(\cdot)$. According to Eq. (16), texture part $f(x)_T$ and structure part $f(x)_S$ after dehazing can be fused into final dehazing image $u(x)$. Flow chart of dehazing method in this paper is shown in Fig. 2.

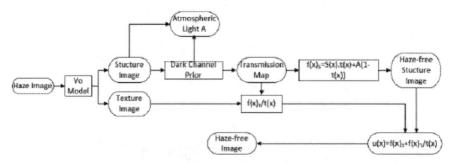

Fig. 2. Flowchart of the proposed dehazing method

4 Experimental Results

In this paper, Matlab (R2014a) is used to carry out simulation experiments on the proposed method. The computer is configured with Intel(R) Core(TM) i5-4590 CPU @3.30 ghz and 4 GB RAM. A large number of hazy images from the LIVE Image Defogging database [33] are tested in this paper. To evaluate the performance of dehazing, this paper first considered a large number of comparative experiments with He [31], and then compared with several other typical methods, simply denoted as Fattal [11], Gibson [1], Nishino [23], Kratz [24], Tan [10], Tarel [12], Aucti [25], Kopf [26], DEFADE [27].

4.1 Dehazing Performance

(1) Compared with He [31] methods. Since the method in this paper is based on He [31], a lot of experiments has been done to compare with dehazing effect of dark channel prior [31]. Figure 3 is the dehazing comparison of He [31] and our method.

Fig. 3. Representative image dehazing results (Building, Tian'anmen, TATTOO)

According to Fig. 3, it can be clearly seen that our dehazing effect is cleaner and more thorough than that of He [31]. For example, in the red frame of the building, windows of the Building are clearer and brighter. In the Tian'anmen square, no matter the trees in the distance, the vegetation nearby or the roof, texture details are more prominent. In the TATTOO image, trademarks in the distance are also clearer.

(2) Compared with other representative methods. Typical dehazing performances were shown in Fig. 4, Fig. 5, Fig. 6, respectively.

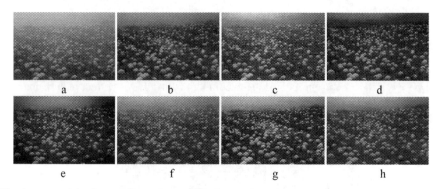

Fig. 4. Image dehazing results on the pumpkin image. (a) hazy image. (b)–(g) the results of Fattal et al. [11], Gibson et al. [1], Nishi et al. [23], Kratz et al. [24], He et al. [31], Tan et al. [10] and ours.

Fig. 5. Image dehazing results on the road image. (a) hazy image. (b)–(g) the results of Fattal et al. [11], Gibson et al. [1], Nishi et al. [23], Kratz et al. [24], He et al. [31], Tan et al. [10] and ours.

By contrast, the proposed method yielded very comparative and even better results than other competing techniques. It is to be noted that the proposed method not only significantly improved the visibility of distant views, but also enhanced texture details and simultaneously maintained original color appearances. Also, halo artifacts within our recovered results were quite small.

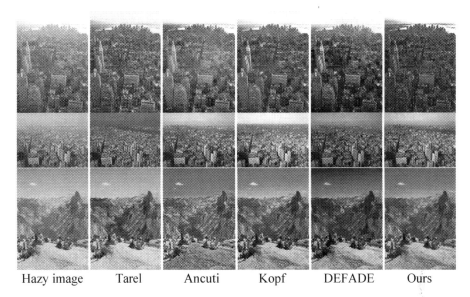

| Hazy image | Tarel | Ancuti | Kopf | DEFADE | Ours |

Fig. 6. Representative image dehazing results (Tower, River and Mountain)

4.2 Quantitative Comparison

The measure of Hautiere et al. [34] provides a quantitative evaluation of a dehazing algorithm using three metrics which are based on the ratio between gradients of hazy image and corresponding dehazed image. e represents rate of new visible edges in dehazed image against hazy image, while σ denotes the percentage of pixels that become black or white following defogging. A higher positive value of e and a value of σ closer to zero imply better performance. $\bar{r}$ denotes mean ratio of gradient norms before and after dehazing. A higher value of $\bar{r}$ represents stronger restoration of local contrast. The perceptual haze density D delivered by FADE is a no-reference method that does not require original hazy image. A lower value of D implies better dehazing performance [27]. Table 1, Table 2, and Table 3 are quantitative comparisons of the methods in Fig. 3, Fig. 4, Fig. 5 and Fig. 6, respectively.

Many experiments have been done in this paper. Table 1 shows that in the three pictures, whether it is e value or $\bar{r}$ value, this method is larger than He method [31], especially building image, e of two algorithms differ by 2.65, $\bar{r}$ of two methods differ by 2.25, It can be seen that texture preservation effect of this method is much better than He [31]. And D, this method is much smaller than He [31].

As shown in Table 2, this method has a larger e, $\bar{r}$, and smaller σ than Fattal [11], Gibson [1], and Nishino [23] methods; Kratz [24] method produces a larger e, $\bar{r}$, but dehazing effect is darker, produced a larger σ; Although Tan method achieves the greatest reduction of perceptual haze density after restoration, most dehazed images produced by that method lose visible edges yielding higher values of metrics due to oversaturation.

As shown in Table 3, Tarel [12], DEFADE [27], ours has a lower perceived haze density D than other methods, but Tarel [12] method produces a negative value, which is due to the fact that dehazing effect is too dark to cause some details to be lost. DEFADE

Table 1. Quantitative comparisons with different measurements

Algorithm	Building				Tian'anmen				TATTOO			
	e	$\bar{r}$	σ	D	e	$\bar{r}$	σ	D	e	$\bar{r}$	σ	D
He	4.65	1.97	0.00	0.98	0.65	1.48	0.00	0.69	0.42	2.17	0.00	0.25
Ours	7.30	4.22	0.00	0.48	0.99	3.48	0.00	0.32	0.42	2.44	0.01	0.13

Table 2. Quantitative comparisons with different measurements

Algorithm	Pumpkin				Road			
	e	$\bar{r}$	σ	D	e	$\bar{r}$	σ	D
Fattal	0.37	3.41	0.00	0.17	0.62	1.99	0.02	0.15
Gibson	0.14	1.74	0.02	0.17	0.46	1.95	0.01	0.15
Nishino	0.26	3.32	0.05	0.09	0.56	2.51	0.01	0.13
Kratz	0.32	3.57	0.03	0.12	0.89	2.39	0.03	0.13
He	0.36	3.55	0.01	0.14	0.72	2.43	0.01	0.19
Tan	0.17	2.34	0.06	0.07	0.59	2.01	0.00	0.09
Ours	0.54	3.99	0.01	0.11	0.73	2.51	0.00	0.13

Table 3. Quantitative comparisons with different measurements

Algorithm	Tower				River				Mountain			
	e	$\bar{r}$	σ	D	e	$\bar{r}$	σ	D	e	$\bar{r}$	σ	D
Tarel	0.07	1.88	0.00	0.24	0.01	1.87	0.00	0.24	−0.01	2.01	0.00	0.25
Ancuti	0.02	1.49	0.00	0.31	0.12	1.54	0.00	0.30	0.18	1.46	0.01	0.33
Kopf	0.05	1.41	0.00	0.48	0.01	1.62	0.01	0.41	−0.01	1.34	0.00	0.48
DEFADE	0.09	1.56	0.00	0.20	0.03	1.49	0.00	0.19	0.15	1.44	0.03	0.24
Ours	0.09	1.72	0.00	0.21	0.03	1.83	0.00	0.17	0.21	1.67	0.01	0.26

method, although the natural color is produced after dehazing, the visible edge is smaller than this method, and texture retention effect is not as good as this article.

That is, our proposed method was able to preserve more texture details. Remarkably, the values of σ were almost close to zero. Therefore, our proposed method was able to recover vivid appearance without sacrificing color fidelity visually.

5 Conclusions

In this paper, VO model is first used to decompose hazy image into low-frequency structure part and high-frequency texture part. We use dark channel prior method to dehaze low-frequency structure part. Then fuse the dehazed structure part and texture part so as to get the final haze-free image. Several experiments are carried out in this paper, and experiment results prove that the method proposed in this paper is more effective in removing haze and maintaining texture details.

Nevertheless, hazy image usually contains a certain amount of noise, which belongs to high-frequency component. How to remove noise from texture part of high-frequency component so as to achieve final effect of dehazing and denoising is our future research direction.

References

1. Gibson, K.B., Vo, D.T.: An investigation of dehazing effects on image and video coding. IEEE Trans. Image Process. **21**(2), 662–673 (2012)
2. Hong, D.F., Yokoya, N., Chanussot, J., Zhu, X.X.: CoSpace: common subspace learning from hyperspectral-multispectral correspondences. IEEE Trans. Geosci. Remote Sens. (TGRS) **57**(7), 4349–4359 (2019)
3. Hong, D.F., Yokoya, N., Ge, N., Chanussot, J., Zhu, X.X.: Learnable Manifold Alignment (LeMA): a semi-supervised cross-modality learning framework for land cover and land use classification. ISPRS J. Photogram. Remote Sens. **147**, 193–205 (2019)
4. Liu, Z., Xiang, B., Song, Y.Q., Lu, H., Liu, Q.F.: An improved unsupervised image segmentation method based on multi-objective particle. Swarm Optim. Clustering Algorithm Comput. Mater. Continua **58**(2), 451–461 (2019)
5. Hong, D.F., Yokoya, N., Chanussot, J., Zhu, X.X.: An augmented linear mixing model to address spectral variability for hyperspectral unmixing. IEEE Trans. Image Process. (TIP) **28**(4), 1923–1938 (2019)
6. Hou, G.J., Pan, Z.K., Wang, G.D., Yang, H.: An efficient nonlocal variational method with application to underwater image restoration. Neurocomputing **369**, 106–121 (2019)
7. Wang, N.B., He, M., Sun, J.G., Wang, H.B., Zhou, L.K., Chu, C.: ia-PNCC: noise processing method for underwater target recognition convolutional neural network. Comput. Mater. Continua **58**(1), 169–181 (2019)
8. Narasimhan, S.G., Nayar, S.K.: Contrast restoration of weather degraded images. IEEE Trans. Pattern Anal. Mach. Intell. **25**(6), 713–724 (2003)
9. Hautiére, N., Tarel, J.P., Lavenant, J., et al.: Automatic fog detection and estimation of visibility distance through use of an onboard camera. Mach. Vis. Appl. **17**(1), 8–20 (2006)
10. Tan, R.: Visibility in bad weather from a single image. In: Proceeding of IEEE Conference on Computer Vision and Pattern Recognition, Washington, DC, pp. 2347–2354. IEEE Computer Society (2008)
11. Fattal, R.: Single image dehazing. ACM Trans. Graph. (TOG) **27**(3), 1–9 (2008)
12. Tarel, J.P., Hautière, N.: Fast visibility restoration from a single color or gray level image. In: 2009 IEEE 12th International Conference on Computer Vision, pp. 2201–2208. IEEE (2009)
13. He, K., Sun, J., Tang, X.: Single image haze removal using dark channel prior. In: IEEE Conference on Computer Vision and Pattern Recognition, CVPR 2009, pp. 1956–1963. IEEE (2011)
14. Wang, Z., Hou, G.J., Pan, Z.K., Wang, G.D.: Single image dehazing and denoising combining dark channel prior and variational models. IET Comput. Vis. **12**(4), 303–402 (2018)
15. Hou, G.J., Li, J.M., Wang, G.D., Yang, H., Huang, B.X., Pan, Z.K.: A novel dark channel prior guided variational framework for underwater image restoration. J. Vis. Commun. Image Represent. **66**, 102732 (2019)

16. Guo, Y.C., Li, C., Liu, Q.: R2N: a novel deep learning architecture for rain removal from single image. Comput. Mater. Continua **58**(3), 829–843 (2019)
17. Xu, Y., Wen, J., Fei, L., Zhang, Z.: Review of video and image defogging algorithms and related studies on image restoration and enhancement. IEEE Access **4**, 165–188 (2016)
18. Li, Y., You, S., Brown, M.S., Tan, R.T.: Haze visibility enhancement: a survey and quantitative benchmarking. Comput. Vis. Image Underst. **165**, 1–16 (2017)
19. Du, Y., Guindon, B., Cihlar, J.: Haze detection and removal in high resolution satellite image with wavelet analysis. IEEE Trans. Geocsi. Remote Sens. **40**(1), 210–217 (2002)
20. Liu, X., Zhang, H., Cheung, Y.M., et al.: Efficient single image dehazing and denosing: an efficient multiscale correlated wavelet approach. Comput. Vis. Image Underst. **162**, 23–33 (2017)
21. Su, J., Sheng, Z., Xie, L., Li, G., Liu, A.: Fast splitting based tag identification algorithm for anti-collision in UHF RFID system. IEEE Trans. Commun. **67**(3), 2527–2538 (2019)
22. Su, J., Sheng, Z., Liu, A., Han, Y., Chen, Y.: A group-based binary splitting algorithm for UHF RFID anti-collision systems. IEEE Trans. Commun. **68**, 1–14 (2019)
23. Nishino, K., Kratz, L., Lombardi, S.: Bayesian defogging. Int. J. Comput. Vis. **98**(3), 263–278 (2012)
24. Kratz, L., Nishino, K.: Factorizing scene albedo and depth from a single foggy image. In: Proceedings of the IEEE International Conference on Computer Vision, pp. 1701–1708 (2009)
25. Ancuti, C.O., Ancuti, C.: Single image dehazing by multi-scale fusion. IEEE Trans. Image Process. **22**(8), 3271–3282 (2013)
26. Kopf, J., Neubert, B., Chen, B., Cohen, M., Cohen-Or, D., Deussen, O., Uyttendaele, M., Lischinski, D.: Deep photo: model-based photograph enhancement and viewing. In: ACMSIGGRAPH Asia Papers, pp. 116:1–116:10. ACM Press, New York (2008)
27. Choi, L., You, J., Alan, C.: Referenceless prediction of perceptual fog density and perceptual image defogging. IEEE Trans. Image Process. **24**(11), 3888–3901 (2015)
28. Meyer, Y.: Oscillating patterns in image processing and nonlinear evolution equations. University Lecture Series, American Mathematical Society, Boston, USA (2001)
29. Gilles, J., Osher, S.: Bregman implementation of Meyer's G-Norm for Cartoon +textures decomposition. In: UCLA CAM, pp. 11–73 (2011)
30. Schechner, Y.Y., Narasimhan, S.G., Nayar, S.K.: Instant dehazing of images using polarization. In: Proceedings of the 2001 IEEE Computer Society Conference on Computer Vision and Pattern Recognition, CVPR 2001, pp. 325–332. IEEE (2001)
31. He, K., Sun, J., Tang, X.: Guided image filtering. In: Daniilidis, K., Maragos, P., Paragios, N. (eds.) ECCV 2010. LNCS, vol. 6311, pp. 1–14. Springer, Heidelberg (2010). https://doi.org/10.1007/978-3-642-15549-9_1
32. Yang, A.P., Wang, N.: Night image dehazing algorithm based on structure-texture layering. Prog. Laser Optoelectron. **55** No. 629(06), 101–108 (2018). (in Chinese)
33. Choi, L.K.: LIVE image defogging database [OL] (2015). http://live.ece.utexas.edu/research/fog/fade_defade.html
34. Hautière, N., Tarel, J.-P., Aubert, D., Dumont, É.: Blind contrast enhancement assessment by gradient ratioing at visible edges. J. Image Anal. Stereol. **27**(2), 87–95 (2008)

Traffic-Behavioral Anomaly Detection of Endhosts Based on Community Discovery

Mingda Wang$^{(\boxtimes)}$ ⓘ, Xuemeng Zhai, Hangyu Hu ⓘ, and Guangmin Hu

University of Electronic Science and Technology of China, Chengdu 611731, China
wangmingdauestc@gmail.com, ZhaiXM@std.uestc.edu.cn,
huhangyuuestc@gmail.com, hgm@uestc.edu.cn

Abstract. Traffic-behavioral anomaly detection of endhosts can discover rarely-observed behaviors exhibited by endhosts based on traffic analysis, and it is a powerful tool for better network management. However, the related researches are still limited. In this paper, a novel method of endhost-traffic-behavioral anomaly detection based on community discovery is proposed. In this approach, a traffic-behavioral graph is generated for each endhost based on its traffic-behavioral features in different time intervals. After this, the endhost-traffic-behavioral anomalies can be detected by discovering anomalous communities from the graph. The experiment results show that the proposed method has a high accuracy performance, and the coverage can be effectively enhanced with proper parameter control.

Keywords: Endhost · Traffic behavior · Anomaly detection · Community discovery · Feature extraction

1 Introduction

As an efficient provider tool of high-quality information for network monitoring, endhost profiling [1, 2] utilizes descriptive information extracted from network traffic to construct thorough profiles for endhosts. Deep Packet Inspection (DPI) techniques [3] can effectively extract such information from the payload of the network packets, and it is used as the fundamental information-mining tools for most of the existing endhost profiling researches. However, DPI is rendered less effective under the prevailing of payload-encrypted communication (such as HTTPS), as well as those endhost profiling techniques based on DPI [4]. Though, there is still a vault of endhost information hiding inside the header of a packet, which can support less descriptive but similarly crucial monitoring task, to wit, the detection of the traffic-behavioral anomaly on endhosts.

The detection of traffic-behavioral anomalies of certain endhosts can reveal key information such as whether the endhost is running normally, whether there is a user replacement, whether the endhost is under attack and so on, thus it is considered a critical mission in endhost monitoring.

To the best of our knowledge, the research about the traffic-behavioral anomaly detection of endhosts based on traffic analysis is still a rarely-touched area. Though,

© Springer Nature Switzerland AG 2020
X. Sun et al. (Eds.): ICAIS 2020, LNCS 12239, pp. 751–762, 2020.
https://doi.org/10.1007/978-3-030-57884-8_66

related works mainly lie in network traffic anomaly detection [5] and host behavior analysis [2, 6, 7]. The research of network traffic anomaly detection has become a hot spot worldwide in the past decade. Nevertheless, nearly all of the network traffic anomaly detection schemes focus on detecting backbone traffic anomalies of a distributed network, where their primary goals stay close to finding large-scale anomalous activities, such as the distributive denial of service attacks (DDoS) [8]. While to the existing traffic behavior researches, most of them have endeavoured to classify different endhosts [6], to cluster endhosts [7], to profile endhosts [2] and so on.

In this paper, we try to explore the possibility to detect endhost anomalies based on traffic-behavior analysis. The traffic of a given endhost is firstly split into non-overlapping segments based on the time of capture, and a traffic-behavioral feature vector is extracted from each traffic segment. The traffic-behavioral feature vectors are deemed as the nodes of a traffic behavior graph, and a binarized similarity is adopted to generate the edge between a given node pair in the graph. With the traffic-behavior graph as input, a community discovery algorithm is adopted to split the graph into communities. Then the anomalous community is separated from the normal communities according to the community size.

The contributions of our work can be summarized into three aspects. The first contribution is that we propose a traffic-behavior-based anomaly detection scheme for single endhost analysis. Secondly, we consider various feature information for an endhost, a number of different kinds of features are extracted to guarantee a thorough depiction of the traffic behavior for an endhost. Finally, a novel framework of endhost anomaly detection based on community discovery are proposed.

2 Related Works

Network traffic anomaly detection has attracted lots of attention during the last decade. Apart from wireless-scenario researches [9–11], we focus on the survey of researches concerning fixed networks. In the survey presented by Chandola [12], the author has given a general discussion and a thorough summary on the existing traffic-anomaly-detection works. In this survey, the author claims that the traffic anomaly is a certain kind of traffic pattern which deviates from well-defined normal behaviors.

Lakhina [13] and his research team has proposed a decomposition-based method to deal with the traffic anomaly detection problem in a backbone network scale. The author argues that the traffic anomaly would stay stealth in a backbone network scale because of the existence of heavy noise, and by leverage a principal component analysis (PCA) method the traffic anomaly can be successfully revealed in certain higher-order subspaces. Andreas [14] has proposed a histogram-based traffic anomaly detection scheme. This method detects traffic anomalies by finding deviations from well-modeled feature distributions.

As far as we know, still there is no published survey paper on network endhost profiling researches. Though, there are several typical techniques for endhost profiling, which are presented as following. Karagiannis [2] has proposed a graph-based endhost profiling technique. This technique leverages endhost communication patterns (such as how many endhost it is communicating with, the port numbers and the protocols) to construct an endhost communication profile. After that, the graphlet profile is extracted based on the learning of a large number of communication graphs.

Tao [15] and his research crew has proposed a user-behavior profiling scheme based on multi-layered information, which was applied to a user cluster mining work. Xia [1] has proposed a user profiling technique based on multi-source information. This technique filters out the user-specific-traffic blocks according to the online-social-network (OSN) information, the corresponding uniqueness and consistency measurement. With the user-specific-traffic blocks, a user profile named with "Mosaic" is constructed based on the detailed OSN information. The work also reveal that the user information contained in the network traffic is far more abundant and dynamic than that obtained from OSN crawling.

Although both the academia and the industrial has put a lot of efforts into the researching of network-wide anomaly detection and the analysis of enhost behavior, the passive detection of traffic-behavioral anomalies on a single host is still unseen, let alone a systematic solution to this problem. By modeling the detection of endhost-traffic-behavioral anomalies into a community discovery problem, together with the proposal of a set of traffic-behavioral features, our work gives an unprecedented exploration into the passive detection of the traffic-behavioral anomaly on single endhosts, which together builds up the uniqueness of this paper.

3   Methodological Description

3.1   Problem Statement and Hypothesis Explanation

In this paper, an endhost-traffic-behavioral anomaly refers to a certain kind of traffic behavior which the endhost does not usually exhibit. The key hypothesis behind this research is that: When given a proper behavioral representation (to wit, use a set of behavioral features to stand for a traffic behavior of an end host), the normal behavior will be distinguishably different from the anomalous behavior, while the behaviors belonging to the same behavioral category should have smaller differences between each other. This hypothesis sheds light upon the idea to leverage community discovery algorithm as the solution to this anomaly detection problem.

Community discovery techniques [16] are widely applied in graph-based data analysis. The main goal of almost every community-discovery method is to divide a given graph into closely inter-connected communities, where the nodes inside a community have strong connectivity between each other, while nodes belonging to different communities have relatively looser connectivity. The homogeneousness that bonds the hypothesis of this paper and the goal of community discovery methods can be stated as: Both of them try to separate internally-close-related groups from the wholeness, based on the closeness measurement of each node pair. This homogeneousness is the main motive that propel us to choose the community discovery method for countering the traffic-behavioral anomaly detection problem.

After modeling the objective as a community-discovery problem, the smaller-sized communities stand exactly for the rarely-observed behaviors. In this paper, we claim that the smaller-sized communities are the anomalous communities, and the nodes inside these communities are the anomalies we try to detect. Thus, detecting the anomalies is equivalent to finding smaller-sized communities of a given graph. Based on this conception, an example drawn from real dataset is presented in Fig. 1, where two large communities are identified as normal communities, and the other smaller ones are detected as being anomalous.

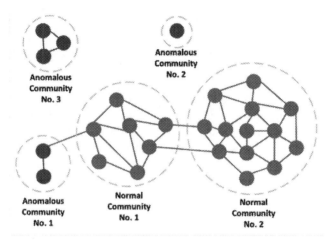

Fig. 1. An illustration on the detection of endhost-traffic-behavioral anomaly under community-discovery modeling. This graph is generated from real data.

The problem now falls into how to bridge the gap between our hypothesis and a typical community discovery problem. The large picture is: Denoting traffic-behavior samples of a given endhost as the nodes, and the similarities between samples as the edges, the traffic-behavioral graph of an endhost can thus come into form, after which the community discovery method can be applied. A more detailed explanation about this process is provided in the next section.

3.2 Methodology Explanation

The Representation Layer. The representation layer is designed to filter out unwanted traffic, and transform the traffic into proper data representations for later use. As is shown in Fig. 2, in the traffic sorting module the raw network traffic is firstly split and sorted according to the identifiers of the interested endhosts in order to obtain the network traffic for each endhost. In the segmentation module, the traffic is segmented into different time bins according to the time of capture and the time granularity needed, e.g., if the time granularity is set as 1 h, the traffic will be divided into non-overlapping segments with 1-h length. In the feature extraction module, a set of traffic-behavioral features are extracted for each traffic segment of each endhost. The features for use are shown in Table 1.

The extracted features are organized into a vector, which is later referred as the traffic-behavioral feature vector of the endhost. The traffic-behavioral feature vectors serve as the inputs of the detection layer, which will be explained in the next subsection.

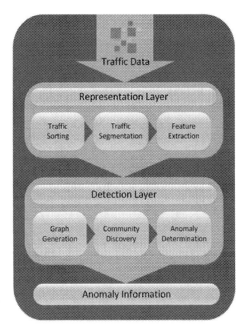

Fig. 2. The work flow and module illustration.

Table 1. .

Feature name	Feature explanation
dst. IP count	The number of IP addresses that the endhost has communicated with
srt. port count	The number of ports that the endhost has used for communications
traffic volume statistics	The flow count, the packet count and the byte count of the traffic, outbound and inbound respectively
TCP flag count	The number of TCP packets transported with the flag fields set to 1, 8 kinds of flags respectively
application usage statistics	The byte count and flow count of the application traffic. The applications include: Web, P2Pdownload, P2Pstreaming, Services, Files and Streaming

The Detection Layer. The main goal of the detection layer is to discover the traffic segments which contain traffic-behavioral anomalies, and the community discovery method is adopted in order to achieve this goal. To be specific, the detection layer takes feature vectors generated by the representation layer as its input, and the traffic segments as its output. This is achieved through three submodules, as is shown in Fig. 2.

In the Graph Generation module, the feature vectors are treated as the nodes of the traffic-behavioral graph of the endhost. A certain similarity measurement is adopted to compute the similarities between samples. A threshold is chosen to binarize the similarity measurements and thus to decide whether there is an edge connecting those two nodes. If the similarity goes beyond the threshold, there will be an edge. If the similarity is less than the threshold, there will be no edge connecting those two nodes. The nodes and the edges then compose the traffic-behavioral graph of the endhost.

After obtaining the traffic-behavioral graph of an endhost, a community discovery method is applied to the graph in the Community Discovery Module. The community discovery method to be used in our scheme will be the Blondel-Guillaume-Lambiotte-Lefebvre's (BGLL) algorithm [17].

The BGLL algorithm finds the best community partition of a given graph through a hierarchical maximization of the modularity measurement [18]. The modularity is an effective indicator of how good is a community partition for a given graph. The larger the modularity, the more closely connected are the nodes inside each community, and the more distant are the nodes belonging to different communities, thus the better the community partition. For a given graph with n nodes, m edges and C communities, Let Δ_{vw} denote a sign function, where $\Delta_{vw} = 1$ if node v and node w are connected by an edge, and $\Delta_{vw} = 0$ if not. Let e_{ij} denote the fraction of the edges with one end attached to a node inside community c_i and the other end attached to a node inside community c_j, so e_{ij} can be computed as:

$$e_{ij} = \sum_{v \in c_i, w \in c_j} \frac{\delta_{vw}}{2m} \tag{1}$$

Let a_i be the fraction of edges with at least one end attached to a node inside community c_i, so a_i can be computed as:

$$a_i = \sum_j^C e_{ij} \tag{2}$$

With a_i and e_{ij} defined, the modularity Q of this graph with this community partition can be computed with the equation below:

$$Q = \sum_i^C \left(e_{ii} - a_i^2 \right) \tag{3}$$

At the start-up of BGLL algorithm, each node of the graph is treated as a community, and the best community partition in this step is found by merging any pair of communities into a new community if the resulting modularity gain is positive and maximized (for all possible community pairs). Denoting c_x as the x th community and c_y as the y th community in the current community partition, denoting the new community merged with c_x and c_y as c_z, the new modularity as Q' and the new community set as C', the modularity gain ΔQ is computed using the equation below:

$$\Delta Q = Q' - Q \tag{4}$$

The equation above can be further simplified as:

$$\Delta Q = \sum_i^{C'} \left(e_{ii} - a_i^2 \right) - \sum_j^C \left(e_{jj} - a_j^2 \right) \tag{5}$$

Noting that the only difference between the two summation terms in the equation above is the difference between the modularities of c_z, c_x and c_y, thus the equation above can be simplified as:

$$\Delta Q = \left[e_{zz} - \left(e_{xx} + e_{yy} \right) \right] - \left[a_z^2 - \left(a_x^2 + a_y^2 \right) \right] \tag{6}$$

Since c_z is merged from c_x and c_y, the term $e_{zz} - \left(e_{xx} + e_{yy} \right)$ is equal to the fraction of edges connecting c_x and c_y, thus $e_{zz} - \left(e_{xx} + e_{yy} \right) = e_{xy}$. So the final form of the equation to compute ΔQ can be derived as below:

$$\Delta Q = e_{xy} - a_z^2 + a_x^2 + a_y^2 \tag{7}$$

Once the maximized modularity gain is found, BGLL algorithm enters a merging process: each community is treated as a new node of a new graph. As is to the edges connecting the new node, a new self-loop edge will replace all the edges inside the original community (to wit, the intra-connecting edges) with the weight equal to the sum of all the weights of the original edges in the community. As is to the inter-connecting edges between different communities, the new edge connecting the two corresponding new nodes will have a weight equal to the sum of all the weights of the original edges connecting these two communities. An example is given in Fig. 3 to clarify this merging process.

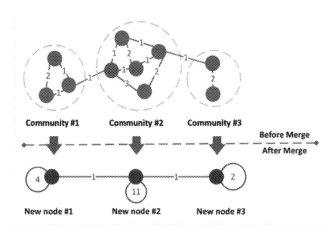

Fig. 3. An example of the community-to-node merge process in BGLL algorithm.

After the merging process, the modularity maximization process is again applied to the merged graph, and again the merging process follows. These interlaced processes are repeated until convergence (to wit, there is no more modularity gain), then the resulting community partition is the final output of BGLL algorithm. With the BGLL algorithm, the communities inside each graph can thus be discovered.

In the Anomaly Determination module, the partitioned communities are evaluated based on their relative sizes. Given an anomaly-determination threshold, if the relative size of a community is less than the threshold, this community will be determined as the anomalous community, and the nodes inside the community will be determined as the traffic-behavioral anomalies of the endhost. On the contrary, the larger-sized communities will be filtered out by this thresholding method, since they stand for the frequent traffic behavior of the endhost. The traffic-behavioral anomalies are exactly the final outputs of our method.

4 Experiments and Analysis

4.1 Experiment Configuration

The traffic dataset used for experiments is collected from the gateway of our networking laboratory, where there are 48 desktop computers inside the network. For the traffic, both inbound and outbound traffic is captured. For each packet in the traffic, the full length of the packet is captured. The overall time duration of the dataset spans over two months, during which the user of each desktop computer was kept unchanged. Noting that there is no artificial generated traffic throughout the whole monitoring period, the traffic is captured under the daily usage of laboratorial computers. Besides, in order to keep the experiment truthful and factual, all the anomalies involved in the experiment are natural occurrences during the monitoring, without any manual intervention.

The ground truth generation of this data is based on three sources. The first source is a process monitor installed on every desktop computer inside the laboratory. The process monitor collected and maintained a complete list of the life cycle for each process which has ever used the network for data communication during the monitoring period. The second source is an event list manually maintained by each PC user, which keeps track of the description and the life cycle of the activities. For example, if a user has done much more downloading someday than they would usually do, he/she is required to log the downloading behavior into an event timeline, which contains detailed explanation towards the downloading, such as the file size, the start time, the end time and so on. The third source is a manual inspection into the endhost traffic, where the intention is to label anomalies which cannot be easily perceived by the first two sources, such as port scans and congestions. For instance, if multiple endhosts in the monitored network receive or send TCP packets with ECN or CWR flags, which most likely indicate an ongoing congestion anomaly, all the related endhost samples will be tagged as being anomalous.

The anomaly categories adopted for evaluation in our experiment are chosen based on two major principles. Firstly, the anomaly should be able to raise enough interest for network monitoring purpose. For example, the anomaly should indicate some real-world event such as a switch of device users, or it should represent the incoming of certain network attacks, so on so forth. Secondly, the anomaly should be detectable through traffic analysis, which is to say, the anomaly should be able to bring perceptible traffic behavior changes so that the method could separate it from the normal behaviors. Based on these principles, 3 major categories are defined for our anomaly detection experiment. The name, the corresponding description and some examples are given in Table 2. As is to the performance-evaluation metrics, the accuracy and the coverage are both adopted. The accuracy is computed as the percentage of the real anomaly instances in all the instances which are determined by our method as anomalies. The coverage is the percentage of the real anomaly instances detected by our method in all the existing anomaly instances.

Table 2. Anomaly category.

Category	Description	Example
Discovery anomaly	A discovery anomaly is exhibited when the endhost tries to access new content never seen before, or the endhost overuse certain applications	App. Overuse
Damage anomaly	A damage anomaly will occur when the endhost participates in a network attack event, neither as a victim or as an attacker	Port scanning
Malfunction anomaly	The endhost exhibits malfunction anomaly when it experiences some network malfunctions	network outage

4.2   Experiment Results and Analysis

In this subsection, the results of the experiments on the key parameters affecting the performance of our method are discussed. The results for edge-formation coefficient evaluation are shown in Fig. 4 where the anomaly-determination coefficient is set to 0.1.

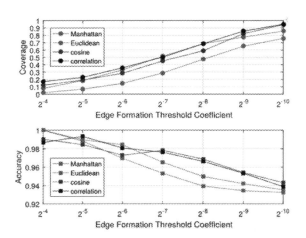

Fig. 4. Experiment results of the edge-formation coefficient rotation.

In Fig. 4 we can see that the coverage rises when the edge-formation coefficient decreases. This is mainly because that as the coefficient decreases, nodes belonging to different traffic behavior tend to have fewer edges interconnecting each other, and community count will increase. During this process, the edges between the anomalies and the normal behavioral instances will break apart at some point since they are different from each other. Thus, the anomalies who were formerly trapped inside a larger normal community get separated out, and they form into small-sized anomalous communities which can be determined correctly through our method, which explain the increase in the coverage.

However, when the edge formation threshold coefficient decreases, not only more anomaly instances get detected, but more normal instances are also separated out and forms small-sized communities. In this way, more false positives are generated, leading to a decrease in accuracy. Fortunately, as is shown in Fig. 4, the decrease in accuracy is overall far less significant than the increase of coverage, where the lowest accuracy value still stays beyond 0.9. Considering the overall performance of this method, we suggest that a relatively small (such as 2^{-10} or smaller) value of edge formation threshold coefficient should be selected in this method.

In Fig. 5 the results for anomaly-determination coefficient experiments are illustrated, and the edge-formation coefficient is set to 2^{-10} during the experiments.

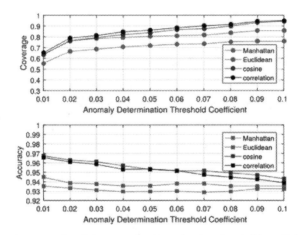

Fig. 5. Experiment results of the anomaly-determination-threshold coefficient rotation.

As is to the rotation of anomaly-determination coefficient shown in Fig. 5, the coverage increases as the coefficient rises, and the accuracy deceases on the contrary. The reason behind this phenomenon is that the larger the anomaly-determination coefficient, the more communities, whether anomalous or not, will be determined as anomalous. Thus, the coverage will rise since more anomalous instances are detected, and accuracy will decrease because more false positives are introduced. According to this analysis, we suggest a relatively large value of anomaly-determination coefficient should be chosen for this method, for example, a value larger than 0.09, where both the coverage and the accuracy can exceed 0.9 according to Fig. 5.

The choice of similarity measurement does stand a role in the performance of this method, as is shown in both Fig. 4 and Fig. 5. It is obvious that the Euclidean distance has the lowest coverage values over both edge-formation and anomaly-determination coefficient settings, while correlation strikes overall the best coverage values in comparison to the other three measurements do. For the accuracies, in the anomaly-determination coefficient evaluation there is a fixed superiority order that cosine and correlation are better than the other two, but no fixed order can be drawn when the edge-formation coefficient is relatively large (larger than 2^{-7}). Though, as the edge formation threshold keeps on decreasing thereafter, the inferiority-superiority order stabilizes as: Cosine, correlation, Euclidean and Manhattan, in an accuracy-decreasing manner. To sum up, the correlation and the cosine similarities are better choices for this method.

5 Conclusion

In this paper, a novel method to detect endhost-traffic-behavioral anomaly based on community discovery is proposed. This method leverages BGLL algorithm to separate anomalous-traffic-behavioral instances from normal ones inside a traffic-behavioral graph of a giving endhost, thus fulfil the task of endhost anomaly detection. According to the experimental results, a small value of edge-formation coefficient (2^{-10} or smaller), a large value of anomaly-detection coefficient (0.09 or greater) and correlation or cosine distance for similarity measurement are suggested empirically. Through all these experiments and analysis, we have confidence that this method can achieve good results in more application scenarios.

Acknowledgment. This research is carried out under grants from National Natural Science Foundation of China (No. 61471101 and No. 61571094). M.W. and G.H. gave the idea, M.W. did the experiments, M.W. and X.Z. interpreted the results, M.W. and H.H. wrote the paper.

References

1. Xia, N., et al.: Mosaic: quantifying privacy leakage in mobile networks. ACM SIGCOMM Comput. Commun. Rev. **43**(4), 279–290 (2013)
2. Karagiannis, T., Papagiannaki, K., Taft, N., Faloutsos, M.: Profiling the end host. In: Uhlig, S., Papagiannaki, K., Bonaventure, O. (eds.) PAM 2007. LNCS, vol. 4427, pp. 186–196. Springer, Heidelberg (2007). https://doi.org/10.1007/978-3-540-71617-4_19
3. Dharmapurikar, S., Krishnamurthy, P., Sproull, T., Lockwood, J.: Deep packet inspection using parallel bloom filters. In: High Performance Interconnects 2003, vol. 20, pp. 44–51. IEEE, New York (2003)
4. Shen, M., Wei, M., Zhu, L., Wang, M.: Classification of encrypted traffic with second-order Markov chains and application attribute bigrams. IEEE Trans. Inf. Forensics Secur. **12**(8), 1830–1843 (2017)
5. Bhuyan, M.H., Bhattacharyya, D.K., Kalita, J.K.: Network anomaly detection: methods, systems and tools. IEEE Commun. Surv. Tutorials **16**(1), 303–336 (2014)
6. Joumblatt, D., Teixeira, R., Chandrashekar, J., Taft, N.: Perspectives on tracing end-hosts: a survey summary. ACM SIGCOMM Comput. Commun. Rev. **40**(2), 51–55 (2010)
7. Jakalan, A., Gong, J., Su, Q., Hu, X., Abdelgder, A.M.: Social relationship discovery of IP addresses in the managed IP networks by observing traffic at network boundary. Comput. Netw. **100**, 12–27 (2016)

8. Mirkovic, J., Reiher, P.: A taxonomy of DDoS attack and DDoS defense mechanisms. ACM SIGCOMM Comput. Commun. Rev. **34**(2), 39–53 (2004)
9. Jian, S., Dan-feng, H., Jun-lin, T., Hai-peng, C.: An efficient anti-collision algorithm based on improved collision detection scheme. IEICE Trans. Commun. **99-B**(2), 465–469 (2016)
10. Jian, S., Xue-feng, Z., Zhong-qiang, L., Hai-peng, C.: Q-value fine-grained adjustment based RFID anti-collision algorithm. IEICE Trans. Commun. **99-B**(7), 1593–1598 (2016)
11. Jian, S., Guang-jun, W., Dan-feng, H.: A new RFID anti-collision algorithm based on the Q-ary search scheme. Chin. J. Electron. **24**(4), 679–683 (2015)
12. Chandola, V., Banerjee, A., Kumar, V.: Anomaly detection: a survey. ACM Comput. Surv. **41**(3), 15 (2009)
13. Lakhina, A., Crovella, M., Diot, C.: Diagnosing network-wide traffic anomalies. ACM SIGCOMM Comput. Commun. Rev. **34**(4), 219–230 (2004)
14. Kind, A., Stoecklin, M.P., Dimitropoulos, X.: Histogram-based traffic anomaly detection. IEEE Trans. Netw. Serv. Manage. **6**(2), 110–121 (2009)
15. Qin, T., Guan, X., Wang, C., Liu, Z.: MUCM: multilevel user cluster mining based on behavior profiles for network monitoring. IEEE Syst. J. **9**(4), 1322–1333 (2015)
16. Coscia, M., Giannotti, F., Pedreschi, D.: A classification for community discovery methods in complex networks. Stat. Anal. Data Min. ASA Data Sci. J. **4**(5), 512–546 (2011)
17. Blondel, V.D., Guillaume, J.L., Lambiotte, R., Lefebvre, E.: Fast unfolding of communities in large networks. J. Stat. Mech. Theory Exp. **2008**(10), 10008 (2008)
18. Newman, M.E.: Modularity and community structure in networks. Proc. Nat. Acad. Sci. **103**(23), 8577–8582 (2006)

A Lightweight Indoor Location Method Based on Image Fingerprint

Ran Gao and Yanchao Zhao[(✉)]

Collage of Computer Science and Technology, Nanjing University of Aeronautics
and Astronautics, Nanjing, China
rangao_nuaa@163.com, yczhao@nuaa.edu.cn

Abstract. Image-based localization is featured with low deployment
cost and universal accessibility with smartphones, thus attracts great
attentions from both academy and industry. The fingerprints from the
image data, which mostly consists of the image features, are promising
for indoor localization. However, constructing such fingerprint database
is usually time-consuming and labor-intensive. Moreover the high computational and storage cost render this localization scheme highly relies
on the cloud-based solution. Meanwhile, this solution suffers from the
transmission delay and query contentions in the cloud. To this end, we
propose a smartphone-based indoor localization method with enhanced
image fingerprint extraction and edge computing paradigm. Basically,
we extract image feature vectors with MobileNet and offload part of the
database to the smartphone. Then, we query image through the second-
level retrieval and verification of the smartphone orientation information
to finds out six most similar images. After this, we obtain the most similar image by carrying out query expansion. In the phase of collecting
image database, video and samrtphone sensors data are collaboratively
used to build image database. We have implement the prototype system
on the Android platform, and conduct real experiments to verify the
feasibility of the positioning method. Experiment results show that our
method has good accuracy (90% location error is within 1.5 m) and high
real-time performance (average location delay is about 1.11 s) .

Keywords: Image fingerprint · Smartphone-based localization ·
Indoor location

1 Introduction

With the rapid improvement of mobile device performance and the increase
of location-based applications, location sensing plays an increasingly important
role. No matter in indoor and outdoor environments, location information can
bring better experience for users in the Location based services.

According to different signal sources, indoor location can be generally divided
into image, RIFD, WIFI, Bluetooth and UWB location and so on. Except image-
based location, the others require to deploy infrastructures into the indoor environments [3,9,20]. Image-based location, featured with infrastructure-free, has

© Springer Nature Switzerland AG 2020
X. Sun et al. (Eds.): ICAIS 2020, LNCS 12239, pp. 763–774, 2020.
https://doi.org/10.1007/978-3-030-57884-8_67

got a lot of attentions recently. Landmark-based location [7,12] uses objects that are easily distinguished from the surrounding environment, such as store logos or wall paintings. However, this method is not suitable for some environments with sparse landmarks while the accuracy of position is high related to the identification of landmarks. In contrary, the image fingerprint based method does not need to model for the environment, thus is more suitable for complex indoor environment. The location accuracy is usually related to the distance between adjacent images in database and the accuracy of image retrieval. In [14,18], they generate vision vectors by the SIFT descriptor and the vocabulary trained in advance. Its time consumption is related to SIFT feature detection and the number of visual vocabulary. In [1,21], the image features are extracted by CNN neural network and fine-grained location is carried out by bundle adjustment or polar geometry. Compared with the paper [14,18], it achieves higher accuracy and less time cost. But the fine-grained location uses bundle adjustment and polar geometry, the application scenarios are limited. In addition, mast RCNN and VGG16 network is not suitable for smartphone. In [6], the author fuses WIFI signal, orientation and image signal to improve the location performance. But its time complexity is high, due to using Multiple Level Image Description (MLID) method to describe images. In summary, these methods are not suitable for smartphone because of computing and storage consumptions. In addition, It is very difficult to collect image database based on image fingerprint in the large scene.

To this end, we propose a lightweight image fingerprint location method based on smartphone. Our method can achieve accurate location with very low delay or even the smartphone is offline. The basic idea is we extract the image feature by MobileNet transplanted smartphone. Specifically, first, we divide the image database into several parts, and then select some representative images as the index of the image database. In order to reduce the calculation and time consumption and improve accuracy of image query, we obtain the location estimation through two-level query and orientation verification. In order to reduce the cost of time and labor, we use video stream to build image library with sensor data. We only store the feature vectors of the image database in the smartphone, and feature vectors database is 6.1MB in total, which is suitable for mobile terminals with limited storage resources. The main contributions of our method are as follows:

– Among the existing image fingerprint localization methods, the application scenarios are relatively limited and need to rely on the sever. In contrast, our method is lightweight and achieves positioning on the smartphone. Firstly, We extract image feature thought mobilenet which is especially designed for mobile devices. Then we only store feature vectors of partial image database.
– Compared with the above positioning method of multimodal fusion and image fingerprint in the range of acceptable location accuracy, it has a good performance in real time.
– In our location method, we improve the speed and accuracy of image retrieval by the orientation of mobile devices and index of image database and we do not need to rely on the surrounding infrastructures.

The rest of this paper is organized as follows. We review the related work in Sect. 2, followed by system overview in Sect. 3. The important parts of the location method based on smartphone are described in Sect. 4. We present illustrative experimental results in Sect. 5. Finally, we conclude in Sect. 6.

2   Related Work

Since GPS signal can not penetrate indoor, researchers have explored many other RF signals for indoor location, such as WIFI, bluetooth and and Radio Frequency Identification(RFID) [3,9,20]. These methods are very promising, but there are still some limitations. Firstly, because of deploying location anchors, they need additional infrastructure. The accuracy of echo schemes is high, but the dense sampling points prevents them from continuous positioning in a large-scale deployment. In paper [8,11,17], they leverages the channel state information (CSI) to build a propagation model and a fingerprinting system at the receiver. They achieves the median accuracy of 0.65 m, but they also needs to sample a large number of dense training points, thus making it difficult for large-scale applications. With the popularity of smartphones and the improvement of camera sensors performance, many scholars use image-based indoor location to solve the shortages of RF signal indoor location. Because it does not extra infrastructure and can achieve competitive accuracy that meets the requirements of indoor location service. In general, there are image-based localization methods related to our method as follows. The first kind of image-based location method is by landmarks [7,12,15]. SweepLoc [7] selects store logo as landmark and then users locate themselves by taking a short video clip (about 6 to 8 s) of his/her surroundings by sweeping the camera. It can select key frames (where potential landmarks are centered) and subsequently reduces the decision error on landmarks. With identified landmarks, SweepLoc formulates an optimization problem to locate the user, taking compass noise and floor map constraint into account. Lost Shopping [15] achieves localization by taking a Phone and then jointly reasons about text detection (localizing shop names in the image with precise bounding boxes), shop facade segmentation, as well as camera rotation and translation within the entire shopping mall. The Ocrapose [12] takes advantage of OCR to read the text/numbers and provide a rough estimate by using the floor plan. Next, it performs OCR-aided stereo feature matching to refine the estimate by solving a PnP problem. The second kind of image-based positioning method is to use image database [1,14,18,21]. Its main technology is image retrieval, which the accuracy of positioning is highly related to. In order to improve accuracy of the location, they use BA(bundle adjustment) or PNP for fine-grained location after image retrieval. The paper [14] uses Laplacian feature extraction and SIFT feature descriptor to extract image feature vectors and generates vocabulary vectors by clustering and pre-trained vocabulary in the offline stage. In the online stage, firstly, rough location estimation is obtained by matching the feature vectors of the query image with the database feature vectors. Then, carrying out by voting improves the accuracy of the query. Similar to the paper [14], the paper [18] adopts a new classification algorithm using KNN and support vector machine to improve the accuracy of the query. The paper [1,21] also use image retrieval, differing from them in the method of extracting image feature and improving location accuracy. They use deep CNN to extract image feature and BA(bundle adjustment) or PNP for fine-grained location.

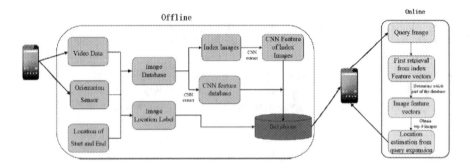

Fig. 1. System framework of location method.

Smartphone has rich sensors, so some researchers try to achieve indoor location via multi-modal sense on smartphone [4,16,19]. Now as the WIFI signal is very universal, they firstly locate themselves roughly by WIFI signal and then refine location through smartphone sensors. Although their methods can achieve better accuracy and reduce the time and labor costs of building image database, they must rely on servers and real time is not very good. In paper [6], it uses GPU to extract image feature and takes about 5s for each positioning.

3   System Overview

In this section, we mainly introduce the workflow of the positioning system and main modules of the positioning system, including the construction of image database and the retrieval based on CNN neural network and query expansion. The system architecture is showed by Fig. 1. It consists of two phases: an offline phase and an online phase. In the offline stage, it is mainly to establish location database. Firstly, the investigator follows the planned routes in advance with the hand holding smartphone to acquire the video data. Meanwhile, the smartphone automatically acquires and records the orientation. Then the image database is constructed and the image position labels are marked by the image selection algorithm. At last, we extract the image features by mobilenet. In the online stage, it mainly achieves location estimation. Firstly, users download image feature vectors database and image location labels to the smartphone. Secondly, the user obtains the image with location tag and extracts the image feature vector using the mobilenet network. Finally users obtain the location estimation by two-level retrieving and query expansion.

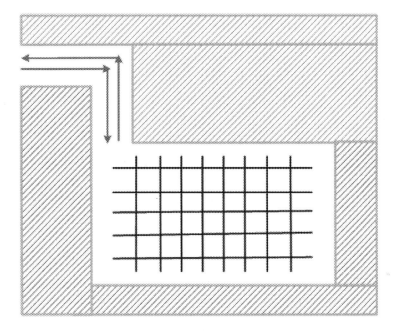

Fig. 2. An example of survey paths.

4   Indoor Location Method Based on Smartphone

4.1   Database Construction

In this section, we introduce the construction of image database in detail. It's very difficult to build image database from photos. In order to reduce the difficulty of building image database and cut down consumption of the labor and time, we use video to construct image database and automatically mark the location label of the image according to the start and end location and the orientation of the smartphone. In the experimental scenario, we plan the path in advance and the experimenter walks along it. When collecting video data, we assume that the experimenter walks at a constant speed. Our experimental scene is on the first floor of the computer college, including the hall and the corridor of the office area. In the hall we divide it into grids because it is relatively broad. We collect the video along the grid line and then build the image database using the video data. We collected 4 images at each grid intersection. In the corridor, one-dimensional position is enough considering that it is long and narrow. We collect video along a straight line. We take two photos at each point. An example is shown in Fig. 2. Red and black lines are the path we set. After collecting video data, we use algorithm 1 to build image database from video frames. In the algorithm, we firstly use smoothing filter to process the orientation sensor data so as to segment the video correctly. Then we use blur detection and similarity detection to remove the blurred image and with the unavailable image caused by turning and shake.

Algorithm 1. the method of building picture database from video

---

Input: video data; orientation sensor data; distance interval image sampling; The length of the path;

Output: image database; image location label;

1: Process direction sensor data by smoothing filter;
2: Segment video according to sensor;
3: Detect and remove blurred image in video frame;
4: remove the unavailable image through similarity detection;
5: Select database pictures and calculate the position coordinates of each image according to the picture sampling interval and path length

---

4.2   Extract Image Features

There are many methods of image feature extraction, among which SIFT (scale invariant feature transform) and CNN are commonly used. The SIFT is a local feature extraction method which uses a Gaussian blur along with a scale-invariant matching of local extremum extrema to find feature points. In image retrieval, most methods based on SIFT feature rely on BOW model. In recent years, with the development of deep neural network, the performance of CNN based feature extraction method is getting better. It is mainly through a lot of filters, constantly extracting features, from the local features to the overall features, so as to carry out image recognition and other functions. Neural network architectures, such as VGG, Google Net, ResNet and inception, have very good performance in feature extraction and are often used in target recognition or image retrieval tasks [10]. However, they are not applicable to the mobile end because the computing and storage resources of the mobile end are limited. Lightweight neural networks, such as squeezenet [5], mobilenet [13] and shufflenet [22], are designed to reduce the storage consumption of models and the speed of problem prediction. Generally, the lightweight neural network architecture compresses the feature map by using 1 * 1 convolution kernel. In order to achieve location independent of the server, we select mobilenet neural network to extract image features. Mobilenet network has smaller volume, less computation and higher precision. It has great advantages in lightweight neural network. We use the output of the average pooling layer as the feature vectors of the image. Image features are compressed in a vector of fixed length 1028. We extract the feature vectors of index image set and image database as $v_i{}^{index}$ and $v_i{}^d$. And the feature vector of the query image is marked as v_q.

4.3   Image Retrieval from Database of Image Feature Vectors

In order to improve the speed of image retrieval, we divide the image database into four parts, and select several representative images as the index of the four image databases. We extract the feature vector(v_q) of the queried image through mobilenet. We compare similarities between images by calculating the Euclidean distance of the feature vectors. In this paper [13], the accuracy of mobilenet's image retrieval top-1 is only 71.7% which is a little low for us. So we select the six images that are the most similar to the query images. And we compare the orientation of query image and image database so as to reduce image retrieval errors.

Algorithm 2. query image retrieval strategy

Input: v_q;v_i^{index} ;v_i^d;label of index image;Orientation of each image
Output: v_s
1: Calculate the Euclidean distance of V_i^{index} and V_q
2: Sort and select the image with the smallest Euclidean distance and similar direction(direction difference less than 20 degrees) to the query image
3: Calculate the Euclidean distance of V_q and V_i^d
4: Sort and select six the image with the smallest Euclidean distance and similar direction(direction difference less than 20 degrees) to the query image.

4.4 Position Estimation

We use average query expansion to get the most similar geo-tagged image related to the query image. And its coordination of the reference point will be taken as the estimation of the query image's position. Total Recall [2] proposes several query expansion methods, including transitive closure expansion, average query expansion and recursive average query expansion. Average query expansion method is the simplest method and its performance is quite good. We get 6 candidate images through image retrieval strategy and then a new query is constructed by averaging their feature vectors. A new query Q_{avg} is then formed by taking the average of the original query Q_0 and the m results.

$$d_{avg} = \frac{1}{m+1}(d_0 + \sum_{i=0}^{m} d_i) \tag{1}$$

where d_0 is feature vector of the query image and d_0 is feature vector of 6 images. Again we requery once, and the results of Q_{avg} are appended to those (top m) of Q_0.

5 Experiment Result

5.1 Experiment Set

In the database preparation stage, the computer is Intel Core i7-4510U CPU @ 2.00 GHz 2. 60 GHz, memories 4.00 GB, 64 bits Win10 OS. Data collection and test equipment is Google nexus 5. We develop an app on Android platform to

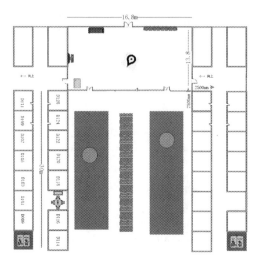

Fig. 3. Maps of the localization scenarios.

collect images database and achieve positioning. Our experimental scene is on the first floor of the computer college, including the hall and the corridor of the office area, which is showed by Fig. 3. In the hall, we divide the hall into grids. The interval between each grid point is 0.6 m. We collect 4 images for each grid point. There are 1120 images in the image database. In order to improve the retrieval efficiency, we divide the image database into four parts and each part has about 280 images. Then we select 10 images as database index. As for the test dataset, 44 query images are randomly taken in the location scene and their true positions are recorded in the form of the two-dimensional coordinate. In the corridor, one-dimensional position is enough considering that it is long and narrow. We collect video along a straight line and 2 images for each point. There are 160 images in the image database. As for the test dataset, 18 query images are randomly taken in the location scene and their true positions are recorded in the form of the one-dimensional coordinate.

5.2   Experiment Result and Evaluation

We evaluate the feasibility of the location method by the time consumption of establishing image feature database, location accuracy, positioning time consumption and so on. In addition, we also verify the feasibility of using video to build the image database by comparing location accuracy of different databases established by photos and videos.

When collecting the image database through video, we assume that the experimenter walks at a constant speed, which is uncontrollable in fact. So the image location labels are an estimate value. And the quality of the video is not as high as that of the photo due to walking. Therefore to verify the feasibility of method of building image database, we compare the location accuracy based on image database built by photo and video. The result is as shown in the Fig. 4. From the figure, we can see that about 79% and 97% of the location errors are within 0.6 m and 1.5 m respectively with the image-based method, while about 56% and 90% of the location errors are within 0.6 m and 1.5 m respectively with the video-based method. The location error of video method is larger than that of image method, but the location error based on video is still within the acceptable range. Compared with the photo method, the time and labor consumption of video method to collect image database is far less than the image-based method. Generally speaking, it is feasible to build database based on video. After the smartphone collects the data, we use the computer to select the database image, to mark the position labels of the images and to extract the image feature vectors. The time consumption is shown in Table 1:

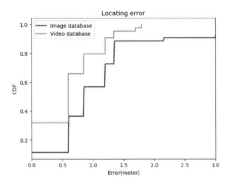

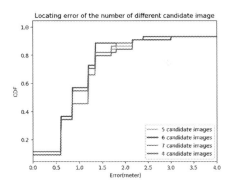

Fig. 4. location error distribution map of image databases builded by photo and video.

Fig. 5. Locating error distribution of the number of different candidate image.

Table 1. Offline database preparation time

	Select images and mark label	Extract the image feature vectors	Total time
The hall	18.511	134.5194	153.03
The corridor	3.263	15.1607	18.423

In the database preparation stage, we don't need to train the model and its time consumption only needs 153.03 and 18.423 s respectively in the hall and corridor, which is very small. And the size of our image feature vectors is 6.1 MB. So it is very suitable for smartphone location.

If the number of candidate images is too large, it may bring big error. However, if it is too small, it may miss the best matching image. In order to explore the number of candidate images in the retrieval phase, we select 4–7 candidate images and draw their cumulative distribution of location error as shown in the Fig. 5. From the cumulative error distribution, when there are 6 candidate images, the positioning performance is the best. We evaluated the positioning accuracy of top-1 during the retrieval phase and the positioning accuracy of retrieval plus query expansion in the hall and corridor, as shown in the Fig. 6 and 7. We can see that the effect of the two methods is almost the same when the location error is less than 1 m. However, 90% of the location error of the combined query expansion is within 1.5 m, while only 70% of location error is within 1.5 m. Obviously, adding query expansion can improve positioning accuracy and robustness. In the corridor 95% of positioning error is within 2m.

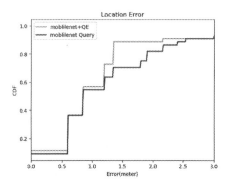

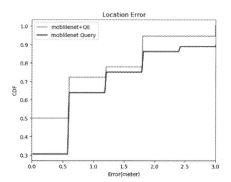

Fig. 6. Location error distribution in the hall

Fig. 7. Location error distribution in the corridor

Finally, we test the time consumption of locating on the computer and smartphone in the hall and corridor. At the same time, we test the time consumption of using secondary query on the smartphone. The time consumption is shown in the Table 2. Because the image database is relatively small in the corridor, we did not use two-level query. We also didn't use secondary query on the computer. The average time consumed for each location is 0.72 s on the computer. It takes 1.32 s to locate on the smartphone. After using the two-level query, the average time of each location is 1.11 s. Positioning time consumption is less than that in the hall due to the small image database. Obviously, the query efficiency is much higher after using the two-level query strategy. And it doesn't need extra infrastructures by using images as index of image database.

Table 2. Comparison of average time in positioning process

	Computer	Smartphone	Smartphone+secondary query
The hall	0.72	1.32	1.11
The corridor	0.68	1.07	-

6 Conclusion

In this paper, we propose a lightweight indoor location method based on smartphone. We use mobilenet neural network to extract image features and our feature vectors database is only 6.1MB, which makes localization smartphone based possible. In addition, to reduce the time consumption of query, we use two-level retrieve to find the best matching images. We improve the positioning accuracy by using query expansion and orientation signal. We propose a method of video to build database and verify its feasibility, which greatly reduces the time and labor consumption of image database building. Experimental results show that 90% positioning error is within 1.5 m and our average positioning time is 1.11 s. In general, our method has high real-time performance and positioning accuracy.

References

1. Chen, Y., Ruizhi, C., Mengyun, L., Aoran, X., Dewen, W., Shuheng, Z.: Indoor visual positioning aided by CNN-based image retrieval: training-free, 3D modeling-free. Sensors **18**(8), 2692 (2018)
2. Chum, O., Philbin, J., Sivic, J., Isard, M., Zisserman, A.: Total recall: automatic query expansion with a generative feature model for object retrieval. In: 2007 IEEE 11th International Conference on Computer Vision (2007)
3. Guo, Y., Yang, L., Li, B., Liu, T., Liu, Y.: Rollcaller: user-friendly indoor navigation system using human-item spatial relation. In: IEEE INFOCOM 2014-IEEE Conference on Computer Communications (2014)
4. Han, X., Zheng, Y., Zhou, Z., Shangguan, L., Ke, Y., Liu, Y.: Indoor localization via multi-modal sensing on smartphones (2016)
5. Iandola, F.N., Han, S., Moskewicz, M.W., Ashraf, K., Dally, W.J., Keutzer, K.: Squeezenet: Alexnet-level accuracy with 50x fewer parameters and <0.5mb model size (2016)
6. Jiang, W., Yin, Z.: Indoor localization with a signal tree. Multimed. Tools Appl. **76**(19), 20317–20339 (2017). https://doi.org/10.1007/s11042-017-4779-6
7. Li, M., Liu, N., Niu, Q., Liu, C., Chan, S.G., Gao, C.: Sweeploc: automatic video-based indoor localization by camera sweeping. IMWUT **2**(3), 120:1–120:25 (2018). https://doi.org/10.1145/3264930
8. Wang, L., Liu, H., Liu, W., Jing, N., Adnan, A., Wu, C.: Leveraging logical anchor into topology optimization for indoor wireless fingerprinting. Comput. Mater. Continua **58**(2), 437–449 (2019). https://doi.org/10.32604/cmc.2019.03814. http://www.techscience.com/cmc/v58n2/23019
9. Liu, K., Liu, X., Xie, L., Li, X.: Towards accurate acoustic localization on a smartphone. In: 2013 Proceedings IEEE INFOCOM, pp. 495–499. IEEE (2013)

10. Wang, M., Niu, S., Gao, Z.: A novel scene text recognition method based on deep learning. Comput. Mater. Continua **60**(2), 781–794 (2019). https://doi.org/10.32604/cmc.2019.05595. http://www.techscience.com/cmc/v60n2/23061

11. Noelia, H., Manuel, O., Jose, A., Euntai, K.: Continuous space estimation: increasing wifi-based indoor localization resolution without increasing the site-survey effort. Sensors **17**(12), 147 (2017)

12. Sadeghi, H., Valaee, S., Shirani, S.: Ocrapose: an indoor positioning system using smartphone/tablet cameras and OCR-aided stereo feature matching (2015)

13. Sandler, M., Howard, A., Zhu, M., Zhmoginov, A., Chen, L.C.: Inverted residuals and linear bottlenecks: mobile networks for classification, detection and segmentation. arXiv preprint arXiv:1801.04381 (2018)

14. Wang, J., Zha, H., Cipolla, R.: Coarse-to-fine vision-based localization by indexing scale-invariant features. IEEE Trans. Syst. Man Cybern. Part B **36**(2), 413–422 (2006)

15. Wang, S., Fidler, S., Urtasun, R.: Lost shopping! monocular localization in large indoor spaces. In: 2015 IEEE International Conference on Computer Vision (ICCV) (2015)

16. Werner, M., Kessel, M.: Indoor positioning using smartphone camera (2011)

17. Wu, K., Jiang, X., Yi, Y., Chen, D., Ni, L.M.: Csi-based indoor localization. IEEE Trans. Parallel Distrib. Syst. **24**(7), 1300–1309 (2013)

18. Xia, Y., Xiu, C., Yang, D.: Visual indoor positioning method using image database, pp. 1–8 (2018). https://doi.org/10.1109/UPINLBS.2018.8559714

19. Wei, Y., Wang, Z., Guo, D., Yu, F.R.: Deep q-learning based computation offloading strategy for mobile edge computing. Comput. Mater. Continua **59**(1), 89–104 (2019). https://doi.org/10.32604/cmc.2019.04836. http://www.techscience.com/cmc/v59n1/27923

20. Zhang, G., Rao, S.V.: Position localization with impulse ultra wide band (2005)

21. Zhang, W., Liu, G., Tian, G.: A coarse to fine indoor visual localization method using environmental semantic information. IEEE Access **7**, 21963–21970 (2019)

22. Zhang, X., Zhou, X., Lin, M., Sun, J.: Shufflenet: an extremely efficient convolutional neural network for mobile devices (2018)

Performance Analysis and Optimization for Multiple Carrier NB-IoT Networks

Sijia Lou[1,2], En Tong[2], and Fei Ding[1]([⊠])

[1] Jiangsu Key Laboratory of Broadband Wireless Communication and Internet of Things,
Nanjing University of Posts and Telecommunications, Nanjing, China
dingfei@njupt.edu.cn
[2] China Mobile Group Jiangsu Co., Ltd., Nanjing, China

Abstract. With the increase of NB-IoT traffic, throughput and concurrent users, the resource of 200 K in a single NB-IoT cell becomes more and more limited. In this paper, we use a PRB of FDD cell to enable in-band operation mode of NB-IoT, which compose a multi-carrier cell with the original SA cell. Traffic equalization and Narrowband IoT Load Distribution are utilized to improve the system capacity by ensuring the absorption of traffic in IB cell. At last, we verify the network optimization we proposed has a positive effect on the system performance through a practical deployment of a commercial case.

Keywords: Internet of Things · NB-IoT · Multi-carrier cell · Traffic equalization

1 Introduction

With the continuous development of mobile communication technology, the interconnection of everything has become an inevitable trend [1]. The application scenario of next generation communication networks includes high rate service and low rate service, among which the low rate applications, such as internet of things (IoT), have not yet been well supported by mature cellular technology. By 2021, the number of IoT terminals will be more than twice that of Internet terminals, while maintaining an annual compound growth rate of more than 20% [2]. Therefore, the demand for low-speed communication services is very large. In order to adapt the characteristics and needs of low rate businesses, narrowband Internet of things (NB-IoT) was proposed. NB-IoT is a burgeoning radio access technology based on cellular network defined by 3GPP, which works in authorized frequency band [3–6]. The uplink and downlink transmission rates supported by NB-IoT are within 250 kbits. Its channel bandwidth is limited to 180 kHz. NB-IoT has the characteristics of strong anti-interference, high reliability, wide coverage and so on. Consequently, it works well for low-rate communication businesses [7].

At present, the research on NB-IoT is mainly focused on latency, security, availability, coverage area and energy consumption. Most of the existing literatures give the performance evaluation of NB-IoT by means of measurement or simulation [8–15]. The result is different because of the different scenes, and the research on the design of NB-IoT optimization is even less. In reference [16, 17], the coverage performance of NB-IoT

© Springer Nature Switzerland AG 2020
X. Sun et al. (Eds.): ICAIS 2020, LNCS 12239, pp. 775–785, 2020.
https://doi.org/10.1007/978-3-030-57884-8_68

network deployed by operators is tested on the current network, in which, reference [16] gives the configuration suggestions for the number of preamble repetition, in reference [17], the coverage performance and capacity performance of LTE-M and NB-IoT are simulated and measured both in typical urban and rural channel propagation models. Furthermore, due to the massive data volume in NB-IoT, requirement for latency and capacity of communication networks are increasingly stringent. When numerous NB-IoT terminals are connecting to the network, it is fatal to improve the network capacity and random access performance of system. In this paper, we focus on improving NB-IoT network performance through multi-carrier networking in commercial environment without increasing the consumption of network resources. Our main contributions can be summarized as follows:

- We construct a network coverage model which combining SA and IB cells. For comparison, we also construct a single SA cell. We use key performance indicators (KPIs) of network optimization to compare the performance of multi-carrier scenarios with that of single carrier scenarios.
- For the condition that massive NB-IoT devices access simultaneously, we utilize traffic balance adjustment and Narrowband IoT Load Distribution (NILD) to make sure that the IB cell can effectively share traffic load. Then, we compare KPIs of system before and after optimization.
- Furthermore, we implement the proposed scheme in a limited area of cellular networks, instead of simulate on a computer. We analyze the system performance such as proportion of online terminals, communication success rate, and concurrent access performance. It is proved that the multi-carrier scenario can effectively improve the capacity of NB cell and realize load balancing between different cells.

In Sect. 2, we describe NB-IoT operation modes, system model and Multi-carrier parameters configuration. In Sect. 3, we analyze performance of multi-carrier cell, and propose two optimization methods. Then we analyze system performance of a commercial multi-carrier case in Sect. 4. Finally, we conclude this paper in Sect. 5.

2 Related Work and System Model

2.1 NB-IoT Operation Modes

NB-IoT networks occupies 180 kHz bandwidth, which is the same as the bandwidth of a resource block in LTE frame structure. Therefore, there are three operation modes as shown in Fig. 1.

Stand-alone (SA) Operation: It can be completely decoupled from LTE, which is suitable for re-cultivating GSM frequency band. The bandwidth of GSM is 200 kHz, which just makes room for 180 kHz NB-IoT bandwidth, and two 10 kHz protection intervals.

Guard-band (GB) Operation: The 180 kHz unused bandwidth in LTE protection band is used for NB-IoT networks, whose advantage is no additional resource occupied.

In-band (IB) Operation: In this mode, NB-IoT networks occupy 1 physical resource block (PRB) resource of LTE.

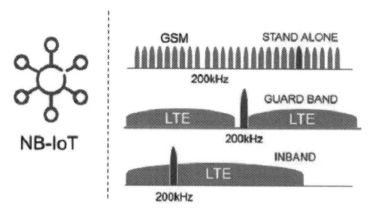

Fig. 1. The NB-IoT operation modes.

2.2 System Model

With the increase of NB-IoT traffic, throughput and concurrent users, the resource of 200 K in single NB-IoT cell becomes more and more limited. We attempt to enable a NB-IoT IB cell by using a PRB resource on the same station of FDD 900 MHz. The IB cell and the original SA cell compose a multi-carrier cell, that is, the multi-carrier technology improves capacity of the cell by using multiple 200 K spectrum resources at the same time.

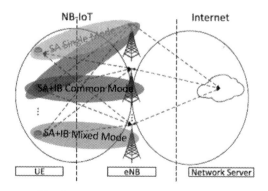

Fig. 2. Three mode of NB-IoT networks.

As shown in Fig. 2, we construct a set of NB-IoT cells, whose configurations are completely same with the commercial network. The system model contains narrowband user equipment (UE), eNodeB, Evolved Packet Core (EPC), and applications on Internet. There are three scenarios we analyzed in this paper with different operation modes.

SA Single Scenario. NB-IoT networks Occupy one 200 kHz bandwidth, which is used as reference in this paper. Limited by spectrum resources, communication failure is easy to occur when a large number of devices access simultaneously.

SA+IB Mixed Scenario. As shown in Fig. 3, the area is covered by multi-carrier network which is composed by SA and IB cells on the basis of existing hardware. The capacity of single cell improves by using multiple 200 K spectrum resources at the same time.

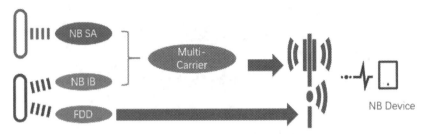

Fig. 3. SA+IB mixed scenario

SA+IB Common Scenario. This Area is covered by the main service cell which worked on SA mode, meanwhile covered by one IB network from another cell. The research of this scenario focuses on the effect of IB cells on neighborhood performance.

2.3 Multi-carrier Parameters Configuration

In this section, we introduce the network parameter configuration of a typical SA cell, and the parameters we need to modify to construct a SA+IB mixed scenario. The parameters of original SA cell are set as follows:

Table 1. Typical configuration of SA cell.

Indicator	Value
Reference signal power	32 dbm
UE Maximum transmitting power	23 dbm
Paging cycle	2.56 s
Paging density	T/64
earfcn	3734, 3736, 3738
RRC connection establishment timer	20 s
TAC	Re-planning
Minimum access level	−130 dbm

The modified parameters involved in opening in band cell on FDD 900 MHz are set as follows:

Other parameters follow the reference value of NB-IoT SA cell. The parameters configuration of NB-IoT original SA cell does not make any changes (Tables 1 and 2).

Table 2. Network configuration of IB cell.

Parameters	Value	Notes
nbIotCellType	1	NB Operation Mode set to 1 (NBIOT_INBAND)
allocatedPrbIndex	19	Set the downlink PRB of NB-IoT
allocatedUlPrb	0	Set the Uplink PRB of NB-IoT
earfcn	3611	Frequency point
physicalLayerCellId	Same as FDD Cell	PCI is set as same as FDD cell
TAC	Re-planning	Cannot be same as the FDD cell
eutranCellRef	FDD cell in same station	Specify FDD cell as the Host cell

3   Analysis and Optimization of Multi-carrier NB Cell

Facing the increasing capacity demand of NB-IoT, we firstly analyze the capacity and access success rate of three scenarios mentioned above. Then we utilize traffic balance adjustment and narrowband IoT load distribution to make sure that the IB cell can effectively balance traffic load, what tends to increase overall capacity.

3.1   Performance Analysis of Multi-carrier Cell

Comparison of NB-IoT Indicators Between Before and After Using Multi-carrier. Using original SA cell as a reference, the network indicators are significantly improved after IB cell is enabled, which expressed as the following four aspects.

1. IB cell has positive effect on network performance RRC success rate increased from 93.03% to 95.28%, meanwhile paging success rate increased from 87% to 97.58%. That means the network can support more terminals to access at the same time.
2. Because IB cell shares part of the businesses, the maximum RRC connections in the cell reduced from 56 to 44, and the uplink and downlink traffic also has different degrees of reduction. The utilization of NPDCCH, NPDSCH and NPUSCH decreased by 3.46%, 2.89% and 3.35% respectively.
3. The average background noise reduced by about 2 dbm.
4. After enabling IB cell, the general trend of performance in SA+IB scenario is consistent with that in SA scenario (Table 3).

Comparison of LTE Indicator of Related Cells. In order to avoid the NB-IoT multi-carrier we utilized influencing the existing LTE networks, we compared the KPIs of the related cells before and after enabling multi-carrier. As the result shows in Table 4, the main indicators of FDD cells remain stable and unaffected by IB deployment.

Table 3. NB-IoT indicators of three scenarios.

Indicator	SA cell without IB	SA cell with IB	SA + IB Mixed Cell
RRC success rate	93.03%	95.28%	96.00%
Maximum RRC connections	56	44	62
Background noise of the cell	−100.41 dbm	−102.01 dbm	−102.19 dbm
Paging success rate	87%	97.58%	98.65%
Uplink traffic	2007.55 KB	1737.18 KB	2102.66 KB
Downlink traffic	1125.48 KB	1121.18 KB	1345.78 KB
NPDCCH average resource utilization	19.69%	16.23%	11.25%
NPDSCH average resource utilization	17.01%	14.18%	10.83%
NPUSCH average resource utilization	11.32%	7.97%	6.61%

Table 4. Comparison of LTE indicators.

Indicator	Before IB deployment	After IB deployment
Connection rate	99.84%	99.86%
Switch success rate	99.36%	99.35%
Uplink traffic	2.26 GB	2.20 GB
Downlink traffic	31.11 GB	31.76 GB

3.2 Parameters Optimization of Multi-carrier Cell

Multicarrier scheme is proved to improve the capacity of NB-IoT networks by using multiple 200 K spectrum resources. However, the coverage of IB cell is relatively weaker than SA cell. From the view of KPIs, it can also be seen that the traffic absorption of IB cell is about 15% of the whole traffic, which fails to effectively realize traffic equalization of SA cell. In this section, two optimization schemes are considered.

Adjustment of Traffic Equalization. We attempt to enhance the traffic absorption ability of IB cell and balance the traffic between IB cell and SA cell by adjustment of reselection threshold and transmission power.

Parameter adjustment strategy:

1. the power ratio between SA cell and IB cell is set to 1:8.
2. The reselection threshold of IB cell is set to −96 dm for both same/different frequency network.

3. Frequency measurement of different frequency in IB cell only add frequency points of SA cells in the same sector.
4. Measurement frequency points in the neighborhood with a radius of 800 m add IB cell frequency points.

As shown in the table, the traffic proportion of IB cell is 15%–23% of that of SA cell after parameter adjustment, which means the proportion of absorbed traffic in IB cell is still low (Table 5).

Table 5. NB-IoT indicators of three scenarios.

Indicator	SA cell without IB	SA cell with IB	IB Cell
pmRrcConnEstabAtt	2519	1538	187
RRC success rate	97.78%	98.63%	100%
Uplink traffic	485.5 KB	295.4 KB	37.9 KB
Downlink traffic	325.9 KB	200.2 KB	26.2 KB

Fig. 4. Process of NB-IoT load distribution

Narrowband IoT Load Distribution. Compared with SA carrier, the power of IB carrier is significantly weaker, even if the power of FDD cell what IB carrier belongs to is increased to 40 W, the power of corresponding SA cell is decreased to 5 W. Reference signal powers are 24.2 dBm(IB)/26.2 dBm(SA). Power of SA cell is slightly stronger. We adopt the technique named Narrowband IoT load distribution to increase the traffic absorbed by IB cell. As shown in the Fig. 4, narrowband IOT load distribution will require UE to re-direct according to the pre-configured frequency point list and the proportion of terminals requiring load distribution during RRC release.

We configure 65% of terminals using load distribution at 5 frequency points, and the proportion of 5 frequency points is 10%, 15%, 12%, 25%, 3%. Then the algorithm will randomly select frequency points for UE according to the configured proportion. If one of the five configured frequency points is not supported by UE, the algorithm will allocate the remaining four frequency points to 65% of UE.

The current configuration makes the terminal approximately redirect to the IB cell. The daily average traffics of uplink and downlink in IB cell are increased by about 5 times and 4 times respectively. The application of narrowband IoT load distribution greatly improves the absorption of traffic in IB cell.

Multi-carrier scheme with narrowband IoT load distribution can decrease NPDCCH utilization of SA cell by 15%, while the KPIs of co-sector FDD cell remains on normal level. It has a positive effect on the traffic balance of the original coverage SA cell and reduces the utilization of the original SA cell (Table 6).

Table 6. Traffic of IB cell after NILD enabled.

Status	Round	pmRrcConnEstabAtt	RRC success ratio	Maximum RRC connections	Uplink traffic	Downlink traffic
NILD unable	Round 1	1627	99.08%	8	305.375	216.625
	Round 2	1654	99.70%	12	353.125	221.375
	Round 3	1773	99.55%	11	324.875	237.5
NILD enabled	Round 1	3933	99.36%	39	859.625	533.25
	Round 2	6333	99.69%	47	1389.875	878.5
	Round 3	7201	99.64%	43	1562.75	992.125

4   Performance Analysis

In this section, we analyze the performance of a commercial NB-IoT system in practical application scenarios. We have deployed over 30000 street lamps in this case, most of them in urban areas. Each cell participating in this performance analysis contains about 60 terminals. ALL the UEs attempt to access simultaneously when it is getting dark, and keep transmitting every 30 min. This kind of communication model has heavy load on NB-IoT networks. There are performance bottlenecks in the existing operation mode. So we deploy multi-carrier cell to solve the dilemma.

Figures 5 and 6 shows the UE online rate before and after optimization mentioned above. The comparison is shown in the table. In the construction stage of the project, we found that the UE online rate and communication success rate were lower than average in Xintang north road. Then SA+IB multi-carrier cell and NILD were used in this area. As the results shows, the UE online rate increased by 24 pp after optimization (Table 7).

Fig. 5. UE online rate before optimization.

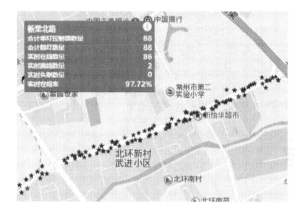

Fig. 6. UE online rate after optimization.

Table 7. Comparison of business indicators.

Indicator	Before optimization	After optimization
Number of UE	59	88
Number of online	43	86
Number of offline	0	2
Number of lost	16	0
Online rate	72.88%	97.72%

5 Conclusion

In this paper, we construct a network coverage model which combining SA and IB networks, and utilize traffic balance adjustment and Narrowband IoT Load Distribution to make sure that the added IB cell have a positive impact on performance indicators, such as cell capacity, load balancing and communication success rate. This scheme is applied to Changzhou smart lamp commercial project. The experimental results prove our point of view.

Acknowledgements. This work is partially supported by the National Natural Science Foundation of China (No. 61427801); the Ministry of Education–China Mobile Research Foundation, China (No. MCM20170205); the Communication Science Research Project of Ministry of Industry and Information Technology, China (No. 2019-R-26); the Six Talent Peaks Project of Jiangsu Province, China (No. DZXX-008); the Postdoctoral Science Foundation, China (Nos. 2019M661900 and 2019K026); the NUPTSF (Nos. NY217146 and NY220028); the STITP of NUPT (No. SYB2019027).)

References

1. Akpakwu, G.A., Silva, B.J., Hancke, G.P., Abu-Mahfouz, A.M.: A survey on 5G networks for the internet of things: communication technologies and challenges. IEEE Access **6**, 3619–3647 (2018)
2. Mattisson, S.: An overview of 5G requirements and future wireless networks: accommodating scaling technology. IEEE Solid-State Circuits Mag. **10**(3), 54–60 (2018)
3. Overall Description: Stage 2, document 3GPP TS 36.300 v13.4.0 Release 13, 3GPP, June 2016
4. Cellular System Support for Ultra Low Complexity and low Throughput Internet of Things, document 3GPP TR 4S.820 v1.3.1 Release 13, 3GPP (2015)
5. Oh, S.-M., Jung, K.-R., Bae, M.S., Shin, J.: Performance analysis for the battery consumption of the 3GPP NB-IoT device. In: Information and Communications Technology Council (ICTC), pp. 981–983 (2017)
6. Ratasuk, R., Mangalvedhe, N., Zhang, Y.: Overview of narrowband IoT in LTE Rel-13. In: 2016 IEEE Conference on Standards for Communications and Networking, pp. 1–7 (2016)
7. Zayas, A.D., Merino, P.: The 3GPP NB-IoT system architecture for the Internet of Things. In: IEEE International Conference on Communications, pp. 277–282 (2017)
8. Ratasuk, R., Vejlgaard, B., Mangalvedhe, N., et al.: NB-IoT system for M2M communication. In: 2016 IEEE Wireless Communications and Networking Conference, pp. 1–5 (2016)
9. He, X.D., Song, L.: NB-Iot uplink coverage capability considering rate requirement. Telecommun. Sci. **S1**, 149–156 (2016)
10. Huang, Y., Tang, Y.: Discussion on networking and coverage capability for NB-IoT. Mob. Commun. **41**(18), 11–15 (2017)
11. Li, J.: Research on NB-IoT networking scheme. Mob. Commun. **41**(6), 14–18 (2017)
12. Chen, W.: Simulation analysis and verification of single station coverage based on NB-IoT technology. Comput. Telecommun. **8**, 36–38 (2017)
13. Zhao, Y., Zhang, L.F., Xing, Y.L.: Research on technology evaluation and networking scheme of NB-IoT. Designing Tech. Posts Telecommun. **8**, 40–45 (2017)

14. Lauridsen, M., Kovács, I.Z., Mogensen, P., et al.: Coverage and capacity analysis of LTE-M and NB-IoT in a rural area. In: 2016 IEEE 84th Vehicular Technology Conference, pp. 1–5 (2016)
15. Adhikary, A., Lin, X., Wang, Y.P.E.: Performance evaluation of NB-IoT coverage. In: 2016 IEEE 84th Vehicular Technology Conference, pp. 1–5 (2016)
16. Zhao, Y., Zhang, L., Xing, Y.: Research on technology evaluation and networking scheme of NB-IoT. Designing Tech. Posts Telecommun., 40–45 (2017)
17. Lauridsen, M., Kovacs, I.Z., Mogensenp, P.: Coverage and capacity analysis of LTE-M and NB-IoT in a rural area. In: 2016 IEEE 84th Vehicular Technology Conference, pp. 1–5 (2016)
18. Xiao, D., Liang, J., Ma, Q., Xiang, Y., Zhang, Y.: High capacity data hiding in encrypted image based on compressive sensing for nonequivalent resources. Comput. Mater. Continua **58**(1), 1–13 (2019)
19. Luo, M., Wang, K., Cai, Z., Liu, A., Li, Y., Cheang, C.F.: Using imbalanced triangle synthetic data for machine learning anomaly detection. Comput. Mater. Continua **58**(1), 15–26 (2019)
20. Deng, Z., Ren, Y., Liu, Y., Yin, X., Shen, Z., Kim, H.-J.: Blockchain-based trusted electronic records preservation in cloud storage. Comput. Mater. Continua **58**(1), 135–151 (2019)

A Short Flows Fast Transmission Algorithm Based on MPTCP Congestion Control

Hua Zhong, PingPing Dong$^{(\boxtimes)}$, WenSheng Tang, Bo Yang, and JingYun Xie

Hunan Provincial Key Laboratory of Intelligent Computing and Language Information Processing, Hunan Normal University, Changsha 410081, China
ppdong@csu.edu.cn

Abstract. With the development of cloud computing and big data technology, traffic in the network presents new features. Real-time online interactive applications are increasing, mainly including web services, instant messaging, online commerce, advertising systems. They have very stringent latency requirements. However, the data traffic generated by data backup, movie download, data mining require high bandwidth. When large and small flows are transmitted concurrently, a series of problems will arise. MPTCP treats long and short flows equally. In this way, packets of long flows will preempt bandwidth at the bottleneck link, fill the router's whole queue buffer. This will cause the short flows transmission timeout. To solve these problems, we propose a novel algorithm named MPTCP-SPTA (MPTCP short flows fast transmission algorithm). The algorithm give high priority to short flows based on the congestion control algorithm, then introduce the Explicit Congestion Notification mechanism(ECN) in MPTCP. Through extensive network simulator (NS-3) simulations, we demonstrate that MPTCP-SPTA outperforms TCP and MPTCP in three typical network topology, which can significantly reduce the completion time of the small flow and effectively improving the total throughput of the network.

Keywords: Small flow priority · MPTCP · ECN · Congestion control

1 Introduction

With the rapid development of cloud computing and big data technologies, the traffic in the network takes on the characteristics of long and short flows transmitted concurrently [1, 2]. The network traffic is composed of numerous latency-sensitive small flows which are consisted of only several packets and a few throughput-sensitive long flows which occupy more than 80% of data flow [3, 4]. For example, real-time applications like web search and online retail generate short flows, whereas background traffic from

This work was supported in part by the National Natural Science Foundation of China under Grant 61602171 and Grant U1636106, in part by Natural Science Foundation of Hunan Province under Grant 2017JJ3223, in part by the Science and Technology Project of Hunan Province Department of Education under 16B1179, and in part by Hunan Province's Strategic and Emerging Industrial Projects under Grant 2018GK4035.

© Springer Nature Switzerland AG 2020
X. Sun et al. (Eds.): ICAIS 2020, LNCS 12239, pp. 786–797, 2020.
https://doi.org/10.1007/978-3-030-57884-8_69

data migration and backup generate long flow [5, 6]. When long and short flows are transmitted concurrently, the two serious problems are brought about in the network. First, the large flows will aggress bandwidth of small flows, causing its transmission time to be increased at the bottleneck link. Furthermore, a few large flows will exhaust the limited buffer in queue, causing small flows packets to be dropped in the router. These questions will be more serious with MPTCP. MPTCP is a set of extensions to TCP, which distributes TCP traffic among sub-flows to improve the end-to-end throughput [9]. Prior work has shown that MPTCP is not good at handling short flows, when long and shot flows are transmitted over multiple paths [10–12]. The reason lies in that. The router ports share the buffer queue pattern, and its buffer is very limited [13]. When multiple sub-flows are aggregated on the router, packets of long flows will occupy the queue in the router quickly, making small flows packets to be dropped easily [14, 15]. To solve this problem, some researchers have done much positive research.

Mohammad Alizadeh et al. proposed the DCTCP algorithm which uses the ECN (Explicit Congestion Notification) mechanism to effectively suppress the long flows in the queue buffer of the router. The DCTCP can deal with the tail drop of the small flows because it keeps the average queue at a lower value [3]. However, DCTCP is not suitable for MPTCP. YuCao et al. proposed the XPM algorithm which suppresses links buffer occupancy in MPTCP by controlling the transmission rate of long flows. But it also reduces the transmission of the small flows [15]. Morteza Kheirkhah et al. proposed the AMP algorithm in MPTCP to mitigates buffer inflation and achieve high fairness by turning off some sub-flows with poor performance [14]. However, it does not consider the small flows fast transmission problem when long and short flows are transmitted simultaneously. To solve the small flows fast transmission problem, Wenfei Wu et al. proposed scheme that is designed to steal bandwidth from large flows over time and reallocate to small flows. The long flows gain more transmission time because the short flows can be transmitted quickly. However, the ATCP does not consider the queue buffer problem of the router in the network [19].

This observation motivates us to give high weight to short flows, which control a flow's rate and prefer small flows to large flows at bottleneck link, then introduce the ECN mechanism to suppress the rapid growth of queue in router. Based on the above observations, we propose a short flows optimized transmission algorithm, namely MPTCP-SPTA. It can well solve the problem that long flows invades resources of small flows in the network when the long and short flows transmit concurrently. In this paper, we make three significant contributions.

- We propose a mechanism which give high priority to short flows based on the MPTCP congestion control. It can allocate sufficient bandwidth to the small flows at the bottleneck links to ensure short flows fast transmission while increasing the throughput of long flows.
- Then, MPTCP-SPTA is proposed by introducing the ECN(Explicit congestion notification) into the new congestion control mechanism mentioned above. It forms the second contribution of this paper. It can keep the average queue of the router in the lower value to reduce latency and avoid packet loss while improving network performance.

• We validate the effectiveness of MPTCP-SPTA by conducting extensive simulations experiments in NS-3. We find that MPTCP-SPTA can reduce the completion time of small flows while improving the throughput of long flow. Thus delay-sensitive applications see the most significant improvements. The results demonstrate that MPTCP-SPTA improves 71% of the total throughput, deduces 87% of the completion time of small flows compared with MPTCP.

2   Related Work

There have been significant studies to reduce the completion time of the short flows. These studies can mainly summarize two types of explicit congestion control and implicit congestion control. We briefly review these relevant works in this section.

Explicit congestion control means that through the cooperation of the sender and receiver intermediate router, the congestion of the intermediate path is perceived. Then it actively slows down the transmission rate of TCP to avoid early congestion. DCTCP use the ECN mechanism to react to the extent of congestion estimated from the fraction of marked packets [3]. TDCTCP improves on the existing DCTCP protocol, which designed to increase throughput without significantly prolonging end-to-end latency [2]. However, it needs to modify the sender, receiver, and router in the network. TDCTCP affects the end-to-end delay of the packet when the queue length of TDCTCP is higher than DCTCP. DT-DCTCP improves the DCTCP algorithm by introducing a new marking mechanism for detecting network congestion [17]. It uses two thresholds to share the load of a single threshold K used in DCTCP. These algorithms reduce the completion time of small flows from different aspects. But they are designed specifically for TCP rather than for the MPTCP. XPM and AMP algorithms are using explicit congestion control in MPTCP [18]. XPM consists of the BOX and Trash algorithm [15]. BOX can suppress the long flows consuming link queue buffer. Trash can transfer traffic from the congestion path to the non-congested path, thus reducing the completion time of the short flows and achieving high throughput. However, XPM can't deal effectively with questions of TCP Incest and Minimum Window Syndrome. AMP can quickly detect the onset of these problems and switch from multiple paths flows to a single-path flow [14].

In the implicit congestion mode, we mainly summarize from two aspects of TCP and MPTCP. ATCP reduces the small flows completion time by controlling the flow rate of TCP [19]. Then short flows of ATCP are given high priority according to the size of the data flow in TCP. DMPTCP proposes the analytical model to predict the transmission time of the data flow [20]. It can predict the amount of data transmitted between the long flows and the small flows, then selects the best sub-flow to transfer data. MPTCP-TOASF firstly uses an optimal path set to transmit small flows, then switches to MPTCP for transmission of long flows [21]. MMPTCP runs in two phases. Initially, MMPTCP randomly selects one of path in all available paths to transmit the small flows then runs MPTCP mode to improve the throughput of long flows [22]. MPTCP-SF sorts the path according to the RTT of sub-flows [23]. If the data traffic is less than 100 K, MPTCP-SF selects the smallest sub-flow. For data traffic between 100 kB and 400 KB, the second-best sub-flow is selected. If the data traffic is more than 400 KB, all sub-flows are selected.

3 Motivation

Nowadays, there is a wide variety of applications on the web, which also introduce flows with very different properties [24]. On the one hand, long flows want to achieve as high throughput as possible and thus occupancy the entire bandwidth in bottleneck link and queue buffer in router. Small flows, on the other hand, are much more concerned with latency, because they have only tens of kilobytes for transfer. More importantly, small flows have strict deadlines [20, 21]. Therefore the long and short flows inevitably compete with each other for network resources, thus affecting the completion time of small flows. These problems will get worse in the MPTCP.

To illustrate the issues, we conduct our experiments with the topology shown in Fig. 1 that the client S1 and server R1 are connected through router R1. There are two paths that share with the bottleneck link. Figure 2 shows that the completion time of the small flows with the background flows of 10 MB is higher than the completion time of the small flows without background flows. Figure 3 shows the completion time of the small flows under the MPTCP and TCP, when the small flows changes from 10 KB to 200 KB in 10 MB background flow. The completion time of the small flows in MPTCP is significantly worse than completion time of the small flows in TCP. The main reasons for this phenomenon are as follows. First, MPTCP leads to increase delays of small flows because the more sub-flows are, the bigger the probability of long flows preempt short flows. Second, the congestion window of small flows may be tiny in its lifetime. A single lost packet can force an entire connection to wait for an RTO to be triggered.

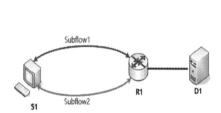

Fig. 1. The topology of shared links.

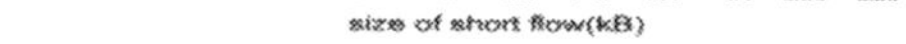

Fig. 2. The completion time of the small flows in the case of background flow and no-background flow in the topology of shared links.

To further analyze the root cause of this phenomenon, we test the instantaneous buffering and number of re-transmission in the topology of shared links. In Fig. 4 the small flows is set to 200 kB, the long flow is 10 MB, and the link buffer is 120 packets. Figure 4 shows the instantaneous buffer and number of timeout between MPTCP and TCP. Their instantaneous buffer and timeout reach the maximum at 2 s. With the triggering of the congestion control mechanism, the instantaneous buffer and timeout will gradually decrease, then these values reach the maximum again. This observation motivates us to design a novel algorithm which allocates more bandwidth to small flows in bottleneck link and suppress the rapid growth of the buffer queue in the router. In the rest of this paper, we present MPTCP-SPTA as well as its performance validation with extensive experiments.

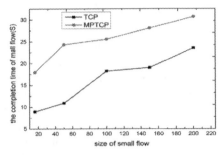

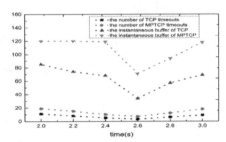

Fig. 3. The completion time of the small flows in the case of TCP and MPTCP in the topology of shared links.

Fig. 4. The instantaneous buffer and the number of timeout in Shared links topology under with background flow(10 MB).

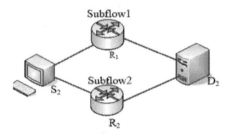

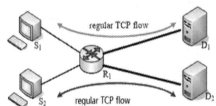

Fig. 5. The topology of no-shared links.

Fig. 6. The topology of the background with the regular TCP flow.

4 Algorithm Design

The goal of the MPTCP-SPTA algorithm is to solve the problem of long flows preempting small flows resources in bottleneck link and router. In the section, we firstly give high priority to short flows in the OLIA algorithm then introduce the ECN mechanism.

There are various congestion control mechanisms in MPTCP, such as EWTCP [21], COUPLED [22], LIA [23], OLIA [24]. EWTCP is to solve the problem of TCP and MPTCP competition bottleneck link bandwidth. COUPLED can migrate data flow from the congested path to the other path, however it does not take into account the fact that the RTT changes frequently. From the perspective of bottleneck link equity, LIA designs an α according to RTT to control window changes, however it does not meet Pareto-optimization' requirements [25]. OLIA improves LIA from the perspective of fairness and optimal resource pools to maximize resource utilization. So, we apply the weight of short flow priority to OLIA.

OLIA firstly defines two sets. M(t) is the set of paths with the maximum congestion window(CWND) at time t. B(t) indicate the set of optimal paths at time t.

$$M(t) = \left\{ i(t) | i(t) = \arg \max_{p \in Ru} Wp(t) \right\} \tag{1}$$

$$B(t) = \left\{ j(t) | j(t) = \arg \max_{p \in Ru} \frac{ep(t)}{rttp(t)^2} \right\} \tag{2}$$

The $W_p(t)$ and $RTT_p(t)$ respectively are CWND and round trip time of path p at time t. R_u is the set of paths available to user U, and $r \in Ru$ is a path. $E_p(t) = \max\{E_{1r}(t), E_{2r}(t)\}$. $E_{1r}(t)$ is the number of bits that were successfully transmitted by user U over path P between the last two packet losses. $E_{2r}(t)$ is the number of bits that successfully transmitted over P after the last loss.

According to ATCP, the weight of short flows priority can be calculated by Eq. 3 [19]. It can prioritize short flows bandwidth allocation and make them complete faster than the long flows when long and short flows are contending.

$$Q_r(t) = \begin{cases} q_H & S < T \\ q_L & S \geq T \end{cases} \tag{3}$$

Let S denote the number of bytes sent by the flow, T is the threshold between the larger and short flows. By the short flows priority weight, the flows that are smaller than T will obtain by the high weight, vice versa. If the long flows is sending in low weight, the short flow will be more aggressive. The weight of short flows priority does not result in long flow starvation. Therefore, a new OLIA mechanism is as follows.

- ☐For each ACK on path P, increase Wr(t+1) by:

$$\frac{w_r(t)/rtt_r^2}{\left(\sum_{p \in Ru} w_p/rtt_p\right)^2} * Q_r(t) + \frac{a_r(t)}{w_r(t)} \tag{4}$$

- For each loss on path P, decrease Wr(t+1) by.☐

$$\frac{w_r(t)}{2} \tag{5}$$

Where $\alpha_r(t)$ calculated as follows.

$$ar = \begin{cases} \frac{1/|R_u|}{|B/M|} & r \in B\backslash M \\ -\frac{1/|R_u|}{|M|} & r \in M \cap B\backslash M \neq \theta \\ 0 & otherwise \end{cases} \tag{6}$$

$a_r(t)$ is the weight of the optimal path relative to the growth of the window. The OLIA with $Q_r(t)$ can dynamically adjust the CWND of each sub-flow based on data flow and path quality. If the data traffic is short flows, the path will takes high weight qH that its CWND grows faster. If the data traffic is long flows, the path will obtain low weight qL that its CWND grows slowly.

The competition between long and short flows occurs not only in bottleneck links but also in routers in the network. The ECN mechanism can solve the problem of long flows preempt resources of small flows in routers when long and small flows transmitted simultaneously [3].

The ECN performs the Exponentially Weighted Moving Average(EWMV) algorithm to estimate average queue length. It assigns a probability of marking the arriving packet with the Congestion Experienced (CE) code-point in the IP header. When receiving a

CE signal, the receiver feeds it back to the sender by setting the ECN Echo (ECE) code-point in the TCP header of ACK. The sender reduces its congestion window(CWND) in response to ECE signals. Meanwhile, it sets the congestion window reduced (CWR) code-point in the TCP header to inform the receiver of ceasing sending ECE signals. The MPTCP-SPTA combines the ECN with new mechanism of OLIA. It allows buffer occupancy to keep a lower value, meanwhile throttle the large flows to invade the bandwidth of short flows in the bottleneck link. It can reduce the packet loss of the short flows, while still achieving high throughput. The pseudo-code of MPTCP-SPTA is given by Algorithm 1.

Algorithm 1: The pseudo-code of MPTCP-SPTA
Initialization s=0
S calculates the data sent by the socket
/*At receiving a new ACK over subflow r*/
Increase CWND(r, $w_r(t+1)$)

$$w_r(t+1) \leftarrow \frac{w_r(t)/rtt_r^2}{\left(\sum_{p \in R_u} w_p/rtt_p\right)^2} * q_r(t) + \frac{a_r(t)}{w_r(t)}$$

/* *Respond to ECN over subflow r* */
$W_r(t+1) \leftarrow W_r(1/\beta)$
/* *Response to duplicate ACKs over sub-flow r* */
Decrease CWND(r)
$W_r(t+1) \leftarrow Wr/2$

If the sender receives an ACK, the congestion window of the sub-flow grows following "Increase CWND(r, Wr(t+1))" in the Algorithm 1. When the sender detects the ECE code-point of ACK, the congestion window of sub-flow decreases by $1/\beta$ in response to the ECN signal. For duplicate ACKs, MPTCP-SPTA reduces the congestion window in the normal way.

5   Evaluation

In this section, we explore the performance by performing extensive experiments based on NS-3.14 under three topology [26]. The Google MPTCP group provides the MPTCP NS-3.14 code for our experiment. We select the completion time of short flows, large flows throughput, and total throughput as the primary evaluation metrics of the simulation experiment.

We examine MPTCP-SPTA under three typical topology, shared links network topology, no-shared links network topology, and topology of the background with the regular TCP flow. To validate the effectiveness of our algorithm, we conduct a simulation experiment under the three topologis.

5.1   Parameter Settings

The parameters are qH, qL and T in $Q_r(t)$. The Christo Wilson et al. mentions that the traffic consists of the short flows of less than 1 MB and long flows of between 1 MB and

100 MB [21]. So we set the parameter T to be 1 MB. We determine the value of both q_H and q_L base on the formula $q_H + q_L$ and $q_H : q_L$ [15]. $q_H : q_L$ plays a vital role in bandwidth allocation, then $q_H + q_L$ influences the oscillation of total throughput. When the value of the formula 7 is equal to 4, the results of q_H and q_L are ideal [15]. So the values of q_H and q_L are 3.2 and 0.8 respectively.

$$q_H + q_L = 4$$
$$q_H \div q_L = 4 \tag{7}$$

The ECN can maintain a lower buffer queue value using a constant factor β to reduce the CWND if the average queue threshold K set to an appropriate value [15]. Yu Cao et al. presents the right solution: the K should satisfy $(K + BDP)/\beta \leq K$. So we have the following formula.

$$k \geq \frac{BDP}{\beta - 1} \quad \beta \geq 2 \tag{8}$$

Considering CWND changes with packet granularity in our experiment, it is sufficient to choose 4 as an appropriate β. In our experiment, the bandwidth of the link is 5 Mbps, and the round-trip time is 30us, So the BDP is 18 packets. By formula 8, we figure out that K must be bigger than 6. Through experiments, we determine an appropriate value of K. According to Fig. 7, we can find that the total throughput is best at 30 packets, but the completion time of small flows is not minimal. When K is 40 packets, the total throughput in the network is the second-largest, and the completion time of the short flows is minimum, So K in our experiments is set to 40.

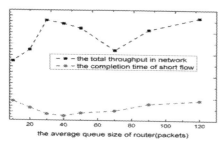

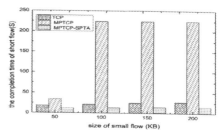

Fig. 7. The completion time of the short flow and the total throughput in the network take corresponding values.

Fig. 8. The completion time of small flow in Shared links topology.

5.2 Experimental Results

Whether our algorithm can preferentially handle small flows in the bottleneck link and buffer queue in router, we build the topology of shared links and router shown in Fig. 1. In the topology, the long flow size is set to 10 MB, and the small flows size changes from 50 KB to 200 KB. Figure 8 shows that the completion time of MPTCP-SPTA is the smallest in the topology of the shared link and router, however the completion time of TCP is better than the one of MPTCP. Since there is a bottleneck link in the shared mode, the MPTCP-SPTA allocates more bandwidth to short flows at bottleneck link. So the completion time of the small flows of MPTCP-SPTA is the shortest. The completion time of MPTCP is the largest, becomes MPTCP is harmful to latency-sensitive small flows. As shown in Fig. 9 that the throughput of MPTCP-SPTA is the largest because large flows can get time compensation after the small flows complete more quickly. Based on analysis above, the total throughput is also improved as showed Fig. 10.

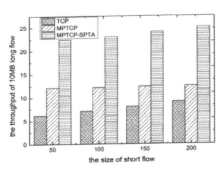

Fig. 9. The throughput of 10 MB in share links topology.

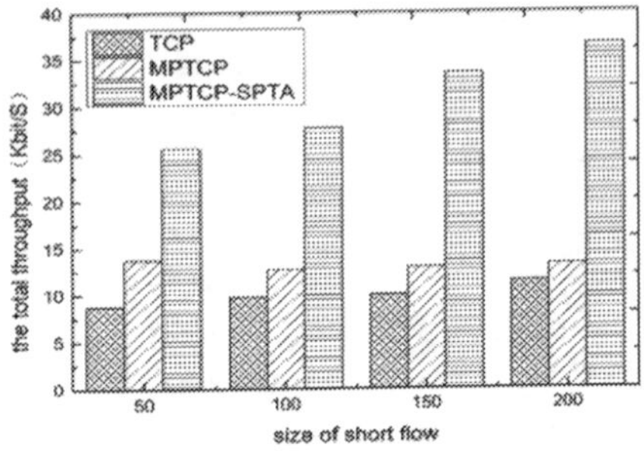

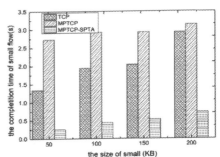

Fig. 10. The total throughput in the network in share links topology.

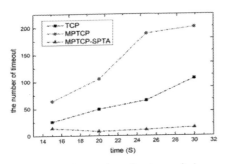

Fig. 11. It is the number of re-transmission under the scenario where concurrent traffic with only one long flow and one short flow.

Fig. 12. The completion time of small flow in no-Share links topology.

The path heterogeneity often exists in MPTCP. To simulate this topology, we build a non-shared topology, as shown in Fig. 5. In this topology, the client S2 and server R2 connected through router R1 and R2 separately. The transmission rates of the two sub-flows are 10Mbps and 1Mbps, and their RTT is 10us, 100us respectively. Figure 12 shows the same result as Fig. 8. However there are heterogeneous sub-flows in the non-shared link topology, so the completion time of MPTCP is worse than the completion time of TCP. But MPTCP-SPTA can reduce Head-of-Line blocking caused by heterogeneous path, therefor its completion time is the shortest. Figures 13 and 14 shows long flows and total throughput. The MPTCP-SPTA inherits the advantages of OLIA's load balancing, meanwhile it reduces the packet loss of the small flows in the router. So that the router queue can better forward the large flows. Therefore, its performance is better than MPTCP and TCP.

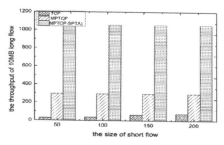

Fig. 13. The throughput of 10 MB in no-shared links topology.

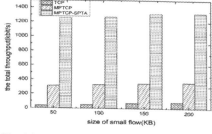

Fig. 14. The total throughput in the network in no-share links topology.

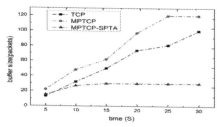

Fig. 15. The instantaneous buffer size under traffic model with only one long flow and one short flow

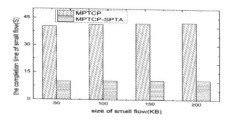

Fig. 16. The completion time of the short flow under the topology of the background with the regular TCP flow.

TCP flow is still the majority of network traffic. So we build a competing network topology as shown in Fig. 6. Figure 16 shows that the completion time of MPTCP-SPTA is reduced by 28% compared to MPTCP. MPTCP-SPTA obtain more bandwidth because it can give the short flows height weight. The ECN in MPTCP-SPTA makes short flows to get a much buffer in the router to reduce the number of lost packets and shorten the short flows completion time. As shown in Figs. 17 and 18, the long flows and total throughput of MPTCP-SPTA is better than MPTCP. Since the completion time of the short flows is reduced, this is advantageous for accelerating the transmission of the long flows. At the same time, the router's buffer maintains a low value to minimize packet loss and re-transmission caused by queue buildup and buffer pressure. As shown in Figs. 11 and 15, the number of the timeout and the instantaneous buffer of the MPTCP-SPTA are minimal.

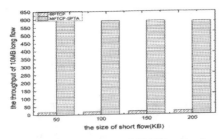

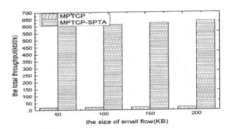

Fig. 17. The throughput of 10 MB under the topology of the competition with the regular TCP flow.

Fig. 18. The total throughput under the topology of the competition with the regular TCP flow.

6   Conclusions

In this paper, we aim to reduce the completion time of short flow and enhance the throughput of the long flows by modifying MPTCP in the network. We propose an MPTCP-SPTA algorithm that satisfies the above requirements. MPTCP-SPTA consists of the adaptive data flow category's MPTCP congestion control for short flow fast transmission and ECN mechanism. MPTCP-SPTA can keep queue buffer of the router at a lower value in responding to the bursty characteristic of the short flows and the tail drop policy. It ensures that more bandwidth allocates to the short flows in the bottleneck links. The simulation experiment fully shows that MPTCP-SPTA can reduce the completion time of the short flows, improve the total throughput of the network in three typical topologies.

References

1. Liu, Y., Yang, Z., Yan, X., et al.: A novel multi-hop algorithm for wireless network with unevenly distributed nodes. Comput. Mater. Continua **58**(1), 79–100 (2019)
2. Das, T., Sivalingam, K.M.: TCP improvements for data center networks. In: 2013 Fifth International Conference on Communication Systems and Networks (COMSNETS), pp. 1–10 (2013)
3. Alizadeh, M., Greenberg, A., Maltz, D.A., et al.: Data center tcp (dctcp). ACM SIGCOMM Comput. Commun. Rev. **41**(4), 63–74 (2011)
4. Munir, A., Qazi, I.A., Qaisar, S.B.: On achieving low latency in data centers. In: 2013 IEEE International Conference on Communications (ICC), pp. 3721–3725. IEEE (2013)
5. Urase, K., Toomey, M.W., Fang, K., et al.: Double neutron star mergers and short Gamma-ray bursts: long-lasting high-energy signatures and remnant dichotomy. Astrophys. J. **854**(1), 60 (2018)
6. Schliwa Bertling, P., Eriksson, M., Purushothama, R., et al.: MPTCP scheduling: U.S. Patent 10,143,001, pp. 11–27 (2018)
7. Li, M., Lukyanenko, A., Ou, Z., et al.: Multipath transmission for the internet: A survey. IEEE Commun. Surv. Tutorials **18**(4), 2887–2925 (2016)
8. Yang, W., Dong, P., et al.: A MPTCP scheduler for web transfer. Comput. Mater. Continua **57**(2), 205–222 (2018)

9. Viernickel, T., Froemmgen, A., Rizk, A., et al.: Multipath quic: a deployable multipath transport protocol. In: 2018 IEEE International Conference on Communications (ICC), pp. 1–7. IEEE (2018)
10. Dong, P., et al.: Performance enhancement of multipath TCP for wireless communications with multiple radio interfaces. IEEE Trans. Commun. **64**(8), 3456–3466 (2016)
11. Hamdi, M.M., Rashid, S.A.J., Ismail, M., et al.: Performance evaluation of active queue management algorithms in large network. In: IEEE 4th International Symposium on Telecommunication Technologies (ISTT), pp. 1–6. IEEE (2018)
12. Kheirkhah, M., Lee, M.: AMP: a better multipath TCP for data center networks (2017). arXiv preprint arXiv:1707.00322
13. Cao, Y., Xu, M., Fu, X., et al.: Explicit multipath congestion control for data center networks. In: Proceedings of the Ninth ACM Conference on Emerging Networking Experiments and Technologies, pp. 73–84. ACM (2013)
14. Dong, P., Yang, W., Tang, W., et al.: Reducing transport latency for short flows with multipath TCP. J. Netw. Comput. Appl. **108**, 20–36 (2018)
15. Kheirkhah, M., Lee, M.: AMP: an adaptive multipath TCP for data center networks. In: 2019 IFIP Networking Conference (IFIP Networking), pp. 1–9. IEEE (2019)
16. Wu, W., Chen, Y., Durairajan, R., et al.: Adaptive data transmission in the cloud. In: 2013 IEEE/ACM 21st International Symposium on Quality of Service (IWQoS), pp. 1–10. IEEE (2013)
17. Vamanan, B., Hasan, J., Vijaykumar, T.N.: Deadline-aware datacenter TCP (d2tcp). ACM SIGCOMM Comput. Commun. Rev. **42**(4), 115–126 (2012)
18. Tang, W., Fu, Y., Dong, P., et al.: An MPTCP scheduler combined with congestion control for short flow delivery in signal transmission. IEEE Access **7**, 116195–116206 (2019)
19. Joy, S., Nayak, A.: Improving flow completion time for short flows in datacenter networks. In: 2015 IFIP/IEEE International Symposium on Integrated Network Management (IM), pp. 700–705. IEEE (2015)
20. Dong, P., Yang, W., Tang, W., et al.: Reducing transport latency for short flows with multipath TCP. J. Netw. Comput. Appl. **108**, 20–36 (2018)
21. Raiciu, C., Handley, M., Wischik, D.: Coupled congestion control for multipath transport protocols. IETF RFC 6356, October 2011
22. Dong, E., Xu, M., Fu, X., et al.: A loss aware MPTCP scheduler for highly lossy networks. Comput. Netw. **157**, 146–158 (2019)
23. Khalili, R., Gast, N., Popovic, M., et al.: MPTCP is not Pareto-optimal: performance issues and a possible solution. IEEE/ACM Trans. Networking (ToN) **21**(5), 1651–1665 (2013)
24. Kheirkhah, M., Wakeman, I., Parisis, G.: Multipath-TCP in ns-3 (2015). arXiv preprint arXiv:1510.07721
25. Melki, R., Mansour, M.M., Chehab, A.: A fairness-based congestion control algorithm for multipath TCP. In: 2018 IEEE Wireless Communications and Networking Conference (WCNC), pp. 1–6. IEEE (2018)
26. Hu, B., Xiang, F., et al.: Research on time synchronization method under arbitrary network delay in wireless sensor networks. Comput. Mater. Continua **61**(3), 1323–1344 (2019)

Analysis of Public Transport System Efficiency Based on Multi-layer Network

Lu Xu$^{(\boxtimes)}$, Funing Yang$^{(\boxtimes)}$, Zhezhou Yu$^{(\boxtimes)}$, Qiuyang Huang$^{(\boxtimes)}$, Xuan Ji$^{(\boxtimes)}$, and Yuanbo Xu$^{(\boxtimes)}$

Jilin University, 2699 Qianjin Street, Chaoyang District, Changchun, China
xlu17@mails.jlu.edu.cn, {yfn,yuzz}@jlu.edu.cn,
huangqyjlu@foxmail.com, 729468722@qq.com, 174728098@qq.com

Abstract. Public transportation plays a pivotal part in urban traffic systems, where enhancing the efficiency of public transport is one of the essential strategies for solving the traffic congestion. In this paper, we propose a novel method to define the weights of edges which connected neighbor stations, by considering multi-layer structural and functional properties. To set the structural weights, a three-layer network is constructed to show the importance of links in the whole public transport system rather than a single traffic mode. Besides, the functional weight represents transport capacity of bus lines, which helps to improve the efficiency of public transport. In order to prove the practicality and reliability of our proposed model, the experiment is conducted by comparing with the traditional index. The experimental results demonstrate that our proposed model is superior to the state-of-the-art models. Moreover, we propose an optimization algorithm to improve single bus line's efficiency without losing the performance of the whole network.

Keywords: Public transport · Multi-layer network · System efficiency

1 Introduction

Urban public transport refers to bus, rail transit, public bicycle, ferry, ropeway and other transport modes which are operated within the whole city. Improving the travel sharing rate of public transport is of great significance to promote green travel and alleviate traffic congestion. In China, many local governments have clear requirements for the public transport travel share index during the 13th Five-Year Plan period, that the sharing ratio of public transport needs to be increased to more than 60% in cities with a population of more than 1 million [1]. In order to achieve this goal, it is a major issue for urgent solutions what makes travelers choose public transport as much as possible and raises public transportation efficiency. In other words, enhancing the efficiency of public transport is one of the essential decisions to creating a harmonious urban traffic environment.

The most existing research focuses on the analysis of the topological structure and network characteristics of urban public transport, which only explains phenomena occurred. Few works take into account the factors such as passenger volume and

X. Sun et al. (Eds.): ICAIS 2020, LNCS 12239, pp. 798–809, 2020.
https://doi.org/10.1007/978-3-030-57884-8_70

road conditions. With regard to multiple transportation, previous papers just discussed simply either one single mode separately or merged all modes in an integrated network. However, for urban public transport system, the relationship between various modes of transportation cannot be simply expressed by the integrated single-layer network. That is to say, the traditional representation has been unable to benefit in a more realistic network analysis of public transport efficiency. Therefore, we need a new network model to solve these problems.

In this paper, we start by establishing the concentration points in bus layer which aim to reflect a more accurate structure. Then we propose a three-layer network structure to abstract the real public transportation system. To connect the bus system with the rail system, the connected layer is added as an intermediate layer. In a such multi-layer model, the weights of edges are defined, in which not only represent the importance of structural networks, but also reflect the transmission capacity between two sites, that is, the importance of functional networks. Finally, we choose some line which has poor transmission capacity according to the weight. A bus line optimization algorithm is proposed to increase public transport efficiency.

The remainder of the paper is organized as follows. We briefly discuss related research in Sect. 2, our three-layers public transportation network model is described in Sect. 3. Link weight definition and bus lines optimization algorithm are relatively presented in Sect. 4 and Sect. 5, and conclusions in Sect. 6.

2 Related Work

Urban transportation has been a classical research field in recent years which contains many problems such as planning, construction, operation, management and so on. Since the study of complex network has made remarkable progress, topological and dynamical properties of networked systems provide useful insights to reveal principals in urban public transportation networks. Based on L-space representation, Anderson [2] can identify the probabilistic congestion areas in cities according to some indicators of public transport network. In addition, weighted links make congestion measurement more accurate. Pijaya [3] selected five commonly used network indicators, namely, network diameter, gamma index, degree centrality, closeness centrality and betweenness centrality, to analyze the evolution of the topological structure of the rail transit network in Bangkok. Xin Zhao [4] analyses the importance of nodes in public transport network, considering the actual factors in various transportation systems, and establishes a weighted model. Finally, a simulation experiment results show on effectiveness of this method. In Michele's [5] paper, a new angle is put forward, which is the destructive effect of natural disasters on the public transport system. The weak connection of Singaporean public transport network, that is, the route that can't meet the people's travel needs, is found out through the test experiment. This work can enable the public transport system to cope with future unexpected stress events.

Although these studies have analyzed the urban public transport network from different perspectives and methods, they are basically based on a single mode of transport, or different modes of transport separately, which does not clearly reflect the interdependence and competition among various modes of transport. Riccardo [6] presented coupling relationship among multiple traffic modes. But he only used the model of multi-layer network to connect different traffic modes, and did not explore the relationship between different layers in depth. The application of multilayer networks is still relatively rare [7–9]. In our approach, a three-layer network structure is proposed, which links the bus layer and the rail layer with the middle layer. Interlayer edges represent transfers within distances. The weight of the inner edge of the bus layer not only reflects the transmission capacity between the two stations, but also represents the position of the edge in the topological network structure.

In terms of operation efficiency, Yansha Wei [10] illustrates a load distribution model to adjust the bus operation efficiency, and the full load rate and other indicators are used as evaluation criteria to verify the validity of the model. Darshan [11] develops a GPS real-time monitoring system, so that passengers can arrange travel more reasonably and avoid waiting time. It is also a way to improve travel efficiency. Xinxin Yang [12] establishes a two-tier weighted network to reveal the relationship between bus and subway, and the travel efficiency under the two-tier traffic mode is analyzed. Singh researches multimodal GIS data and improves inefficient bus lines to efficient ones. New ideas and methods are put forward in the field of traffic planning.

Different from those methods, we take the total travel time as the general goal, and through a series of constraints, we redesign the original route based on the road network in the candidate route set, and find out the new route with less time and high coverage. Although the result of this algorithm is the optimization of a single bus line, it actually optimizes the performance of the whole network.

3   Multi-layer Network Structure Representation

We consider each mode of transportation as a layer. The metro network as a core network which has the advantages of high transportation capacity, fast speed and great convenience, but its construction cost is higher. Because of wide coverage and low construction cost, bus network is considered as the basic network. Moreover, we will add an intermediate layer called connected layer, which to communicate among two separate independent layers.

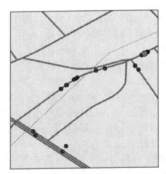

Fig. 1. An example of the concentration point. Red points are bus stops located in a shopping mall in Changchun, Jilin province of China. Green point is the concentration point of the red points. (Color figure online)

We started abstracting the original network that each bus stop will be a node and there will be an edge if two next stops belong to the same bus line in tradition. However, bus stops are distributed densely and miscellaneously in cities, for instance, there are several bus stations which have been set up in different locations around the railway station or large shopping malls (as shown in the Fig. 1). These bus stops are located around the same POI, therefore have similar functions, so we will introduce the new node representation Concentration point.

The structure of the Concentration point not only reflects the original network with fewer nodes, but also integrates the smashed hot stations into a concentrated node, enlarging its role in the topological structure. A graph consisting of Concentration points compared with the original bus map is shown in the Fig. 2. For example, there are n stations around the original railway station, passengers with different needs choose the most suitable station, that is to say, n stations divert traffic of the railway station. It is of positive significance to the actual bus system, since it effectively alleviates the problem of large passenger flow and passenger detention in railway stations, but such station layout has a negative impact on our analysis of the topological structure. Traffic hubs with high degree are diluted into n different stations here. Introducing the Concentration point well solved problem of discrepancy of node definition between the real network and the abstract topological network, and also provides powerful information for our following efficiency analysis of public transport. Because of the farther distance between two near rail transit stations, the above problems do not exist in the Metro layer, it is represented by the traditional graph method, e.g. a node represents a subway stop and an edge represents the link of two next stops.

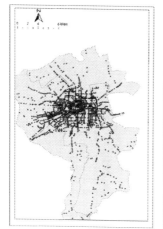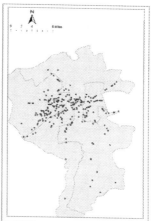

Fig. 2. The left picture is the original bus stops distribution which the red points represent the bus stops. The right picture is the concentration points distribution. As you can see, the whole distribution has been streamlined a lot. (Color figure online)

A mid layer is set between bus transit system and rail transit system to make connections and establish an interactive model. As the most important connection is transfer, we regard these bus stops set as the mid layer which are within miles of the rail stations. Usually, 500 m is an acceptable transfer distance for most passengers [13]. However, for urban centers with dense bus stops, passengers' requirements for transfer are also increasing gradually. In other words, they need a shorter transfer distance. As for sub-urban areas, due to fewer bus stops, the transfer distance of more than 500 m has to be chosen. In this case about setting the radius, with the decreasing density of urban public transport, we will increase the tolerance of passenger transfer distance, that is, the radius is increasing with the decreasing of the urban bus density. We calculate radius as

$$R = \alpha * 500, \tag{1}$$

where α is an adjustment factor and the range of values is from 1 to 2. More concretely, in this work we regard the People's Square as the downtown point, with a value of 1 within five kilometers, 1.6 within 10 km and 2 outside 10 km. That is to say, people in the downtown area tolerate walking distance of 500 and on the outskirts tolerate 800, but who lives in the suburbs can acceptable 1000 m. Subregional considerations are certainly more meaningful than setting uniform values globally (Fig. 3).

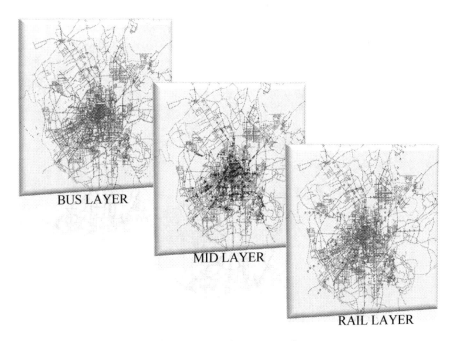

Fig. 3. A three-layers network.

In this way, bus stops in the range of R can be linked to the rail station. Note that the physical meaning of the interlayer edge here is transfer. The transfer station determines the connection relationship of each line in the network, and decides the skeleton structure of the whole network. The transfer of different modes of transportation can quickly relieve a large number of passengers and provide more choices for citizens to travel. That is to say, when we study the structural characteristics of bus stops and links, we no longer only consider the influence in the single-layer, but also consider the importance of the whole public transport system. This structure expresses the mixed cooperative relationship among the various modes of urban public transport system, as well as embody the unique topological structure and functional characteristics of each model. It provides an advanced model for urban public transport, which is more conducive to subsequent analysis.

4 Link Weight Definition

Setting link weight needs us to concentrate mainly on two levels here: topological structure and functional characteristics. We consider the position of bus edges in three-tier networks, instead of using simple definitions in single-layer complex networks traditionally. Since the three-layers network can better reflect its role in the whole public transport system, known as comprehensiveness. Functional characteristics, that is the unique characteristics of bus transport network, which are distinct from abstract network anymore, namely passenger flow and speed. Here we mainly consider two factors, as they can best reflect the transmission capacity of the bus link, in other words, reflect efficiency.

Assign a functional weight to each link

$$w_{\theta (func)} = F/v, \tag{2}$$

where w_{func} is part of link weight of link θ about functional characteristics, F is the passenger flow of link θ and v is the average speed of the bus.

First of all, we need to make clear what kind of bus lines are inefficient, that is, high passenger flow, which plays an important role in carrying capacity, but the speed is relatively slow. Especially in the rush hour, if the speed of the bus carrying relatively more passengers can be increased, the gathered traffic will be quickly dredged and the congestion will be eased as soon as possible. Although increasing the number of bus shifts is also a solution, the practice provides pressure for congested sections and requires a higher cost. Therefore, improving the speed works well with lower cost and better effect.

Before giving the topology weight definition, let's discuss the characteristics of the nodes in the network. In complex networks, the characteristics of nodes, such as degree and betweenness centrality [14], reflect the structural characteristics of networks. In the abstract network, the definitions of these characteristics are relatively abstract. In this paper, we no longer use the abstract definitions in the traditional sense in the unweighted network, but tailor-made for the multi-layer urban public transport network.

As the most representative index to report the importance of nodes in complex networks, degree measures the impact of neighbor nodes on local nodes. The higher the degree, the more important the nodes are in networks. The degree of node i is defined as the sum of the number of neighbor nodes. We give degree a more reasonable definition in the urban public transport network as

$$d_i = \sum_{j \in N} F_{ij}, \tag{3}$$

where d_i is the sum of the traffic flow that the node i connects to all its neighbors. N is the set of neighbor nodes of node i and F_{ij} is the passengers flow of link between node i and node j. With the new understanding of degree, the following definition of edge can be more accurate. For example, there are four routes through station i, but the total passenger flow is not as large as that of one line through the station j. It can be seen that the number of edges can not accurately represent the transmission capacity of the line between the two stations, d_i can. Similarly, the degree of link θ is the sum of its neighbors' passenger flow.

Edge betweenness is one of the most reliable measures in complex networks, so it's the main body of formula of structural weight. By multiplying a multiplex participation coefficient [15], the edges that play an essential role in the multi-layer network become more important. Here we give the weight of the structure as

$$w_{\theta(stru)} = P_\theta * B_\theta, \tag{4}$$

$$P_\theta = \frac{M}{M-1} \left[1 - \sum_{\alpha=1}^{M} \left(\frac{d_\theta^{[\alpha]}}{\sum_{\alpha=1}^{M} d_\theta^{[\alpha]}} \right)^2 \right], \tag{5}$$

$$B_\theta = 2 \sum_g^n \sum_k^n \frac{L(g,k,\theta)}{L(g,k)} / n(n-1), \tag{6}$$

where w_{stru} is part of link weight of link θ about structural characteristics, B_θ is link betweenness which is defined as the proportion of the number of the link θ passing through the shortest path in the network to the total number of shortest paths. It has done well in single-layer networks, but not so well in explaining the importance of multi-layer networks. Therefore, we use a coefficient to enhance its performance in multi-layer networks. The definition of the multiplex participation coefficient P_θ is introduced in Refs. [16, 17] to detect the link distribution in each layer. The range of values is 0 to 1. Generally, the larger value of P_θ means the more evenly link-distribution in each layer.

In this way, the link weight is denoted as

$$w_\theta = w_{\theta(func)} + w_{\theta(stru)}, \tag{7}$$

where w_θ is what we define the link weight finally. Note that, we take the normalization processing measures to the weight values of two aspects.

5 Bus Lines Optimization

The process of bus route optimization is quite complicated, and it cannot be solved by the only abstract bus layer, because the buses drive on the road network. When discussing the structural characteristics of the bus network, we can only connect the two stations without considering the road network. But in the actual route optimization, we must take the road network as the basic layer to optimize the route. In this problem, we use a two-layers network structure, the basic layer is the road network, and the operation layer is the abstract bus network. The cost of optimizing the overall network structure is very high, so we propose a single-line optimization algorithm. It increases the bus line efficiency and adds the global network characteristics to the constraint conditions, which ensures overall system performance.

First, some of bus lines should be selected. Here questions regarding link selection are primarily asked about (i) What kind of lines need to be optimized? As described in above section, the links with high weight values, in other words, those lines which play an important role in the network structure but have poor transport capacity, might have required optimization. Specifically, we adjust the route on the road network and it will not affect the importance of adjusted links in the bus layer. That's what we need for a two-layers structure. (ii) how to properly determine the number of the adjusted links. In the other hand, the scale of a city's ground public transport system must be controlled. The increase of stations and the expansion of coverage not only increase the cost, but also may lead to the reduction of efficiency [18]. Therefore, the number of routes adjusted is not the more the better, and road resources are certain, the increase of bus efficiency will inevitably affect the traffic volume of other motor vehicles. We rank the links in descending order according to the weight. Every five edge is updated, we calculate the overall efficiency. The speed of improving the efficiency will be slower. When it gradually becomes flat, we think that adjusting more lines will not improve the overall efficiency more, but also increase the cost of resources. In this way, we can investigate the number of the adjusted links.

In general, there is a considerable complexity of getting the optimal path based on Global Searching. Hence, we introduce the candidate path set. Note that by default the optimal solution is in the candidate path set. The candidate path set includes the first k paths with the least time, the first k paths with the shortest distance and the first k paths with the largest coverage. Here coverage refers to the number of POIs around the station. If the lines in the candidate path set cannot be connected to a connected network, we add some auxiliary sections artificially to make the candidate path set become a local network. Based on this local network, we optimize a single line as shown Fig. 4.

Fig. 4. An example of candidate path set. The blue lines are the basis road network and the red lines are the local network consisting of the candidate routes. (Color figure online)

As we mentioned above, bus route optimization is a complicated problem with many factors to be considered. For this multi-objective problem, we can't express all the factors as a mathematical formula. Then we regard the efficiency of bus as the objective function which is the most important factor. Then other factors are expressed by constraints to obtain feasible solutions and complete multi-objective decision analysis.

Next, we give the overall optimization objective function

$$min = \sum_{\theta}^{N} F_{\theta} * t_{\theta}, \tag{8}$$

$$s.t. \{\theta \in R | F_{\theta} > thres_{\theta}\}, \tag{9}$$

$$\{\theta \in R | \theta \notin B\}, \tag{10}$$

$$\{\theta \in R | dis(\theta, \lambda) \leq 500\,\text{m}, \lambda \in O\}, \tag{11}$$

where F_{θ} is the passenger flow and t_{θ} is the time spent between two bus stops. That is to say, it's to minimize the total travel time. R is the result route set, B is the original bus lines set and O is the original route. Equation (9) (10) (11) represents three different constraints, which is guaranteeing bus flow, trying not to use the existing bus lines to ensure the coverage of the whole bus network for the city and adjustment as far as possible on the basis of the original line.

The goal is to find as direct as possible and with a larger coverage, which ensures efficiency and takes traffic into account, so as to greatly improve the transport capacity of bus line. However, the planning of bus routes is a complex problem, and we need to consider more factors, which of course will be reflected in our constraints.

We use Changchun bus IC data as traffic data, and the GPS traces data describing the bus route to calculate the average speed of a day. Firstly, according to the method of generating candidate routes set, the first K routes with the shortest time, the largest coverage and the nearest distance are formed into a local road network, as shown in the Fig. 4.

For parameter k, we select 2 in this paper. The value of K depends on the final generated road network. Maybe the routes between the two stations are dense. Then the value of K should be larger so that the final local road network is a connected graph. If the distance between stations is relatively close and the road network between them is sparse, the large value will lead to the excessive deviation of the final generated route from the original one. K is generally taken as [1, 5].

According to the algorithm (Table 1), the genetic algorithm is used to find the optimal route (an example shown in Fig. 5) within the constraints in the local network, and iterates continuously until all routes that need modified are updated. Under the condition of updating the same number of lines, the global efficiency of the network $E = \frac{1}{n^2} \sum_{j=1}^{N} \sum_{i=1}^{N} t_{ij}$, as research target, and the weight model of this paper is compared with the traditional weight model. The results are shown in Table 2. From Table 2, the traditional structural characteristics are worse for improving the efficiency of the whole network than considering the functional characteristics only, and the weight of comprehensive consideration is the best. We can see that under the analysis model, the efficiency of the whole network after updating the routes selected according to the weight model of this paper is higher than that under the traditional model. This is because our weights take into account both structural and functional characteristics. For the purpose of improving efficiency, these edges play a much more important role in the network than those that only consider structural characteristics. Corresponding to the real traffic system, if we can improve the efficiency of buses passing through traffic fortresses and densely populated areas, it will play a key role in alleviating traffic congestion and quickly dredging dense traffic. Therefore, under the same number of routes, the network efficiency will be improved more. This is also consistent with the actual urban network characteristics. Through the optimization results of public transport lines, we can prove that the weight setting in this paper is better than the index considering only the topological structure and other characteristics without considering multi-layer factors for public transport network.

Table 1. Algorithmic description.

Algorithm: BUS LINES OPTIMIZATION

---

1. Input: the links need optimized set A in descending order by weight, the parameter k
2. Output: the Optimization set R
3. R=list[]
4. for each $\theta \in$ A,
5.   get the origin V and destination W of the θ
6.   generates the heap (Q=V(G)) %generate the candidate set Q
7.   while(!isEmpty(Q))
8.   v=EXTRACT-MIN(Q)
9.   for each vertex v_adj belongs to Adj[v]
10.   update v_adj %the adjacent point of v
11.   Q' = Q satisfies constraint condition
12.   t = 0; %genetic algorithm
13.   Initialize P (t);
14.   while(P(t) is fit enough or enough time) do
15.   t = t+1;
16.   Select P (t) from P (t-1);
17.   do if R $\cup$ {P(t) $satisfy\ minf\,(P(t)$}
18.   then R$\leftarrow$ R $\cup$ {θ}
19. return R

Table 2. Comparison of experimental results of different indicators for efficiency improvement. Increase rate is the radio of the difference between the value of optimized network and the value of original efficiency and the value of original efficiency. Degree is defined as the sum of the number of neighbor edge and d_i is what Eq. (3) describes. Betweenness and w_θ is separately described in Eq. (6) and Eq. (7)

Symbol	Degree	d_i	Betweenness	w_θ
E(h)	2.58	2.01	2.30	1.57
Increase rate	9.79%	29.72%	19.58%	45.1%

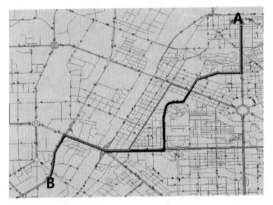

Fig. 5. The red line is the new routes between stop A and B generated by algorithms and the black one is the original line. (Color figure online)

6 Conclusion

In this paper, an effective weight representation method is proposed for the links of public transport network. It not only takes into account the unique functional characteristics of urban public transport network compared with other specific networks such as social network, communication network and power grid, which can reflect the transmission capacity, but also adds the multi-layer topological structure characteristics. That is to say, a multi-level factor is multiplied by a well-performing edge index in a single-layer network to reflect the relationship between different modes of transport in the public transport system. In order to define the structure characteristics, we also propose a three-layer network structure. For the bus layer, we also propose a method to represent named concentration point, which can not only simplify the structure of the bus layer with very dense stations, but also make the stations with large passenger flow and important status more prominent. Finally, the correctness of weights is verified by optimizing urban bus routes, and some specific proposals and schemes are provided for urban planning and later work.

References

1. China's Main Urban Traffic Analysis Report (2018)
2. De Bona, A.: Congestion potential – a new way to analyze public transportation based on complex networks. In: IEEE International Smart Cities Conference (ISC2) (2018)
3. Bangxang, P.N.: Topological evolution of public transportation network: a case study of Bangkok rail transit network. In: International Conference on Industrial Engineering and Applications (ICIEA) (2018)
4. Zhao, X.: Research and simulation of node importance in transport network based on weighted model. In: IEEE 2nd Advanced Information Technology, Electronic and Automation Control Conference (IAEAC) (2017)
5. Ferretti, M., Barlacchi, G., Pappalardo, L., Lucchini, L., Lepri, B.: Weak nodes detection in urban transport systems: planning for resilience in Singapore. In: IEEE 5th International Conference on Data Science and Advanced Analytics (DSAA) (2018)
6. Gallotti, R., Barthelemy, M.: The multilayer temporal network of public transport in Great Britain. Sci. Data **2**, 140056 (2015)
7. Bin, S., et al.: Collaborative filtering recommendation algorithm based on multi-relationship social network. Comput. Mater. Contin. **60**(2), 659–674 (2019)
8. Wang, Y., Zhang, X., Zhang, Y.: Joint spectrum partition and performance analysis of full-duplex D2D communications in multi-tier wireless networks. Comput. Mater. Contin. **61**(1), 171–184 (2019)
9. Sun, W., Hongji, D., Nie, S., He, X.: Traffic sign recognition method integrating multi-layer features and kernel extreme learning machine classifier. Comput. Mater. Contin. **60**(1), 147–161 (2019)
10. Wei, Y.: The evaluation and adjustment of bus operation efficiency based on load profile. In: IEEE 4th International Conference on Cloud Computing and Big Data Analysis (ICCCBDA) (2019)
11. Yang, X., Li, Y., Meng, F., Song, Y.: A model of bus-metro bi-level weighted network with travel efficiency. In: 37th Chinese Control Conference (CCC) (2018)
12. Pitam, S., Kumar, S.A., Priyamvada, S.: Multimodal data modeling for efficiency assessment of social priority based urban bus route transportation system using GIS and data envelopment analysis. Multimedia Tools Appl. **78**(17), 23897–23915 (2018). https://doi.org/10.1007/s11 042-018-6147-6
13. Zhang, Y.: Discussion on spatial match between land use and rail transit: a case study of Shenzhen subway line 2. Urban Plan. (2017)
14. Newman, M.E.J.: The structure and function of complex networks. SIAM Rev. **45**(2), 167–256 (2003)
15. Battiston, F., Nicosia, V., Latora, V.: Structural measures for multiplex networks. Phys. Rev. E **89**(3), 032804 (2014)
16. Guimerà, R., Amaral, L.A.: Cartography of complex networks: modules and universal roles. J. Stat. Mech. **2005**(P02001), P02001I–P0200113 (2005). nihpa35573
17. Guimerà, R., Amaral, L.A.N.: Functional cartography of complex metabolic networks. Nature **433**(7028), 895 (2005)
18. Wei, W.: Urban Public Transportation System Planning Method and Management Technology. Science Press (2002)

Fast Video Classification with CNNs in Compressed Domain

Lorxayxang Kai, Yang Wu, Xiaodong Dai, and Ming Ma[✉]

College of Computer Science and Engineering, Inner Mongolian University, Hohhot, China
csmaming@imu.edu.cn

Abstract. In recent years, Convolutional Neural Network (CNN) has proven to be one of the most successful tools for video-based action recognition. As the most popular method for video action recognition, the two-stream CNNs method, utilizing the optical flows, is not available for real-time applications because of the high computation requirement. In this paper, we present that accelerating the CNN architecture by replacing optical flows with the Motion Vector (MVs) can achieve a faster process speed that can be used in real-time applications. The MVs is designed to extracted information directly from the compressed video bitstreams. We explored how the proposed video classification method gives the very impressive result. First, using motion vector via taking raw video bitstream as input to directly predict action classes without explicitly computing optical flow. Secondly, we demonstrate a strong base-line two-stream ConvNet using pre-train models and transfer learning for our both spatial stream and temporal stream. The finding of our approach proves to be significantly faster than the original two-stream approaches, and achieves high accuracy and satisfies real-time requirement.

Keywords: Deep learning · Video classification · Computer vision · Compressed domain

1 Introduction

In recent years, a number of books and scholarly journals published on human action recognition and image classification very success on computer vision, and Convolutional Neural Networks (CNNs), and Deep learning. Human action recognition [14–17] is due to its great potential in a real application and dynamic action recognition systems applied to video surveillance, video retrieval, video streaming, human computer interaction, and traffic sign classification [31]. The existence of large action datasets and worldwide competitions have shown the effect of performance on the standard benchmarks such as UCF-101 [19], YouTube videos classification, HMDB5 [18], have been researching in this area. Early approaches in this area utilize Convolutional Neural Networks which mainly consist of feature extraction, feature encoding, classification, recognition, detection steps. This method shows that performance increases more than handcrafted features. Numerous researchers exploit Deep Convolutional Neural Networks (Deep Convnets) to solve images processing problems from image classification,

© Springer Nature Switzerland AG 2020
X. Sun et al. (Eds.): ICAIS 2020, LNCS 12239, pp. 810–821, 2020.
https://doi.org/10.1007/978-3-030-57884-8_71

object recognition, object detection, to video classification. Once successful, the two-stream architecture [25] includes both RGB CNNs and optical flow CNNs will be used for classification and receiving performance on several large action datasets. However, the two-stream CNNs not available for processing video in real-times. Optical flow is time consuming for calculation of processing speed (Fig. 1).

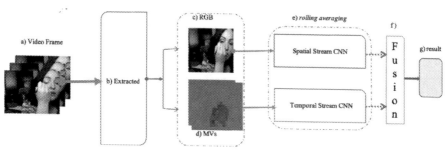

Fig. 1. structure of video classifies. The input is a sequence of video frames (a). Extracting the motion vectors (MVs) and RGB images from the video frames for training (b). RGB images and MVs are separately trained on two-stream CNNs (c)(d). RGB images are processed on spatial CNN and MVs are processed on temporal CNN, and get each predicted score using fine-tuned network. We call this network rolling averaging (e). Finally, fusing the best score of each stream (f), and predicting our two-stream rolling averaging (TSRA) score (g).

The main objective of this paper is to address the video classification method with the high-performance base and accuracy via a two-stream CNNs. Optical flow is computationally expensive, optical flow can be conducted at the speed of 16.7 frames per second(fps) with K40 GPU [27], which leads to inferior recognition accuracy. To circumvent this problem, we present the MV representations obtained from the video codec. MVs are extracted from bitstreams directly from certain compressed macroblocks (MBs) [26], the information utilized in video codec and can be directly extracted almost without extra computation.

MV is originally proposed for video coding. MV is designed to exploit the motion information from the reference frame to the current frames and the corresponding image blocks to reduce bit rat of video. Early research [27] conducted showed that motion vector can be used for video classifier link optical flow, the motion vectors represents movement patterns of images block which resemble optical flows in terms of describing local motions. Although, the MV is very powerful for video coded but MV contains noisy and imprecise movement information. Giving the accuracy similar to optical flows.

To solve the issue of accuracy, we proposed a various method that using the main point of images classifiers to train an RGB [25] and MV [27] to learn motion representation of the action and other vision tasks in the video. To address these challenges, we first train the RGB via pre-trained and transfer-learning on VGG and ResNet models for our spatial stream. Seconded using motion vector to takes raw video frames as input and learned the activities in video to avoid the heavy computational cost of time-consuming and significantly improves the performance.

We face many changes of Network architecture:

1. First, take a scalpel and cut off the final set of fully connected layers from a pre-trained Neural Networks using TF-Slim (TensorFlow-Slim) pre-train models.
2. Then, replace the head with a new set of fully connected layers with random initializations start from pre-train model and fine-tune it.
3. Finally, train the network using a very small learning rate so the new set of fully connected layers can learn patterns from the previously learned CONV layers earlier in the network this process is called allowing the FC layers to "warm up".

2   Related Work

In the success of CNNs has made the number of research focus to the field of computer vision, especially in human Action recognition, and images recognition has been widely explored in an active field of research with many algorithms [30]. Recent Deep Convolutional Networks (Deep ConvNet) have made significant progress in Action recognition in videos [3–5]. Deep ConvNet have been initially extended to be applied for the action in video and activity recognition achieved the state-of-the-art performance [5, 8, 9, 19].

Video-based algorithms have been proven to be powerful for computer vision and using Convolutional Neural Network to extract information from video data. With the success in the static of image in the 2D Convolutional Neural network (2D CNNs) [1, 2], and in many instances, frame-level feature is combined in some form of ConvNet. The early work Due Tran [6] proposed a simple but effective approach for spatiotemporal feature using deep 3-dimensional Convolutional Networks (3D ConvNets) trained on large scale supervised video dataset and learn the futures while exploring different fusion approach on the early fusion and show fusion. Another proposed method, further to incorporate the appearance and motion information, Feichtenhofer et al. [5], effect of fusing Convolutional layer to learn both spatially and temporally feature. They found a spatial and temporal network can be fused at a convolution layer without loss of performance, but with a substantial saving in parameters.

The idea of a Two-stream CNNs is not a new. Liu et al. [8, 19] proposed an architecture that using Convolutional 3D Network (T-C3D), to learns video action representations din a hierarchical multi-granularity manner. Specifically, they combine a residual 3D ConvNets which captures complementary information on the appearance of a single frame and the motion between consecutive frames with a new temporal encoding method to explore the temporal dynamics of the whole video. To avoid heavy calculations when doing the inference. Ma et al. [11], propose the two different networks to further integrate spatiotemporal information extracted from RGB and optical flow vector for person reidentification, the temporal segment Recurrent Neural Networks (RNN), and Inception-style Temporal-ConvNet were proposed. This demonstrated that using both RNNs and Temporal-ConvNets on spatiotemporal feature matrices are able to exploit spatiotemporal dynamics to improve the overall performance.

To improve the method of video information extraction, the MV has been put forward. Simonyan et al. [9] proposed a two-stream CNNs architecture which incorporates spatial and temporal networks to capture the complementary information on appearance from still frames and motion between frames. And then, Chadha et al. [2] proposed that selective access to macroblock MV for selective, and high speed in comparison to conventional full-frame decoding followed by optical flow estimation. Replacing optical flow with MV can be obtained directly from compressed videos without extra calculation. However, MV lacks fine structures and it contains noisy and inaccurate motion patterns, leading to the evident degradation of recognition performance. Zhang et al. [10]'s method is that optical flow and motion vector are inherent correlated. Exploited transferring the knowledge learned with optical flow CNNs to MV CNNs can significantly boost the performance of the latter. video-based action recognition with Deep ConvNets has shown remarkable progress. However, most of the existing approaches are too computationally expensive due to the complex network architecture. It's still improving to optimize these methods. Ruohan Meng et al. [32] propose the current steganography method into the mainly of spatial domain and temporal domain, that divides an image into several matrices to improving the method preferment.

3 Methodology

We propose a CNN-based video classification method for a sequence of video frame. The main idea is that video is a hierarchical data structure composed of events, scenes, shots, super-frames and frames. First, in Sect. 3.1, we propose the networks for training our spatial stream CNNs and temporal stream CNNs. We use ImageNet pre-trained models and transfer learning to train Resnet50 and VGG-16 with our datasets, then fine-tune the top layer of spatial stream CNNs and temporal stream CNNs, and use 10 x-channels and 10 y-channels for each motion vector. we call it rolling averaging in Sect. 3.2. Then we train the spatial stream and temporal stream network upon rolling averaging. Finally, we combine the spatial stream and temporal stream.

3.1 Two-Stream Architecture

Two steam CNNs is consisting of two-part spatial stream and temporal stream. spatial stream is taking the RGB images as input. Temporal stream is the motion vector images as input [25]. A great literature shows that using deep ConvNets can improve overall efficiency for two methods. Especially, the performance from VGG-16 [4], GoogleNet [20], and BN-Inception [4]. We chose a Resnet50 as our network for spatial and temporal. Feicht-enhofer et al. [5, 10] experiments how to fusion stage of spatial stream and temporal stream. Their results indicate that fusion can achieve the best performance can be fusion at the last stage. From this idea, we aim at exploring feature fusion using the last layer from both spatial-stream and temporal-stream CNNs.

Spatial Stream. We using the single RGB from the spatial stream shown that stacking RGB frames channel-wise and ingesting such volumes into a 2D CNNs does not necessarily improve performance [4, 5]. The spatial-stream ConvNet is pre-trained on ImageNet and fine-tuned on RGB images extracted from the UCF101 dataset. Using TensorFlow-slim pre-trained models.

Temporal Stream. For our temporal stream, we stack MV images for the temporal stream which has been considered as a standard for two-stream ConvNets for action recognition [4, 6, 21, 23, 24]. We follow the standard to show how each of our framework design and training practices can improve the classification accuracy by using a fine-tuning.

Fuse the Network. For the fusion of two-streams of network, we fine-tuned those networks with batch size of 20 and a learning late starting from 10^{-4}, and 0.85 of the dropout rates for both of VGG and ResNet. In our experiments, we use the pre-training and fine-tuned models of TensorFlow-slim (TF-slim) for both of spatial stream and temporal stream. We only fuse between layers with the output resolution by average identity matrices to sum the two networks of RGB and motion vectors stream and fine-tuned the model. This is chosen because in preliminary experiments, it provided better results.

3.2   Fine-Tuned Network

Deep Learning task, one that involves training a Convolutional Neural Networks on a dataset of images, first instinct would be to train the network from scratch. However, in practice, deep neural networks like Convolutional Neural Networks has a huge number of parameters, often in the range of millions. Training a Convolutional Neural Network on a small dataset (one that is smaller than the number of parameters) greatly affects the Convolutional Neural Networks ability to generalize, often result in overfitting. Therefore, more often in practice, one would fine-tune existing networks that are trained on a large dataset like the ImageNet1.2 M labeled images by continue training it on the smaller dataset we have. Provided that our dataset is not drastically different in context to the original dataset the pre-trained model will already have learned features that are relevant to own classification problems.

Fine-tuning. is the type of transfer learning. We apply fine-tuning to deep learning model that have already been trained on given dataset. Typically, these networks are state-of-the-art architecture such as VGG-16, Resnet50, and inception that have been trained on the ImageNet dataset. This method we name rolling averaging requires to perform network surgery. First, we take a scalpel and cut off the final set of fully connected layers from a pre-trained Convolutional Neural Network like VGG, Resnet, and inception. We then replace the head with a new set of fully connected layers with random initialization. From the all layer's head are frozen so their weights cannot be updated. We then train the network using very small learning rate so the new set of FC can be to start to learn pattern from the previously learning Convolutional Neural Networks earlier in the network. Applying the fine-tuning to allow apply pre-trained network to recognize class that they not the originally train on show on the Fig. 2. (a). the layers of the VGG16 network. The final set of layers are fully-connected layers along with the Softmax classifier. (b). When performing, we actually remove the head from the networks, as in future extraction. (c). however, unlike future extraction, when we fine-tuning actually build a new fully connected head and replace it on the top of originals architecture.

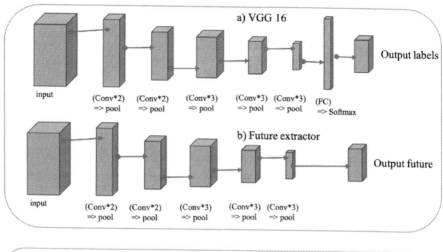

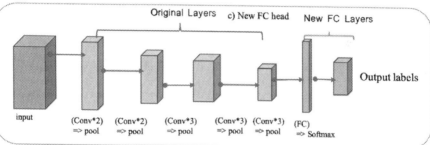

Fig. 2. a). The original VGG16 network architecture. b). Removing the FC layers from VGG16 and treating the final POOL layer as a feature extractor. c). Removing the original FC Layers and replacing them with a brand-new FC head. These FC layers can then be fine-tuned to a specific dataset the old FC Layers are no longer to used.

Training. Our network is train with the standard back-propagation algorithm, this permits sequential training on a single GPU. First, using the mini-batch of stochastic gradient descents (SGD) algorithms to optimizer the cross-entropy error function. Then, forward propagated through the networks as normally. However, the backpropagation is stop after FC layers, to allow these layers to start to learn patterns from the highly discriminative Convolutional Neural Networks layers. After the FC head have been started to learn pattern in our dataset, pause training, and unfreeze the body. Then continues the training. Fine-tuning (rolling averaging) is very powerful method to obtain video classification from pre-train Convolutional Neural Network on the custom datasets. Even more powerful then future extraction.

Data Augmentation. Data augmentation has been very helpful especially when the training data are limited. As the name suggests, data augmentation randomly jitters our training data by applying a series of random translations, rotations, shears, and flips. Applying these simple transformations does not change the class label of the input image; however, each augmented image can be considered a "new" image that the training algorithm has not seen before. Therefore, our training algorithm is being constantly presented with new training samples, allowing it to learn more robust and discriminative patterns.

Hyper-parameter Optimization. The learning rate for all our stream is initially set to 1×10^{-4}. The weight decay is set to be 1×10^{-4}, pre-trained is set with ResRet-50, VGG-16 and momentum is 0.9. The batch sizes trained with SGD optimizer. Set Batch-size by default we use $b = 15$ for temporal, $b = 10$ for spatial, and 5 for **TSRA** (Two Steam Rolling Averaging). Set dropout rate is 0.85 for VGG-16, and set 0.5 for Resnet50 by default, and run the back-propagation algorithm for 500 iteration. All experiments are performed under Intel corei7 of CPU, 32 GB of RAM, Nvidia GeForce GTX 2080Ti of GPU, and CUDA 10.0. for benchmarking the computing power. Our framework is implemented with TensorFlow, and report the inference speed.

Testing. We fuse the two streams by averaging the softmax scores, and uniformly sample a number of frames in each video and the video level prediction is the voting result of all frame-level predictions. We pick the starting frame among those early enough to guarantee a desired number of frames. For shorter videos, we looped the video as many times as necessary to satisfy each model's input interface.

4   Experiments

4.1   Datasets and Evaluation Protocol

We evaluate the performance of our proposed method in the UCF-101 [19] and HMDB51 [18] datasets. The dataset consists of 13,320 videos from 101 classes of deferments human action categories, HMDB51 consists of 6766 videos from 51 classes of human action categories. The videos in 101 human action categories are grouped into 25 groups, where each group consists of 4–7 videos of an action.

Fig. 3. The category datasets of 101 deferent human action classes from videos in the wild.

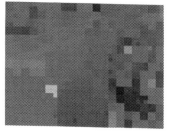

Fig. 4. The MVs activities map corresponding action classes

4.2 Baseline Two-Stream ConvNet

In this section, we report the experiment result of our spatial stream, temporal stream and two-stream CNNs on the UCF-101 datasets. Our two-stream performance is obtained by taking from spatial stream and temporal stream ConvNet. The method that we proposed to leverage the baseline two-stream ConvNet and show significant improvement by the RGB and MV modeling in Table 1.

Table 1. Performance from spatial and temporal-stream ConvNets, and two-stream ConvNet on the UCF101 datasets

Method	Dataset	Accuracy (%)
Spatial stream	UCF-101	74.3%
Temporal	UCF-101	64.6%
Two streams	UCF-101	78.8%

4.3 Performance of Rolling Averaging

In this section, we discuss various experiments and individual performance with new architecture design. Using the new high-level API from Tensorflow (TF-Slim) for images classification models. It also contains pre-trained models and fine-tuning network. However, we can train on very large datasets, because training models can be very computationally intensive process. We use the ImageNet pre-trained models and fine-tuning with a very deep neural network on our datasets. ResNet50 can obtain 87.8 for spatial and 77.2 for temporal. VGG-16 can obtain 86.7 for spatial and 78.5 for temporal both stream on the split1 in UCF-101 datasets. our experiments show in Table 2.

Table 2. Exploration of different very deep ConvNet architecture on the UCF-101 dataset.

Training setting	Spatial	Temporal	Two streams
VGG16	86.8%	78.5%	87.81%
ResNet50	87.8%	77.2%	88.5%

4.4 Final Performance

The result from the two streams of our spatial and temporal stream both on UCF-101. We name of final stream is **TSRA** (Two Steam Rolling Averaging). Out **TSRA** concatenate achieved significant segments of each contribution from both spatial and temporal. We propose temporal feature to take an action information from a MV images in each training step which can achieve better accuracy on UCF101 (see Table 3, Figures 3 and 4).

Our **TSRA** is using only 300 frames per video in each class of UCF-101 dataset. The baseline two-stream ConvNet shows the accuracy of 88.5% for ResNet and 87.81 for VGG, the spatial rises from 74.3% to 87.8% and the temporal rises from 64.6% to 78.5%. This verifies that the methods we proposed can express the temporal information by using very motion vectors (see Tables 3 and 4).

Table 3. The performance illustration of the last three show on the UCF-101 dataset.

Method	Split1	Split2
Spatial	87.8%	86.8%
Temporal	77.2%	78.5%
TSRA	88.5%	87.81%

Table 4. State-of-the-art video classification on the UCF-101 datasets.

Method	Accuracy
MV+FV [27]	78.5%
C3D (1 net) [28]	82.3%
C3D (3 net) [28]	85.2%
iDT+FV [29]	85.9%
Two-steam CNNs [9]	88.9%
TSRA (our)	88.5%

In the experiments, we compare our proposed method with several state-of-the- art methods. The result is summarized in Table 5, our approach is faster than iDT+FV, classical Two-stream CNNs respectively. And it also achieves higher performance than iDT+FV and achieves impressive on UCF-101 datasets and can be conducted in real time. For optical flow-based approach, TSRA is slightly better than iDT+CNN and shows comparable performance with DTMV+RGB-CNN. However, iDT+CNN need to calculate the optical flow, and DTMV+RGB-CNN need to transfer the knowledge of optical flow to MVs wasting time to calculate optical flow.

Table 5. Comparison accuracy and speed with state-of-the-art on UCF-101

Method	Accuracy	FPS
MV+FV [27]	78.5%	132.8
C3D (1 net) [28]	82.3%	313.9
C3D (3 net) [28]	85.2%	–
iDT+FV [29]	85.9%	2.1
Two-steam CNNs [9]	88.9%	14.3
DTMV+RGB-CNN [25]	87.5%	390.7
TSRA (our)	88.5%	401.5

5   Conclusions and Discussion

Two-stream ConvNets have been widely used in video understanding, especially for human action recognition. In this paper, we thoroughly explored the motion vector CNN to accelerate the processing speed of deep learning for real-time application. The motion vector is designed for video encoded stream and can be directly extracted without extra computation. Our findings using convolutional neural network fine-tuning is a very powerful method obtained to train the big datasets and achieve high processing speed. The caution must be given to account for strengths and weaknesses in each approach. However, to boot the performance of real-time video classifiers and train the model, we replace the optical flow with motion vector. TSRA is proposed to apply for the new challenging datasets and verify the effectiveness of our training approach which shows significantly better performance. We recommit to fine-tuning because this method can lead to higher accuracy than transfer learning via feature extraction. On the other hand, it is clear that convolutional networks can exploit information as well. In the future, we plan to pursue these directions on larger datasets as well as investigate how to better regularize.

References

1. Ciregan, D., Meier, U., Schmidhuber, J.: Multi-column deep neural networks for image classification (2012). Arxiv preprint arXiv:1202.2745

2. Ballas, N., Yao, L., Pal, C., Courville, A.: Delving deeper into convolutional networks for learning video representations. In: International Conference of Learning Representations (ICLR), November 2016

3. Karpathy, A.: Large-scale video classification with convolutional neural networks. In: Proceedings of the IEEE conference on Computer Vision and Pattern Recognition (CVPR), pp. 1725–1732 (2014)

4. Simonyan, K., Zisserman, A.: Two-stream convolutional networks for action recognition in videos. In: Proceedings of the Advances in Neural Information Processing Systems (NIPS), pp. 568–576 (2014)

5. Feichtenhofer, C., Pinz, A., Zisserman, A.: Convolutional two-stream network fusion for video action recognition (2016). arXiv preprint arXiv:1604.06573

6. Tran, D.: Learning Spatiotemporal Features with 3D Convolutional Networks (2015)

7. Chadha, A., Abbas, A., Andreopoulos, Y.: Video classification with CNNs: using the codeca as a spatio-temporal activity sensor. IEEE Trans. Circ. Syst. Video Technol. **29**(2), 475–485 (2017)

8. Liu, K., Liu, W., Gan, C., Tan, M., Ma, H.: T-C3D: temporal convolutional 3D network for real-time action recognition. In: AAAI, pp. 7138–7145 (2018)

9. Simonyan, K., Zisserman, A.: Two-stream convolutional networks for action recognition in videos. In: Advances in Neural Information Processing Systems, pp. 568–576 (2014)

10. Zhang, B., et al.: Real-time Recognition with Enhanced Motion Vector CNNs. In: Proceedings of the IEEE Conference on Computer Vision and Pattern Recognition, pp. 2718–2726 (2016)

11. Ma, C.Y., et al.: TS-LSTM and temporal-inception: exploiting spatiotemporal dynamics for activity recognition. Sig. Process. Image Commun. **71**, 76–87 (2019)

12. Sun, L., Jia, K., Yeung, D.-Y., Shi, B.E.: Human action recognition using factorized spatial-temporal convolutional networks. In: Proceedings of the IEEE International Conference on Computer Vision, pp. 4597–4605 (2015)

13. Wang, L., Qiao, Y., Tang, X.: MoFAP: a multi-level representation for action recognition. Int. J. Comput. Vis. **119**(3), 254–271 (2016)

14. Tran, D., Bourdev, L., Fergus, R., Torresani, L., Paluri, M.: Learning spatiotemporal features with 3D convolutional networks. In: Proceedings of the IEEE International Conference on Computer Vision (ICCV), December 2015, pp. 4489–4497 (2015)

15. Wang, L., Qiao, Y., Tang, X.: Action recognition with trajectorypooled deep-convolutional descriptors. In: Proceedings of CVPR, Boston, MA, USA, June 2015, pp. 4305–4314 (2015)

16. Soomro, K., Zamir, A.R., Shah, M.: UCF101: a dataset of 101 human actions classes from videos in the wild. In: CoRR, pp. 1–7, November 2012

17. Kuehne,H., Jhuang, H., Garrote, E., Poggio, T., Serre, T.: HMDB: a large video database for human motion recognition. In: Proceedings of the IEEE International Conference on Computer Vision (ICCV), pp. 2556–2563 (2011)

18. Ioffe, S., Szegedy, C.: Batch normalization: accelerating deep network training by reducing internal covariate shift. In: Blei, D., Bach, F. (eds.) Proceedings of the 32nd International Conference on Machine Learning (ICML-15), pp. 448–456. JMLR Workshop and Conference Proceedings (2015)

19. Szegedy, C., Liu, W., Jia, Y., Sermanet, P., Reed, S., Anguelov, D., Erhan, D., Vanhoucke, V., Rabinovich, A.: Going deeper with convolutions. In: Proceedings of the IEEE Conference on Computer Vision and Pattern Recognition, pp. 1–9 (2015)

20. Wang, L., Xiong, Y., Wang, Z., Qiao, Y., Lin, D., Tang, X., Van Gool, L.: Temporal segment networks: towards good practices for deep action recognition. In: ECCV (2016)

21. Wang, X., Farhadi, A., Gupta, A.: Actions transformations. In: The IEEE Conference on Computer Vision and Pattern Recognition (CVPR), June 2016

22. Wang, Y., Song, J, Wang, L., Van Gool, L., Hilliges, O.: Two-stream sr-cnns for action recognition in videos. In: BMVC (2016)

23. Yue-Hei Ng, J., Hausknecht, M., Vijayanarasimhan, S., Vinyals, O., Monga, R., Toderici, G.: Beyond short snippets: deep networks for video classification. In: Proceedings of the IEEE Conference on Computer Vision and Pattern Recognition, pp. 4694–4702 (2015)
24. Chadha, A., Abbas, A., Andreopoulos, Y.: Video classification with CNNs: using the codec as a spatio-temporal activity sensor. In: IEEE International Conference (2017)
25. Zheng, B., Wang, L., Wang, Z., Qiao, Y.: Real-time action recognition with deeply transferred motion vector CNNs. In: Proceedings of the IEEE Proceedings of the International Conference on Computer Vision (ICCV), pp. 1–13 (2018)
26. Kantorov, V., Laptev, I.: Efficient feature extraction, encoding, and classification for action recognition. In: Proceedings of the CVPR, Columbus, OH, USA, September 2014, pp. 2593–2600 (2014)
27. Tran, D., Bourdev, L.D., Fergus, R., Torresani, L., Paluri, M.: C3D: generic features for video analysis (2014). CoRR, abs/1412.0767
28. Wang, H., Schmid, C.: Action recognition with improved trajectories. In: ICCV'13, pp. 3551–3558 (2013)
29. VGG and ResNet. https://github.com/tensorflow/models/tree/master/research/slim
30. Fang, W., Zhang, F., Sheng, V.S., Ding, Y.: A method for improving CNN-based image recognition using DCGAN. Comput. Mater. Continua 57(1), 167–178 (2018)
31. Zhou, S., Liang, W., Li, J., Kim, J.-U.: Improved VGG model for road traffic sign recognition. Comput. Mater. Continua 57(1), 11–24 (2018)
32. Meng, R., Rice, S.G., Wang, J., Sun, X.: A fusion steganographic algorithm based on faster R-CNN. Comput. Mater. Continua 55(1), 001–016 (2018)

Author Index

Printed in the United States
By Bookmasters